physica status solidi

SUBJECT AND AUTHOR INDEX
VOLUMES 16 TO 30 · 1966 TO 1968

AKADEMIE-VERLAG · BERLIN

THEORIA CUM PRAXI

Classification Scheme

1. Structure of Solids
   1.1 Alloys. Metallurgy
   1.2 Solid-State Phase Transformations
   1.3 Surfaces
   1.4 Films

2. Non-Crystalline State

3. Crystallography
   3.1 Crystal Growth
   3.2 Interatomic Forces

4. Microstructure of Solids

5. Perfectly Periodic Structures

6. Lattice Mechanics. Phonons
   6.1 Mössbauer Investigations

7. Acoustic Properties of Solids

8. Thermal Properties of Solids

9. Diffusion in Solids

10. Defect Properties of Solids (Irradiation Defects See 11)
    10.1 Defect Properties of Metals
    10.2 Photochemical Reactions. Colour Centres

11. Irradiation Effects in Solids

12. Mechanical Properties of Solids (Plastic Deformations See 10)
    12.1 Mechanical Properties of Metals (Plastic Deformations See 10.1)

13. Electron States in Solids
    13.1 Band Structure. Fermi Surfaces
    13.2 Excitons
    13.3 Surface States
    13.4 Impurity and Defect States

14. Electrical Properties of Solids. Transport Phenomena
    14.1 Metals. Conductors
    14.2 Superconductivity. Superconducting Materials and Devices
    14.3 Semiconductors
        14.3.1 Semiconducting Films
        14.3.2 Semiconducting Devices. Junctions (Contact Problems See 14.4.1)
    14.4 Dielectrics
        14.4.1 High Field Phenomena, Space Charge Effects, Inhomogeneities, Injected Carriers (Electroluminescence See 20.3; Junctions See 14.3.2)
        14.4.2 Ferroelectric Materials and Phenomena

15. Thermoelectric and Thermomagnetic Properties of Solids

16. Photoconductivity. Photovoltaic Effects

17. Emission of Electrons and Ions from Solids

(Continued on cover three)

physica status solidi

SUBJECT AND AUTHOR INDEX

Volumes 16—30, 1966—1968

Edited by

W. BORCHARDT

AKADEMIE-VERLAG · BERLIN

Subscriptions and orders for this Subject Index or single copies of physica status solidi should be addressed to

AKADEMIE-VERLAG GmbH, 108 Berlin, Leipziger Straße 3 – 4 or to Buchhandlung KUNST UND WISSEN, Erich Bieber, 7 Stuttgart 1, Wilhelmstr. 4 – 6 or to Deutsche Buch-Export und -Import GmbH, 701 Leipzig, Postschließfach 160

Herausgeber und verantwortlich für den Inhalt des Registers: Dr. W. Borchardt, 102 Berlin, Neue Schönhauser Str. 20. Verlag: Akademie-Verlag GmbH, 108 Berlin, Leipziger Str. 3–4, Fernruf: 22 04 41, Telex-Nr. 112 020, Postscheckkonto: Berlin 350 21. — Bestellnummer des Registers 1068/R 2. Gesamtherstellung: VEB Druckerei „Thomas Müntzer" Bad Langensalza. — Veröffentlicht unter der Lizenznummer 1310 des Presseamtes beim Vorsitzenden des Ministerrates der Deutschen Demokratischen Republik

Preface

After the "Subject and Author Index Volumes 1 to 15, 1961 to 1966" has found general assent, in future we shall comprise every 1500 to 2000 papers in a Subject and Author Index, thus complying with the numerous wishes of our readers. The good development of physica status solidi made it possible to comply with the demand for the next Index Volume by the present "Subject and Author Index Volumes 16 to 30, 1966 to 1968" already $2^1/_2$ years later and thus increase its actuality. Whereas the first Index comprised about 1700 papers having appeared in physica status solidi within the first five years, the second Index comprises about 1850 papers having appeared from July 1966 to December 1968.

The general scheme of the first Index Volume was maintained. Meeting with many wishes, the classification scheme was completed by adding the Substance Classification which was introduced in January 1967 in physica status solidi. The basis for quoting the respective papers is the subject classification given from January 1967 for every paper in each issue. Before taking over these subject classifications into the Index Volume they were once more carefully revised and in some cases improved and completed in order to get a most uniform and clear classification for all papers. The average number of entries per paper was increased to 3.1 of which an average of 1.1 entries refers to the substance classification (compared with 1.8 entries per paper in the first Index Volume).

I readily avail myself of the opportunity of thanking the readers for their interest and numerous letters and I would be very grateful for critical remarks and proposals for improvement. I am indebted to Prof. Dr. P. Görlich, Dr. E. Gutsche, and Dr. S. Oberländer for their interest and valuable remarks. In the compilation of the manuscript I was again actively supported by Mrs. E. Schwan to whom I am much indebted. Especially sincere thanks are due to my wife for her unresting help in proof reading.

Berlin, April 1969

W. Borchardt

Some Comments on the Use of the Subject Classification and the Alphabetic Author Index

The classification system[1]) comprises 20 broad sections each devoted to a major field of solid state physics and 3 broad sections devoted to the main groups of substances. A section may contain a number of sub-sections dealing with aspects of this field or special substance, respectively, which tend to receive particular attention. A paper dealing with this field will either be classified under one or more of the sub-sections or, if none of these are relevant, it will just appear under the general section heading (i. e. sections and sub-sections are equivalent in this sense). For example, a paper dealing with problems of band structure would appear under 13.1 but not 13, whereas a paper on polaron states would be quoted under 13; a paper dealing with GaP would appear under 22.2.2, not under 22 or 22.2, whereas a paper on AlN would be quoted under 22.2, and a paper on a II—V or a III—VI compound would be quoted under 22.

All papers will be quoted under each relevant section. There is made a distinction between one or two principal entries (referring to the main subject matter of the paper) and secondary entries (referring to subsiduary subject matter and substance, respectively). In the principal entries the title of the paper is given in its original language and in English. In addition, every principal entry is marked by * in front of the title of the paper. In the secondary entries the title is only given in English.

Each entry gives the volume, the number of the issue, the initial and final page numbers, and the year in which the paper appeared. The letter "K" before the page numbers indicates a short note published in the off-set part of the journal. Each entry is followed by references to the other entries. These references give the subject classifications and the pages in the index where the other entries appear. The references to the principal entries are given in italics.

Review articles appear at the head of the list of entries under a given subject and are given the sub-title 'Review Articles'. These articles are listed in the chronological order of their publication. The subject classification for a review article consisting of several parts is based on the article as a whole rather than on the individual parts.

The entries for original papers and short notes are made together in a single list for each subject and this follows the entries for review articles. These are again ordered according to date of publication. Original papers are sometimes published in several parts under the same principal title (occasionally with special sub-titles). The various parts of papers of this type are treated separately and each part is classified according to its particular contents.

---

[1]) See cover, pages 2 und 3. For the case of losing the cover by bookbinding we repeat the classification scheme on pages 642 and 643.

In using the subject index attention should be given to the following details:

Subject	quoted in	remarks
representations of space groups	3 and 13.1	
plastic deformations	10 or 10.1	not in 12 or 12.1
colour centres	10.2	not in 20.1, X-ray coloration not in 11
plasmons, Alfvén waves, helicon waves	13 and 20	
hot electrons	13 and 14.3	possibly another sub-section of 14 instead of 14.3
transport phenomena	14 or sub-section of 14	if not clearly to be assigned to 8 or 15
galvanomagnetic effects	14 or sub-section of 14	not in 18
electro-optical effects	20 and 14 or sub-section of 14	generally not in 20.1
piezoelectricity, electrostriction	14 or sub-section of 14	not in 12 or 12.1; different electromechanical effects are quoted in 12 or 12.1
ionic conduction	14.4	not in 14.3
magneto-optical effects	18 and 20	possibly in additional sections, e.g. Faraday effect in 13
X-ray spectra	20	but not X-ray studies of structure

Substance	quoted in	remarks
II—V and III—VI compounds	22	
mixed crystals of III—V compounds, II—VI compounds, and Halides, respectively	22.8 and 22.2 (22.4, 22.5, respectively) or sub-sections	e.g. ZnS_xSe_{1-x} in 22.4.1, 22.4.2, and and 22.8; $RbMnF_3$ in 22.5 and 22.8
MIS and MOS structures	the section assigned to the semiconductor	not in the section assigned to the oxide or the metal
MIM structures	the section assigned to the insulator	generally not in the section assigned to the metal
epitaxial layers	the section assigned to the layer	not in the section assigned to the substrate
alnico and permalloy	21.1.1	not supplementary in 21 and 21.1

The author index gives the names of the authors in alphabetical order. Each name is followed by the bibliographic details of the paper and the numbers of the pages on which the paper appears in the subject index. The principal entry is again given in italics. When several papers have been published by the same author they are arranged according to the dates of publication.

Subject Index

1. Structure of Solids

Original Papers and Short Notes

J. Spyridelis, P. Delavignette, and S. Amelinckx
*On the Superstructures of Ta_2O_5 and Nb_2O_5
Vol. **19**, No. 2, p. 683—704 (1967).
See also 4, p. 40; 10, p. 98; 22.6, p. 537.

G. Kunze
Refraction-Modified Bragg Diffraction in the Case of Asymmetric Back-Reflection
Vol. **20**, No. 1, p. 179—195 (1967).
See also *4*, p. *41*.

V. A. Bokov, N. A. Grigoryan, and M. F. Bryzhina
*X-Ray Diffraction and Magnetic Studies of Solid Solutions $Bi_{1-x}Ca_xMnO_3$
Vol. **20**, No. 2, p. 745—754 (1967).
See also 1.2, p. 10; 18.2, p. 306; 18.3, p. 317; 18.4, p. 322; 22.8, p. 549.

B. Ray and P. M. Spencer
*Phase Diagram of the Alloy System HgTe–CdTe
Vol. **22**, No. 2, p. 371—372 (1967).
See also *1.2*, p. *11*; 22.4.3, p. 509; 22.8, p. 551.

F. E. Roberts
*Evaporation of As from GaAs
Vol. **22**, No. 2, p. 373—376 (1967).
See also 22.2.1, p. 484.

1.1 Alloys. Metallurgy

Original Papers and Short Notes

E. Kovács-Csetényi, C. R. Vassel, and I. Kovács
*Determination of the Solubility of Si in Dilute Al–Si Alloys by Low-Temperature Resistivity Measurements
Vol. **17**, No. 1, p. K123–K126 (1966).
See also 14.1, p. 220; 21, p. 375.

D. H. Howling
*Identification and Morphology of Oxide Particles in Internally Oxidized Cu–Mn Alloys
Vol. **18**, No. 2, p. 579–591 (1966).
See also *4*, p. *39*; 21.1, p. 394.

A. van den Beuckel, P. C. J. Coremans, and M. M. A. Vrijhoef
*On the Kinetics of Short-Range Ordering in AuAg (50,50) and CuAl (85,15)
Vol. **19**, No. 1, p. 177–183 (1967).
See also 10.1, p. 120; 21.1, p. 395; 21.6, p. 432.

V. I. Syutkina and E. S. Yakovleva
The Mechanism of Deformation of the Ordered CuAu Alloy
Vol. **21**, No. 2, p. 465–480 (1967).
See also 4, p.; 42; *10.1*, p. *123*; 21.1, p. 398; 21.6, p. 433.

B. Roessler and P. A. Flinn
A Theory of X-Ray Diffraction from an Ordered Alloy Containing Antiphase Domains
Vol. **24**, No. 1, p. 263–274 (1967).
See also *4*, p. *43*; 21.1, p. 401; 21.6, p. 434.

L. Delaey
Electron Microscopic Investigation of Lamellar Mixtures of Two Close-Packed Phases
Vol. **25**, No. 2, p. 697–710 (1968).
See also *4*, p. *45*; 21.1, p. 403.

A. S. Sastri, M. J. Marcinkowski, and D. Koskimaki
Nature of the Ni–Ti Martensite Transformation
Vol. **25**, No. 2, p. K67–K69 (1968).
See also *1.2*, p. *11*; 21, p. 385; 21.1, p. 403.

H.-J. Bunge
The Orientation Distribution Function of the Crystallites in Cold-Rolled and Annealed Low-Carbon Steel Sheets.
Vol. **26**, No. 1, p. 167–172 (1968).
See also *4*, p. *45*; 21.1.1, p. 420.

S. Radelaar
*The Kinetics of Short-Range Order in Some Ag–Au Alloys
Vol. **27**, No. 1, p. K63–K67 (1968).
See also 9, p. 91; 10.1, p. 131; 21.6, p. 436.

E. Ruedl, P. Delavignette, and S. Amelinckx
*Electron Diffraction and Electron Microscopic Study of Long- and Short-Range Order in Ni_4Mo and of the Substructure Resulting from Ordering
Vol. **28**, No. 1, p. 305–328 (1968).
See also 4, p. 46; 21, p. 386; 21.1, p. 405.

1.2 Solid-State Phase Transformations

Original Papers and Short Notes

S. W. Kennedy, G. F. Taylor, and J. H. Patterson
*Misfit and Hysteresis at Structure Transformations in Rubidium and Cesium Nitrates; and Thermal Expansions
Vol. **16**, No. 2, p. K175–K178 (1966).
See also 3, p. 26; 8, p. 79; 22.8, p. 547.

M. Balarin und M. Schenk
*Zur Verschiebung von Umwandlungspunkten zwischen kristallinen Phasen durch Gitterdefekte
(On the Defect-Induced Shift of Transition Points between Crystalline Phases)
Vol. **17**, No. 1, p. 91–99 (1966).
See also 10, p. 94; 11, p. 146; 14.4.2, p. 266.

A. J. Morton and C. M. Wayman
*On the Feasibility of a "Composite" Martensite Model for the $(225)_f$ Austenite–Martensite Transformation
Vol. **17**, No. 1, p. 101–103 (1966).
See also 3, p. 26; 21.1.1, p. 412.

E. V. Chisler and M. S. Shur
Raman Spectrum and Phase Transition in the Ferroelectric Crystal $NaNO_2$
Vol. **17**, No. 1, p. 173–176 (1966).
See also *6*, p. *57*; 14.4.2, p. 266; *20.1*, p. *345*.

P. F. Lindquist, M. L. Hammond, and R. H. Bragg
Comments on the Paper "On the Interpretation of Electron Diffraction Patterns from 'Amorphous' Boron"
Vol. **17**, No. 1, p. K25–K31 (1966).
See also *3*, p. *26*; 4, p. 38; 22.1, p. 456.

K. Ozawa and S. Anzai
*Effect of Pressure on the α-Transition Point of Iron Monosulphide
Vol. **17**, No. 2, p. 697–700 (1966).
See also 12, p. 156; 14.3, p. 233; 22, p. 441.

B. Staliński, Z. Biegański, and R. Troć
Low-Temperature Heat Capacity and Thermodynamical Functions of Ferromagnetic Uranium Phosphide U_3P_4
Vol. **17**, No. 2, p. 837–841 (1966).
See also *8*, p. *80*; 18.2, p. 303; 22, p. 441.

H. Gärtner and E. Schelzke
Structure of Thin Cobalt Layers Prepared by Cathodic Sputtering
Vol. **17**, No. 2, p. K241–K244 (1966).
See also *1.4*, p. *17*; 21.1, p. 393.

Hj. Matzke
Radiation Damage and Xenon Release in Quartz and Fused Silica Following Ion Bombardment
Vol. **18**, No. 1, p. 285–295 (1966).
See also 2, p. 22; *11*, p. *147*; 22.6, p. 536.

K. S. Aleksandrov, L. M. Reshchikova, and B. V. Beznosikov
Behaviour of the Elastic Constants of $KMnF_3$ Single Crystals near the Transition of Puckering
Vol. **18**, No. 1, p. K17—K20 (1966).
See also *12*, p. *156*; 22.5, p. 512; 22.8, p. 548.

O. Hauser und M. Sckenk
*Strahleninduzierte Phasenumwandlungen einiger Substanzen des Perowskit-Gittertyps und ihre thermodynamische Behandlung
(Radiation-Induced Phase Transitions of Some Perovskite-Type Substances and Their Thermodynamic Treatment)
Vol. **18**, No. 2, p. 547—555 (1966).
See also 8, p. 80; 11, p. 147; 14.4.2, p. 266.

B. Dorner and H. H. Stiller
Internal Dynamics of the Low-Temperature Phases of CH_4 Molecular Crystals
Vol. **18**, No. 2, p. 795—806 (1966).
See also *6*, p. *58*; 23, p. 561.

R. Mănăilă and P. Păuşescu
*The Tetragonal to Cubic Transition in the System $Cd_xMg_{1-x}Mn_2O_4$
Vol. **19**, No. 1, p. K19—K22 (1967).
See also 22.8, p. 549.

E. I. Gavrilitsa and S. I. Radautsan
*Influence of Ordering on the Properties of Solid Solutions of the System $(HgSe)_{3x}$–$(In_2Se_3)_{1-x}$
Vol. **19**, No. 2, p. 609—611 (1967).
See also 8, p. 81; 14.3, p. 235; 22.8, p. 549.

H. M. Otte, H. A. Lipsitt, P. F. Lindquist, M. L. Hammond, and R. H. Bragg
Further Comments on the Paper "On the Interpretation of Electron Diffraction Patterns from 'Amorphous' Boron"
Vol. **19**, No. 2, p. K99—K101 (1967).
See also *3*, p. *26*; 4, p. 40; 22.1, p. 457.

L. Zsoldos
Lattice Parameter Change of FeRh Alloys due to Antiferromagnetic–Ferromagnetic Transformation
Vol. **20**, No. 1, p. K25—K28 (1967).
See also 5, p. 51; *18.2*, p. *306*; *18.4*, p. *322*; 21.1.1, p. 416.

V. A. Bokov, N. A. Grigoryan, and M. F. Bryzhina
X-Ray Diffraction and Magnetic Studies of Solid Solutions $Bi_{1-x}Ca_xMnO_3$
Vol. **20**, No. 2, p. 745—754 (1967).
See also *1*, p. *7*; 18.2, p. 306; 18.3, p. 317; 18.4, p. 322; 22.8, p. 549.

T. Sakata, K. Sakata, and I. Nishida
*Study of Phase Transition in NbO_2
Vol. **20**, No. 2, p. K155—K157 (1967).
See also 22.6, p. 537.

N. Kristoffel and P. Konsin
*Pseudo-Jahn-Teller Effect and Second Order Phase Transitions in Crystals
Vol. **21**, No. 1, p. K39—K43 (1967).
See also 6, p. 61; *13*, p. *166*; 14.4.2, p. 267.

E. J. Kramer and Ch. L. Bauer
Variations of Young's Modulus throughout the Mixed State in Niobium
Vol. **22**, No. 1, p. 199–202 (1967).
See also *10.1*, p. *124*; 14.2, p. 229; 18, p. 292; 21, p. 380.

L. A. Shuvalov and N. R. Ivanov
The Second Ferroelectric Phase Transition in $NaH_3(SeO_3)_2$ Crystals and Their Physical Properties
Vol. **22**, No. 1, p. 279–287 (1967).
See also *14.4.2*, p. *267*.

B. Ray and P. M. Spencer
*Phase Diagram of the Alloy System HgTe–CdTe
Vol. **22**, No. 2, p. 371–372 (1967).
See also *1*, p. *7*; 22.4.3, p. 509; 22.8, p. 551.

I. Dézsi, L. Keszthelyi, D. Kulgawczuk, B. Molnár, and N. A. Eissa
Mössbauer Study of β- and δ-FeOOH and Their Disintegration Products
Vol. **22**, No. 2, p. 617–629 (1967).
See also *6.1*, p. *72*; 18.3, p. 318; 18.4, p. 323; 22.8, p. 551.

E. Nagy and G. Petö
*Ordering in the Alloy Cu_3Au–V
Vol. **23**, No. 1, p. K49–K52 (1967).
See also 16, p. 280; 21.1, p. 400.

I. Dima and D. Borşan
The Polymorphism of CdTe Thin Films
Vol. **23**, No. 2, p. K113–K115 (1967).
See also *1.4*, p. *19*; 22.4.3, p. 510.

S. Toschev and I. Gutzow
Derivation of Nucleation Kinetics in Solids by Examination of Plane Sections through the Samples
Vol. **24**, No. 1, p. 349–359 (1967).
See also *3.1*, p. *32*; 22, p. 448.

B. I. Kuliev and G. M. Nedlin
The Types of Magnetic Ordering in Crystals Having the Structure of $RbNiF_3$
Vol. **24**, No. 1, p. K25–K28 (1967).
See also 3.2, p. 35; *18*, p. *293*; 22.5, p. 513, 22.8, p. 552.

K. H. J. Buschow, J. F. Fast, A. M. van Diepen, and H. W. de Wijn
Magnetic Dilution of Rare-Earth Aluminium Cubic Laves Phases RAl_2
Vol. **24**, No. 2, p. 715–720 (1967).
See also *18.2*, p. *309*; 21, p. 383.

A. G. Fitzgerald and G. Thomas
*Electron Microscope Observations of Twinning and Phase Transformations in Indium Sulfide Crystals
Vol. **25**, No. 1, p. 263–271 (1968).
See also *4*, p. *44*; 22, p. 449.

A. S. Sastri, M. J. Marcinkowski, and D. Koskimaki
*Nature of the Ni–Ti Martensite Transformation
Vol. **25**, No. 2, p. K67–K69 (1968).
See also 1.1, p. 8; 21, p. 385; 21.1, p. 403.

G. B. ABDULLAEV, Z. A. ALIYAROVA, E. H. ZAMANOVA, and G. A. ASADOV
Investigation of the Electric Properties of Cu_2S Single Crystals
Vol. **26**, No. 1, p. 65–68 (1968).
See also *14.3*, p. *240*; 15, p. 273; 22, p. 450.

E. O. OSMANOV, YU. V. RUD, and M. E. STRYALKOVSKII
*High-Temperature Modifications of the Semiconductor Compounds $CdSnAs_2$ and $CdGeAs_2$
Vol. **26**, No. 1, p. 85–90 (1968).
See also 2, p. 24; 8, p. 84; 12, p. 158; 14.3, p. 240; 22.7, p. 546; 22.8, p. 553.

K. WETZIG
Some Investigations Concerning the Position and the Melting Point of TaFe Lamellas in the System TaFe–$TaFe_2$
Vol. **27**, No. 1, p. K7–K9 (1968).
See also *4*, p. *46*; 21, p. 386; 21.1.1, p. 421.

B. A. STRUKOV, M. AMIN, and V. A. KOPCHIK
Comparative Investigation of the Specific Heat of KH_2PO_4 (KDP) and KD_2PO_4 (DKDP) Single Crystals
Vol. **27**, No. 2, p. 741–749 (1968).
See also *8*, p. *85*; 14.4.2, p. 268.

W. BITTERLICH and R. J. JELITTO
*Computational Evidence for the Transition Point of a Square Ising Lattice with First and Second Neighbour Interactions
Vol. **28**, No. 1, p. 365–372 (1968).
See also 18.2, p. 312; 18.4, p. 324.

C. MUNTONI and A. SERPI
*On an Allotropic Transformation of ZnS after Pressure Treatment
Vol. **28**, No. 1, p. K43–K45 (1968).
See also 13.1, p. 187; 20.1, p. 356; 22.4.1, p. 503.

N. KRISTOFFEL and P. KONSIN
*Displacive Vibronic Phase Transitions in Narrow-Gap Semiconductors
Vol. **28**, No. 2, p. 731–739 (1968).
See also *6*, p. *67*; 14.4.2, p. 269; 22, p. 453.

G. M. DE'MUNARI, F. GIUSIANO, and G. MAMBRIANI
*A Phase Transition and the Photoemissive Quantum Yield at Low Temperatures in Cs_3Sb Films
Vol. **29**, No. 1, p. 341–356 (1968).
See also 3, p. 27; 13.1, p. 188; 17, p. 289; 22.7, p. 546.

H. WEIJMA and B. R. KNOTTNERUS
Ionic Conductivity of Pure and Ba^{2+}-Doped CsCl
Vol. **29**, No. 1, p. K43–K45 (1968).
See also 10, p. 110; *14.4*, p. *254*; 22.5.2, p. 529.

V. PLAKHTY and W. COCHRAN
*X-Ray Study of the Phase Transition and Lattice Vibrations of Lanthanum Aluminate
Vol. **29**, No. 1, p. K81–K83 (1968).
See also *6*, p. *68*; 22.8, p. 555.

R. C. Thiel and C. B. van den Berg
Temperature Dependence of Hyperfine Interactions in Near-Stoichiometric FeS. (I): Experiment
Vol. **29**, No. 2, p. 837—846 (1968).
See also *6.1*, p. *74*; 18.4, p. 325; 22, p. 454.

P. P. Salhotra, E. C. Subbarao, and P. Venkateswarlu
*Polymorphism of Rubidium Nitrate
Vol. **29**, No. 2, p. 859—864 (1968).
See also *3*, p. *28*; 8, p. 86; 22.8, p. 556.

R. Hsiang-Tao Yeh
Possible Proximity Effect in Ferroelectric Films
Vol. **30**, No. 1, p. K45—K47 (1968).
See also *14.4.2*, p. *269*.

R. E. Hummel and J. W. Koger
*Effect of Neutron Irradiation on the Martensitic Transformation in β_1-Copper–Zinc
Vol. **30**, No. 2, p. 515—523 (1968).
See also *11*, p. *155*; 14.1, p. 227; 21.1, p. 409.

L. E. Tanner
*The Ordering of Ni_3V
Vol. **30**, No. 2, p. 685—701 (1968).
See also 3.1, p. 33; 4, p. 48; 21, p. 388.

W. Cochran
Neutron Scattering and Dielectric Properties of Perovskite-Type Crystals
Vol. **30**, No. 2, p. K157—K160 (1968).
See also *6*, p. *70*; 14.4.2, p. 269.

1.3 Surfaces

Original Papers and Short Notes

CHR. LEHMANN and P. SIGMUND
On the Mechanism of Sputtering
Vol. **16**, No. 2, p. 507—511 (1966).
See also *11*, p. *145*; 21.1, p. 390.

CHR. NANEV
*Elektronenmikroskopische Untersuchung der Wachstumsmorphologie von Zink- und Kadmiumeinkristallen
(Electron-Microscopic Study of the Growth Morphology of Zinc and Cadmium Single Crystals)
Vol. **16**, No. 2, p. 777—787 (1966).
See also 3.1, p. 29; 21, p. 374; 21.1, p. 391.

J. PFEIFER
Epitaxial Growth of Germanium on Germanium Substrates Cleaned and Heated by Electron Bombardment
Vol. **17**, No. 1, p. K15—K18 (1966).
See also *3.1*, p. *29*; 22.1.1, p. 459.

L. VALENTA and A. SUKIENNICKI
*Order–Disorder Phenomena at the Surface of Solids
Vol. **17**, No. 2, p. 903—909 (1966).
See also *3.2*, p. *34*; 21.1.1, p. 413.

L. ERNST
*Field-Emission Microscopy of Germanium
Vol. **19**, No. 1, p. 89—94 (1967).
See also 22.1.1, p. 461.

A. G. REVESZ
*Thermal Oxidation of Silicon: Growth Mechanism and Interface Properties
Vol. **19**, No. 1, p. 193—202 (1967).
See also 13.3, p. 197; 14.3.2, p. 248; 22.1.2, p. 470.

E. IGRAS and T. WARMIŃSKI
*Electron Mirror Observation of Lithium Precipitation at Scratch-Induced Dislocations in Silicon
Vol. **19**, No. 2, p. K67—K69 (1967).
See also 4, p. 40; 10, p. 98; 22.1.2, p. 470.

K. WETZIG
*Investigations of Surfaces of the Binary System Ta–Fe by Means of an Emission Electron Microscope
Vol. **19**, No. 2, p. K71—K73 (1967).
See also 4, p. 40; 21, p. 377; 21.1.1, p. 415.

E. IGRAS and T. WARMIŃSKI
Electron Mirror Microscopic Investigation of Surfa ceDiffusion of Lithium in Silicon
Vol. **20**, No. 1, p. K5—K8 (1967).
See also *9*, p. *88*; 22.1.2, p. 471.

H. P. UTECH
*Interpretation of de Grinberg's Experiments on Lateral Surface Fringes in Sodium Chloride
Vol. **20**, No. 1, p. K41–K43 (1967).
See also 22.5.2, p. 521.

R. VANSELOW
*Veränderungen an reinen Platinoberflächen durch gelöste Fremdstoffe: Beobachtungen mit dem Feldelektronenmikroskop
(Changes due to Solved Impurities at Pure Platinum-Surfaces: Observations with the Field Electron-Microscope)
Vol. **21**, No. 1, p. 69–76 (1967).
See also 10.1, p. 122; 21.6, p. 433.

I. L. SOKOLSKAYA, V. G. IVANOV, and G. N. FURSEY
Study of Barium Adsorption on Germanium by Field-Emission Microscopy
Vol. **21**, No. 2, p. 789–795 (1967).
See also *13.3*, p. *197*; 17, p. 287; 22.1.1, p. 462.

J. PUNZEL
*Temperature Dependence of the Surface Structure of Ion-Bombarded Si Single Crystals
Vol. **24**, No. 1, p. K1–K2 (1967).
See also 11, p. 150; 22.1.2, p. 472.

WILLIAM R. SAVAGE
*Field Electron Microscope Study of Niobium Surfaces
Vol. **25**, No. 1, p. 131–138 (1968).
See also 17, p. 288; 21, p. 383.

E. IGRAS and T. WARMIŃSKI
*On Electron Mirror Microscopic Observations of the Details of p–n Junction Regions in Si
Vol. **27**, No. 1, p. 69–76 (1968).
See also *14.3.2*, p. *250*; 22.1.2, p. 473.

D. MICHELL and A. P. SMITH
The Nature of Oxide Layers on Single Crystals of Cadmium and Magnesium
Vol. **27**, No. 1, p. 291–300 (1968).
See also *1.4*, p. *20*; 10, p. 107; 22.6, p. 541.

P. LUKÁŠ, M. KLESNIL, and J. KREJČÍ
Dislocations and Persistent Slip Bands in Copper Single Crystals Fatigued at Low Stress Amplitude
Vol. **27**, No. 2, p. 545–558 (1968).
See also *10.1*, p. *131*; 21.1, p. 405.

1.4 Films

Review Articles

L. E. MURR
*Calibration and Use of an Electron Microscope for Precision Micromeasurements in Thin Film Materials
Vol. **19**, No. 1, p. 7—34 (1967).
See also **4**, p. 37; 10, p. 93; 21.1, p. 390; 21.1.1, p. 410.

Original Papers and Short Notes

M. J. STOWELL and T. J. LAW
*The Migration of Double Positioning Boundaries during the Growth of Epitaxial (111) Gold Films
Vol. **16**, No. 1, p. 117—125 (1966).
See also *3.1*, p. *29*; 21.6, p. 430.

P. ESCUDIER, F. BIRAGNET, J. DEVENYI, and A. YELON
*Structural and Magnetic Properties of Epitaxial Permalloy Films
Vol. **16**, No. 1, p. 295—300 (1966).
See also *18.2*, p. *299*; 21.1.1, p. 410.

A. CORCIOVEI and G. COSTACHE
Influence of the Perpendicular and Parallel Anisotropy on the Ferromagnetic Properties of Thin Films
Vol. **16**, No. 1, p. 329—336 (1966).
See also *18.2*, p. *300*.

M. J. STOWELL and T. J. LAW
*Interrupted Vapour Deposition of Epitaxial Gold Films Grown Inside an Electron Microscope
Vol. **16**, No. 2, p. 479—485 (1966).
See also **3.1**, p. 29; 21.6, p. 430.

K. W. BÖER, J. W. FEITKNECHT, and D. G. KANNENBERG
*Properties of Recrystallized Evaporated CdS Layers
Vol. **16**, No. 2, p. 697—709 (1966).
See also **13.4**, p. 201; 16, p. 276; 22.4.1, p. 497.

P. WEBER and W. RUSKE
Temperature Dependence of Oblique-Incidence Anisotropy in Permalloy Films
Vol. **17**, No. 1, p. 185—190 (1966).
See also *18.2*, p. *301*; 21.1.1, p. 412.

L. V. KIRENSKII, V. G. PYNKO, N. I. SIVKOV, G. P. PYNKO, R. V. SUKHANOVA, and M. A. OVSYANNIKOV
Growth of Single-Crystal Films of Iron, Nickel, Cobalt, and Their Alloys
Vol. **17**, No. 1, p. 243—248 (1966).
See also *3.1*, p. *29*; 21.1, p. 392; 21.1.1, p. 412.

W. METZDORF and H. E. WIEHL
Negative Oblique-Incidence Anistropy in Magnetostriction-Free Permalloy Films
Vol. **17**, No. 1, p. 285—294 (1966).
See also *18.2*, p. *302*; 21.1.1, p. 412.

Z. KAMBERSKÁ and V. KAMBERSKÝ
Oblique-Incidence Anisotropy in Cobalt Films
Vol. **17**, No. 1, p. 411—415 (1966).
See also *18.2*, p. *302*; 21.1, p. 392.

I. B. PUCHALSKA and A. SUKIENNICKI
Temperature Dependence of Oblique-Incidence Anisotropy Observed by Lorentz Microscopy
Vol. **17**, No. 1, p. K1 (1966).
See also *18.2*, p. *302*; 21.1.1, p. 412.

W. D. DOYLE
Epitaxial Influence of the Substrates on the Anisotropy in Thin Electrodeposited Ni–Fe Cylindrical Films
Vol. **17**, No. 1, p. K67—K69 (1966).
See also *18.2*, p. *302*; 21.1.1, p. 412.

H. HOFFMANN
Principal Considerations about a Non-Linear Theory of the Initial Suspectibility of Uniaxial Ferromagnetic Films
Vol. **17**, No. 1, p. K105—K107 (1966).
See also *18.2*, p. *303*.

L. S. PALATNIK, M. YA. FUKS, L. I. LUKASHENKO, A. G. RAVLIK und A. A. KOZMA
*Struktur und magnetische Eigenschaften von kondensierten ferromagnetischen Schichten
(Structure and Magnetic Properties of Condensed Ferromagnetic Films)
Vol. **17**, No. 2, p. 543—554 (1966).
See also **3.1**, p. 30; 10.1, p. 116; *18.2*, p. *303*; 21.1, p. 392; 21.1.1, p. 413.

H. JAHRREISS und H.-J. ISKEN
*Zum Epitaxieverhalten dünner Silberaufdampfschichten
(Epitaxial Behaviour of Thin Evaporated Silver Layers)
Vol. **17**, No. 2, p. 619—629 (1966).
See also *3.1*, p. *30*; 21.6, p. 431.

H. GÄRTNER und E. SCHELZKE
*Die Struktur dünner Kobalt-Schichten aus der Kathodenzerstäubung
(Structure of Thin Cobalt Layers Prepared by Cathodic Sputtering)
Vol. **17**, No. 2, p. K241—K244 (1966).
See also 1.2, p. 9; 21.1, p. 393.

K. M. KUNZ, A. K. GREEN, and E. BAUER
*On the Formation of Single Crystal Films of F.C.C. Metals on Alkali Halide Cleavage Planes in Ultrahigh Vacuum
Vol. **18**, No. 1, p. 441—457 (1966).
See also *3.1*, p. *30*; 21, p. 376; 21.6, p. 431.

I. SPÎNULESCU-CARNARU
*The Crystalline Structure of ZnTe Thin Films
Vol. **18**, No. 2, p. 769—778 (1966).
See also **3.1**, p. 30; 5, p. 50; 22.4.3, p. 509.

R. H. WADE and J. SILCOX
*Small Angle Electron Scattering from Vacuum Condensed Metallic Films. (I): Theory
Vol. **19**, No. 1, p. 57—62 (1967).
See also **4**, p. 39; 21, p. 376.

R. H. WADE and J. SILCOX
*Small Angle Electron Scattering from Vacuum Condensed Metallic Films. (II): Experimental Results
Vol. **19**, No. 1, p. 63–76 (1967).
See also 4, p. 40; 21.1, p. 395; 21.6, p. 432.

W. A. JESSER and D. KUHLMANN-WILSDORF
On the Theory of Interfacial Energy and Elastic Strain of Epitaxial Overgrowths in Parallel Alignment on Single Crystal Substrates
Vol. **19**, No. 1, p. 95–105 (1967).
See also *10*, p. *97*.

G. A. JONES
*Preferred Orientation and Domain Structure of Iron Films Deposited on (100) Copper Substrates
Vol. **19**, No. 2, p. 811–821 (1967).
See also 18.2, p. 305; 21.1.1, p. 415.

D. J. D. THOMAS
*The Effect of Heat Treatment on Silicon Nitride Layers on Silicon
Vol. **20**, No. 1, p. 131–142 (1967).
See also 22, p. 443.

J. AULEYTNER, K. GODWOD, J. LITWIN, and Z. WOŁOSZYN
*Application of X-Ray Methods to the Study of the Real Structure of Monocrystalline Epitaxial Layers
Vol. **20**, No. 2, p. K77–K79 (1967).
See also 22.1.1, p. 461; 22.2.1, p. 483.

G. A. JONES
*Structural and Magnetic Properties of Epitaxial Films of Iron
Vol. **20**, No. 2, p. K95–K97 (1967).
See also *18.2*, p. *306*; 21.1.1, p. 416.

I. B. PUCHALSKA, A. SUKIENNICKI, Z. PETYKIEWICZ, and T. TYMOSZ
*The Influence of Annealing on the Structure of Polycrystalline Permalloy Thin Films
Vol. **21**, No. 1, p. 183–189 (1967).
See also 18.2, p. 306; 21.1.1, p. 417.

J. E. KNAPPETT and S. J. T. OWEN
Observation of the Growth Process of Gallium Arsenide on Tungsten
Vol. **21**, No. 2, p. K99–K101 (1967).
See also *3.1*, p. *31*; 14.3.1, p. 245; 22.2.1, p. 484.

N. PASCHOFF
*Wachstum und Struktur dünner, aus schmelzflüssigen Filmen auskristallisierter Germanium-Kristalle
(Growth and Structure of Thin Germanium Crystals Crystallized from Molten Films)
Vol. **22**, No. 1, p. 83–91 (1967).
See also *3.1*, p. *31*; 22.1.1, p. 462.

N. PASCHOFF and J. HEYDENREICH
Dislocation Arrangement in Thin Germanium Foils Prepared by Crystallization
Vol. **22**, No. 1, p. 93–101 (1967).
See also *10*, p. *101*; 22.1.1, p. 462.

K. Otsuka and C. M. Wayman
Epitaxial Growth of Vacuum-Evaporated Co on NaCl. (I): Conditions Leading to Epitaxy
Vol. 22, No. 2, p. 559—578 (1967).
See also *3.1*, p. *32*; 21.1, p. 399.

K. Otsuka and C. M. Wayman
Epitaxial Growth of Vacuum-Evaporated Co on NaCl. (II): Analysis of Diffraction Patterns
Vol. 22, No. 2, p. 579—592 (1967).
See also *3.1*, p. *32*; 21.1, p. 399.

M. V. Coleman and D. J. D. Thomas
*The Structure of Silicon Oxide Films
Vol. 22, No. 2, p. 593—602 (1967).
See also 22.1.2, p. 471; 22.6, p. 539.

B. Persson and H. Norden
Epitaxial Growth of Gold on Cadmium Iodide
Vol. 22, No. 2, p. 631—647 (1967).
See also *3.1*, p. *32*; 22.6, p. 539.

I. Spînulescu-Carnaru
*The Orientation of Crystallites in ZnTe Thin Films
Vol. 23, No. 1, p. 157—163 (1967).
See also 4, p. 43; 22.4.3, p. 510.

R. Ludeke and W. Paul
Growth and Optical Properties of Wurtzite and Sphalerite CdSe Epitaxial Thin Films
Vol. 23, No. 1, p. 413—418 (1967).
See also *3.1*, p. *32*; 13.1, p. 181; *20.1*, p. *350*; 22.4.2, p. 507.

I. Dima and D. Borşan
*The Polymorphism of CdTe Thin Films
Vol. 23, No. 2, p. K113—K115 (1967).
See also 1.2, p. 11; 22.4.3, p. 510.

A. G. Revesz
*Noncrystalline Structure and Electronic Conduction of Silicon Dioxide Films
Vol. 24, No. 1, p. 115—126 (1967).
See also 2, p. 23; *14.3.1*, p. *246*; 14.3.2, p. 249; 22.6, p. 539.

L. E. Murr
*Study of Erbium Thin Film Oxidation in the Electron Microscope
Vol. 24, No. 1, p. 135—148 (1967).
See also 5, p. 52; 22.6, p. 539.

M. V. Coleman and D. J. D. Thomas
*The Structure of Amorphous Silicon Films
Vol. 24, No. 2, p. K 111—K 115 (1967).
See also 2, p. 23; 22.1.2, p. 472.

M. J. Stowell and T. J. Law
*The Growth and Defect Structure of Gold Films Formed on Molybdenite in Ultra-High Vacuum
Vol. 25, No. 1, p. 139—146 (1968).
See also *3.1*, p. *33*; 10.1, p. 127; 21.6, p. 435.

M. V. COLEMAN and D. J. D. THOMAS
*The Structure of Amorphous Silicon Nitride Films
Vol. 25, No. 1, p. 241–252 (1968).
See also 2, p. 23; 22, p. 449.

W. SCHÜZ
Forbidden Reflexes in Electron Diffraction from Tellurium Films Evaporated in Vacuum
Vol. 25, No. 1, p. 253–261 (1968).
See also *5*, p. *53*; 22.1.3, p. 478.

L. E. MURR
*Investigation of Grain Boundary Energetics in Er_2O_3 Thin Foils by Transmission Electron Microscopy
Vol. 25, No. 2, p. 629–640 (1968).
See also *4*, p. *44*; 22.6, p. 540.

R. GEMPERLE and A. GEMPERLE
Dependence of the Ferromagnetic Domain Width on the Thickness of Cobalt Foils
Vol. 26, No. 1, p. 207–217 (1968).
See also *18.2*, p. *310*; 21.1, p. 404.

D. MICHELL and A. P. SMITH
*The Nature of Oxide Layers on Single Crystals of Cadmium and Magnesium
Vol. 27, No. 1, p. 291–300 (1968).
See also 1.3, p. 15; 10, p. 107; 22.6, p. 541.

U. GRADMANN and J. MÜLLER
*Flat Ferromagnetic, Epitaxial 48Ni/52Fe(111) Films of Few Atomic Layers
Vol. 27, No. 1, p. 313–324 (1968).
See also *18.2*, p. *311*; 21.1, p. 405; 21.1.1, p. 421.

R. J. MURPHY and J. L. CRAWFORD
Electron Microscope Image Profiles of Paired and Triple Dislocations
Vol. 27, No. 2, p. 559–572 (1968).
See also *10*, p. *108*; *10.1*, p. *131*.

C. JECH
Depth Distribution Measurements Using Bombardement-Enhanced Solubility
Vol. 27, No. 2, p. 573–577 (1968).
See also *11*, p. *152*; 22, p. 452; 22.1.2, p. 473.

R. P. FERRIER and I. B. PUCHALSKA
360° Walls and Strong Stripe Domains in Permalloy Films
Vol. 28, No. 1, p. 335–347 (1968).
See also *18.2*, p. *312*; 21.1.1, p. 422.

P. GUYOT
*Theory of Orientation Contrast from Inclusions
Vol. 28, No. 1, p. 349–354 (1968).
See also 4, p. 46.

D. J. CRAIK and M. J. WOOD
Thermally Activated Domain Wall Movement in Nickel–Iron Films
Vol. 28, No. 1, p. 407–414 (1968).
See also *18.2*, p. *313*; 21.1, p. 405; 21.1.1, p. 422.

B. F. Bilenkii, A. G. Mikolaichuk und D. M. Freik
*Struktur und optische Eigenschaften von epitaxialen SnTe-, SnSe- und SnS-Schichten (Structure and Optical Properties of Epitaxial SnTe, SnSe, and SnS Films)
Vol. **28**, No. 1, p. K5—K7 (1968).
See also *20.1*, p. *355*; 22, p. 452.

P. Ehrhart, W. Rupp, and R. Sizmann
Stopping Power of 0.5 to 3.5 MeV α-Particles in Cu–Au Alloys
Vol. **28**, No. 1, p. K35—K37 (1968).
See also *11*, p. *153*; 21.1, p. 405; 21.6, p. 436.

H. Berger, W. Kahle, and G. Jäniche
Thickness Dependence of Conductivity due to the Polycrystalline Structure in Evaporated CdS Thin Films
Vol. **28**, No. 2, p. K97—K100 (1968).
See also *14.3.1*, p. *247*; 22.4.1, p. 504.

K. Kleinhenz and P. Runow
Preparation of Thin Boron Foils and Electron Microscopic Observation of Their Lattice Faults
Vol. **29**, No. 2, p. 627—637 (1968).
See also *10*, p. *110*; 22.1, p. 457.

E. E. Galloni and R. Zimmerman
*Double Diffraction in Epitaxially Grown Nickel Polycrystalline Films
Vol. **29**, No. 2, p. K91—K94 (1968).
See also 21.1, p. 407.

S. E. Bronisz and D. L. Douglass
*Grain Growth in Thin Films of ThO_2, NpO_2, and PuO_2
Vol. **29**, No. 2, p. K95—K97 (1968).
See also 3.1, p. 33; 22.6, p. 543.

2. Non-Crystalline State

Original Papers and Short Notes

L. I. van Torne
Pumping Induced Imperfections in Glass:Nd^{3+} Lasers
Vol. **16**, No. 1, p. 171—182 (1966).
See also 4, p. 37; *20.2*, p. *360*; 22.6, p. 535.

G. B. Abdullaev, S. I. Mekhtieva, G. M. Aliev, D. Sh. Abdinov, and T. G. Kerimova
Determination of the Oxygen Content in Selenium
Vol. **16**, No. 1, p. K31—K34 (1966).
See also 10, p. 94; *20.1*, p. *344*; 22.1.3, p. 476.

R. Grigorovici, N. Croitoru, and A. Dévényi
Thermoelectric Power in Amorphous Germanium
Vol. **16**, No. 2, p. K143—K145 (1966).
See also 14.3.1, p. 245; *15*, p. *270*; 22.1.1, p. 459.

A. I. Gubanov
*The LCAO Method for an Arbitrary Mixture of Atoms
Vol. **17**, No. 2, p. 807—814 (1966).
See also *13*, p. *163*; 21, p. 375; 22, p. 441.

Hj. Matzke
Radiation Damage and Xenon Release in Quartz and Fused Silica Following Ion Bombardment
Vol. **18**, No. 1, p. 285—295 (1966).
See also 1.2, p. 9; *11*, p. *147*; 22.6, p. 536.

G. B. Abdullaev, A. U. Malsagov, and V. M. Glazov
Thermoelectric Power of $A^{I}B^{III}C_2^{VI}$ Type Compounds in the Solid and Liquid State
Vol. **18**, No. 1, p. K29—K31 (1966).
See also *15*, p. *271*; 22.8, p. 548.

W. van der Lugt and S. B. van der Molen
Nuclear Magnetic Resonance in Liquid Gallium Alloys
Vol. **19**, No. 1, p. 327—331 (1967).
See also *19*, p. *327*; 21, p. 377.

V. V. Tarassov
Heat Capacity and Structure of Vitreous Silica and Diamond-Like Lattices (I)
Vol **20**, No. 1, p. 37—43 (1967).
See also *6*, p. *59*; 8, p. 81; 22, p. 443.

V. V. Tarassov
Heat Capacity and Structure of Vitreous Silica and Diamond-Like Lattices (II)
Vol. **20**, No. 1, p. 45—57 (1967).
See also *6*, p. *59*; 8, p. 81; 22, p. 443.

B. Eckstein
*A Disorder Model of Melting and Melts
Vol. **20**, No. 1, p. 83—93 (1967).
See also *10.1*, p. *121*.

S. U. Dzhalilov and K. I. Rzaev
*On the Phenomenon of Selenium Vitrification
Vol. **20**, No. 1, p. 261—266 (1967).
See also 22.1.3, p. 477.

L. L. HENCH and D. A. JENKINS
AC Conductivity of a Glass Semiconductor
Vol. **20**, No. 1, p. 327—330 (1967).
See also *14.3*, p. *235*; 22.6, p. 537; 22.8, p. 549.

G. HOLTMEYER and J. KALUS
Effect of the Short-Range Order around Magnesium Impurities in Bismuth Melts on the Yield of Nuclear Resonance Fluorescence
Vol. **21**, No. 1, p. 297—302 (1967).
See also 6.1, p. 72; *11*, p. *148*; 21.7, p. 438.

N. F. MOTT and R. S. ALLGAIER
Localized States in Disordered Lattices
Vol. **21**, No. 1, p. 343—356 (1967).
See also *13.4*, p. *206*; 21, p. 378; 22, p. 445.

L. MÜLLER and Y. MIHAI
Absorption of Amorphous Se Thin Layers Doped with Bi and Zn
Vol. **21**, No. 1, p. K65—K68 (1967).
See also 13.4, p. 207; *20.1*, p. *348*; 22.1.3, p. 477.

G. GETOV, P. SIMIDTCHIEVA, M. NIKIFOROVA, and R. ANDREYTCHIN
Effect of Some Elements on the Optical Absorption Edge of Vitreous As_2S_3
Vol. **21**, No. 1, p. K87—K89 (1967).
See also 13.4, p. 207; *20.1*, p. *348*; 22, p. 445.

S. LUNGU
Point Defects in Neutron-Damaged Silica Glass
Vol. **23**, No. 1, p. 147—156 (1967).
See also 10, p. 102; *11*, p. *150*; 22.8, p. 552.

R. GRIGOROVICI, N. CROITORU, and A. DÉVÉNYI
Thermoelectric Power in Amorphous Silicon
Vol. **23**, No. 2, p. 621—625 (1967).
See also 13.1, p. 182; 14.3.1, p. 246; *15*,p. *273*; 22.1.2, p. 472.

R. GRIGOROVICI, N. CROITORU, and A. DÉVÉNYI
Photoconductivity in Amorphous Germanium
Vol. **23**, No. 2, p. 627—631 (1967).
See also *16*, p. *280*; 22.1.1, p. 463.

A. G. REVESZ
Noncrystalline Structure and Electronic Conduction of Silicon Dioxide Films
Vol. **24**, No. 1, p. 115—126 (1967).
See also *1.4*, p. *19*; *14.3.1*, p. *246*; 14.3.2, p. 249; 22.6, p. 539.

M. V. COLEMAN and D. J. D. THOMAS
The Structure of Amorphous Silicon Films
Vol. **24**, No. 2, p. K111—K115 (1967).
See also *1.4*, p. *19*; 22.1.2, p. 472.

M. V. COLEMAN and D. J. D. THOMAS
The Structure of Amorphous Silicon Nitride Films
Vol. **25**, No. 1, p. 241—252 (1968).
See also *1.4*, p. *20*; 22, p. 449.

J. VIŠČAKAS, P. MAČKUS, and A. SMILGA
Space-Charge-Limited Currents and High Electric Field Effect in Vitreous Selenium Films
Vol. **25**, No. 1, p. 331—335 (1968).
See also 13.4, p. 211; 14.3.1, p. 246; *14.4.1*, p. *261*; 22.1.3, p. 478.

E. O. Osmanov, Yu. V. Rud, and M. E. Stryalkovskii
High-Temperature Modifications of the Semiconductor Compounds $CdSnAs_2$ and $CdGeAs_2$
Vol. **26**, No. 1, p. 85–90 (1968).
See also *1.2*, p. *12*; 8, p. 84; 12, p. 158; 14.3, p. 240; 22.7, p. 546; 22.8, p. 553.

R. H. Tredgold, R. H. Williams, G. Kavalyauskene, and R. Keezer
Photoemission Studies on Amorphous Layers and Single Crystals of Selenium
Vol. **26**, No. 1, p. K5–K8 (1968).
See also *17*, p. *288*; 22.1.3, p. 479.

Kh. M. Khalilov, B. B. Kuliev, and F. A. Gambarov
*Effect of Phosphorus Admixtures on Viscoelastic Properties of Selenium
Vol. **26**, No. 2, p. K113–K117 (1968).
See also 10, p. 106; *12*, p. *158*; 22.1.3, p. 479.

L. Ask
Effect of Mechanical Stress on the I-U Characteristic of a GaP p–n Junction
Vol. **26**, No 2, p. K143–K145 (1968).
See also 12, p. 158; *14.3.2*, p. *250*; 22.2.2, p. 488.

B. T. Kolomiets, T. N. Mamontova, and V. V. Negreskul
*On the Recombination Radiation of Vitreous Semiconductors
Vol. **27**, No. 1, p. K15–K16 (1968).
See also *20.3*, p. *369*; 22.8, p. 553.

O. Brümmer and G. Dräger
On the Interpretation of the Fine Structure of X-Ray K-Absorption Spectra Far from the Edge
Vol. **27**, No. 2, p. 513–516 (1968).
See also *20*, p. *341*; 22.1.1, p. 465; 22.6, p. 541; 22.8, p. 554.

V. K. Jain and P. K. C. Pillai
Characteristics of Plastic Electrets
Vol. **28**, No. 1, p. 415–419 (1968).
See also *14.4*, p. *254*; 22.9, p. 560.

N. A. Goryunova, S. M. Ryvkin, G. P. Shpenikov, I. I. Tichina, and V. G. Fedotov
Investigations of Some Properties of Vitreous and Crystalline $CdGeP_2$
Vol. **28**, No. 2, p. 489–494 (1968).
See also 14.3, p. 242; *20.3*, p. *370*; 22.8, p. 554.

F. B. Gadzhiev, Ch. M. Askerov, and G. M. Aliev
Effect of Heat Treatment and Natrium Admixtures on the Temperature Dependence of Electric Conductivity of Amorphous Selenium
Vol. **29**, No. 1, p. K47–K50 (1968).
See also *14.3*, p. *243*; 22.1.3, p. 480.

R. A. Burley
An Infrared Spectrophotometric Study of Vitreous Selenium Doped with Selenium Dioxide
Vol. **29**, No. 2, p. 551–558 (1968).
See also *10*, p. *110*; 20.1, p. 357; 22.1.3, p. 480.

I. Dézsi, N. A. Eissa, L. Keszthelyi, B. Molnár, and D. L. Nagy
Mössbauer Study of $SnCl_2$ and $Dy(ClO_4)_3$ in Ice
Vol. **30**, No. 1, p. 215–218 (1968).
See also *6.1*, p. *74*; 22.8, p. 556.

G. J. Morgan
Green Functions for "One-Body" Equations Describing the Disordered State
Vol. **30**, No. 1, p. 247–250 (1968).
See also 6, p. 69; *13*, p. *173*.

H. Imagawa
ESR Studies of Cupric Ion in Various Oxide Glasses
Vol. **30**, No. 2, p. 469–478 (1968).
See also 3.2, p. 36; *19*, p. *336*; 22.8, p. 557.

Nguyen van Dong and Pham Ngu Tung
Transport Properties of Silver Telluride in the Solid and Liquid States
Vol. **30**, No. 2, p. 557–567 (1968).
See also *14.3*, p. *244*; 14.4, p. 254; *15*, p. *275*; 22, p. 455.

3. Crystallography

Review Articles

P. Krishna and Ajit Ram Verma
*Crystal-Polymorphism in One Dimension
Vol. **17**, No. 2, p. 437–477 (1966).
See also *5*, p. *49*; 10, p. 93; 22.3, p. 494; 22.4.1, p. 497; 22.5, p. 512.

Original Papers and Short Notes

W. L. McCubbin
*Symmetry Orbitals and UV Selection Rules for the Polymethylene Chain
Vol. **16**, No. 1, p. 289–293 (1966).
See also *13.1*, p. *174*; 20, p. 337; 22.9, p. 558.

H. M. Otte and A. G. Crocker
Crystallographic Formulae for Hexagonal Lattices — Simplifications
Vol. **16**, No. 1, p. K25–K26 (1966).
See also *5*, p. *49*.

K. H. J. Buschow and J. F. Fast
*Magnetic and Structural Characteristics of Some Equiatomic Rare-Earth Germanides
Vol. **16**, No. 2, p. 467–473 (1966).
See also 5, p. 49; 18.1, p. 296; 18.2, p. 300; 18.3, p. 316; 18.4, p. 321; 22.7, p. 545.

S. W. Kennedy, G. F. Taylor, and J. H. Patterson
Misfit and Hysteresis at Structure Transformations in Rubidium and Cesium Nitrates; and Thermal Expansions
Vol. **16**, No. 2, p. K175–K178 (1966).
See also *1.2*, p. *9*; 8, p. 79, 22.8, p. 547.

A. J. Morton and C. M. Wayman
On the Feasibility of a "Composite" Martensite Model for the $(225)_f$ Austenite–Martensite Transformation
Vol. **17**, No. 1, p. 101–103 (1966).
See also *1.2*, p. *9*; 21.1.1, p. 412.

P. F. Lindquist, M. L. Hammond, and R. H. Bragg
*Comments on the Paper "On the Interpretation of Electron Diffraction Patterns from 'Amorphous' Boron"
Vol. **17**, No. 1, p. K25–K31 (1966).
See also 1.2, p. 9; 4, p. 38; 22.1, p. 456.

P. Neumann
Simplification of Crystallographic Calculations for Hexagonal Lattices by Consequently Using the Quadruaxial Hexagonal Coordinate-System
Vol. **17**, No. 1, p. K71–K74 (1966).
See also *5*, p. *49*.

H. M. Otte, H. A. Lipsitt, P. F. Lindquist, M. L. Hammond, and R. H. Bragg
*Further Comments on the Paper "On the Interpretation of Electron Diffraction Patterns from 'Amorphous' Boron"
Vol. **19**, No. 2, p. K99–K101 (1967).
See also 1.2, p. 10; 4, p. 40; 22.1, p. 457.

M. Bevis and N. Swindells
The Determination of the Orientation of Micro-Crystals Using a Back-Reflection Kossel Technique and an Electron Probe Microanalyser
Vol. **20**, No. 1, p. 197–212 (1967).
See also *4*, p. *41*; 21.1.1, p. 415.

I. I. Pinchuk
Thermomagnetic Effects in Anisotropic Media
Vol. **20**, No. 2, p. 537–543 (1967).
See also *15*, p. *271*; 21.7, p. 438.

P. R. Okamoto and G. Thomas
*On the Four-Axis Hexagonal Reciprocal Lattice and Its Use in the Indexing of Transmission Electron Diffraction Patterns
Vol. **25**, No. 1, p. 81–91 (1968).
See also 5, p. 53.

J. G. Allpress, J. V. Sanders, and A. D. Wadsley
Electron Microscopy of High-Temperature Nb_2O_5 and Related Phases
Vol. **25**, No. 2, p. 541–550 (1968).
See also *5*, p. *53*; 22.6, p. 540.

M. Hart and A. D. Milne
Direct Observations of Plane Wave and Spherical Wave Pendellösung Fringes
Vol. **26**, No. 1, p. 185–189 (1968).
See also *5*, p. *53*; 22.1.2, p. 473.

P. Rowlands and M. Bevis
A Rapid Method for Orientating Crystals from Kossel Patterns
Vol. **26**, No. 1, p. K25–K28 (1968).
See also *4*, p. *45*; 21.1.1, p. 420.

R. F. Egorov, B. I. Reser, and V. P. Shirokovskii
Consistent Treatment of Symmetry in the Tight Binding Approximation
Vol. **26**, No. 2, p. 391–408 (1968).
See also *13.1*, p. *185*.

H. Overhof and U. Rössler
*Magnetic Space Groups and Their Irreducible Representations
Vol. **26**, No. 2, p. 461–468 (1968).
See also 13.1, p. 185; 18, p. 294.

T. P. Srinivasan and S. D. Nigam
Invariant Elastic Constants for Crystals
Vol. **28**, No. 1, p. K71–K73 (1968).
See also *12*, p. *158*; *12.1*, p. *160*.

J.-U. Arnold and P. Grosse
*The Temperature Dependence of the Atom Coordinates in the Tellurium Lattice
Vol. **28**, No. 1, p. K93–K96 (1968).
See also 5, p. 54; 8, p. 85; 22.1.3, p. 479.

G. M. de'Munari, F. Giusiano, and G. Mambriani
A Phase Transition and the Photoemissive Quantum Yield at Low Temperatures in Cs_3Sb Films
Vol. **29**, No. 1, p. 341–356 (1968).
See also *1.2*, p. *12*; 13.1, p. 188; 17, p. 289; 22.7, p. 546.

P. P. Salhotra, E. C. Subbarao, and P. Venkateswarlu
*Polymorphism of Rubidium Nitrate
Vol. **29**, No. 2, p. 859–864 (1968).
See also *1.2*, p. *13*; 8, p. 86; 22.8, p. 556.

S. P. Taneja and S. P. Puri
Crystal Field Calculation of Principal Magnetic Susceptibilities and Anisotropy Variation with Temperature of Anhydrous Neodymium Chloride
Vol. **30**, No. 1, p. 193–197 (1968).
See also *18.1*, p. *298*; 22.5, p. 514.

A. Gemperle
Crystallography of Antiphase Boundaries in Fe–Si Alloys
Vol. **30**, No. 2, p. 541–550 (1968).
See also *4*, p. *48*; 21.1.1, p. 423.

A. T. Davidson and A. D. Yoffe
Hopping Electical Conduction and Thermal Breakdown in Natural and Synthetic Mica
Vol. **30**, No. 2, p. 741–754 (1968).
See also 10, p. 113; *14.4.1*, p. *265*; 20.1, p. 359; 22, p. 455.

J. Przystawa
Magnetic Symmetry of Uranium Compounds of UX_2 Type
Vol. **30**, No. 2, p. K115–K117 (1968).
See also *18.4*, p. *325*; 22, p. 455.

3.1 Crystal Growth

Review Articles

J. R. DALE
Alloyed Semiconductor Heterojunctions
Vol. **16**, No. 2, p. 351–387 (1966).
See also 5, p. 49; 13.4, p. 200; *14.3.2*, p. *248*; 16, p. 276; 22.1.1, p. 458; 22.1.2, p. 468; 22.2.1, p. 482; 22.2.2, p. 487; 22.2.3, p. 489.

A. S. BORSHCHEVSKII, N. A. GORYUNOVA, F. P. KESAMANLY, and D. N. NASLEDOV
Semiconducting $A^{II}B^{IV}C_2^{V}$ Compounds
Vol. **21**, No. 1, p. 9–55 (1967).
See also 3.2, p. 34; 5, p. 49; 8, p. 79; 10, p. 93; 13.1, p. 174; *14.3*, p. *231*; 15, p. 270; 16, p. 276; 17, p. 286; 18, p. 291; 20.1, p. 344; 22.8, p. 547.

Original Papers and Short Notes

M. J. STOWELL and T. J. LAW
*The Migration of Double Positioning Boundaries during the Growth of Epitaxial (111) Gold Films
Vol. **16**, No. 1, p. 117–125 (1966).
See also *1.4*, p. *16*; 21.6, p. 430.

M. J. STOWELL and T. J. LAW
Interrupted Vapour Deposition of Epitaxial Gold Films Grown Inside an Electron Microscope
Vol. **16**, No. 2, p. 479–485 (1966).
See also *1.4*, p. *16*; 21.6, p. 430.

R. T. DELVES
*The Theory of the Stability of the Solid-Liquid Interface under Constitutional Supercooling (I)
Vol. **16**, No. 2, p. 621–632 (1966).

CHR. NANEV
Electron-Microscopic Study of the Growth Morphology of Zinc and Cadmium Single Crystals
Vol. **16**, No. 2, p. 777–787 (1966).
See also *1.3*, p. *14*; 21, p. 374; 21.1, p. 391.

R. T. DELVES
*The Theory of the Stability of the Solid-Liquid Interface under Constitutional Supercooling (II)
Vol. **17**, No. 1, p. 119–130 (1966).

L. V. KIRENSKII, V. G. PYNKO, N. I. SIVKOV, G. P. PYNKO, R. V. SUKHANOVA und M. A. OVSYANNIKOV
*Züchtung von Einkristallschichten aus Eisen, Nickel, Kobalt und deren Legierungen (Growth of Single-Crystal Films of Iron, Nickel, Cobalt, and Their Alloys)
Vol. **17**, No. 1, p. 243–248 (1966).
See also 1.4, p. 16; 21.1, p. 392; 21.1.1, p. 412.

J. PFEIFER
*Epitaxial Growth of Germanium on Germanium Substrates Cleaned and Heated by Electron Bombardment
Vol. **17**, No. 1, p. K15–K18 (1966).
See also 1.3, p. 14; 22.1.1, p. 459.

L. S. PALATNIK, M. YA. FUKS, L. I. LUKASHENKO, A. G. RAVLIK, and A. A. KOZMA
Structure and Magnetic Properties of Condensed Ferromagnetic Films
Vol. **17**, No. 2, p. 543–554 (1966).
See also *1.4*, p. *17*; 10.1, p. 116; *18.2*, p. *303*; 21.1, p. 392; 21.1.1, p. 413.

P. BENNEMA
*Technique for Measuring the Rate of Growth of Crystals from Solutions in Dependence on the Degree of Supersaturation at Low Supersaturations
Vol. **17**, No. 2, p. 555–562 (1966).
See also 22.8, p. 547.

P. BENNEMA
*Crystal Growth Measurements on Potassium Aluminium Alum and Sodium Chlorate from Slightly Supersaturated Solutions
Vol. **17**, No. 2, p. 563–570 (1966).
See also 22.8, p. 548.

D. NENOW
*Über die phänomenologische Theorie der Gleichgewichtsformen von Kristallen (Phenomenological Theory of the Equilibrium Shapes of Crystals)
Vol. **17**, No. 2, p. 575–579 (1966).

H. JAHRREISS und H.-J. ISKEN
*Zum Epitaxieverhalten dünner Silberaufdampfschichten (Epitaxial Behaviour of Thin Evaporated Silver Layers)
Vol. **17**, No. 2, p. 619–629 (1966).
See also *1.4*, p. *17*; 21.6, p. 431.

A. G. LESNIK, G. I. LEVIN, and V. M. NEDOSTUP
The Importance of Vacancies and Interstitial Atoms in the Development of Induced Anisotropy
Vol. **17**, No. 2, p. 745–751 (1966).
See also 10.1, p. 116; *18.2*, p. *303*; 21.1.1, p. 413.

H. P. D. LANYON and E. F. HOCKINGS
*The Selenium-Tellurium System
Vol. **17**, No. 2, p. K185–K186 (1966).
See also 22.1.3, p. 476.

G. B. GIBBS
Vacancy Condensation on Precipitate Particles
Vol. **17**, No. 2, p. K191–K194 (1966).
See also *10*, p. *96*.

K. M. KUNZ, A. K. GREEN, and E. BAUER
*On the Formation of Single Crystal Films of F.C.C. Metals on Alkali Halide Cleavage Planes in Ultrahigh Vacuum
Vol. **18**, No. 1, p. 441–457 (1966).
See also *1.4*, p. *17*; 21, p. 376; 21.6, p. 431.

I. SPÎNULESCU-CARNARU
The Crystalline Structure of ZnTe Thin Films
Vol. **18**, No. 2, p. 769–778 (1966).
See also *1.4*, p. *17*; 5, p. 50; 22.4.3, p. 509.

P. Bennema, R. Kern, and B. Simon
*Interpretation of Deviating Points in the Relation between the Rate of Crystal Growth and the Relative Supersaturation
Vol. **19**, No. 1, p. 211–216 (1967).
See also 22.8, p. 549.

N. A. Pangarov
*Calculation of the Specific Surface Energies of Crystal Planes by Investigating the Orientation of Crystal Nuclei on Inert Substrates
Vol. **20**, No. 1, p. 365–369 (1967).
See also 3.2, p. 35; 21.6, p. 432.

N. A. Pangarov
*Twinning Processes in the Electrocrystallization of Face-Centred Cubic Metals
Vol. **20**, No. 1, p. 371–377 (1967).
See also 4, p. 41; 21, p. 377.

R. T. Delves
*The Theory of Stability during Temperature Gradient Zone Melting
Vol. **20**, No. 2, p. 693–704 (1967).

W. P. R. George, C. H. L. Goodman, H. F. Sterling, and R. W. Warren
A Possible New Group of Semiconducting Compounds
Vol. **21**, No. 1, p. 205–210 (1967).
See also 3.2, p. 35; *14.3*, p. *236*; 21.7, p. 438; 22.7, p. 545.

G. B. Abdullaev, Z. A. Aliyarova, and G. A. Asadov
Preparation of Cu_2Se Single Crystals and Investigation of Their Electrical Properties
Vol. **21**, No. 2, p. 461–464 (1967).
See also *14.3*, p. *236*; 15, p. 272; 22, p. 445.

W. A. Jesser and D. Kuhlmann-Wilsdorf
*The Geometry and Energy of a Twist Boundary between Crystals with Unequal Lattice Parameters
Vol. **21**, No. 2, p. 533–544 (1967).
See also 10, p. 100; 10.1, p. 123; 21, p. 379; 22, p. 445.

S. Toschev and I. Gutzow
*Time Lag in Heterogenous Nucleation due to Nonstationary Effects
Vol. **21**, No. 2, p. 683–691 (1967).

J. E. Knappett and S. J. T. Owen
*Observation of the Growth Process of Gallium Arsenide on Tungsten
Vol. **21**, No. 2, p. K99–K101 (1967).
See also 1.4, p. 18; 14.3.1, p. 245; 22.2.1, p. 484.

N. Paschoff
*Wachstum und Struktur dünner, aus schmelzflüssigen Filmen auskristallisierter Germanium-Kristalle
(Growth an Structure of Thin Germanium Crystals Crystallized from Molten Films)
Vol. **22**, No. 1, p. 83–91 (1967).
See also *1.4*, p. *18*; 22.1.1, p. 462.

M. Paunov
*Feldelektronenmikroskopische Untersuchungen der Quecksilberkeimbildung
(Field Electronmicroscopic Study of the Nucleation of Mercury)
Vol. **22**, No. 1, p. 103–112 (1967).
See also 21, p. 380.

A. GEORGIEVA and D. NENOW
Spontaneous Solid–Vapour Transformation of Negative Crystals in Diphenyl
Vol. 22, No. 2, p. 415–421 (1967).
See also *4*, p. *42*; 22.9, p. 559.

P. W. GRAVES
*Single Crystal Growth of Calcium Pyroniobate
Vol. 22, No. 2, p. 499–504 (1967).
See also 22.6, p. 539; 22.8, p. 551.

K. OTSUKA and C. M. WAYMAN
*Epitaxial Growth of Vacuum-Evaporated Co on NaCl. (I): Conditions Leading to Epitaxy
Vol. 22, No. 2, p. 559–578 (1967).
See also 1.4, p. 19; 21.1, p. 399.

K. OTSUKA and C. M. WAYMAN
*Epitaxial Growth of Vaccuum-Evaporated Co on NaCl. (II): Analysis of Diffraction Patterns
Vol. 22, No. 2, p. 579–592 (1967).
See also 1.4, p. 19; 21.1, p. 399.

B. PERSSON and H. NORDEN
*Epitaxial Growth of Gold on Cadmium Iodide
Vol. 22, No. 2, p. 631–647 (1967).
See also 1.4, p. 19; 21.6, p. 434.

H. OPPERMANN, G. WEISE, J. BARTHEL und K.-H. BERTHEL
*Molybdäneinkristalle mit sehr kleinen Restwiderstandsverhältnissen (Molybdenum Single Crystals with Very Small Residual-Resistivity Ratios)
Vol. 22, No. 2, p. K151–K153 (1967).
See also 14.1, p. 223; 21, p. 381.

R. LUDEKE and W. PAUL
*Growth and Optical Properties of Wurtzite and Sphalerite CdSe Epitaxial Thin Films
Vol. 23, No. 1, p. 413–418 (1967).
See also 1.4, p. 19; 13.1, p. 181; *20.1*, p. *350*; 22.4.2, p. 507.

R. D. GRETZ
*Nucleation of Zinc Crystals in Multilayer Adsorption
Vol. 23, No. 2, p. 453–460 (1967).
See also 21.1, p. 400.

CHR. NANEV and D. IWANOV
*Growth of Zinc and Cadmium Whiskers from the Vapour Phase on a Single Crystal Substrate of the Same Material
Vol. 23, No. 2, p. 663–673 (1967).
See also 21, p. 382; 21.1, p. 400.

S. TOSCHEV and I. GUTZOW
*Derivation of Nucleation Kinetics in Solids by Examination of Plane Sections through the Samples
Vol. 24, No. 1, p. 349–359 (1967).
See also 1.2, p. 11; 22, p. 448.

R. BAUBINAS, A. SAKALAS, A. SMILGA, and J. VIŠČAKAS
p-Type CdSe Single Crystals
Vol. **24**, No. 1, p. K91–K93 (1967).
See also *14.3*, p. *239*; *16*, p. *280*; 22.4.2, p. 507.

M. J. STOWELL and T. J. LAW
*The Growth and Defect Structure of Gold Films Formed on Molybdenite in Ultra-High Vacuum
Vol. **25**, No. 1, p. 139–146 (1968).
See also *1.4*, p. *19*; 10.1, p. 127; 21.6, p. 435.

J. H. EVANS and B. L. EYRE
*The Effect of Tungsten Content on the Residual Resistivity of Zone Refined Molybdenum
Vol. **25**, No. 1, p. K39–K41 (1968).
See also *14.1*, p. *224*; 21, p. 384.

N. A. GORYUNOVA, V. M. ORLOV, V. I. SOKOLOVA, G. P. SHPENKOV, and E. V. TSVETKOVA
*Preparation and Some Properties of CuP_2 Single Crystals
Vol. **25**, No. 2, p. 513–519 (1968).
See also 5, p. 53; 14.3, p. 240; p. 15, p. 273; 20.1, p. 351; 22, p. 450.

I. SPÎNULESCU-CARNARU and I. DRĂGHICI
On Electrical Properties of PbTe Thin Films
Vol. **26**, No. 1, p. K39–K42 (1968).
See also *14.3.1*, p. *246*; 22, p. 450.

R. KRISHNAN
*Preparation and Resonance Properties of CuO Doped YIG Single Crystals
Vol. **26**, No. 1, p. K47–K50 (1968).
See also 18.3, p. 319; *19*, p. *333*.

J. T. CALOW, S. J. T. OWEN, and P. W. WEBB
*The Growth and Electrical Characteristics of Epitaxial Layers of Zinc Selenide on p-Type Germanium
Vol. **28**, No. 1, p. 295–303 (1968).
See also *14.3.1*, p. *247*; 14.3.2, p. 251; 22.4.2, p. 508.

P. REIMERS and W. RUPPEL
*The Preparation of CdS–CdSe Graded Single Crystals
Vol. **29**, No. 1, p. K31–K33 (1968).
See also 5, p. 54; 20.1, p. 357; 22.4.1, p. 504; 22.4.2, p. 508; 22.8, p. 555.

P. FLÖGEL, E. NEBAUER, H.-J. ULLRICH, S. DÄBRITZ, and F. ZIMMER
Electron Microprobe and Scanning Electron Microscope Investigations on Plate-Shaped CdS Single Crystals
Vol. **29**, No. 2, p. K89–K90 (1968).
See also *4*, p. *47*; 22.4.1, p. 505.

S. E. BRONISZ and D. L. DOUGLASS
Grain Growth in Thin Films of ThO_2, NpO_2, and PuO_2
Vol. **29**, No. 2, p. K95–K97 (1968).
See also *1.4*, p. *21*; 22.6, p. 543.

L. E. TANNER
The Ordering of Ni_3V
Vol. **30**, No. 2, p. 685–701 (1968).
See also *1.2*, p. *13*; 4, p. 48; 21, p. 388.

3.2 Interatomic Forces

Review Articles

A. S. Borshchevskii, N. A. Goryunova, F. P. Kesamanly, and D. N. Nasledov
Semiconducting $A^{II}B^{IV}C_2^V$ Compounds
Vol. **21**, No. 1, p. 9—55 (1967).
See also 3.1, p. 29; 5, p. 49; 8, p. 79; 10, p. 93; 13.1, p. 174; *14.3*, p. *231*; 15, p. 270; 16, p. 276; 17, p. 286; 18, p. 291; 20.1, p. 344; 22.8, p. 547.

Original Papers and Short Notes

Z. Šroubek
*On the Theory of the Superexchange Interaction in Ionic Crystals: $KNiF_3$
Vol. **16**, No. 2, p. 405—412 (1966).
See also 13, p. 163; 18.4, p. 321; 22.5, p. 512; 22.8, p. 547.

V. K. Nikulin
*Cohesive Energy of Noble Metals
Vol. **16**, No. 2, p. K125—K127 (1966).
See also 5, p. 49; 21.1, p. 391; 21.6, p. 430.

F. Kubec
Crystal Field Calculation in the $CdWO_4$ Lattice
Vol. **17**, No. 2, p. 571—574 (1966).
See also *6*, p. *57*; 13.4, p. 202; 22.8, p. 548.

A. I. Gubanov and V. K. Nikulin
*The Theory of the Cohesion of Noble Metals
Vol. **17**, No. 2, p. 815—823 (1966).
See also 5, p. 50; 12.1, p. 159; 21.1, p. 393; 21.6, p. 431.

L. Valenta and A. Sukiennicki
*Order–Disorder Phenomena at the Surface of Solids
Vol. **17**, No. 2, p. 903—909 (1966).
See also *1.3*, p. *14*; 21.1.1, p. 413.

R. Wehner
Second-Order Dipole Moment of the F.C.C. and Diamond Lattices
Vol. **17**, No. 2, p. K179—K183 (1966).
See also *6*, p. *57*.

R. C. Lincoln, K. M. Koliwad, and P. B. Ghate
Elastic Constants of Some NaCl Type Alkali Halides
Vol. **18**, No. 1, p. 265—277 (1966).
See also *12*, p. *156*; 22.5.2, p. 519.

N. P. Gupta and B. Dayal
Effect of Zero-Point Energy on the Specific Heats of Solidified Neon and Xenon
Vol. **18**, No. 2, p. 731—736 (1966).
See also *6*, p. *58*; *8*, p. *80*; 23, p. 561.

R. I. Anishchenko
*Calculation of the Mean Inner Potential of a Crystal in the Statistical Theory
Vol. **18**, No. 2, p. 923—928 (1966).
See also 13, p. 164; 21, p. 376; 21.1, p. 394; 21.1.1, p. 414; 21.2, p. 424; 21.3, p. 426; 21.4, p. 428; 21.6, p. 432; 21.7, p. 438; 22.1.1, p. 460.

J. D. Keys and H. P. Dibbs
*On the Bonding in Bismuth Telluride
Vol. **19**, No. 1, p. K11–K13 (1967).
See also 14.3, p. 235; 22, p. 442.

N. A. Pangarov
Calculation of the Specific Surface Energies of Crystal Planes by Investigating the Orientation of Crystal Nuclei on Inert Substrates
Vol. **20**, No. 1, p. 365–369 (1967).
See also *3.1*, p. *31*; 21.6, p. 432.

E. Jäger and R. Perthel
Some Remarks to the Crystal Field Theory of Spinel Lattices
Vol. **20**, No. 2, p. 433–441 (1967).
See also *13*, p. *165*; 18.3, p. 317.

J. Przystawa and W. Suski
Magnetic Ordering in Uranium Compounds with Crystal Structures of Anti-Cu_2Sb and PbFCl Types. Molecular Field Approximation
Vol. **20**, No. 2, p. 451–459 (1967).
See also *18.2*, p. *306*; *18.4*, p. *322*; 22, p. 444.

E. A. Metzbower
Noncentral Force Model for Hexagonal Close-Packed Crystal Lattices
Vol. **20**, No. 2, p. 681–692 (1967).
See also *6*, p. *60*; 8, p. 81; 12.1, p. 160; 21.1, p. 397; 21.3, p. 426.

W. P. R. George, C. H. L. Goodman, H. F. Sterling, and R. W. Warren
A Possible New Group of Semiconducting Compounds
Vol. **21**, No. 1, p. 205–210 (1967).
See also 3.1, p. 31; *14.3*, p. *236*; 21.7, p. 438; 22.7, p. 545.

A. Weiss and K. Zohner
Nuclear Quadrupole Interaction of ^{23}Na and ^{11}B and the Crystal Structure of $NaBF_4$
Vol. **21**, No. 1, p. 257–270 (1967).
See also 5, p. 51; *19*, p. *329*; 22.5, p. 513; 22.8, p. 550.

B. P. Kotrubenko, V. N. Lange, and M. M. Markus
*Chemical Bond and Structure of the Ternary Compound InAsTe
Vol. **22**, No. 1, p. K15–K17 (1967).
See also *5*, p. *52*; 22.8, p. 551.

S. V. Vonsovskii, V. I. Cherepanov, A. N. Men, and A. E. Nikiforov
Group-Theoretical Classification of Electronic States in Crystals Composed of Pairs and Complexes of Bound Impurity Ions
Vol. **24**, No. 1, p. 51–63 (1967).
See also *13.4*, p. *210*; 22.6, p. 539.

B. I. Kuliev and G. M. Nedlin
The Types of Magnetic Ordering in Crystals Having the Structure of $RbNiF_3$
Vol. **24**, No. 1, p. K25–K28 (1967).
See also 1.2, p. 11, *18*, p. *293*; 22.5, p. 513; 22.8, p. 552.

J. B. Page, Jr. and D. Strauch
Shell Model Treatment of the Vibrations of Crystals Containing Substitutional Defects. (I): U-Center Localized Modes
Vol. **24**, No. 2, p. 469–479 (1967).
See also *6*, p. *64*; 10.2, p. 141; 22.5.2, p. 524.

M. F. Rose and R. T. Ramsey
Higher Order Elastic Constants in H.C.P. Crystals
Vol. **25**, No. 1, p. 103–108 (1968).
See also *12.1*, p. *160*; 21, p. 383; 21.3, p. 426.

E. A. Metzbower
*Noncentral Force Model for Hexagonal Close-Packed Crystal Lattices (II)
Vol. **25**, No. 1, p. 403–416 (1968).
See also 6, p. 64; 8, p. 84; 12.1, p. 160; 21.1, p. 402.

M. Jaroš
*On the Covalency in Transition-Metal Impurity Covalent Complexes
Vol. **27**, No. 1, p. K21–K24 (1968).
See also 19, p. 333; 22.4.3, p. 510.

S. Haussühl
Elastic and Thermoelastic Behaviour of MgF_2 and MnF_2
Vol. **28**, No. 1, p. 127–130 (1968).
See also *12*, p. *158*; 22.5, p. 514; 22.5.3, p. 534.

K. J. Rao and C. N. R. Rao
Schottky Defects in Alkali Halides: Their Creation, Interaction, and Migration
Vol. **28**, No. 1, p. 157–169 (1968).
See also *10*, p. *108*; 22.5.2, p. 527.

V. K. Nikulin and M. B. Trzhaskovskaya
*The Theory of Cohesion of Noble Metals (II)
Vol. **28**, No. 2, p. 801–806 (1968).
See also 13.1, p. 187; 21.1, p. 406; 21.6, p. 437.

A. R. Ruffa
The Valence Bond Approximation in Crystals — Application to an Analysis of the Ultraviolet Spectrum in Quartz
Vol. **29**, No. 2, p. 605–616 (1968).
See also *13.1*, p. *189*; 20.1, p. 357; 22.6, p. 543.

K. H. Timmesfeld
Anharmonic Broadening of Resonant Modes
Vol. **30**, No. 1, p. 73–85 (1968).
See also *6*, p. *69*; *20.1*, p. *358*; 22.5.2, p. 530.

A. Anderman and W. G. Gehman
Xenon–Vacancy Defect Clusters in Copper
Vol. **30**, No. 1, p. 283–289 (1968).
See also *10.1*, p. *135*; 21.1, p. 408.

M. A. Kumakhov
Formation of Radiation Defects by Channeling Particles
Vol. **30**, No. 1, p. 379–385 (1968).
See also *11*, p. *155*; 21.1, p. 408; 22.1.2, p. 474.

H. Imagawa
ESR Studies of Cupric Ion in Various Oxide Glasses
Vol. **30**, No. 2, p. 469–478 (1968).
See also 2, p. 25; *19*, p. *336*; 22.8, p. 557.

P. G. Baranov, R. A. Zhitnikov, and N. I. Melnikov
Theoretical Study of Properties of Copper and Silver Atoms Stabilized in Alkali Halide Crystals
Vol. **30**, No. 2, p. 859–866 (1968).
See also 10.2, p. 144; 13.4, p. 218; *19*, p. *336*; 22.5.2, p. 531.

4. Microstructure of Solids

Review Articles

A. Aharoni
Magnetization Curling
Vol. **16**, No. 1, p. 3—42 (1966).
See also *18.2*, p. *299*; 21.1, p. 390; 21.1.1, p. 410.

L. E. Murr
Calibration and Use of an Electron Microscope for Precision Micromeasurements in Thin Film Materials
Vol. **19**, No. 1, p. 7—34 (1967).
See also *1.4*, p. *16*; 10, p. 93; 21.1, p. 390; 21.1.1, p. 410.

Original Papers and Short Notes

L. I. van Torne
Pumping Induced Imperfections in Glass:Nd^{3+} Lasers
Vol. **16**, No. 1, p. 171—182 (1966).
See also 2, p. 22; *20 2*, p. *360*; 22.6, p. 535.

R. Straubel and D. Robaschik
Calculation of the Interaction in an Aggregate of Aligned Magnetic Single-Domain Particles
Vol. **16**, No. 1, p. 237—246 (1966).
See also *18.2*, p. *299*; 18.3, p. 316.

H. Lachowicz
Ramp Field Switching in Thin Magnetic Films
Vol. **16**, No. 1, p. 313—320 (1966).
See also *18.2*, p. *299*; 21.1.1, p. 410.

D. J. Craik and M. J. Wood
Thermally Activated Domain Wall Movement
Vol. **16**, No. 1, p. 321—328 (1966).
See also *18.2*, p. *300*; 21.1.1, p. 410.

H. Spreen and Ch. Schwink
Complete Solution of the Magnetoelastic Problem for Domain Structures with Cyclic Magnetization
Vol. **16**, No. 1, p. K45—K49 (1966).
See also *18.2*, p. *300*; 21.1, p. 390.

M. Attardo and J. M. Galligan
A Field Ion Microscope Study of Neutron Irradiated Tungsten
Vol. **16**, No. 2, p. 449—457 (1966).
See also *11*, p. *145*; 17, p. 286; 21, p. 374.

D. J. Barber
Electron Microscopy of Irradiation-Induced Defect Clusters in Magnesium Fluoride
Vol. **16**, No. 2, p. 531—547 (1966).
See also *10*, p. *94*; *11*, p. *145*; 22.5.3, p. 532.

F. Biragnet, J. Devenyi, G. Clerc, O. Massenet, R. Montmory, and A. Yelon
Interactions between Domain Walls in Coupled Films
Vol. **16**, No. 2, p. 569—576 (1966).
See also *18.2*, p. *300*; 21.1.1, p. 411.

J. van Landuyt
*Determination of the Displacement Vector at the Anti-Phase Boundaries in Rutile by Contrast Experiments in the Electron Microscope
Vol. **16**, No. 2, p. 585–590 (1966).
See also 10, p. 94; 22.6, p. 535.

J. A. Cundall and A. P. King
Comparison of Magnetic Measurements and Theoretical Predictions for Nickel–Iron Films over a Wide Composition Range
Vol. **16**, No. 2, p. 613–619 (1966).
See also *18.2*, p. *300*, 21.1.1, p. 411.

V. A. Buravikhin and V. I. Popov
Structure and Polarity of Domain Boundaries in Ferromagnetic Films
Vol. **16**, No. 2, p. 657–662 (1966).
See also 12.1, p. 159; *18.2*, p. *301*; 21.1.1, p. 411.

V. Gerold and H. Haberkorn
On the Critical Resolved Shear Stress of Solid Solutions Containing Coherent Precipitates
Vol. **16**, No. 2, p. 675–684 (1966).
See also *10.1*, p. *114*; 21.1, p. 391.

A. A. Glazer, A. P. Potapov, R. I. Tagirov, and Ya. S. Shur
On Particularities of Magnetic Properties of Thin Films with Exchange Anisotropy
Vol. **16**, No. 2, p. 745–750 (1966).
See also *18.2*, p. *301*; 21.1.1, p. 411.

L. V. Kirenskii, V. G. Pynko, G. P. Pynko, R. V. Sukhanova, N. I. Sivkov, G. I. Rusov, and I. S. Edelman
Magnetic Properties of Single-Crystal Films of Iron, Nickel, Cobalt, and Their Alloys
Vol. **17**, No. 1, p. 249–257 (1966).
See also *18.2*, p. *302*; 21.1, p. 392; 21.1.1, p. 412.

P. F. Lindquist, M. L. Hammond, and R. H. Bragg
Comments on the Paper "On the Interpretation of Electron Diffraction Patterns from 'Amorphous' Boron"
Vol. **17**, No. 1, p. K25–K31 (1966).
See also 1.2, p. 9; *3*, p. *26*; 22.1, p. 456.

G. J. van Gurp
Superconducting Fluxline Pinning by Twin Boundaries
Vol. **17**, No. 1, p. K135–K137 (1966).
See also *14.2*, p. *228*.

K. Godwod, R. Kowalczyk, and M. Lefeld-Sosnowska
*The Influence of the Primary X-Ray Beam Divergence on the X-Ray Wave Field Absorption Coefficients Measured by Means of the Three-Crystal X-Ray Spektrometer
Vol. **17**, No. 2, p. 869–874 (1966).
See also 20, p. 338; 22.1.2, p. 469.

H. C. Bolton, L. A. Bursill, A. C. McLaren, and R. G. Turner
On the Origin of the Colour of Labradorite
Vol. **18**, No. 1, p. 221–230 (1966).
See also *20.1*, p. *346*; 22.8, p. 548.

R. GEVERS, J. VAN LANDUYT, and S. AMELINCKX
*On a Simple Derivation of the Amplitudes of the Electron Beams Transmitted and Scattered by a Crystal Containing Planar Interfaces — Images of Subgrain Boundaries
Vol. 18, No. 1, p. 325—342 (1966).
See also 22.6, p. 536.

R. GEVERS, J. VAN LANDUYT, and S. AMELINCKX
The Fine Structure of Spots in Electron Diffraction Resulting from the Presence of Planar Interfaces and Dislocations. (I): General Theory and Its Application to Stacking Faults and Anti-Phase Boundaries
Vol. 18, No. 1, p. 343—361 (1966).
See also *10*, p. *96*; *10.1*, p. *118*.

J. VAN LANDUYT, R. GEVERS, and S. AMELINCKX
The Fine Structure of Spots in Electron Diffraction Resulting from the Presence of Planar Interfaces and Dislocations. (II): Observations on Crystals Containing Stacking Faults.
Vol. 18, No. 1, p. 363—378 (1966).
See also *10*, p. *96*; *10.1*, p. *118*; 21.1.1, p. 414; 22.6, p. 536.

K. C. A. BLASDALE, R. KING, and K. E. PUTTICK
The Microstructure of Tensile Kinks in Cadmium Crystals
Vol. 18, No. 2, p. 491—503 (1966).
See also *10.1*, p. *118*; 21, p. 376.

D. H. HOWLING
*Identification and Morphology of Oxide Particles in Internally Oxidized Cu–Mn Alloys
Vol. 18, No. 2, p. 579—591 (1966).
See also *1.1*, p. *8*; 21.1, p. 394.

J. H. C. HOGG and W. J. DUFFIN
*Twinning in InS
Vol. 18, No. 2, p. 755—760 (1966).
See also 5, p. 50; 22, p. 441.

H. TRÄUBLE and U. ESSMANN
High-Resolution Technique for Studying Magnetic Structures of Superconductors
Vol. 18, No. 2, p. 813—828 (1966).
See also *14.2*, p. *228*.

V. I. STARTSEV, V. P. SOLDATOV, and M. M. BRODSKY
*The Rate of Twin Layer Growth in Bismuth Single Crystals
Vol. 18, No. 2, p. 863—871 (1966).
See also 21.7, p. 438.

H. REHME
*Elektronenmikroskopischer Nachweis von Sperrschichten in Bariumtitanat-Kaltleiterkeramik
(Electron-Microscopic Detection of Barrier Layers in Barium Titanate Ceramics)
Vol. 18, No. 2, p. K101—K102 (1966).
See also 14.4.2, p. 266.

R. H. WADE and J. SILCOX
Small Angle Electron Scattering from Vacuum Condensed Metallic Films. (I): Theory
Vol. 19, No. 1, p. 57—62 (1967).
See also *1.4*, p. *17*; 21, p. 376.

R. H. WADE and J. SILCOX
Small Angle Electron Scattering from Vacuum Condensed Metallic Films. (II): Experimental Results
Vol. 19, No. 1, p. 63–76 (1967).
See also *1.4*, p. *18*; 21.1, p. 395; 21.6, p. 432.

CH. SCHWINK and O. GRÜTER
Studies of the Domain Structure of Nickel Single Crystals. (II): Measurements and Observations on Prismatic Crystals with an ⟨110⟩ Axis
Vol. 19, No. 1, p. 217–229 (1967).
See also *18.2*, p. *304*; 21.1, p. 395.

A. GEMPERLE and J. KOČÍK
Transmission Electron Microscopic Observation of Lattice Deformation in Antiphase Boundaries of Ordered Fe–Si Alloys
Vol. 19, No. 1, p. 333–338 (1967).
See also *5*, p. *50*; 21.1.1, p. 415.

D. D. MISHIN and R. M. GRETCHISHKIN
Electron Microscopic Examination of High-Coercive Cobalt–Platinum Alloys
Vol. 19, No. 1, p. K1–K3 (1967).
See also *18.2*, p. *305*; 21.1, p. 395; 21.6, p. 432.

A. C. McLAREN, J. A. RETCHFORD, D. T. GRIGGS, and J. M. CHRISTIE
Transmission Electron Microscope Study of Brazil Twins and Dislocations Experimentally Produced in Natural Quartz
Vol. 19, No. 2, p. 631–644 (1967).
See also *10*, p. *98*; 22.6, p. 536.

J. SPYRIDELIS, P. DELAVIGNETTE, and S. AMELINCKX
On the Superstructures of Ta_2O_5 and Nb_2O_5
Vol. 19, No. 2, p. 683–704 (1967).
See also *1*, p. *7*; 10, p. 98; 22.6, p. 537.

R. H. WADE
Electron Diffraction from a Magnetic Phase Grating
Vol. 19, No. 2, p. 847–854 (1967).
See also *18.2*, p. *305*; 21.1, p. 396; 21.1.1, p. 415.

E. IGRAS and T. WARMIŃSKI
Electron Mirror Observation of Lithium Precipitation at Scratch-Induced Dislocations in Silicon
Vol. 19. No. 2, p. K67–K69 (1967).
See also *1.3*, p. *14*; 10, p. 98; 22.1.2, p. 470.

K. WETZIG
Investigations of Surfaces of the Binary System Ta–Fe by Means of an Emission Electron Microscope
Vol. 19, No. 2, p. K71–K73 (1967).
See also *1.3*, p. *14*; 21, p. 377; 21.1.1, p. 415.

H. M. OTTE, H. A. LIPSITT, P. F. LINDQUIST, M. L. HAMMOND, and R. H. BRAGG
Further Comments on the Paper "On the Interpretation of Electron Diffraction Patterns from 'Amorphous' Boron"
Vol. 19, No. 2, p. K99–K101 (1967).
See also 1.2, p. 10; *3*, p. *26*; 22.1, p. 457.

S. Lele, T. R. Anantharaman, and Ch. A. Johnson
*X-Ray Diffraction by Hexagonal Close-Packed Crystals Containing Extrinsic Stacking Faults
Vol. **20**, No. 1, p. 59–68 (1967).

G. Kunze
*Refraktionsmodifizierte Braggsche Beugung bei asymmetrischer Rückstrahl-Reflexion
(Refraction-Modified Bragg Diffraction in the Case of Asymmetric Back-Reflection)
Vol. **20**, No. 1, p. 179–195 (1967).
See also **1**, p. 7.

M. Bevis and N. Swindells
*The Determination of the Orientation of Micro-Crystals Using a Back-Reflection Kossel Technique and an Electron Probe Microanalyser
Vol. **20**, No. 1, p. 197–212 (1967).
See also **3**, p. 27; 21.1.1, p. 415.

N. A. Pangarov
Twinning Processes in the Electrocrystallization of Face-Centred Cubic Metals
Vol. **20**, No. 1, p. 371–377 (1967).
See also *3.1*, p. *31*; 21, p. 377.

E. P. Lautenschlager, T. C. Tisone, and J. O. Brittain
*Electron Transmission Microscopy of NiAl
Vol. **20**, No. 2, p. 443–450 (1967).
See also 10.1, p. 122; 21.1, p. 396.

P. M. Robinson and H. G. Scott
Tilt Boundaries in Deformed Anthracene Single Crystals
Vol. **20**, No. 2, p. 461–471 (1967).
See also *10*, p. *99*; 22.9, p. 559.

G. Remaut, R. Gevers, and S. Amelinckx
*Wavy Fringes at Domain Boundaries in Barium Titanate Observed in the Electron Microscope
Vol. **20**, No. 2, p. 613–621 (1967).
See also 14.4.2, p. 266; 22.8, p. 549.

B. Roessler
Observations of Ferromagnetic Domains by Anomalous Transmission of X-Rays. The Effects of Variation in the Reflecting Plane
Vol. **20**, No. 2, p. 713–723 (1967).
See also *18.2*, p. *306*; 21.1.1, p. 416.

D. Vaughan and J. M. Silcock
*The Orientation and Shape of θ Precipitates Formed in an Al–Cu Alloy
Vol. **20**, No. 2, p. 725–736 (1967).
See also 21, p. 378.

P. Asadi
*X-Ray Topography in Coloured and Uncoloured Zones of Antozonite–Fluorite
Vol. **20**, No. 2, p. K71–K72 (1967).
See also 10.2, p. 139; 22.8, p. 550.

P. Asadi
*On the Existence of Piezoelectric Textures in Single Crystals of Antozonite–Fluorite
Vol. **20**, No. 2, p. K73–K75 (1967).
See also 14.4, p. 252; 22.8, p. 550.

S. Libovický
*Antiphase Boundaries in Silicon Iron Single Crystals Revealed by Etching
Vol. **20**, No. 2, p. K85–K87 (1967).
See also 21.1.1, p. 416.

B. A. Greenberg
Effect of Segregation at Dislocations on the Shape of the NMR Line
Vol. **20**, No. 2, p. K103–K106 (1967).
See also 10, p. 99; *19*, p. *328*.

J. Kobayashi
Optical Observation of Domain Motion in Ferroelectric $BaTiO_3$
Vol. **21**, No. 1, p. 151–158 (1967).
See also *14.4.2*, p. *267*.

M. Rosenberg, C. Tănăsoiu, and V. Florescu
The Influence of the Magnetic Field on the Domain Structure of Ferrimagnetic Oxides with High Anisotropy Field
Vol. **21**, No. 1, p. 197–204 (1967).
See also *18.3*, p. *317*; 22.6, p. 537.

Y. T. Chou and D. M. Barnett
Stacked Screw Dislocation Arrays in an Anisotropic Two-Phase Medium
Vol. **21**, No. 1, p. 239–246 (1967).
See also *10*, p. *99*.

J. A. M. Salter and P. Charsley
The Effect of Grain Size on the Lattice Thermal Conductivity of Copper Aluminium Alloys
Vol. **21**, No. 1, p. 357–368 (1967).
See also 6, p. 61; *8*, p. *82*; 21, p. 379; 21.1, p. 397.

I. B. Puchalska and A. Sukiennicki
On the Temperature Dependence of the Domain Structure
Vol. **21**, No. 1, p. K57–K58 (1967).
See also *18.2*, p. *307*; 21.1.1, p. 417.

V. I. Syutkina and E. S. Yakovleva
The Mechanism of Deformation of the Ordered CuAu Alloy
Vol. **21**, No. 2, p. 465–480 (1967).
See also 1.1, p. 8; *10.1*, p. *123*; 21.1, p. 398; 21.6, p. 433.

L. Fiermans and J. Vennik
*Microprobe Investigations of Copper Precipitates in Silicon Single Crystals
Vol. **21**, No. 2, p. 627–634 (1967).
See also 22.1.2, p. 471.

M. Polcarová and J. Kaczér
X-Ray Diffraction Contrast on Ferromagnetic Domain Walls in Fe–Si Single Crystals
Vol. **21**, No. 2, p. 635–642 (1967).
See also *18.2*, p. *307*; 21.1.1, p. 417.

A. Georgieva and D. Nenow
*Spontaneous Solid–Vapour Transformation of Negative Crystals in Diphenyl
Vol. **22**, No. 2, p. 415–421 (1967).
See also 3.1, p. 32; 22.9, p. 559.

L. Fiermans and J. Vennik
*The Influence of Carbon on Precipitation of Copper in Silicon Single Crystals
Vol. **22**, No. 2, p. 463–471 (1967).
See also 22.1.2, p. 471.

K.-P. Chik, M. Wilkens und M. Rühle
*Die Interpretation des elektronenmikroskopischen Beugungskontrastes von Einschlüssen nahe einer kräftefreien Folienoberfläche
(Interpretation of the Electron-Microscopic Diffraction Contrast of Inclusions near a Traction-Free Foil Surface)
Vol. **23**, No. 1, p. 113–126 (1967).

J. Arends and J. F. Verwey
ESR on UV Irradiated Lead Halides at 80 °K
Vol. **23**, No. 1, p. 137–145 (1967).
See also 10.2, p. 140; *19*, p. *330*; 22.5, p. 513.

I. Spînulescu-Carnaru
The Orientation of Crystallites in ZnTe Thin Films
Vol. **23**, No 1, p. 157–163 (1967).
See also *1.4*, p. *19*; 22.4.3, p. 510.

F. Eichhorn, D. Sippel, and K. Kleinstück
*Influence of Oxygen Segregations in Silicon Single Crystals on the Halfwidth of the Double-Crystal Rocking Curve of Thermal Neutrons
Vol. **23**, No. 1, p. 237–240 (1967).
See also 22.1.2, p. 472.

R. Gevers, J. van Landuyt, and S. Amelinckx
*Fine Structure of Weak Kinematical Electron Diffraction Beams for a Foil Containing Stacking Fault or Anti-Phase Boundaries — Two-Beam Case
Vol. **23**, No. 2, p. 549–562 (1967).
See also *10.1*, p. *125*; 21.1.1, p. 419.

A. G. Khachaturyan and M. P. Usikov
*A New Type of X-Ray and Electron Diffuse Scattering and Corresponding Strain Contrast on Electron Microscopic Images
Vol. **23**, No. 2, p. 745–754 (1967).
See also 21.1, p. 400; 22.6, p. 539.

A. Fourdeux, R. Gevers, and S. Amelinckx
Electron Microscopic Contrast Effects at the Interface between Two Different Polytypes. (I): Interface Cubic-Hexagonal
Vol. **24**, No. 1, p. 195–206 (1967).
See also *10*, p. *103*; *10.1*, p. *126*; 21.1, p. 401.

B. Roessler and P. A. Flinn
*A Theory of X-Ray Diffraction from an Ordered Alloy Containing Antiphase Domains
Vol. **24**, No. 1, p. 263–274 (1967).
See also 1.1, p. 8; 21.1, p. 401; 21.6, p. 434.

A. J. Springthorpe
The Rhombohedral Distortion in NiO
Vol. **24**, No. 1, p. K3–K4 (1967).
See also *18.4*, p. *323*; 22.6, p. 539.

H. Spreen

Study of the Domain Structure in Nickel Single Crystals. (III): The Closure Domain Structure of Cylindrical Crystals with an $\langle 110 \rangle$ Axis
Vol. **24**, No. 2, p. 413–429 (1967).
See also *18.2*, p. *309*; 21.1, p. 401.

H. Spreen

Study of the Domain Structure in Nickel Single Crystals. (IV): The Basic Structure of Cylindrical Crystals with an $\langle 100 \rangle$ Axis
Vol. **24**, No. 2, p. 431–441 (1967).
See also *18.2*, p. *309*; 21.1, p. 401.

H. Diepers

Electron-Microscopic Study of Defect Agglomerates in Ion-Irradiated Copper Foils. (II): Formation of Agglomerates
Vol. **24**, No. 2, p. 623–633 (1967).
See also 10.1, p. 127; *11*, p. *150*; 21.1, p. 402.

A. Hubert

*Zur Theorie der zweiphasigen Domänenstrukturen in Supraleitern und Ferromagneten
(On the Theory of Two-Phase Domain Structures in Superconductors and Ferromagnets)
Vol. **24**, No. 2, p. 669–682 (1967).
See also 14.2, p. 229; 18.2, p. 309; 21.1, p. 402.

L. Delaey

*Anomalous Spots in Electron Diffraction Patterns of Lamellar Mixture of Two Closed-Packed Phases
Vol. **24**, No. 2, p. K103–K105 (1967).
See also 21.1, p. 402.

T. E. Volin and R. W. Balluffi

Annealing Kinetics of Voids and the Self-Diffusion Coefficient in Aluminium
Vol. **25**, No. 1, p. 163–173 (1968).
See also *9*, p. *91*; 21, p. 384.

A. G. Fitzgerald and G. Thomas

*Electron Microscope Observations of Twinning and Phase Transformations in Indium Sulfide Crystals
Vol. **25**, No. 1, p. 263–271 (1968).
See also *1.2*, p. *11*; 22, p. 449.

A. Holz

Theoretical Study of Nucleation in Ferromagnetics
Vol. **25**, No. 2, p. 567–582 (1968).
See also *18.2*, p. *310*; 21.1, p. 403; 21.1.1, p. 420.

L. E. Murr

*Investigation of Grain Boundary Energetics in Er_2O_3 Thin Foils by Transmission Electron Microscopy
Vol. **25**, No. 2, p. 629–640 (1968).
See also *1.4*, p. *20*; 22.6, p. 540.

M. Kuriyama and G. M. McManus
*X-Ray Interference Fringes and Domain Arrangements in Fe + 3 wt% Si Single Crystals
Vol. **25**, No. 2, p. 667–677 (1968).
See also 18.2, p. 310; 21.1.1, p. 420.

L. Delaey
*Electron Microscopic Investigation of Lamellar Mixtures of Two Close-Packed Phases
Vol. **25**, No. 2, p. 697–710 (1968).
See also 1.1, p. 8; 21.1, p. 403.

H.-J. Bunge
*The Orientation Distribution Function of the Crystallites in Cold-Rolled and Annealed Low-Carbon Steel Sheets
Vol. **26**, No. 1, p. 167–172 (1968).
See also 1.1, p. 8; 21.1.1, p. 420.

J. Kobayashi, H. Schmid, and E. Ascher
Optical Study on the Ferroelectric Orthorhombic Phase of Fe-I-Boracite
Vol. **26**, No. 1, p. 277–283 (1968).
See also *14.4.2*, p. *268*; 20, p. 340.

H. Beyer and R. Sizmann
Recrystallization of Cold-Worked Gold during α-Irradiation
Vol. **26**, No. 1, p. 359–370 (1968).
See also 10.1, p. 129; *11*, p. *151*; 14.1, p. 225; 21.6, p. 435.

H. Rehme
Electron-Microscopic Study of the Mechanism of Barium Titanate Ceramics
Vol. **26**, No. 1, p. K1–K3 (1968).
See also *14.4.1*, p. *262*; 14.4.2, p. 268.

P. Rowlands and M. Bevis
*A Rapid Method for Orientating Crystals from Kossel Patterns
Vol. **26**, No. 1, p. K25–K28 (1968).
See also 3, p. 27; 21.1.1, p. 420.

R. P. Harrison
*The Effect of Particle Size on the Fracture Strength of Sodium Chloride Single Crystals
Vol. **26**, No. 1, p. K65–K67 (1968).
See also 10, p. 105;. 12, p. 158; 22.5.2, p. 526.

A. Szytuła, A. Burewicz, Ž. Dimitrijević, S. Kraśnicki, H. Rżany, J. Todorović, A. Wanic, and W. Wolski
*Neutron Diffraction Studies of α-FeOOH
Vol. **26**, No. 2, p. 429–434 (1968).
See also 18.3, p. 319; 18.4, p. 324; 22.8, p. 553.

A. Authier, A. D. Milne, and M. Sauvage
*X-Ray Dynamical Contrast of a Planar Defect
Vol. **26**, No. 2, p. 469–484 (1968).

B. J. Wicks and M. H. Lewis
*Direct Observations of Ferroelectric Domains in Lithium Niobate
Vol. **26**, No. 2, p. 571–576 (1968).
See also 10, p. 105; *14.4.2*, p. *268*.

R. GEVERS, J. VAN LANDUYT, and S. AMELINCKX
*The Fine Structure of Spots in Electron Diffraction Resulting from the Presence of Planar Interfaces and Dislocations. (III): Domain Boundaries
Vol. **26**, No. 2, p. 577–590 (1968).
See also 10, p. 105.

A. HOLZ
Magnetization Distribution at the Basal Faces of Cylindrical Ferromagnetics with Application to Nucleation
Vol. **26**, No. 2, p. 751–764 (1968).
See also *18.2*, p. *311*; 21.1.1, p. 421.

J. M. PANKRATZ and M. L. RUDEE
An Application of Stereo Electron Microscopy to Anomalously Wide Images in Neutron Irradiated Silicon
Vol. **26**, No. 2, p. K97–K98 (1968).
See also *10*, p. *106*; 22.1.2, p. 473.

E. FELDTKELLER
Magnetic Domain Wall Dynamics
Vol. **27**, No. 1, p. 161–170 (1968).
See also *18.2*, p. *311*; 21.1, p. 404; 21.1.1, p. 421.

K. WETZIG
*Some Investigations Concerning the Position and the Melting Point of TaFe Lamellas in the System TaFe-$TaFe_2$
Vol. **27**, No. 1, p. K7–K9 (1968).
See also 1.2, p. 12; 21, p. 386; 21.1.1, p. 421.

D. G. COATES
*Improvements in the Orientation Effect Observed by Scanning Electron Microscopy
Vol. **27**, No. 1, p. K11–K13 (1968).
See also 5, p. 53; 22.2.1, p. 485.

G. F. ALFREY and D. B. MEADOWCROFT
The Influence of Structure on the Electroluminescence of Zinc Sulphide
Vol. **27**, No. 2, p. 541–544 (1968).
See also 10, p. 107; *20.3*, p. *370*; 22.4.1, p. 502.

M. VICTORIA and A. E. VIDOZ
Tensile Behavior upon Ordering of Ni_3Fe Single Crystals
Vol. **28**, No. 1, p. 131–144 (1968).
See also *10.1*, p. *131*; 21.1, p. 405; 21.1.1, p. 422.

A. E. VIDOZ
On the Work Hardening of Superstructures
Vol. **28**, No. 1, p. 145–156 (1968).
See also *10.1*, p. *131*; 21.1, p. 405; 21.1.1; p. 422; 21.6, p. 436.

E. RUEDL, P. DELAVIGNETTE, and S. AMELINCKX
Electron Diffraction and Electron Microscopic Study of Long- and Short-Range Order in Ni_4Mo and of the Substructure Resulting from Ordering
Vol. **28**, No. 1, p. 305–328 (1968).
See also *1.1*, p. *8*; 21, p. 386; 21.1, p. 405.

P. GUYOT
Theory of Orientation Contrast from Inclusions
Vol. **28**, No. 1, p. 349–354 (1968).
See also *1.4*, p. *20*.

A. Farwer and F. Matossi†
Internal Optical Reflection in Cubic ZnS with One-Dimensional Stacking Faults
Vol. 28, No. 1, p. 373–383 (1968).
See also 10, p. 109; *20.1*, p. *355*; 22.4.1, p. 503.

E. Zielińska-Rohozińska
*Observations of the Effects of X-Ray Polarization on Pendellösung Fringes with the Three-Crystal X-Ray Spectrometer
Vol. 28, No. 2, p. K139–K142 (1968).
See also 5, p. 54; 22.1.2, p. 474.

E. Taglauer
Small Angle Scattering of Subthermal Neutrons from Deformed Polycrystalline Copper
Vol. 29, No. 1, p. 259–268 (1968).
See also *10.1*, p. *133*; 21.1, p. 407.

A. Bourret and D. Dautreppe
Study of Agglomerated Defects in Irradiated Pure Nickel by Electron Microscopy
Vol. 29, No. 1, p. 283–298 (1968).
See also *10.1*, p. *133*; 11, p. 154; 21.1, p. 407.

A. Sukiennicki
On a New Type of Stripe Domains
Vol. 29, No. 1, p. 417–421 (1968).
See also *18.2*, p. *314*; 21.1.1, p. 422.

A. A. Vaipolin, N. A. Goryunova, L. I. Kleshchinskii, G. V. Loshakova, and E. O. Osmanov
The Structure and Properties of the Semiconducting Compound $ZnSnP_2$
Vol. 29, No. 1, p. 435–442 (1968).
See also *5*, p. *54*; 6, p. 68; 10, p. 110; 14.3, p. 243; 22.8, p. 555.

E. S. Meieran and I. A. Blech
*Contrast Asymmetries in Lang Topographs of Crystals Strained by Thin Films
Vol. 29, No. 2, p. 653–667 (1968).
See also *10*, p. *111*; 22.1.2, p. 474; 22.2.1, p. 486.

J. P. Guigay and R. H. Wade
*Mainly on the Fresnel Mode in Lorentz Microscopy
Vol. 29, No. 2, p. 799–805 (1968).
See also 18.2, p. 314.

A. Jostsons and P. G. McDougall
*Fault Structures in Ti_2O
Vol. 29, No. 2, p. 873–889 (1968).
See also 10, p. 111; 22.6, p. 543.

P. Flögel, E. Nebauer, H.-J. Ullrich, S. Däbritz, and F. Zimmer
*Electron Microprobe and Scanning Electron Microscope Investigations on Plate-Shaped CdS Single Crystals
Vol. 29, No. 2, p. K89–K90 (1968).
See also 3.1, p. 33; 22.4.1, p. 505.

H. R. Vydyanath, D. H. Sastry, and K. I. Vasu
*On the Kinetics of Clustering in a Magnesium-3% Zinc Alloy
Vol. 29, No. 2, p. K137–K140 (1968).
See also 10.1, p. 134; 14.1, p. 226; 21.1, p. 408; 21.3, p. 427.

I. GAAL and L. URAY
*Investigation of Non-Conductive Second Phases by Means of the Shape Factor of Electrical Resistance
Vol. **29**, No. 2, p. K163–K166 (1968).
See also 14.1, p. 226; 21, p. 388.

J. LECIEJEWICZ, A. MURASIK, and R. TROĆ
The Antiferromagnetic Ordering in Uranium Monoarsenide and Monoantimonide
Vol. **30**, No. 1, p. 157–162 (1968).
See also *18.4*, p. *325*; 22, p. 454.

F. SECCO D'ARAGONA, R. C. RAU, and R. L. LADD
*Domain Boundaries in Vanadium
Vol. **30**, No. 1, p. K1–K3 (1968).
See also 10.1, p. 135; 21.1, p. 408.

A. GEMPERLE
*Crystallography of Antiphase Boundaries in Fe–Si Alloys
Vol. **30**, No. 2, p. 541–550 (1968).
See also 3, p. 28; 21.1.1, p. 423.

CH. SCHWINK and O. SCHÄRPF
Electron-Optic Investigation of the Magnetic Stray Field above Bloch Walls in Cylindric Nickel Crystals
Vol. **30**, No. 2, p. 637–644 (1968).
See also *18.2*, p. *315*; 21.1, p. 409.

L. E. TANNER
The Ordering of Ni_3V
Vol. **30**, No. 2, p. 685–701 (1968).
See also *1.2*, p. *13*; 3.1, p. 33; 21, p. 388.

R. DE RIDDER, J. VAN LANDUYT, R. GEVERS, and S. AMELINCKX
*The Fine Structure of Spots in Electron Diffraction Resulting from the Presence of Planar Interfaces and Dislocations. (IV): Wedge Crystals
Vol. **30**, No. 2, p. 797–815 (1968).
See also 10, p. 113; 10.1, p. 136; 21.1.1, p. 423; 22.6, p. 544.

TH. RICKER
Domain Walls in Thin Ferromagnetic Films Investigated by Low-Angle Electron Deflection
Vol. **30**, No. 2, p. K93–K95 (1968).
See also *18.2*, p. *315*; 21.1, p. 409; 21.1.1, p. 423.

5. Perfectly Periodic Structures

Review Articles

J. R. Dale
Alloyed Semiconductor Heterojunctions
Vol. **16**, No. 2, p. 351–387 (1966).
See also 3.1, p. 29; 13.4, p. 200; *14.3.2*, p. *248*; 16, p. 276; 22.1.1, p. 458; 22.1.2, p. 468; 22.2.1, p. 482; 22.2.2, p. 487; 22.2.3, p. 489.

P. Krishna and Ajit Ram Verma
*Crystal-Polymorphism in One Dimension
Vol. **17**, No. 2, p. 437–477 (1966).
See also *3*, p. *26*; 10, p. 93; 22.3, p. 494; 22.4.1, p. 497; 22.5, p. 512.

A. S. Borshchevskii, N. A. Goryunova, F. P. Kesamanly, and D. N. Nasledov
Semiconducting $A^{II}B^{IV}C_2^V$ Compounds
Vol. **21**, No. 1, p. 9–55 (1967).
See also 3.1, p. 29; 3.2, p. 34; 8, p. 79; 10, p. 93; 13.1, p. 174; *14.3*, p. *231*; 15, p. 270; 16, p. 276; 17, p. 286; 18, p. 291; 20.1, p. 344; 22.8, p. 547.

Original Papers and Short Notes

H. M. Otte and A. G. Crocker
*Crystallographic Formulae for Hexagonal Lattices — Simplifications
Vol. **16**, No. 1, p. K25–K26 (1966).
See also **3**, p. 26.

K. H. J. Buschow and J. F. Fast
Magnetic and Structural Characteristics of Some Equiatomic Rare-Earth Germanides
Vol. **16**, No. 2, p. 467–473 (1966).
See also *3*, p. *26*; 18.1, p. 296; 18.2, p. 300; 18.3, p. 316; 18.4, p. 321; 22.7, p. 545.

J. Treusch and R. Sandrock
Energy Band Structures of Selenium and Tellurium (Kohn-Rostoker Method)
Vol. **16**, No. 2, p. 487–497 (1966).
See also *13.1*, p. *175*; 22.1.3, p. 476.

V. K. Nikulin
Cohesive Energy of Noble Metals
Vol. **16**, No. 2, p. K125–K127 (1966).
See also *3.2*, p. *34*; 21.1, p. 391; 21.6, p. 430.

E. V. Chisler and M. S. Shur
The Raman Spectrum of $NaNO_2$ in the Ferroelectric Phase
Vol. **17**, No. 1, p. 163–171 (1966).
See also *6*, p. *56*; 14.4.2, p. 266; *20.1*, p. *345*.

O. N. Efimov and E. K. Kovev
*Integral Characteristics of Anomalous Transmission of X-Rays
Vol. **17**, No. 1, p. 397–410 (1966).
See also 10, p. 94; 22.1.1, p. 459.

P. Neumann
*Vereinfachung kristallographischer Rechnungen in hexagonalen Gittern durch konsequente Benutzung des vierachsigen hexagonalen Koordinatensystems (Simplification of Crystallographic Calculations for Hexagonal Lattices by Consequently Using the Quadruaxial Hexagonal Coordinate-System)
Vol. **17**, No. 1, p. K71–K74 (1966).
See also **3**, p. 26.

U. RÖSSLER and M. LIETZ
Band Structure of Cubic and Hexagonal ZnS (APW Method)
Vol. **17**, No. 2, p. 597–604 (1966).
See also *13.1*, p. *175*; 20.1, p. 346; 22.4.1, p. 498.

A. I. GUBANOV and V. K. NIKULIN
The Theory of the Cohesion of Noble Metals
Vol. **17**, No. 2, p. 815–823 (1966).
See also *3.2*, p. *34*; 12.1, p. 159; 21.1, p. 393; 21.6, p. 431.

G. V. DAVYDOV
*Dependence of the Geometry of the Formation of X-Ray Diffraction Lines on the Instrumental Factors in X-Ray Investigations of Polycrystals
Vol. **18**, No. 1, p. 41–48 (1966).

R. COLELLA and A. MERLINI
*A Study of the (222) "Forbidden" Reflection in Germanium and Silicon
Vol. **18**, No. 1, p. 157–166 (1966).
See also 22.1.1, p. 460; 22.1.2, p. 469.

J. H. C. HOGG and W. J. DUFFIN
Twinning in InS
Vol. **18**, No. 2, p. 755–760 (1966).
See also *4*, p. *39*; 22, p. 441.

I. SPÎNULESCU-CARNARU
The Crystalline Structure of ZnTe Thin Films
Vol. **18**, No. 2, p. 769–778 (1966).
See also *1.4*, p. *17*; 3.1, p. 30; 22.4.3, p. 509.

G. D. GUSEINOV, M. Z. ISMAILOV, and A. G. TALYBOV
*On the New Semiconducting Compound $HgTlS_2$
Vol. **18**, No. 2, p. 929–935 (1966).
See also 8, p. 80; 13.4, p. 204; 14.3, p. 234; 16, p. 278; 22.8, p. 548.

C. P. KEMPTER
*Vegard's "Law"
Vol. **18**, No. 2, p. K117–K118 (1966).
See also 21.1, p. 394; 21.6, p. 432.

G. V. DAVYDOV
*Dependence of the Geometry of the Diffractional Line Formation on the Instrumental and Other Physical Factors at X-Ray Investigation of the Polycrystalline Fine Structure
Vol. **18**, No. 2, p. K145–K147 (1966).

A. GEMPERLE and J. KOČÍK
*Transmission Electron Microscopic Observation of Lattice Deformation in Antiphase Boundaries of Ordered Fe–Si Alloys
Vol. **19**, No. 1, p. 333–338 (1967).
See also 4, p. 40; 21.1.1, p. 415.

A. CZACHOR and A. PINDOR
Dependence of Calculated Elastic Constants on the c/a Ratio of Hexagonal Close-Packed Metals
Vol. **20**, No. 1, p. K17–K20 (1967).
See also *12.1*, p. *159*; 21, p, 377; 21.1, p. 396; 21.3, p. 426.

L. Zsoldos
Lattice Parameter Change of FeRh Alloys due to Antiferromagnetic–Ferromagnetic Transformation
Vol. **20**, No. 1, p. K25–K28 (1967).
See also 1.2, p. 10; *18.2*, p. *306*; *18.4*, p. *322*; 21.1.1, p. 416.

H.-J. Ullrich
*Precision Lattice Parameter Measurements by Interferences from Lattice Sources (Kossel Lines) and Divergent Beam X-Ray Diffraction (Pseudo-Kossel Lines) in Back Reflection
Vol. **20**, No. 2, p. K113–K117 (1967).
See also 21, p. 378; 22, p. 444.

H. Goretzki
*Neutron Diffraction Studies on Titanium–Carbon and Zirconium–Carbon Alloys
Vol. **20**, No. 2, p. K141–K143 (1967).
See also 21, p. 378.

M. Järvinen and O. Inkinen
An X-Ray Diffraction Study of Rubidium Chloride
Vol. **21**, No. 1, p. 127–135 (1967).
See also *13*, p. *165*; 22.5.2, p. 522.

S. S. Sheinin
*The Effect of Systematic Reflections on the (110) Extinction Distance in Molybdenum
Vol. **21**, No. 1, p. 247–256 (1967).
See also 21, p. 378.

A. Weiss and K. Zohner
Nuclear Quadrupole Interaction of ^{23}Na and ^{11}B and the Crystal Structure of $NaBF_4$
Vol. **21**, No. 1, p. 257–270 (1967).
See also **3.2**, p. 35; *19*, p. *329*; 22.5, p. 513; 22.8, p. 550.

J. S. Ziomek and P. B. Pickar
Optical Studies of β-SiC
Vol. **21**, No. 1, p. 271–278 (1967).
See also 6, p. 61; 13.1, p. 178; *20.1*, p. *348*; 22.3, p. 494.

J. Tobisch and W. Hase
*Structure Investigations of Two Uranium Nitrides U_2N_{3+x}
Vol. **21**, No. 1, p. K11–K12 (1967).
See also 22, p. 445.

K. H. J. Buschow and J. F. Fast
*Crystal Structure and Magnetic Properties of Some Rare Earth Germanides
Vol. **21**, No. 2, p. 593–600 (1967).
See also 18.1, p. 296; 18.3, p. 318; 18.4, p. 322; 22.7, p. 545.

Ya. A. Iosilevskii
*Some Model Problems on the Dynamics of Two-Component Crystals
Vol. **22**, No. 1, p. 255–266 (1967).
See also 6, p. 62; 10, p. 101.

O. N. Efimov
*Contribution of Thermal Vibrations to the Anomalous Transmission of X-Rays
Vol. **22**, No. 1, p. 297–308 (1967).
See also 6, p. 62; 22.1.1, p. 462.

B. P. Kotrubenko, V. N. Lange, and M. M. Markus
*Chemical Bond and Structure of the Ternary Compound InAsTe
Vol. **22**, No. 1, p. K15—K17 (1967).
See also *3.2*, p. *35*; 22.8, p. 551.

J. Leciejewicz, R. Troć, A. Murasik, and A. Zygmunt
Neutron-Diffraction Study of Antiferromagnetism in USb_2 and UBi_2
Vol. **22**, No. 2, p. 517—526 (1967).
See also *18.4*, p. *323*; 22, p. 447.

H. P. Singh and B. Dayal
X-Ray Determination of the Thermal Expansion of Zinc Selenide
Vol. **23**, No. 1, p. K93—K95 (1967).
See also *8*, p. *83*; 22.4.2, p. 507.

W. Brückner, K. Kleinstück, and G. E. R. Schulze
*Atomic Arrangement in the Homogeneity Range of the Laves Phases $ZrFe_2$ and $TiFe_2$
Vol. **23**, No. 2, p. 475—480 (1967).
See also 21.1.1, p. 419.

B. W. Batterman and G. Hildebrandt
*Observation of X-Ray Pendellösung Fringes in Darwin Reflection
Vol. **23**, No. 2, p. K147—K149 (1967).
See also 22.1.2, p. 472.

L. E. Murr
Study of Erbium Thin Film Oxidation in the Electron Microscope
Vol. **24**, No. 1, p. 135—148 (1967).
See also *1.4*, p. *19*; 22.6, p. 539.

G. Hildebrandt
*Röntgenwellenfelder in einem Dreistrahlfall
(X-Ray Wave Fields in a Three-Beam Case)
Vol. **24**, No. 1, p. 245—261 (1967).

V. A. Ivanova, D. Sh. Abdinov, and G. M. Aliev
Some Structural, Electrical, and Thermal Parameters of VCr_2Se_4, $FeCr_2Se_4$, and $NiCr_2Se_4$ Ternary Compounds
Vol. **24**, No. 1, p. K23—K24 (1967).
See also 8, p. 83; *14.3*, p. *239*; 15, p. 273; 22.8, p. 552.

M. Kuriyama
*Distortion Correction in Anomalous Absorption Coefficients
Vol. **24**, No. 2, p. 743—748 (1967).

V. A. Ivanova, D. Sh. Abdinov, and G. M. Aliev
On Some Characteristics of Chromium Selenides
Vol. **24**, No. 2, p. K145—K147 (1967).
See also *8*, p. *84*; 14.3, p. 239; *15*, p. *273*; 22, p. 449.

T. O. Baldwin
The Temperature Dependence of the Borrmann Effect in Copper
Vol. **25**, No. 1, p. 71—74 (1968).
See also *6*, p. *64*; 21.1, p. 402.

P. R. Okamoto and G. Thomas
On the Four-Axis Hexagonal Reciprocal Lattice and Its Use in the Indexing of Transmission Electron Diffraction Patterns
Vol. **25**, No. 1, p. 81–91 (1968).
See also *3*, p. *27*.

W. Schüz
*Über verbotene Reflexe bei Elektronenbeugung an vakuumaufgedampften Tellurschichten
(Forbidden Reflexes in Electron Diffraction from Tellurium Films Evaporated in Vacuum)
Vol. **25**, No. 1, p. 253–261 (1968).
See also 1.4, p. 20; 22.1.**3**, p. 478.

E. E. Anderson
Concerning the Uniqueness of a Molecular Field Solution for Two Magnetic Sublattices
Vol. **25**, No. 1, p. K31–K34 (1968).
See also *18*, p. *293*.

E. Jäger
Jahn-Teller Effect at Lattice Sites of Symmetry D_{3d}
Vol. **25**, No. 1, p. K43–K46 (1968).
See also *13*, p. *169*; 18.3, p. 319.

N. A. Goryunova, V. M. Orlov, V. I. Sokolova, G. P. Shpenkov, and E. V. Tsvetkova
Preparation and Some Properties of CuP_2 Single Crystals
Vol. **25**, No. 2, p. 513–519 (1968).
See also *3.1*, p. *33*; 14.3, p. 240; 15, p. 273; 20.1, p. 351; 22, p. 450.

J. G. Allpress, J. V. Sanders, and A. D. Wadsley
*Electron Microscopy of High-Temperature Nb_2O_5 and Related Phases
Vol. **25**, No. 2, p. 541–550 (1968).
See also 3, p. 27; 22.6, p. 540.

M. Hart and A. D. Milne
*Direct Observations of Plane Wave and Spherical Wave Pendellösung Fringes
Vol. **26**, No. 1, p. 185–189 (1968).
See also 3, p. 27; 22.1.2, p. 473.

D. Kordes
*Vanadiummonocarbid mit geordneten Leerstellen
(Vanadium Monocarbide Containing Ordered Vacancies)
Vol. **26**, No. 2, p. K103–K105 (1968).
See also 10, p. 106; 22.3, p. 494.

D. G. Coates
Improvements in the Orientation Effect Observed by Scanning Electron Microscopy
Vol. **27**, No. 1, p. K11–K13 (1968).
See also *4*, p. *46*; 22.2.1, p. 485.

J. S. Tiwari, A. Mehra, and K. G. Srivastava
Electronic Absorption Spectrum of K_2CoF_4
Vol. **28**, No. 1, p. 61–63 (1968).
See also 13, p. 171; *20.1*, p. *355*; 22.5, p. 514; 22.8, p. 554.

S. S. Sheinin and C. D. Cann
*Effect of Absorption on Multi-Beam Intensity Profiles of Wedge Crystals
Vol. **28**, No. 1, p. 187–192 (1968).
See also 21, p. 386; 22.6, p. 542.

A. M. Afanasev, Yu. Kagan, and F. N. Chukhovskii
Dynamical Treatment of the Thermal Diffuse Scattering of X-Rays
Vol. 28, No. 1, p. 287—294 (1968).
See also *6*, p. *66*.

J.-U. Arnold and P. Grosse
The Temperature Dependence of the Atom Coordinates in the Tellurium Lattice
Vol. 28, No. 1, p. K93—K96 (1968).
See also *3*, p. *27*; 8, p. 85; 22.1.3, p. 479.

T. Paakkari and P. Suortti
*Absolute Atomic Scattering Factors of Tungsten
Vol. 28, No. 2, p. 539—545 (1968).
See also 21, p. 387.

V. A. Bokov, N. A. Grigoryan, M. F. Bryzhina, and V. V. Tikhonov
Effect of Lattice Distortions on the Magnetic Behaviour of Perovskite-Type Manganites
Vol. 28, No. 2, p. 835—847 (1968).
See also *18.2*, p. *313*; *18.4*, p. *324*; 22.8, p. 554.

L. Śniadower, M. Psoda, and R. R. Gałązka
*X-Ray Measurements of Lattice Dilatation in HgTe
Vol. 28, No. 2, p. K121—K123 (1968).
See also *8*, p. *85*; 22.4.3, p. 511.

E. Zielińska-Rohozińska
Observations of the Effects of X-Ray Polarization on Pendellösung Fringes with the Three-Crystal X-Ray Spectrometer
Vol. 28, No. 2, p. K139—K142 (1968).
See also *4*, p. *47*; 22.1.2, p. 474.

P. B. Pickar and H. D. Tiller
Optical Energy Gap in TlSe
Vol. 29, No. 1, p. 153—158 (1968).
See also 13.1, p. 188; 16, p. 283; *20.1*, p. *356*; 22, p. 453.

H. Gamari-Seale and P. Papamantellos
*A Neutron Diffraction Study on Gallium-Substituted Magnetite
Vol. 29, No. 1, p. 323—331 (1968).
See also 18.3, p. 320.

A. A. Vaipolin, N. A. Goryunova, L. I. Kleshchinskii, G. V. Loshakova, and E. O. Osmanov
*The Structure and Properties of the Semiconducting Compound $ZnSnP_2$
Vol. 29, No. 1, p. 435—442 (1968).
See also 4, p. 47; 6, p. 68; 10, p. 110; 14.3, p. 243; 22.8, p. 555.

P. Reimers and W. Ruppel
The Preparation of CdS–CdSe Graded Single Crystals
Vol. 29, No. 1, p. K31—K33 (1968).
See also *3.1*, p. *33*; 20.1, p. 357; 22.4.1, p. 504; 22.4.2, p. 508; 22.8, p. 555.

B. Chalupa, R. Michalec, V. Petržílka, J. Tichý, and J. Zelenka
*Diffraction of Neutrons on a Vibrating Quartz Crystal
Vol. 29, No. 1, p. K51—K54 (1968).
See also 14.4, p. 254; 22.6, p. 543.

M. Lefeld-Sosnowska and C. Malgrange
*Observation of Oscillations in Rocking Curves of the Laue Reflected and Refracted Beams from Thin Si Single Crystals
Vol. **30**, No. 1, p. K23–K25 (1968).
See also 22.1.2, p. 474.

C. D. Cann and S. S. Sheinin
*The Dependence of Extinction Distance on Deviation from the Bragg Condition in Silicon
Vol. **30**, No. 2, p. 791–796 (1968).
See also 22.1.2, p. 475.

6. Lattice Mechanics. Phonons

Review Article

D. P. Morgan
Helicon Waves in Solids
Vol. **24**, No. 1, p. 9—36 (1967).
See also *13*, p. *162*; 14.1, p. 220; 14.3, p. 231; 19, p. 326; *20*, p. *337*; 21, p. 373; 21.1, p. 390; 21.2, p. 424; 21.6, p. 430; 22.1.1, p. 458; 22.2.1, p. 482; 22.2.3, p. 489.

V. L. Bonch-Bruevich and E. G. Landsberg
Recombination Mechanisms
Vol. **29**, No. 1, p. 9—43 (1968).
See also 13, p. 162; 13.2, p. 191; *13.4*, p. *200*; *14.3*, p. *231*; 20, p. 337; 20.3, p. 363; 2.1.1, p. 458; 22.1.2, p. 468.

Original Papers and Short Notes

E. Kamieniecki
Distribution Function of Hot Carriers
Vol. **16**, No. 1, p. 57—62 (1966).
See also *13*, p. *162*; *14.3*, p. *231*; 22.1.1, p. 459; 22.1.2, p. 468.

A. I. Bobrysheva
The Interaction of Two Excitons in a Crystal
Vol. **16**, No. 1, p. 337—348 (1966).
See also *13.2*, p. *191*; 22, p. 439.

A. G. Samoilovich, M. V. Nitsovich, and V. M. Nitsovich
On the Theory of Anisotropic Thermoelectric Power in Semiconductors
Vol. **16**, No. 2, p. 459—465 (1966).
See also 13.1, p. 174; *15*, p. *270*; 22, p. 440.

M. M. Shukla and B. Dayal
*Lattice Vibrations of Aluminium on the Basis of Krebs's Model
Vol. **16**, No. 2, p. 513—516 (1966).
See also 8, p. 79; 21, p. 374.

R. Granger
Anisotropic Electron Distribution in Pure Semiconductors near Breakdown
Vol. **16**, No. 2, p. 599—612 (1966).
See also 13, p. 163; *14.4.1*, p. *255*; 22, p. 440; 22.2.3, p. 489.

G. Kluge
Acoustic Activity
Vol. **17**, No. 1, p. 109—118 (1966).
See also *7*, p. *76*; 22.6, p. 535.

Y. Kazumata, T. Kamikawa, and K. Ozawa
*Zero-Phonon Lines of Color Centers in LiF and NaF
Vol. **17**, No. 1, p. 131—138 (1966).
See also *10.2*, p. *137*; 22.5.2, p. 518.

L. Elstner
Temperature Dependence of the Hole Mobility in Silicon Crystals
Vol. **17**, No. 1, p. 139—150 (1966).
See also *14.3*, p. *232*; 22.1.2, p. 469.

E. V. Chisler and M. S. Shur
*The Raman Spectrum of $NaNO_2$ in the Ferroelectric Phase
Vol. **17**, No. 1, p. 163—171 (1966).
See also 5, p. 49; 14.4.2, p. 266; *20.1*, p. *345*.

E. V. CHISLER and M. S. SHUR
*Raman Spectrum and Phase Transition in the Ferroelectric Crystal $NaNO_2$
Vol. **17**, No. 1, p. 173–176 (1966).
See also 1.2, p. 9; 14.4.2, p. 266; *20.1*, p. *345*.

R. B. LAIBOWITZ and H. S. SACK
Luminescence and Photoconductivity in Silver Chloride with Laser Excitation
Vol. **17**, No. 1, p. 353–360 (1966).
See also 13.4, p. 202; *16*, p. *277*; *20.3*, p. *364*; 22.5.1, p. 515.

M. E. ERTL
Thermal Resistivity due to Mass Fluctuation Scattering in Bismuth–Antimony Alloys
Vol. **17**, No. 1, p. K63–K65 (1966).
See also *8*, p. *79*; 21.7, p. 438.

F. KUBEC
*Crystal Field Calculation in the $CdWO_4$ Lattice
Vol. **17**, No. 2, p. 571–574 (1966).
See also 3.2, p. 34; 13.4, p. 202; 22.8, p. 548.

R. GLEISS
*Berechnung des Debye-Waller-Faktors von Nickel
(Calculation of the Debye-Waller Factor of Nickel)
Vol. **17**, No. 2, p. 761–767 (1966).
See also 8, p. 79; 21.1, p. 393.

G. B. ABDULLAEV, M. I. ALIEV, and G. I. SAFARALIEV
About Phonon Scattering in Doped GaSb Crystals
Vol. **17**, No. 2, p. 777–781 (1966).
See also *8*, p. *79*; 22.2.3, p. 490.

O. GOEDE and E. GUTSCHE
Edge Emission in CdS as Internal Radiative Transition in Donor–Acceptor Associates
Vol. **17**, No. 2, p. 911–923 (1966).
See also 13.4, p. 203; *20.3*, p. *364*; 22.4.1, p. 498.

R. WEHNER
*Second-Order Dipole Moment of the F.C.C. and Diamond Lattices
Vol. **17**, No. 2, p. K179–K183 (1966).
See also 3.2, p. 34.

R. P. GUPTA, P. K. SHARMA, and S. PAL
Temperature Variation of the Grüneisen Parameter of Copper
Vol. **18**, No. 1, p. 119–124 (1966).
See also *8*, p. *80*; 21.1, p. 393.

W. NEUMANN
Thermal Conductivity of Yellow PbO between 0 and 500 °C
Vol. **18**, No. 1, p. 151–156 (1966).
See also *8*, p. *80*; 22.6, p. 535.

S. A. MOSKALENKO and P. I. KHADSHI
Infrared Absorption by Excitons due to Photoionization and Intraband Lattice Scattering
Vol. **18**, No. 1, p. 379–390 (1966).
See also *13.2*, p. *191*; 20.1, p. 346; 22.1.1, p. 460; 22.2.2, p. 487; 22.2.3, p. 490; 22.6, p. 536.

B. M. Askerov and F. M. Gashimzade
Quantum Theory of the Electrical Conductivity of Semiconductors with a Non-Standard Energy Band
Vol. **18**, No. 2, p. 667–675 (1966).
See also 13.1, p. 176; *14.3*, p. *233*; 22.2, p. 481; 22.2.3, p. 490.

N. P. Gupta and B. Dayal
*Effect of Zero-Point Energy on the Specific Heats of Solidified Neon and Xenon
Vol. **18**, No. 2, p. 731–736 (1966).
See also 3.2, p. 34; *8*, p. *80*; 23, p. 561.

B. Dorner und H. H. Stiller
*Die innere Dynamik der Tieftemperaturphasen des Molekülkristalls CH_4
(Internal Dynamics of the Low-Temperature Phases of CH_4 Molecular Crystals)
Vol. **18**, No. 2, p. 795–806 (1966).
See also 1.2, p. 10; 23, p. 561.

A. F. Lubchenko and S. I. Dudkin
Effect of Concentration on Light Absorption and Dispersion by Impurity Centres in the Case of Weak Electron–Phonon Coupling
Vol. **18**, No. 2, p. 853–862 (1966).
See also 13.4, p. 204; *20.1*, p. *346*.

G. D. Guseinov and A. I. Rasulov
Heat Conductivity Study of GaSe Monocrystals
Vol. **18**, No. 2, p. 911–922 (1966).
See also *8*, p. *80*; 22, p. 442.

V. V. Paranjape
Distribution Function of Hot Electrons
Vol. **18**, No. 2, p. K103–K104 (1966).
See also *13*, p. *164*; *14.3*, p. *234*; 22, p. 442.

E. Kamieniecki
Remark on Solving Boltzmann Equation for Hot Carriers
Vol. **18**, No. 2, p. K105 (1966).
See also *13*, p. *164*; *14.3*, p. *234*; 22, p. 442.

I. V. Ioffe
*On the Theory of the Lattice Oscillations of Dielectric Crystals in an External Electric Field
Vol. **19**, No. 1, p. 51–56 (1967).
See also 18.2, p. 304; 18.4, p. 321; 19, p. 327; 22, p. 442.

W. Ludwig and H. Pick
*Scattering of Phonons at Defect Planes in the Simple Cubic Lattice
Vol. **19**, No. 1, p. 313–326 (1967).
See also 10, p. 97.

V. L. Broude, E. I. Rashba, and E. F. Sheka
*A New Approach to the Vibronic Spectra of Molecular Crystals
Vol. **19**, No. 1, p. 395–406 (1967).
See also 20.1, p. 347; 22.9, p. 558.

E. S. Mashkova, V. A. Molchanov, and V. Soshka
*Influence of the Thermal Vibrations of the Crystal Lattice upon the Energy Distribution of Scattered Ions
Vol. **19**, No. 1, p. 425–428 (1967).
See also 21.1, p. 395.

B. Velický
Polaron Effect on the Optical Absorption Edge of Semiconductors
Vol. **19**, No. 1, p. K39–K42 (1967).
See also 13, p. 164; 13.1, p. 176; *20.1*, p. *347*; 22, p. 442.

A. Axmann und W. Gissler
*Streuung langsamer Neutronen an polykristallinem Selen und Tellur
(Scattering of Slow Neutrons on Polycrystalline Selenium and Tellurium)
Vol. **19**, No. 2, p. 721–727 (1967).
See also 22.1.3, p. 477.

M. M. Shukla and B. Dayal
*The Lattice Vibrations of Nickel in Krebs's Model
Vol. **19**, No. 2, p. 729–733 (1967).
See also 21.1, p. 396.

M. P. Verma and B. Dayal
*Lattice Dynamics of MgO
Vol. **19**, No. 2, p. 751–756 (1967).
See also 22.6, p. 537.

R. Stockmeyer and H. H. Stiller
*Lattice Dynamics of Solid Adamantane
Vol. **19**, No. 2, p. 781–785 (1967).
See also 22.9, p. 558.

A. M. Stoneham
Interference Effects in Spin Lattice Relaxation by Two-Phonon Processes
Vol. **19**, No. 2, p. 787–792 (1967).
See also *13*, p. *165*; 22.6, p. 537.

G. Björkman and G. Grimvall
*Polarization Vectors for Lattice Vibrations in Sodium
Vol. **19**, No. 2, p. 863–865 (1967).
See also 21.2, p. 424.

V. V. Tarassov
*Heat Capacity and Structure of Vitreous Silica and Diamond-Like Lattices (I)
Vol. **20**, No. 1, p. 37–43 (1967).
See also 2, p. 22; 8, p. 81; 22, p. 443.

V. V. Tarassov
*Heat Capacity and Structure of Vitreous Silica and Diamond-Like Lattices (II)
Vol. **20**, No. 1, p. 45–57 (1967).
See also 2, p. 22; 8, p. 81; 22, p. 443.

A. S. Davydov
*On the Operator of Exciton–Phonon Interaction
Vol. **20**, No. 1, p. 143–151 (1967).
See also *13.2*, p. *192*; 22, p. 443.

R. P. Gupta
Lattice Specific Heats of Thallium and Yttrium
Vol. **20**, No. 1, p. 291–294 (1967).
See also *8*, p. *81*; 21, p. 377.

G. S. Pawley
*A Model for the Lattice Dynamics of Naphthalene and Anthracene
Vol. **20**, No. 1, p. 347–360 (1967).
See also 22.9, p. 559.

R. E. DeWames and W. F. Hall
*A Relation between the Energy Loss Spectrum of Incoherently Scattered Electrons and the Phonon Density of States in Crystals
Vol. **20**, No. 1, p. K11—K15 (1967).
See also 22.5.2, p. 521.

A. F. Lubchenko and S. I. Dudkin
Effect of Concentration on Light Absorption and Dispersion by Impurity Centres in the Case of Weak Elektron–Phonon Coupling (II)
Vol. **20**, No. 2, p. 427—432 (1967).
See also 13.4, p. 205; *20.1*, p. *347*.

T. N. Casselman and H. N. Spector
*Interaction of Optical Phonons with Spin Waves in Ionic Crystals
Vol. **20**, No. 2, p. 605—612 (1967).
See also 13, p. 165; 18, p. 292; *19*, p. *328*; 22, p. 444.

E. A. Metzbower
*Noncentral Force Model for Hexagonal Close-Packed Crystal Lattices
Vol. **20**, No. 2, p. 681—692 (1967).
See also 3.2, p. 35; 8, p. 81; 12.1, p. 160; 21.1, p. 397; 21.3, p. 426.

N. A. Chernoplekov, G. Kh. Panova, M. G. Zemlyanov, B. N. Samoilov, and V. I. Kutaitsev
*Investigation of the Quasi-Local Level in the Vibrational Spectrum of the Ti Lattice with Heavy Impurity Atoms
Vol. **20**, No. 2, p. 767—770 (1967).
See also 8, p. 81; 21.1, p. 397.

A. A. Abdurakhmanova and M. I. Aliev
On the Mechanism of Thermal Conductivity in $(GaSb)_{3x}$–$(Ga_2Te_3)_{1-x}$ and $(GaSb)_x$–$(GaTe)_{1-x}$ Solid Solutions
Vol. **20**, No. 2, p. 777—781 (1967).
See also *8*, p. *82*; 22.8, p. 550.

H. Bilz, R. Zeyher, and R. K. Wehner
*Anharmonic Sideband Effects of the U-Center Local Mode
Vol. **20**, No. 2, p. K167—K171 (1967).
See also 10.2, p. 139; 20.1, p. 348; 22.5.2, p. 522.

G. O. Müller, H. Peibst, E. Schnürer, and H. Thiel
*The Influence of High Phonon Flux Densities on the X-Ray Reflectivity of Nearly Ideal CdS Crystals
Vol. **20**, No. 2, p. K173—K175 (1967).
See also 7, p. 76; 22.4.1, p. 499.

M. Ali and P. Nagels
*Evaluation of the Debye Temperature of Thorium Dioxide
Vol. **21**, No. 1, p. 113—116 (1967).
See also 8, p. 82; 22.6, p. 537.

K. Ishikawa
*On the Quasi-Harmonic Theory of Phonons
Vol. **21**, No. 1, p. 137—144 (1967).

M. F. Rose
Higher Order Elastic Constants and the Equation of State of F.C.C. Metals
Vol. **21**, No. 1, p. 235—238 (1967).
See also 8, p. 82; *12.1*, p. *160*; 21, p. 378; 21.1, p. 397; 21.6, p. 433.

J. S. ZIOMEK and P. B. PICKAR
Optical Studies of β-SiC
Vol. **21**, No. 1, p. 271—278 (1967).
See also 5, p. 51; 13.1, p. 178; *20.1*, p. *348*; 22.3, p. 494.

K. MIKA
*Die Richtungsverteilung der Frequenzen von Gitterschwingungen einer festen Wellenzahl
(Directional Distribution of Lattice Vibrational Frequencies with Fixed Wave-Number)
Vol. **21**, No. 1, p. 279—290 (1967).

V. NÜSSLEIN and U. SCHRÖDER
*Calculations of Dispersion Curves and Specific Heat for LiF and NaCl Using the Breathing Shell Model
Vol. **21**, No. 1, p. 309—314 (1967).
See also 8, p. 82; 22.5.2, p. 522.

J. A. M. SALTER and P. CHARSLEY
The Effect of Grain Size on the Lattice Thermal Conductivity of Copper Aluminium Alloys
Vol. **21**, No. 1, p. 357—368 (1967).
See also 4, p. 42; *8*, p. *82*; 21, p. 379; 21.1, p. 397.

N. KRISTOFFEL and P. KONSIN
Pseudo-Jahn-Teller Effect and Second Order Phase Transitions in Crystals
Vol. **21**, No. 1, p. K39—K43 (1967).
See also *1.2*, p. *10*; *13*, p. *166*; 14.4.2, p. 267.

Z. A. DEMIDENKO and V. S. MASHKEVICH
Indirect Optical Transitions at the Centre of the Brillouin Zone
Vol. **21**, No. 1, p. K59—K63 (1967).
See also *20.1*, p. *348*; 22, p. 445.

M. HULIN
*Selection Rules for Two-Phonon Absorption Processes
Vol. **21**, No. 2, p. 607—618 (1967).
See also *20.1*, p. *348*.

D. G. ARASLI and M. I. ALIEV
Influence of Defects and of the Interaction between Them on Phonon Scattering in Heavily Doped Ge and Si Crystals
Vol. **21**, No. 2, p. 643—648 (1967).
See also *8*, p. *82*; 10, p. 100; 22.1.1, p. 462; 22.1.2, p. 471.

N. P. GUPTA and B. DAYAL
Thermal Expansion and Compressibility of Solid Neon and Xenon
Vol. **21**, No. 2, p. 661—664 (1967).
See also *8*, p. *82*; *12*, p. *157*; 23, p. 561.

V. P. KALASHNIKOV
*Theory of Spin-Lattice Relaxation of Conduction Electrons of Ionic Semiconductors at Extremely High Magnetic Fields
Vol. **21**, No. 2, p. 775—784 (1967).
See also 13, p. 166; 14.4, p. 253; 18, p. 292; 22, p. 446.

B. PEGEL
On the Isotope Effect in Interstitial Diffusion
Vol. **22**, No. 1, p. 223—228 (1967).
See also *9*, p. *89*.

YA. A. IOSILEVSKII
Some Model Problems on the Dynamics of Two-Component Crystals
Vol. 22, No. 1, p. 255—266 (1967).
See also *5*, p. *51*; 10, p. 101.

O. N. EFIMOV
Contribution of Thermal Vibrations to the Anomalous Transmission of X-Rays
Vol. 22, No. 1, p. 297—308 (1967).
See also *5*, p. *51*; 22.1.1, p. 462.

B. PEGEL
On the Isotope Effect in Interstitial Diffusion. (II): Extended Model
Vol. 22, No. 1, p. K45—K48 (1967).
See also *9*, p. *89*.

R. K. WEHNER
*Phonon Green Functions by Functional Methods
Vol. 22, No. 2, p. 527—535 (1967).

H.-G. KAHLE, H. KALBFLEISCH, and E. STEIN
*Paramagnetic Relaxation and Observation of a Phonon Bottleneck in Terbium Ethyl Sulfate
Vol. 22, No. 2, p. 537—540 (1967).
See also *18.1*, p. *296*; 19, p. 330; 22.8, p. 551.

B. MUKHERJEE and A. J. F. BOYLE
*On the Debye Characteristic Temperature of Diamond
Vol. 22, No. 2, p. K131—K134 (1967).
See also 8, p. 83; 12, p. 157; 22.1, p. 457.

J. B. GRUN, A. MYSYROWICZ, F. RAGA, A. BIVAS, R. LEVY, and S. NIKITINE
Photon-Assisted Emission in CdSe
Vol. 22, No. 2, p. K155—K156 (1967).
See also 13.2, p. 193; *20.3*, p. *367*; 22.4.2, p. 507.

B. PEGEL and D. LEPSKI
On the Isotope Effect in Interstitial Diffusion. (III): Quantum Corrections
Vol. 23, No. 1, p. 335—339 (1967).
See also *9*, p. *90*.

S. PAL and P. K. SHARMA
*On Lindemann's Melting Criterion
Vol. 23, No. 1, p. 361—364 (1967).
See also 21, p. 381; 21.1, p. 400; 21.1.1, p. 419; 21.2, p. 424; 21.6, p. 434.

M. PORSCH and J. RÖSELER
Recoil Effects in the Polaron Problem
Vol. 23, No. 1, p. 365—376 (1967).
See also *13*, p. *167*; 13.1, p. 181; 22, p. 447.

P. H. DEDERICHS
Dynamical Scattering Theory for Crystals with Point Defects
Vol. 23, No. 1, p. 377—386 (1967).
See also *10*, p. *102*; *10.1*, p. *125*.

J. KOŁODZIEJCZAK
Effect of Strong Electric Fields on Magneto-Resistivity and Hall Coefficient in Semiconductors
Vol. 23, No. 1, p. 423—427 (1967).
See also 13.1, p. 181; *14.3*, p. *238*; 14.4.1, p. 260; 22, p. 447; 22.2.3, p. 491.

I. P. Ipatova, A. A. Klochichin, and A. V. Subashiev
*The Influence of Localized and Resonance Modes of Lattice Vibrations on the Radiative Recombination in Semiconductors
Vol. **23**, No. 2, p. 467–473 (1967).
See also 13.4, p. 209; *20.3*, p. *367*; 22.1.2, p. 472.

P. S. Ho and A. L. Ruoff
*A Quasi-Harmonic Calculation of Lattice Dynamics for Na
Vol. **23**, No. 2, p. 489–499 (1967).
See also 13.1, p. 182; 14.1, p. 224; 21.2, p. 425.

A. E. Lord, Jr.
Coupling of Magnons and Phonons at High Frequencies
Vol. **23**, No. 2, p. 521–526 (1967).
See also *13*, p. *167*; 18.2, p. 308.

R. G. J. Grisar, K. P. Reiners, K. F. Renk, and L. Genzel
*Impurity-Induced Far-Infrared Absorption in the Phonon Gap Regions of KI and KBr
Vol. **23**, No. 2, p. 613–620 (1967).
See also *20.1*, p. *350*; 22.5.2, p. 524.

G. G. Kovalevskaya, D. N. Nasledov, and S. V. Slobodchikov
*Photoconductivity Oscillations in InP
Vol. **23**, No. 2, p. 755–760 (1967).
See also 13.1, p. 182; *16*, p. *280*; 22.2.2, p. 488.

M. Bitter, W. Gissler, and T. Springer
*Lattice Dynamics of Solid Helium at 2.9 °K and 125 atm by Neutron Scattering
Vol. **23**, No. 2, p. K155–K158 (1967).
See also 23, p. 561.

L. Wojtczak
The Melting Point of Thin Films
Vol. **23**, No. 2, p. K163–K166 (1967).
See also *8*, p. *83*; 21, p. 382; 21.1, p. 400; 21.6, p. 434.

E. R. Fitzgerald and T. W. Wrigth
Invariance of Sound Velocity Sums in Crystals
Vol. **24**, No. 1, p. 37–44 (1967).
See also *7*, p. *77*.

H. Böttger
Splitting of ESR Lines by Spin–Phonon Interaction for Ions with Spin 1
Vol. **24**, No. 1, p. 65–75 (1967).
See also 13.4, p. 210; *19*, p. *331*; 22.6, p. 539.

R. Geick, U. Schröder, and J. Stuke
*Lattice Vibrational Properties of Trigonal Selenium
Vol. **24**, No. 1, p. 99–108 (1967).
See also 20.1, p. 351; 22.1.3, p. 478.

D. Bäuerle and B. Fritz
*Infrared Vibrational Absorption by U-Centers in NaI
Vol. **24**, No. 1, p. 207–215 (1967).
See also 10.2, p. 141; *20.1*, p. *351*; 22.5.2, p. 524.

R. Zeyfang
Lattice Thermal Resistivity due to Dislocations in Alloys at Low Temperatures
Vol. **24**, No. 1, p. 221–234 (1967).
See also *8*, p. *83*; 10.1, p. 126; 21.1, p. 401.

H. G. Reik and D. Heese
The Small Polaron Problem with an Application to Optical and DC Data of Reduced Barium Titanate
Vol. 24, No. 1, p. 281—287 (1967).
See also *13*, p. *168*; 14.4.2, p. 267; 20.1, p. 351.

C. N. Plăviţu
On the Phonon Drag Effect Contribution to Thermomagnetic Phenomena
Vol. 24, No. 1, p. 361—372 (1967).
See also 13, p. 168; 14.3, p. 239; *15*, p. *273*; 22, p. 448.

J. B. Page, Jr. and D. Strauch
*Shell Model Treatment of the Vibrations of Crystals Containing Substitutional Defects. (I): U-Center Localized Modes
Vol. 24, No. 2, p. 469—479 (1967).
See also 3.2, p. 35; 10.2, p. 141; 22.5.2, p. 524.

B. Pietrass
*Application of the S.C.F. Method of Lattice Dynamics to the Coulomb Lattice
Vol. 24, No. 2, p. 571—581 (1967).

Ya. A. Iosilevskii
Mössbauer Effect in Complex Multicomponent Systems
Vol. 24, No. 2, p. 749—755 (1967).
See also *6.1*, p. *73*.

V. A. Besfamilnaya, V. F. Elesin, I. A. Kurova, N. N. Ormont, and V. V. Ostroborodova
On the Nature of Oscillations in Impurity Photoconductivity Spectra of Ge
Vol. 24, No. 2, p. 757—762 (1967).
See also 13, p. 168; *16*, p. *281*; 22.1.1, p. 464.

T. O. Baldwin
*The Temperature Dependence of the Borrmann Effect in Copper
Vol. 25, No. 1, p. 71—74 (1968).
See also 5, p. 52; 21.1, p. 402.

L. Merten
*Zur Richtungsdispersion der optischen Gitterschwingungen des tetragonalen $BaTiO_3$ bei Zimmertemperatur
(Orientational Dispersion of Optical Lattice Vibrations of Tetragonal $BaTiO_3$ at Room Temperature)
Vol. 25, No. 1, p. 125—130 (1968).
See also 14.4.2, p. 268.

W. Cochran and A. Zia
*Structure and Dynamics of Perovskite-Type Crystals
Vol. 25 No. 1, p. 273—283 (1968).
See also 14.4.2, p. 268.

J. Röseler
*A New Variational Ansatz in the Polaron Theory
Vol. 25, No. 1, p. 311—316 (1968).
See also *13*, p. *168*.

E. A. Metzbower
Noncentral Force Model for Hexagonal Close-Packed Crystal Lattices (II)
Vol. 25, No. 1, p. 403—416 (1968).
See also *3.2*, p. *36*; 8, p. 84; 12.1, p. 160; 21.1, p. 402.

P. Görlich, H. Karras, G. Kötitz, and R. Rauch
Phonon-Assisted Colour Centre Fluorescence of Additively Coloured Alkali Earth Fluoride Crystals
Vol. **25**, No. 1, p. K15–K18 (1968).
See also *10.2*, p. *141*; *20.3*, p. *368*; 22.5.3, p. 533.

P. Charsley, J. A. M. Salter, and A. D. W. Leaver
*An Experimental Investigation of Dislocation-Phonon Scattering in Some Copper Alloys
Vol. **25**, No. 2, p. 531–540 (1968).
See also 8, p. 84; *10.1*, p. *128*; 21, p. 384; 21.1, p. 402.

M. Luukkala
*On the Hysteresis of Fractional Phonons
Vol. **25**, No. 2, p. K99–K101 (1968).
See also 21, p. 385.

A. Hadni, J. Claudel, and P. Strimer
Far Infrared Absorption and Reflection Spectra of ZnSe, ZnTe, and CdSe at Low Temperature
Vol. **26**, No. 1, p. 241–252 (1968).
See also *20.1*, p. *352*; 22.4.2, p. 507; 22.4.3, p. 510.

H. Böttger
Jahn-Teller Splitting of an ESR Line
Vol. **26**, No. 2, p. 681–692 (1968).
See also *13.4*, p. *212*; 19, p. 333; 22.6, p. 541.

A. E. Lord, Jr.
Sound Wave Velocity Change due to the Magnon–Phonon Interaction
Vol. **26**, No. 2, p. 717–724 (1968).
See also *7*, p. *77*; 13, p. 169; 18.2, p. 311; 22.8, p. 553.

J. Donecker
*Phonon Scattering by Point-Defect Aggregates in AgCl Crystals
Vol. **26**, No. 2, p. K131–K134 (1968).
See also 8, p. 84; 10, p. 106; 22.5.1, p. 516.

L. N. Kurbatov, E. F. Gross, V. E. Mashenko, B. S. Razbirin, N. N. Mochalkin, and A. D. Britov
Coherent Emission of Free Excitons in CdSe Crystals
Vol. **26**, No. 2, p. K139–K142 (1968).
See also *13.2*, p. *194*; *20.2*, p. *361*; 22.4.2, p. 507.

I. N. Krupskii, V. G. Manzhely, and L. A. Koloskova
Thermal Conductivity of Solid Ammonia
Vol. **27**, No. 1, p. 263–268 (1968).
See also *8*, p. *84*; 23, p. 561.

R. Stockmeyer and H. Stiller
*Phonons, Torsons, and Rotational Diffusion in Adamantane
Vol. **27**, No. 1, p. 269–280 (1968).
See also 22.9, p. 560.

S. I. Masharov
*Effect of Phonon Drag on the Kinetic Properties of Alloys
Vol. **27**, No. 1, p. 455–461 (1968).
See also 8, p. 84; 13, p. 170; 14.1, p. 225; 15, p. 274; 21, p. 386.

T. M. Fitzgerald and B. D. Silverman
Ultrasonic Attenuation in Unirradiated and Neutron Irradiated Quartz
Vol. 27, No. 2, p. 473–478 (1968).
See also *7*, p. *77*; 11, p. 152; 22.6, p. 541.

J. Sak
On the Theory of Shallow Impurity States
Vol. 27, No. 2, p. 521–527 (1968).
See also 13.1, p. 186; *13.4*, p. *213*; 22, p. 452.

V. Fedoseev and V. Hizhnyakov
On the Theory of Low-Temperature Exciton Absorption Spectra
Vol. 27, No. 2, p. 751–755 (1968).
See also *13.2*, p. *194*; 20.1, p. 355; 22, p. 452.

W. Haubenreisser
Nonlinear Two-Magnon-One-Phonon Confluence Process in BVC Garnet
Vol. 27, No. 2, p. K145–K147 (1968).
See also 13, p. 170; *18.3*, p. *320*.

Yu. E. Perlin and Sh. N. Gifeisman
The Theory of Shallow F-Centres
Vol. 28, No. 1, p. 45–60 (1968).
See also *10.2*, p. *143*; 13, p. 171; 14.4, p. 254; 18, p. 294; 22.2.1, p. 485; 22.2.2, p. 488; 22.2.3, p. 493; 22.4, p. 495; 22.4.1, p. 503; 22.4.2, p. 508; 22.4.3, p. 511; 22.5.1, p. 517; 22.5.2, p. 527.

A. A. Lipnik and M. A. Ruvinskii
Attenuation and Amplification of Hypersound by Mott Excitons in Semiconductors with Piezoelectric and Deformational Electron–Phonon Interactions
Vol. 28, No. 1, p. 75–81 (1968).
See also *7*, p. *78*; 13.1, p. 187; 13.2, p. 194; 22, p. 452.

V. P. Kalashnikov and A. M. Zlobin
On the Theory of Thermogalvanomagnetic Phenomena in Semiconductors at High Magnetic Fields
Vol. 28, No. 1, p 105–109 (1968).
See also *8*, p. *85*; *14.3*, p. *242*; 22, p. 452.

L. Merten
*Directional Dispersion of Extraordinary Optical Phonons of α-Quartz
Vol. 28, No. 1, p. 111–119 (1968).
See also 20.1, p. 355; 22.6, p. 542.

A. M. Afanasev, Yu. Kagan, and F. N. Chukhovskii
*Dynamical Treatment of the Thermal Diffuse Scattering of X-Rays
Vol. 28, No. 1, p. 287–294 (1968).
See also 5, p. 54.

F. Kelemen, A. Néda, E. Cruceanu, and D. Niculescu
On the Thermal Conductivity of $Zn_xHg_{1-x}Te$ Solid Solutions
Vol. 28, No. 1, p. 421–428 (1968).
See also *8*, p. *85*; 22.4.3, p. 511; 22.8, p. 554.

S. M. Atakishiev, D. Sh. Abdinov, and G. A. Akhundov
Heat Conductivity of Indium Selenide
Vol. 28, No. 1, p. K47–K50 (1968).
See also *8*, p. *85*; 22, p. 453.

D. K. Roy
On the Separation of the Phonon-Assisted Current Components in Germanium Tunnel Diodes
Vol. **28**, No. 1, p. K75–K79 (1968).
See also *14.3.2*, p. *251*; 22.1.1, p. 465.

H. Böttger
*Multiphononabsorption vòn Kristallen mit Punktdefekten im starken Strahlungsfeld (Multiphonon Absorption of Crystals with Point Defects in a Strong Radiation Field)
Vol. **28**, No. 2, p. 611–621 (1968).
See also *10*, p. *109*; 22.5.2, p. 528.

J. Schnakenberg
*Polaronic Impurity Hopping Conduction
Vol. **28**, No. 2, p. 623–633 (1968).
See also *13*, p. *171*; 14.3, p. 242; 22, p. 453.

A. K. Singh and P. K. Sharma
*Thermal Expansion of Cubic Metals in Krebs's Model
Vol. **28**, No. 2, p. 643–648 (1968).
See also *8*, p. *85*; 21, p. 387; 21.1, p. 406; 21.2, p. 425; 21.6, p. 436.

I. F. Chang, S. S. Mitra, J. N. Plendl, and L. C. Mansur
*Long-Wavelength Longitudinal Phonons of Multi-Mode Crystals
Vol. **28**, No. 2, p. 663–673 (1968).
See also 20.1, p. 356; 22.4.1, p. 504; 22.4.2, p. 508; 22.6, p. 542; 22.8, p. 554.

N. Kristoffel and P. Konsin
*Displacive Vibronic Phase Transitions in Narrow-Gap Semiconductors
Vol. **28**, No. 2, p. 731–739 (1968).
See also *1.2*, p. *12*; 14.4.2, p. 269; 22, p. 453.

O. G. Vendik
*Calculation of the Point Lattice Potential of a Bounded Crystal
Vol. **28**, No. 2, p. 789–796 (1968).
See also 22.5.2, p. 528.

Yu. B. Rosenfeld, B. G. Vekhter, and B. S. Tsukerblat
Optical Bands in Electron-Vibrational Systems with Degeneracy
Vol. **28**, No. 2, p. K179–K182 (1968).
See also *20.1*, p. *356*.

W. Kress, H. Borik, and R. K. Wehner
*Infrared Lattice Absorption of Silicon and Germanium
Vol. **29**, No. 1, p. 133–143 (1968).
See also 20.1, p. 356; 22.1.1, p. 466; 22.1.2, p. 474.

K. Pfeuffer
Determination of the Electron Spin–Lattice Relaxation Time T_1 in the Direct Process
Vol. **29**, No. 1, p. 171–177 (1968).
See also *18.1*, p. *297*; 19, p. 335; 22.8, p. 555.

A. N. Basu and S. Sengupta
*A Deformable Shell Model for the Alkali Halides
Vol. **29**, No. 1, p. 367–375 (1968).
See also 22.5.2, p. 528.

M. Luukkala
On Coherent Phonon Fusion in Quartz Transducers
Vol. **29**, No. 1, p. 377–382 (1968).
See also 7, p. *78*; 22.6, p. 543.

A. Czachor
*On the Lattice Dynamics of Hexagonal Structure Metals
Vol. **29**, No. 1, p. 423–434 (1968).
See also 21.3, p. 427.

A. A. Vaipolin, N. A. Goryunova, L. I. Kleshchinskii, G. V. Loshakova, and E. O. Osmanov
The Structure and Properties of the Semiconducting Compound $ZnSnP_2$
Vol. **29**, No. 1, p. 435–442 (1968).
See also 4, p. 47; *5*, p. *54*; 10, p. 110; 14.3, p. 243; 22.8, p. 555.

M. S. R. Chari and N. S. Natarajan
The Phonon–Electron Scattering Coefficients in Dilute Silver–Manganese Alloys
Vol. **29**, No. 1, p. K69–K72 (1968).
See also *8*, p. *85*; 13, p. 172; 21.6, p. 437.

V. Plakhty and W. Cochran
*X-Ray Study of the Phase Transition and Lattice Vibrations of Lanthanum Aluminate
Vol. **29**, No. 1, p. K81–K83 (1968).
See also *1.2*, p. *12*; 22.8, p. 555.

B. G. Dick
One-Phonon Relaxation of Hydroxyl Ions in Alkali Halide Crystals
Vol. **29**, No. 2, p. 587–596 (1968).
See also *10.2*, p. *144*; 22.5.2, p. 529.

D. I. Bower, E. Claridge, and I. S. T. Tsong
Low-Temperature Elastic Constants and Specific Heats of F.C.C. Nickel–Iron Alloys
Vol. **29**, No. 2, p. 617–625 (1968).
See also 8, p. 86; *12.1*, p. *160*; 13, p. 172; 18.2, p. 314; 21.1, p. 407; 21.1.1, p. 422.

D. Bäuerle and B. Fritz
*Vibrational Spectra of U_1-Centres in KI
Vol. **29**, No. 2, p. 639–643 (1968).
See also 10.2, p. 144; *20.1*, p. *357*; 22.5.2, p. 529.

R. G. Howell and D. J. Newman
High Resolution Study of Crystal Field Lines in the Far Infra-Red
Vol. **29**, No. 2, p. 697–706 (1968).
See also 18.1, p. 297; *20.1*, p. *357*; 22.8, p. 555.

J. Dillinger, Č. Koňák, V. Prosser, J. Sak, and M. Zvára
Phonon-Assisted Exciton Transitions in $A^{II}B^{VI}$ Semiconductors
Vol. **29**, No. 2, p. 707–714 (1968).
See also *13.2*, p. *195*; 20.1, p. 357; 22.4.1, p. 505; 22.4.2, p. 508; 22.4.3, p. 511.

K. N. Shrivastava
*Phonon-Induced Hyperfine Coupling of V^{2+} Ion in Cubic Crystalline Fields
Vol. **29**, No. 2, p. 737–742 (1968).
See also 19, p. 335; 22.6, p. 543.

J. E. Lewis, H. Rodot, and P. Haen
The Low-Temperature Thermoelectric Power and Thermal Conductivity of GeTe and of Some GeTe-MnTe Alloys
Vol. **29**, No. 2, p. 743–754 (1968).
See also *8*, p. *86*; 13, p. 173; *15*, p. *274*; 22, p. 454; 22.8, p. 555.

A. Mehra
Optical Absorption of Mn^{2+}-Doped Alkali Halides
Vol. **29**, No. 2, p. 847–857 (1968).
See also *10*, p. *111*; 13.4, p. 215; *20.1*, p. *357*; 22.5.2, p. 529.

R. Ramji Rao and R. Srinivasan
Calculation of the Generalized Grüneisen Parameters for Acoustic Waves in Uniaxial Crystals from the Third Order Elastic Constants
Vol. **29**, No. 2, p. 865–871 (1968).
See also *7*, p. *78*; *8*, p. *86*.

B. Pegel
*Self-Vibrations of Parallel Edge Dislocations
Vol. **29**, No. 2, p. K133–K136 (1968).
See also 10.1, p. 134.

H. Vogt and H. Happ
*The Temperature Dependence of Absorption of Sodium Nitrite in the Far Infrared
Vol. **30**, No. 1, p. 67–72 (1968).
See also 14.4.2, p. 269; *20.1*, p. *358*.

K. H. Timmesfeld
*Anharmonic Broadening of Resonant Modes
Vol. **30**, No. 1, p. 73–85 (1968).
See also 3.2, p. 36; *20.1*, p. *358*; 22.5.2, p. 530.

V. V. Tarassov and A. F. Demidenko
*Heat Capacity and Quasi-Chain Dynamics of Diamond-Like Structures
Vol. **30**, No. 1, p. 147–155 (1968).
See also *8*, p. *86*; 13.1, p. 189; 22.1.1, p. 466; 22.1.2, p. 474; 22.2.1, p. 486; 22.2.2, p. 488; 22.2.3, p. 493.

M. Sodha and S. Sharma
Some Transport Coefficients in Cadmium Sulphide
Vol. **30**, No. 1, p. 239–245 (1968).
See also *14.3*, p. *244*; 22.4.1, p. 505.

G. J. Morgan
Green Functions for "One-Body" Equations Describing the Disordered State
Vol. **30**, No. 1, p. 247–250 (1968).
See also 2, p. 25; *13*, p. *173*.

F. J. Schedewie
Paramagnetic Relaxation in Neodymium-Ethylsulfate: Phonon Re-Absorption and Spectral Spin Diffusion
Vol. **30**, No. 1, p. 251–259 (1968).
See also 13, p. 173; *18.1*, p. *298*; 19, p. 336; 22.8, p. 556.

A. S. Davydov
The Radiationless Transfer of Energy of Electronic Excitation between Impurity Molecules in Crystals
Vol. **30**, No. 1, p. 357–366 (1968).
See also *13.4*, p. *216*; 20.1, p. 358; 20.3, p. 371; 22, p. 454.

R. K. Gupta and N. P. Gupta
*Anharmonic Potential Parameters of the Buckingham Potential for Solid Neon
Vol. **30**, No. 1, p. K41–K43 (1968).
See also 23, p. 562.

L. Merten
*Polariton Dispersion in Biaxial and Uniaxial Crystals
Vol. **30**, No. 2, p. 449—454 (1968).
See also *20*, p. *343*, 22.6, p. 544.

K. Peuker and E. D. Trifonov
Theory of Light Emission by Impurity Centres from Non-Equilibrium Vibrational States
Vol. **30**, No. 2, p. 479—484 (1968).
See also 13.4, p. 217; *20.3*, p. *371*; 22, p. 455.

D. Strauch
*Shell Model Treatment of the Vibrations of Crystals Containing Substitutional Defects. (II): On the Theory of Defect-Induced One-Phonon Absorption
Vol. **30**, No. 2, p. 495—505 (1968).
See also 10, p. 112; *20.1*, p. *359*; 22.1, p. 457; 22.1.1, p. 466; 22.1.2, p. 475.

G. Baldini and M. Guzzi
*Zero-Phonon Transitions of Eu^{++}- and Sm^{++}-Doped Alkali Halides
Vol. **30**, No. 2, p. 601—612 (1968).
See also 10, p. 112; 20.1, p. 359; 20.3, p. 372; 22.5.2, p. 531.

R. Siems
Mechanical Interactions of Point Defects
Vol. **30**, No. 2, p. 645—658 (1968).
See also *10*, p. *112*; *10.1*, p. *135*.

J. Voigt and F. Spiegelberg
On the Dynamical Behaviour of Excitons in CdS Single Crystals. (I): Correlation between Intrinsic Excitonic Absorption and Emission
Vol. **30**, No. 2, p. 659—670 (1968).
See also *13.2*, p. *195*; 20.1, p. 359; 20.3, p. 372; 22.4.1, p. 505.

R. Kishore
*Green Function Theory of Transport Properties of Anharmonic Metals
Vol. **30**, No. 2, p. 817—826 (1968).
See also 8, p. 86; *13*, p. *173*; 14.1, p. 227; 21, p. 389.

H. Böttger
On the Spin–Phonon Interaction in External Electric Field
Vol. **30**, No. 2, p. K119—K123 (1968).
See also 13, p. 173; *19*, p. *336*; 22.6, p. 544.

W. Cochran
*Neutron Scattering and Dielectric Properties of Perovskite-Type Crystals
Vol. **30**, No. 2, p. K157—K160 (1968).
See also 1.2, p. 13; 14.4.2, p. 269.

6.1 Mössbauer Investigations

Original Papers and Short Notes

E. Wieser, W. Meisel und K. Kleinstück
*Mößbauer-Untersuchungen an Mn- und Mg-Mn-Ferriten
(Mössbauer Studies of Mn and Mg-Mn Ferrites)
Vol. **16**, No. 1, p. 127–135 (1966).
See also 18.3, p. 316.

Ya. A. Iosilevskii
*Mössbauer Effect in Some Polyatomic Crystals
Vol. **16**, No 2, p. 633–649 (1966).
See also 22, p. 440.

F. van der Woude
*Mössbauer Effect in α-Fe_2O_3
Vol. **17**, No. 1, p. 417–432 (1966).
See also 18.4, p. 321; 22.6, p. 535.

J. Bara, H. U. Hrynkiewicz, A. Z. Hrynkiewicz, M. Karapandžić, and T. Matlak
*Magnetic Hyperfine Structure of the ^{57}Fe Impurity Nucleus in Non-Magnetic Materials
Vol. **17**, No. 1, p. K53–K55 (1966).
See also 21, p. 374.

A. Szytuła, A. Burewicz, K. Dyrek, A. Hrynkiewicz, D. Kulgawczuk, Z. Obuszko, H. Rżany, and A. Wanic
Susceptibility Measurements of the Antiferromagnetic A-Goethite
Vol. **17**, No. 2, p. K195–K197 (1966).
See also *18.4*, p. *321*; 22, p. 441.

W. Hase und W. Meisel
*Mößbauer-Effekt von Fe^{57} in Sesquioxiden des D 5_3-Strukturtyps
(Mössbauer Effect of Fe^{57} in Sesquioxides of the D 5_3 Structure)
Vol. **18**, No. 1, p. K41–K43 (1966).
See also 22.6, p. 536.

B. C. van Zorge, W. J. Caspers, and A. J. Dekker
*Note on Relaxation Effects in Mössbauer Spectra
Vol. **18**, No. 2, p. 761–768 (1966).

F. Borsa, R. G. Barnes, and R. A. Reese
*Nuclear Magnetic Resonance and Mössbauer Effect Study of Sn^{119} in Rare Earth–Tin Intermetallic Compounds
Vol. **19**, No. 1, p. 359–371 (1967).
See also *19*, p. *328*; 22.7, p. 545.

R. H. Nussbaum and J. G. Dash
*Relaxation Narrowing and Thermal Shift of Mössbauer Lines
Vol. **19**, No. 1, p. K31–K34 (1967).

L. Cser, J. Ostanevich, and L. Pál
*Mössbauer Effect in Iron–Aluminium Alloys (I)
Vol. **20**, No. 2, p. 581–589 (1967).
See also 18, p. 292; 21.1.1, p. 416.

L. CSER, J. OSTANEVICH, and L. PÁL
*Mössbauer Effect in Iron–Aluminium Alloys (II)
Vol. **20**, No. 2, p. 591–596 (1967).
See also 18, p. 292; 21.1.1, p. 416.

B. C. DUTTA, G. HOLTMEYER, and J. KALUS
Study of the Anisotropic Slowing-Down of Low-Energy Magnesium Atoms in an NaCl Single Crystal at Various Temperatures
Vol. **21**, No. 1. p. 291–295 (1967).
See also *11*, p. *148*; 22.5.2, p. 522.

G. HOLTMEYER and J. KALUS
Effect of the Short-Range Order around Magnesium Impurities in Bismuth Melts on the Yield of Nuclear Resonance Fluorescence
Vol. **21**, No. 1, p. 297–302 (1967).
See also 2, p. 23; *11*, p. *148*; 21.7, p. 438.

U. GONSER and R. W. GRANT
*Determination of Spin Directions and Electric Field Gradient Axes in Vivianite by Polarized Recoil-Free γ-Rays
Vol. **21**, No. 1, p. 331–342 (1967).
See also 13, p. 165; 18.1, p. 296; 18.4, p. 322; 22.8, p. 550.

A. BUREWICZ, D. KULGAWCZUK, and A. SZYTUŁA
*Internal Magnetic Field in α-FeOOD
Vol. **21**, No. 1, p. K85–K86 (1967).
See also 18.4, p. 322; 22.8, p. 550.

I. DÉZSI, GY. ERLAKI, and L. KESZTHELYI
*Mössbauer Effect Studies on the $x(Rh_2O_3)1-x(Fe_2O_3)$ System
Vol. **21**, No. 2, p. K121–K125 (1967).
See also 18.4, p. 322; 22.6, p. 538; 22.8, p. 551.

U. BERTELSEN, J. M. KNUDSEN, and H. KROGH
*Mössbauer Effect in FeF_3
Vol. **22**, No. 1, p. 59–64 (1967).
See also 18.4, p. 322; 22.5, p. 513.

G. YA. SELYUTIN
*On the Shift Intensity of the Mössbauer Line in Metals
Vol. **22**, No. 1, p. K23–K26 (1967).
See also 21, p. 380.

G. P. HUFFMAN, P. R. ERRINGTON, and R. M. FISHER
*Mössbauer Study of the Fe-Mn Carbides $(Fe_{1-x}Mn_x)_3C$ and $(Fe_{1.1}Mn_{3.9})C_2$
Vol. **22**, No. 2, p. 473–481 (1967).
See also 18.2, p. 308; 22.8, p. 551.

I. DÉZSI, L. KESZTHELYI, D. KULGAWCZUK, B. MOLNÁR, and N. A. EISSA
*Mössbauer Study of β- and δ-FeOOH and Their Disintegration Products
Vol. **22**, No. 2, p. 617–629 (1967).
See also 1.2, p. 11; 18.3, p. 318; 18.4, p. 323; 22.8, p. 551.

H. GABRIEL
*Mössbauer Spectra in the Presence of Electron Spin Relaxation (I)
Vol. **23**, No. 1, p. 195–205 (1967).
See also 13, p. 167.

E. F. Makarov, V. A. Povitskii, E. B. Granovskii, and A. A. Fridman
*Study of Alnico 8 by Mössbauer Spectroscopy
Vol. **24**, No. 1, p. 45–50 (1967).
See also 18.2, p. 308; 21.1.1, p. 419.

J. M. Trooster and A. Dymanus
*Mössbauer Effect in $Ga_{2-x}Fe_xO_3$ and Related Compounds
Vol. **24**, No. 2, p. 487–499 (1967).
See also 18.3, p. 318; 22.6, p. 540; 22.8, p. 552.

Ya. A. Iosilevskii
*Mössbauer Effect in Complex Multicomponent Systems
Vol. **24**, No. 2, p. 749–755 (1967).
See also 6, p. 64.

E. Wieser, V. A. Povitskii, E. F. Makarov, and K. Kleinstück
*Mössbauer Investigations of Mg, Mn, and Mg-Mn Ferrites in External Magnetic Fields
Vol. **25**, No. 2, p. 607–611 (1968).
See also 18.3, p. 319.

L. Valenta, W. Haubenreisser, and W. Brodkorb
A Contribution to the Problem of the Spatial Distribution of the Spontaneous Magnetization in Thin Films
Vol. **26**, No. 1, p. 191–196 (1968).
See also *18.2*, p. *310*.

Sh. Sh. Bashkirov and G. Ya. Selyutin
*Mössbauer Spectrum Shift Depending on the Degree of Spin-System Ordering
Vol. **26**, No. 1, p. 253–259 (1968).
See also 13, p. 169; 18, p. 294.

F. van der Woude
*On the Interpretation of Broadened and Shifted Lines in Mössbauer Spectra
Vol. **26**, No. 1, p. K37–K38 (1968).
See also 18.4, p. 324; 22.5, p. 513.

S. J. Lewis and P. A. Flinn
*Disappearance of Quadrupole Splitting of the Mössbauer Resonance in Iron–Carbon Austenite at Elevated Temperatures
Vol. **26**, No. 1, p. K51–K54 (1968).
See also 21.1.1, p. 420.

I. Y. Dekhtyar, E. G. Madatova, P. S. Nisin, and R. G. Fedchenko
Plastic Deformation Effect on the Electronic Properties of Solid Solutions with Localized Magnetic Moments
Vol. **26**, No. 2, p. K151–K154 (1968).
See also *10.1*, p. *130*, 14.1, p. 225; 18.1, p. 297; 21.6, p. 436.

L. Cser, I. Dézsi, I. Gladkih, L. Keszthelyi, D. Kulgawczuk, N. A. Eissa, and E. Sterk
*Mössbauer Study of Hyperfine Fields in Mn-Zn Ferrites
Vol. **27**, No. 1, p. 131–138 (1968).
See also 18.3, p. 319.

H. Gabriel, J. Bosse, and K. Rander
*Mössbauer Spectra in the Presence of Electron Spin Relaxation (II)
Vol. **27**, No. 1, p. 301–312 (1968).
See also 13, p. 170.

B. H. ZIMMERMANN, H. JENA, G. ISCHENKO, H. KILIAN, and D. SEYBOTH
*Mössbauer Experiments with Coulomb-Excited ^{73}Ge after Coulomb Recoil Implantation
Vol. **27**, No. 2, p. 639–651 (1968).
See also 22.1.1, p. 465.

K. HENNIG
*Study of Iron Ions in ^{57}Fe-Doped AgCl Using the Mössbauer Effect
Vol. **27**, No. 2, p. K115–K116 (1968).
See also 10, p. 108; 22.5.1, p. 517.

K. HENNIG, KIM YUNG, and B. S. SKORCHEV
On States of Iron Impurity Ions in AgCl and NaCl Investigated by the Mössbauer Method
Vol. **27**, No. 2, p. K161–K164 (1968).
See also *10*, p. *108*; 22.5.1, p. 517; 22.5.2, p. 527.

I. BUNGET
*Note on Mössbauer Effect in $MnFe_2O_4$
Vol. **28**, No. 1, p. K39–K42 (1968).
See also 18.3, p. 320.

V. V. NEMOSHKALENKO, O. N. RASUMOV, and V. V. GORSKII
*Investigation of the Mössbauer Effect in Some Fe-Al Alloys
Vol. **29**, No. 1, p. 45–48 (1968).
See also 13, p. 172; 21.1.1, p. 422.

S. E. GUKASYAN and V. S. SHPINEL
*Mössbauer Effect in Antimony-Organic Compounds
Vol. **29**, No. 1, p. 49–52 (1968).
See also 22.9, p. 560.

M. A. KRIVOGLAZ and L. B. KVASHNINA
*On the Spin Broadening of Mössbauer Lines of Impurities in Ferromagnetics and of Atoms in Antiferromagnetics
Vol. **29**, No. 1, p. 61–73 (1968).
See also 13, p. 172; 18.2, p. 314; 18.4, p. 324.

V. BECKMANN, W. BRUCKNER, W. FUCHS, G. RITTER, and H. WEGENER
The Measurement of the Anisotropy Constant of Antiferromagnetic $FePO_4$ by Means of the Mössbauer Effect
Vol. **29**, No. 2, p. 781–791 (1968).
See also *18.4*, p. *325*; 22.8, p. 555.

R. C. THIEL and C. B. VAN DEN BERG
*Temperature Dependence of Hyperfine Interactions in Near-Stoichiometric FeS. (I): Experiment
Vol. **29**, No. 2, p. 837–846 (1968).
See also 1.2, p. 13; 18.4, p. 325; 22, p. 454.

J. PIEKOSZEWSKI, J. SUWALSKI, and S. LIGENZA
*Mössbauer Effect Study in Chalcopyrite
Vol. **29**, No. 2, p. K99–K101 (1968).
See also 22.8, p. 556.

I. DÉZSI, N. A. EISSA, L. KESZTHELYI, B. MOLNÁR, and D. L. NAGY
*Mössbauer Study of $SnCl_2$ and $Dy(ClO_4)_3$ in Ice
Vol. **30**, No. 1, p. 215–218 (1968).
See also 2, p. 24; 22.8, p. 556.

G. Albanese and M. Musci
*Mössbauer Resonant Absorption in $B8_2$ Germanium Compounds
Vol. **30**, No. 1, p. K37—K40 (1968).
See also 22.8, p. 557.

J. Celý
*Mössbauer Effect in Simple Cubic Lattice with Line Defect
Vol. **30**, No. 2, p. 671—683 (1968).
See also 10, p. 112; 10.1, p. 135.

7. Acoustic Properties of Solids

Review Article

J. Mertsching
*Theory of Electromagnetic Waves in Metals and Their Interaction with Ultrasonic Waves (II)
Vol. **26**, No. 1, p. 9–64 (1968).
See also 13, p. 162; 19, p. 326; 20, p. 337; 21, p. 373.

Original Papers and Short Notes

J. Mertsching
*Zur Bestimmung der freien Weglänge der Elektronen aus Messungen der Ultraschall-absorption in Metallen
(On the Determination of the Electron Free Path from Measurements of Ultrasonic Absorption in Metals)
Vol. **16**, No. 1, p. 267–275 (1966).
See also 13, p. 162; 21, p. 373.

G. Kluge
*Akustische Aktivität
(Acoustic Activity)
Vol. **17**, No. 1, p. 109–118 (1966).
See also 6, p. 56; 22.6, p. 535.

R. Laiho, P. Ketolainen, and E. Mäntysalo
Ultrasonic Modulation of F-Band in Additively Colored KBr
Vol. **18**, No. 2, p. K143–K144 (1966).
See also *10.2*, p. *138*; 22.5.2, p. 520.

H. Markert
*Decrease of Number and Size of Barkhausen Jumps during Ultrasonic Irradiation
Vol. **20**, No. 1, p. K67–K70 (1967).
See also 18.2, p. 306; 18.3, p. 317; 21.1.1, p. 416.

G. O. Müller, H. Peibst, E. Schnürer, and H. Thiel
The Influence of High Phonon Flux Densities on the X-Ray Reflectivity of Nearly Ideal CdS Crystals
Vol. **20**, No. 2, p. K173–K175 (1967).
See also *6*, p. *60*; 22.4.1, p. 499.

K. M. Koliwad, P. B. Ghate, and A. L. Ruoff
Pressure Derivatives of the Elastic Constants of NaBr and KF
Vol. **21**, No. 2, p. 507–516 (1967).
See also *12*, p. *157*; 22.5.2, p. 522.

J.-P. Ropé
Experimental Study of the Dynamics of High-Field Domains in Electron–Phonon Coupling
Vol. **21**, No. 2, p. 517–524 (1967).
See also *14.4.1*, p. *258*; 22.2.1, p. 484.

A. Bonnot
Relaxation of High-Field Domains and Light Emission Associated with Current Oscillations in GaAs
Vol. **21**, No. 2, p. 525–531 (1967).
See also *14.4.1*, p. *258*; *20.3*, p. *366*; 22.2.1, p. 484.

E. A. Kaner and V. L. Falko
Magnetoacoustic Resonance Effects in Metals in Inclined Magnetic Fields
Vol. **22**, No. 1, p. 319–331 (1967).
See also 13. 1, p. 179; *19*, p. *329*; 21, p. 380.

Ľ. Hrivnák and J. Kovár
*Absorption of Ultrasound in Indium Antimonide
Vol. **23**, No. 1, p. 189–193 (1967).
See also 8, p. 83; 22.2.3, p. 491.

S. G. Kalashnikov, A. I. Morosov, and V. V. Proklov
*Some Data on Acousto-Electrical Domains in Photoconductive Cadmium Sulphide
Vol. **23**, No. 1, p. K9–K12 (1967).
See also *14.4.1*, p. *260*; 22.4.1, p. 501.

E. R. Fitzgerald and T. W. Wrigth
*Invariance of Sound Velocity Sums in Crystals
Vol. **24**, No. 1, p. 37–44 (1967).
See also 6, p. 63.

B. K. Jones
*Measurement of the Magnitude of the Electronic Attenuation of Sound in Metals
Vol. **24**, No. 1, p. 297–300 (1967).
See also 13.1, p. 182; 21, p. 382; 21.1, p. 401; 21.2, p. 425.

I. V. Ioffe and A. L. Kazakov
*On the Theory of Ferroacoustic Resonance
Vol. **24**, No. 1, p. K9–K10 (1967).
See also *19*, p. *331*.

B. Autin
Transient Acousto-Electric Phenomena in Semiconducting Piezo-Electric Crystals
Vol. **24**, No. 2, p. 501–508 (1967).
See also *14.4.1*, p. *261*; 22, p. 448.

H. L. Hartnagel
*Some Ideas on Electroacoustic Domains
Vol. **24**, No. 2, p. K133–K135 (1967).
See also *14.4.1*, p. *261*; 22, p. 449.

R. H. T. Yeh
*Ultrasonic Attenuation of Anisotropic Superconductor in $ql \ll 1$ Limit
Vol. **25**, No. 2, p. 551–555 (1968).
See also *14.2*, p. 230.

A. A. Grinberg and A. A. Kastalskii
*The Ultrasound Generation under the Gunn Effect in a Piezoelectric Material
Vol. **26**, No. 1, p. 219–224 (1968).
See also *14.4.1*, p. *262*; 22.2.1, p. 485.

A. E. Lord, Jr.
*Sound Wave Velocity Change due to the Magnon-Phonon Interaction
Vol. **26**, No. 2, p. 717–724 (1968).
See also 6, p. 65; 13, p. 169; 18.2, p. 311; 22.8, p. 553.

T. M. Fitzgerald and B. D. Silverman
*Ultrasonic Attenuation in Unirradiated and Neutron Irradiated Quartz
Vol. **27**, No. 2, p. 473–478 (1968).
See also 6, p. 66; 11, p. 152; 22.6, p. 541.

S. A. MIRONOV, K. V. SHEVLYAGIN, A. G. GUREVICH, and B. M. LEBED
Excitation of Magnetostatic, Magnetoelastic, and Acoustic Waves in Ca-Bi-V-Fe Garnet
Vol. **27**, No. 2, p. K85—K87 (1968).
See also *18.3*, p. *320*.

A. A. LIPNIK and M. A. RUVINSKII
*Attenuation and Amplification of Hypersound by Mott Excitons in Semiconductors with Piezoelectric and Deformational Electron–Phonon Interactions
Vol. **28**, No. 1, p. 75—81 (1968).
See also 6, p. 66; 13.1, p. 187; 13.2, p. 194; 22, p. 452.

M. LUUKKALA
*On Coherent Phonon Fusion in Quartz Transducers
Vol. **29**, No. 1, p. 377—382 (1968).
See also 6, p. 67; 22.6, p. 543.

R. RAMJI RAO and R. SRINIVASAN
*Calculation of the Generalized Grüneisen Parameters for Acoustic Waves in Uniaxial Crystals from the Third Order Elastic Constants
Vol. **29**, No. 2, p. 865—871 (1968).
See also 6, p. 69; *8*, p. *86*.

D. BOCCON-GIBOD
Modulation of Near Infrared Radiation by Acoustoelectric Domains in n-Type GaAs at Room Temperature
Vol. **30**, No. 2, p. 525—531 (1968).
See also *14.4.1*, p. *264*; *20.1*, p. *359*; 22.2.1, p. 486.

N. E. BYER and H. S. SACK
*Ultrasonic Velocity and Attenuation in Alkali Halides Containing CN^-. (I): Experimental Results
Vol. **30**, No. 2, p. 569—577 (1968).
See also 10, p. 112; 13.4, p. 217; 22.5.2, p. 530.

N. E. BYER and H. S. SACK
*Ultrasonic Velocity and Attenuation in Alkali Halides Containing CN^-. (II): Theoretical Interpretation
Vol. **30**, No. 2, p. 579—586 (1968).
See also 10, p. 112; 13.4, p. 217; 22.5.2, p. 531.

R. K. L. GAY and H. L. HARTNAGEL
The Spatial and Time Variation of Electric Fields in Semiconducting CdS
Vol. **30**, No. 2, p. 755—758 (1968).
See also *14.4.1*, p. *265*; 22.4.1, p. 505.

8. Thermal Properties of Solids

Review Articles

A. S. Borshchevskii, N. A. Goryunova, F. P. Kesamanly, and D. N. Nasledov
Semiconducting $A^{II}B^{IV}C_2^V$ Compounds
Vol. **21**, No. 1, p. 9—55 (1967).
See also 3.1, p. 29; 3.2, p. 34; 5, p. 49; 10, p. 93; 13.1, p. 174; *14.3*, p. *231*; 15, p. 270; 16, p. 276; 17, p. 286; 18, p. 291; 20.1, p. 344; 22.8, p. 547.

Original Papers and Short Notes

C. N. Plǎviţu
Some Kinetic Phenomena in Impure Ionic Semiconductors
Vol. **16**, No. 1, p. 69—78 (1966).
See also 13, p. 162; *14.4*, p. *252*; 15, p. 270; 22, p. 439.

J. S. Rajput and A. K. Gupta
*On the Specific Heat Anomaly in Superconducting Lead
Vol. **16**, No. 1, p. K51—K53 (1966).
See also *14.2*, p. *228*; 21, p. 373.

M. M. Shukla and B. Dayal
Lattice Vibrations of Aluminium on the Basis of Krebs's Model
Vol. **16**, No. 2, p. 513—516 (1966).
See also *6*, p. *56*; 21, p. 374.

S. W. Kennedy, G. F. Taylor, and J. H. Patterson
Misfit andHysteresis at Structure Transformations in Rubidium and CesiumNitrates; and Thermal Expansions
Vol. **16**, No. 2, p. K175—K178 (1966).
See also *1.2*, p. *9*; 3, p. 26; 22.8, p. 547.

K. N. Pathak and B. Deo
*Thermal Expansion of a Linear Lattice
Vol. **17**, No. 1, p. 77—81 (1966).

M. E. Ertl
*Thermal Resistivity due to Mass Fluctuation Scattering in Bismuth-Antimony Alloys
Vol. **17**, No. 1, p. K63—K65 (1966).
See also 6, p. 57; 21.7, p. 438.

V. G. Manzhelii, V. G. Gavrilko, and E. I. Voitovich
*Thermal Expansion of Solidified Rare Gases
Vol. **17**, No. 1, p. K139—K141 (1966).
See also 23, p. 561.

R. Gleiss
Calculation of the Debye-Waller Factor of Nickel
Vol. **17**, No. 2, p. 761—767 (1966).
See also *6*, p. *57*; 21.1, p. 393.

G. B. Abdullaev, M. I. Aliev, and G. I. Safaraliev
*About Phonon Scattering in Doped GaSb Crystals
Vol. **17**, No. 2, p. 777—781 (1966).
See also 6, p. 57; 22.2.3, p. 490.

B. STALIŃSKI, Z. BIEGAŃSKI, and R. TROĆ
*Low-Temperature Heat Capacity and Thermodynamical Functions of Ferromagnetic Uranium Phosphide U_3P_4
Vol. **17**, No. 2, p. 837–841 (1966).
See also 1.2, p. 9; 18.2, p. 303; 22, p. 441.

R. P. GUPTA, P. K. SHARMA, and S. PAL
*Temperature Variation of the Grüneisen Parameter of Copper
Vol. **18**, No. 1, p. 119–124 (1966).
See also 6, p. 57; 21.1, p. 393.

W. NEUMANN
*Die Wärmeleitfähigkeit des gelben PbO zwischen 0 und 500 °C
(Thermal Conductivity of Yellow PbO between 0 and 500 °C)
Vol. **18**, No. 1, p. 151–156 (1966).
See also 6, p. 57; 22.6, p. 535.

V. A. KUTASOV und I. A. SMIRNOV
*Der Einfluß von überschüssigem antistrukturellem Bi auf die Wärmeleitfähigkeit des Kristallgitters von Bi_2Te_3, Bi_2Te_3-Sb_2Te_3 und Bi_2Te_3-Bi_2S_3 mit Jodbeimischung
(Effect of Excess Antistructural Bi on the Lattice Thermal Conductivity of Bi_2Te_3, Bi_2Te_3-Sb_2Te_3, and Bi_2Te_3-Bi_2S_3 Containing Iodine)
Vol. **18**, No. 1, p. 479–488 (1966).
See also *10*, p. *96*; 14.3, p. 233; 15, p. 271; 22, p. 441; 22.8, p. 548.

O. HAUSER and M. SCHENK
Radiation-Induced Phase Transitions of Some Perovskite-Type Substances and Their Thermodynamic Treatment
Vol. **18**, No. 2, p. 547–555 (1966).
See also *1.2*, p. *10*; 11, p. 147; 14.4.2, p. 266.

N. P. GUPTA and B. DAYAL
*Effect of Zero-Point Energy on the Specific Heats of Solidified Neon and Xenon
Vol. **18**, No. 2, p. 731–736 (1966).
See also 3.2, p. 34; *6*, p. *58*; 23, p. 561.

G. D. GUSEINOV and A. I. RASULOV
*Heat Conductivity Study of GaSe Monocrystals
Vol. **18**, No. 2, p. 911–922 (1966).
See also 6, p. 58; 22, p. 442.

G. D. GUSEINOV, M. Z. ISMAILOV, and A. G. TALYBOV
On the New Semiconducting Compound $HgTlS_2$
Vol. **18**, No. 2, p. 929–935 (1966).
See also *5*, p. *50*; 13.4, p. 204; 14.3, p. 234; 16, p. 278; 22.8, p. 548.

G. ALBRECHT and G. WOLF
Electron Specific Heat and λ-Anomaly of $CrH_{0.84}$
Vol. **18**, No. 2, p. K119–K122 (1966).
See also *13*, p. *164*; 22, p. 442.

G. S. NIKOLSKII and V. V. EREMENKO
*The Effect of a Magnetic Field on the Heat Conductivity of Er Polycrystals
Vol. **18**, No. 2, p. K123–K125 (1966).
See also 14.1, p. 222; 21.4, p. 428.

R. DE BATIST, R. GEVERS, and M. VERSCHUEREN
*Magnon Contribution to the Low-Temperature Specific Heat of UO_2
Vol. **19**, No. 1, p. 77–88 (1967).
See also 18.4, p. 322; 22.6, p. 536.

M. S. R. Chari
The Electrical and Thermal Resistivities of Iron
Vol. **19**, No. 1, p. 169—172 (1967).
See also *14.1*, p. *222*; 21.1.1, p. 415.

G. D. Guseinov, A. I. Rasulov, E. M. Kerimova, and M. Z. Ismailov
*On Heat Conductivity of $A^{III}B^{VI}$-Type Semiconductors
Vol. **19**, No. 1, p. K7—K10 (1967).
See also 22, p. 442.

P. Charsley and J. A. M. Salter
*Further Comments on the Analysis of Measurements of the Lattice Thermal Conductivity
Vol. **19**, No. 1, p. K63—K65 (1967).
See also 21, p. 377; 21.1, p. 396.

E. I. Gavrilitsa and S. I. Radautsan
Influence of Ordering on the Properties of Solid Solutions of the System $(HgSe)_{3x}$-$(In_2Se_3)_{1-x}$
Vol. **19**, No. 2, p. 609—611 (1967).
See also *1.2*, p. *10*; 14.3, p. 235; 22.8, p. 549.

V. V. Tarassov
Heat Capacity and Structure of Vitreous Silica and Diamond-Like Lattices (I)
Vol. **20**, No. 1, p. 37—43 (1967).
See also 2, p. 22; *6*, p. *59*; 22, p. 443.

V. V. Tarassov
Heat Capacity and Structure of Vitreous Silica and Diamond-Like Lattices (II)
Vol. **20**, No. 1, p. 45—57 (1967).
See also 2, p. 22; *6*, p. *59*; 22, p. 443.

J. Šedivý and H. Šíchová
*Röntgenographische Messung der charakteristischen Debye-Temperatur bei vielkristallinem Silber mit Anwendung der Korrektur auf primäre Extinktion
(X-Ray Measurement of the Characteristic Debye Temperature of Polycrystalline Silver Correcting for Primary Extinction)
Vol. **20**, No. 1, p. 249—254 (1967).
See also 21.6, p. 432.

R. P. Gupta
*Lattice Specific Heats of Thallium and Yttrium
Vol. **20**, No. 1, p. 291—294 (1967).
See also 6, p. 59; 21, p. 377.

N. P. Gupta and B. Dayal
*Equation of State of Solid Argon and Krypton
Vol. **20**, No. 1, p. 321—325 (1967).
See also 23, p. 561.

E. A. Metzbower
Noncentral Force Model for Hexagonal Close-Packed Crystal Lattices
Vol. **20**, No. 2, p. 681—692 (1967).
See also 3.2, p. 35; *6*, p. *60*; 12.1, p. 160; 21.1, p. 397; 21.3, p. 426.

N. A. Chernoplekov, G. Kh. Panova, M. G. Zemlyanov, B. N. Samoilov, and V. I. Kutaitsev
Investigation of the Quasi-Local Level in the Vibrational Spectrum of the Ti Lattice with Heavy Impurity Atoms
Vol. **20**, No. 2, p. 767—770 (1967).
See also *6*, p. *60*; 21.1, p. 397.

A. A. Abdurakhmanova and M. I. Aliev
*On the Mechanism of Thermal Conductivity in $(GaSb)_{3x}$-$(Ga_2Te_3)_{1-x}$ and $(GaSb)_x$-$(GaTe)_{1-x}$ Solid Solutions
Vol. **20**, No. 2, p. 777—781 (1967).
See also 6, p. 60; 22.8, p. 550.

M. Ali and P. Nagels
Evaluation of the Debye Temperature of Thorium Dioxide
Vol. **21**, No. 1, p. 113—116 (1967).
See also *6*, p. *60*; 22.6, p. 537.

O. Budke and J. Krempaský
*Measurement of Thermal Parameters of Thin Films and Plane-Parallel Plates Located on a Thermally Conductive Base
Vol. **21**, No. 1, p. 175—182 (1967).
See also 22.2.3, p. 491; 22.5.2, p. 522.

M. F. Rose
Higher Order Elastic Constants and the Equation of State of F.C.C. Metals
Vol. **21**, No. 1, p. 235—238 (1967).
See also 6, p. 60; *12.1*, p. *160*; 21, p. 378; 21.1, p. 297; 21.6, p. 433.

V. Nüsslein and U. Schröder
Calculations of Dispersion Curves and Specific Heat for LiF and NaCl Using the Breathing Shell Model
Vol. **21**, No. 1, p. 309—314 (1967).
See also *6*, p. *61*; 22.5.2, p. 522.

J. A. M. Salter and P. Charsley
*The Effect of Grain Size on the Lattice Thermal Conductivity of Copper Aluminium Alloys
Vol. **21**, No. 1, p. 357—368 (1967).
See also 4, p. 42; 6, p. 61; 21, p. 379; 21.1, p. 397.

F. Kelemen, A. Néda, D. Niculescu, and E. Cruceanu
*On the Thermal Conductivity of CdTe and Some $CdTe_{1-x}S_x$ and $CdTe_{1-x}Se_x$ Solid Solutions
Vol. **21**, No. 2, p. 557—565 (1967).
See also 22.4.1, p. 500; 22.4.2, p. 506; 22.4.3, p. 509; 22.8, p. 550.

D. G. Arasli and M. I. Aliev
*Influence of Defects and of the Interation between Them on Phonon Scattering in Heavily Doped Ge and Si Crystals
Vol. **21**, No. 2, p. 643—648 (1967).
See also 6, p. 61; 10, p. 100; 22.1.1, p. 462; 22.1.2, p. 471.

N. P. Gupta and B. Dayal
*Thermal Expansion and Compressibility of Solid Neon and Xenon
Vol. **21**, No. 2, p. 661—664 (1967).
See also 6, p. 61; *12*, p. *157*; 23, p. 561.

N. S. Natarajan and M. S. R. Chari
*The Anomalous Thermal Conductivity of Dilute Ag-Mn Alloys at Helium Temperatures
Vol. **21**, No. 2, p. K127—K130 (1967).
See also 21.6, p. 433.

A. M. de Goer and B. Dreyfus
*γ-Irradiation Effect on the Thermal Conductivity of Al_2O_3 with Cr or Mn Impurities at Low Temperature
Vol. **22**, No. 1, p. 77–82 (1967).
See also 10, p. 101; *11*, p. *149*; 22.6, p. 538.

S. A. Zeinalov and M. I. Aliev
*Investigation of the Thermal Conductivity of InSb-GaSb Solid Solutions
Vol. **22**, No. 1, p. 153–156 (1967).
See also 22.2.3, p. 491; 22.8, p. 551.

B. Mukherjee and A. J. F. Boyle
On the Debye Characteristic Temperature of Diamond
Vol. **22**, No. 2, p. K131–K134 (1967).
See also *6*, p. *62*; 12, p. 157; 22.1, p. 457.

Ľ. Hrivnák and J. Kovár
Absorption of Ultrasound in Indium Antimonide
Vol. **23**, No. 1, p. 189–193 (1967).
See also *7*, p. *77*; 22.2.3, p. 491.

J. P. Peverley and P. H. E. Meijer
*Entropy and Susceptibility of a Pure Dipole-Dipole Substance
Vol. **23**, No. 1, p. 353–360 (1967).
See also *18*, p. *293*; 22.8, p. 552.

H. P. Singh and B. Dayal
*X-Ray Determination of the Thermal Expansion of Zinc Selenide
Vol. **23**, No. 1, p. K93–K95 (1967).
See also 5, p. 52; 22.4.2, p. 507.

V. A. Finkel, Y. N. Smirnov und V. V. Vorobjov
*Zum Problem der thermischen Ausdehnung von Gadolinium
(On the Problem of the Thermal Expansion of Gadolinium)
Vol. **23**, No. 1, p. K97–K98 (1967).
See also 21.4, p. 428.

A. R. Adams, F. Baumann, and J. Stuke
*Thermal Conductivity of Selenium and Tellurium Single Crystals and Phonon Drag of Tellurium
Vol. **23**, No. 1, p. K99–K104 (1967).
See also 13, p. 167; 15, p. 273; 22.1.3, p. 477.

L. Wojtczak
*The Melting Point of Thin Films
Vol. **23**, No. 2, p. K163–K166 (1967).
See also 6, p. 63; 21, p. 382; 21.1, p. 400; 21.6, p. 434.

R. Zeyfang
*Der Gitterwärmewiderstand durch Versetzungen in Legierungen bei tiefen Temperaturen
(Lattice Thermal Resistivity due to Dislocations in Alloys at Low Temperatures)
Vol. **24**, No. 1, p. 221–234 (1967).
See also 6, p. 63; 10.1, p. 126; 21.1, p. 401.

V. A. Ivanova, D. Sh. Abdinov, and G. M. Aliev
Some Structural, Electrical, and Thermal Parameters of VCr_2Se_4, $FeCr_2Se_4$, and $NiCr_2Se_4$ Ternary Compounds
Vol. **24**, No. 1, p. K23–K24 (1967).
See also 5, p. 52; *14.3*, p. *239*; 15, p. 273; 22.8, p. 552.

I. N. KRUPSKII and V. G. MANZHELY
*Thermal Conductivity of Solid Argon
Vol. **24**, No. 1, p. K53–K56 (1967).
See also 23, p. 561.

V. A. IVANOVA, D. SH. ABDINOV, and G. M. ALIEV
*On Some Characteristics of Chromium Selenides
Vol. **24**, No. 2, p. K145–K147 (1967).
See also 5, p. 52; 14.3, p. 239; *15*, p. *273*; 22, p. 449.

J. SAK
Temperature Dependence of the Band Structure
Vol. **25**, No. 1, p. 155–161 (1968).
See also *13.1* p. *183*.

E. A. METZBOWER
Noncentral Force Model for Hexagonal Close-Packed Crystal Lattices (II)
Vol. **25**, No. 1, p. 403–416 (1968).
See also *3.2*, p. *36*; 6, p. 64; 12.1, p. 160; 21.1, p. 402.

J. SÓLYOM and A. ZAWADOWSKI
*On the Specific Heat of Dilute Magnetic Alloys
Vol. **25**, No. 1, p. 473–483 (1968).
See also 18, p. 293.

P. CHARSLEY, J. A. M. SALTER, and A. D. W. LEAVER
An Experimental Investigation of Dislocation-Phonon Scattering in Some Copper Alloys
Vol. **25**, No. 2, p. 531–540 (1968).
See also *6*, p. *65*; *10.1*, p. *128*; 21, p. 384; 21.1, p. 402.

E. O. OSMANOV, YU. V. RUD, and M. E. STRYALKOVSKII
High-Temperature Modifications of the Semiconductor Compounds $CdSnAs_2$ and $CdGeAs_2$
Vol. **26**, No. 1, p. 85–90 (1968).
See also *1.2*, p. *12*; 2, p. 24; 12, p. 158; 14.3, p. 240; 22.7, p. 546; 22.8, p. 553.

M. I. BAGATSKII, V. A. KUCHERYAVY, V. G. MANZHELII, and V. A. POPOV
*Thermal Capacity of Solid Nitrogen
Vol. **26**, No. 2, p. 453–460 (1968).
See also 23, p. 561.

J. DONECKER
Phonon Scattering by Point-Defect Aggregates in AgCl Crystals
Vol. **26**, No. 2, p. K131–K134 (1968).
See also *6*, p. *65*; 10, p. 106; 22.5.1, p. 516.

I. N. KRUPSKII, V. G. MANZHELY, and L. A. KOLOSKOVA
*Thermal Conductivity of Solid Ammonia
Vol. **27**, No. 1, p. 263–268 (1968).
See also 6, p. 65; 23, p. 561.

S. I. MASHAROV
Effect of Phonon Drag on the Kinetic Properties of Alloys
Vol. **27**, No. 1, p. 455–461 (1968).
See also *6*, p. *65*; 13, p. 170; 14.1, p. 225; 15, p. 274; 21, p. 386.

B. A. Strukov, M. Amin, and V. A. Kopchik
*Comparative Investigation of the Specific Heat of KH_2PO_4 (KDP) and KD_2PO_4 (DKDP) Single Crystals
Vol. 27, No. 2, p. 741—749 (1968).
See also 1.2, p. 12; 14.4.2, p. 268.

A. I. Mitsek, N. P. Kolmakova, V. A. Feigin, and P. F. Gaidanskii
Spin-Wave Distribution Function and Thermodynamics of the Heisenberg Ferromagnetic
Vol. 28, No. 1, p. 83—96 (1968).
See also 13, p. 171; *18.2*, p. *312*.

V. P. Kalashnikov and A. M. Zlobin
*On the Theory of Thermogalvanomagnetic Phenomena in Semiconductors at High Magnetic Fields
Vol. 28, No. 1, p. 105—109 (1968).
See also 6, p. 66; *14.3*, p. *242*; 22, p. 452.

F. Kelemen, A. Néda, E. Cruceanu, and D. Niculescu
*On the Thermal Conductivity of $Zn_xHg_{1-x}Te$ Solid Solutions
Vol. 28, No. 1, p. 421—428 (1968).
See also 6, p. 66; 22.4.3, p. 511; 22.8, p. 554.

S. M. Atakishiev, D. Sh. Abdinov, and G. A. Akhundov
*Heat Conductivity of Indium Selenide
Vol. 28, No. 1, p. K47—K50 (1968).
See also 6, p. 66; 22, p. 453.

J.-U. Arnold and P. Grosse
The Temperature Dependence of the Atom Coordinates in the Tellurium Lattice
Vol. 28, No. 1, p. K93—K96 (1968).
See also *3*, p. *27*; 5, p. 54; 22.1.3, p. 479.

A. van den Beukel
*Specific Heat of Ice near the "Ferroelectric" Transition Temperature
Vol. 28, No. 2, p. 565—568 (1968).
See also 14.4.2, p. 269.

A. K. Singh and P. K. Sharma
*Thermal Expansion of Cubic Metals in Krebs's Model
Vol. 28, No. 2, p. 643—648 (1968).
See also *6*, p. *67*; 21, p. 387; 21.1, p. 406; 21.2, p. 425; 21.6, p. 436.

Ya. S. Feiziev, Ya. N. Nasirov, and O. N. Novruzov
*Thermal Properties of Tin Sulphide Single Crystals
Vol. 28, No. 2, p. K113—K115 (1968).
See also 22, p. 453.

L. Śniadower, M. Psoda, and R. R. Gałązka
*X-Ray Measurements of Lattice Dilatation in HgTe
Vol. 28, No. 2, p. K121—K123 (1968).
See also *5*, p. *54*; 22.4.3, p. 511.

M. S. R. Chari and N. S. Natarajan
*The Phonon–Electron Scattering Coefficients in Dilute Silver–Manganese Alloys
Vol. 29, No. 1, p. K69—K72 (1968).
See also 6, p. 68; 13, p. 172; 21.6, p. 437.

D. I. Bower, E. Claridge, and I. S. T. Tsong
Low-Temperature Elastic Constants and Specific Heats of F.C.C. Nickel–Iron Alloys
Vol. **29**, No. 2, p. 617–625 (1968).
See also 6, p. 68; *12.1*, p. *160*; 13, p. 172; 18.2, p. 314; 21.1, p. 407; 21.1.1, p. 422.

J. E. Lewis, H. Rodot, and P. Haen
*The Low-Temperature Thermoelectric Power and Thermal Conductivity of GeTe and of Some GeTe–MnTe Alloys
Vol. **29**, No. 2, p. 743–754 (1968).
See also 6, p. 68; 13, p. 173; *15*, p. *274*; 22, p. 454; 22.8, p. 555.

P. P. Salhotra, E. C. Subbarao, and P. Venkateswarlu
Polymorphism of Rubidium Nitrate
Vol. **29**, No. 2, p. 859–864 (1968).
See also *1.2*, p. *13*; *3*, p. *28*; 22.8, p. 556.

R. Ramji Rao and R. Srinivasan
*Calculation of the Generalized Grüneisen Parameters for Acoustic Waves in Uniaxial Crystals from the Third Order Elastic Constants
Vol. **29**, No. 2, p. 865–871 (1968).
See also 6, p. 69; *7*, p. *78*.

V. V. Tarassov and A. F. Demidenko
*Heat Capacity and Quasi-Chain Dynamics of Diamond-Like Structures
Vol. **30**, No. 1, p. 147–155 (1968).
See also *6*, p. *69*; 13.1, p. 189; 22.1.1, p. 466; 22.1.2, p. 474; 22.2.1, p. 486; 22.2.2, p. 488; 22.2.3, p. 493.

R. Kishore
Green Function Theory of Transport Properties of Anharmonic Metals
Vol. **30**, No. 2, p. 817–826 (1968).
See also *6*, p. *70*; *13*, p. *173*; 14.1, p. 227; 21, p. 389.

9. Diffusion in Solids

Review Articles

P. Süptitz and J. Teltow
*Transport of Matter in Simple Ionic Crystals (Cubic Halides)
Vol. **23**, No. 1, p. 9–56 (1967).
See also *10*, p. *93*; 14.4, p. 252; 15, p. 270; 22.5.1, p. 515; 22.5.2, p. 518; 22.5.3, p. 532.

A. Seeger and K. P. Chik
*Diffusion Mechanisms and Point Defects in Silicon and Germanium
Vol. **29**, No. 2, p. 455–542 (1968).
See also *10*, p. *93*; 11, p. 145; 13.4, p. 200; 19, p. 327; 22.1.1, p. 458; 22.1.2, p. 468.

Original Papers and Short Notes

G. B. Gibbs
*Grain Boundary Impurity Diffusion
Vol. **16**, No. 1, p. K27–K29 (1966).
See also 10, p. 94.

J. Askill
*Environmental Effects on the Diffusion of Ta^{182} in B.C.C. Titanium
Vol. **16**, No. 1, p. K63–K65 (1966).
See also 21.1, p. 390.

S. Ceresara, T. Federighi, and F. Pieragostini
*Determination of Diffusion Coefficients in Metals by Resistometric Method — Application to the Diffusion of Zn in Al
Vol. **16**, No. 2, p. 439–447 (1966).
See also 21, p. 373.

N. K. Archipova, S. M. Klotsman, A. N. Timofeev, and I. Sh. Trakhtenberg
*Effect of a DC Field on the Lattice Diffusion of Silver-110 in Copper and Gold
Vol. **16**, No. 2, p. 729–736 (1966).
See also 21.1, p. 391; 21.6, p. 430.

D. Bergner und W. Lange
*Diffusion von Sn und Zn in Sn und in einer homogenen Sn–Zn-Legierung
(Diffusion of Sn and Zn in Sn and in a Homogeneous Sn–Zn Alloy)
Vol. **18**, No. 1, p. 67–74 (1966).
See also 21, p. 375.

D. Bergner und W. Lange
*Volumen- und Korngrenzendiffusion von Sn und Zn in heterogenen Sn–Zn-Legierungen
(Volume- and Grain-Boundary Diffusion of Sn and Zn in Heterogeneous Sn–Zn Alloys)
Vol. **18**, No. 1, p. 75–84 (1966).
See also 21, p. 375.

G. B. Gibbs
*Solvent Diffusivity in a Dilute B.C.C. Binary Alloy (Weak Perturbation Case)
Vol. **18**, No. 1, p. 85–88 (1966).
See also 21, p. 375.

H. Wever
*Eine Methode zur Bestimmung partieller Diffusionskoeffizienten aus dem Kirkendall-Effekt im gesamten Konzentrationsbereich
(Method for Determining Partial Diffusion-Coefficients from the Kirkendall Effect in the Whole Concentration Range)
Vol. 18, No. 1, p. K25–K27 (1966).

S. M. Klotsman, A. N. Timofeev, and I. Sh. Trakhtenberg
*On the Mechanism of Lattice Electromigration in Metals
Vol. 18, No. 2, p. 847–852 (1966).
See also 14.1, p. 222; 21.1, p. 394; 21.6, p. 432.

R. N. Ghoshtagore
*Diffusion from a Thin Film into a Dissociating or Evaporating Solid
Vol. 19, No. 1, p. 123–127 (1967).

B. I. Boltaks and T. D. Dzhafarov
*The Effect of Applied Electric Field on Diffusion of Impurities in Gallium Arsenide
Vol. 19, No. 2, p. 705–714 (1967).
See also 22.2.1, p. 483.

R. Böttcher, W. Windsch, and W. Lüdke
EPR Investigations on Mn^{2+}-Doped LiF Single Crystals
Vol. 20, No. 1, p. 121–126 (1967).
See also 13.4, p. 205; *19*, p. *328*; 22.5.2, p. 521.

P. Schmeling
*Diffusion of Xenon in Barium Fluoride
Vol. 20, No. 1, p. 127–129 (1967).
See also 22.5.3, p. 532.

T. Geszti
On the Theory of Thermally Activated Processes
Vol. 20, No. 1, p. 165–177 (1967).
See also *10*, p. *98*; 12, p. 157; 14.4, p. 252.

B. Ya. Pines, I. P. Grebennik, R. I. Bazyura, and G. V. Zozulya
*Surface and Volume Diffusion in Thin Films of the System Ag-Se
Vol. 20, No. 1, p. 213–223 (1967).
See also 21.6, p. 432; 22, p. 443; 22.1.3, p. 477.

E. Igras and T. Warmiński
*Electron Mirror Microscopic Investigation of Surface Diffusion of Lithium in Silicon
Vol. 20, No. 1, p. K5–K8 (1967).
See also 1.3, p. 14; 22.1.2, p. 471.

K. J. Richards and K. R. Cramer
*Diffusion Measurements Involving Discontinuous Interfacial Contact
Vol. 20, No. 1, p. K21–K23 (1967).

H. P. Bonzel
*Diffusion of Nickel in Silicon
Vol. 20, No. 2, p. 493–504 (1967).
See also 10, p. 99; 22.1.2, p. 471.

R. F. Peart
*Effect of Pressure on the Diffusion of Fe in Ti and Ti + 10% Fe
Vol. 20, No. 2, p. 545–550 (1967).
See also 12.1, p. 160; 21, p. 378; 21.1.1, p. 416.

P. Schmeling
*Non-Ideal Diffusion of Argon in Potassium Chloride
Vol. **20**, No. 2, p. 597–603 (1967).
See also 11, p. 148; 22.5.2, p. 521.

R. N. Ghoshtagore
*On the Mechanism of Substitutional Diffusion in Silicon
Vol. **20**, No. 2, p. K89–K94 (1967).
See also 10, p. 99; 22.1.2, p. 471.

E. D. Hondros
*Herring-Nabarrro Creep Studies on the Self-Diffusion of Iron in the Presence of Solutes
Vol. **21**, No. 1, p. 375–382 (1967).
See also 10.1, p. 122; 21.1.1, p. 417.

J. Levasseur et J. Philibert
*Détermination des coefficients de diffusion intrinsèques par mesure de l'effet Kirkendall
(Determination of Intrinsic Diffusion Coefficients by Measuring the Kirkendall Effect)
Vol. **21**, No. 1, p. K1–K4 (1967).
See also 21.1, p. 397; 21.1.1, p. 417.

J. Bougnot, E. Monteil, and C. Llinares
*Diffusion and Solubility in Gallium Antimonide
Vol. **21**, No. 1, p. K31–K34 (1967).
See also 10, p. 100; 14.3, p. 236; 22.2.3, p. 491.

R. Kelly
*Bubble Diffusion and the Motion of Point Defects near Surfaces
Vol. **21**, No. 2, p. 451–459 (1967).
See also 11, p. 148.

A. Lodding
*Thermotransport and the Associated Isotope Effect in Solids
Vol. **22**, No. 1, p. 157–161 (1967).
See also 21.2, p. 424; 22.8, p. 551.

B. Pegel
*On the Isotope Effect in Interstitial Diffusion
Vol. **22**, No. 1, p. 223–228 (1967).
See also 6, p. 61.

P. Reimers
*Änderung der Eindringkurven bei Diffusionsmessungen durch Verdampfung des Diffusionsmaterials
(Change of the Penetration Curves in Diffusion Measurements due to the Evaporation of the Diffusion Material)
Vol. **22**, No. 1, p. K27–K31 (1967).
See also 21.6, p. 433.

B. Pegel
*On the Isotope Effect in Interstitial Diffusion. (II): Extended Model
Vol. **22**, No. 1, p. K45–K48 (1967).
See also 6, p. 62.

P. Kordoš
*Heat Treatment of InSb Crystals with Different Dislocation Densities
Vol. **22**, No. 1, p. K59–K61 (1967).
See also 10, p. 101; 14.3, p. 237; 22.2.3, p. 491.

J. Pelleg
*A Note on Nickel Diffusion in Copper
Vol. 22, No. 2, p. K83—K85 (1967).
See also 21.1, p. 400.

M. Matyáš und L. Švob
*Bemerkungen zur Weiserschen Theorie der Diffusion von Zwischengitterionen im Diamantgitter
(Remarks on Weiser's Theory of Interstitial Diffusion in the Diamond Lattice)
Vol. 22, No. 2, p. K141—K145 (1967).
See also 21.2, p. 424.

A. van den Beukel
*On the Relation between Self-Diffusion, Melting, and Vacancy-Impurity Binding in Dilute Alloys
Vol. 23, No. 1, p. 165—170 (1967).
See also *10.1*, p. *125*; 21, p. 381.

G. Martin, D. A. Blackburn et Y. Adda
*Autodiffusion au joint de grains de bicristaux d'argent soumis à une pression hydrostatique
(Self-Diffusion in the Grain Boundaries of Silver Bicrystals under Hydrostatic Pressure)
Vol. 23, No. 1, p. 223—228 (1967).
See also 10.1, p. 125; 21.6, p. 434.

R. F. Peart and J. Askill
*The Mechanism of Diffusion in B.C.C. Transition Metals
Vol. 23, No. 1, p. 263—275 (1967).
See also 10, p. 102; 21, p. 381; 21.1, p. 400; 21.1.1, p. 419.

B. Pegel and D. Lepski
*On the Isotope Effect in Interstitial Diffusion. (III): Quantum Corrections
Vol. 23, No. 1, p. 335—339 (1967).
See also 6, p. 62.

J. Askill
*Tracer Diffusion of Tungsten in Molybdenum
Vol. 23, No. 1, p. K21—K23 (1967).
See also 21, p. 381.

D. Heitkamp
*On the Thermodynamics of Thermal Diffusion in Substitutional Alloys
Vol. 24, No. 1, p. 341—348 (1967).
See also 10.1, p. 126; 21, p. 382.

A. S. Nowick and G. J. Dienes
*On the Inherent Curvature of the Arrhenius Plot in Diffusion Experiments
Vol. 24, No. 2, p. 461—467 (1967).
See also 21.1, p. 401; 21.6, p. 435.

A. Glasner, R. Reisfeld, and A. Linenberg
*Diffusion of Tl^+ in Doped and Pretreated KCl
Vol. 24, No. 2, p. 695—706 (1967).
See also 22.5.2, p. 525.

M. H. van Maaren
*Note on the Diffusion and Solubility of Lithium in Gallium Antimonide
Vol. 24, No. 2, p. K125—K128 (1967).
See also 10, p. 103; 22.2.3, p. 492.

H. Dübler und H. Wever
*Thermo- und Elektrotransport in β-Titan und β-Zirkonium
(Thermo- and Electrotransport in β-Titanium and β-Zirconium)
Vol. **25**, No. 1, p. 109–118 (1968).
See also 10.1, p. 127; 21, p. 383.

T. E. Volin and R. W. Balluffi
*Annealing Kinetics of Voids and the Self-Diffusion Coefficient in Aluminium
Vol. **25**, No. 1, p. 163–173 (1968).
See also 4, p. 44; 21, p. 384.

H. P. Bonzel and N. A. Gjostein
*Surface Self-Diffusion Measurements on Copper
Vol. **25**, No. 1, p. 209–222 (1968).
See also 21.1, p. 402.

D. Schumacher, A. Seeger, and O. Härlin
Vacancies, Divacancies, and Self-Diffusion in Platinum
Vol. **25**, No. 1, p. 359–371 (1968).
See also *10.1*, p. *128*; 21.6, p. 435.

R. Kelly, Č. Jech, and Hj. Matzke
*Diffusion in Inert-Gas Bombarded KCl, KBr, and KI
Vol. **25**, No. 2, p. 641–653 (1968).
See also 11, p. 151; 22.5.2, p. 525.

K. R. Cramer and K. J. Richards
*Diffusion Concentration Distribution near an Idealized Contact Spot
Vol. **26**, No. 1, p. 97–105 (1968).

A. Sobaszek
*Experimental Comparison of the Grain Boundary Diffusion Models
Vol. **26**, No. 1, p. K59–K63 (1968).
See also 21.6, p. 435.

J. Bougnot, L. Szepessy, and S. F. Da Cunha
*Diffusion of Cadmium in Gallium Antimonide
Vol. **26**, No. 2, p. K127–K129 (1968).
See also 22.2.3, p. 492.

J. Biersack and W. Diez
*Motion of Markers and Bubbles in Solids by Self-Diffusion in a Temperature Gradient
Vol. **27**, No. 1, p. 139–144 (1968).
See also 21, p. 385; 22, p. 451.

S. Radelaar
The Kinetics of Short-Range Order in Some Ag–Au Alloys
Vol. **27**, No. 1, p. K63–K67 (1968).
See also *1.1*, p. *8*; 10.1, p. 131; 21.6, p. 436.

S. Ceresara
*A Step Annealing Procedure for the Determination of Diffusion Coefficients in Metals by the Resistometric Method — Application to the Diffusion of Cu in Al
Vol. **27**, No. 2, p. 517–520 (1968).
See also 21, p. 386.

F. W. Felix
*Rare-Gas Diffusion in Neutron-Irradiated Rubidium Halides
Vol. **27**, No. 2, p. 529–534 (1968).
See also 10, p. 107; 11, p. 152; 22.5.2, p. 526.

P. Süptitz and R. Weidmann
*Diffusion of Mn^{2+} in AgCl and AgBr Crystals
Vol. **27**, No. 2, p. 631—638 (1968).
See also 10, p. 108; 22.5.1, p. 517.

H. Bakker
*A Curvature in the ln D versus $1/T$ Plot for Self-Diffusion in Nickel at Temperatures from 980 to 1400 °C
Vol. **28**, No. 2, p. 569—576 (1968).
See also 21.1, p. 406.

A. Seeger and H. Mehrer
*Interpretation of Self-Diffusion and Vacancy Properties in Gold
Vol. **29**, No. 1, p. 231—243 (1968).
See also *10.1*, p. *133*; 21.6, p. 437.

E. Nebauer
*Vapour-Pressure Investigations of Impurity Diffusion and Solubility in A^{II}-B^{VI} Compounds Demonstrated for the System CdS:Au:S_2
Vol. **29**, No. 1, p. 269—281 (1968).
See also 10, p. 110; 22.4.1, p. 504.

T. J. Turner, C. M. Dozier, and G. P. Williams, Jr.
Activation Energies for Zener Relaxation in α-AgCd
Vol. **30**, No. 1, p. 87—91 (1968).
See also *10.1*, p. *134*; *12.1*, p. *161*; 21, p. 388; 21.6, p. 437.

A. B. Gardner, R. L. Sanders, and L. M. Slifkin
*Self-Diffusion in α-Silver–Cadmium
Vol. **30**, No. 1, p. 93—96 (1968).
See also 10.1, p. 134; 21, p. 388; 21.6, p. 437.

M. C. Naik, M. S. Anand, A. R. Paul, and R. P. Agarwala
*Impurity Diffusion in Face-Centred Cubic Metals
Vol. **30**, No. 1, p. 319—328 (1968).
See also 21, p. 388; 21.1, p. 408; 21.6, p. 437.

N. P. Kowalenko und W. Ebeling
*Zur statistischen Theorie der Diffusionsprozesse in kondensierten Systemen (On the Statistical Theory of Diffusion Processes in Condensed Systems)
Vol. **30**, No. 2, p. 533—540 (1968).

10. Defect Properties of Solids (Irradiation Defects See 11)

Review Articles

P. Krishna and Ajit Ram Verma
Crystal-Polymorphism in One Dimension
Vol. **17**, No. 2, p. 437–477 (1966).
See also *3*, p. *26*; *5*, p. *49*; 22.3, p. 494; 22.4.1, p. 497; 22.5, p. 512.

L. E. Murr
Calibration and Use of an Electron Microscope for Precision Micromeasurements in Thin Film Materials
Vol. **19**, No. 1, p. 7–34 (1967).
See also *1.4*, p. *16*; 4, p. 37; 21.1, p. 390; 21.1.1, p. 410.

A. S. Borshchevskii, N. A. Goryunova, F. P. Kesamanly, and D. N. Nasledov
Semiconducting $A^{II}B^{IV}C_2^{V}$ Compounds
Vol. **21**, No. 1, p. 9–55 (1967).
See also 3.1, p. 29; 3.2, p. 34; 5, p. 49; 8, p. 79; 13.1, p. 174; *14.3*, p. *231*; 15, p. 270; 16, p. 276; 17, p. 286; 18, p. 291; 20.1, p. 344; 22.8, p. 547.

P. Stüptitz and J. Teltow
*Transport of Matter in Simple Ionic Crystals (Cubic Halides)
Vol. **23**, No. 1, p. 9–56 (1967).
See also *9*, p. *87*; 14.4, p. 252; 15, p. 270; 22.5.1, p. 515; 22.5.2, p. 518; 22.5.3, p. 532.

F. Williams
Donor–Acceptor Pairs in Semiconductors
Vol. **25**, No. 2, p. 493–512 (1968).
See also *13.4*, p. *200*; 20.3, p. 363; 22.1, p. 456; 22.1.1, p. 458; 22.1.2, p. 468; 22.2.2, p. 487; 22.4.1, p. 497.

K. N. Subramanian
*Cyclic Stressing in Crystals of Rock-Salt Structure
Vol. **28**, No. 1, p. 9–37 (1968).
See also 10.1, p. 114; 21, p. 373; 21.1, p. 390; 22.5.1, p. 515; 22.5.2, p. 518; 22.6, p. 535.

A. Seeger and K. P. Chik
*Diffusion Mechanisms and Point Defects in Silicon and Germanium
Vol. **29**, No. 2, p. 455–542 (1968).
See also *9*, p. *87*; 11, p. 145; 13.4, p. 200; 19, p. 326; 22.1.1, p. 458; 22.1.2, p. 468.

Original Papers and Short Notes

I. B. Bersuker and B. G. Vekhter
Jahn-Teller Distortions and Inversion Splitting for Transition Metal Complexes and Local Centres in Crystals with Degenerate Electronic T-Terms
Vol. **16**, No. 1, p. 63–68 (1966).
See also *13.4*, p. *200*; 22, p. 433.

P. Sauer, O. Schirmer, and J. Schneider
Comments on Paraelectric Resonance
Vol. **16**, No. 1, p. 79–88 (1966).
See also *13.4*, p. *201*; 22.5.2, p. 518.

B. I. Smirnov and B. A. Efimov
*Development of Glide Bands during Plastic Deformation of LiF Crystals
Vol. **16**, No. 1, p. 191–196 (1966).
See also 22.5.2, p. 518.

G. B. Gibbs
Grain Boundary Impurity Diffusion
Vol. **16**, No. 1, p. K27—K29 (1966).
See also *9*, p. *87*.

G. B. Abdullaev, S. I. Mekhtieva, G. M. Aliev, D. Sh. Abdinov, and T. G. Kerimova
Determination of the Oxygen Content in Selenium
Vol. **16**, No. 1, p. K31—K34 (1966).
See also 2, p. 22; *20.1*, p. *344*; 22.1.3, p. 476.

D. J. Barber
*Electron Microscopy of Irradiation-Induced Defect Clusters in Magnesium Fluoride
Vol. **16**, No. 2, p. 531—547 (1966).
See also 4, p. 37; *11*, p. *145*; 22.5.3, p. 532.

J. van Landuyt
Determination of the Displacement Vector at the Anti-Phase Boundaries in Rutile by Contrast Experiments in the Electron Microscope
Vol. **16**, No. 2, p. 585—590 (1966).
See also *4*, p. *38*; 22.6, p. 535.

M. Musa
*Some Results Concerning the Co^{++} Site Symmetry in Alkali Halide Crystals
Vol. **16**, No. 2, p. 771—776 (1966).
See also 20.1, p. 345; 22.5.2, p. 518.

G. B. Abdullaev, N. I. Ibragimov, Sh. V. Mamedov, and T. Ch. Zhuvarly
ESR Study of the Behaviour of Oxygen in Selenium
Vol. **16**, No. 2, p. K113—K116 (1966).
See also *19*, p. *326*; 22.1.3, p. 476.

G. A. Bogomolova, A. A. Kaminskii, and V. A. Timofeeva
Nd^{3+} Optical Centres in $Y_3Al_5O_{12}$ Crystals
Vol. **16**, No. 2, p. K165—K166 (1966).
See also *20.1*, p. *345*; 20.2, p. 360; 22.8, p. 547.

B. Pödör
Electron Mobility in Plastically Deformed Germanium
Vol. **16**, No. 2, p. K167—K170 (1966).
See also *14.3*, p. *232*; 22.1.1, p. 459.

M. Balarin and M. Schenk
On the Defect-Induced Shift of Transition Points between Crystalline Phases
Vol. **17**, No. 1, p. 91—99 (1966).
See also *1.2*, p. *9*; 11, p. 146; 14.4.2, p. 266.

P. Delavignette, R. Trivedi, R. Gevers, and S. Amelinckx
*Electron Microscopic Images of Dislocation Ribbons
Vol. **17**, No. 1, p. 221—241 (1966).
See also 22.1, p. 456.

J. Nihoul and L. Stals
*On the Reaction Order of Bimolecular Recovery Processes in Solids
Vol. **17**, No. 1, p. 295—305 (1966).

O. N. Efimov and E. K. Kovev
Integral Characteristics of Anomalous Transmission of X-Rays
Vol. **17**, No. 1, p. 397—410 (1966).
See also *5*, p. *49*; 22.1.1, p. 459.

J. F. Petroff et A. Authier
*Franges d'égale déformation dans les topographies aux rayons X
(Fringes of Equal Deformation in X-Ray Topographs)
Vol. 17, No. 1, p. K3—K5 (1966).
See also 22.9, p. 558.

J. H. Beaumont and P. W. M. Jacobs
Effect of Gaseous Impurities on the Structure-Sensitive Conductance of Potassium Chloride Single Crystals
Vol. 17, No. 1, p. K45—K48 (1966).
See also *14.4*, p. *252*; 22.5.2, p. 519.

G. Völkel and W. Windsch
EPR Studies of Copper-Doped Rochelle-Salt Single Crystals
Vol. 17, No. 1, p. K75—K78 (1966).
See also 14.4.2, p. 266; *19*, p. *326*.

S. S. Sheinin and C. D. Cann
*On the Measurement of Two Beam Values of Extinction Distance
Vol. 17, No. 1, p. K143—K145 (1966).
See also 22.1.2, p. 469.

Y. T. Chou
*Screw Dislocations in and near Lamellar Inclusions
Vol. 17, No. 2, p. 509—516 (1966).

M. Hagino and T. Takahashi
Photoelectric Properties of Cs–Sb Films Having Various Compositions
Vol. 17, No. 2, p. 609—618 (1966).
See also 14.3.1, p. 245; 16, p. 277; *17*, p. *286*; 22.7, p. 545.

B. A. Greenberg
The Effect of Dislocation Dipoles on the Shape of the Nuclear Magnetic Resonance Line
Vol. 17, No. 2, p. 673—681 (1966).
See also *19*, p. *327*; 22, p. 441.

A. Seeger und G. Wobser
*Anordnungen paralleler Versetzungen in anisotropen Medien
(Parallel Dislocation Arrays in Anisotropic Media)
Vol. 17, No. 2, p. 709—724 (1966).
See also *10.1*, p. *116*; 21, p. 375; 21.1, p. 392; 21.3, p. 426; 21.6, p. 431; 22.1.1, p. 460; 22.1.2, p. 469.

F. Rueda and W. Dekeyser
*Charge of Dislocations in Pure γ-Irradiated LiF
Vol. 17, No. 2, p. 739—744 (1966).
See also 10.2, p. 138; 22.5.2, p. 519.

G. Hauret et M. Girard-Nottin
*Étude de la diffusion de la lumière par des cristaux de chlorure de potassium dopés au calcium
(Study of Light Scattering in Potassium Chloride Crystals Doped with Calcium)
Vol. 17, No. 2, p. 881—888 (1966).
See also 20, p. 338; 22.5.2, p. 519.

C. Willaime, P. Delavignette, R. Gevers, and S. Amelinckx
*The Effect of the Presence of Dislocations on the Shape of Diffraction Spots in Electron Diffraction
Vol. **17**, No. 2, p. K173—K174 (1966).
See also 22.8, p. 548.

R. B. Clough
*The Activation Energy for Plastic Flow in Stress Relaxation
Vol. **17**, No. 2, p. K175—K178 (1966).
See also *10.1*, p. *117*.

G. B. Gibbs
*Vacancy Condensation on Precipitate Particles
Vol. **17**, No. 2, p. K191—K194 (1966).
See also 3.1, p. 30.

T. R. Duncan and D. Kuhlmann-Wilsdorf
*The Stress Fields of Dislocation Tilt Boundaries in Anisotropic Cubic and Diamond-Type Crystals
Vol. **18**, No. 1, p. 231—239 (1966).
See also *10.1*, p. *117*; 21, p. 375; 21.1, p. 393; 21.5, p. 429; 21.6, p. 431; 22.1, p. 456; 22.1.1, p. 460; 22.1.2, p. 470.

R. Gevers, J. van Landuyt, and S. Amelinckx
*The Fine Structure of Spots in Electron Diffraction Resulting from the Presence of Planar Interfaces and Dislocations. (I): General Theory and Its Application to Stacking Faults and Anti-Phase Boundaries
Vol. **18**, No. 1, p. 343—361 (1966).
See also 4, p. 39; *10.1*, p. *118*.

J. van Landuyt, R. Gevers, and S. Amelinckx
*The Fine Structure of Spots in Electron Diffraction Resulting from the Presence of Planar Interfaces and Dislocations. (II): Observations on Crystals Containing Stacking Faults
Vol. **18**, No. 1, p. 363—378 (1966).
See also 4, p. 39; *10.1*, p. *118*; 21.1.1, p. 414; 22.6, p. 536.

W. A. Brantley and Ch. L. Bauer
*The Geometry of Charged Dislocations in the NaCl Structure
Vol. **18**, No. 1, p. 465—478 (1966).

V. A. Kutasov und I. A. Smirnov
*Der Einfluß von überschüssigem antistrukturellem Bi auf die Wärmeleitfähigkeit des Kristallgitters von Bi_2Te_3, Bi_2Te_3-Sb_2Te_3 und Bi_2Te_3-Bi_2S_3 mit Jodbeimischung (Effect of Excess Antistructural Bi on the Lattice Thermal Conductivity of Bi_2Te_3, Bi_2Te_3-Sb_2Te_3, and Bi_2Te_3-Bi_2S_3 Containing Iodine)
Vol. **18**, No. 1, p. 479—488 (1966).
See also *8*, p. *80*; 14.3, p. 233; 15, p. 271; 22, p. 441; 22.8, p. 548.

G. C. T. Liu and J. C. M. Li
*Energy of Polygonal and Elliptical Dislocation Loops
Vol. **18**, No. 2, p. 517—525 (1966).
See also *10.1*, p. *118*.

G. C. T. Liu and J. C. M. Li
*Energy of Hexagonal Dislocation Loops
Vol. **18**, No. 2, p. 527—539 (1966).
See also *10.1*, p. *118*; 21, p. 376; 22, p. 441; 22.1.2, p. 470.

M. Asche, O. G. Sarbej, and V. M. Vasetskii
Piezoresistance of p-Germanium
Vol. **18**, No. 2, p. 749–754 (1966).
See also 12, p. 156; *14.3*, p. *234*; 22.1.1, p. 460.

D. Schwarz
Extrinsic Absorption of Chalcogen Doped AgBr Crystals in the Case of Variable Cation-Impurity Concentration
Vol. **18**, No. 2, p. 787–794 (1966).
See also *20.1*, p. *346*; 22.5.1, p. 515.

G. B. Abdullaev, F. B. Gadiev, G. M. Aliev, S. I. Mekhtieva, and D. Sh. Abdinov
*Interaction of Sodium Admixtures and Oxygen in Selenium
Vol. **18**, No. 2, p. K153–K156 (1966).
See also 14.3, p. 234; 22.1.3, p. 477.

V. L. Vinetskii and G. A. Kholodar
On the Electric Conductivity of Semiconductors Caused by the Ionization of Thermal Lattice Defects
Vol. **19**, No. 1, p. 41–49 (1967).
See also 13.4, p. 204; *14.3*, p. *234*; 22.6, p. 536.

W. A. Jesser and D. Kuhlmann-Wilsdorf
*On the Theory of Interfacial Energy and Elastic Strain of Epitaxial Overgrowths in Parallel Alignment on Single Crystal Substrates
Vol. **19**, No. 1, p. 95–105 (1967).
See also 1.4, p. 18.

S. Schäfer
*Messung von Versetzungsgeschwindigkeiten in Germanium
(Measurement of Dislocation Velocities in Germanium)
Vol. **19**, No. 1, p. 297–312 (1967).
See also 22.1.1, p. 461.

W. Ludwig and H. Pick
Scattering of Phonons at Defect Planes in the Simple Cubic Lattice
Vol. **19**, No. 1, p. 313–326 (1967).
See also *6*, p. *58*.

R. P. Harrison
*The Effect of Cadmium Additions on the Fracture Strength of Sodium Chloride Single Crystals
Vol. **19**, No. 1, p. 353–357 (1967).
See also 12, p. 156; 22.5.2, p. 520.

Yu. S. Boyarskaya, R. P. Zhitaru, and V. G. Koloskova
*Effect of Different Lattice Defects on the Mobility of Dislocations in Alkali Halide Crystals
Vol. **19**, No. 1, p. 441–451 (1967).
See also 11, p. 147; 22.5.2, p. 520.

V. B. Pariiskii
*Effect of Stress State on the Shape of the Dislocation Rosette near an Indentation in Some Alkali Halide Crystals
Vol. **19**, No. 2, p. 525–532 (1967).
See also 22.5.2, p. 521.

S. N. Komnik, V. Z. Bengus, and E. D. Lyak
*On the Kinetics of Plastic Deformation of Some Alkali Halide Crystals
Vol. **19**, No. 2, p. 533—541 (1967).
See also 22.5.2, p. 521.

A. C. McLaren, J. A. Retchford, D. T. Griggs, and J. M. Christie
*Transmission Electron Microscope Study of Brazil Twins and Dislocations Experimentally Produced in Natural Quartz
Vol. **19**, No. 2, p. 631—644 (1967).
See also 4, p. 40; 22.6, p. 536.

R. C. Rau and G. A. Chase
*Clustering of Defects in Neutron-Irradiated Beryllium Oxide
Vol. **19**, No. 2, p. 645—654 (1967).
See also 11, p. 148; 22.6, p. 536.

J. Spyridelis, P. Delavignette, and S. Amelinckx
On the Superstructures of Ta_2O_5 and Nb_2O_5
Vol. **19**, No. 2, p. 683—704 (1967).
See also *1*, p. *7*; 4, p. 40; 22.6, p. 537.

E. Igras and T. Warmiński
Electron Mirror Observation of Lithium Precipitation at Scratch-Induced Dislocations in Silicon
Vol. **19**, No. 2, p. K67—K69 (1967).
See also *1.3*, p. *14*; 4, p. 40; 22.1.2, p. 470.

M. Arndt and A. F. Rudolph
*Density Changes due to Dislocations in Silicon Single Crystals
Vol. **19**, No. 2, p. K79—K82 (1967).
See also 22.1.2, p. 470.

J. H. P. van Weeren, R. Struikmans, and J. Blok
Test of the Read Model Concerning the Electron Mobility in Plastically Deformed n-Type Ge
Vol. **19**, No. 2, p. K107—K110 (1967).
See also *14.3*, p. *235*; 22.1.1, p. 461.

B. Reppich
*Plastische Verformung von Eisen-II-Oxid
(Plastic Deformation of Iron-II-Oxide)
Vol. **20**, No. 1, p. 69—82 (1967).
See also 22.6, p. 537.

T. Geszti
*On the Theory of Thermally Activated Processes
Vol. **20**, No. 1, p. 165—177 (1967).
See also 9, p. 88; 12, p. 157; 14.4, p. 252.

Y. T. Chou
*On Stacked Screw Dislocation Arrays in an Anisotropic Medium
Vol. **20**, No. 1, p. 285—290 (1967).

P. Asadi
On the Existence of a Spontaneous Polarization in Fluorite Single Crystals Doped with Lime
Vol. **20**, No. 1, p. K55—K58 (1967).
See also *14.4.1*, p. *258*; 22.5.3, p. 532.

P. Asadi
On the Electrical Conductivity of Fluorite Single Crystals Doped with Uranium Dioxide
Vol. **20**, No. 1, p. K59–K61 (1967).
See also *14.4*, p. *252*; 22.5.3, p. 533.

P. M. Robinson and H. G. Scott
*Tilt Boundaries in Deformed Anthracene Single Crystals
Vol. **20**, No. 2, p. 461–471 (1967).
See also 4, p. 41; 22.9, p. 559.

H. P. Bonzel
Diffusion of Nickel in Silicon
Vol. **20**, No. 2, p. 493–504 (1967).
See also *9*, p. *88*; 22.1.2, p. 471.

R. de Wit
*Some Relations for Straight Dislocations
Vol. **20**, No. 2, p. 567–573 (1967).

R. de Wit
*The Self-Energy of Dislocation Configurations Made up of Straight Segments
Vol. **20**, No. 2, p. 575–580 (1967).

W. Blum und B. Ilschner
*Über das Kriechverhalten von NaCl-Einkristallen
(Creep Behaviour of NaCl Single Crystals)
Vol. **20**, No. 2, p. 629–642 (1967).
See also 22.5.2, p. 521.

R. N. Ghoshtagore
On the Mechanism of Substitutional Diffusion in Silicon
Vol. **20**, No. 2, p. K89–K94 (1967).
See also *9*, p. *89*; 22.1.2, p. 471.

B. A. Greenberg
Effect of Segregation at Dislocations on the Shape of the NMR Line
Vol. **20**, No. 2, p. K103–K106 (1967).
See also 4, p. 42; *19*, p. *328*.

J. Blinowski and M. Grynberg
On the Direct Interband Magnetoabsorption in Uniaxially Stressed Germanium
Vol. **20**, No. 2, p. K107–K111 (1967).
See also *18*, p. *292*; *20*, p. *339*; 22.1.1, p. 461.

W. Schröter
Electric Properties of Dislocations in Germanium
Vol. **21**, No. 1, p. 211–224 (1967).
See also 13.4, p. 206; *14.3*, p. *236*; 22.1.1, p. 461.

Y. T. Chou and D. M. Barnett
*Stacked Screw Dislocation Arrays in an Anisotropic Two-Phase Medium
Vol. **21**, No. 1, p. 239–246 (1967).
See also 4, p. 42.

R. Gevers, J. van Landuyt, and S. Amelinckx
*Two Beam Kinematical Theory for the Diffraction of Electrons by Crystals with Stacking Faults
Vol. **21**, No. 1, p. 393–411 (1967).
See also *10.1*, p. *122*; 21.1.1, p. 417; 22.6, p. 538.

F. BALIBAR et A. AUTHIER
*Etude théorique et expérimentale du contraste des images de dislocations
(Experimental and Theoretical Study of the Contrast of Dislocation Images)
Vol. 21, No. 1, p. 413—422 (1967).
See also 22.1.2, p. 471.

R. KRISHNAN
*Ionic Redistribution and Anisotropy in Heat Treated NiO-Doped YIG Single Crystals
Vol. 21, No.1, p. K5—K7 (1967).
See also 18.3, p. 317; 19, p. 329.

A. BEHR, H. PEISL, and W. WAIDELICH
X-Ray Production of Vacancies in KCl at Low Temperatures
Vol. 21, No. 1, p. K9—K10 (1967).
See also *11*, p. *148*; 22.5.2, p. 522.

J. BOUGNOT, E. MONTEIL, and C. LLINARES
Diffusion and Solubility in Gallium Antimonide
Vol. 21, No. 1, p. K31—K34 (1967).
See also *9*, p. *89*; 14.3, p. 236; 22.2.3, p. 491.

J. HESSE
*Fließspannung und Versetzungsdichte Kalzium-dotierter NaCl-Einkristalle
(Yield Stress and Dislocation Density in Calcium-Doped NaCl Single Crystals)
Vol. 21, No. 2, p. 495—505 (1967).
See also 22.5.2, p. 522.

W. A. JESSER and D. KUHLMANN-WILSDORF
The Geometry and Energy of a Twist Boundary between Crystals with Unequal Lattice Parameters
Vol. 21, No. 2, p. 533—544 (1967).
See also *3.1*, p. *31*; 10.1, p. 123; 21, p. 379; 22, p. 445.

D. G. ARASLI and M. I. ALIEV
Influence of Defects and of the Interaction between Them on Phonon Scattering in Heavily Doped Ge and Si Crystals
Vol. 21, No. 2, p. 643—648 (1967).
See also 6, p. 61; *8*, p. *82*; 22.1.1, p. 462; 22.1.2, p. 471.

P. MÜLLER
*Zur Berechnung der Fehlordnung in Silberhalogeniden
(On the Calculation of Disorder in Silver Halides)
Vol. 21, No. 2, p. 693—701 (1967).
See also 22.5.1, p. 516.

L. P. KHIZNICHENKO, P. F. KROMER, D. K. KAIPNAZAROV, E. OTENYAZOV, D. YUSUPOVA, and L. G. ZOTOVA
The Properties of the Low-Temperature Internal Friction Peak in Silicon
Vol. 21, No. 2, p. 805—810 (1967).
See also 11, p. 149; *12*, p. *157*; 22.1.2, p. 471.

P. F. KROMER and L. P. KHIZNICHENKO
On the Low-Temperature Internal Friction in Silicon
Vol. 21, No. 2, p. 811—818 (1967).
See also *12*, p. *157*; 22.1.2, p. 471.

M. Peřinová
*Origin of Perfect Dislocations and Glide Elements in $NaNO_3$ Single Crystals
Vol. **21**, No. 2, p. K103—K106 (1967).
See also 22.8, p. 550.

H. Pink
*The Determination of Chromium Distribution in Laser-Rubies with Neutron Activation
Vol. **21**, No. 2, p. K111—K113 (1967).
See also *20.2*, p. *361*; 22.6, p. 538.

A. M. de Goer and B. Dreyfus
γ-Irradiation Effect on the Thermal Conductivity of Al_2O_3 with Cr or Mn Impurities at Low Temperature
Vol. **22**, No. 1, p. 77—82 (1967).
See also *8*, p. *83*; *11*, p. *149*; 22.6, p. 538.

N. Paschoff und J. Heydenreich
*Zur Versetzungsanordnung in dünnen, durch Kristallisation gewonnenen Germanium-Folien
(Dislocation Arrangement in Thin Germanium Foils Prepared by Crystallization)
Vol. **22**, No. 1, p. 93—101 (1967).
See also 1.4, p. 18; 22.1.1, p. 462.

M. Balarin, R. Rattke und A. Zetzsche
*Zur Auswertung isochroner Ausheilkurven
(On the Evaluation of Isochronal Annealing Curves)
Vol. **22**, No. 1, p. 123—129 (1967).

K. Lal
Dielectric Properties of Interstitial Impurity Cations in Alkali Halide Crystals
Vol. **22**, No. 1, p. 141—146 (1967).
See also *14.4*, p. *253*; 22.5.2, p. 523.

Ya. A. Iosilevskii
Some Model Problems on the Dynamics of Two-Component Crystals
Vol. **22**, No. 1, p. 255—266 (1967).
See also *5*, p. *51*; 6, p. 62.

P. Kordoš
Heat Treatment of InSb Crystals with Different Dislocation Densities
Vol. **22**, No. 1, p. K59—K61 (1967).
See also *9*, p. *89*; 14.3, p. 237; 22.2.3, p. 491.

J. L. Martin, B. Jouffrey, and P. Costa
*Stacking Faults in a Non-Stoichiometric Face-Centred Cubic TaC
Vol. **22**, No. 2, p. 349—354 (1967).
See also 22.3, p. 494.

J. E. Caffyn, T. L. Goodfellow, E. M. Olymbios, and D. J. Perrin
Dielectric and Optical Absorption Measurements on CaF_2 and CaF_2:Y
Vol. **22**, No. 2, p. 549—557 (1967).
See also 13.4, p. 209; *14.4*, p. *253*; *20.1*, p. *350*; 22.5.3, p. 533.

R. Yu. Khansevarov, T. V. Mashovets, and N. A. Vitovskii
Damage Formation Rates in Ge Subject to Thermal Treatment
Vol. **22**, No. 2, p. K103—K107 (1967).
See also *11*, p. *150*; 14.3, p. 237; 22.1.1, p. 462.

S. Lungu
Point Defects in Neutron-Damaged Silica Glass
Vol. **23**, No. 1, p. 147–156 (1967).
See also 2, p. 23; *11*, p. *150*; 22.8, p. 552.

Z. Knésl
*The Equilibrium Position of an Orientation Junction of an Edge Dislocation Dipole
Vol. **23**, No. 1, p. 185–188 (1967).

R. F. Peart and J. Askill
The Mechanism of Diffusion in B.C.C. Transition Metals
Vol. **23**, No. 1, p. 263–275 (1967).
See also *9*, p. *90*; 21, p. 381; 21.1, p. 400; 21.1.1, p. 419.

H. Teichler
*Berechnung des Peierls-Potentials im Diamantgitter mit Hilfe der Pseudopotentialmethode
(Calculation of the Peierls Potential in the Diamond Lattice by the Pseudopotential Method)
Vol. **23**, No. 1, p. 341–352 (1967).
See also 22.1.1, p. 463; 22.1.2, p. 472.

P. H. Dederichs
*Dynamical Scattering Theory for Crystals with Point Defects
Vol. **23**, No. 1, p. 377–386 (1967).
See also 6, p. 62; *10.1*, p. *125*.

H. Peisl, H. Spalt, and W. Waidelich
*X-Ray Diffuse Scattering in γ-Irradiated LiF
Vol. **23**, No. 1, p. K75–K78 (1967).
See also 10.2, p. 140; 22.5.2, p. 524.

G. D. Guseinov and A. M. Ramazanzade
*Visualization of Dislocations on Basal Planes of GaSe Single Crystals
Vol. **23**, No. 2, p. 461–465 (1967).
See also 22, p. 448.

D. J. Bacon
*A Method for Describing a Flexible Dislocation
Vol. **23**, No. 2, p. 527–538 (1967).
See also *10.1*, p. *125*.

E. Yu. Gutmanas and E. M. Nadgornyi
*Estimation of the Peierls Stress in Alkali Halides
Vol. **23**, No. 2, p. 571–576 (1967).
See also 22.5.2, p. 524.

W. Hörstel und G. Kretzschmar
*Akzeptorbildende Gitterdefekte im Tellur
(Lattice Defects Forming Acceptors in Tellurium)
Vol. **23**, No. 2, p. 639–647 (1967).
See also 14.3, p. 238; 22.1.3, p. 478.

F. Vávra
*A Note on Etching of Dislocations in AgCl Crystals
Vol. **23**, No. 2, p. K121–K122 (1967).
See also 22.5.1, p. 516.

M. J. MARCINKOWSKI and H. J. LEAMY
*Analysis of Dislocation Loops in Superlattices
Vol. 24, No. 1, p. 149–162 (1967).
See also *10.1*, p. *126*; 21.1.1, p. 419.

A. FOURDEUX, R. GEVERS, and S. AMELINCKX
*Electron Microscopic Contrast Effects at the Interface between Two Different Polytypes. (I): Interface Cubic-Hexagonal
Vol. 24, No. 1, p. 195–206 (1967).
See also 4, p. 43; *10.1*, p. *126*; 21.1, p. 401.

G. M. BARTENEV and E. G. KORYAK-DORONENKO
*The Theoretical Strength of NaCl Crystals under Tension along Its Body and Face Diagonals
Vol. 24, No. 2, p. 443–449 (1967).
See also 12, p. 158; 22.5.2, p. 524.

J. STANKOWSKI
EPR Study of Cu^{2+} Ion in Triglycine Fluoroberyllate
Vol. 24, No. 2, p. 451–456 (1967).
See also 14.4.2, p. 268; *19*, p. *331*.

F. FRÖHLICH and G. HENSEL
Low Temperature Conductivity of Ca-Doped KCl Crystals Influenced by Segregation Processes
Vol. 24, No. 2, p. 535–541 (1967).
See also *14.4*, p. *253*; 22.5.2, p. 524.

M. K. SHEINKMAN, V. A. TYAGAI, O. V. SNITKO, I. B. ERMOLOVICH, G. L. BELENKY, and V. N. BONDARENKO
Studies on the Nature of CdS Single Crystal Sensitization by Etching
Vol. 24, No. 2, p. 543–549 (1967).
See also *13.4*, p. *210*; 16, p. 281; 20.3, p. 368; 22.4.1, p. 501.

M. HÖHNE and M. STASIW
ESR Studies of Different Se_2^{3-} Centres in Silver Halides
Vol. 24, No. 2, p. 591–604 (1967).
See also *19*, p. *331*; 22.5.1, p. 516.

J. AULEYTNER, E. F. GROSS, A. A. KRALINA, and B. S. RAZBIRIN
*Investigation of the Real Structure of CdS Crystals in Connection with the Shift of Exciton Lines
Vol. 24, No. 2, p. K107–K110 (1967).
See also *13.2*, p. *193*; 20.1, p. 351; 22.4.1, p. 501.

T. V. NORDSTROM and S. AMELINCKX
*On the So-Called "Anomalous" Contrast at the Central Partial of Triple Ribbons
Vol. 24, No. 2, p. K121–K123 (1967).
See also *10.1*, p. *127*.

M. H. VAN MAAREN
Note on the Diffusion and Solubility of Lithium in Gallium Antimonide
Vol. 24, No. 2, p. K125–K128 (1967).
See also *9*, p. *90*; 22.2.3, p. 492.

J. ARENDS
*Defect Formation Energies and Migration Energy Barriers in Cesium Halides
Vol. 24, No. 2, p. K129–K131 (1967).
See also 22.5.2, p. 525.

W. Bollmann and A. J. Perry
*The Sequence of Partials in the F.C.C. Structure
Vol. **24**, No. 2, p. K141—K144 (1967).
See also *10.1*, p. *127*.

A. H. Scholz
*Computer Calculations for Vacancies in Alkali Halides with NaCl Structure
Vol. **25**, No. 1, p. 285—301 (1968).
See also 22.5.2, p. 525.

F. Fröhlich and P. Seifert
*Potential Differences in NaCl Crystals due to Plastic Deformation by Indentation Experiments between RT and 20 °K
Vol. **25**, No. 1, p. 303—310 (1968).
See also 22.5.2, p. 525.

F. Bailly
*Energies of Formation of Metal Vacancies in II-VI Semiconducting Tellurides (HgTe, CdTe, ZnTe)
Vol. **25**, No. 1, p. 317—322 (1968).
See also 22.4.3, p. 510.

J. Grabmaier and C. Watson
*Plastic Deformation in Single Crystals of Mg-Al-Spinel: Slip Parallel to the {111} Plane
Vol. **25**, No. 1, p. K7—K10 (1968).
See also 22.6, p. 540.

K. N. Shrivastava
Temperature Dependence of Paramagnetic Resonance Transition Probabilities: Mn^{2+} in MgO
Vol. **25**, No. 1, p. K47—K50 (1968).
See also *19*, p. *332*; 22.6, p. 540.

W. A. Coghlan and W. D. Nix
*Contributions to the Kinematical Theory of Diffraction Contrast from Screw Dislocations
Vol. **25**, No. 2, p. 679—689 (1968).
See also *10.1*, p. *128*.

U. Kümmel and G. Zollfrank
Phenomenological Study of Brightness Distributions in Electroluminescent Hexagonal ZnS Crystals
Vol. **25**, No. 2, p. 711—720 (1968).
See also *20.3*, p. *368*; 22.4.1, p. 502.

P. Feltham and R. Ghosh
On Colour Centres in Plastically Deformed Alkaline Earth Fluorides
Vol. **25**, No. 2, p. K71—K73 (1968).
See also *10.2*, p. *142*; 19, p. 332; 22.5.3, p. 533.

D. K. Smith and H. W. Newkirk
*Irradiation Decoration of Defects in Single-Crystal BeO
Vol. **25**, No. 2, p. K79—K82 (1968).
See also 11, p. 151; 22.6, p. 540.

Z. Smetana
The Influence of Mn^{3+} Ion Clustering on Ferromagnetic Resonance Linewidth of Manganese-Rich Mn-Fe Spinels
Vol. **25**, No. 2, p. K87—K89 (1968).
See also 18.3, p. 319; *19*, p. *332*.

J. T. MOORE, T. R. DUNCAN, and D. KUHLMANN-WILSDORF
*Corrected Data for the Shear Stresses of Infinite Tilt Boundaries Composed of {111}, 1/2 ⟨1̄10⟩ Dislocations in Elastically Anisotropic F.C.C. Metals
Vol. **25**, No. 2, p. K105–K106 (1968).
See also *10.1*, p. *129*; 21, p. 385; 21.1, p. 403; 21.5, p. 429; 21.6, p. 435; 22.1, p. 457; 22.1.1, p. 464; 22.1.2, p. 473.

R. G. BESSENT and P. FELTHAM
Formation of F-Centres in Alkaline Earth Oxides by Plastic Deformation
Vol. **25**, No. 2, p. K107–K110 (1968).
See also *10.2*, p. *142*; 19, p. 333; 22.6, p. 541.

B. N. OSHCHERIN
*On Enthalpy Calculation of Vacancy Formation in Inorganic Substances
Vol. **25**, No. 2, p. K123–K125 (1968).
See also *10.1*, p. *129*; 21, p. 385; 22, p. 450.

S. DASH and M. L. JOSHI
*Climb of Extrinsic Faults in Silicon
Vol. **26**, No. 1, p. 159–166 (1968).
See also 22.1.2, p. 473.

K.-H. MATUCHA
*Elektronenmikroskopie von Gleitlinien auf verformtem NaCl
(Electron Microscopic Study of Slip Lines on Deformed NaCl)
Vol. **26**, No. 1, p. 291–310 (1968).
See also 22.5.2, p. 526.

R. A. ZHITNIKOV, V. B. KOLTSOV, and N. I. MELNIKOV
Paramagnetic Resonance of Copper Atoms Stabilized in Alkali-Halide Crystals
Vol. **26**, No. 1, p. 371–380 (1968).
See also *19*, p. *333*; 22.5.2, p. 526.

R. P. HARRISON
The Effect of Particle Size on the Fracture Strength of Sodium Chloride Single Crystals
Vol. **26**, No. 1, p. K65–K67 (1968).
See also *4*, p. *45*; 12, p. 158; 22.5.2, p. 526.

J. ŁAGOWSKI
*On Surface Recombination Changes due to Edge Dislocations
Vol. **26**, No. 1, p. K69–K71 (1968).
See also *13.3*, p. *198*; 22.1.1, p. 464.

J. ARENDS and H. NIJBOER
The Ionic Conductivity of CsCl Doped with K^+, Rb^+, and Br^-
Vol. **26**, No. 2, p. 537–542 (1968).
See also *14.4*, p. *254*; 22.5.2, p. 526.

B. J. WICKS and M. H. LEWIS
Direct Observations of Ferroelectric Domains in Lithium Niobate
Vol. **26**, No. 2, p. 571–576 (1968).
See also *4*, p. *45*; *14.4.2*, p. *268*.

R. GEVERS, J. VAN LANDUYT, and S. AMELINCKX
The Fine Structure of Spots in Electron Diffraction Resulting from the Presence of Planar Interfaces and Dislocations. (III): Domain Boundaries
Vol. **26**, No. 2, p. 577–590 (1968).
See also *4*, p. *46*.

R. J. BELTZ, T. L. DAVIS, and K. MALÉN
*Some Unifying Relations for Moving Dislocations
Vol. 26, No. 2, p. 621–637 (1968).
See also *10.1*, p. *129*.

U. MESSERSCHMIDT
*Influence of Impurities on Dislocation Velocities Controlled by Jog-Dragging
Vol. 26, No. 2, p. 701–708 (1968).
See also 22.5.2, p. 526.

H. ALEXANDER
*Elektronenmikroskopie eingefrorener Versetzungen. (I): Grundsätzliches zur Methode und Ergebnisse am Verformungsbeginn von Germaniumkristallen
(Electron Microscopic Study of Frozen-in Dislocations. (I): Basic Considerations on the Method and Results at the Deformation Onset of Germanium Crystals)
Vol. 26, No. 2, p. 725–741 (1968).
See also 22.1.1, p. 464.

D. MICHELL, P. M. ROBINSON, and A. P. SMITH
*Direct Observation of Dislocations in Anthracene
Vol. 26, No. 2, p. K93–K95 (1968).
See also 22.9, p. 560.

J. M. PANKRATZ and M. L. RUDEE
*An Application of Stereo Electron Microscopy to Anomalously Wide Images in Neutron Irradiated Silicon
Vol. 26, No. 2, p. K97–K98 (1968).
See also 4, p. 46; 22.1.2, p. 473.

D. KORDES
Vanadium Monocarbide Containing Ordered Vacancies
Vol. 26, No. 2, p. K103–K105 (1968).
See also *5*, p. *53*; 22.3, p. 494.

KH. M. KHALILOV, B. B. KULIEV, and F. A. GAMBAROV
Effect of Phosphorus Admixtures on Viscoelastic Properties of Selenium
Vol. 26, No. 2, p. K113–K117 (1968).
See also *2*, p. *24*; *12*, p. *158*; 22.1.3, p. 479.

J. DONECKER
Phonon Scattering by Point-Defect Aggregates in AgCl Crystals
Vol. 26, No. 2, p. K131–K134 (1968).
See also *6*, p. *65*; 8, p. 84; 22.5.1, p. 516.

A. AUTHIER
*Contrast of a Stacking Fault on X-Ray Topographs
Vol. 27, No. 1, p. 77–93 (1968).
See also 22.1.2, p. 473.

P. GÖRLICH, H. KARRAS, G. KÖTITZ, and R. RAUCH
Polarized Luminescence of X-Irradiated CaF_2:Y and SrF_2:Y Crystals and the Structure of Their Luminescence Centres
Vol. 27, No. 1, p. 109–115 (1968).
See also *20.3* p. *369*; 22.5.3, p. 534.

J. H. P. VAN WEEREN, G. KOOPMANS, and J. BLOK
The Position of the Dislocation Acceptor Level in n-Type Ge
Vol. 27, No. 1, p. 219–224 (1968).
See also *13.4*, p. *213*; 14.3, p. 241; 22.1.1, p. 465.

J. H. P. van Weeren, R. Struikmans, G. Koopmans, and J. Blok
Low-Field Magnetoresistance and Magnetoconductivity in n-Type Ge with Dislocations
Vol. **27**, No 1, p. 225–235 (1968).
See also *14.3*, p. *241*; 22.1.1, p. 465.

D. Michell and A. P. Smith
The Nature of Oxide Layers on Single Crystals of Cadmium and Magnesium
Vol. **27**, No. 1, p. 291–300 (1968).
See also 1.3, p. 15; *1.4*, p. *20*; 22.6, p. 541.

W. Ulrici
*Untersuchung dreiwertiger Übergangsmetallionen in Silberhalogeniden. (I): AgCl:V
(Study of Trivalent Transition-Metal Ions in Silver Halides. (I): AgCl:V)
Vol. **27**, No. 1, p. 333–344 (1968).
See also 13.4, p. 213; 14.4, p. 254; 20.1, p. 354; 22.5.1, p. 516.

H. Alexander
*Elektronenmikroskopie eingefrorener Versetzungen. (II): Ergebnisse in den Bereichen I und II der Verfestigungskurve
(Electron Microscopic Study of Frozen-in Dislocations. (II): Results Obtained for Stages I and II of the Work-Hardening Curve)
Vol. **27**, No. 1, p. 391–412 (1968).
See also 22.1.1, p. 465.

A. F. Lubchenko
Theory of Urbach's Rule for Impurity Centres
Vol. **27**, No. 1, p. K73–K76 (1968).
See also 13.4, p. 213; *20.1*, p. *354*; 22, p. 452.

D. Schwarz
ESR Study of Cu^{2+} in AgCl Crystals Additionally Doped with Se or S
Vol. **27**, No. 1, p. K77–K80 (1968).
See also 13.4, p. 213; *19*, p. *333*; 22.5.1, p. 516.

W. Ulrici
*Untersuchung dreiwertiger Übergangsmetallionen in Silberhalogeniden. (II): Chrom und Titan
(Study of Trivalent Transition-Metal Ions in Silver Halides. (II): Chromium and Titanium)
Vol. **27**, No. 2, p. 489–500 (1968).
See also 13.4, p. 213; 14.4, p. 254; 20.1, p. 354; 22.5.1, p. 517.

F. W. Felix
Rare-Gas Diffusion in Neutron-Irradiated Rubidium Halides
Vol. **27**, No. 2, p. 529–534 (1968).
See also *9*, p. *91*; 11, p. 152; 22.5.2, p. 526.

B. K. Daniels and D. B. Meadowcroft
*Twist Boundaries and Electroluminescence
Vol. **27**, No. 2, p. 535–539 (1968).
See also *20.3*, p. *369*; 22.4.1, p. 502.

G. F. Alfrey and D. B. Meadowcroft
The Influence of Structure on the Electroluminescence of Zinc Sulphide
Vol. **27**, No. 2, p. 541–544 (1968).
See also 4, p. 46; *20.3*, p. *370*; 22.4.1, p. 502.

R. J. Murphy and J. L. Crawford
*Electron Microscope Image Profiles of Paired and Triple Dislocations
Vol. 27, No. 2, p. 559–572 (1968).
See also 1.4, p. 20; *10.1*, p. *131*.

T. Jøssang
*The Self-Energy of a Dislocation and the Elastic Energy of Piecewise Straight Dislocation Configurations
Vol. 27, No. 2, p. 579–592 (1968).

P. Süptitz and R. Weidmann
Diffusion of Mn^{2+} in AgCl and AgBr Crystals
Vol. 27, No. 2, p. 631–638 (1968).
See also *9*, p. *92*; 22.5.1, p. 517.

K. Hennig
Study of Iron Ions in ^{57}Fe-Doped AgCl Using the Mössbauer Effect
Vol. 27, No. 2, p. K115–K116 (1968).
See also *6.1*, p. *74*; 22.5.1, p. 517.

R. P. Harrison
*The Yield Strength of Sodium Chloride Containing Manganese
Vol. 27, No. 2, p. K153–K156 (1968).
See also 22.5.2, p. 527.

K. Hennig, Kim Yung, and B. S. Skorchev
*On States of Iron Impurity Ions in AgCl and NaCl Investigated by the Mössbauer Method
Vol. 27, No. 2, p. K161–K164 (1968).
See also 6.1, p. 74; 22.5.1, p. 517; 22.5.2, p. 527.

R. Balzer, H. Peisl, and W. Waidelich
*Evidence for Frenkel Defect Recombination during Thermal Annealing of Low-Temperature X-Irradiated KCl
Vol. 27, No. 2, p. K165–K167 (1968).
See also 11, p. 153; 22.5.2, p. 527.

V. I. Cherepanov and A. E. Nikiforov
*Consideration of Polarization and Covalency Corrections to the Quasi-Coulomb Shift of Optical Lines of Ion Pairs in Crystals
Vol. 28, No. 1, p. 65–73 (1968).
See also 13.4, p. 214; 20.3, p. 370; 22.6, p. 542.

K. J. Rao and C. N. R. Rao
*Schottky Defects in Alkali Halides: Their Creation, Interaction, and Migration
Vol. 28, No. 1, p. 157–169 (1968).
See also 3.2, p. 36; 22.5.2, p. 527.

K. Ždánský
Vibrational Resonance Mode and the Temperature Dependence of Hyperfine Coupling of Mn^{2+} in BaO
Vol. 28, No. 1, p. 181–186 (1968).
See also 13.4, p. 214 ; *19*, p. *334*; 22.6, p. 542.

R. Balzer, H. Peisl, and W. Waidelich
*Lattice Disorder in Alkali Halides X-Irradiated at Room Temperature
Vol. 28, No. 1, p. 207–210 (1968).
See also 11, p. 153; 22.5.2, p. 527.

J. Hatz
Some Effects of Material Inhomogeneities on the Near-Field Pattern of GaAs Diode Lasers
Vol. **28**, No. 1, p. 233–245 (1968).
See also 14.3.2, p. 251; *20.2*, p. *362*; 22.2.1, p. 485.

Yu. A. Sherstkov, V. I. Nepsha, V. A. Vagenin, A. E. Nikiforov, and A. I. Krotkii
Effect of Applied Electric Field on EPR Spectra of Crystal Impurity Centres
Vol. **28**, No. 1, p. 269–275 (1968).
See also 13, p. 171; *19*, p. *334*; 22.6, p. 542.

A. Farwer and F. Matossi†
Internal Optical Reflection in Cubic ZnS with One-Dimensional Stacking Faults
Vol. **28**, No. 1, p. 373–383 (1968).
See also 4, p. 47; *20.1*, p. *355*; 22.4.1, p. 503.

T. V. Samoilova, B. I. Smirnov, and E. E. Chernov
*Electron Microscopy Observation of Dislocation Multipoles in Slip Bands of LiF Crystals
Vol. **28**, No. 1, p. K1–K3 (1968).
See also 22.5.2, p. 527.

T. Okada, H. Kawai, S. Mukai, and T. Suita
*Hardening of KCl Crystals by F_A-Centers
Vol. **28**, No. 1, p. K19–K21 (1968).
See also *10.2*, p. *143*; 22.5.2, p. 527.

H. A. Bahr, W. Pompe, and H.-G. Schöpf
*On the Dynamics of Dislocations
Vol. **28**, No. 1, p. K23–K26 (1968).
See also *10.1*, p. *132*.

B. Escaig
*L'activation thermique des déviations sous faibles contraintes dans les structures h.c. et c.c.
(Thermal Activation of Deviations at Low Stresses in H.C. and C.C. Structures)
Vol. **28**, No. 2, p. 463–474 (1968).
See also *10.1*, p. *132*.

P. O. Szomor
*Elastic Constants and Tensile Strength of NaCl-Type Alkali Halides
Vol. **28**, No. 2, p. 529–537 (1968).
See also *12*, p. *158*; 22.5.2, p. 527.

H. Böttger
*Multiphononabsorption von Kristallen mit Punktdefekten im starken Strahlungsfeld
(Multiphonon Absorption of Crystals with Point Defects in a Strong Radiation Field)
Vol. **28**, No. 2, p. 611–621 (1968).
See also *6*, p. *67*; 22.5.2, p. 528.

O. Goede
Investigation of the Emission of CdS Single Crystals in the 2.0 eV Region
Vol. **28**, No. 2, p. K167–K170 (1968).
See also *20.3*, p. *370*; 22.4.1, p. 504.

R. C. Whelan and D. Shaw
Evidence of a Doubly Ionized Native Donor in CdTe
Vol. **29**, No. 1, p. 145–152 (1968).
See also *13.4*, p. *215*; 14.3, p. 243; 22.4.3, p. 511.

E. Nebauer
Vapour-Pressure Investigations of Impurity Diffusion and Solubility in A^{II}-B^{VI} Compounds Demonstrated for the System CdS:Au:S_2
Vol. **29**, No. 1, p. 269—281 (1968).
See also *9*, p. *92*; 22.4.1, p. 504.

R. Boyn
Optical Absorption due to Intrinsic Defects in CdS Single Crystals
Vol. **29**, No. 1, p. 307—321 (1968).
See also *13.4*, p. *215*; *20.1*, p. *356*; 22.4.1, p. 504.

W. Frank
*Die kritische Schubspannung mit zweiwertigen Kationen dotierter Alkalihalogenide. (I): Nicht-agglomerierte Fremdion–Leerstellen-Paare
(The Critical Shear Stress of Alkali Halides Doped with Divalent Cations. (I): Non-Associated Impurity–Cation Vacancy Pairs)
Vol. **29**, No. 1, p. 391—402 (1968).
See also 22.5.2, p. 528.

A. A. Vaipolin, N. A. Goryunova, L. I. Kleshchinskii, G. V. Loshakova, and E. O. Osmanov
The Structure and Properties of the Semiconducting Compound $ZnSnP_2$
Vol. **29**, No. 1, p. 435—442 (1968).
See also 4, p. 47; *5*, p. *54*; 6, p. 68; 14.3, p. 243; 22.8, p. 555.

Š. Kupča
*Hardness of NaCl Single Crystals Prepared with Various Concentrations of $BaCl_2$ in the Melt
Vol. **29**, No. 1, p. K15—K17 (1968).
See also 14.4, p. 254; 22.5.2, p. 529.

H. Weijma and B. R. Knottnerus
Ionic Conductivity of Pure and Ba^{2+}-Doped CsCl
Vol. **29**, No. 1, p. K43—K45 (1968).
See also 1.2, p. 12; *14.4*, p. *254*; 22.5.2, p. 529.

B. von Guérard, H. Peisl, und W. Waidelich
*Equilibrium Vacancy Concentration in KCl
Vol. **29**, No. 1, p. K59—K61 (1968).
See also 22.5.2, p. 529.

D. J. Bacon
*Stability of the Dislocation Dipole in Half-Space
Vol. **29**, No. 2, p. 543—549 (1968).

R. A. Burley
*An Infrared Spectrophotometric Study of Vitreous Selenium Doped with Selenium Dioxide
Vol. **29**, No. 2, p. 551—558 (1968).
See also 2, p. 24; 20.1, p. 357; 22.1.3, p. 480.

K. Kleinhenz und P. Runow
*Herstellung dünner Bor-Folien und elektronenmikroskopische Beobachtung ihrer Baufehler
(Preparation of Thin Boron Foils and Electron Microscopic Observation of Their Lattice Faults)
Vol. **29**, No. 2, p. 627—637 (1968).
See also 1.4, p. 21; 22.1, p. 457.

E. S. Meieran and I. A. Blech
*Contrast Asymmetries in Lang Topographs of Crystals Strained by Thin Films
Vol. **29**, No. 2, p. 653—667 (1968).
See also *4*, p. *47*; 22.1.2, p. 474; 22.2.1, p. 486.

M. Sauvage
*Observations de sources et de réactions entre dislocations partielles de macle sur des topographies aux rayons X
(Observation of Sources and Reactions between Partial Twinning Dislocations on X-Ray Topographs)
Vol. **29**, No. 2, p. 725—736 (1968).
See also 22.8, p. 555.

W. Frank
*Die kritische Schubspannung mit zweiwertigen Kationen dotierter Alkalihalogenide. (II): Fremdion–Leerstellen-Dipol-Agglomerate und fremdionenreiche Ausscheidungen
(The Critical Shear Stress of Alkali Halides Doped with Divalent Cations. (II): Associated Impurity–Vacancy Dipoles and Impurity-Rich Precipitations)
Vol. **29**, No. 2, p. 767—780 (1968).
See also 22.5.2, p. 529.

A. Mehra
*Optical Absorption of Mn^{2+}-Doped Alkali Halides
Vol. **29**, No. 2, p. 847—857 (1968).
See also 6, p. 69; 13.4, p. 215; *20.1*, p. *357*; 22.5.2, p. 529.

A. Jostsons and P. G. McDougall
Fault Structures in Ti_2O
Vol. **29**, No. 2, p. 873—889 (1968).
See also *4*, p. *47*; 22.6, p. 543.

E. V. R. Sastry and T. M. Srinivasan
Growth and Bleach of F-Centers in KCl with Interstitial Cobalt
Vol. **29**, No. 2, p. K107—K109 (1968).
See also *10.2*, p. *144*; 22.5.2, p. 529.

W. W. Walker and L. J. Demer
*Indentation Creep in Crystals
Vol. **29**, No. 2, p. K141—K143 (1968).
See also 22.5.1, p. 517; 22.5.2, p. 529; 22.5.3, p. 534; 22.6, p. 543.

E. P. Berdnikov, V. A. Drozdov, and L. A. Mozgovaya
*On the Relation between Microhardness and Point Defects of Crystals
Vol. **29**, No. 2, p. K145—K147 (1968).
See also 12, p. 158; 15, p. 274; 22.1.2, p. 474.

G. Chanussot and H. Arend
Thermocurrents in Paraelectric $BaTiO_3$
Vol. **29**, No. 2, p. K149—K151 (1968).
See also *14.4.2*, p. *269*.

Yu. A. Osipiyan and I. S. Smirnova
*Perfect Dislocations in the Wurtzite Lattice
Vol. **30**, No. 1, p. 19—29 (1968).
See also 22.4, p. 496.

P. Feltham
*A Stochastic Model of Creep
Vol. **30**, No. 1, p. 135—146 (1968).
See also *10.1*, p. *134*; 21, p. 388; 22.5.2, p. 530.

P. Humble
*Electron Diffraction Contrast from Three Parallel Overlapping Stacking Faults
Vol. **30**, No. 1, p. 183–192 (1968).
See also *10.1*, p. *135*.

N. Itoh
Degradation of Low-Energy Electrons in Alkali Halides
Vol. **30**, No. 1, p. 199–207 (1968).
See also *11*, p. *154*; 22.5.2, p. 530.

L. I. Datsenko, M. Ya. Skorokhod, and A. S. Vasilkovskii
*The Effect of Dislocations on the Intensity Jumps at the Interferential Transmission of X-Rays near the K-Edge of Absorption in Ge
Vol. **30**, No. 1, p. 231–237 (1968).
See also 20, p. 342; 22.1.1, p. 466.

T. Okada and K. Ozawa
*Anomalous Ductility of LiF Crystals at Low Temperatures
Vol. **30**, No. 1, p. K69–K71 (1968).
See also 11, p. 155; 22.5.2, p. 530.

D. Strauch
Shell Model Treatment of the Vibrations of Crystals Containing Substitutional Defects. (II): On the Theory of Defect-Induced One-Phonon Absorption
Vol. **30**, No. 2, p. 495–505 (1968).
See also *6*, p. *70*; *20.1*, p. *359*; 22.1, p. 457; 22.1.1, p. 466; 22.1.2, p. 475.

N. E. Byer and H. S. Sack
Ultrasonic Velocity and Attenuation in Alkali Halides Containing CN^-. (I): Experimental Results
Vol. **30**, No. 2, p. 569–577 (1968).
See also *7*, p. *78*; 13.4, p. 217; 22.5.2, p. 530.

N. E. Byer and H. S. Sack
Ultrasonic Velocity and Attenuation in Alkali Halides Containing CN^-. (II): Theoretical Interpretation
Vol. **30**, No. 2, p. 579–586 (1968).
See also *7*, p. *78*; 13.4, p. 217; 22.5.2, p. 531.

F. Granzer, G. Wagner und J. Eisenblätter
*Atomistische Berechnung der Core-Struktur, Core-Energie und Peierls-Spannung einer Stufenversetzung in NaCl
(Atomistic Calculation of Core Structure, Core Energy, and Peierls Stress of an Edge Dislocation in NaCl)
Vol. **30**, No. 2, p. 587–600 (1968).
See also 22.5.2, p. 531.

G. Baldini and M. Guzzi
Zero-Phonon Transitions of Eu^{++}- and Sm^{++}-Doped Alkali Halides
Vol. **30**, No. 2, p. 601–612 (1968).
See also *6*, p. *70*; 20.1, p. 359; 20.3, p. 372; 22.5.2, p. 531.

R. Siems
*Mechanical Interactions of Point Defects
Vol. **30**, No. 2, p. 645–658 (1968).
See also 6, p. 70; *10.1*, p. *135*.

J. Celý
Mössbauer Effect in Simple Cubic Lattice with Line Defect
Vol. **30**, No. 2, p. 671–683 (1968).
See also *6.1*, p. *75*; 10.1, p. 135.

A. T. DAVIDSON and A. D. YOFFE
Hopping Electrical Conduction and Thermal Breakdown in Natural and Synthetic Mica
Vol. **30**, No. 2, p. 741–754 (1968).
See also 3, p. 28; *14.4.1*, p. *265*; 20.1, p. 359; 22, p. 455.

R. DE RIDDER, J. VAN LANDUYT, R. GEVERS, and S. AMELINCKX
The Fine Structure of Spots in Electron Diffraction Resulting from the Presence of Planar Interfaces and Dislocations. (IV): Wedge Crystals
Vol. **30**, No. 2, p. 797–815 (1968).
See also *4*, p. *48*; 10.1, p. 136; 21.1.1, p. 423; 22.6, p. 544.

H. BETHGE and U. MESSERSCHMIDT
*Nonconservative Processes during Dislocation Glide at Elevated Temperatures
Vol. **30**, No. 2, p. K87–K89 (1968).
See also 22.5.2, p. 531.

G. HENSEL and J. ZILLER
Precipitation Influenced Low Temperature Conductivity of Ca Doped NaCl Crystals
Vol. **30**, No. 2, p. K165–K168 (1968).
See also *14.4*, p. *254*, 22.5.2, p. 531.

10.1 Defect Properties of Metals

Review Articles

K. N. Subramanian
Cyclic Stressing in Crystals of Rock-Salt Structure
Vol. **28**, No. 1, p. 9–37 (1968).
See also *10*, p. *93*; 21, p. 373; 21.1, p. 390; 22.5.1, p. 515; 22.5.2, p. 518; 22.6, p. 535.

Original Papers and Short Notes

B. Šesták and S. Kadečková
*Etch Pits at Dislocations in High-Purity Iron Single Crystals
Vol. **16**, No. 1, p. 89–94 (1966).
See also 21.1.1, p. 410.

P. C. J. Gallagher
*An Absolute Determination of the Extrinsic and Intrinsic Stacking Fault Energies in Ag–In Alloys
Vol. **16**, No. 1, p. 95–115 (1966).
See also 21.6, p. 430.

J. Melngailis
*A Model for the Change in Elastic Modulus of a Solid Caused by Point Defects
Vol. **16**, No. 1, p. 247–252 (1966).
See also 12.1, p. 159; 21.1, p. 390.

P. Peretto, J. L. Oddou, C. Minier-Cassayre, D. Dautreppe et P. Moser
*Étude et interprétation du stade I dans le nickel irradié aux électrons à 20 °K
(Investigation and Interpretation of Stage I of Nickel Electron Irradiated at 20°K)
Vol. **16**, No. 1, p. 281–287 (1966).
See also 14.1, p. 220; 18.2, p. 299; 21.1, p. 390.

D. Rühlicke
*Zum Gleitmechanismus plastisch verformter Magnesiumeinkristalle oberhalb Raumtemperatur
(Slip Mechanism of Plastically Deformed Magnesium Single Crystals above Room Temperature)
Vol. **16**, No. 1, p. K1–K4 (1966).
See also 21.3, p. 426.

V. Gerold and H. Haberkorn
*On the Critical Resolved Shear Stress of Solid Solutions Containing Coherent Precipitates
Vol. **16**, No. 2, p. 675–684 (1966).
See also 4, p. 38, 21.1, p. 391.

K.-P. Chik
*Electron Diffraction Contrast of Small Stacking Fault Tetrahedra with Applications to Quenched Gold
Vol. **16**, No. 2, p. 685–696 (1966).
See also 21.6, p. 430

R. Dietz and W. Hellenthal
*Recovery Processes in Nickel Thin Films between 20 and 300 °C.
Vol. **16**, No. 2, p. 789–794 (1966).
See also 14.1, p. 220; 21.1, p. 391.

M. RÜHLE und M. WILKENS
*Die Natur der elektronenmikroskopisch beobachtbaren Defekte in neutronenbestrahltem Kupfer
(Nature of the Electron-Microscopically Observable Defects in Neutron-Irradiated Copper)
Vol. **16**, No. 2, p. K105—K108 (1966).
See also 11, p. 145; 21.1, p. 391.

H. DIEPERS und J. DIEHL
*Nachweis und Entstehung von Zwischengitteratom-Agglomeraten in ionenbestrahlten Kupferfolien
(Detection and Formation of Interstitial Agglomerates in Ion-Irradiated Copper Foils)
Vol. **16**, No. 2, p. K109—K112 (1966).
See also 11, p. 145; 21.1, p. 391.

D. A. BLACKBURN
*Concentration de défauts au voisinage des interfaces dans un métal soumis à des gradients de potentiel électrique ou de température
(Concentration of Defects near Interfaces in a Metal in the Presence of Electric Potential and Temperature Gradients)
Vol. **17**, No. 1, p. 3—9 (1966).
See also 21, p. 374.

T. J. KOPPENAAL and R. J. ARSENAULT
*Some Notes on the Strengthening Mechanism in Neutron Irradiated Copper Single Crystals
Vol. **17**, No. 1, p. 27—33 (1966).
See also 11, p. 145; 21.1, p. 391.

CH. SCHWINK and H. NEUHÄUSER
*On the Strengthening Mechanism in Neutron Irradiated Copper Single Crystals. (I): On Experiments Concerning Strain Rate Changes
Vol. **17**, No. 1, p. 35—42 (1966).
See also 11, p. 145; 21.1, p. 391.

J. DIEHL and G. P. SEIDEL
*On the Strengthening Mechanism in Neutron Irradiated Copper Single Crystals. (II): Analysis of the Thermally Activated Processes
Vol. **17**, No. 1, p. 43—48 (1966).
See also 11, p. 146; 21.1, p. 392.

W. ANDRÄ and K. STEENBECK
Magnetic Anisotropy Induced in Thin Films at Low Temperatures
Vol. **17**, No. 1, p. 191—195 (1966).
See also *18.2*, p. *302*; 21.1, p. 392.

K. ISEBECK, F. RAU, W. SCHILLING, K. SONNENBERG, P. TISCHER, and H. WENZL
Stored Energy, Volume, and Resistivity Change in Neutron Irratiated Aluminium
Vol. **17**, No. 1, p. 259—268 (1966).
See also *11*, p. *146*; 14.1, p. 220; 21, p. 374.

D. AST und J. DIEHL
*Untersuchung des Zerfalls der Leerstellenagglomerate in neutronenbestrahltem Nickel mit Hilfe der kritischen Schubspannung
(Study of the Decay of Vacancy Agglomerates in Neutron-Irradiated Nickel by Measurements of Critical Shear Stress)
Vol. **17**, No. 1, p. 269—284 (1966).
See also 11, p. 146; 21.1, p. 392.

P. Vigier, C. Minier-Cassayre, P. Moser et V. Hivert
*Etude de défauts ponctuels créés par bombardement neutronique dans le fer contenant certaines impuretés
(Study of Point Defects Produced by Neutron Irradition in Doped Iron)
Vol. **17**, No. 1, p. 317–321 (1966).
See also 11, p. 146; 14.1, p. 220; 18.2, p. 302; 21.1.1, p. 412.

R. von Jan
*The Interpretation of Recovery Stages in Irradiated Metals
Vol. **17**, No. 1, p. 361–373 (1966).
See also 11, p. 146; 12, p. 374; 21.1, p. 392; 21.6, p. 430.

R. Pichon, C. Minier-Cassayre, V. Hivert, and P. Moser
Study of the Recovery of Resistivity of Niobium Electron Irradiated at 20 °K
Vol. **17**, No. 1, p. K33–K35 (1966).
See also *14.1*, p. *220*; 21, p. 374.

P. Vigier, V. Hivert, A. Bourret et P. Moser
*Précipitation et rémise en solution du carbone dans le fer
(Precipitation and Re-Solution of Carbon in Iron
Vol. **17**, No. 1, p. K37–K39 (1966).
See also 18.2, p. 302; 21.1.1, p. 412.

P. Kratochvíl and B. Smola
*The Effect of Vacuum Annealing on the Strain Behaviour of Cadmium Single Crystals
Vol. **17**, No. 1, p. K91–K94 (1966).
See also 21, p. 374.

D. N. Seidman and R. W. Balluffi
*On the Efficiency of Dislocation Climb in Gold
Vol. **17**, No. 2, p. 531–541 (1966).
See also 21.6, p. 431.

L. S. Palatnik, M. Ya. Fuks, L. I. Lukashenko, A. G. Ravlik, and A. A. Kozma
Structure and Magnetic Properties of Condensed Ferromagnetic Films
Vol. **17**, No. 2, p. 543–554 (1966).
See also *1.4*, p. *17*; 3.1, p. 30; *18.2*, p. *303*; 21.1, p. 392; 21.1.1, p. 413.

A. Seeger und G. Wobser
*Anordnungen paralleler Versetzungen in anisotropen Medien
(Parallel Dislocation Arrays in Anisotropic Media)
Vol. **17**, No. 2, p. 709–724 (1966).
See also *10*, p. *95*; 21, p. 375; 21.1, p. 392; 21.3, p. 426; 21.6, p. 431; 22.1.1, p. 460; 22.1.2, p. 469.

U. Essmann
*Elektronenmikroskopische Untersuchung der Versetzungsanordnung verformter Kupfereinkristalle. (III): Bestimmung der Versetzungsdichte
(Electron-Microscopic Study of the Dislocation Arrangement in Deformed Copper Single Crystals. (III): Determination of the Dislocation Density)
Vol. **17**, No. 2, p. 725–737 (1966).
See also 21.1, p. 392.

A. G. Lesnik, G. I. Levin, and V. M. Nedostup
The Importance of Vacancies and Interstitial Atoms in the Development of Induced Anisotropy
Vol. **17**, No. 2, p. 745–751 (1966).
See also 3.1, p. 30; *18.2*, p. *303*; 21.1.1, p. 413.

A. G. Lesnik and L. M. Sandler
Effect of Tensile Stress on the Magnetic Properties of Permalloy Films
Vol. **17**, No. 2, p. 753–760 (1966).
See also *18.2*, p. *303*; 19, p. 327; 21.1.1, p. 413.

J. L. Oddou, C. Minier-Cassayre et P. Moser
*Effet des impurités sur les stades de revenu de la résistivité électrique du nickel bombardé à basse température par des électrons
(Effect of Impurities on the Recovery Stages of the Resistivity of Nickel Electron Irradiated at Low Temperatures)
Vol. **17**, No. 2, p. 793–798 (1966).
See also *11*, p. *146*; 14.1, p. 220; 21.1, p. 393.

R. B. Clough
*The Activation Energy for Plastic Flow in Stress Relaxation
Vol. **17**, No. 2, p. K175–K178 (1966).
See also *10*, p. *96*.

S. B. McRickard
*Twinning Deformation in Iron at 4 °K
Vol. **18**, No. 1, p. 89–97 (1966).
See also 21.1.1, p. 413.

R. Chang and L. J. Graham
*Edge Dislocation Core Structure and the Peierls Barrier in Body-Centered Cubic Iron
Vol. **18**, No. 1, p. 99–103 (1966).
See also 21.1.1, p. 413.

J. van Landuyt, R. Gevers, and S. Amelinckx
*On the Determination of the Nature of Stacking Faults in F.C.C. Metals from the Bright Field Image
Vol. **18**, No. 1, p. 167–172 (1966).
See also 21.1.1, p. 413.

K. L. Merkle
Radiation-Induced Point Defect Clusters in Copper and Gold. (I): Clusters Produced in Energetic Dislpacement Cascades
Vol. **18**, No. 1, p. 173–188 (1966).
See also *11*, p. *146*; 21.1, p. 393; 21.6, p. 431.

A. Seeger und G. Wobser
*Stapelfehlerdipole in kubisch-flächenzentrierten Metallen
(Stacking Fault Dipoles in Face-Centered Cubic Metals)
Vol. **18**, No. 1, p. 189–206 (1966).
See also 21.1, p. 393; 21.6, p. 431.

H. Bilger
Temperature and Deformation Dependence of the Coercivc Field Strength in Iron Single Crystals (II)
Vol. **18**, No. 1, p. 207–219 (1966).
See also *18.2*, p. *303*; 21.1.1, p. 414.

T. R. Duncan and D. Kuhlmann-Wilsdorf
*The Stress Fields of Dislocation Tilt Boundaries in Anisotropic Cubic and Diamond-Type Crystals
Vol. **18**, No. 1, p. 231–239 (1966).
See also *10*, p. *96*; 21, p. 375; 21.1, p. 393; 21.5, p. 429; 21.6, p. 431; 22.1, p. 456; 22.1.1, p. 460; 22.1.2, p. 470.

R. Gevers, J. van Landuyt, and S. Amelinckx
*The Fine Structure of Spots in Electron Diffraction Resulting from the Presence of Planar Interfaces and Dislocations. (I): General Theory and Its Application to Stacking Faults and Anti-Phase Boundaries
Vol. **18**, No. 1, p. 343—361 (1966).
See also **4**, p. 39; *10*, p. *96*.

J. van Landuyt, R. Gevers, and S. Amelinckx
*The Fine Structure of Spots in Electron Diffraction Resulting from the Presence of Planar Interfaces and Dislocations. (II): Observations on Crystals Containing Stacking Faults
Vol. **18**, No. 1, p. 363—378 (1966).
See also **4**, p. 39; *10*, p. *96*; **21.1.1**, p. 414; **22.6**, p. 536.

K. Isebeck, R. Müller, W. Schilling, and H. Wenzl
*Reaction Kinetics of Stage III Recovery in Aluminium after Neutron Irradiation
Vol. **18**, No. 1, p. 427—439 (1966).
See also **11**, p. 147; **14.1**, p. 221; **21**, p. 375.

W. Frank
*Die kritische Schubspannung kubischer Kristalle mit Fehlstellen tetragonaler Symmetrie. (III): Elektronen- und neutronenbestrahltesAluminium sowie elektronenbestrahltes Kupfer
(Critical Shear Stress of Cubic. Crystals Containing Defects of Tetragonal Symmetry. (III): Electron- and Neutron-Irradiated Aluminium and Electron-Irradiated Copper)
Vol. **18**, No. 1, p. 459—464 (1966).
See also **11**, p. 147; **21**, p. 376; **21.1**, p. 394.

B. F. Rothenstein, C. Anghel, and M. Lupulescu
*Influence of Hydrogen on the Magnetomechanical Damping in Nickel
Vol. **18**, No. 1, p. K13—K16 (1966).
See also *18.2*, p. *304*; **21.1**, p. 394.

C. Varotto and A. E. Vidoz
On the Recovery and Ordering Processes of Cold Worked Ni–Fe Alloys
Vol. **18**, No. 1, p. K49—K52 (1966).
See also *14.1*, p. *221*; **21.1.1**, p. 414.

K. C. A. Blasdale, R. King, and K. E. Puttick
*The Microstructure of Tensile Kinks in Cadmium Crystals
Vol. **18**, No. 2, p. 491—503 (1966).
See also **4**, p. 39; **21**, p. 376.

G. C. T. Liu and J. C. M. Li
*Energy of Polygonal and Elliptical Dislocation Loops
Vol. **18**, No. 2, p. 517—525 (1966).
See also *10*, p. *96*.

G. C. T. Liu and J. C. M. Li
*Energy of Hexagonal Dislocation Loops
Vol. **18**, No. 2, p. 527—539 (1966).
See also *10*, p. *96*; **21**, p. 376; **22**, p. 441; **22.1.2**, p. 470.

Ch. Schwink
*Über den Begriff „aktives Gleitvolumen“ und dessen Bedeutung
(On the Concept of "Active Slip-Volume" and Its Meaning)
Vol. **18**, No. 2, p. 557—567 (1966).
See also **21.1**, p. 394.

F. Häussermann und M. Wilkens
*Bestimmung der Stapelfehlerenergie kubisch-flächenzentrierter Metalle aus der Analyse des elektronenmikroskopischen Beugungskontrastes von Stapelfehlerdipolen (Determination of the Stacking Fault Energy of Face-Centered Cubic Metals from the Electron-Microscopic Diffraction Contrast of Stacking Fault Dipoles)
Vol. **18**, No. 2, p. 609–624 (1966).
See also 21.1, p. 394; 21.6, p. 431.

H. Yoshinaga
Measurements of the Anisotropy of the Dislocation Resistivity in Au, Ag, and Cu
Vol. **18**, No. 2, p. 625–636 (1966).
See also *14.1*, p. *221*; 21.1, p. 394; 21.6, p. 431.

H. Bilger
*Elektronenmikroskopische Oberflächen- und Durchstrahlungsuntersuchungen von Eisen-Einkristallen sowie der Einfluß der Orientierung auf deren Verfestigungsverhalten
(Electron-Microscopic Surface and Transmission Studies of Iron Single Crystals and the Effect of Orientation on Their Strengthening Behaviour)
Vol. **18**, No. 2, p. 637–651 (1966).
See also 21.1.1, p. 414.

V. Vítek
*Thermally Activated Motion of Screw Dislocations in B.C.C. Metals
Vol. **18**, No. 2, p. 687–701 (1966).
See also 21, p. 376.

V. Vítek and F. Kroupa
*Dislocation Theory of Slip Geometry and Temperature Dependence of Flow Stress in B.C.C. Metals
Vol. **18**, No. 2, p. 703–713 (1966).
See also 21.1.1, p. 414.

B. Harris
*Solution Hardening in Niobium
Vol. **18**, No. 2, p. 715–730 (1966).
See also 21, p. 376.

L. I. van Torne
Point Defect Electron Scattering in Beryllium
Vol. **18**, No. 2, p. 737–742 (1966).
See also *14.1*, p. *221*; 21.3, p. 426.

G. Schmid
*Kriechexperimente an hochreinen Kupfer-Einkristallen
(Creep Experiments on High-Purity Copper Single Crystals)
Vol. **18**, No. 2, p. 829–845 (1966).
See also 21.1, p. 394.

V. F. Moiseev and V. I. Trefilov
*Change of the Deformation Mechanism (Slip ⇄ Twinning) in Polycrystalline α-Iron
Vol. **18**, No. 2, p. 881–895 (1966).
See also 21.1.1, p. 414.

W. Heye and G. Wassermann
*Mechanical Twinning in Cold-Rolled Silver Single Crystals
Vol. **18**, No. 2, p. K107–K111 (1966).
See also 21.6, p. 432.

H. BILGER

*On the Recovery of Plastically Deformed Iron Single Crystals above Room Temperature
Vol. 18, No. 2, p. K135–K138 (1966).
See also 18.2, p. 304; 21.1.1, p. 414.

IR. BOESONO, G. J. ERNST, M. C. LEMMENS, M. J. VAN LANGEN, and G. DE VRIES

A Zener-Damping Peak in Fe–4 Wt% Si
Vol. 19, No. 1, p. 107–110 (1967).
See also *12.1*, p. *159*; 21.1.1, p. 414.

E. KÖSTER

Deformation and Temperature Dependence of Magnetization Processes in Nickel Single Crystals. (I): Bloch Wall Displacements in ⟨110⟩ and ⟨111⟩ Nickel Single Crystals
Vol. 19, No. 1, p. 153–168 (1967).
See also *18.2*, p. *304*; 21.1, p. 395.

A. VAN DEN BEUKEL, P. C. J. COREMANS, and M. M. A. VRIJHOEF

On the Kinetics of Short-Range Ordering in AuAg (50,50) and CuAl (85,15)
Vol. 19, No. 1, p. 177–183 (1967).
See also *1.1*, p. *8*; 21.1, p. 395; 21.6, p. 432.

A. K. HEAD

*Unstable Dislocations in Anisotropic Crystals
Vol. 19, No. 1, p. 185–192 (1967).
See also 21, p. 376; 21.1, p. 395; 21.1.1, p. 415.

W. FRANK

*Die kritische Schubspannung kubischer Kristalle mit Fehlstellen tetragonaler Symmetrie. (IV): Die untere Streckgrenze von α-Eisen
(Critical Shear Stress of Cubic Crystals Containing Defects of Tetragonal Symmetry. (IV): The Lower Field Stress of α-Iron)
Vol. 19, No. 1, p. 239–250 (1967).
See also 21.1.1, p. 415.

H. MUGHRABI and A. SEEGER

*The Study of Defects in Quenched Nickel
Vol. 19, No. 1, p. 251–262 (1967).
See also 21.1, p. 395.

M. RÜHLE

*Elektronenmikroskopie kleiner Fehlstellenagglomerate in bestrahlten Metallen. (I): Theorie des Kontrastes und experimentelle Methoden zur Ermittlung des Defekttyps
(Electron-Microscopic Studies of Small Defect-Agglomerates in Irradiated Metals. (I): Theory of Contrast and Experimental Methods for Determining the Defect Type)
Vol. 19, No. 1, p. 263–278 (1967).
See also 11, p. 147; 21.1, p. 395.

M. RÜHLE

*Elektronenmikroskopie kleiner Fehlstellenagglomerate in bestrahlten Metallen. (II): Untersuchungen an neutronen- und ionenbestrahltem Kupfer sowie neutronenbestrahltem Nickel
(Electron-Miroscopic Studies of Small Defect-Agglomerates in Irradiated Metals.(II): Investigations on Neutron- and Ion-Irradiated Cu and Neutron-Irradiated Ni)
Vol. 19, No. 1, p. 279–295 (1967).
See also 11, p. 147; 21.1, p. 395.

F. Guiu
*Temperature and Strain Rate Dependence of the Flow Stress in Molybdenum
Vol. **19**, No. 1, p. 339–351 (1967).
See also 21, p. 377.

P. Lukáč
*Spannungsrelaxation in Cd-Einkristallen
(Stress Relaxation in Cd Single Crystals)
Vol. **19**, No. 1, p. K47–K50 (1967).
See also 21, p. 377.

J. Man, M. Holzmann, and B. Vlach
*Microstrain Region and Transition to Macrostrain in 99.9% Polycrystalline Copper
Vol. **19**, No. 2, p. 543–553 (1967).
See also 21.1, p. 396.

E. Köster
Deformation and Temperature Dependence of the Magnetization Processes in Nickel Single Crystals. (II): Bloch Wall Displacements and Rotational Processes of the Spontaneous Magnetization in ⟨001⟩ Nickel Single Crystals
Vol. **19**, No. 2, p. 655–671 (1967).
See also *18.2*, p. *305*; 21.1, p. 396.

K.-H. Pfeffer
*Wechselwirkung zwischen Versetzungen und ebenen Blochwänden mit starrem Magnetisierungsverlauf
(Interaction between Dislocations and Plane Bloch-Walls with a Rigid Magnetization Curve)
Vol. **19**, No. 2, p. 735–750 (1967).
See also *18.2*, p. *305*; 21.1, p. 396; 21.1.1, p. 415.

O. Vöhringer and E. Macherauch
*The Yield Point of α-Copper–Tin Alloys
Vol. **19**, No. 2, p. 793–803 (1967).
See also 21.1, p. 396.

L. I. van Torne
Electron Charge Distribution about Zinc Ions in Aluminium
Vol. **19**, No. 2, p. 855–862 (1967).
See also *13*, p. *165*; 14.1, p. 222; 21, p. 377.

P. Guyot
*Camel-Humps Peierls Hills in Iron
Vol. **19**, No. 2, p. K95–K98 (1967).
See also 21.1.1, p. 415.

B. Eckstein
*A Disorder Model of Melting and Melts
Vol. **20**, No. 1, p. 83–93 (1967).
See also *2*, p. *22*.

K.-H. Pfeffer
Micromagnetic Treatment of the Interaction between Dislocations and Plane Bloch-Walls. (I): General Theory
Vol. **20**, No. 1, p. 395–411 (1967).
See also *18.2*, p. *305*; 21.1, p. 396; 21.1.1, p. 416.

E. P. LAUTENSCHLAGER, T. C. TISONE, and J. O. BRITTAIN
Electron Transmission Microscopy of NiAl
Vol. **20**, No. 2, p. 443–450 (1967).
See also *4*, p. *41*; 21.1, p. 396.

A. K. HEAD, M. H. LORETTO, and P. HUMBLE
*The Influence of Large Elastic Anisotropy on the Determination of Burgers Vectors of Dislocations in β-Brass by Electron Microscopy
Vol. **20**, No. 2, p. 505–519 (1967).
See also 12.1, p. 159; 21.1, p. 397.

A. K. HEAD, M. H. LORETTO, and P. HUMBLE
*The Identification of Burgers Vectors and Unstable Directions of Dislocations in β-Brass
Vol. **20**, No. 2, p. 521–536 (1967).
See also 12.1, p. 159; 21.1, p. 397.

F. J. WORZALA
*Hardening in Irradiated Copper Crystals
Vol. **20**, No. 2, p. K81–K83 (1967).
See also 11, p. 148; 21.1, p. 397.

R. VANSELOW
Changes due to Solved Impurities at Pure Platinum-Surfaces: Observations with the Field Electron-Microscope
Vol. **21**, No. 1, p. 69–76 (1967).
See also *1.3*, p. *15*; 21.6, p. 433.

L. M. CLAREBROUGH, D. MICHELL, and A. P. SMITH
*A Source of Dislocations in Copper Crystals
Vol. **21**, No. 1, p. 369–373 (1967).
See also 21.1, p. 397.

E. D. HONDROS
Herring-Nabarro Creep Studies on the Self-Diffusion of Iron in the Presence of Solutes
Vol. **21**, No. 1, p. 375–382 (1967).
See also *9*, p. *89*; 21.1.1, p. 417.

R. GEVERS, J. VAN LANDUYT, and S. AMELINCKX
*Two Beam Kinematical Theory for the Diffraction of Electrons by Crystals with Stacking Faults
Vol. **21**, No. 1, p. 393–411 (1967).
See also *10*, p. *99*; 21.1.1, p. 417; 22.6, p. 538.

J. STEINERT
Induced Uniaxial Magnetic Anisotropy of Fe–Al Alloys at Low Concentrations
Vol. **21**, No. 1, p. K13–K15 (1967).
See also *18.2*, p. *307*; 21, p. 379; 21.1.1, p. 417.

A. KÖTHE and F. SCHLÄT
*Recovery of Deformed Niobium and Tantalum Degassed in Ultra-High Vacuum
Vol. **21**, No. 1, p. K73–K76 (1967).
See also 14.1, p. 222; 21, p. 379.

A. A. ADAMENKO and I. J. DEKHTYAR
Effect of Plastic Deformation on the Absolute Thermopower of α-Brass
Vol. **21**, No. 1, p. K81–K84 (1967).
See also *15*, p. *272*; 21.1, p. 398.

V. I. SYUTKINA and E. S. YAKOVLEVA
*The Mechanism of Deformation of the Ordered CuAu Alloy
Vol. **21**, No. 2, p. 465–480 (1967).
See also 1.1, p. 8; 4, p. 42; 21.1, p. 398; 21.6, p. 433.

W. A. JESSER and D. KUHLMANN-WILSDORF
The Geometry and Energy of a Twist Boundary between Crystals with Unequal Lattice Parameters
Vol. **21**, No. 2, p. 533–544 (1967).
See also *3.1*, p. *31*; 10, p. 100; 21, p. 379; 22, p. 445.

J. A. SIGLER and D. KUHLMANN-WILSDORF
*Calculations on the Mechanical Energy of Vacancy Condensation Loops, Stacking Fault Tetrahedra, and Voids
Vol. **21**, No. 2, p. 545–556 (1967).
See also 21, p. 379; 21.1, p. 398; 21.6, p. 433.

J. POLÁK
*Kinetics of Quenched-in Vacancies in Pure Platinum
Vol. **21**, No. 2, p. 581–591 (1967).
See also 21.6, p. 433.

K. C. A. BLASDALE
*Arrays of Dislocation Dipoles in Cadmium
Vol. **21**, No. 2, p. 649–660 (1967).
See also 21, p. 379.

P. LUKÁŠ and M. KLESNIL
*Fatigue Hardening of Cu–Zn Alloys
Vol. **21**, No. 2, p. 717–731 (1967).
See also 21.1, p. 398.

P. HUMBLE
*A Comparison of Electron Microscope Images of Dislocations Based on Isotropic and Anisotropic Elasticity Theory
Vol. **21**, No. 2, p. 733–740 (1967).
See also 21, p. 379; 21.1, p. 398; 21.1.1, p. 417; 21.6, p. 433.

K.-H. PFEFFER
Micromagnetic Treatment of the Interaction between Dislocations and Plane Bloch-Walls. (II): Applications
Vol. **21**, No. 2, p. 837–856 (1967).
See also *18.2*, p. *307*; 21.1, p. 398.

K.-H. PFEFFER
On the Theory of Coercive Field Strength and Initial Susceptibility
Vol. **21**, No. 2, p. 857–872 (1967).
See also *18.2*, p. *307*; 21.1, p. 398; 21.1.1, p. 418.

A. CLEMENT, N. CLEMENT et P. COULOMB
*Paires de défauts intrinsèque et extrinsèque dans un acier inoxydable et dans un alliage cuivre–silicium
(Intrinsic and Extrinsic Defect Pairs in a Nonoxidizable Steel and a Copper–Silicon Alloy)
Vol. **21**, No. 2, p. K97–K98 (1967).
See also 21.1, p. 398; 21.1.1, p. 418.

F. J. Kedves
*On Quenching Experiments in Gold
Vol. **21**, No. 2, p. K183—K186 (1967).
See also 14.1, p. 223; 21.6, p. 433.

K. R. Evans and W. F. Flanagan
*An Analysis of the Cottrell-Stokes Law
Vol. **22**, No. 1, p. 195—198 (1967).
See also 21.1, p. 398.

E. J. Kramer and Ch. L. Bauer
*Variations of Young's Modulus throughout the Mixed State in Niobium
Vol. **22**, No. 1, p. 199—202 (1967).
See also 1.2, p. 11; 14.2, p. 229; 18, p. 292; 21, p. 380.

J. L. Routbort and H. S. Sack
*Low Temperature Internal Friction of Aluminium at 1 Hz
Vol. **22**, No. 1, p. 203—212 (1967).
See also 21, p. 380.

J. Demny
*Stacking Faults in Tungsten
Vol. **22**, No. 1, p. K1—K2 (1967).
See also 21, p. 380.

V. Vítek
*The Effect of the Type of Splitting of Screw Dislocations on the Slip Geometry in B.C.C. Metals
Vol. **22**, No. 2, p. 453—462 (1967).
See also 21, p. 381; 21.1.1, p. 418.

K. R. Evans, D. J. Bailey, and W. F. Flanagan
*An Interpretation of the Dislocation Velocity-Stress Exponent
Vol. **22**, No. 2, p. 607—616 (1967).
See also 21.1, p. 399; 21.1.1, p. 418.

L. Gerward
Magnetic After-Effect in Deformed Cobalt Single Crystals
Vol. **22**, No. 2, p. 659—666 (1967).
See also *18.2*, p. *308*; 21.1, p. 399.

H. Bilger und H. Kronmüller
*Das Erholungsspektrum von Kobalt
(Recovery Spectrum of Cobalt)
Vol. **22**, No. 2, p. 683—688 (1967).
See also 14.1, p. 223; 21.1, p. 399.

M. Wilkens, M. Rühle, and F. Häussermann
*On the Nature of the Long-Range Dislocation Contrast in Electron Transmission Micrographs
Vol. **22**, No. 2, p. 689—708 (1967).
See also 21.1, p. 399.

M. O. Speidel
*Interaction of Dislocations with Coherent Particles and Stress Corrosion Cracking of High-Strength Aluminium Alloys
Vol. **22**, No. 2, p. K71—K74 (1967).
See also 21, p. 381.

V. G. Kurdyumov, L. M. Morgulis, and L. M. Utevskii
*Slip Band Structure of Deformed Silicon Iron Single Crystals
Vol. **22**, No. 2, p. K75—K78 (1967).
See also 21.1.1, p. 418.

A. van den Beukel
*On the Relation between Self-Diffusion, Melting, and Vacancy–Impurity Binding in Dilute Alloys
Vol. **23**, No. 1, p. 165—170 (1967).
See also *9*, p. *90*; 21, p. 381.

D. A. Blackburn
*Non-Equilibrium Vacancy Concentrations in Metals Subject to Thermal Gradients and Electric Fields
Vol. **23**, No. 1, p. 177—183 (1967).
See also 21, p. 381; 21.6, p. 434.

G. Martin, D. A. Blackburn, and Y. Adda
Self-Diffusion in the Grain Boundaries of Silver Bicrystals under Hydrostatic Pressure
Vol. **23**, No. 1, p. 223—228 (1967).
See also *9*, p. *90*; 21.6, p. 434.

P. H. Dederichs
*Dynamical Scattering Theory for Crystals with Point Defects
Vol. **23**, No. 1, p. 377—386 (1967).
See also 6, p. 62; *10*, p. *102*.

G. Duesing, H. Hemmerich, D. Meissner, and W. Schilling
The Influence of Vacancies and Impurities on the Damage Production in Platinum Electron-Irradiated at 90 °K
Vol. **23**, No. 2, p. 481—488 (1967).
See also *11*, p. *150*; 14.1, p. 223; 21.6, p. 434.

D. J. Bacon
*A Method for Describing a Flexible Dislocation
Vol. **23**, No. 2, p. 527—538 (1967).
See also *10*, p. *102*.

I. M. Bernstein and J. C. M. Li
*Thermally Activated Deformation of Potassium
Vol. **23**, No. 2, p. 539—548 (1967).
See also 21.2, p. 425.

R. Gevers, J. van Landuyt, and S. Amelinckx
*Fine Structure of Weak Kinematical Electron Diffraction Beams for a Foil Containing Stacking Fault or Anti-Phase Boundaries — Two-Beam Case
Vol. **23**, No. 2, p. 549—562 (1967).
See also *4*, p. *43*; 21.1.1, p. 419.

M. L. Swanson
The Effects of Defect Doping on Low Temperature Neutron-Irradiation Damage and Recovery in Aluminium and Platinum
Vol. **23**, No. 2, p. 649—661 (1967).
See also *11*, p. *150*; 14.1, p. 224; 21, p. 381; 21.6, p. 434.

P. Regnier et J. M. Dupouy
*Glissement prismatique de monocristaux de béryllium
(Prismatic Glide of Beryllium Monocrystals)
Vol. **23**, No. 2, p. K109—K111 (1967).
See also 21.3, p. 426.

J. L. DAVIDSON and J. M. GALLIGAN
*Kinetics of Vacancy Annihilation in Quenched Copper
Vol. 23, No. 2, p. K139–K141 (1967).
See also 21.1, p. 400.

F. KROUPA
*A Note on Work Hardening in B.C.C. Metals
Vol. 23, No. 2, p. K143–K145 (1967).
See also 21, p. 382.

P. PAUFLER and G. E. R. SCHULZE
*Plastic Deformation of the Intermetallic Compound $MgZn_2$
Vol. 24, No. 1, p. 77–87 (1967).
See also 21.1, p. 401; 21.3, p. 426.

M. J. MARCINKOWSKI and H. J. LEAMY
*Analysis of Dislocation Loops in Superlattices
Vol. 24, No. 1, p. 149–162 (1967).
See also *10*, p. *103*; 21.1.1, p. 419.

F. J. KRINGS
*Ferromagnetische Hochtemperaturnachwirkung an Eisen–Silizium-Legierungen bei gleichzeitiger plastischer Verformung
(High-Temperature Ferromagnetic After-Effect in Iron–Silicon Alloys during Plastic Deformation)
Vol. 24, No. 1, p. 163–176 (1967).
See also 18.2, p. 309; 21.1.1, p. 419.

A. FOURDEUX, R. GEVERS, and S. AMELINCKX
*Electron Microscopic Contrast Effects at the Interface between Two Different Polytypes. (I): Interface Cubic-Hexagonal
Vol. 24, No. 1, p. 195–206 (1967).
See also 4, p. 43; *10*, p. *103*; 21.1, p. 401.

R. ZEYFANG
Lattice Thermal Resistivity due to Dislocations in Alloys at Low Temperatures
Vol. 24, No. 1, p. 221–234 (1967).
See also 6 p. 63; *8*, p. *83*; 21.1, p. 401.

H. DIEPERS
Electron-Microscopic Study of Defect Agglomerates in Ion-Irradiated Copper Foils. (I): Analysis of Agglomerate Type and Configuration
Vol. 24, No. 1, p. 235–244 (1967).
See also *11*, p. *150*; 21.1, p. 401.

T. O. OGURTANI and R. A. HUGGINS
*Theory of Electric Field Gradient due to Conduction Electron Charge Density Redistribution around Screw Dislocations in Metals
Vol. 24, No. 1, p. 301–311 (1967).
See also 13.4, p. 210; 21.1, p. 401.

D. HEITKAMP
On the Thermodynamics of Thermal Diffusion in Substitutional Alloys
Vol. 24, No. 1, p. 341–348 (1967).
See also *9*, p. *90*; 21, p. 382.

B. SPRUŠIL and P. VOSTRÝ
*Reply to "On Quenching Experiments in Gold" by F. J. Kedves
Vol. 24, No. 1, p. K41–K42 (1967).
See also 14.1, p. 224; 21.6, p. 434.

R. Brugière and P. Lucasson
*On the Stage III Recovery of Electron Irradiated Aluminium
Vol. **24**, No. 1, p. K77–K81 (1967).
See also 11, p. 150; 14.1, p. 224; 21, p. 383.

H. Diepers
Electron-Microscopic Study of Defect Agglomerates in Ion-Irradiated Copper Foils. (II): Formation of Agglomerates
Vol. **24**, No. 2, p. 623–633 (1967).
See also 4, p. 44; *11*, p. *150*; 21.1, p. 402.

H. P. van de Braak and W. J. Caspers
*Point Defects in Ferromagnets
Vol. **24**, No. 2, p. 733–741 (1967).
See also *18.2*, p. *310*.

T. V. Nordstrom and S. Amelinckx
*On the So-Called "Anomalous" Contrast at the Central Partial of Triple Ribbons
Vol. **24**, No. 2, p. K121–K123 (1967).
See also *10*, p. *103*.

W. Bollmann and A. J. Perry
*The Sequence of Partials in the F.C.C. Structure
Vol. **24**, No. 2, p. K141–K144 (1967).
See also *10*, p. *104*.

J. Perez and P. Gobin
*Room Temperature Peak of Internal Friction of Aluminium
Vol. **24**, No. 2, p. K167–K172 (1967).
See also 21, p. 383.

A. Bourret and D. Dautreppe
Type of Small Dislocation Loops in Neutron Irradiated Copper
Vol. **24**, No. 2, p. K173–K174 (1967).
See also *11*, p. *151*; 21.1, p. 402.

H. Dübler and H. Wever
Thermo- and Electrotransport in β-Titanium and β-Zirconium
Vol. **25**, No. 1, p. 109–118 (1968).
See also *9*, p. *91*; 21, p. 383.

M. J. Stowell and T. J. Law
The Growth and Defect Structure of Gold Films Formed on Molybdenite in Ultra-High Vacuum
Vol. **25**, No. 1, p. 139–146 (1968).
See also *1.4*, p. *19*; *3.1*, p. *33*; 21.6, p. 435.

G. P. Seidel
*Einfluß von Neutronenbestrahlung auf die plastische Verformung von Eisen-Einkristallen mit unterschiedlichem Kohlenstoffgehalt. (I): Die Fließspannung am Verformungsbeginn
(Effect of Neutron Irradiation on the Plastic Deformation of Iron Single Crystals with Different Carbon Concentrations. (I): The Field Stress at the Deformation Onset)
Vol. **25**, No. 1, p. 175–188 (1968).
See also *11*, p. *151*; 21.1.1, p. 420.

F. Guiu
*Deformation of Molybdenum Single Crystals at Slow Rates of Strain
Vol. **25**, No. 1, p. 189–202 (1968).
See also 21, p. 384.

F. Guiu
*The Stress Dependence of the Dislocation Velocity in Molybdenum
Vol. **25**, No. 1, p. 203–207 (1968).
See also 21, p. 384.

T. Leffers
*Computer Simulation of the Plastic Deformation in Face-Centred Cubic Polycrystals and the Rolling Texture Derived
Vol. **25**, No. 1, p. 337–344 (1968).
See also 21.1, p. 402.

D. Schumacher, A. Seeger, and O. Härlin
*Vacancies, Divacancies, and Self-Diffusion in Platinum
Vol. **25**, No. 1, p. 359–371 (1968).
See also **9**, p. 91; 21.6, p. 435.

H. Träuble and U. Essmann
Defects in the Flux Thread Lattice in Type II Superconductors
Vol. **25**, No. 1, p. 373–393 (1968).
See also *14.2*, p. *229*; 21, p. 384.

H. Träuble and U. Essmann
Direct Evidence for Motive of Flux Threads during Current Passage in Superconductors
Vol. **25**, No. 1, p. 395–402 (1968).
See also *14.2*, p. *230*; 21, p. 384.

J. C. Williams and M. J. Blackburn
*The Identification of a Non-Basal Slip Vector in Titanium and Titanium–Aluminium Alloys
Vol. **25**, No. 1, p. K1–K3 (1968).
See also 21, p. 384.

P. Charsley, J. A. M. Salter, and A. D. W. Leaver
*An Experimental Investigation of Dislocation–Phonon Scattering in Some Copper Alloys
Vol. **25**, No. 2, p. 531–540 (1968).
See also *6*, p. *65*; 8, p. 84; 21, p. 384; 21.1, p. 402.

H. Neuhäuser
*Untersuchungen zur aktiven Kristallänge im Fließbereich neutronenbestrahlter Kupfereinkristalle. (I): Die Ausbildung der Gleitbänder
(Study of the Active Crystal-Length in the Yield Range of Neutron-Irradiated Copper Single Crystals. (I): Slip Band Formation)
Vol. **25**, No. 2, p. 593–606 (1968).
See also 11, p. 151; 21.1, p. 403.

W. A. Coghlan and W. D. Nix
*Contributions to the Kinematical Theory of Diffraction Contrast from Screw Dislocations
Vol. **25**, No. 2, p. 679–689 (1968).
See also *10*, p. *104*.

A. Schweighofer and F. W. Marsh
*Deformation of Zinc Single Crystals (99.999% Purity) at High Strain Rates
Vol. **25**, No. 2, p. K91–K94 (1968).
See also 21.1, p. 403.

J. T. MOORE, T. R. DUNCAN, and D. KUHLMANN-WILSDORF
*Corrected Data for the Shear Stresses of Infinite Tilt Boundaries Composed of {111}, ½ ⟨1$\bar{1}$0⟩ Dislocations in Elastically Anisotropic F.C.C. Metals
Vol. **25**, No. 2, p. K105—K106 (1968).
See also *10*, p. *105*; 21, p. 385; 21.1, p. 403; 21.5, p. 429; 21.6, p. 435; 22.1, p. 457; 22.1.1, p. 464; 22.1.2, p. 473.

B. N. OSHCHERIN
*On Enthalpy Calculation of Vacancy Formation in Inorganic Substances
Vol. **25**, No. 2, p. K123—K125 (1968).
See also *10*, p. *105*; 21, p. 385; 22, p. 450.

W. FRANK
*Thermally Activated Dislocation Motion in a Solid Containing a Multiple Spectrum of Dislocation Obstacles
Vol. **26**, No. 1, p. 197—206 (1968).
See also 21.1, p. 403.

K. DIFFERT
*Versetzungsecken in kubisch-flächenzentrierten Kristallen. (I): Die Geometrie der Versetzungsecken
(Dislocation Corners in Face Centered Cubic Crystals. (I): Geometry of Dislocation Corners)
Vol. **26**, No. 1, p. 327—344 (1968).
See also 21.1, p. 404.

J. L. DAVIDSON and J. M. GALLIGAN
*A Study of Quench Hardening in Copper
Vol. **26**, No. 1, p. 345—358 (1968).
See also 21.1, p. 404.

H. BEYER and R. SIZMANN
Recrystallization of Cold-Worked Gold during α-Irradiation
Vol. **26**, No. 1, p. 359—370 (1968).
See also 4, p. 45; *11*, p. *151*; 14.1, p. 225; 21.6, p. 435.

M. E. DE MORTON
*Atomic Relaxations in a Face-Centred-Tetragonal Alloy
Vol. **26**, No. 1, p. K73—K75 (1968).
See also 12.1, p. 160; 21, p. 385.

L. KAUN, A. LUFT, J. RICHTER, and D. SCHULZE
*Slip Line Pattern and Active Slip Systems of Tungsten and Molybdenum Single Crystals Weakly Deformed in Tension at Room Temperature
Vol. **26**, No. 2, p. 485—499 (1968).
See also 21, p. 385.

C. W. CHEN, L. R. EDWARDS, and S. LEGVOLD
Resistance Anomaly in Palladium–Silver Alloys Containing Rare Earth Impurities
Vol. **26**, No. 2, p. 611—619 (1968).
See also *14.1*, p. *225*; 21, p. 385; 21.6, p. 435.

R. J. BELTZ, T. L. DAVIS, and K. MALÉN
*Some Unifying Relations for Moving Dislocations
Vol. **26**, No. 2, p. 621—637 (1968).
See also *10*, p. *106*.

G. P. WILLIAMS, JR. and T. J. TURNER
*Zener Relaxation in Silver Solid Solutions
Vol. **26**, No. 2, p. 645–651 (1968).
See also *12.1*, p. *160*; 21.6, p. 435.

Y. HAVEN
*Relation between Relaxation Time and Jump Frequencies for Zener Relaxation in Disordered F.C.C. Alloys
Vol. **26**, No. 2, p. 653–660 (1968).
See also *12.1*, p. *160*; 21, p. 385.

M. RÜHLE
*Radiation Hardening of Neutron-Irradiated Copper Single Crystals. (I): Experimental Results
Vol. **26**, No. 2, p. 661–670 (1968).
See also *11*, p. *151*; 21.1, p. 404; 21.6, p. 435.

W. FRANK, M. RÜHLE, and M. SAXLOVÁ
*Radiation Hardening of Neutron-Irradiated Copper Single Crystals. (II): Interpretation
Vol. **26**, No. 2. p. 671–680 (1968).
See also *11*, p. *151*; 21.1, p. 404.

G. F. BOLLING, A. ARROTT, and R. H. RICHMAN
A Study of Curie Temperatures and the Effect of Carbon in Face-Centered Cubic Iron–Nickel Alloys
Vol. **26**, No. 2, p. 743–750 (1968).
See also *18.2*, p. *311*; 21.1.1, p. 421.

I. Y. DEKHTYAR, E. G. MADATOVA, P. S. NISIN, and R. G. FEDCHENKO
*Plastic Deformation Effect on the Electronic Properties of Solid Solutions with Localized Magnetic Moments
Vol. **26**, No. 2, p. K151–K154 (1968).
See also 6.1, p. 73; 14.1, p. 225; 18.1, p. 297; 21.6, p. 436.

L. K. IVES and A. W. RUFF, JR.
*Studies of the Effect of Annealing on Extended Dislocation Nodes in Silver–Tin Alloys
Vol. **27**, No. 1, p. 117–123 (1968).
See also 21.1, p. 404; 21.6, p. 436.

T. C. TISONE, J. O. BRITTAIN, and M. MESHII
*Stacking Faults in a Cu–15 at% Al Alloy. (I): The Short Range Order and Temperature Dependence of the Stacking Fault Energy
Vol. **27**, No. 1, p. 185–194 (1968).
See also 21, p. 385; 21.1, p. 404.

T. SCHOBER and R. W. BALLUFFI
*Direct Evidence for Dechanneling of 40 kV Gold Ions at Planar Defects in Silver and Gold
Vol. **27**, No. 1, p. 195–200 (1968).
See also *11*, p. *152*; 21.6, p. 436.

H. NEUHÄUSER
*Untersuchungen zur aktiven Kristallänge im Fließbereich neutronenbestrahlter Kupfereinkristalle. (II): Die aktive Kristallänge
(Study of the Active Crystal-Length in the Yield Range of Neutron-Irradiated Copper Single Crystals. (II): The Active Crystal-Length)
Vol. **27**, No. 1, p. 281–290 (1968).
See also 11, p. 152; 21.1, p. 404.

A. SEEGER and H. KRONMÜLLER
The Micromagnetic Equations of a Superconductor
Vol. 27, No. 1, p. 371—382 (1968).
See also *14.2*, p. *230*.

J. M. GALLIGAN and M. J. ATTARDO
Radiation Hardening in Platinum
Vol. 27, No. 1, p. 383—390 (1968).
See also *11*, p. *152*; 21.6, p. 436.

J. KINEL, L. KOZŁOWSKI, and J. W. MOROŃ
Diffusional Relaxation Effects in Hydrogen Charged Silicon Iron
Vol. 27, No. 1, p. K37—K39 (1968).
See also *12.1*, p. *160*; *18.2*, p. *312*; 21.1.1, p. 421.

D. D. MISHIN
*The Effect of Dislocations on the Properties of Silicon Iron in Alternating Magnetic Fields
Vol. 27, No. 1, p. K49—K50 (1968).
See also 18.2, p. 312; 21.1.1, p. 421.

S. RADELAAR
The Kinetics of Short-Range Order in Some Ag–Au Alloys
Vol. 27, No. 1, p. K63—K67 (1968).
See also *1.1*, p. *8*; 9, p. 91; 21.6, p. 436.

F. P. BULLEN and S. McK. COUSLAND
*The Temperature Dependence of the Flow Stress of Copper Single Crystals
Vol. 27, No. 2, p. 501—512 (1968).
See also 21.1, p. 405.

P. LUKÁŠ, M. KLESNIL, and J. KREJČÍ
*Dislocations and Persistent Slip Bands in Copper Single Crystals Fatigued at Low Stress Amplitude
Vol. 27, No. 2, p. 545—558 (1968).
See also 1.3, p. 15; 21.1, p. 405.

R. J. MURPHY and J. L. CRAWFORD
*Electron Microscope Image Profiles of Paired and Triple Dislocations
Vol. 27, No. 2, p. 559—572 (1968).
See also 1.4, p. 20; *10*, p. *108*.

M. WUTTIG, J. T. STANLEY, and H. K. BIRNBAUM
*Interstitial Solute Trapping in Irradiated and Quenched Iron
Vol. 27, No. 2, p. 701—712 (1968).
See also 21.1.1, p. 421.

J. L. BRIMHALL and B. MASTEL
*Defect Clusters in Neutron Irradiated Rhenium
Vol. 27, No. 2, p. K89—K92 (1968).
See also 11, p. 152; 21, p. 386.

M. VICTORIA and A. E. VIDOZ
*Tensile Behavior upon Ordering of Ni_3Fe Single Crystals
Vol. 28, No. 1, p. 131—144 (1968).
See also 4, p. 46; 21.1, p. 405; 21.1.1, p. 422.

A. E. VIDOZ
*On the Work Hardening of Superstructures
Vol. 28, No. 1, p. 145—156 (1968).
See also 4, p. 46, 21.1, p. 405; 21.1.1, p. 422; 21.6, p. 436.

J. Van der Planken and A. Deruyttere
*Twinning and Fracture along $\{10\bar{1}1\}$ Planes in a Magnesium-Tin Alloy
Vol. 28, No. 1, p. K9–K10 (1968).
See also 21, p. 387; 21.3, p. 427.

H. A. Bahr, W. Pompe, and H.-G. Schöpf
*On the Dynamics of Dislocations
Vol. 28, No. 1, p. K23–K26 (1968).
See also *10*, p. *109*.

P. Regnier et J. M. Dupouy
*Dissociation et glissement des dislocations sur le plan prismatique des métaux hexagonaux
(Dissociation and Glide of Dislocations on a Prismatic Plane of Hexagonal Metals)
Vol. 28, No. 1, p. K55–K57 (1968).
See also 21, p. 387.

B. Escaig
*L'activation thermique des déviations sous faibles contraintes dans les structures h.c. et c.c.
(Thermal Activation of Derivations at Low Stresses in H.C. and C.C. Structures)
Vol. 28, No. 2, p. 463–474 (1968).
See also *10*, p. *109*.

R. Conte, P. Groh et B. Escaig
*Étude de l'écoulement plastique de monocristaux filamentaires de fer entre 300 et 4 °K
(Study of Plastic Flow of Iron Whiskers between 300 and 4 °K)
Vol. 28, No. 2, p. 475–488 (1968).
See also 21.1.1, p. 422.

B. Pegel
*Stacking Faults on {110} Planes in the B.C.C. Lattice
Vol. 28, No. 2, p. 603–609 (1968).
See also 21, p. 387; 21.1, p. 406; 21.1.1, p. 422.

H. H. Neely, D. W. Keefer, and A. Sosin
Electron Irradiation and and Recovery of Tungsten
Vol. 28, No. 2, p. 675–682 (1968).
See also *11*, p. *153*; 14.1, p. 226; 21, p. 387.

A. Holz and H. Kronmüller
Contribution of Dislocations to the Anisotropy of the Magnetic Saturation Curve in Nickel
Vol. 28, No. 2, p. 755–771 (1968).
See also *18.2*, p. *313* ; 21.1, p. 406.

J. Polák
*Thermal Quasi-Equilibrium of Vacancies, Divacancies, and Impurities in Platinum
Vol 28, No. 2, p. 773–782 (1968).
See also 14.1, p. 226; 21.6, p. 436.

F. Dworschak, H. Schuster, H. Wollenberger, and J. Wurm
*Analysis of Point Defect States in Copper. (I): The Influence of the Point Defect State on the Irradiation Damage Rate at 80 °K
Vol. 29, No. 1, p. 75–80 (1968).
See also 11, p. 153; 21.1, p. 406.

F. Dworschak, H. Schuster, H. Wollenberger, and J. Wurm
*Analysis of Point Defect States in Copper. (II): The Effects of Stage I and Stage III Annealing on Electron-Irradiation-Induced Frenkel Pairs
Vol. 29, No. 1, p. 81–89 (1968).
See also 11, p. 153; 21.1, p. 406.

P. Humble and C. T. Forwood
*A Dynamical Treatment of the Stress-Induced Dissociation of Triangular Frank Dislocation Loops in F.C.C. Metals
Vol. 29, No. 1, p. 99–106 (1968).
See also 21.1, p. 407; 21.6, p. 437.

A. Seeger and H. Mehrer
*Interpretation of Self-Diffusion and Vacancy Properties in Gold
Vol. 29, No. 1, p. 231–243 (1968).
See also *9*, p. *92*; 21.6, p. 437.

F. Walz
Investigation of the Magnetic After-Effect in Nickel after Neutron Irradiation and Plastic Deformation
Vol. 29, No. 1, p. 245–258 (1968).
See also 11, p. 153; *18.2*, p. *314*; 21.1, p. 407.

E. Taglauer
*Small Angle Scattering of Subthermal Neutrons from Deformed Polycrystalline Copper
Vol. 29, No. 1, p. 259–268 (1968).
See also 4, p. 47; 21.1, p. 407.

A. Bourret and D. Dautreppe
*Study of Agglomerated Defects in Irradiated Pure Nickel by Electron Microscopy
Vol. 29, No. 1, p. 283–298 (1968).
See also 4, p. 47; 11, p. 154; 21.1, p. 407.

R. T. Pascoe and C. W. A. Newey
*Deformation Modes of the Intermediate Phase NiAl
Vol. 29, No. 1, p. 357–366 (1968).
See also 21.1, p. 407.

B. Harris
*Solution-Hardening and Deformation of Niobium–Tungsten Alloys
Vol. 29, No. 1, p. 383–390 (1968).
See also 21, p. 387.

F. F. Lavrentev, O. P. Salita, and V. L. Vladimirova
*Dislocation Mobility in the $\{11\bar{2}2\}$ $\langle 11\bar{2}3\rangle$ Slip System of Zinc Single Crystals
Vol. 29, No. 2, p. 569–574 (1968).
See also 21.1, p. 407.

D. Veselý
*The Study of Deformation of Thin Foils of Mo under the Electron Microscope
Vol. 29, No. 2, p. 675–683 (1968).
See also 21, p. 388.

D. Veselý
*The Study of Slip Bands on the Surface of Mo Single Crystals
Vol. 29, No. 2, p. 685–696 (1968).
See also 21, p. 388.

J. Delaplace, J. C. Nicoud, D. Schumacher, and G. Vogl
Low-Temperature Neutron Radiation Damage and Recovery in Beryllium
Vol. **29**, No. 2, p. 819–824 (1968).
See also *11*, p. *154*; 14.1, p. 226; 21.3, p. 427.

G. Roth and V. Naundorf
Measurement of the Activation Energies of Dislocation Pinning in Copper after Electron Irradiation at 78 °K
Vol. **29**, No. 2, p. K123–K125 (1968).
See also *11*, p. *154*; 12.1, p. 161; 21.1, p. 408.

B. Pegel
Self-Vibrations of Parallel Edge Dislocations
Vol. **29**, No. 2, p. K133–K136 (1968).
See also *6*, p. *69*.

H. R. Vydyanath, D. H. Sastry, and K. I. Vasu
On the Kinetics of Clustering in a Magnesium–3% Zinc Alloy
Vol. **29**, No. 2, p. K137–K140 (1968).
See also *4*, p. *47*; 14.1, p. 226; 21.1, p. 408; 21.3, p. 427.

J. L. Leveque, T. Anagnostopoulos, H. Bilger, and P. Moser
*Point Defects in Iron–Carbon Solid Solution Studied by Magnetic After-Effect Measurements
Vol. **29**, No. 2, p. K179–K181 (1968).
See also *11*, p. *154*; 18.2, p. 315; 21.1.1, p. 423.

D. D. Mishin
*Dislocation Theory of Susceptibility and Coervice Force of Silicon Iron
Vol. **30**, No. 1, p. 31–35 (1968).
See also *18.2*, p. *315*; 21.1.1, p. 423.

T. J. Turner, C. M. Dozier, and G. P. Williams, Jr.
*Activation Energies for Zener Relaxation in α-AgCd
Vol. **30**, No. 1, p. 87–91 (1968).
See also 9, p. 92; *12.1*, p. *161*; 21, p. 388; 21.6, p. 437.

A. B. Gardner, R. L. Sanders, and L. M. Slifkin
Self-Diffusion in α-Silver–Cadmium
Vol. **30**, No. 1, p. 93–96 (1968).
See also *9*, p. *92*; 21, p. 388; 21.6, p. 437.

J. Delaplace, J. Hillairet, J. C. Nicoud, D. Schumacher, and G. Vogl
Low Temperature Neutron Radiation Damage and Recovery in Magnesium
Vol. **30**, No. 1, p. 119–126 (1968).
See also *11*, p. *154*; 14.1, p. 226; 21.3, p. 427.

O. Kanert
*Calculation of Quadrupole Broadening from (111) $\langle 110 \rangle$ Dislocations in Cubic Crystals
Vol. **30**, No. 1, p. 127–134 (1968).
See also 19, p. 335; 21, p. 388; 21.1, p. 408.

P. Feltham
*A Stochastic Model of Creep
Vol. **30**, No. 1, p. 135–146 (1968).
See also *10*, p. *111*; 21, p. 388; 22.5.2, p. 530.

P. Humble
*Electron Diffraction Contrast from Three Parallel Overlapping Stacking Faults
Vol. **30**, No. 1, p. 183–192 (1968).
See also *10*, p. *112*.

D. Becker, F. Dworschak, C. Lehmann, K. T. Rie, H. Schuster, H. Wollenberger, and J. Wurm
Determination of the Average Number of Replacement Collisions in an Electron-Irradiated F. C. C. Crystal (Ni_3Mn)
Vol. **30**, No. 1, p. 219–229 (1968).
See also *11*, p. *155*; 14.1, p. 227; 18.2, p. 315; 21.1, p. 408.

J. Gemperlová
*Force on Dislocations in Anisotropic Bicrystals and Half-Spaces
Vol. **30**, No. 1, p. 261–274 (1968).
See also 21, p. 388; 21.1, p. 408; 21.1.1, p. 423; 21.2, p. 425.

A. Anderman and W. G. Gehman
*Xenon–Vacancy Defect Clusters in Copper
Vol. **30**, No. 1, p. 283–289 (1968).
See also 3.2, p. 36; 21.1, p. 408.

Ch. Schwink and H. Traub
*The Active Slip Volume and the Temperature-Dependent Processes in Stage I of Single Crystals of Cu Alloys
Vol. **30**, No. 1, p. 387–397 (1968).
See also 21.1, p. 408.

F. Secco D'Aragona, R. C. Rau, and R. L. Ladd
Domain Boundaries in Vanadium
Vol. **30**, No. 1, p. K1–K3 (1968).
See also *4*, p. *48*; 21.1, p. 408.

T. Skośkiewicz and B. Baranowski
Investigation of the Electrical Resistance Anomaly in the Palladium–Hydrogen System
Vol. **30**, No. 1, p. K33–K35 (1968).
See also *14.1*, p. *227*; 21, p. 388.

E. Köster
Magnetization and Magnetostriction of Hexagonal Cobalt Single Crystals with Internal Stresses
Vol. **30**, No. 2, p. 455–460 (1968).
See also *18.2*, p. *315*; 21.1, p. 409.

R. Siems
*Mechanical Interactions of Point Defects
Vol. **30**, No. 2, p. 645–658 (1968).
See also 6, p. 70; *10*, p. *112*.

J. Celý
Mössbauer Effect in Simple Cubic Lattice with Line Defect
Vol. **30**, No. 2, p. 671–683 (1968).
See also *6.1*, p. *75*; 10, p. 112.

Y. Quéré
*Dechannelling Cylinder of Dislocations
Vol. **30**, No. 2, p. 713–722 (1968).
See also *11*, p. *155*; 21, p. 389.

R. DE RIDDER, J. VAN LANDUYT, R. GEVERS, and S. AMELINCKX
The Fine Structure of Spots in Electron Diffraction Resulting from the Presence of Planar Interfaces and Dislocations. (IV): Wedge Crystals
Vol. **30**, No. 2, p. 797–815 (1968).
See also *4*, p. *48*; 10, p. 113; 21.1.1, p. 423; 22.6, p. 544.

P. GASSENMEIER und M. WILKENS
*Röntgenographische Untersuchung des Übergangs vom Bereich I zum Bereich II der Verfestigungskurve kubisch-flächenzentrierter Kupfer–Gallium-Einkristalle (X-Ray Investigation of the Transition between Stages I and II of the Work Hardening Curve of Cubic Face Centered Copper–Gallium Single Crystals)
Vol. **30**, No. 2, p. 833–843 (1968).
See also 21, p. 389; 21.1, p. 409.

A. SEEGER and H. KRONMÜLLER
Reversible Magnetization Curve and Magnetostriction of Uniaxial Ferromagnetic Crystals with Internal Stresses
Vol. **30**, No. 2, p. K101–K106 (1968).
See also *18.2*, p. *315*; 21.1, p. 409.

R. BRUGIÈRE and P. LUCASSON
*On the Stage III Recovery of Electron Irradiated Copper
Vol. **30**, No. 2, p. K139–K143 (1968).
See also 11, p. 155; 14.1, p. 227; 21.1, p. 409.

10.2 Photochemical Reactions. Colour Centres

Original Papers and Short Notes

V. Topa and S. Haragea
*Colour Centres in X-Irradiated NaCl:Ag and NaCl:(Ag + Ca) Crystals
Vol. **16**, No. 1, p. 137–145 (1966).
See also 22.5.2, p. 518.

W. Dreybrodt and D. Silber
EPR Measurements on Diatomic Halogen Centers in Allkali Halides
Vol. **16**, No. 1, p. 215–223 (1966).
See also 13.4, p. 201; *19*, p. *326*; 22.5.2, p. 518.

V. V. Ratnam
*On the Electrical and Magnetic Properties of Coloured and Uncoloured Calcium Fluoride Crystals
Vol. **16**, No. 2, p. 549–557 (1966).
See also 13.4, p. 201; 14.4, p. 252; 18.1, p. 296; 22.5.3, p. 532.

V. V. Ratnam
*On the Colour Centres and X-Ray Luminescence of Calcium Fluoride Crystals
Vol. **16**, No. 2, p. 559–567 (1966).
See also *20.3*, p. *364*; 22.5.3, p. 532.

M. L. Mukherjee and H. N. Bose
*Nonradiative Destruction of F-Centres in X-Irradiated KCl Doped with Anionic Impurities
Vol. **16**, No. 2, p. 591–598 (1966).
See also 20.3. p. 364; 22.5.2, p. 518.

O. Adametz
*Umwandlung von U-Zentren in OH-Zentren bei KCl-Kristallen
(Conversion of U-Centres in KCl Crystals into OH-Centres)
Vol. **16**, No. 2, p. K161–K164 (1966).
See also 22.5.2, p. 518.

Y. Kazumata, T. Kamikawa, and K. Ozawa
*Zero-Phonon Lines of Color Centers in LiF and NaF
Vol. **17**, No. 1, p. 131–138 (1966).
See also *6*, p. *56*; 22.5.2, p. 518.

F. Van de Wiele and G. Jacobs
*Optical and Thermal Properties of Color Centers in CsBr Single Crystals Irradiated at Liquid-Helium Temperature
Vol. **17**, No. 1, p. 177–184 (1966).
See also 22.5.2, p. 519.

H. D. Koswig
"Optical Bleaching" of Ni^{2+}-Doped AgCl Crystals
Vol. **17**, No. 1, p. K83–K87 (1966).
See also *20.1*, p. *345*; 22.5.1, p. 515.

T. Kamikawa, Y. Kazumata, and K. Ozawa
*New Infrared Absorption in Neutron-Irradiated Lithium Fluoride
Vol. **17**, No. 2, p. 645–655 (1966).
See also 11, p. 146; 20.1, p. 346; 22.5.2, p. 519.

F. RUEDA and W. DEKEYSER
Charge of Dislocations in Pure and γ-Irradiated LiF
Vol. 17, No. 2, p. 739–744 (1966).
See also *10*, p. *95*; 22.5.2, p. 519.

W. DEGRIECK and G. JACOBS
*Properties of the B-Centre in Additively Coloured Ag^+ Doped KCl Crystals
Vol. 18, No. 1, p. 279–284 (1966).
See also 22.5.2, p. 519.

P. PETRESCU and V. GHEORDĂNESCU
Anisotropy of the Photoemission from Some Coloured Alkali Halides
Vol. 18, No. 2, p. 511–515 (1966).
See also *17*, p. *287*; 22.5.2, p. 520.

L. BEN-DOR, A. GLASNER, and S. ZOLOTOV
*Colour Centres in KCl Doped with Zn^{+2}, Cd^{+2}, and Hg^{+2}
Vol. 18, No. 2, p. 593–599 (1966).
See also 22.5.2, p. 520.

D. SCHMID
The *g*-Factor of F-Centres in Alkali Halide Crystals
Vol. 18, No. 2, p. 653–666 (1966).
See also *19*, p. *327*; 22.5.2, p. 520.

K. K. SHVARTS, A. YA. VITOL, A. V. PODIN, D. O. KALNIN, and YU. A. EKMANIS
Radiation Effects in Pile-Irradiated LiF Crytals
Vol. 18, No. 2, p. 897–909 (1966).
See also *11*, p. *147*; 19, p. 327; 20.1, p. 346; 22.5.2, p. 520.

P. V. SASTRY
*On the Photo- and Thermo-Chemical Reactions in Additively Colored Sodium Chloride
Vol. 18, No. 2, p. K127–K130 (1966).
See also 22.5.2, p. 520.

R. J. FLEMING
*Potential Energy Curve for F-Centre Electrons in Calcium Fluoride
Vol. 18, No. 2, p. K139–K142 (1966).
See also 13.4, p. 204; 22.5.3, p. 532.

R. LAIHO, P. KETOLAINEN, and E. MÄNTYSALO
*Ultrasonic Modulation of F-Band in Additively Colored KBr
Vol. 18, No. 2, p. K143–K144 (1966).
See also 7, p. 76; 22.5.2, p. 520.

P. GÖRLICH, H. KARRAS, E. MEISEL, and H. MOTHES
*IR Absorption of Uncoloured and Additively Coloured Synthetic Crystals CaF_2 and CaF_2/Na Crystals
Vol. 19, No. 1, p. K23–K25 (1967).
See also 22.5.3, p. 532.

YU. A. LOGATCHOV
Model Potential Approximation for Z_1-Centres in Ionic Crystals
Vol. 19, No. 1, p. K35–K38 (1967).
See also *13.4*, p. *204*; 22.5.2, p. 520.

W. BARTH and B. FRITZ
*Local Mode Absorption by U-Centres in Alkali-Halide Mixed Crystals
Vol. 19, No. 2, p. 515–524 (1967).
See also 22.5.2, p. 520.

O. E. Facey and P. W. M. Jacobs
*Photochemical Properties of SH^- Ions in KCl and KBr Crystals
Vol. **19**, No. 2, p. 565—576 (1967).
See also 22.5.2, p. 521.

F. T. Goldstein
*F-Center Formation by Fundamental Absorption in KI
Vol. **20**, No. 1, p. 379—393 (1967).
See also 20.3, p. 365; 22.5.2, p. 521.

M. Höhne and M. Stasiw
Te_2^{3-} and Se_2^{3-} Centres in AgBr. (I): The *g*-, Superhyperfine, and Hyperfine Tensors
Vol. **20**, No. 2, p. 657—666 (1967).
See also 13.4, p. 206; *19*, p. *328*; 22.5.1, p. 515.

M. Höhne and M. Stasiw
Te_2^{3-} and Se_2^{3-} Centres in AgBr. (II): Electronic Structure
Vol. **20**, No. 2, p. 667—673 (1967).
See also 13.4, p. 206; *19*, p. *328*; 22.5.1, p. 515.

P. Feltham
*F-Centres in Alkaline Earth Fluorides
Vol. **20**, No. 2, p. 675—680 (1967).
See also 13.4, p. 206; 22.5.3, p. 533.

P. Asadi
X-Ray Topography in Coloured and Uncoloured Zones of Antozonite–Fluorite
Vol. **20**, No. 2, p. K71—K72 (1967).
See also *4*, p. *41*; 22.8, p. 550.

H. Bilz, R. Zeyher, and R. K. Wehner
Anharmonic Sideband Effects of the U-Center Local Mode
Vol. **20**, No. 2, p. K167—K171 (1967).
See also *6*, p. *60*; 20.1, p. 348; 22.5.2, p. 522.

H. Kühnert
Polarization of F-Center Luminescence in KCl due to Electric Fields
Vol. **21**, No. 2, p. K171—K173 (1967).
See also *20.3*, p. *366*; 22.5.2, p. 522.

R. Sittig
Stark Effect Measurements of the OH^- Absorption Band in KCl
Vol. **21**, No. 2, p. K175—K177 (1967).
See also 14.4.1, p. 259; *20.1*, p. *349*; 22.5.2, p. 523.

H. W. den Hartog and J. Arends
*F-Centers in SrF_2
Vol. **22**, No. 1, p. 131—140 (1967).
See also 13.4, p. 208; 19, p. 329; 22.5.3, p. 533.

W. Degrieck and G. Jacobs
Luminescence and Photoconductivity of the B-Centre in Additively Coloured KCl: Ag Crystals
Vol. **22**, No. 1, p. 177—184 (1967).
See also *16*, p. *279*; *20.3*, p. *366*; 22.5.2, p. 523.

G. Wolfram and V. Witt
*Kinetics of the Thermal Mutual Conversion of F and Z_2 Color Centers in KCl:Sr Crystals
Vol. **22**, No. 1, p. 245—253 (1967).
See also 22.5.2, p. 523.

G. E. Arkhangelskii, Z. L. Morgenshtern, and V. B. Neustruev
*Colour Centres in Ruby Crystals
Vol. 22, No. 1, p. 289–295 (1967).
See also 20.1, p. 349; 20.2, p. 361; 20.3, p. 366; 22.6, p. 538.

A. Mukerji, G. A. Tanton, and J. E. Williams
*X-Ray Induced F-Centers in CaF_2
Vol. 22, No. 1, p. K19–K22 (1967).
See also 11, p. 149; 22.5.3, p. 533.

F. Agulló-López
*Role of γ-Radiation Intensity on the First Stage of the Room-Temperature F-Colouring in NaCl
Vol. 22, No. 2, p. 483–490 (1967).
See also 11, p. 149; 22.5.2, p. 523.

J. Arends and J. F. Verwey
ESR on UV Irradiated Lead Halides at 80 °K
Vol. 23, No. 1, p. 137–145 (1967).
See also 4, p. 43; *19*, p. *330*; 22.5, p. 513.

J. H. Crawford, Jr., W. A. Sibley, and E. Sonder
*The Influence of Electron-Trapping Impurity on Defect Creation and Bleaching in KCl Irradiated at 78 °K
Vol. 23, No. 1, p. 301–306 (1967).
See also 11, p. 150; 13.4, p. 209; 22.5.2, p. 523.

S. Wang
Optical Absorption of an Interstitial Alkali-Earth Cation an in Alkali-Halide Crystal
Vol. 23, No. 1, p. 387–399 (1967).
See also 13.2, p. 193; *13.4*, p. *209*; 22.5.2, p. 523.

S. Wang
Electronic States of the M_A-Centers in KCl
Vol. 23, No. 1, p. 401–412 (1967).
See also *13.4*, p. *209*; 22.5.2, p. 523.

R. Voszka, T. Horváth, and A. Watterich
*A Possible Model for V-Centres Stable at Room Temperature in X-Irradiated KCl(Ca) Crystals
Vol. 23, No. 1, p. K71–K73 (1967).
See also 22.5.2, p. 523.

H. Peisl, H. Spalt, and W. Waidelich
X-Ray Diffuse Scattering in γ-Irradiated LiF
Vol. 23, No. 1, p. K75–K78 (1967).
See also *10*, p. *102*; 22.5.2, p. 524.

H. W. den Hartog and J. Arends
Electronic Structure of F-Centers in Alkaline Earth Fluorides
Vol. 23, No. 2, p. 713–719 (1967).
See also *13.4*, p. *210*; 19, p. 330; 22.5.3, p. 533.

P. Petrescu
*L-Bands Detected in Photoemission Spectra of NaCl, LiCl, and RbBr
Vol. 23, No. 2, p. K105–K107 (1967).
See also 17, p. 287; 22.5.2, p. 524.

D. Bäuerle and B. Fritz
Infrared Vibrational Absorption by U-Centers in NaI
Vol. 24, No. 1, p. 207—215 (1967).
See also *6*, p. *63*; *20.1*, p. *351*; 22.5.2, p. 524.

T. A. T. Cowell and J. Woods
Thermally Stimulated Currents and Infra-Red Luminescence in CdS Crystals
Vol. 24, No. 1, p. K37—K40 (1967).
See also *13.4*, p. *210*; 14.3, p. 239; 20.3, p. 367; 22.4.1, p. 501.

J. B. Page, Jr. and D. Strauch
Shell Model Treatment of the Vibrations of Crystals Containing Substitutional Defects. (I): U-Center Localized Modes
Vol. 24, No. 2, p. 469—479 (1967).
See also 3.2, p. 35; *6*, p. *64*; 22.5.2, p. 524.

T. H. Cheung and S. Wang
Theoretical Studies on Z_1-Bands in Alkali Halide Crystals
Vol. 24, No. 2, p. 509—523 (1967).
See also *13.4*, p. *210*; 22.5.2, p. 524.

F. Fröhlich und O. Adametz
*Umwandlung von OH-Zentren in U-Zentren bei KCl-Kristallen durch elektrolytische Verfärbung
(Conversion of OH-Centres, in KCl Crystals into U-Centres by Electrolytic Coloration)
Vol. 24, No. 2, p. 583—590 (1967).
See also 22.5.2, p. 524.

H. W. den Hartog
Fourth Shell Isotropic H. F. S.-Constants of F-Centers in CaF_2
Vol. 24, No. 2, p. K117—K120 (1967).
See also *13.4*, p. *211*; 19, p. 331; 22.5.3, p. 533.

P. Görlich, H. Karras, Ch. Symanowski, and P. Ullmann
*The Colour Centre Absorption of X-Ray Coloured Alkaline Earth Fluoride Crystals
Vol. 25, No. 1, p. 93—101 (1968).
See also 22.5.3, p. 533.

H. L. Arora, S. Kim, and S. Wang
*Optical Absorption of the Impurity Ag and Models for the E-Band in an Alkali Halide
Vol. 25, No. 1, p. 223—232 (1968).
See also 13.4, p. 211; 22.5.2, p. 525.

G. Giuliani and E. Reguzzoni
*X-Ray Production and Thermal Annealing of Frenkel Pairs in KCl
Vol. 25, No. 1, p. 437—448 (1968).
See also *11*, p. *151*; 22.5.2, p. 525.

P. Görlich, H. Karras, G. Kötitz, and R. Rauch
*Phonon-Assisted Colour Centre Fluorescence of Additively Coloured Alkali Earth Fluoride Crystals
Vol. 25, No. 1, p. K15—K18 (1968).
See also 6, p. 65; *20.3*, p. *368*; 22.5.3, p. 533.

M. Höhne and M. Stasiw
Detection of Ag^{2+} in Doped AgCl Crystals by ESR
Vol. 25, No. 1, p. K55—K57 (1968).
See also *19*, p. *332*; 22.5.1, p. 516.

Y. Kazumata
*Atomic ^{3}H Center in Neutron-Irradiated LiF
Vol. **25**, No. 2, p. 563—566 (1968).
See also 11, p. 151; 19, p. 332; 22.5.2, p. 525.

L. Pelsers and G. Jacobs
Photoconductivity of CsBr Single Crystals Containing F- and M-Centres
Vol. **25**, No. 2, p. 621—628 (1968).
See also *16*, p. *281*; 22.5.2, p. 525.

P. Feltham and R. Ghosh
*On Colour Centres in Plastically Deformed Alkaline Earth Fluorides
Vol. **25**, No. 2, p. K71—K73 (1968).
See also 10, p. 104; 19, p. 332; 22.5.3, p. 533.

R. G. Bessent and P. Feltham
*Formation of F-Centres in Alkine Earth Oxides by Plastic Deformation
Vol. **25**, No. 2, p. K107—K110 (1968).
See also 10, p. 105; 19, p. 333; 22.6, p. 541.

H. W. den Hartog
*F-Centers in BaF_2
Vol. **25**, No. 2, p. K111—K114 (1968).
See also 13.4, p. 211; 19, p. 333; 22.5.3, p. 533.

J. Hingsammer and H. Jodl
*Z_1- and Z_2-Centres in Mixed Crystals of Alkali Halides Doped with Sr^{++}
Vol. **26**, No. 1, p. 261—266 (1968).
See also 22.5.2, p. 526; 22.8, p. 553.

A. Buroff and J. Malinowski
Charge Compensation of Trapped Holes in Silver Bromide
Vol. **26**, No. 1, p. 267—276 (1968).
See also 13.4, p. 212; *16*, p. *282*; 22.5.1, p. 516.

Y. Kazumata
Self-Trapped Holes in Neutron Irradiated CaF_2 and BaF_2 Crystals
Vol. **26**, No. 2, p. K119—K122 (1968).
See also 11, p. 152; *19*, p. *333*; 22.5.3, p. 534.

B. Henderson and R. D. King
*Colour Centres in Neutron Irradiated Calcium Oxide
Vol. **26**, No. 2, p. K147—K149 (1968).
See also 11, p. 152; 22.6, p. 541.

R. A. Shatas and J. D. Stettler
Mollwo-Ivey Relation in Aggregate Defect Centers
Vol. **26**, No. 2, p. K155—K157 (1968).
See also *13.4*, p. *212*; 22.5.2, p. 526; 22.6, p. 541.

M. Georgiev and E. Petrova
*Investigation of the Drift of Photoelectrons in AgBr at Moderate Temperatures
Vol. **27**, No. 1, p. K33—K36 (1968).
See also 22.5.1, p. 516.

S. D. McLaughlan and H. W. Evans
Production of Colloidal Calcium by Electron Irradiation of CaF_2 Crystals
Vol. **27**, No. 2, p. 695—700 (1968).
See also *11*, p. *152*; 19, p. 333; 20.1, p. 354; 22.5.3, p. 534.

K. MAIER and W. GEBHARDT
*Stress-Induced Inhomogeneous Effects of the K-Absorption in RbCl
Vol. 27, No. 2, p. 713–721 (1968).
See also 13.4, p. 214; 22.5.2, p. 526.

YU. E. PERLIN and SH. N. GIFEISMAN
*The Theory of Shallow F-Centres
Vol. 28, No. 1, p. 45–60 (1968).
See also 6, p. 66; 13, p. 171; 14.4, p. 254; 18, p. 294; 22.2.1, p. 485; 22.2.2, p. 488; 22.2.3, p. 493; 22.4, p. 495; 22.4.1, p. 503; 22.4.2, p. 508; 22.4.3, p. 511; 22.5.1, p. 517; 22.5.2, p. 527.

B. G. DICK
*Calculation of Properties of Vacancy-Centered Z_1-Center Models in Alkali Halides
Vol. 28, No. 1, p. 223–232 (1968).
See also *13.4*, p. *214*; 22.5.2, p. 527.

M. HÖHNE and M. STASIW
ESR Detection of Self-Trapped Holes in AgCl
Vol. 28, No. 1, p. 247–253 (1968).
See also *19*, p. *334*; 20.1, p. 355; 22.5.1, p. 517.

G. WOLFRAM
Photoconductivity of the Z_2-Center in KCl:Sr Crystals
Vol. 28, No. 1, p. K11–K13 (1968).
See also 13.4, p. 214; *16*, p. *283*; 22.5.2, p. 527.

T. OKADA, H. KAWAI, S. MUKAI, and T. SUITA
*Hardening of KCl Crystals by F_A-Centers
Vol. 28, No. 1, p. K19–K21 (1968).
See also *10*, p. *109*; 22.5.2, p. 527.

G. K. BORN, R. J. GRASSER, and A. O. SCHARMANN
EPR and Thermoluminescence Measurements on $CaWO_4$ Crystals
Vol. 28, No. 2, p. 583–588 (1968).
See also *11*, p. *153*; 19, p. 334; 20.3, p. 370; 22.8, p. 554.

M. GEORGIEV, E. ZAGORSKA, and J. MALINOWSKI
*Temperature Dependence of the Diffusion of Photoexcited Holes in AgBr
Vol. 28, No. 2, p. 725–730 (1968).
See also 16, p. 283; 22.5.1, p. 517.

R. A. SHATAS, G. A. TANTON, and A. MUKERJI
*Mollwo-Ivey Relation of H- and F-Bands in Alkaline Earth Fluorides
Vol. 28, No. 2, p. K143–K145 (1968).
See also 22.5.3, p. 534.

C. SÁNCHEZ and F. AGULLÓ-LÓPEZ
*Transient Effects in the Room-Temperature F-Colouring of NaCl Irradiated with X- or γ-Rays
Vol. 29, No. 1, p. 217–230 (1968).
See also 11, p. 153; 22.5.2, p. 528.

P. PETRESCU
On the L-Bands and the First Exciton Bands in the Photoemission Spectra of Alkali Iodides
Vol. 29, No. 1, p. 333–340 (1968).
See also 13.2, p. 195; *17*, p. *289*; 22.5.2, p. 528.

P. Petrescu
On the Photoemission Spectrum of LiF:Mg
Vol. **29**, No. 1, p. K1—K2 (1968).
See also *17*, p. *289*; 22.5.2, p. 528.

P. Petrescu
Evidence for L-Bands in Photoemission Spectra of KF and KCl
Vol. **29**, No. 1, p. K3—K5 (1968).
See also *17*, p. *289*; 22.5.2, p. 529.

R. J. Fleming
*Evaluation of 1s–2p Optical Transition Energy for F-Centres in CaF_2
Vol. **29**, No. 1, p. K77—K80 (1968).
See also 13.4, p. 215; 22.5.3, p. 534.

R. Kersten
Hyperfine and Quadrupole Interactions of the F-Centre in KCl as Measured by ENDOR
Vol. **29**, No. 2, p. 575—586 (1968).
See also *19*, p. *335*; 22.5.2, p. 529.

B. G. Dick
*One-Phonon Relaxation of Hydroxyl Ions in Alkali Halide Crystals
Vol. **29**, No. 2, p. 587—596 (1968).
See also 6, p. 68; 22.5.2, p. 529.

D. Bäuerle and B. Fritz
Vibrational Spectra of U_1-Centres in KI
Vol. **29**, No. 2, p. 639—643 (1968).
See also *6*, p. *68*; *20.1*, p. *357*, 22.5.2, p. 529.

G. E. Arkhangelskii, Z. L. Morgenshtern, and V. B. Neustruev
*On the Nature of the Colour Centres of Ruby
Vol. **29**, No. 2, p. 831—836 (1968).
See also 13.4, p. 215; *20.3*, p. *371*; 22.6, p. 543.

E. V. R. Sastry and T. M. Srinivasan
*Growth and Bleach of F-Centers in KCl with Interstitial Cobalt
Vol. **29**, No. 2, p. K107—K109 (1968).
See also 10, p. 111; 22.5.2, p. 529.

R. L. Gilbert
*Empirical Relationships for the Emission Energies and Lifetimes of the F-Center
Vol. **30**, No. 1, p. 173—182 (1968).
See also 13.4, p. 216; 20.3, p. 371; 22.5.2, p. 530.

P. G. Baranov, R. A. Zhitnikov, and N. I. Melnikov
Paramagnetic Resonance of Silver Atoms Stabilized in Alkali Halide Crystals
Vol. **30**, No. 2, p. 851—858 (1968).
See also *19*, p. *336*; 22.5.2, p. 531.

P. G. Baranov, R. A. Zhitnikov, and N. I. Melnikov
Theoretical Study of Properties of Copper and Silver Atoms Stabilized in Alkali Halide Crystals
Vol. **30**, No. 2, p. 859—866 (1968).
See also 3.2, p. 36; 13.4, p. 218; *19*, p. *336*; 22.5.2, p. 531.

K.-J. Berg and K.-P. Meyer
*Thermoluminescence of Z_1-Centers in Ca Doped KCl Crystals after X-Irradiation
Vol. **30**, No. 2, p. K161—K164 (1968).
See also *20.3*, p. *372*: 22.5.2, p. 531.

11. Irradiation Effects in Solids

Review Articles

A. Seeger and K. P. Chik
Diffusion Mechanisms and Point Defects in Silicon and Germanium
Vol. **29**, No. 2, p. 455–542 (1968).
See also *9*, p. *87*; *10*, p. *93*; 13.4, p. 200; 19, p. 326; 22.1.1, p. 458; 22.1.2, p. 468.

Original Papers and Short Notes

P. C. Euthymiou and A. L. Tanenbaum
*Effect of Neutron and γ-Ray Irradiation on the Carrier Lifetime and Detectivity of InSb
Vol. **16**, No. 1, p. 183–190 (1966).
See also 14.3, p. 231; 22.2.3, p. 489

M. Attardo and J. M. Galligan
*A Field Ion Microscope Study of Neutron Irradiated Tungsten
Vol. **16**, No. 2, p. 449–457 (1966).
See also 4, p. 37; 17, p. 286; 21, p. 374.

Chr. Lehmann and P. Sigmund
*On the Mechanism of Sputtering
Vol. **16**, No. 2, p. 507–511 (1966).
See also 1.3, p. 14; 21.1, p. 390.

D. J. Barber
*Electron Microscopy of Irradiation-Induced Defect Clusters in Magnesium Fluoride
Vol. **16**, No. 2, p. 531–547 (1966).
See also 4, p. 37; *10*, p. *94*, 22.5.3, p. 532.

M. Rühle and M. Wilkens
Nature of the Electron-Microscopically Observable Defects in Neutron-Irradiated Copper
Vol. **16**, No. 2, p. K105–K108 (1966).
See also *10.1*, p. *115*; 21.1, p. 391.

H. Diepers and J. Diehl
Detection and Formation of Interstitial Agglomerates in Ion-Irradiated Copper Foils
Vol. **16**, No. 2, p. K109–K112 (1966).
See also *10.1*, p. *115*; 21.1, p. 391.

T. J. Koppenaal and R. J. Arsenault
Some Notes on the Strengthening Mechanism in Neutron Irradiated Copper Single Crystals
Vol. **17**, No. 1, p. 27–33 (1966).
See also *10.1*, p. *115*; 21.1, p. 391.

Ch. Schwink and H. Neuhäuser
On the Strengthening Mechanism in Neutron Irradiated Copper Single Crystals. (I): On Experiments Concerning Strain Rate Changes
Vol. **17**, No. 1, p. 35–42 (1966).
See also *10.1*, p. *115*; 21.1, p. 391.

J. DIEHL and G. P. SEIDEL
On the Strengthening Mechanism in Neutron Irradiated Copper Single Crystals. (II): Analysis of the Thermally Activated Processes
Vol. **17**, No. 1, p. 43–48 (1966).
See also *10.1*, p. *115*; 21.1, p. 392.

M. G. ASCH, M. C. FÉLIX, and M. R. ONGARO
Electrization of Polyethylene and Polystyrene after γ-Irradiation
Vol. **17**, No. 1, p. 49–57 (1966).
See also *13.3*, p. *196*; 22.9, p. 558.

M. BALARIN and M. SCHENK
On the Defect-Induced Shift of Transition Points between Crystalline Phases
Vol. **17**, No. 1, p. 91–99 (1966).
See also *1.2*, p. *9*; 10, p. 94; 14.4.2, p. 266.

K. ISEBECK, F. RAU, W. SCHILLING, K. SONNENBERG, P. TISCHER, and H. WENZL
*Stored Energy, Volume, and Resistivity Change in Neutron Irradiated Aluminium
Vol. **17**, No. 1, p. 259–268 (1966).
See also 10.1, p. 115; 14.1, p. 220; 21, p. 374.

D. AST and J. DIEHL
Study of the Decay of Vacancy Agglomerates in Neutron-Irradiated Nickel by Measurements of Critical Shear Stress
Vol. **17**, No. 1, p. 269–284 (1966).
See also *10.1*, p. *115*; 21.1, p. 392.

P. VIGIER, C. MINIER-CASSAYRE, P. MOSER, and V. HIVERT
Study of Point Defects Produced by Neutron Irradiation in Doped Iron
Vol. **17**, No. 1, p. 317–321 (1966).
See also *10.1*, p. *116*; 14.1, p. 220; 18.2, p. 302; 21.1.1, p. 412.

R. VON JAN
The Interpretation of Recovery Stages in Irradiated Metals
Vol. **17**, No. 1, p. 361–373 (1966).
See also *10.1*, p. *116*; 21, p. 374; 21.1, p. 392; 21.6, p. 430.

T. KAMIKAWA, Y. KAZUMATA, and K. OZAWA
New Infrared Absorption in Neutron-Irradiated Lithium Fluoride
Vol. **17**, No. 2, p. 645–655 (1966).
See also *10.2*, p. *137*; 20.1, p. 346; 22.5.2, p. 519.

G. SCHOFFA
Electron Spin Resonance of Radiation-Produced Radicals in 2,6-Butylhydroxytoluene Single Crystals
Vol. **17**, No. 2, p. 683–689 (1966).
See also *19*, p. *327*; 22.9, p. 558.

J. L. ODDOU, C. MINIER-CASSAYRE et P. MOSER
*Effet des impuretés sur les stades de revenu de la résistivité électrique du nickel bombardé à basse température par des électrons
(Effect of Impurities on the Recovery Stages of the Resistivity of Nickel Electron Irradiated at Low Temperatures)
Vol. **17**, No. 2, p. 793–798 (1966).
See also *10.1*, p. *117*; 14.1, p. 220; 21.1, p. 393.

K. L. MERKLE
*Radiation-Induced Point Defect Clusters in Copper and Gold. (I): Clusters Produced in Energetic Displacement Casades
Vol. **18**, No. 1, p. 173–188 (1966).
See also 10.1, p. 117; 21.1, p. 393; 21.6, p. 431.

HJ. MATZKE
*Radiation Damage and Xenon Release in Quartz and Fused Silica Following Ion Bombardment
Vol. **18**, No. 1, p. 285–295 (1966).
See also 1.2, p. 9; 2, p. 22; 22.6, p. 536.

HJ. MATZKE
Diffusion of Tritium from Lithium Fluoride
Vol. **18**, No. 1, p. 317–324 (1966).
See also *17*, p. *286*; 22.5.2, p. 519.

K. ISEBECK, R. MÜLLER, W. SCHILLING, and H. WENZL
Reaction Kinetics of Stage III Recovery in Aluminium after Neutron Irradiation
Vol. **18**, No. 1, p. 427–439 (1966).
See also *10.1*, p. *118*; 14.1, p. 221; 21, p. 375.

W. FRANK
Critical Shear Stress of Cubic Crystals Containing Defects of Tetragonal Symmetry. (III): Electron- and Neutron-Irradiated Aluminium and Electron-Irradiated Copper
Vol. **18**, No. 1, p. 459–464 (1966).
See also *10.1*, p. *118*; 21, p. 376; 21.1, p. 394.

K. F. LIDER, B. V. NOVIKOV, and S. A. PERMOGOROV
Application of Bound-Exciton Optical Spectra in the Study of Radiation Damage in Crystals
Vol. **18**, No. 1, p. K1–K4 (1966).
See also *13.2*, p. *191*; *20.3*, p. *364*; 22.4.1, p. 499.

O. HAUSER and M. SCHENK
Radiation-Induced Phase Transitions of Some Perovskite-Type Substances and Their Thermodynamic Treatment
Vol. **18**, No. 2, p. 547–555 (1966).
See also *1.2*, p. *10*; 8, p. 80; 14.4.2, p. 266.

K. K. SHVARTS, A. YA. VITOL, A. V. PODIN, D. O. KALNIN, and YU. A. EKMANIS
*Radiation Effects in Pile-Irradiated LiF Crystals
Vol. **18**, No. 2, p. 897–909 (1966).
See also 10.2, p. 138; 19, p. 327; 20.1, p. 346; 22.5.2, p. 520.

M. RÜHLE
Electron-Microscopic Studies of Small Defect-Agglomerates in Irradiated Metals. (I): Theory of Contrast and Experimental Methods for Determining the Defect Type
Vol. **19**, No. 1, p. 263–278 (1967).
See also *10.1*, p. *120*; 21.1, p. 395.

M. RÜHLE
Electron-Microscopic Studies of Small Defect-Agglomerates in Irradiated Metals. (II): Investigations on Neutron- and Ion-Irradiated Cu and Neutron-Irradiated Ni
Vol. **19**, No. 1, p. 279–295 (1967).
See also *10.1*, p. *120*; 21.1, p. 395.

YU. S. BOYARSKAYA, R. P. ZHITARU, and V. G. KOLOSKOVA
Effect of Different Lattice Defects on the Mobility of Dislocations in Alkali Halide Crystals
Vol. **19**, No. 1, p. 441–451 (1967).
See also *10*, p. *97*; 22.5.2, p. 520.

R. C. Rau and G. A. Chase
Clustering of Defects in Neutron-Irradiated Beryllium Oxide
Vol. **19**, No. 2, p. 645–654 (1967).
See also *10*, p. *98*; 22.6, p. 536.

S. Y. Chuang and B. G. Hogg
Positron Annihilation in Alnico and $(La_{0.7}Pb_{0.3})MnO_3$
Vol. **20**, No. 1, p. 331–335 (1967).
See also *13*, p. *165*; 20, p. 338; 21.1.1, p. 415; 22.8, p. 549.

P. Schmeling
Non-Ideal Diffusion of Argon in Potassium Chloride
Vol. **20**, No. 2, p. 597–603 (1967).
See also *9*, p. *89*; 22.5.2, p. 521.

F. J. Worzala
Hardening in Irradiated Copper Crystals
Vol. **20**, No. 2, p. K81–K83 (1967).
See also *10.1*, p. *122*; 21.1, p. 397.

R. Lallement et D. A. Wigley
*Étude de la variation de longueur du plutonium α par autoirradiation à 4 °K
(Study of the Variation of Lenght of α-Plutonium by Self-Irradiation at 4 °K)
Vol. **21**, No. 1, p. 145–150 (1967).
See also 12, p. 157; 14.1, p. 222; 21.5, p. 429.

B. C. Dutta, G. Holtmeyer und J. Kalus
*Untersuchung der anisotropen Abbremsung von niederenergetischen Magnesiumatomen in einem NaCl-Einkristall bei verschiedenen Temperaturen
(Study of the Anisotropic Slowing-Down of Low-Energy Magnesium Atoms in an NaCl Single Crystal at Various Temperatures)
Vol. **21**, No. 1, p. 291–295 (1967).
See also 6.1, p. 72; 22.5.2, p. 522.

G. Holtmeyer und J. Kalus
*Einfluß der Nahordnung um Magnesiumverunreinigungen in Wismutschmelzen auf die Kernresonanzfluoreszenz-Ausbeute
(Effect of the Short-Range Order around Magnesium Impurities in Bismuth Melts on the Yield of Nuclear Resonance Fluorescence)
Vol. **21**, No. 1, p. 297–302 (1967).
See also 2, p. 23; 6.1, p. 72; 21.7, p. 438.

F. J. Bryant and E. Webster
*Threshold Energy for Atomic Displacement in Cadmium Telluride
Vol. **21**, No. 1, p. 315–321 (1967).
See also *20.3*, p. *365*; 22.4.3, p. 509.

A. Behr, H. Peisl, and W. Waidelich
*X-Ray Production of Vacancies in KCl at Low Temperatures
Vol. **21**, No. 1, p. K9–K10 (1967).
See also 10, p. 100; 22.5.2, p. 522.

R. C. Rau and K. Lacefield
Microhardness of Irradiated Beryllium Oxide
Vol. **21**, No. 1, p. K69–K71 (1967).
See also *12*, p. *157*; 22.6, p. 538.

R. Kelly
Bubble Diffusion and the Motion of Point Defects near Surfaces
Vol. **21**, No. 2, p. 451–459 (1967).
See also *9*, p. *89*.

Č. Jech
*Ion-Bombardment Enhanced Solubility in Solids
Vol. **21**, No. 2, p. 481–486 (1967).
See also 22, p. 445; 22.6, p. 538.

F. Dworschak, H. Schuster, H. Wollenberger, and J. Wurm
*The Influence of the Size Effect on Electrical Resistivity Measurements in Irradiated Metals
Vol. **21**, No. 2, p. 741–745 (1967).
See also 14.1, p. 223; 21, p. 379.

L. P. Khiznichenko, P. F. Kromer, D. K. Kaipnazarov, E. Otenyazov, D. Yusupova, and L. H. Zotova
Properties of the Low-Temperature Internal Friction Peak in Silicon
Vol. **21**, No. 2, p. 805–810 (1967).
See also 10, p. 100; *12*, p. *157*; 22.1.2, p. 471.

A. M. de Goer and B. Dreyfus
*γ-Irradiation Effect on the Thermal Conductivity of Al_2O_3 with Cr or Mn Impurrities at Low Temperature
Vol. **22**, No. 1, p. 77–82 (1967).
See also *8*, p. *83*; 10, p. 101; 22.6, p. 538.

A. Mukerji, G. A. Tanton, and J. E. Williams
X-Ray Induced F-Centers in CaF_2
Vol. **22**, No. 1, p. K19–K22 (1967).
See also *10.2*, p. *140*; 22.5.3, p. 533.

L. N. Cojocaru
*Effects of Fast Neutrons on Electric Properties of NiO Single Crystals
Vol. **22**, No. 2, p. 361–364 (1967).
See also 13.4, p. 208; 14.3, p. 237; 15, p. 272; 22.6, p. 539.

K. Dettmann, G. Leibfried, and K. Schroeder
*Spontaneous Recombination of Frenkel Pairs for Electron Irradiation
Vol. **22**, No. 2, p. 423–432 (1967).
See also 21.1, p. 399.

K. Dettmann, G. Leibfried, and K. Schroeder
*Spontaneous Recombination of Frenkel Pairs in Neutron-Irradiated Solids
Vol. **22**, No. 2, p. 433–440 (1967).
See also 21, p. 380; 21.1, p. 399; 21.6, p. 434.

F. Agulló-López
Role of γ-Radiation Intensity on the First Stage of the Room-Temperature F-Colouring in NaCl
Vol. **22**, No. 2, p. 483–490 (1967).
See also *10.2*, p. *140*; 22.5.2, p. 523.

L. Elstner and W. Kamprath
Quenched-in Levels in p-Type Silicon
Vol. **22**, No. 2, p. 541–547 (1967).
See also *13.4*, p. *208*; 14.3, p. 237; 22.1.2, p. 471.

R. Yu. Khansevarov, T. V. Mashovets, N. A. Vitovskii, and B. Chelustka
*On the Nature of Damages in Indium Antimonide Produced by γ-and X-Irradiation
Vol. **22**, No. 2, p. K95–K101 (1967).
See also 14.3, p. 237; 22.2.3, p. 491.

R. Yu. Khansevarov, T. V. Mashovets, and N. A. Vitovskii
*Damage Formation Rates in Ge Subject to Thermal Treatment
Vol. **22**, No. 2, p. K103—K107 (1967).
See also 10, p. 101; 14.3, p. 237; 22.1.1, p. 462.

S. Lungu
*Point Defects in Neutron-Damaged Silica Glass
Vol. **23**, No. 1, p. 147—156 (1967).
See also 2, p. 23; 10, p. 102; 22.8, p. 552.

J. H. Crawford, Jr., W. A. Sibley, and E. Sonder
The Influence of Electron-Trapping Impurity on Defect Creation and Bleaching in KCl Irradiated at 78 °K
Vol. **23**, No. 1, p. 301—306 (1967).
See also *10.2*, p. *140*; 13.4, p. 209; 22.5.2, p. 523.

G. Duesing, H. Hemmerich, D. Meissner, and W. Schilling
*The Influence of Vacancies and Impurities on the Damage Production in Platinum Electron-Irradiated at 90 °K
Vol. **23**, No. 2, p. 481—488 (1967).
See also 10.1, p. 125; 14.1, p. 223; 21.6, p. 434.

M. L. Swanson
*The Effects of Defect Doping on Low Temperature Neutron-Irradiation Damage and Recovery in Aluminium and Platinum
Vol. **23**, No. 2, p. 649—661 (1967)
See also 10.1, p. 125; 14.1, p. 224; 21, p. 381; 21.6, p. 434.

H. H. Neely and D. W. Keefer
*Resistivity Studies of Electron Irradiated Iron
Vol. **24**, No. 1, p. 217—220 (1967).
See also 14.1, p. 224; 21.1.1, p. 419.

H. Diepers
*Elektronenmikroskopische Untersuchung von Fehlstellenagglomeraten in ionenbestrahlten Kupferfolien. (I): Analyse von Agglomerattyp und -konfiguration (Electron-Microscopic Study of Defect Agglomerates in Ion-Irradiated Copper Foils. (I): Analysis of Agglomerate Type and Configuration)
Vol. **24**, No. 1, p. 235—244 (1967).
See also 10.1, p. 126; 21.1, p. 401.

J. Punzel
Temperature Dependence of the Surface Structure of Ion-Bombarded Si Single Crystals
Vol. **24**, No. 1. p. K1—K2 (1967).
See also *1.3*, p. *15*; 22.1.2, p. 472.

R. Brugière and P. Lucasson
On the Stage III Recovery of Electron Irradiated Aluminium
Vol. **24**, No. 1, p. K77—K81 (1967).
See also *10.1*, p. *127*; 14.1, p. 224; 21, p. 383.

H. Diepers
*Elektronenmikroskopische Untersuchung von Fehlstellenagglomeraten in ionenbestrahlten Kupferfolien. (II): Entstehung der Agglomerate (Electron-Microscopic Study of Defect Agglomerates in Ion-Irradiated Copper Foils. (II): Formation of Agglomerates)
Vol. **24**, No. 2, p. 623—633 (1967).
See also 4, p. 44; 10.1, p. 127; 21.1, p. 402.

A. BOURRET and D. DAUTREPPE
*Type of Small Dislocation Loops in Neutron Irradiated Copper
Vol. **24**, No. 2, p. K173–K174 (1967).
See also 10.1, p. 127; 21.1, p. 402.

G. P. SEIDEL
*Einfluß von Neutronenbestrahlung auf die plastische Verformung von Eisen-Einkristallen mit unterschiedlichem Kohlenstoffgehalt. (I): Die Fließspannung am Verformungsbeginn
(Effect of Neutron Irradiation on the Plastic Deformation of Iron Single Crystals with Different Carbon Concentrations. (I): The Field Stress at the Deformation Onset)
Vol. **25**, No. 1, p. 175–188 (1968).
See also *10.1*, p. *127*; 21.1.1, p. 420.

G. GIULIANI and E. REGUZZONI
*X-Ray Production and Thermal Annealing of Frenkel Pairs in KCl
Vol. **25**, No. 1, p. 437–448 (1968).
See also *10.2*, p. *141*; 22.5.2, p. 525.

Y. KAZUMATA
Atomic ^{3}H Center in Neutron-Irradiated LiF
Vol. **25**, No. 2, p. 563–566 (1968).
See also *10.2*, p. *142*; 19, p. 332; 22.5.2, p. 525.

H. NEUHÄUSER
Study of the Active Crystal-Lenght in the Yield Range of Neutron-Irradiated Copper Single Crystals. (I): Slip Band Formation
Vol. **25**, No. 2, p. 593–606 (1968).
See also *10.1*, p. *128*; 21.1, p. 403.

R. KELLY, Č. JECH, and HJ. MATZKE
Diffusion in Inert-Gas Bombarded KCl, KBr, and KI
Vol. **25**, No. 2, p. 641–653 (1968).
See also *9*, p. *91*; 22.5.2, p. 525.

D. K. SMITH and H. W. NEWKIRK
Irradiation Decoration of Defects in Single-Crystal BeO
Vol. **25**, No. 2, p. K79–K82 (1968).
See also *10*, p. *104*; 22.6, p. 540.

H. BEYER and R. SIZMANN
*Recrystallization of Cold-Worked Gold during α-Irradiation
Vol. **26**, No. 1, p. 359–370 (1968).
See also 4, p. 45; 10.1, p. 129; 14.1, p. 225; 21.6, p. 435.

M. SOROHAN
*The Influence of Ion Bombardment upon the Induced Magnetic Anisotropy in Thin Nickel Films
Vol. **26**, No. 1, p. K9–K11 (1968).
See also *18.2*, p. *310*; 21.1, p. 404.

M. RÜHLE
*Radiation Hardening of Neutron-Irradiated Copper Single Crystals. (I): Experimental Results
Vol. **26**, No. 2, p. 661–670 (1968).
See also *10.1*, p. *130*; 21.1, p. 404; 21.6, p. 435.

W. FRANK, M. RÜHLE, and M. SAXLOVÁ
*Radiation Hardening of Neutron-Irradiated Copper Single Crystals. (II): Interpretation
Vol. **26**, No. 2, p. 671–680 (1968).
See also *10.1*, p. *130*; 21.1, p. 404.

Y. Kazumata
Self-Trapped Holes in Neutron Irradiated CaF_2 and BaF_2 Crystals
Vol. **26**, No. 2, p. K119–K122 (1968).
See also 10.2, p. 142; *19*, p. *333*; 22.5.3, p. 534.

B. Henderson and R. D. King
Colour Centres in Neutron Irradiated Calcium Oxide
Vol. **26**, No. 2, p. K147–K149 (1968).
See also *10.2*, p. *142*; 22.6, p. 541.

T. Schober and R. W. Balluffi
*Direct Evidence for Dechanneling of 40 kV Gold Ions at Planar Defects in Silver and Gold
Vol. **27**, No. 1, p. 195–200 (1968).
See also *10.1*, p. *130*; 21.6, p. 436.

H. Neuhäuser
Study of the Active Crystal-Lenght in the Yield Range of Neutron-Irradiated Copper Single Crystals. (II): The Active Crystal-Lenght
Vol. **27**, No. 1, p. 281–290 (1968).
See also *10.1*, p. *130*; 21.1, p. 404.

D. N. Bose
Effect of Low Energy Electron and UV Irradiations on the Charge Condition of Etched Germanium Surfaces
Vol. **27**, No. 1, p. 325–331 (1968).
See also *13.3*, p. *198*; 22.1.1, p. 465.

J. M. Galligan and M. J. Attardo
*Radiation Hardening in Platinum
Vol. **27**, No. 1, p. 383–390 (1968).
See also 10.1, p. 131; 21.6, p. 436.

T. M. Fitzgerald and B. D. Silverman
Ultrasonic Attenuation in Unirradiated and Neutron Irradiated Quartz
Vol. **27**, No. 2, p. 473–478 (1968).
See also 6, p. 66; *7*, p. *77*; 22.6, p. 541.

F. W. Felix
Rare-Gas Diffusion in Neutron-Irradiated Rubidium Halides
Vol. **27**, No. 2, p. 529–534 (1968).
See also *9*, p. *91*; 10, p. 107; 22.5.2, p. 526.

Č. Jech
*Depth Distribution Measurements Using Bombardment-Enhanced Solubility
Vol. **27**, No. 2, p. 573–577 (1968).
See also 1.4, p. 20; 22, p. 452; 22.1.2, p. 473.

S. D. McLaughlan and H. W. Evans
*Production of Colloidal Calcium by Electron Irradiation of CaF_2 Crystals
Vol. **27**, No. 2, p. 695–700 (1968).
See also 10.2, p. 142; 19, p. 333; 20.1, p. 354; 22.5.3, p. 534.

J. L. Brimhall and B. Mastel
Defect Clusters in Neutron Irradiated Rhenium
Vol. **27**, No. 2, p. K89–K92 (1968).
See also *10.1*, p. *131*; 21, p. 386.

E. W. Kreutz, H. Pagnia, and W. Waidelich
*X-Ray Induced Conductivity Changes in InSb
Vol. **27**, No. 2, p. K111–K114 (1968).
See also 13.3, p. 198; 14.3, p. 242; 22.2.3, p. 492.

R. Balzer, H. Peisl, and W. Waidelich
Evidence for Frenkel Defect Recombination during Thermal Annealing of Low-Temperature X-Irradiated KCl
Vol. **27**, No. 2, p. K165–K167 (1968).
See also *10*, p. *108*; 22.5.2, p. 527.

R. Balzer, H. Peisl, and W. Waidelich
Lattice Disorder in Alkali Halides X-Irradiated at Room Temperature
Vol. **28**, No. 1, p. 207–210 (1968).
See also *10*, p. *108*; 22.5.2, p. 527.

T. O. Tuomi
*Effect of Reactor Irradiation on the Conductivity and the Photoconductivity of Hexagonal Selenium Single Crystals
Vol. **28**, No. 1, p. 329–334 (1968).
See also 14.3, p. 242; 16, p. 282; 22.1.3, p. 479.

P. Ehrhart, W. Rupp, and R. Sizmann
*Stopping Power of 0.5 to 3.5 MeV α-Particles in Cu-Au Alloys
Vol. **28**, No. 1, p. K35–K37 (1968).
See also 1.4, p. 21; 21.1, p. 405; 21.6, p. 436.

G. K. Born, R. J. Grasser, and A. O. Scharmann
*EPR and Thermoluminescence Measurements on $CaWO_4$ Crystals
Vol. **28**, No. 2, p. 583–588 (1968).
See also 10.2, p. 143; 19, p. 334; 20.3, p. 370; 22.8, p. 554.

H. H. Neely, D. W. Keefer, and A. Sosin
*Electron Irradiation and Recovery of Tungsten
Vol. **28**, No. 2, p. 675–682 (1968).
See also 10.1, p. 132; 14.1, p. 226; 21, p. 387.

F. Dworschak, H. Schuster, H. Wollenberger, and J. Wurm
Analysis of Point Defect States in Copper. (I): The Influence of the Point Defect State on the Irradiation Damage Rate at 80 °K
Vol. **29**, No. 1, p. 75–80 (1968).
See also *10.1*, p. *132*; 21.1, p. 406.

F. Dworschak, H. Schuster, H. Wollenberger, and J. Wurm
Analysis of Point Defect States in Copper. (II): The Effects of Stage I and Stage III Annealing on Electron-Irradiation-Induced Frenkel Pairs
Vol. **29**, No. 1, p. 81–89 (1968).
See also *10.1*, p. *133*; 21.1, p. 406.

C. Sánchez and F. Agulló-López
Transient Effects in the Room-Temperature F-Colouring of NaCl Irradiated with X- or γ-Rays
Vol. **29**, No. 1, p. 217–230 (1968).
See also *10.2*, p. *143*; 22.5.2, p. 528.

F. Walz
Investigation of the Magnetic After-Effect in Nickel after Neutron Irradiation and Plastic Deformation
Vol. **29**, No. 1, p. 245–258 (1968).
See also 10.1, p. 133; *18.2*, p. *314*; 21.1, p. 407.

A. Bourret and D. Dautreppe
Study of Agglomerated Defects in Irradiated Pure Nickel by Electron Microscopy
Vol. **29**, No. 1, p. 283–298 (1968).
See also **4**, p. 47; *10.1*, p. *133*; 21.1, p. 407.

K. Ibel and R. Sizmann
*Energy Loss of ⟨110⟩ Channelled α-Recoil Atoms in Gold
Vol. **29**, No. 1, p. 403–415 (1968).
See also 21.6, p. 437.

R. von Jan
*Scattering Effects with Channelled Ions
Vol. **29**, No. 2, p. 755–759 (1968).

J. Delaplace, J. C. Nicoud, D. Schumacher, and G. Vogl
*Low-Temperature Neutron Radiation Damage and Recovery in Beryllium
Vol. **29**, No. 2, p. 819–824 (1968).
See also 10.1, p. 134; 14.1, p. 226; 21.3, p. 427.

L. N. Cojocaru
*Effects of Fission Fragments on Electric Properties of the UO_2–SiO_2 System
Vol. **29**, No. 2, p. K119–K122 (1968).
See also 14.3, p. 244; 15, p. 274; 22.6, p. 543; 22.8, p. 556.

G. Roth und V. Naundorf
*Messung der Aktivierungsenergien der Versetzungsverankerung in Kupfer nach Elektronenbestrahlungen bei 78 °K
(Measurement of the Activation Energies of Dislocation Pinning in Copper after Electron Irradiation at 78 °K)
Vol. **29**, No. 2, p. K123–K125 (1968).
See also 10.1, p. 134; 12.1, p. 161; 21.1, p. 408.

J. C. Jousset
*Défauts d'irradiation dans l'uranium α et règle de Matthiessen
(Irradiation Defects in α-Uranium and Matthiessen's Rule)
Vol. **29**, No. 2, p. K127–K131 (1968).
See also 14.1, p. 226; 21.5, p. 429.

J. L. Leveque, T. Anagnostopoulos, H. Bilger, and P. Moser
*Point Defects in Iron–Carbon Solid Solution Studied by Magnetic After-Effect Measurements
Vol. **29**, No. 2, p. K179–K181 (1968).
See also *10.1*, p. *134*; 18.2, p. 315; 21.1.1, p. 423.

R. Kelly
*On the Problem of Depth Distributions in Oxides Following High-Dose Labelings
Vol. **30**, No. 1, p. 37–48 (1968).
See also 22.6, p. 544.

J. Delaplace, J. Hillairet, J. C. Nicoud, D. Schumacher, and G. Vogl
*Low Temperature Neutron Radiation Damage and Recovery in Magnesium
Vol. **30**, No. 1, p. 119–126 (1968).
See also 10.1, p. 134; 14.1, p. 226; 21.3, p. 427.

N. Itoh
*Degradation of Low-Energy Electrons in Alkali Halides
Vol. **30**, No. 1, p. 199–207 (1968).
See also 10, p. 112; 22.5.2, p. 530.

D. Becker, F. Dworschak, C. Lehmann, K. T. Rie, H. Schuster, H. Wollenberger, and J. Wurm
*Determination of the Average Number of Replacement Collisions in an Electron-Irradiated F.C.C. Crystal (Ni_3Mn)
Vol. **30**, No. 1, p. 219—229 (1968).
See also 10.1, p. 135; 14.1, p. 227; 18.2, p. 315; 21.1, p. 408.

W. Stössel and W. Zimmermann
*Photoconduction in CdS Crystals Generated by a Modulated Electron Beam
Vol. **30**, No. 1, p. 311—318 (1968).
See also 13.4, p. 216; *16*, p. *284*; 22.4.1, p. 505.

M. A. Kumakhov
*Formation of Radiation Defects by Channeling Particles
Vol. **30**, No. 1, p. 379—385 (1968).
See also 3.2, p. 36; 21.1, p. 408; 22.1.2, p. 474.

T. Okada and K. Ozawa
Anomalous Ductility of LiF Crystals at Low Temperatures
Vol. **30**, No. 1, p. K69—K71 (1968).
See also *10*, p. *112*; 22.5.2, p. 530.

R. E. Hummel and J. W. Koger
*Effect of Neutron Irradiation on the Martensitic Transformation in β_1-Copper-Zinc
Vol. **30**, No. 2, p. 515—523 (1968).
See also *1.2*, p. *13*; 14.1, p. 227; 21.1, p. 409.

Y. Quéré
*Dechanneling Cylinder of Dislocations
Vol. **30**, No. 2, p. 713—722 (1968).
See also *10.1*, p. *135*; 21, p. 389.

R. Brugière and P. Lucasson
On the Stage III Recovery of Electron Irradiated Copper
Vol. **30**, No. 2, p. K139—K143 (1968).
See also *10.1*, p. *136*; 14.1, p. 227; 21.1, p. 409.

12. Mechanical Properties of Solids (Plastic Deformations See 10)

Original Papers and Short Notes

E. I. Rashba, V. A. Romanov, I. I. Boiko, and I. P. Zhadko
Electrical Pinch in Elastically Deformed Germanium
Vol. **16**, No. 1, p. 43–56 (1966).
See also *14.3*, p. *231*; 14.4.1, p. 255; 22.1.1, p. 459.

V. A. Romanov, I. P. Zhadko, and I. I. Boiko
Photoconductivity of Anisotropically Deformed Intrinsic Germanium
Vol. **17**, No. 1, p. 389–396 (1966).
See also *16*, p. *277*; 22.1.1, p. 459.

K. Ozawa and S. Anzai
Effect of Pressure on the α-Transition Point of Iron Monosulphide
Vol. **17**, No. 2, p. 697–700 (1966).
See also *1.2*, p. *9*; 14.3, p. 233; 22, p. 441.

G. P. Upit and S. A. Varchenya
*Microhardness of Alkali Halide Crystals
Vol. **17**, No. 2, p. 831–835 (1966).
See also 22.5.2, p. 519.

R. C. Lincoln, K. M. Koliwad, and P. B. Ghate
*Elastic Constants of Some NaCl Type Alkali Halides
Vol. **18**, No. 1, p. 265–277 (1966).
See also 3.2, p. 34; 22.5.2, p. 519.

K. S. Aleksandrov, L. M. Reshchikova, and B. V. Beznosikov
*Behaviour of the Elastic Constants of $KMnF_3$ Single Crystals near the Transition of Puckering
Vol. **18**, No. 1, p. K17–K20 (1966).
See also 1.2, p. 10; 22.5, p. 512; 22.8, p. 548.

H. L. Brown and C. P. Kempter
*Elastic Properties of Zirconium Carbide
Vol. **18**, No. 1, p. K21–K23 (1966).
See also 22.3, p. 494.

M. Asche, O. G. Sarbej, and V. M. Vasetskii
Piezoresistance of p-Germanium
Vol. **18**, No. 2, p. 749–754 (1966).
See also 10, p. 97; *14.3*, p. *234*; 22.1.1, p. 460.

G. R. Barsch
*Adiabatic, Isothermal, and Intermediate Pressure Derivatives of the Elastic Constants for Cubic Symmetry. (I): Basic Formulae
Vol. **19**, No. 1, p. 129–138 (1967).

G. R. Barsch and Z. P. Chang
*Adiabatic, Isothermal, and Intermediate Pressure Derivatives of the Elastic Constants for Cubic Symmetry. (II): Numerical Results for 25 Materials
Vol. **19**, No. 1, p. 139–151 (1967).
See also 21, p. 376; 22, p. 442.

R. P. Harrison
The Effect of Cadmium Additions on the Fracture Strength of Sodium Chloride Single Crystals
Vol. **19**, No. 1, p. 353–357 (1967).
See also *10*, p. *97*; 22.5.2, p. 520.

T. Geszti
On the Theory of Thermally Activated Processes
Vol. **20**, No. 1, p. 165–177 (1967).
See also 9, p. 88; *10*, p. *98*; 14.4, p. 252.

R. Lallement and D. A. Wigley
Study of the Variation of Length of α-Plutonium by Self-Irradiation at 4 °K
Vol. **21**, No. 1, p. 145–150 (1967).
See also *11*, p. *148*; 14.1, p. 222; 21.5, p. 429.

R. C. Rau and K. Lacefield
*Microhardness of Irradiated Beryllium Oxide
Vol. **21**, No. 1, p. K69–K71 (1967).
See also 11, p. 148; 22.6, p. 538.

K. M. Koliwad, P. B. Ghate, and A. L. Ruoff
*Pressure Derivatives of the Elastic Constants of NaBr and KF
Vol. **21**, No. 2, p. 507–516 (1967).
See also 7, p. 76; 22.5.2, p. 522.

N. P. Gupta and B. Dayal
*Thermal Expansion and Compressibility of Solid Neon and Xenon
Vol. **21**, No. 2, p. 661–664 (1967).
See also 6, p. 61; *8*, p. *82*; 23, p. 561.

L. P. Khiznichenko, P. F. Kromer, D. K. Kaipnazarov, E. Otenyazov, D. Yusupova, and L. G. Zotova
*The Properties of the Low-Temperature Internal Friction Peak in Silicon
Vol. **21**, No. 2, p. 805–810 (1967).
See also 10, p. 100; 11, p. 149; 22.1.2, p. 471.

P. F. Kromer and L. P. Khiznichenko
*On the Low-Temperature Internal Friction in Silicon
Vol. **21**, No. 2, p. 811–818 (1967).
See also 10, p. 100; 22.1.2, p. 471.

H. L. Jetter and H. C. Wolf
On the Davydov Splitting of the First Excited Singulet-State of Anthracene Crystals
Vol. **22**, No. 1, p. K39–K43 (1967).
See also 13.2, p. 193; *20.1*, p. *349*; 22.9, p. 559.

V. A. Drozdov and L. A. Mozgovaya
*Infrared Quenching of the Photomechanical Effect in Cadmium Sulphide
Vol. **22**, No. 2, p. K109–K112 (1967).
See also 16, p. 280; 22.4.1, p. 500.

B. Mukherjee and A. J. F. Boyle
On the Debye Characteristic Temperature of Diamond
Vol. **22**, No. 2, p. K131–K134 (1967).
See also *6*, p. *62*; 8, p. 83; 22.1, p. 457.

Z. P. Chang, G. R. Barsch, and D. L. Miller
*Pressure Dependence of Elastic Constants of Cesium Halides
Vol. **23**, No. 2, p. 577–586 (1967).
See also 22.5.2, p. 524.

G. K. Afanaseva, K. S. Aleksandrov, and A. I. Kitaigorodskii
*Elastic Constants of Anthracene
Vol. **24**, No. 1, p. K61–K63 (1967).
See also 22.9, p. 559.

G. M. Bartenev and E. G. Koryak-Doronenko
The Theoretical Strength of NaCl Crystals under Tension along Its Body and Face Diagonals
Vol. **24**, No. 2, p. 443—449 (1967).
See also *10*, p. *103*; 22.5.2, p. 524.

E. O. Osmanov, Yu. V. Rud, and M. E. Stryalkovskii
High-Temperature Modifications of the Semiconductor Compounds $CdSnAs_2$ and $CdGeAs_2$
Vol. **26**, No. 1, p. 85—90 (1968).
See also *1.2*, p. *12*; 2, p. 24; 8, p. 84; 14.3, p. 240; 22.7, p. 546; 22.8, p. 553.

R. P. Harrison
The Effect of Particle Size on the Fracture Strength of Sodium Chloride Single Crystals
Vol. **26**, No. 1, p. K65—K67 (1968).
See also *4*, p. *45*; 10, p. 105; 22.5.2, p. 526.

Kh. M. Khalilov, B. B. Kuliev, and F. A. Gambarov
*Effect of Phosphorus Admixtures on Viscoelastic Properties of Selenium
Vol. **26**, No. 2, p. K113—K117 (1968).
See also *2*, p. *24*; 10, p. 106; 22.1.3, p. 479.

L. Ask
Effect of Mechanical Stress on the I-U Characteristic of a GaP p-n Junction
Vol. **26**, No. 2, p. K143—K145 (1968).
See also 2, p. 24; *14.3.2*, p. *250*; 22.2.2, p. 488.

S. Haussühl
*Elastisches und thermoelastisches Verhalten von MgF_2 und MnF_2
(Elastic and Thermoelastic Behaviour of MgF_2 and MnF_2)
Vol. **28**, No. 1, p. 127—130 (1968).
See also 3.2, p. 36; 22.5, p. 514; 22.5.3, p. 534.

T. P. Srinivasan and S. D. Nigam
*Invariant Elastic Constants for Crystals
Vol. **28**, No. 1, p. K71—K73 (1968).
See also 3, p. 27; *12.1*, p. *160*.

P. O. Szomor
*Elastic Constants and Tensile Strength of NaCl-Type Alkali Halides
Vol. **28**, No. 2, p. 529—537 (1968).
See also *10*, p. *109*; 22.5.2, p. 527.

J. H. Gieske and G. R. Barsch
*Pressure Dependence of the Elastic Constants of Single Crystalline Aluminum Oxide
Vol. **29**, No. 1, p. 121—131 (1968).
See also 22.6, p. 542.

E. P. Berdnikov, V. A. Drozdov, and L. A. Mozgovaya
On the Relation between Microhardness and Point Defects of Crystals
Vol. **29**, No. 2, p. K145—K147 (1968).
See also *10*, p. *111*; 15, p. 274; 22.1.2, p. 474.

S. Haussühl
*Piezoelektrisches und elektrisches Verhalten von Lithiumjodat
(Piezoelectrical and Electrical Behaviour of Lithiumjodate)
Vol. **29**, No. 2, p. K159—K162 (1968).
See also *14.3*, p. *244*; 22.8, p. 556.

12.1 Mechanical Properties of Metals (Plastic Deformations See 10.1)

Original Papers and Short Notes

J. Melngailis
A Model for the Change in Elastic Modulus of a Solid Caused by Point Defects
Vol. **16**, No. 1, p. 247–252 (1966).
See also *10.1*, p. *114*; 21.1, p. 390.

V. A. Buravikhin, V. G. Kazakov, V. S. Khristosenko, and P. I. Krukover
Effect of Uniaxial Elastic Stress on the Remagnetization Processes in Thin Permalloy Films
Vol. **16**, No. 2, p. 651–656 (1966).
See also *18.2*, p. *301*; 21.1.1, p. 411.

V. A. Buravikhin and V. I. Popov
Structure and Polarity of Domain Boundaries in Ferromagnetic Films
Vol. **16**, No. 2, p. 657–662 (1966).
See also 4, p. 38; *18.2*, p. *301*; 21.1.1, p. 411.

A. I. Gubanov and V. K. Nikulin
The Theory of the Cohesion of Noble Metals
Vol. **17**, No. 2, p. 815–823 (1966).
See also *3.2*, p. *34*; 5, p. 50; 21.1, p. 393; 21.6, p. 431.

M. F. Rose
*Pressure Dependence of the Second Order Elastic Constants
Vol. **17**, No. 2, p. K199–K201 (1966).
See also 21, p. 375; 21.1, p. 393; 21.6, p. 431.

Ir. Boesono, G. J. Ernst, M. C. Lemmens, M. J. van Langen, and G. de Vries
*A Zener-Damping Peak in Fe–4 Wt% Si
Vol. **18**, No. 1, p. 107–110 (1967).
See also 10.1, p. 120; 21.1.1, p. 414.

R. Labusch
Elastic Constants of the Flux Line Lattice in Type II Superconductors
Vol. **19**, No. 2, p. 715–719 (1967).
See also *14.2*, p. *228*.

A. Czachor and A. Pindor
*Dependence of Calculated Elastic Constants on the c/a Ratio of Hexagonal Close-Packed Metals
Vol. **20**, No. 1, p. K17–K20 (1967).
See also 5, p. 50; 21, p. 377; 21.1, p. 396; 21.3, p. 426.

A. K. Head, M. H. Loretto, and P. Humble
The Influence of Large Elastic Anisotropy on the Determination of Burgers Vectors of Dislocations in β-Brass by Electron Microscopy
Vol. **20**, No. 2, p. 505–519 (1967).
See also *10.1*, p. *122*; 21.1, p. 397.

A. K. Head, M. H. Loretto, and P. Humble
The Identification of Burgers Vectors and Unstable Directions of Dislocations in β-Brass
Vol. **20**, No. 2, p. 521–536 (1967).
See also *10.1*, p. *122*; 21.1, p. 397.

R. F. Peart
Effect of Pressure on the Diffusion of Fe in Ti and Ti + 10% Fe
Vol. **20**, No. 2, p. 545—550 (1967).
See also *9*, p. *88*; 21, p. 378; 21.1.1, p. 416.

E. A. Metzbower
Noncentral Force Model for Hexagonal Close-Packed Crystal Lattices
Vol. **20**, No. 2, p. 681—692 (1967).
See also 3.2, p. 35; *6*, p. *60*; 8, p. 81; 21.1, p. 397; 21.3, p. 426.

M. F. Rose
*Higher Order Elastic Constants and the Equation of State of F.C.C. Metals
Vol. **21**, No. 1, p. 235—238 (1967).
See also 6, p. 60; 8, p. 82; 21, p. 378; 21.1, p. 397; 21.6, p. 433.

M. F. Rose and R. T. Ramsey
*Higher Order Elastic Constants in H.C.P. Crystals
Vol. **25**, No. 1, p. 103—108 (1968).
See also 3.2, p. 36; 21, p. 383; 21.3, p. 426.

E. A. Metzbower
Noncentral Force Model for Hexagonal Close-Packed Crystal Lattices (II)
Vol. **25**, No. 1, p. 403—416 (1968).
See also *3.2*, p. *36*; 6, p. 64; 8, p. 84; 21.1, p. 402.

M. E. de Morton
Atomic Relaxations in a Face-Centred-Tetragonal Alloy
Vol. **26**, No. 1, p. K73—K75 (1968).
See also *10.1*, p. *129*; 21, p. 385.

G. P. Williams, Jr. and T. J. Turner
*Zener Relaxation in Silver Solid Solutions
Vol. **26**, No. 2, p. 645—651 (1968).
See also *10.1*, p. *130*; 21.6, p. 435.

Y. Haven
*Relation between Relaxation Time and Jump Frequencies for Zener Relaxation in Disordered F.C.C. Alloys
Vol. **26**, No. 2, p. 653—660 (1968).
See also *10.1*, p. *130*; 21, p. 385.

J. Kinel, L. Kozłowski, and J. W. Moroń
*Diffusional Relaxation Effects in Hydrogen Charged Silicon Iron
Vol. **27**, No. 1, p. K37—K39 (1968).
See also 10.1, p. 131; *18.2*, p. *312*; 21.1.1, p. 421.

T. P. Srinivasan and S. D. Nigam
*Invariant Elastic Constants for Crystals
Vol. **28**, No. 1, p. K71—K73 (1968).
See also 3, p. 27; *12*, p. *158*.

D. I. Bower, E. Claridge, and I. S. T. Tsong
*Low-Temperature Elastic Constants and Specific Heats of F.C.C. Nickel–Iron Alloys
Vol. **29**, No. 2, p. 617—625 (1968).
See also 6, p. 68; 8, p. 86; 13, p. 172; 18.2, p. 314; 21.1, p. 407; 21.1.1, p. 422.

G. Roth and V. Naundorf
Measurement of the Activation Energies of Dislocation Pinning in Copper after Electron Irradiation at 78 °K
Vol. **29**, No. 2, p. K123–K125 (1968).
See also 10.1, p. 134; *11*, p. *154*; 21.1, p. 408.

T. J. Turner, C. M. Dozier, and G. P. Williams, Jr.
*Acitvation Energies for Zener Relaxation in α-AgCd
Vol. **30**, No. 1, p. 87–91 (1968).
See also 9, p. 92; *10.1*, p. *134*; 21, p. 388; 21.6, p. 437.

13. Electron States in Solids

Review Articles

L. Eckertová
Field Cathodes with Thin Insulating Films
Vol. **18**, No. 1, p. 3–40 (1966).
See also 13.3, p. 196; 13.4, p. 200; 14.4.1, p. 255; *17*, p. *286*; 22.6, p. 535.

D. P. Morgan
*Helicon Waves in Solids
Vol. **24**, No. 1, p. 9–36 (1967).
See also 6, p. 56; 14.1, p. 220; 14.3, p. 231; 19, p. 326; *20*, p. *337*; 21, p. 373; 21.1, p. 390; 21.2, p. 424; 21.6, p. 430; 22.1.1, p. 458; 22.2.1, p. 482; 22.2.3, p. 489.

J. Mertsching
Theory of Electromagnetic Waves in Metals and Their Interaction with Ultrasonic Waves (II)
Vol. **26**, No. 1, p. 9–64 (1968).
See also *7*, p. *76*; 19, p. 326; 20, p. 337; 21, p. 373.

W. Steinmann
*Optical Plasma Resonances in Solids
Vol. **28**, No. 2, p. 437–462 (1968).
See also 13.3, p. 196; *20*, p. *337*; 20.1, p. 344; 21, p. 373; 21.2, p. 424; 21.6, p. 430.

V. L. Bonch-Bruevich and E. G. Landsberg
Recombination Mechanisms
Vol. **29**, No. 1, p. 9–43 (1968).
See also 6, p. 56; 13.2, p. 191; *13.4*, p. *200*; *14.3*, p. *231*; 20, p. 337; 20.3, p. 363; 22.1.1, p. 458; 22.1.2, p. 468.

Original Papers and Short Notes

E. Kamieniecki
*Distribution Function of Hot Carriers
Vol. **16**, No. 1, p. 57–62 (1966).
See also 6, p. 56; *14.3*, p. *231*; 22.1.1, p. 459; 22.1.2, p. 468.

C. N. Plăviţu
Some Kinetic Phenomena in Impure Ionic Semiconductors
Vol. **16**, No. 1, p. 69–78 (1966).
See also 8, p. 79; *14.4*, p. *252*; 15, p. 270; 22, p. 439.

H. Kronmüller and A. Seeger
Magnetostrictive and Magnetocrystalline Residual Electric Resistivity in Ferromagnetics
Vol. **16**, No. 1, p. 253–266 (1966).
See also *14.1*, p. *220*; *18.2*, p. *299*; 21, p. 373.

J. Mertsching
On the Determination of the Electron Free Path from Measurements of Ultrasonic Absorption in Metals
Vol. **16**, No. 1, p. 267–275 (1966).
See also *7*, p. *76*; 21, p. 373.

M. Zvára, F. Žaloudek, and V. Prosser
Interband Faraday Effect in CdTe Single Crystals
Vol. **16**, No. 1, p. K21–K24 (1966).
See also *18*, p. *291*; *20*, p. *337*; 22.4.3, p. 509.

V. P. KALASHNIKOV
*Influence of Spatial Inhomogeneities on the Nuclear Polarization due to Direct Current in Semiconductors
Vol. **16**, No. 1, p. K35—K37 (1966).
See also *14.3*, p. *231*; 22, p. 439.

M. DUBEJKO and S. OLSZEWSKI
Quantum-Statistical Calculations on the Work Function of Metals
Vol. **16**, No. 2, p. 399—404 (1966).
See also *17*, p. *286*; 21, p. 373.

Z. ŠROUBEK
On the Theory of the Superexchange Interaction in Ionic Crystals: $KNiF_3$
Vol. **16**, No. 2, p. 405—412 (1966).
See also *3.2*, p. *34*; 18.4, p. 321; 22.5, p. 512; 22.8, p. 547.

R. GRANGER
Anisotropic Electron Distribution in Pure Semiconductors near Breakdown
Vol. **16**, No. 2, p. 599—612 (1966).
See also 6, p. 56; *14.4.1*, p. *255*; 22, p. 440; 22.2.3, p. 489.

TH. KLUPSCH and W. HAUBENREISSER
*A New Nonlinear Mechanism for Establishing the Saturation Susceptibility during Spin Wave Parallel Pumping
Vol. **16**, No. 2, p. K147—K151 (1966).
See also *18*, p. *291*.

B. LEMAIRE and H. LE GALL
*Spin-Wave Instability in Parallel Pumping with Long Duration Pulses and Steady State
Vol. **17**, No. 1, p. K41—K43 (1966).
See also *18.3*, p. *316*.

N. SINGH and M. D. KARKHANAVALA
Studies on the Electrical Properties of Uranium Oxides. (II): Non-stoichiometric U_3O_8
Vol. **17**, No. 2, p. 501—508 (1966).
See also *14.3*, p. *232*; 22.6, p. 535.

V. DIENYS and J. POZHELA
Anisotropy of the Conductivity of n-Type Germanium at High Electric Fields
Vol. **17**, No. 2, p. 769—776 (1966).
See also *14.4.1*, p. *256*; 22.1.1, p. 460.

J. MERTSCHING
*Theorie des Dielektrizitätstensors in Relaxationszeit-Approximation
(Theory of the Dielectric Tensor in the Relaxation Time Approximation)
Vol. **17**, No. 2, p. 799—806 (1966).
See also *20*, p. *337*.

A. I. GUBANOV
*The LCAO Method for an Arbitrary Mixture of Atoms
Vol. **17**, No. 2, p. 807—814 (1966).
See also *2*, p. *22*; 21, p. 375; 22, p. 441.

J. DEVREESE, R. DE CONINCK, and H. POLLAK
On the Conduction Mechanism in Uranium Dioxide
Vol. **17**, No. 2, p. 825—829 (1966).
See also *14.3*, p. *233*; 15, p. 270; 22.6, p. 535

T. L. TANSLEY
Forward Bias Current-Voltage Characteristics for a Heterojunction in which Tunnelling Dominates
Vol. 18, No. 1, p. 105–112 (1966).
See also *14.3.2*, p. *248*; 22.2.1, p. 483.

H. STOLZ
*Contribution to the Theory of Nonlinear Response
Vol. 18, No. 1, p. 251–264 (1966).

H. FLIETNER und N. KEMPE
*Untersuchung der Schwingungseigenschaften des Oszillistors
(Study of the Oscillatory Properties of the Oscillistor)
Vol. 18, No. 1, p. 415–426 (1966).
See also *20*, p. *338*; 22.1.1, p. 460.

I. LICEA
On the Variation of the Electron Temperature in Polar Semiconductors by an Electric Field
Vol. 18, No. 1, p. K9–K12 (1966).
See also *14.3*, p. *233*; 22.2.3, p. 490.

H. MAYER and J. HÖLZL
Experimental Determination of the Maximum Emission-Depths of Monoenergetic Secondary Electrons
Vol. 18, No. 2, p. 779–785 (1966).
See also *17*, p. *287*; 21.2, p. 424.

R. I. ANISHCHENKO
Calculation of the Mean Inner Potential of a Crystal in the Statistical Theory
Vol. 18, No. 2, p. 923–928 (1966).
See also *3.2*, p. *34*; 21, p. 376; 21.1, p. 394; 21.1.1, p. 414; 21.2, p. 424; 21.3, p. 426; 21.4, p. 428; 21.6, p. 432; 21.7, p. 438; 22.1.1, p. 460.

V. V. PARANJAPE
*Distribution Function of Hot Electrons
Vol. 18, No. 2, p. K103–K104 (1966).
See also 6, p. 58; *14.3*, p. *234*; 22, p. 442.

E. KAMIENIECKI
*Remark on Solving Boltzmann Equation for Hot Carriers
Vol. 18, No. 2, p. K105 (1966).
See also 6, p. 58; *14.3*, p. *234*; 22, p. 442.

G. ALBRECHT und G. WOLF
*Spezifische Elektronenwärme und λ-Anomalie von $CrH_{0,84}$
(Electron Specific Heat and λ-Anomaly of $CrH_{0.84}$)
Vol. 18, No. 2, p. K119–K122 (1966).
See also 8, p. 80; 22, p. 442.

H. P. VAN DE BRAAK and W. J. CASPERS
Disorder Resistivity and Spin Depolarization in Rare-Earth Alloys
Vol. 19, No. 1, p. 385–393 (1967).
See also *14.1*, p. *222*; 18.2, p. 305; 21.4, p. 428.

B. VELICKÝ
Polaron Effect on the Optical Absorption Edge of Semiconductors
Vol. 19, No. 1, p. K39–K42 (1967).
See also 6, p. 59; 13.1, p. 176; *20.1*, p. *347*; 22, p. 442.

W. FRANZ und P. MIOSGA
*Verteilungsfunktion und Driftgeschwindigkeit warmer Elektronen in Germanium
(Distribution Function and Drift Velocity of Warm Electrons in Germanium)
Vol. 19, No. 2, p. 597–602 (1967).
See also 14.3, p. 235; 22.1.1, p. 461.

J. Villain
*Oscillations collectives de spins et ordre à courte distance
(Collective Spin Oscillations and Short-Range Order)
Vol. **19**, No. 2, p. 767–776 (1967).
See also *18*, p. *291*.

A. M. Stoneham
*Interference Effects in Spin Lattice Relaxation by Two-Phonon Processes
Vol. **19**, No. 2, p. 787–792 (1967).
See also 6, p. 59; 22.6, p. 537.

L. I. van Torne
*Electron Charge Distribution about Zinc Ions in Aluminium
Vol **19**, No. 2, p. 855–862 (1967).
See also 10.1, p. 121; 14.1, p. 222; 21, p. 377.

F. Brouers
*The Pair Distribution Function in the R.P.A. and in the Hubbard Approximation
Vol. **19**, No. 2, p. 867–871 (1967).

R. E. Newnham and R. P. Santoro
*Magnetic and Optical Properties of Dioptase
Vol. **19**, No. 2, p. K87–K90 (1967).
See also 18, p. 291; *20*, p. *338*; 22.8, p. 549.

S. Y. Chuang and B. G. Hogg
*Positron Annihilation in Alnico and $(La_{0.7}Pb_{0.3})MnO_3$
Vol. **20**, No. 1, p. 331–335 (1967).
See also 11, p. 148; 20, p. 338; 21.1.1, p. 415; 22.8, p. 549.

E. Jäger and R. Perthel
*Some Remarks to the Crystal Field Theory of Spinel Lattices
Vol. **20**, No. 2, p. 433–441 (1967).
See also 3.2, p. 35; 18.3, p. 317.

T. N. Casselman and H. N. Spector
Interaction of Optical Phonons with Spin Waves in Ionic Crystals
Vol. **20**, No. 2, p. 605–612 (1967).
See also *6*, p. *60*; 18, p. 292; 19, p. 328; 22, p. 444.

G. Jones, G. Smith, and A. R. Beattie
*The Drifted Maxwellian Distribution Function in InSb
Vol. **20**, No. 2, p. K135–K139 (1967).
See also 14.3, p. 236; 22.2.3, p. 490.

H. Stolz
Second Order Conductivity and Nonlinear Plasma Excitation of an Isotropic Electron Gas
Vol. **21**, No. 1, p. 77–88 (1967).
See also *14*, p. *219*; 20, p. 339.

M. Järvinen and O. Inkinen
*An X-Ray Diffraction Study of Rubidium Chloride
Vol. **21**, No. 1, p. 127–135 (1967).
See also 5, p. 51; 22.5.2, p. 522.

U. Gonser and R. W. Grant
Determination of Spin Directions and Electric Field Gradient Axes in Vivianite by Polarized Recoil-Free γ-Rays
Vol. **21**, No. 1, p. 331–342 (1967).
See also *6.1*, p. *72*; 18.1, p. 296; 18.4, p. 322; 22.8, p. 550.

N. KRISTOFFEL and P. KONSIN
*Pseudo-Jahn-Teller Effect and Second Order Phase Transitions in Crystals
Vol. 21, No. 1, p. K39–K43 (1967).
See also *1.2*, p. *10*; 6, p. 61; 14.4.2, p. 267.

V. P. KALASHNIKOV
Theory of Spin-Lattice Relaxation of Conduction Electrons of Ionic Semiconductors at Extremely High Magnetic Fields
Vol. 21, No. 2, p. 775–784 (1967).
See also *6*, p. *61*; 14.4, p. 253; 18, p. 292; 22, p. 446.

G. LAUTZ und M. SCHULZ
*Schraubenförmige Dichtewellen und Oszillistor-Effekt im Elektron-Loch-Plasma des n-Germaniums
(Helical Density-Waves and Oscillistor Effect in the Electron-Hole Plasma of n-Germanium)
Vol. 21, No. 2, p. K139–K143 (1967).
See also 14.4.1, p. 259; *20*, p. *339*; 22.1.1, p. 462.

Ž. DIMITRIJEVIĆ, S. KRAŚNICKI, H. RŻANY, J. TODOROVIĆ, A. WANIC, H. CURIEN, and A. MILOJEVIĆ
*Anisotropy of Magnon Dispersion Relation in Haematite
Vol. 21, No. 2, p. K163–K166 (1967).
See also *18*, p. *292*; 22.6, p. 538.

I. LICEA
Electrical Conduction of Semiconducting Thin Films in Strong Electric Fields
Vol. 22, No. 1, p. 147–152 (1967).
See also *14.3.1*, p. *245*; 22, p. 446.

F. BROUERS
*Plasmon Satellites of Soft X-Ray Spectra
Vol. 22, No. 1, p. 213–221 (1967).
See also *20*, p. *339*; 21, p. 380; 21.2, p. 424.

V. L. BONCH-BRUEVICH
On the Theory of Electrical Domains in Hot Electron Semiconductors
Vol. 22, No. 1, p. 267–278 (1967).
See also 14.3, p. 237; *14.4.1*, p. *259*; 22, p. 446; 22.1.1, p. 462; 22.2.1, p. 484.

YU. P. IRKHIN, A. N. VOLOSHINSKII, and SH. SH. ABELSKII
On the Theory of Spontaneous Hall Effect at Low Temperatures
Vol. 22, No. 1, p. 309–318 (1967).
See also *14.1*, p. *223*; 18.2, p. 308.

E. A. KANER and V. G. SKOBOV
*On the Possibility of Existence of Quantum Electromagnetic Waves in Metals
Vol. 22, No. 1, p. 333–340 (1967).
See also 13.1 p. 179; *20*, p. *339*; 21, p. 380.

H. PUSZKARSKI
*Surface Spin Waves in Hexagonal Cobalt Thin Films
Vol. 22, No. 2, p. 355–359 (1967).
See also 13.3, p. 197; 19, p. 329; 21.1, p. 399.

A. OTTO
*Theory of Plasmon Excitation in Thin Films by Electrons of Non-Normal Incidence
Vol. 22, No. 2, p. 401–406 (1967).
See also 13.3, p. 197; 20 p. 339.

VU DINH KY
Planar Hall and Nernst Effect in Ferromagnetic Metals
Vol. **22**, No. 2, p. 729–736 (1967).
See also *14.1*, p. *223*; *15*, p. *272*; 18.2, p. 308; 21.1, p. 400; 21.1.1, p. 418.

J. B. SWAN, A. OTTO, and H. FELLENZER
*Observed Retardation Effects on the Energy of the ω_+-Surface Plasmons in Thin Aluminium Foils
Vol. **23**, No. 1, p. 171–176 (1967).
See also 13.3, p. 198; *20*, p. *340*; 21, p. 381.

H. GABRIEL
Mössbauer Spectra in the Presence of Electron Spin Relaxation (I)
Vol. **23**, No. 1, p. 195–205 (1967).
See also *6.1*, p. *72*.

A. G. SAMOILOVICH, I. S. BUDA, and I. V. DAKHOVSKII
Anisotropy of Thermomagnetic Effects in n-Ge
Vol. **23**, No. 1, p. 229–236 (1967).
See also 14.3, p. 238; *15*, p. *272*; 22.1.1, p. 463.

H. P. MÜLLER, P. THOMA, and G. VAUBEL
The Phosphorescence of Anthracene Single Crystals and Its Spectrum
Vol. **23**, No. 1, p. 253–262 (1967).
See also *20.3*, p. *367*; 22.9, p. 559.

M. PORSCH and J. RÖSELER
*Recoil Effects in the Polaron Problem
Vol. **23**, No. 1, p. 365–376 (1967).
See also 6, p. 62; 13.1, p. 181; 22, p. 447.

W. GÜTH and H. BLEHER
Avalanche Breakdown in Germanium in the Case of Optical Band-Band Excitation
Vol. **23**, No. 1, p. K13–K16 (1967).
See also *14.4.1*, p. *260*; 16, p. 280; 22.1.1, p. 463.

P. BYSZEWSKI, B. KALIŃSKA, and J. KOŁODZIEJCZAK
Interband Faraday Rotation and Ellipticity in Germanium and Gallium Arsenide
Vol. **23**, No. 1, p. K53–K56 (1967).
See also *18*, p. *293*; *20*, p. *340*; 22.1.1, p. 463; 22.2.1, p. 484.

J. KOŁODZIEJCZAK
*Free Carrier Effect on the Generation of Higher Harmonics in Semiconductors
Vol. **23**, No. 1, p. K57–K59 (1967).
See also 13.1, p. 181; 14.3, p. 238; *20*, p. *340*; 22, p. 447.

A. R. ADAMS, F. BAUMANN, and J. STUKE
Thermal Conductivity of Selenium and Tellurium Single Crystals and Phonon Drag of Tellurium
Vol. **23**, No. 1, p. K99–K104 (1967).
See also *8*, p. *83*; 15, p. 273; 22.1.3, p. 477.

B. HEINRICH, D. FRAITOVÁ, and V. KAMBERSKÝ
*The Influence of s–d Exchange on Relaxation of Magnons in Metals
Vol. **23**, No. 2, p. 501–507 (1967).
See also *19*, p. *330*; 21.1.1, p. 419.

A. E. LORD, JR.
*Coupling of Magnons and Phonons at High Frequencies
Vol. **23**, No. 2, p. 521–526 (1967).
See also 6, p. 63; 18.2, p. 308.

F. Voigt
The Orientation Dependence of the Spin Wave Damping in Ferromagnetic Insulators with Spinel-Structure
Vol. 23, No. 2, p. 607–612 (1967).
See also 18.3, p. 318; *19*, p. *330*.

J. Loos
The Nonlinear Excitation of Strongly Coupled Spin Waves
Vol. 23, No. 2, p. 721–728 (1967).
See also *19*, p. *331*.

V. L. Bonch-Bruevich
On the Problem of Static Domains in Hot Electron Semiconductors
Vol. 23, No. 2, p. 761–766 (1967).
See also 14.3, p. 238; *14.4.1*, p. *260*; 22.1.1, p. 463.

H. G. Reik and D. Heese
*The Small Polaron Problem with an Application to Optical and DC Data of Reduced Barium Titanate
Vol. 24, No. 1, p. 281–287 (1967).
See also 6, p. 64; 14.4.2, p. 267; 20.1, p. 351.

C. N. Plăviţu
On the Phonon Drag Effect Contribution to Thermomagnetic Phenomena
Vol. 24, No. 1, p. 361–372 (1967).
See also 6, p. 64; 14.3, p. 239; *15*, p. *273*; 22, p. 448.

J. Geiger
*Inelastic Electron Scattering in Thin Films at Oblique Incidence
Vol. 24, No. 2, p. 457–460 (1967).
See also 20, p. 340; 21, p. 383.

C. Hammar and P. Weissglas
*Hot Electron Polar Scattering in InSb
Vol. 24, No. 2, p. 531–534 (1967).
See also *14.3*, p. *239*; 22.2.3, p. 492.

R. E. Nettleton
*Statistical Perturbation Theory of Order–Disorder Ferroelectrics: Zeroth Approximation
Vol. 24, No. 2, p. 561–570 (1967).
See also 14.4.2, p. 268.

V. A. Besfamilnaya, V. F. Elesin, I. A. Kurova, N. N. Ormont, and V. V. Ostroborodova
On the Nature of Oscillations in Impurity Photoconductivity Spectra of Ge
Vol. 24, No. 2, p. 757–762 (1967).
See also 6, p. 64; *16*, p. *281*; 22.1.1, p. 464.

E. M. Ledovskaya
*On the Adiabatic Potential of a Complex System
Vol. 25, No. 1, p. 63–69 (1968).

J. Röseler
*A New Variational Ansatz in the Polaron Theory
Vol. 25, No. 1, p. 311–316 (1968).
See also *6*, p. *64*.

I. Licea
*Electron Temperature in Polar Semiconductors
Vol. 25, No. 1, p. 461–471 (1968).
See also *14.3*, p. *240*; 22.2, p. 481; 22.4, p. 495.

E. Jäger
*Jahn-Teller Effect at Lattice Sites of Symmetry D_{3d}
Vol. **25**, No. 1, p. K43–K46 (1968).
See also 5, p. 53; 18.3, p. 319.

M. Schulz
*Helical Density Waves and the Oscillistor Effect in the Electron Hole Plasma of the n-Type Ge (Theory)
Vol. **25**, No. 2, p. 521–530 (1968).
See also 14.4.1, p. 261; *20*, p. *340*; 22.1.1, p. 464.

G. Vertogen and W. J. Caspers
*A Remark on the Derivation of the Indirect Interaction between Nuclear or Ionic Spins
Vol. **25**, No. 2, p. 721–728 (1968).

G. Vertogen
*Ground State of a System with Ferromagnetic s–f Interaction
Vol. **25**, No. 2, p. 729–740 (1968).
See also 18.2, p. 310.

R. Fatehally, N. P. Sastry, and R. Nagarajan
*Exchange and Spin–Orbit Coupling in $KFeF_3$
Vol. **26**, No. 1, p. 91–96 (1968).
See also 18.4, p. 323; 22.5, p. 513; 22.8, p. 553.

I. Licea
*Hot Electron Mobility in Polar Semiconductors
Vol **26**, No. 1, p. 115–123 (1968).
See also *14.3*, p. *240*; 22.2.3, p. 492.

Sh. Sh. Bashkirov and G. Ya. Selyutin
Mössbauer Spectrum Shift Depending on the Degree of Spin-System Ordering
Vol. **26**, No. 1, p. 253–259 (1968).
See also *6.1*, p. *73*; 18, p. 294.

P. de Chatel and I. Szabó
*Spin Density Oscillations in a Polarized Free Electron Gas
Vol. **26**, No. 1, p. 319–325 (1968).
See also 18, p. 294.

A. E. Lord, Jr.
Sound Wave Velocity Change due to the Magnon–Phonon Interaction
Vol. **26**, No. 2, p. 717–724 (1968).
See also 6, p. 65; *7*, p. *77*; 18.2, p. 311; 22.8, p. 553.

A. Otto
*Eine neue Methode der Anregung nichtstrahlender Oberflächenplasmaschwingungen
(A New Method for Exciting Non-Radiative Surface Plasma Oscillations)
Vol. **26**, No. 2, p. K99–K101 (1968).
See also *20*, p. *341*; 21.6, p. 436.

K.-H. Müller und M. Wolf
*Magnetisierung und Spinwellensteifigkeit im Heisenbergmodell für Fe-Al-Legierungen
(Magnetization and Stiffness of Spin Waves in the Heisenberg Model for Fe-AlAlloys)
Vol. **26**, No. 2, p. K109–K111 (1968).
See also *18.2*, p. *311*; 21.1.1, p. 421.

H. Gabriel, J. Bosse, and K. Rander
Mössbauer Spectra in the Presence of Electron Spin Relaxation (II)
Vol. 27, No. 1, p. 301–312 (1968).
See also *6.1*, p. *73*.

U. Birkholz and J. Schelm
Mechanism of Electrical Conduction in β-$FeSi_2$
Vol. 27, No. 1, p. 413–425 (1968).
See also 13.1, p. 186; *14.3*, p. *241*; 15, p. 274; 22, p. 452; 22.7, p. 546.

V. K. Saxena and D. L. Bhattacharya
Effect of Electron-Lattice Interactions on the Spin-Paramagnetism of Metals
Vol. 27, No. 1, p. 427–433 (1968).
See also *18.1*, p. *297*; 21, p. 386; 21.2, p. 425.

V. N. Bogomolov and D. N. Mirlin
Optical Absorption by Polarons in Rutile (TiO_2) Single Crystals
Vol. 27, No. 1, p. 443–453 (1968).
See also *20.1*, p. *354*; 22.6, p. 541.

S. I. Masharov
Effect of Phonon Drag on the Kinetic Properties of Alloys
Vol. 27, No. 1, p. 455–461 (1968).
See also *6*, p. *65*; 8, p. 84; 14.1, p. 225; 15, p. 274; 21, p. 386.

M. I. Klinger
*On the Experimental Observation of Small Polarons in Semiconductors
Vol. 27, No. 2, p. 479–488 (1968).
See also 14.3, p. 241; 20.1, p. 354; 22.6, p. 541.

B. S. Tošić and R. B. Žakula
*Kinematic and Dynamic Interaction between Elementary Excitations in Ferromagnetics with Spin $S = 1/2$
Vol. 27, No. 2, p. 623–629 (1968).
See also *18.2*, p. *312*.

J. K. Pozhela and K. K. Repshas
Thermoelectric Force of Hot Carriers
Vol. 27, No. 2, p. 757–762 (1968).
See also 14.4.1, p. 262; *15*, p. *274*; 22.1.1, p. 465.

I. Licea
*Warm Electron Temperature and Mobility in Polar Semiconductors
Vol. 27, No. 2, p. K137–K141 (1968).
See also 13.1, p. 186; *14.3*, p. *242*; 22.2.3, p. 492.

I. Licea
*Hot Electron Temperature and Mobility in Polar Semiconductors
Vol. 27, No. 2, p. K143–K144 (1968).
See also *14.3*, p. *242*; 14.4.1, p. 263; 22.2.3, p. 492.

W. Haubenreisser
Nonlinear Two-Magnon-One-Phonon Confluence Process in BVC Garnet
Vol. 27, No. 2, p. K145–K147 (1968).
See also 6, p. 66; *18.3*, p. *320*.

M. Zvára
Interband Faraday Rotation in Gallium Arsenide
Vol. 27, No. 2, p. K157–K160 (1968).
See also 18, p. 294; *20*, p. *341*; 22.2.1, p. 485.

Yu. E. Perlin and Sh. N. Gifeisman
The Theory of Shallow F-Centres
Vol. 28, No. 1, p. 45—60 (1968).
See also 6, p. 66, *10.2*, p. *143*; 14.4, p. 254; 18, p. 294; 22.2.1, p. 485; 22.2.2, p. 488 22.2.3, p. 493; 22.4, p. 495; 22.4.1, p. 503; 22.4.2, p. 508; 22.4.3, p. 511; 22.5.1, p. 517; 22.5.2, p. 527.

J. S. Tiwari, A. Mehra, and K. G. Srivastava
Electronic Absorption Spectrum of K_2CoF_4
Vol. 28, No. 1, p. 61—63 (1968).
See also 5, p. 53; *20.1*, p. *355*; 22.5, p. 514; 22.8, p. 554.

A. I. Mitsek, N. P. Kolmakova, V. A. Feigin, and P. F. Gaidanskii
Spin-Wave Distribution Function and Thermodynamics of the Heisenberg Ferromagnetic
Vol. 28, No. 1, p. 83—96 (1968).
See also 8, p. 85; *18.2*, p. *312*.

C. Marti, R. Romestain, and R. Visocekas
Forbidden Transitions ($\Delta M = 2$) Induced by an Axial Crystal Field in Electronic Paramagnetic Resonance
Vol. 28, No. 1, p. 97—103 (1968).
See also *19*, p. *334*; 22.4.1, p. 503.

B. W. Holland
Pseudopotentials and the Knight Shift
Vol. 28, No. 1, p. 121—125 (1968).
See also *19*, p. *334*; 21, p. 386; 21.2, p. 425.

Yu. A. Sherstkov, V. I. Nepsha, V. A. Vagenin, A. E. Nikiforov, and A. I. Krotkii
Effect of Applied Electric Field on EPR Spectra of Crystal Impurity Centres
Vol. 28, No. 1, p. 269—275 (1968).
See also 10, p. 109; *19*, p. *334*; 22.6, p. 542.

H. Heinrich, G. Bauer, and D. Kasperkovitz
Spatial Dependence of Hot Carrier Oscillations in n-Type Germanium
Vol. 28, No. 1, p. K51—K53 (1968).
See also *14.4.1*, p. *263*; 22.1.1, p. 465.

L. Ya. Kobelev and O. V. Troshin
*The Gaussian Approximation for Quantum Corrections in the Statistical Model of Atoms
Vol. 28, No. 1, p. K81—K83 (1968).

H. le Gall
Quantum-Mechanical Treatment of the Transient Processes with Three-Magnon Non-Linear Transitions under Perpendicular Pumping
Vol. 28, No. 2, p. 495—505 (1968).
See also 18, p. 294; *19*, p. *334*.

J. Schnakenberg
*Polaronic Impurity Hopping Conduction
Vol. 28, No. 2, p. 623—633 (1968).
See also *6*, p. *67*; 14.3, p. 242; 22, p. 453.

D. I. Lalović, B. S. Tošić, and R. B. Žakula
*On the Problem of Kinematic Interaction in the Quantum Theory of Magnetism
Vol. 28, No. 2, p. 635—642 (1968).
See also *18.2*, p. *313*.

V. L. Bonch-Bruevich
On the Problem of Recombination Autooscillations
Vol. 28, No. 2, p. 687—693 (1968).
See also 13.4, p. 215; *14.4.1*, p. *263*; 22.1.1, p. 466.

P. Brand
*Die Unterscheidung zwischen Leitungs- und Rumpfelektronen bei der feldtheoretischen Behandlung des Vielelektronen-Problems
(Distinction between Conduction and Core Electrons in the Field Theoretical Treatment of the Many-Electron Problem)
Vol. 28, No. 2, p. 741—753 (1968).

E. Praveczki
On the Use of Bose Operators for Ferromagnets at General Temperatures
Vol. 28, No. 2, p. K105—K107 (1968).
See also *18.2*, p. *313*.

V. V. Nemoshkalenko, O. N. Rasumov, and V. V. Gorskii
Investigation of the Mössbauer Effect in Some Fe–Al Alloys
Vol. 29, No. 1, p. 45—48 (1968).
See also *6.1*, p. *74*; 21.1.1, p. 422.

M. A. Krivoglaz and L. B. Kvashnina
*The Correlation Function $S_{zz}(\omega)$ for Weakly Coupled Impurity Atom Spin in a Ferromagnetic
Vol. 29, No. 1, p. 53—59 (1968).
See also 18.2, p. 313; 19, p. 335.

M. A. Krivoglaz and L. B. Kvashnina
On the Spin Broadening of Mössbauer Lines of Impurities in Ferromagnetics and of Atoms in Antiferromagnetics
Vol. 29, No. 1, p. 61—73 (1968).
See also *6.1*, p. *74*; 18.2, p. 314; 18.4, p. 324.

A. M. van Diepen, H. W. de Wijn, and K. H. J. Buschow
Nuclear Magnetic Resonance and Susceptibility of Equiatomic Rare-Earth–Aluminium Compounds
Vol. 29, No. 1, p. 189—192 (1968).
See also 18.1, p. 297; *19*, p. *335*; 21, p. 387; 21.4, p. 428.

B. Garben and H. Seliger
Faraday-Effect on Amorphous Selenium
Vol. 29, No. 1, p. K27—K29 (1968).
See also *18*, p. *294*; *20*, p. *342*; 22.1.3, p. 480.

M. S. R. Chari and N. S. Natarajan
The Phonon–Electron Scattering Coefficients in Dilute Silver-Manganese Alloys
Vol. 29, No. 1, p. K69—K72 (1968).
See also 6, p. 68; *8*, p. *85*; 21.6, p. 437.

W. Schilz
*Anisotropy of High-Field Helicon Propagation in PbTe
Vol. 29, No. 2, p. 559—568 (1968).
See also 14.3, p. 244; *20*, p. *342*; 22, p. 454.

D. I. Bower, E. Claridge, and I. S. T. Tsong
Low-Temperature Elastic Constants and Specific Heats of F.C.C. Nickel-Iron Alloys
Vol. 29, No. 2, p. 617—625 (1968).
See also 6, p. 68; 8, p. 86; *12.1*, p. *160*; 18.2, p. 314; 21.1, p. 407; 21.1.1, p. 422.

J. E. LEWIS, H. RODOT, and P. HAEN
The Low-Temperature Thermoelectric Power and Thermal Conductivity of GeTe and of Some GeTe-MnTe Alloys
Vol. **29**, No. 2, p. 743–754 (1968).
See also 6, p. 68; *8*, p. *86*; *15*, p. *274*; 22, p. 454; 22.8, p. 555.

C. SIMOI, I. HRIANCA, and P. CRĂCIUN
Exoemission of Electrons without Photostimulation
Vol. **29**, No. 2, p. 761–766 (1968).
See also *17*, p. *290*; 21, p. 388.

G. J. MORGAN
*Green Functions for "One-Body" Equations Describing the Disordered State
Vol. **30**, No. 1, p. 247–250 (1968).
See also 2, p. 25; 6, p. 69.

F. J. SCHEDEWIE
Paramagnetic Relaxation in Neodymium-Ethylsulfate: Phonon Re-Absorption and Spectral Spin Diffusion
Vol. **30**, No. 1, p. 251–259 (1968).
See also 6, p. 69; *18.1*, p. *298*; 19, p. 336; 22.8, p. 556.

R. V. PISAREV, I. G. SINY, N. N. NESTEROVA, G. A. SMOLENSKY, and P. P. SYRNIKOV
Faraday Rotation and Magnetic Dichroism in $RbNi(Co)F_3$
Vol. **30**, No. 1, p. 367–372 (1968).
See also *18.3*, p. *320*; *20*, p. *342*; 22.5, p. 514; 22.8, p. 556.

J. HOFMANN and W. STEINMANN
*Plasma Resonance in the Photoemission of Silver
Vol. **30**, No. 1, p. K53–K56 (1968).
See also 17, p. 290; *20*, p. *342*; 21.6, p. 437.

H. W. DE WIJN, K. H. J. BUSCHOW, and A. M. VAN DIEPEN
Experimental Evidence for Interband Mixing in Rare-Earth Intermetallic Compounds
Vol. **30**, No. 2, p. 759–766 (1968).
See also 13.1, p. 190; 18.1, p. 298; *19*, p. *336*; 21, p. 389; 21.4, p. 428.

L. A. ALMASOV and I. M. DYKMAN
*Plasma Oscillations in Semiconductors with Non-Standard Dispersion Law in the Presence of Electric and Magnetic Fields
Vol. **30**, No. 2, p. 767–778 (1968).
See also *20*, p. *343*; 22.1.1, p. 467.

R. KISHORE
*Green Function Theory of Transport Properties of Anharmonic Metals
Vol. **30**, No. 2, p. 817–826 (1968).
See also *6*, p. *70*; 8, p. 86; 14.1, p. 227; 21, p. 389.

E. M. LEDOVSKAYA
*Use of Perturbation Theory in the Calculation of Adiabatic Potential
Vol. **30**, No. 2, p. 867–870 (1968).

H. BÖTTGER
On the Spin–Phonon Interaction in External Electric Field
Vol. **30**, No. 2, p. K119–K123 (1968).
See also 6, p. 70; *19*, p. *336*; 22.6, p. 544.

13.1 Band Structure. Fermi Surfaces

Review Articles

Y. P. VARSHNI
Band-to-Band Radiative Recombination in Groups IV, VI, and III–V Semiconductors (I)
Vol. **19**, No. 2, p. 459–514 (1967).
See also 13.2, p. 191; 13.4, p. 200; 20.1, p. 344; 20.2, p. 360; *20.3*, p. *363*; 22.1, p. 456; 22.1.1, p. 458; 22.1.2, p. 468; 22.1.3, p. 476; 22.2.1, p. 482; 22.2.2, p. 487; 22.2.3, p. 489.

Y. P. VARSHNI
Band-to-Band Radiative Recombination in Groups IV, VI, and III–V Semiconductors (II)
Vol. **20**, No. 1, p. 9–36 (1967).
See also 13.2, p. 191; 13.4, p. 200; 20.1, p. 344; 20.2, p. 360; *20.3*, p. *363*; 22.1, p. 456; 22.1.1, p. 458; 22.1.2, p. 468; 22.1.3, p. 476; 22.2.1, p. 482; 22.2.2, p. 487; 22.2.3, p. 489.

A. S. BORSHCHEVSKII, N. A. GORYUNOVA, F. P. KESAMANLY, and D. N. NASLEDOV
Semiconducting $A^{II}B^{IV}C_2^{V}$ Compounds
Vol. **21**, No. 1, p. 9–55 (1967).
See also 3.1, p. 29; 3.2, p. 34; 5, p. 49; 8, p. 79; 10, p. 93; *14.3*, p. *231*; 15, p. 270; 16, p. 276; 17, p. 286; 18, p. 291; 20. 1, p. 344; 22.8, p. 547.

G. HARBEKE
*On the Band Structure of Anisotropic Crystals
Vol. **27**, No. 1, p. 9–44 (1968).
See also 13.2, p. 191; 14.3, p. 231; 20.1, p. 344; 22, p. 439; 22.1, p. 456; 22.1.1, p. 458; 22.1.2, p. 468; 22.1.3, p. 476; 22.4.1, p. 497; 22.4.2, p. 506; 22.4.3, p. 509; 22.5, p. 512; 22.7, p. 545; 22.8, p. 547.

M. BALKANSKI and E. AMZALLAG
*Band Parameters Determination from Faraday Rotation Measurements
Vol. **30**, No. 2, p. 407–439 (1968).
See also 18, p. 291; 20, p. 337; 22.1.1, p. 458; 22.1.2, p. 468; 22.2.1, p. 482; 22.2.2, p. 487; 22.2.3, p. 489; 22.4, p. 495; 22.4.1, p. 497; 22.4.2, p. 506; 22.4.3, p. 509.

Original Papers and Short Notes

B. VELICKÝ and J. SAK
*Excitonic Effects in the Interband Absorption of Semiconductors
Vol. **16**, No. 1, p. 147–157 (1966).
See also *13.2*, p. *191*; 20.1, p. 344; 22, p. 439.

W. L. MCCUBBIN
*Symmetry Orbitals and UV Selection Rules for the Polymethylene Chain
Vol. **16**, No. 1, p. 289–293 (1966).
See also *3*, p. *26*; 20, p. 337; 22.9, p. 558.

J. KOŁODZIEJCZAK and S. ŻUKOTYŃSKI
*The Effect of Higher Bands on the Effective Masses in InSb and InAs
Vol. **16**, No. 1, p. K55–K58 (1966).
See also 22.2.1, p. 482; 22.2.3, p. 489.

A. G. SAMOILOVICH, M. V. NITSOVICH, and V. M. NITSOVICH
On the Theory of Anisotropic Thermoelectric Power in Semiconductors
Vol. **16**, No. 2, p. 459–465 (1966).
See also 6, p. 56; *15*, p. *270*; 22, p. 440.

J. Treusch and R. Sandrock
*Energy Band Structures of Selenium and Tellurium (Kohn-Rostoker Method)
Vol. **16**, No. 2, p. 487—497 (1966).
See also 5, p. 49; 22.1.3, p. 476.

M. V. Kurik and N. I. Vitrikhovskii
*Optical Properties and Energy Structure of CdS–CdTe Crystals
Vol. **16**, No. 2, p. K139—K142 (1966).
See also *20.1*, p. *345*; 22.4.1, p. 497; 22.4.3, p. 509; 22.8, p. 547.

H. Hora and R. Kantlehner
The Anomalous Vectorial Photoemission from a Cesium–Antimony Layer
Vol. **17**, No. 1, p. 59—66 (1966).
See also *17*, p. *286*; 22.7, p. 545.

H. J. Fischbeck
*Über die Pippardsche Theorie des magnetischen Durchbruchs in Metallen
(On Pippard's Theory of Magnetic Breakdown in Metals)
Vol. **17**, No. 1, p. 197—210 (1966).
See also 18. p. 291; 21, p. 374.

J. Stuke and G. Weiser
Franz-Keldysh Effect in Trigonal and Amorphous Selenium
Vol. **17**, No. 1, p. 343—352 (1966).
See also *14.4.1*, p. *256*; *20*, p. *337*; 22.1.3, p. 476.

A. A. Abdurakhmanova and M. I. Aliev
Electrical Properties of GaSb-Ga_2Te_3 and GaSb-GaTe Solid Solutions
Vol. **17**, No. 1, p. 375—380 (1966).
See also *14.3*, p. *232*; 15, p. 270; 22.8, p. 547.

J. Gyulai, V. K. Subashiev, and G. A. Chalikyan
On the Mechanism of Photoconductivity in High-Resistivity GaP
Vol. **17**, No. 1, p. K49—K51 (1966).
See also *16*, p. *277*; 22.2.2, p. 487.

A. S. Filipchenko, G. G. Kulikov, D. N. Nasledov, and Yu. I. Ukhanov
*Effective Mass of Holes in InSb
Vol. **17**, No. 1, p. K113—K114 (1966).
See also 20.1, p. 345; 22.2.3, p. 490.

A. F. Kravchenko and V. S. Sardaryan
*On the Structure of the Bottom of the Conduction Band in GaAs
Vol. **17**, No. 2, p. 479—488 (1966).
See also 14.3, p. 232; 22.2.1, p. 482.

G. B. Abdullaev, E. S. Guseinova, and B. G. Tagiev
Electrical Conductivity of n-InSe Single Crystals in Strong Electric Fields
Vol. **17**, No. 2, p. 593—596 (1966).
See also *14.4.1*, p. *256*; 22, p. 440.

U. Rössler and M. Lietz
*Band Structure of Cubic and Hexagonal ZnS (APW Method)
Vol. **17**, No. 2, p. 597—604 (1966).
See also 5, p. 50; 20.1, p. 346; 22.4.1, p. 498.

W. Güth and W. Haist
Electronic States of Dislocations in Semiconductors
Vol. **17**, No. 2, p. 691—696 (1966).
See also *13.4*, p. *202*; 22.1, p. 456.

J. Kołodziejczak and H. Stramska
On the Theory of Free Carrier Electro-Magneto-Optical Phenomena
Vol. **17**, No. 2, p. 701–707 (1966).
See also 14.4.1, p. 256; *18*, p. *291*; *20*, p. *337*; 22, p. 441.

C. Verié
*Sur la structure de bandes des alliages HgTe–CdTe. (I): Mesures électriques
(On the Band Structure of HgTe-CdTe. (I): Electric Measurements)
Vol. **17**, No. 2, p. 889–901 (1966).
See also 14.3, p. 233; 22.4.3, p. 509; 22.8, p. 548.

R. Lück
Temperature and Field Dependence of the Galvanomagnetic Properties of Aluminium and Indium
Vol. **18**, No. 1, p. 49–57 (1966).
See also *14.1*, p. *221*; 21, p. 375.

R. Lück
Temperature and Field Dependence of the Galvanomagnetic Properties of Lead and the Effect of Finite Sample-Thickness on These Dependences
Vol. **18**, No. 1, p. 59–66 (1966).
See also *14.1*, p. *221*; 21, p. 375.

B. M. Askerov and F. M. Gashimzade
Quantum Theory of the Electrical Conductivity of Semiconductors with a Non-Standard Energy Band
Vol. **18**, No. 2, p. 667–675 (1966).
See also 6, p. 58; *14.3*, p. *233*., 22.2, p. 481; 22.2.3, p. 490.

I. P. Molodyan, D. N. Nasledov, S. I. Radautsan, and V. G. Sidorov
*The Effective Mass of Electrons in $(InSb)_x \cdot (InTe)_{1-x}$ Crystals
Vol. **18**, No. 2, p. 677–682 (1966).
See also 14.3, p. 234; 15, p. 271; 22.8, p. 548.

G. Richter
Determination of Transport Coefficients for Minority Carriers in InSb
Vol. **18**, No. 2, p. K149–K152 (1966).
See also *14.3*, p. *234*; 16, p. 278; 22.2.3, p. 490.

J. Kołodziejczak
On the Theory of Galvanomagnetic Phenomena in Strong Electric Fields
Vol. **19**, No. 1, p. 373–384 (1967).
See also 14.4.1, p. 257; *15*, p. *271*; 22, p. 442.

B. Velický
Polaron Effect on the Optical Absorption Edge of Semiconductors
Vol. **19**, No. 1, p. K39–K42 (1967).
See also 6, p. 59; 13, p. 164; *20.1*, p. *347*; 22, p. 442.

S. Żukotyński and J. Kołodziejczak
*On the Determination of the Non-Parabolicity of Energy Bands
Vol. **19**, No. 1, p. K51–K54 (1967).
See also 14.3, p. 235; 15, p. 271; 22, p. 443.

J. Kołodziejczak and E. Kierzek-Pecold
Free Carrier Electro-Optical Kerr Effect in Semiconductors
Vol. **19**, No. 1, p. K55–K58 (1967).
See also *14.4.1*, p. *257*; *20*, p. *338*; 22, p. 443.

A. R. Beattie and G. Smith
*Recombination in Semiconductors by a Light Hole Auger Transition
Vol. **19**, No. 2, p. 577–586 (1967).
See also *16*, p. *278*; 22.2.1, p. 483; 22.2.3, p. 490.

J. Treusch
*Green's Function Method for Energy Band Calculations Including Spin-Orbit Coupling
Vol. **19**, No. 2, p. 603–607 (1967).

R. Enderlein and R. Keiper
On the Theory of Electro-Absorption and Electro-Reflectance in Semiconductors
Vol. **19**, No. 2, p. 673–681 (1967).
See also 14.4.1, p. 257; *20*, p. *338*; 22, p. 443.

E. Gutsche and E. Jahne
*Spin-Orbit Splitting of the Valence Band of Wurtzite Type Crystals
Vol. **19**, No. 2, p. 823–832 (1967).
See also 22.4, p. 495.

M. A. Nizametdinova
The Reflection Spectrum of GaSe and GaS Single Crystals near the Fundamental Absorption Edge
Vol. **19**, No. 2, p. K111–K112 (1967).
See also *20.1*, p. *347*; 22, p. 443.

R. R. Gałązka and L. Sosnowski
*Conduction Band Structure of $Cd_{0.1}Hg_{0.9}Te$
Vol. **20**, No. 1, p. 113–120 (1967).
See also 14.3, p. 235; 22.4.3, p. 509; 22.8, p. 549.

R. Enderlein
The Influence of Collisions on the Franz-Keldysh Effect
Vol. **20**, No. 1, p. 295–299 (1967).
See also 14.4.1, p. 257; *20*, p. *338*; 22, p. 444.

G. Coste
Oscillatory Thermomagnetic Effects in Lead Telluride
Vol. **20**, No. 1, p. 361–364 (1967).
See also *15*, p. *271*; 22, p. 444.

Z. Bodó and I. Hevesi
Optical Absorption near the Absorption Edge in V_2O_5 Single Crystals
Vol. **20**, No. 1, p. K45–K49 (1967).
See also *20.1*, p. *347*; 22.6, p. 537.

S. Milošević, and L. Dobrosavljević
*Evalution of the One-Electron Potential for Al Crystals
Vol. **20**, No. 1, p. K63–K66 (1967).
See also 21, p. 378.

I. V. Abarenkov and I. M. Antonova
*The Model Potential for Systems with Closed Shells
Vol. **20**, No. 2, p. 643–651 (1967).
See also 22.5.2, p. 521.

V. P. Naberezhnykh and A. A. Maryakhin
*Radio-Frequency Size Effects in Cadmium in Magnetic Fields Inclined to the Sample Surface
Vol. **20**, No. 2, p. 737–743 (1967).
See also 21, p. 378.

W. Henrion
On an Exciton Structure in the Reflection Spectrum of Trigonal Selenium Single Crystals
Vol. **20**, No. 2, p. K145–K149 (1967).
See also *13.2*, p. *192*; 20.1, p. 347; 22.1.3, p. 477.

E. Jahne and E. Gutsche
*Valence Band Structure of Wurtzite Type Crystals
Vol. **21**, No. 1, p. 57–68 (1967).
See also 22.4, p. 495; 22.4.1, p. 499; 22.4.2, p. 506.

N. Sexer
Classical Galvanomagnetic and Thermomagnetic Effects in Cadmium Arsenide
Vol. **21**, No. 1, p. 225–234 (1967).
See also *14.3*, p. *236*; 15, p. 272; 22.7, p. 545.

J. S. Ziomek and P. B. Pickar
Optical Studies of β-SiC
Vol. **21**, No. 1, p. 271–278 (1967).
See also 5, p. 51; 6, p. 61; *20.1*, p. *348*; 22.3, p. 494.

R. A. French
Type-1 Superconductivity in Molybdenum
Vol. **21**, No. 1, p. K35–K38 (1967).
See also *14.2*, p. *229*; 21, p. 379.

P. Schnupp
Kroning-Penney-Type Calculations for Electron Tunneling through Thin Dielectric Films
Vol. **21**, No. 2, p. 567–573 (1967).
See also *14.3.1*, p. *245*; 22, p. 446.

K. H. Gundlach and G. Heldmann
The Effect of the $E(k)$-Relation on Tunneling through Asymmetric Barriers
Vol. **21**, No. 2, p. 575–579 (1967).
See also *14.3.1*, p. *245*; 14.3.2, p. 249; 22.6, p. 538.

M. Averous and G. Bougnot
Effect of Donor Impurity Concentration on the Piezoresistance of n-Type GaSb
Vol. **21**, No. 2, p. 665–670 (1967).
See also 13.4, p. 207; *14.3*, p. *236*; 22.2.3, p. 491.

R. Herrmann
Oscillations of the Surface Impedance of Tungsten in Weak Magnetic Fields
Vol. **21**, No. 2, p. 703–707 (1967).
See also *14.1*, p. *223*; *19*, p. *329*; 21, p. 379.

D. M. Bercha, A. N. Borets, I. M. Stakhyra, and K. D. Tovstyuk
The Band Edge and the Energy Spectrum of In_2Se
Vol. **21**, No. 2, p. 769–774 (1967).
See also *20.1*, p. *348*; 22, p. 446.

B. M. ASKEROV and F. M. GASHIMZADE
*Determination of Parameters of Degenerate Semiconductors
Vol. **21**, No. 2, p. K155–K158 (1967).
See also 14.3, p. 237; 15, p. 272; 22, p. 446.

A. R. FERCHMIN
*Some Remarks on the Paper "Zur Theorie des Ferromagnetismus dünner Schichten" by H. Hemschik
Vol. **21**, No. 2, p. K179–K182 (1967).
See also *18.2*, p. *307*.

B. TAVGER
Transport Phenomena in Semiconducting Thin Films
Vol. **22**, No. 1, p. 31–38 (1967).
See also *14.3.1*, p. *245*; 15, p. 272; 22, p. 446.

A. YA. BLANK and E. A. KANER
Waves with Discrete Spectrum in Metals in the Vicinity of the Cyclotron Resonance Frequency
Vol. **22**, No. 1, p. 47–58 (1967).
See also *19*, p. *329*; 21, p. 380.

H. N. SPECTOR
Optical Absorption in Electric and Magnetic Fields
Vol. **22**, No. 1, p. 185–193 (1967).
See also 14.4.1, p. 259; 18, p. 292; *20.1*, p. *349*; 22, p. 446.

E. GUTSCHE and H. LANGE
Electroreflectance of CdS and CdSe Single Crystals at the Fundamental Absorption Edge
Vol. **22**, No. 1, p. 229–233 (1967).
See also 13.2, p. 192; *14.4.1*, p. *259*; *20*, p. *339*; 20.1, p. 349; 22.4.1, p. 500; 22.4.2, p. 506.

H. J. FISCHBECK
*Theory of Weakly Bound Bloch Electrons in a Magnetic Field. (I): Nearly Free Approximation
Vol. **22**, No. 1, p. 235–244 (1967).
See also 18, p. 292.

E. A. KANER and V. L. FALKO
Magnetoacoustic Resonance Effects in Metals in Inclined Magnetic Fields
Vol. **22**, No. 1, p. 319–331 (1967).
See also 7, p. 77; *19*, p. *329*; 21; p. 380.

E. A. KANER and V. G. SKOBOV
On the Possibility of Existence of Quantum Electromagnetic Waves in Metals
Vol. **22**, No. 1, p. 333–340 (1967).
See also *13*, p. *166*; *20*, p. *339*; 21, p. 380.

W. HENRION
Reflectivity Measurements on Trigonal Selenium Single Crystals in the Spectral Region between 1.6 and 6.0 eV
Vol. **22**, No. 1, p. K33–K37 (1967).
See also 13.2, p. 192; *20.1*, p. *349*; 22.1.3, p. 477.

E. MOHLER, J. STUKE, and G. ZIMMERER
Some New Features of the Reflectance Spectrum of Trigonal Selenium Single Crystals
Vol. **22**, No. 1, p. K49–K53 (1967).
See also *20.1*, p. *349*; 22.1.3, p. 477.

V. Frei
*On Energy Bands of Crystals with Lowered Point Symmetry
Vol. 22, No. 2, p. 381–390 (1967).
See also 22, p. 446.

J. O. Henningsen
*Cyclotron Resonance in Zinc
Vol. 22, No. 2, p. 441–452 (1967).
See also *19*, p. *330*; 21.1, p. 399.

D. J. Morgan and J. A. Galloway
*Energy Band Structure of Germanium
Vol. 22, No. 2, p. 491–497 (1967).
See also 22.1.1, p. 462.

H. J. Fischbeck
*Theory of Weakly Bound Bloch Electrons in a Magnetic Field. (II): Asymptotic Approximation
Vol. 22, No. 2, p. 649–658 (1967).
See also *18*, p. *292*.

H. Bross und K. H. Anthony
*Eine Verallgemeinerung der Korringa-Kohn-Rostoker-Methode zur Berechnung von Einelektronenzuständen in Kristallen
(Generalized Korringa-Kohn-Rostoker Method for Calculating Electronic States in Crystals)
Vol. 22, No. 2, p. 667–682 (1967).

A. I. Kosinskaya and S. I. Uritsky
On the Classical Theory of the Optical Properties of Free Carriers in Semiconductors
Vol. 23, No. 1, p. 57–68 (1967).
See also 18, p. 293; *20*, p. *339*; 22.1.1, p. 463.

W. Szymańska
*Conduction Band Structure of HgTe from Thermomagnetic Measurements
Vol. 23, No. 1, p. 69–73 (1967).
See also 15, p. 272; 22.4.3, p. 510.

K. D. Tovstyuk and P. I. Voronyuk
Galvanomagnetic Effects in PbTe
Vol. 23, No. 1, p. 75–91 (1967).
See also *14.3*, p. *237*; 22, p. 447.

N. I. Akimenko, P. M. Staryk, and K. D. Tovstyuk
*Magnetic Properties of PbTe
Vol. 23, No. 1, p. 93–96 (1967).
See also 14.3, p. 238; 18, p. 293; 22, p. 447.

D. J. Morgan and J. A. Galloway
*Sum Rules for Effective Masses in Energy Band Theory
Vol. 23, No. 1, p. 97–103 (1967).
See also 22.1.1, p. 463.

R. Enderlein and R. Keiper
On the Dielectric Function of a Semiconductor in an External Electric Field
Vol. 23, No. 1, p. 127–136 (1967).
See also 14.4.1, p. 259; *20*, p. *339*; 22, p. 447.

P. Eckelt
*Energy Band Structures of Cubic ZnS, ZnSe, ZnTe, and CdTe (Korringa-Kohn-Rostoker Method)
Vol. **23**, No. 1, p. 307–312 (1967).
See also 22.4.1, p. 500; 22.4.2, p. 507; 22.4.3, p. 510.

M. Porsch and J. Röseler
Recoil Effects in the Polaron Problem
Vol. **23**, No. 1, p. 365–376 (1967).
See also 6, p. 62; *13*, p. *167*; 22, p. 447.

R. Ludeke and W. Paul
Growth and Optical Properties of Wurtzite and Sphalerite CdSe Epitaxial Thin Films
Vol. **23**, No. 1, p. 413–418 (1967).
See also 1.4, p. 19; *3.1*, p. *32*; *20.1*, p. *350*; 22.4.2, p. 507.

J. Kołodziejczak
Effect of Strong Electric Fields on Magneto-Resistivity and Hall Coefficient in Semiconductors
Vol. **23**, No. 1, p. 423–427 (1967).
See also 6, p. 62; *14.3*, p. *238*; 14.4.1, p. 260; 22, p. 447; 22.2.3, p. 491.

V. G. Lyapin and K. B. Tolpygo
*Partial Account of Interelectron Correlation in the Theory of Band Structure of Crystals with Sphalerite and Diamond Lattices
Vol. **23**, No. 1, p. 429–437 (1967).
See also 22, p. 447.

F. R. Kessler and E. Sutter
Absorption and Photoconductivity of a Series of Selenium-Tellurium Mixed Crystals
Vol. **23**, No. 1, p. K25–K33 (1967).
See also *16*, p. *280*; *20.1*, p. *350*; 22.1.3, p. 477.

R. R. Gałązka and T. Zakrzewski
*Heavy Hole Effective Mass of $Cd_{0.1}Hg_{0.9}Te$
Vol. **23**, No. 1, p. K39–K43 (1967).
See also 14.3, p. 238; 15, p. 272; 22.4.3, p. 510; 22.8, p. 552.

J. Kołodziejczak
Free Carrier Effect on the Generation of Higher Harmonics in Semiconductors
Vol. **23**, No. 1, p. K57–K59 (1967).
See also *13*, p. *167*; 14.3, p. 238; *20*, p. *340*; 22, p. 447.

H. Lange and W. Henrion
Thermoreflectance of CdS and Se Single Crystals
Vol. **23**, No. 1, p. K67–K70 (1967).
See also *20.1*, p. *350*; 22.1.3, p. 477; 22.4.1, p. 501.

D. E. Aspnes
Note on Collision Broadening of Franz-Keldysh Effect
Vol. **23**, No. 1, p. K79–K81 (1967).
See also 14.4.1, p. 260; *20*, p. *340*; 22, p. 447.

W. Schülke and U. Berg
*Measurement of the Compton Band on Beryllium Single Crystals
Vol. **23**, No. 1, p. K87–K91 (1967).
See also 20, p. 340; 21.3, p. 426.

P. S. Ho and A. L. Ruoff
A Quasi-Harmonic Calculation of Lattice Dynamics for Na
Vol. **23**, No. 2, p. 489—499 (1967).
See also *6*, p. *63*; 14.1, p. 224; 21.2, p. 425.

M. Picard and M. Hulin
*A Pseudopotential Approach to the Electron Band Structure of Tellurium
Vol. **23**, No. 2, p. 563—570 (1967).
See also 22.1.3, p. 478.

R. Grigorovici, N. Croitoru, and A. Dévényi
Thermoelectric Power in Amorphous Silicon
Vol. **23**, No. 2, p. 621—625 (1967).
See also 2, p. 23; 14.3.1, p. 246; *15*, p. *273*; 22.1.2, p. 472.

M. D. Stafleu and A. R. de Vroomen
*De Haas-van Alphen Effect in White Tin
Vol. **23**, No. 2, p. 675—682 (1967).
See also 18, p. 293; 21, p. 382.

M. D. Stafleu and A. R. de Vroomen
*Fermi Surface and Pseudopotential Coefficients in White Tin
Vol. **23**, No. 2, p. 683—696 (1967).
See also 21, p. 382.

A. Stella, A. D. Brothers, R. H. Hopkins, and D. W. Lynch
*Pressure Coefficient of the Band Gap in Mg_2Si, Mg_2Ge, and Mg_2Sn
Vol. **23**, No. 2, p. 697—702 (1967).
See also 14.3, p. 238; 20.1, p. 350; 22.7, p. 545.

G. G. Kovalevskaya, D. N. Nasledov, and S. V. Slobodchikov
Photoconductivity Oscillations in InP
Vol. **23**, No. 2, p. 755—760 (1967).
See also *6*, p. *63*; *16*, p. *280*; 22.2.2, p. 488.

A. A. Maryakhin and I. V. Svechkarev
*On the Pseudopotential Coefficients and Spin–Orbit Coupling in Metallic Cadmium
Vol. **23**, No. 2, p. K133—K136 (1967).
See also 21, p. 382.

J. Antula
*A Method of Measuring Effective Electron Mass in Thin Insulating Films
Vol. **24**, No. 1, p. 95—98 (1967).
See also 14.3.2, p. 249; 14.4.1, p. 260; 22.6, p. 539.

H. Mell and J. Stuke
Magnetoconductivity of Trigonal Selenium Single Crystals
Vol. **24**, No. 1, p. 183—193 (1967).
See also *14.3*, p. *238*; 22.1.3, p. 478.

K. H. Herrmann and G. Oelgart
Piezoresistance in Tellurium. (I): Experiments with Uniaxial Stress Parallel to the *c*-Axis
Vol. **24**, No. 1, p. 289—296 (1967).
See also *14.3*, p. *238*; 22.1.3, p. 478.

B. K. Jones
Measurement of the Magnitude of the Electronic Attenuation of Sound in Metals
Vol. **24**, No. 1, p. 297—300 (1967).
See also *7*, p. *77*; 21, p. 382; 21.1, p. 401; 21.2, p. 425.

J. Kołodziejczak
Magnetic Resonance in Higher Harmonics Generated by Free Carriers in Semiconductors
Vol. **24**, No. 1, p. 323–329 (1967).
See also 14.3, p. 238; *19*, p. *331*; 22, p. 448.

P. Guetin
*Note on the Density-of-State Effective Mass in GaAs
Vol. **24**, No. 1, p. K29–K31 (1967).
See also 22.2.1, p. 484.

P. Grosse and M. Selders
*The p-Band of Tellurium at Very Low Temperatures
Vol. **24**, No. 1, p. K33–K36 (1967).
See also 20.1, p. 351; 22.1.3, p. 478.

S. A. Nemnonov and E. Z. Kurmaev
*Band Structure and Superconductivity of A_3B-Type Intermetallic Compounds with β-W Structure
Vol. **24**, No. 1, p. K43–K47 (1967).
See also *14.2*, p. *229*; 21, p. 383.

U. K. Mishra and A. K. Walton
*Infra-Red Faraday Effect in p-Type InAs and Si
Vol. **24**, No. 1, p. K87–K90 (1967).
See also 18, p. 293; 20, p. 340; 22.1.2, p. 472; 22.2.1, p. 485.

B. Pistoulet, J. L. Robert, and M. Marques
*Study of GaSb Valence Band by Magnetoresistance and Hall Effect Measurements at Low Temperature
Vol. **24**, No. 2, p. 481–486 (1967).
See also 14.3, p. 239; 22.2.3, p. 491.

B. Hejda
*The Symmetrized Combinations of Plane Waves of Wurtzite
Vol. **24**, No. 2, p. 659–667 (1967).

J. Sak
*Temperature Dependence of the Band Structure
Vol. **25**, No. 1, p. 155–161 (1968).
See also 8, p. 84.

L. Śniadower and R. Kowalczyk
Magnetoplasma Reflection in Mercury Telluride
Vol. **25**, No. 1, p. 233–240 (1968).
See also *18*, p. *293*; *20*, p. *340*; 20.1, p. 351; 22.4.3, p. 510.

R. Herrmann
*Untersuchung der Fermifläche des Molybdän durch Zyklotronresonanzmessungen (Cyclotron Resonance Study of the Fermi Surface of Molybdenum)
Vol. **25**, No. 1, p. 427–435 (1968).
See also 19, p. 331; 21, p. 384.

P. Schnupp
A Model for Tunneling through a Non-Crystalline Thin Dielectric Film
Vol. **25**, No. 1, p. 455–460 (1968).
See also *14.3.1*, p. *246*; 22, p. 449.

G. JUNKG und H. MENNIGER
*Zum Einfluß der Energiebandstruktur auf Photoeffekte in Silizium
(Influence of Energy Band Structure on Photoeffects in Silicon)
Vol. **25**, No. 1, p. K59—K63 (1968).
See also 16, p. 281; 22.1.2, p. 473.

R. HERRMANN und KARIN HERRMANN
*Zyklotronresonanz im Tellur
(Cyclotron Resonance in Tellurium)
Vol. **25**, No. 2, p. 655—659 (1968).
See also 19, p. 332; 22.1.3, p. 478.

R. HERRMANN
*Zur Fermifläche von Wolfram und Molybdän
(Fermi Surface of Tungsten and Molybdenum)
Vol. **25**, No. 2, p. 661—666 (1968).
See also 19, p. 332; 21, p. 384.

H. P. GESERICH
*Optische und elektrische Messungen an dünnen Thallium(III)-Oxydschichten
(Optical and Electrical Measurements on Thallium(III)Oxide Films)
Vol. **25**, No. 2, p. 741—751 (1968).
See also 14.3.1, p. 246; 20.1, p. 352; 22.6, p. 540.

F. C. GREISEN
*A Systematic Numerical Error in APW Calculations
Vol. **25**, No. 2, p. 753—756 (1968).
See also 21, p. 384; 21.1, p. 403.

J. G. BROERMAN
*Approximate Symmetry Relations for Double Group $\boldsymbol{k} \cdot \boldsymbol{p}$ Expansions
Vol. **25**, No. 2, p. 757—762 (1968).
See also 22.1.1, p. 464.

V. V. NEMOSHKALENKO and M. A. MINDLINA
On the Form of the $K\beta_{2,5}$-Zinc Band
Vol. **25**, No. 2, p. K83—K86 (1968).
See also *20*, p. *340*; 21.1, p. 403.

P. GROSSE und K. WINZER
*Magnetoabsorptionsmessungen an Tellur
(Magnetoabsorption Masurements on Tellurium)
Vol. **26**, No. 1, p. 139—150 (1968).
See also 18, p. 293; *20.1*, p. *352*; 22.1.3, p. 478.

B. KRAMER and P. THOMAS
*Relativistic KKR-Calculations on the Band Structure of Selenium and Tellurium
Vol. **26**, No. 1, p. 151—158 (1968).
See also 22.1.3, p. 479.

W. THIELEMANN
Effect of Doping on the Optical Absorption of Circularly Polarized Waves in n-Type GaAs in a Magnetic Field
Vol. **26**, No. 1, p. K29—K31 (1968).
See also 18, p. 294; *20.1*, p. *352*; 22.2.1, p. 485.

G. A. SIKHARULIDZE and YU. I. UKHANOV
Absorption of Polarized Light in $CdSnAs_2$ Crystals
Vol. **26**, No. 1, p. K33—K36 (1968).
See also *20.1*, p. *353*; 22.7, p. 546; 22.8, p. 553.

R. F. Egorov, B. I. Reser, and V. P. Shirokovskii
*Consistent Treatment of Symmetry in the Tight Binding Approximation
Vol. **26**, No. 2, p. 391–408 (1968).
See also 3, p. 27.

H. Overhof and U. Rössler
Magnetic Space Groups and Their Irreducible Representations
Vol. **26**, No. 2, p. 461–468 (1968).
See also *3*, p. *27*; 18, p. 294.

R. Enderlein
Excitonic Effects in the Electro-Absorption of Semiconductors
Vol. **26**, No. 2, p. 509–517 (1968).
See also 13.2, p. 193; *14.4.1*, p. *262*; *20*, p. *340*; 20.1, p. 353; 22, p. 451.

G. Döhler and K. Hacker
*A Possibility for Band Structure Determination: Intraband and Donor-Band Transitions of Crystal Electrons in High Electric Fields
Vol. **26**, No. 2, p. 551–564 (1968).
See also 13.4, p. 212; 14.4.1, p. 262; 20.1, p. 353; 22, p. 451.

J. Pastrňák and L. Roskovcová
Optical Absorption Edge of AlN Single Crystals
Vol. **26**, No. 2, p. 591–597 (1968).
See also *20.1*, p. *353*; 22.2, p. 481.

G. Vaubel and H. Baessler
*Determination of the Energy of Conduction States in Anthracene Crystals by Photoemission of Electrons from Sodium
Vol. **26**, No. 2, p. 599–606 (1968).
See also 17, p. 288; 22.9, p. 560.

H. Baessler, N. Riehl, and G. Vaubel
*Calculation of the Width of the Second Conduction Band in Anthracene Crystals from Photoemission Data
Vol. **26**, No. 2, p. 607–610 (1968).
See also 17, p. 289; 22.9, p. 560.

L. Rothkirch
Interpretation of the Anisotropy of the Galvanomagnetic Effects in Tellurium by Anisotropic Scattering
Vol. **26**, No. 2, p. 709–715 (1968).
See also *14.3*, p. *241*; 22.1.3, p. 479.

A. R. Ferchmin
*Remarks to the Theory of Ferromagnetism of Thin Films
Vol. **26**, No. 2, p. K107–K108 (1968).
See also *18.2*, p. *311*.

E. Schmidt
Optical Properties of GeSi Alloys in the Energy Region from 1 to 13 eV
Vol. **27**, No. 1, p. 57–67 (1968).
See also *20.1*, p. *353*; 22.1.1, p. 464; 22.1.2, p. 473.

L. I. Anatychuk, V. D. Iskra, O. J. Luste, I. M. Rarenko, and L. I. Zarubin
Anisotropy of Electroductivity in CdSb
Vol. **27**, No. 1, p. 101–107 (1968).
See also *14.3*, p. *241*; 22, p. 451; 22.7, p. 546.

M. M. Akselrod, K. M. Demchuk, I. M. Tsidilkovski, E. L. Broyda, and K. P. Rodionov

*The Effect of Pressure on the Electron Effective Mass in InSb
Vol. **27**, No. 1, p. 249–254 (1968).
See also 14.3, p. 241; 15, p. 274; 22.2.3, p. 492.

M. Grynberg

*Influence of Uniaxial Stress on the Optical Properties of CdSe
Vol. **27**, No. 1, p. 255–261 (1968).
See also 20.1, p. 354; 20.3, p. 369; 22.4.2, p. 507.

H. J. Fischbeck and J. Mertsching

The Theory of Low-Field Oscillations of the Surface Impedance in Metals
Vol. **27**, No. 1, p. 345–357 (1968).
See also 14.1, p. 225; *19*, p. *333*; 21, p. 386.

M. Porsch

Green's Function Calculation of Electro-Optical Effects in Semiconductors. (I): General Formalism and Interband Transitions
Vol. **27**, No. 1, p. 359–370 (1968).
See also 14.4.1, p. 262; *20*, p. *341*; 22, p. 451.

U. Birkholz and J. Schelm

Mechanism of Electrical Conduction in β-$FeSi_2$
Vol. **27**, No. 1, p. 413–425 (1968).
See also 13, p. 170; *14.3*, p. *241*; 15, p. 274; 22, p. 452; 22.7, p. 546.

H. Stramska, Z. Bachan, P. Byszewski, and J. Kołodziejczak

Interband Faraday Rotation and Ellipticity Observed at the Absorption Edge in Silicon
Vol. **27**, No. 1, p. K25–K28 (1968).
See also *18*, p. *294*; *20*, p. *341*; 22.1.2, p. 473.

M. Ya. Valakh, N. I. Vitrikhovskii, and M. P. Lisitsa

*The Electron Effective Mass in CdS_xSe_{1-x} Crystals
Vol. **27**, No. 1, p. K81–K83 (1968).
See also 20.1, p. 354; 22.4.1, p. 502; 22.4.2, p. 507; 22.8, p. 554.

J. Sak

On the Theory of Shallow Impurity States
Vol. **27**, No. 2, p. 521–527 (1968).
See also 6, p. 66; *13.4*, p. *213*; 22, p. 452.

G. Appelt

Fine Structure Measurements in the Energy Angular Distribution of Secondary Electrons from a (110) Face of Copper
Vol. **27**, No. 2, p. 657–669 (1968).
See also *17*, p. *289*; 20.1, p. 354; 21.1, p. 405.

I. Licea

Warm Electron Temperature and Mobility in Polar Semiconductors
Vol. **27**, No. 2, p. K137–K141 (1968).
See also *13*, p. *170*; *14.3*, p. *242*; 22.2.3, p. 492.

I. Ivanov-Omskii, F. P. Kesamanly, B. T. Kolomiets, A. Sh. Mekhtiev, and V. A. Skripkin

*Density of States Effective Mass of Holes in HgTe
Vol. **27**, No. 2, p. K169–K172 (1968).
See also 15, p. 274; 22.4.3, p. 511.

A. A. Lipnik and M. A. Ruvinskii
Attenuation and Amplification of Hypersound by Mott Excitons in Semiconductors with Piezoelectric and Deformational Electron–Phonon Interactions
Vol. 28, No. 1, p. 75–81 (1968).
See also 6, p. 66; *7*, p. *78*; 13.2, p. 194; 22, p. 452.

M. Ali, M. Fridman, M. Denayer, and P. Nagels
Electrical Conduction in Li-Doped MnO Single Crystals
Vol. 28, No. 1, p. 193–206 (1968).
See also 13.4, p. 214; *14.3*, p. *242*; 15, p. 274; 22.6, p. 542.

M. Owner-Petersen and M. R. Samuelsen
Cyclotron Resonance of Holes in Si
Vol. 28, No. 1, p. 211–222 (1968).
See also *19*, p. *334*; 22.1.2, p. 474.

M. V. Andriyashik, M. Yu. Sakhnovskii, V. B. Timofeev, and A. S. Yakimova
Optical Transitions in the Spectra of the Fundamental Absorption and Reflection of InSe Single Crystals
Vol. 28, No. 1, p. 277–285 (1968).
See also 13.2, p. 194; *20.1*, p. *355*; 22, p. 452.

L. K. Thomas
The Normal Spectral Emissivity of Tungsten–Molybdenum Alloys
Vol. 28, No. 1, p. 401–406 (1968).
See also *20.3*, p. *370*; 21, p. 386.

V. V. Nemoshkalenko and V. V. Gorskii
Investigation of the Valence Electron Distribution in Mg–Al Alloys
Vol. 28, No. 1, p. K15–K17 (1968).
See also *20*, p. *341*; 21, p. 387; 21.3, p. 427.

C. Muntoni and A. Serpi
On an Allotropic Transformation of ZnS after Pressure Treatment
Vol. 28, No. 1, p. K43–K45 (1968).
See also *1.2*, p. *12*; 20.1, p. 356; 22.4.1, p. 503.

I. Alstrup and K. Johansen
*Empirical Third Neighbour LCAO Energy Bands of Silicon
Vol. 28, No. 2, p. 555–564 (1968).
See also 22.1.2, p. 474.

K. Maschke and U. Rössler
*The Electronic Structure of CdO. (I): The Energy-Band Structure (APW Method)
Vol. 28, No. 2, p. 577–581 (1968).
See also 22.4, p. 496.

B. M. Askerov and F. M. Gashimzade
The Transverse Magnetic Resistance in Semiconductors with Non-Standard Energy Band
Vol. 28, No. 2, p. 783–788 (1968).
See also *14.3*, p. *243*; 22, p. 453.

V. K. Nikulin and M. B. Trzhaskovskaya
The Theory of Cohesion of Noble Metals (II)
Vol. 28, No. 2, p. 801–806 (1968).
See also *3.2*, p. *36*; 21.1, p. 406; 21.6, p. 437.

T. A. KUDYKINA, K. B. TOLPYGO, and S. M. ZUBKOVA
Exciton Theory in Crystals with Complicated Band Structure with Electron–Phonon Interaction
Vol. **28**, No. 2, p. 807–814 (1968).
See also *13.2*, p. *194*; 22, p. 453; 22.5.2, p. 528.

S. BALLARÒ, A. BALZAROTTI, and V. GRASSO
Electroreflectance at Oblique Incidence in Ge
Vol. **28**, No. 2, p. K109–K112 (1968).
See also *14.4.1*, p. *264*; *20*, p. *341*; 20.1, p. 356; 22.1.1, p. 466.

L. ŚNIADOWER
Temperature Studies of Magnetoplasma Reflection in Mercury Telluride
Vol. **28**, No. 2, p. K125–K130 (1968).
See also *18*, p. *294*; *20.1*, p. *356*; 22.4.3, p. 511.

S. N. SHESTATSKII and V. V. SOBOLEV
Electroreflectance Spectrum of Cuprous Oxide Crystals
Vol. **28**, No. 2, p. K131–K133 (1968).
See also 13.2, p. 194; *14.4.1*, p. *264*; *20*, p. *342*; 20.1, p. 356; 22.6, p. 542.

K. SCHRÖDER and N. P. BAUM
Absolute Thermoelectric Power and Electrical Resistivity of Chromium-Rich Chromium–Nickel Alloys
Vol. **29**, No. 1, p. 107–114 (1968).
See also *14.1*, p. *226*; *15*, p. *274*; 18.4, p. 324; 21.1, p. 407.

A. B. KUNZ
*Electronic Energy Bands for Rubidium Chloride and the Face-Centered Cubic Alkali Bromides
Vol. **29**, No. 1, p. 115–120 (1968).
See also 22.5.2, p. 528.

P. B. PICKAR and H. D. TILLER
Optical Energy Gap in TlSe'
Vol. **29**, No. 1, p. 153–158 (1968).
See also 5, p. 54; 16, p. 283; *20.1*, p. *356*; 22, p. 453.

K. H. HERRMANN and J. PUHLMANN
Piezoresistance of Tellurium. (II): Experiments with Pressures Lowering Symmetry
Vol. **29**, No. 1, p. 193–202 (1968).
See also *14.3*, p. *243*; 22.1.3, p. 480.

M. ALTWEIN, H. FINKENRATH, Č. KOŇÁK, J. STUKE, and G. ZIMMERER
The Electronic Structure of CdO. (II): Spectral Distribution of Optical Constants
Vol. **29**, No. 1, p. 203–209 (1968).
See also 14.3.1, p. 247; *20.1*, p. *356*; 22.4, p. 496.

G. M. DE'MUNARI, F. GIUSIANO, and G. MAMBRIANI
A Phase Transition and the Photoemissive Quantum Yield at Low Temperatures in Cs_3Sb Films
Vol. **29**, No. 1, p. 341–356 (1968).
See also *1.2*, p. *12*; 3, p. 27; 17, p. 289; 22.7, p. 546.

V. P. NABEREZHNYKH
*Radio-Frequency Size Effect on Electrons in Boundary Cross-Sections Having Extremum Displacement during a Cyclotron Period
Vol. **29**, No. 1, p. K23–K26 (1968).
See also 21, p. 387.

K. Platzöder
Temperature Effects on the Vacuum Ultraviolet Reflectance of α-Quartz
Vol. **29**, No. 1, p. K63–K64 (1968).
See also 13.2, p. 195; *20.1*, p. *357*; 22.6, p. 543.

A. R. Ruffa
*The Valence Bond Approximation in Crystals — Application to an Analysis of the Ultraviolet Spectrum in Quartz
Vol. **29**, No. 2, p. 605–616 (1968).
See also 3.2, p. 36; 20.1, p. 357; 22.6, p. 543.

J. Kołodziejczak
Nonlinear Magnetooptical Phenomena in Semiconductors (I)
Vol. **29**, No. 2, p. 645–652 (1968).
See also *18*, p. *295*; *20*, p. *342*; 22, p. 454.

M. Averous, G. Bougnot, and J. Calas
*Conduction Bands of GaSb-InSb Mixed Crystals with a Low Concentration of InSb
Vol. **29**, No. 2, p. 807–812 (1968).
See also 14.3, p. 244; 22.2.3, p. 493; 22.8, p. 556.

E. Z. Dziuba, D. Niculescu, and N. Niculescu
*Energy Band Structure of $Zn_xHg_{1-x}Te$ Solid Solutions
Vol. **29**, No. 2, p. 813–818 (1968).
See also 14.3, p. 244; 20.1, p. 357; 22.4.3, p. 511; 22.8, p. 556.

E. Kierzek-Pecold, J. Kołodziejczak, and I. Pracka
Optical Constants of Crystalline CaB_6
Vol. **29**, No. 2, p. K183–K185 (1968).
See also *20.1*, p. *357*; 22, p. 454.

Yu. G. Litvinenko, V. V. Eremenko, and T. I. Garber
Antiferromagnetic Ordering Effect on the Light Absorption Spectrum by Crystalline Oxygen
Vol. **30**, No. 1, p. 49–60 (1968).
See also *18.4*, p. *325*; 20.1, p. 358; 23, p. 561.

V. V. Tarassov and A. F. Demidenko
Heat Capacity and Quasi-Chain Dynamics of Diamond-Like Structures
Vol. **30**, No. 1, p. 147–155 (1968).
See also *6*, p. *69*; *8*, p. *86*; 22.1.1, p. 466; 22.1.2, p. 474; 22.2.1, p. 486; 22.2.2, p. 488; 22.2.3, p. 493.

V. V. Sobolev, S. D. Shutov, Yu. V. Popov, and S. N. Shestatskii
Reflectivity Spectra of the Rhombohedral Crystals Bi_2Te_3, Bi_2Se_3, and Sb_2Te_3 over the Range from 0.7 to 12.5 eV
Vol. **30**, No. 1, p. 349–355 (1968).
See also *20.1*, p. *358*; 22, p. 454.

D. V. Chepur, D. M. Bercha, I. D. Turyanitsa, and V. Yu. Slivka
*Peculiarities of the Energy Spectrum and Edge Absorption in the Chain Compounds $A^VB^{VI}C^{VII}$
Vol. **30**, No. 2, p. 461–468 (1968).
See also 16, p. 284; 20.1, p. 358; 22.8, p. 557.

V. V. Nemoshkalenko, M. A. Mindlina, and B. P. Mamko
X-Ray Spectra and Electronic Structure of Elements at the End of the First Long Period from ^{29}Cu to ^{34}Se
Vol. **30**, No. 2, p. 703–712 (1968).
See also *20*, p. *343*; 21.1, p. 409; 22.1.1, p. 467; 22.1.3, p. 480.

J. Czerwonko
*On the Thermodynamics and Paramagnetic Properties of Dispersed Metals
Vol. **30**, No. 2, p. 723—734 (1968).
See also 18.1, p. 298; 21, p. 389; 21.1, p. 409; 21.6, p. 437.

H. W. de Wijn, K. H. J. Buschow, and A. M. van Diepen
Experimental Evidence for Interband Mixing in Rare-Earth Intermetallic Compounds
Vol. **30**, No. 2, p. 759—766 (1968).
See also 13, p. 173; 18.1, p. 298; *19*, p. *336*; 21, p. 389; 21.4, p. 428.

H. J. Fischbeck
*General Theory of the Bloch Electron in a Magnetic Field
Vol. **30**, No. 2, p. 779—789 (1968).
See also 18, p. 295.

K. H. Gundlach
Note on the Effect of the Band Structure on Tunnelling in Metal–Semiconductor–Metal Junctions
Vol. **30**, No. 2, p. K135—K138 (1968).
See also 14.3.1, p. 247; *14.3.2*, p. *251*; 22, p. 455.

E. M. Averbakh, V. V. Sobolev, N. N. Syrbu, and Ya. A. Ugai
The Optical Reflectivity Spectra of CdSb and ZnSb Single Crystals
Vol. **30**, No. 2, p. K145—K147 (1968).
See also *20.1*, p. *359*; 22.7, p. 546.

E. Cruceanu, St. Slădaru, and T. Boţilă
Optical Energy Gap of Tl_2Te_3
Vol. **30**, No. 2, p. K149—K151 (1968).
See also *20.1*, p. *359*; 22, p. 455.

13.2 Excitons

Review Articles

Y. P. VARSHNI
Band-to-Band Radiative Recombination in Groups IV, VI, and III–V Semiconductors (I)
Vol. **19**, No. 2, p. 459–514 (1967).
See also 13.1, p. 174; 13.4, p. 200; 20.1, p. 344; 20.2, p. 360; *20.3*, p. *363*; 22.1, p. 456; 22.1.1, p. 458; 22.1.2, p. 468; 22.1.3, p. 476; 22.2.1, p. 482; 22.2.2, p. 487; 22.2.3, p. 489.

Y. P. VARSHNI
Band-to-Band Radiative Recombination in Groups IV, VI, and III–V Semiconductors (II)
Vol. **20**, No. 1, p. 9–36 (1967).
See also 13.1, p. 174; 13.4, p. 200; 20.1, p. 344; 20.2, p. 360; *20.3*, p. *363*; 22.1, p. 456; 22.1.1, p. 458; 22.1.2, p. 468; 22.1.3, p. 476; 22.2.1, p. 482; 22.2.2, p. 487; 22.2.3, p. 489.

G. HARBEKE
On the Band Structure of Anisotropic Crystals
Vol. **27**, No. 1, p. 9–44 (1968).
See also *13.1*, p. *174*; 14.3, p. 231; 20.1, p. 344; 22, p. 439; 22.1, p. 456; 22.1.1, p. 458; 22.1.2, p. 468; 22.1.3, p. 476; 22.4.1, p. 497; 22.4.2, p. 506; 22.4.3, p. 509; 22.5, p. 512; 22.7, p. 545; 22.8, p. 547.

V. L. BONCH-BRUEVICH and E. G. LANDSBERG
Recombination Mechanisms
Vol. **29**, No. 1, p. 9–43 (1968).
See also 6, p. 56; 13, p. 162; *13.4*, p. *200*; *14.3*, p. *231*; 20, p. 337; 20.3, p. 363; 22.1.1, p. 458; 22.1.2, p. 468.

Original Papers and Short Notes

B. VELICKÝ and J. SAK
*Excitonic Effects in the Interband Absorption of Semiconductors
Vol. **16**, No. 1, p. 147–157 (1966).
See also *13.1*, p. *174*; 20.1, p. 344; 22, p. 439.

G. GALLUS and H. C. WOLF
*Direct Measurement of the Diffusion Length of Singulet Excitons in Solid Phenanthrene
Vol. **16**, No. 1, p. 277–280 (1966).
See also 20.3, p. 363; 22.9, p. 558.

A. I. BOBRYSHEVA
*The Interaction of Two Excitons in a Crystal
Vol. **16**, No. 1, p. 337–348 (1966).
See also 6, p. 56; 22, p. 439.

S. A. MOSKALENKO and P. I. KHADSHI
*Infrared Absorption by Excitons due to Photoionization and Intraband Lattice Scattering
Vol. **18**, No. 1, p. 379–390 (1966).
See also 6, p. 57; 20.1, p. 346; 22.1.1, p. 460; 22.2.2, p. 487; 22.2.3, p. 490; 22.6, p. 536.

K. F. LIDER, B. V. NOVIKOV, and S. A. PERMOGOROV
*Application of Bound-Exciton Optical Spectra in the Study of Radiation Damage in Crystals
Vol. **18**, No. 1, p. K1–K4 (1966).
See also 11, p. 147; *20.3*, p. *364*; 22.4.1, p. 499.

B. S. Razbirin, I. Filiński, and B. Wojtowicz-Natanson
*Hot Excitons in Cadmium Selenide
Vol. **18**, No. 1, p. K37–K40 (1966).
See also 20.3, p. 364; 22.4.2, p. 506.

U. Gerhardt and E. Mohler
*Piezooptical Experiments on the Excitons in KBr and KI
Vol. **18**, No. 1, p. K45–K48 (1966).
See also 20.1, p. 346; 22.5.2, p. 519.

A. S. Davydov
*On the Operator of Exciton–Phonon Interaction
Vol. **20**, No. 1, p. 143–151 (1967).
See also *6*, p. *59*; 22, p. 443.

A. S. Davydov and E. N. Myasnikov
*Absorption and Dispersion of Light by Molecular Excitons
Vol. **20**, No. 1, p. 153–163 (1967).
See also *20.1*, p. *347*; 22, p. 443.

J. Spyridelis, J. Stoimenos, and N. Economou
Optical and Photoconductive Phenomena in Cuprous Oxide
Vol. **20**, No. 2, p. 623–628 (1967).
See also 16, p. 279; *20.1*, p. *347*; 22.6, p. 537.

V. D. Egorov
Luminescence of CdS Excited by Electrons of Various Energies
Vol. **20**, No. 2, p. 705–711 (1967).
See also 20.2, p. 361; *20.3*, p. *365*; 22.4.1, p. 499.

V. M. Asnin and A. A. Rogachev
*Exciton Absorption in Doped Germanium
Vol. **20**, No. 2, p. 755–757 (1967).
See also *20.1*, p. *347*; 22.1.1, p. 461.

W. Henrion
*On an Exciton Structure in the Reflection Spectrum of Trigonal Selenium Single Crystals
Vol. **20**, No. 2, p. K145–K149 (1967).
See also 13.1, p. 178; 20.1, p. 347; 22.1.3, p. 477.

E. Majerníková
Faraday Effect and Spin–Orbit Interaction of the Lowest Exciton States in Molecular Crystals
Vol. **22**, No. 1, p. 113–122 (1967).
See also *18*, p. *292*; *20*, p. *339*; 22.9, p. 559.

E. Gutsche and H. Lange
Electroreflectance of CdS and CdSe Single Crystals at the Fundamental Absorption Edge
Vol. **22**, No. 1, p. 229–233 (1967).
See also 13.1, p. 179; *14.4.1*, p. *259*; *20*, p. *339*; 20.1, p. 349; 22.4.1, p. 500; 22.4.2, p. 506.

W. Henrion
Reflectivity Measurements on Trigonal Selenium Single Crystals in the Spectral Region between 1.6 and 6.0 eV
Vol. **22**, No. 1, p. K33–K37 (1967).
See also 13.1, p. 179; *20.1*, p. *349*; 22.1.3, p. 477.

H. L. Jetter and H. C. Wolf
On the Davydov Splitting of the First Excited Singulet-State of Anthracene Crystals
Vol. **22**, No. 1, p. K39–K43 (1967).
See also 12, p. 157; *20.1*, p. *349*; 22.9, p. 559.

J. B. Grun, A. Mysyrowicz, F. Raga, A. Bivas, R. Levy, and S. Nikitine
Photon-Assisted Emission in CdSe
Vol. **22**, No. 2, p. K155–K156 (1967).
See also 6, p. 62; *20.3*, p. *367*; 22.4.2, p. 507.

S. Wang
Optical Absorption of an Interstitial Alkali-Earth Cation in an Alkali-Halide Crystal
Vol. **23**, No. 1, p. 387–399 (1967).
See also 10.2, p. 140; *13.4*, p. *209*; 22.5.2, p. 523.

J.-Y. Roncin and K. Moorjani
On the Absorption Spectrum of Gaseous and Solid Xenon
Vol. **23**, No. 1, p. K1–K3 (1967).
See also *20.1*, p. *350*; 23, p. 561.

D. Haarer, D. Schmid und H. C. Wolf
*Elektronenspin-Resonanz von Triplett-Exzitonen in Anthrazen (Electron Spin Resonance of Triplet Excitons in Anthracene)
Vol. **23**, No. 2, p. 633–638 (1967).
See also *19*, p. *330*; 22.9, p. 559.

V. D. Egorov, G. O. Müller, and H. H. Weber
Cathodoluminescence Yields in Highly Excited CdS
Vol. **23**, No. 2, p. 703–708 (1967).
See also 20.2, p. 361; *20.3*, p. *367*; 22.4.1, p. 501.

J. Auleytner, E. F. Gross, A. A. Kralina, and B. S. Razbirin
*Investigation of the Real Structure of CdS Crystals in Connection with the Shift of Exciton Lines
Vol. **24**, No. 2, p. K107–K110 (1967).
See also *10*, p. *103*; 20.1, p. 351; 22.4.1, p. 501.

L. Ghiţă and E. Ţenescu
On the Exciton Spectrum of PbI_2 on Different Substrates
Vol. **24**, No. 2, p. K175–K178 (1967).
See also *20.1*, p. *351*; 22.5, p. 513.

B. L. Evans and P. A. Young
*Delocalized Excitons in Thin Anisotropic Crystals
Vol. **25**, No. 1, p. 417–425 (1968).
See also 20.1, p. 351; 22, p. 449.

P. G. Harper and J. A. Hilder
*Exciton Spectra in Thin Crystals
Vol. **26**, No. 1, p. 69–76 (1968).
See also 20.1, p. 352; 22, p. 450.

R. Enderlein
Excitonic Effects in the Electro-Absorption of Semiconductors
Vol. **26**, No. 2, p. 509–517 (1968).
See also 13.1, p. 185; *14.4.1*, p. *262*; *20*, p. *340*; 20.1, p. 353; 22, p. 451.

L. N. Kurbatov, E. F. Gross, V. E. Mashenko, B. S. Razbirin, N. N. Mochalkin, and A. D. Britov
*Coherent Emission of Free Excitons in CdSe Crystals
Vol. **26**, No. 2, p. K139–K142 (1968).
See also 6, p. 65; *20.2*, p. *361*; 22.4.2, p. 507.

P. Devaux and M. Schott
Field Effect on Photocarriers in Copper Phthalocyanine Single Crystals
Vol. **27**, No. 1, p. 201–208 (1968).
See also *13.3*, p. *198*; *16*, p. *282*; 22.9, p. 560.

V. M. Agranovich and Yu. V. Konobeev
*Diffusion of Free Excitons in Molecular Crystals
Vol. **27**, No. 1, p. 435–442 (1968).
See also 13.4, p. 213; 22.9, p. 560.

V. Fedoseev and V. Hizhnyakov
*On the Theory of Low-Temperature Exciton Absorption Spectra
Vol. **27**, No. 2, p. 751–755 (1968).
See also 6, p. 66; 20.1, p. 355; 22, p. 452.

Ya. O. Dovhyj
Connection between the Circular Dichroism and Rotatory Dispersion of Gyrotropic Crystals and Its Experimental Verification in the Region of the Excitation of Circular Excitons
Vol. **27**, No. 2, p. K121–K124 (1968).
See also *20*, p. *341*; 22, p. 452.

A. A. Lipnik and M. A. Ruvinskii
Attenuation and Amplification of Hypersound by Mott Excitons in Semiconductors with Piezoelectric and Deformational Electron–Phonon Interactions
Vol. **28**, No. 1, p. 75–81 (1968).
See also 6, p. 66; *7*, p. *78*; 13.1, p. 187; 22, p. 452.

H.-J. Hänsch
Study of the Electroluminescence Mechanism of CdS Single Crystals
Vol. **28**, No. 1, p. 255–268 (1968).
See also 14.4.1, p. 263; *20.3*, p. *370*; 22.4.1, p. 503.

M. V. Andriyashik, M. Yu. Sakhnovskii, V. B. Timofeev, and A. S. Yakimova
Optical Transitions in the Spectra of the Fundamental Absorption and Reflection of InSe Single Crystals
Vol. **28**, No. 1, p. 277–285 (1968).
See also 13.1, p. 187; *20.1*, p. *355*; 22, p. 452.

R. V. Grigoriev, B. V. Novikov, and A. E. Cherednichenko
Influence of Oxygen Adsorption on the Fine Structure of the Spectral Distribution of Photoconductivity in CdS Crystals
Vol. **28**, No. 1, p. K85–K88 (1968).
See also 13.3, p. 198; *16*, p. *283*; 22.4.1, p. 504.

T. A. Kudykina, K. B. Tolpygo, and S. M. Zubkova
*Exciton Theory in Crystals with Complicated Band Structure with Electron–Phonon Interaction
Vol. **28**, No. 2, p. 807–814 (1968).
See also 13.1, p. 188; 22, p. 453; 22.5.2, p. 528.

S. N. Shestatskii and V. V. Sobolev
Electroreflectance Spectrum of Cuprous Oxide Crystals
Vol. **28**, No. 2, p. K131–K133 (1968).
See also 13.1, p. 188; *14.4.1*, p. *264*; *20*, p. *342*; 20.1, p. 356; 22.6, p. 542.

M. R. Tubbs
The Variation of Exciton Peak Position with Lattice Parameter in Metal Iodide Crystals
Vol. **28**, No. 2, p. K135—K137 (1968).
See also *20.1*, p. *356*; 22.5, p. 514; 22.5.2, p. 528; 22.5.3, p. 534.

M. P. Fontana, H. Blume, and W. J. van Sciver
Properties of Exciton States in NaI. (I): The Intrinsic Photoluminescence
Vol. **29**, No. 1, p. 159—166 (1968).
See also 20.3, p. 370; 22.5.2, p. 528.

P. Petrescu
On the L-Bands and the First Exciton Bands in the Photoemission Spectra of Alkali Iodides
Vol. **29**, No. 1, p. 333—340 (1968).
See also 10.2, p. 143; *17*, p. *289*; 22.5.2, p. 528.

V. M. Asnin, G. L. Eristavi, and A. A. Rogachev
Effect of Weak Electric Fields on the Absorption Edge in Doped Germanium
Vol. **29**, No. 1, p. 443—446 (1968).
See also 13.4, p. 215; 14.3, p. 243; *20.1*, p. *356*; 22.1.1, p. 466.

E. Mohler
Linear Electro-Optic Effect of Excitons in CuCl
Vol. **29**, No. 1, p. K55—K57 (1968).
See also *14.3*, p. *243*; *20*, p. *342*; 22.5, p. 514.

K. Platzöder
Temperature Effects on the Vaccuum Ultraviolet Reflectance of α-Quartz
Vol. **29**, No. 1, p. K63—K64 (1968).
See also 13.1, p. 189; *20.1*, p. *357*; 22.6, p. 543.

J. Dillinger, Č. Koňák, V. Prosser, J. Sak, and M. Zvára
*Phonon-Assisted Exciton Transitions in $A^{II}B^{VI}$ Semiconductors
Vol. **29**, No. 2, p. 707—714 (1968).
See also 6, p. 68; 20.1, p. 357; 22.4.1, p. 505; 22.4.2, p. 508; 22.4.3, p. 511.

E. L. Frankevich and B. M. Rumyantsev
Charge-Transfer Excitons and Delayed Fluorescence of Anthracene
Vol. **30**, No. 1, p. 329—340 (1968).
See also *20.3*, p. *371*; 22.9, p. 560.

E. F. Gross, B. S. Razbirin, V. P. Fedorov, and Yu. P. Naumov
*Bound Exciton Complexes in CdSe Crystals
Vol. **30**, No. 2, p. 485—494 (1968).
See also 20.1, p. 358; 20.3, p. 371; 22.4.2, p. 508.

K. Dos, A. Haug und P. Rohner
*Die Berücksichtigung der Austauschwechselwirkung im Zweiteilchenbild der Exzitonen und ihr Einfluß auf die Exzitonenspektren
(Consideration of the Exchange Interaction in the Two-Body Picture of Excitons and Its Influence on the Exciton Spectra)
Vol. **30**, No. 2, p. 619—625 (1968).
See also 20.1, p. 359; 22.4.1, p. 505.

J. Voigt and F. Spiegelberg
*On the Dynamical Behaviour of Excitons in CdS Single Crystals. (I): Correlation between Intrinsic Excitonic Absorption and Emission
Vol. **30**, No. 2, p. 659—670 (1968).
See also 6, p. 70; 20.1, p. 359; 20.3, p. 372; 22.4.1, p. 505.

13.3 Surface States

Review Articles

L. Eckertová
Field Cathodes with Thin Insulating Films
Vol. **18**, No. 1, p. 3—40 (1966).
See also 13, p. 162; 13.4, p. 200; 14.4.1, p. 255; *17*, p. *286*; 22.6, p. 535.

F. R. Brotzen
Emission of Exoelectrons from Metallic Materials
Vol. **22**, No. 1, p. 9—30 (1967).
See also *17*, p. *286*; 21, p. 373.

W. Steinmann
Optical Plasma Resonances in Solids
Vol. **28**, No. 2, p. 437—462 (1968).
See also *13*, p. *162*; *20*, p. *337*; 20.1, p. 344; 21, p. 373; 21.2, p. 424; 21.6, p. 430.

Original Papers and Short Notes

K. W. Böer and R. Schubert
The Influence of Oxygen in the Ultrahigh Vacuum Range on Electrical Properties of CdS
Vol. **16**, No. 1, p. K5—K9 (1966).
See also *16*, p. *276*; 22.4.1, p. 497.

M. Tomášek
*Some Remarks about Electronic Correlation in Crystals Surfaces: The Ideal (100) Surface of a Semi-Infinite Diamond-Like Crystal
Vol. **16**, No. 2, p. 389—397 (1966).
See also 22, p. 440.

V. A. Zuev and V. G. Litovchenko
Theory of the Photovoltaic Effect in Semiconductors
Vol. **16**, No. 2, p. 751—760 (1966).
See also *16*, p. *276*; 22.1.1, p. 459; 22.1.2, p. 469.

R. Schubert
Gas Desorption from Virginal CdS Crystals
Vol. **16**, No. 2, p. K157—K159 (1966).
See also *16*, p. *276*; 22.4.1, p. 497.

M. G. Asch, M. C. Félix et M. R. Ongaro
*Electrisation du polyéthylène et du polystyrène après irradiation γ (Electrization of Polyethylene and Polystyrene after γ-Irradiation)
Vol. **17**, No. 1, p. 49—57 (1966).
See also 11, p. 146; 22.9, p. 558.

E. Gutsche, F. Spiegelberg, and J. Voigt
Transient Photoconductivity Quenching in CdS Single Crystals due to Surface Trapping
Vol. **17**, No. 1, p. K11—K14 (1966).
See also *16*, p. *277*; 22.4.1, p. 498.

F. Forlani, N. Minnaja, and E. Pagiola
*Theory of the Influence of Surface States on the Impedance of a Semiconductor–Insulator Interface
Vol. **17**, No. 2, p. 657—672 (1966).
See also *14.3.2*, p. *248*; 22.1.2, p. 469.

E.-H. Weber
Relationship between Space Charge, Surface Conductivity, and Band Bending for Photoconductors under Uniform Optical Excitation
Vol. **17**, No. 2, p. 843–854 (1966).
See also 13.4, p. 203; 14.4.1, p. 257; *16*, p. *277*; 22.4, p. 495.

V. D. Egorov
Linear Photoconductive Rise in Semiconductors
Vol. **18**, No. 1, p. 145–150 (1966).
See also *16*, p. *278*; 22.1.1, p. 460.

H. Flietner und W. Gräfe
*Großsignal-Feldeffekt zur Oberflächentermbestimmung bei Silizium
(Large-Signal Field Effect for the Determination of the Surface Levels of Silicon)
Vol. **18**, No. 1, p. K5–K7 (1966).
See also 14.3, p. 233; 22.1.2, p. 470.

A. G. Revesz
Thermal Oxidation of Silicon: Growth Mechanism and Interface Properties
Vol. **19**, No. 1, p. 193–202 (1967).
See also *1.3*, p. *14*; 14.3.2, p. 248; 22.1.2, p. 470.

R. D. Levine and A. T. Amos
*On the Theory of Surface and Impurity States
Vol. **19**, No. 2, p. 587–596 (1967).
See also 13.4, p. 205.

M. Henzler
*Leitfähigkeit und Feldeffekt reiner Siliziumspaltflächen
(Conductivity and Field Effect of Pure Cleavage-Surfaces of Silicon)
Vol. **19**, No. 2, p. 833–845 (1967).
See also 14.3, p. 235; 22.1.2, p. 470.

I. A. Akimov, V. M. Bentsa, F. I. Vilesov, and A. N. Terenin
External Photoeffect from Sensitizing Dyes Adsorbed on Semiconductors
Vol. **20**, No. 2, p. 771–776 (1967).
See also 16, p. 279; *17*, p. *287*; 22.4, p. 495; 22.5, p. 512; 22.5.1, p. 515.

S. G. Christov
Unified Theory of Thermionic and Field Emission from Semiconductors
Vol. **21**, No. 1, p. 159–173 (1967).
See also *17*, p. *287*; 22, p. 444.

I. L. Sokolskaya, V. G. Ivanov, and G. N. Fursey
*Study of Barium Adsorption on Germanium by Field-Emission Microscopy
Vol. **21**, No. 2, p. 789–795 (1967).
See also 1.3, p. 15; 17, p. 287; 22.1.1, p. 462.

H. Puszkarski
Surface Spin Waves in Hexagonal Cobalt Thin Films
Vol. **22**, No. 2, p. 355–359 (1967).
See also *13*, p. *166*; 19, p. 329; 21.1, p. 399.

A. Otto
Theory of Plasmon Excitation in Thin Films by Electrons of Non-Normal Incidence
Vol. **22**, No. 2, p. 401–406 (1967).
See also *13*, p. *166*; 20, p. 339.

J. B. Swan, A. Otto, and H. Fellenzer
Observed Retardation Effects on the Energy of the ω_+-Surface Plasmons in Thin Aluminium Foils
Vol. 23, No. 1, p. 171–176 (1967).
See also *13*, p. *167*; *20*, p. *340*; 21, p. 381.

L. Ernst
About the Fowler-Nordheim Plots of Germanium Field Emitters
Vol. 24, No. 1, p. 177–181 (1967).
See also *17*, p. *287*; 22.1.1, p. 463.

D. N. Bose
*Field-Effect Measurements on Silicon Carbide
Vol. 24, No. 2, p. K165–K166 (1967).
See also 14.3, p. 239; 22.3, p. 494.

D. M. Hughes and G. Carter
The Effects of Oxygen Adsorption and Low Energy Ion Bombardment on the Electrical Properties of Cadmium Sulphide Thin Films
Vol. 25, No. 1, p. 449–453 (1968).
See also *14.3.1*, p. *246*; 22.4.1, p. 501.

G. Krenzke, H. Lemke, and G. O. Müller
Properties of Gold Contacts on Clean n-Silicon Surfaces
Vol. 25, No. 2, p. K131–K134 (1968).
See also *14.4.1*, p. *261*; 22.1.2, p. 473.

J. Łagowski
*On Surface Recombination Changes due to Edge Dislocations
Vol. 26, No. 1, p. K69–K71 (1968).
See also *10*, p. *105*; 22.1.1, p. 464.

P. Devaux and M. Schott
*Field Effect on Photocarriers in Copper Phthalocyanine Single Crystals
Vol. 27, No. 1, p. 201–208 (1968).
See also 13.2, p. 194; *16*, p. *282*; 22.9, p. 560.

D. N. Bose
*Effect of Low Energy Electron and UV Irradiations on the Charge Condition of Etched Germanium Surfaces
Vol. 27, No. 1, p. 325–331 (1968).
See also 11, p. 152; 22.1.1, p. 465.

E. W. Kreutz, H. Pagnia, and W. Waidelich
X-Ray Induced Conductivity Changes in InSb
Vol. 27, No. 2, p. K111–K114 (1968).
See also *11*, p. *153*; 14.3, p. 242; 22.2.3, p. 492.

R. V. Grigoriev, B. V. Novikov, and A. E. Cherednichenko
Influence of Oxygen Adsorption on the Fine Structure of the Spectral Distribution of Photoconductivity in CdS Crystals
Vol. 28, No. 1, p. K85–K88 (1968).
See also 13.2, p. 194; *16*, p. *283*; 22.4.1, p. 504.

H. Pagnia
Surface Photoconduction in p-InSb Single Crystals
Vol. 28, No. 1, p. K89–K92 (1968).
See also *16*, p. *283*; 22.2.3, p. 493.

E.-H. WEBER
Surface Photoconductivity of CdS Influenced by Chemisorption and Desorption of Oxygen
Vol. **28**, No. 2, p. 649–662 (1968).
See also *16*, p. *283*; 22.4.1, p. 504.

A. B. FAZAKAS and P. H. HANDEL
*Stability of Surface-Barriers in Semiconductors
Vol **28**, No. 2, p. 699–709 (1968).
See also 14.4.1, p. 263; 22, p. 453.

L. AHMAD
*Change in Electron Affinity of a Semiconductor due to Different Forms of Chemisorption
Vol. **29**, No. 1, p. 179–187 (1968).
See also 22.8, p. 555.

H. WIENHOLD
On Relaxation Effects of Epitaxial Si–MOS Structures
Vol. **30**, No. 1, p. K17–K21 (1968).
See also *14.3.2*, p. *251*; 22.1.2, p. 474.

W. LUDWIG, W. GERICKE, and H. TREPTOW
Electron Tunnelling in MIS-Structures of the Type Al–Al_2O_3–ZnTe
Vol. **30**, No. 1, p. K61–K64 (1968).
See also *14.3.2*, p. *251*; 14.4.1, p. 264; 22.4.3, p. 511.

13.4 Impurity and Defect States

Review Articles

J. R. Dale
Alloyed Semiconductor Heterojunctions
Vol. **16**, No. 2, p. 351—387 (1966).
See also **3.1**, p. **29**; 5, p. **49**; **13.4**, p. 200; *14.3.2*, p. *248*; 16, p. 276; 22.1.1, p. 458; 22.1.2, p. 468; 22.2.1, p. 482; 22.2.2, p. 487; 22.2.3, p. 489.

L. Eckertová
Field Cathodes with Thin Insulating Films
Vol. **18**, No. 1, p. 3—40 (1966).
See also **13**, p. 162; **13.3**, p. 196; 14.4.1, p. 255; *17*, p. *286*; 22.6, p. 535.

Y. P. Varshni
Band-to-Band Radiative Recombination in Groups IV, VI, and III–V Semiconductors (I)
Vol. **19**, No. 2, p. 459—514 (1967).
See also 13.1, p. 174; 13.2, p. 191; 20.1, p. 344; 20.2, p. 360; *20.3*, p. *363*; 22.1, p. 456; 22.1.1, p. 458; 22.1.2, p. 468; 22.1.3, p. 476; 22.2.1, p. 482; 22.2.2, p. 487; 22.2.3, p. 489.

Y. P. Varshni
Band-to-Band Radiative Recombination in Groups IV, VI, and III–V Semiconductors (II)
Vol. **20**, No. 1, p. 9—36 (1967).
See also 13.1, p. 174; 13.2, p. 191; 20.1, p. 344; 20.2, p. 360; *20.3*, p. *363*; 22.1, p. 456; 22.1.1, p. 458; 22.1.2, p. 468; 22.1.3, p. 476; 22.2.1, p. 482; 22.2.2, p. 487; 22.2.3, p. 489.

M. H. Pilkuhn
The Injection Laser
Vol. **25**, No. 1, p. 9—62 (1968).
See also 14.3.2, p. 248; *14.4.1*, p. *255*; *20.2*, p. *360*; 20.3, p. 363; 22.2.1, p. 482; 22.2.2, p. 487; 22.2.3, p. 489.

F. Williams
*Donor–Acceptor Pairs in Semiconductors
Vol. **25**, No. 2, p. 493—512 (1968).
See also 10, p. 93; 20.3, p. 363; 22.1, p. 456; 22.1.1, p. 458; 22.1.2, p. 468; 22.2.2, p. 487; 22.4.1, p. 497.

V. L. Bonch-Bruevich and E. G. Landsberg
*Recombination Mechanisms
Vol. **29**, No. 1, p. 9—43 (1968).
See also 6, p. 56; 13, p. 162; 13.2, p. 191; *14.3*, p. *231*; 20, p. 337; 20.3, p. 363; 22.1.1, p. 458; 22.1.2, p. 468.

A. Seeger and K. P. Chik
Diffusion Mechanisms and Point Defects in Silicon and Germanium
Vol. **29**, No. 2, p. 455—542 (1968).
See also *9*, p. *87*; *10*, p. *93*; 11, p. 145; 19, p. 326; 22.1.1, p. 458; 22.1.2, p. 468.

Original Papers and Short Notes

I. B. Bersuker and B. G. Vekhter
*Jahn-Teller Distortions and Inversion Splitting for Transition Metal Complexes and Local Centres in Crystals with Degenerate Electronic T-Terms
Vol. **16**, No. 1, p. 63—68 (1966).
See also 10, p. 93; 22, p. 439.

P. SAUER, O. SCHIRMER, and J. SCHNEIDER
*Comments on Paraelectric Resonance
Vol. **16**, No. 1, p. 79–88 (1966).
See also 10, p. 93; 22.5.2, p. 518.

N. RIEHL and P. THOMA
Delayed Fluorescence and Thermoluminescence of Anthracene Single Crystals at Very Low Temperatures
Vol. **16**, No. 1, p. 159–170 (1966).
See also *20.3*, p. *363*; 22.9, p. 558.

G. B. ABDULLAEV, E. S. GUSEINOVA, and B. G. TAGIEV
Electrical Conductivity of p-GaSe Single Crystals in Strong Electric Fields
Vol. **16**, No. 1, p. 205–208 (1966).
See also *14.4.1*, p. *255*; 22, p. 439.

W. DREYBRODT and D. SILBER
EPR Measurements on Diatomic Halogen Centers in Alkali Halides
Vol. **16**, No. 1, p. 215–223 (1966).
See also 10.2, p. 137; *19*, p. *326*; 22.5.2, p. 518.

H. LEMKE
On the Time Dependence of Space-Charge Limited Injection Currents in Semiconductors
Vol. **16**, No. 2, p. 413–426 (1966).
See also *14.4.1*, p. *255*; 22, p. 440.

H. LEMKE
Time Dependence of Space-Charge Limited Injection Currents in Fe-Doped p-Silicon
Vol. **16**, No. 2, p. 427–438 (1966).
See also *14.4.1*, p. *255*; 22.1.2, p.469.

M. BANCIE-GRILLOT and P. BOURTAYRE
Visible Photoluminescence Emission of Gadolinium Incorporated in the Crystal Lattice of Zinc Sulphide
Vol. **16**, No. 2, p. 517–523 (1966).
See also *20.3*, p. *363*; 22.4.1, p. 497.

V. V. RATNAM
On the Electrical and Magnetic Properties of Coloured and Uncoloured Calcium Fluoride Crystals
Vol. **16**, No. 2, p. 549–557 (1966).
See also *10.2*, p. *137*; 14.4, p. 252; 18.1, p. 296; 22.5.3, p. 532.

P. THOMA and G. VAUBEL
Electrical Properties of Anthracene Single Crystals at Very Low Temperatures
Vol. **16**, No. 2, p. 663–674 (1966).
See also 14.3, p. 231; 14.4.1, p. 256; *16*, p. *276*; 22.9, p. 558.

K. W. BÖER, J. W. FEITKNECHT, and D. G. KANNENBERG
Properties of Recrystallized Evaporated CdS Layers
Vol. **16**, No. 2, p. 697–709 (1966).
See also *1.4*, p. *16*; 16, p. 276; 22.4.1, p. 497.

L. GOUSKOV, G. LECOY, C. LLINARES et M. SAVELLI
*Études de la durée de vie et du bruit de génération recombinaison controlés par les dislocations dans le germanium
(Study of Lifetime and Generation–Recombination Noise Controlled by Dislocations in Germanium)
Vol. **16**, No. 2, p.721–728 (1966).
See also 14.3, p. 232; 16, p. 276; 22.1.1, p. 459.

H. J. ROHDE
Effect of Shallow Traps on the Current-Voltage Characteristics in the Case of Double Injection
Vol. **16**, No. 2, p. K179–K182 (1966).
See also *14.4.1*, p. *256*; 22.4.1, p. 498.

D. V. EDDOLLS
Electrical Properties of n-Type Epitaxial Gallium Arsenide
Vol. **17**, No. 1, p. 67–76 (1966).
See also *14.3*, p. *232*; 22.2.1, p. 482.

S. A. ALIEV, F. P. KESAMANLY, T. S. LAGUNOVA, and D. N. NASLEDOV
Hall Effect and Magnetoresistance of n-InP Crystals at Low Temperatures
Vol. **17**, No. 1, p. 105–108 (1966).
See also *14.3*, p. *232*; 22.2.2, p. 487.

R. B. LAIBOWITZ and H. S. SACK
Luminescence and Photoconductivity in Silver Chloride with Laser Excitation
Vol. **17**, No. 1, p. 353–360 (1966).
See also 6, p. 57; *16*, p. *277*; *20.3*, p. *364*; 22.5.1, p. 515.

H. D. KOSWIG
*Zur Abhängigkeit der innerelektronischen Abstoßungsparameter der $3d^3$- und $3d^7$-Konfiguration von den Liganden in oktaedrischen Kristallfeldern
(Dependence of the Inner-Electronic Repulsion Parameters of the $3d^3$ and $3d^7$ Configuration on the Ligands in Octahedral Crystal Fields)
Vol. **17**, No. 1, p. K57–K61 (1966).
See also 20.1, p. 345; 22, p. 440.

I. A. AKIMOV, A. M. MESHKOV, and A. N. TERENIN
On Thermal E.M.F. Measurements in Photo-Semiconducting Dyes
Vol. **17**, No. 1, p. K95–K98 (1966).
See also *15*, p. *270*; 16, p. 277; 22, p. 440.

F. KUBEC
Crystal Field Calculation in the $CdWO_4$ Lattice
Vol. **17**, No. 2, p. 571–574 (1966).
See also 3.2, p. 34; *6*, p. *57*; 22.8, p. 548.

K. KIROV
*Impulsmethode zur Untersuchung der Haftstellen in Kristallen vom CdS-Typ
(Pulse Method for Studying Trapping Centres in CdS Type Crystals)
Vol. **17**, No. 2, p. 581–592 (1966).
See also 14.4.1, p. 256; 16, p. 277; 22.4.1, p. 498.

W. GÜTH und W. HAIST
*Elektronenzustände in Halbleiterversetzungen
(Electronic States of Dislocations in Semiconductors)
Vol. **17**, No. 2, p. 691–696 (1966).
See also 13.1, p. 175; 22.1, p. 456.

H.-E. Gumlich, R. Moser, and E. Neumann
Recombination Distribution in the Non-Stationary Phase of the Electroluminescence of ZnS Phosphors Containing Iron-Group Elements
Vol. **17**, No. 2, p. 783—792 (1966).
See also *20.3*, p. *364*; 22.4.1, p. 498.

E.-H. Weber
Relationship between Space Charge, Surface Conductivity, and Band Bending for Photoconductors under Uniform Optical Excitation
Vol. **17**, No. 2, p. 843—854 (1966).
See also 13.3, p. 197; 14.4.1, p. 257; *16*, p. *277*; 22.4, p. 495.

H. Böttger
Temperature Dependence of the Optical Absorption of Transition Element Ions in Silver Chloride
Vol. **17**, No. 2, p. 855—868 (1966).
See also *20.1*, p. *346*; 22.5.1, p. 515.

O. Goede and E. Gutsche
Edge Emission in CdS as Internal Radiative Transition in Donor–Acceptor Associates
Vol. **17**, No. 2, p. 911—923 (1966).
See also 6, p. 57; *20.3*, p. *364*; 22.4.1, p. 498.

F. I. Ismailov, G. A. Akhundov, and O. R. Vernich
Investigation of Electrical Conductivity and Hall Effect in GaSe Single Crystals
Vol. **17**, No. 2, p. K237—K240 (1966).
See also *14.3*, p. *233*; 15, p. 271; 22, p. 441.

A. Räuber and J. Schneider
Localized $^2S_{1/2}$-State Centres in ZnS
Vol. **18**, No. 1, p. 125—132 (1966).
See also *19*, p. *327*; 22.4.1, p. 498.

W. Dreybrodt and K. Fussgaenger
Interpretation of the Temperature Dependence of Ag^+ Dipole Strengths in Alkali Halides
Vol. **18**, No. 1, p. 133—144 (1966).
See also *20.1*, p. *346*; 22.5.2, p. 519.

G. Jungk
*Zur Anregungs–Rekombinationsstatistik in Halbleitern
(On the Generation–Recombination Statistics in Semiconductors)
Vol. **18**, No. 1, p. 297—306 (1966).
See also 14.3, p. 233; 16, p. 278; 22.1.2, p. 470.

P. Bräunlich and A. Scharmann
*Approximate Solution of Schön's Balance Equations for the Thermoluminescence and the Thermally Stimulated Conductivity of Inorganic Photoconducting Crystals
Vol. **18**, No. 1, p. 307—316 (1966).
See also 14.3, p. 233; 16, p. 278; 20.3, p. 364; 22.4, p. 495.

B. P. Kozyrev and T. Boţilă
Some Photoelectrical Properties of KRS-5 Single Crystals
Vol. **18**, No. 1, p. K57—K61 (1966).
See also *16*, p. *278*; 22.5, p. 512; 22.8, p. 548.

B. Tuck
Photoluminescence of $GaAs_{0.7}P_{0.3}$
Vol. 18, No. 2, p. 541–545 (1966).
See also *20.3*, p. *364*; 22.2.1, p. 483; 22.2.2, p. 488; 22.8, p. 548.

G. B. Street and W. D. Gill
Photoconductivity and Drift Mobilities in Single Crystal Realgar (As_4S_4)
Vol. 18, No. 2, p. 601–607 (1966).
See also 14.3, p. 233; *16*, p. *278*; 22, p. 441.

V. V. Eremenko, E. V. Matyushkin, and S. V. Petrov
*Study of Energy Transfer from 3d to 4f Electrons in Antiferromagnetic $MnF_2:Eu^{3+}$ Crystals
Vol. 18, No. 2, p. 683–686 (1966).
See also 18.4, p. 321, *20.3*, p. *364*; 22.5.3, p. 532.

M. Jung
Pseudo-Abrupt Alloyed p-n Junctions in GaP
Vol. 18, No. 2, p. 743–748 (1966).
See also *14.3.2*, p. *248*; 22.2.2, p. 488.

A. F. Lubchenko and S. I. Dudkin
Effect of Concentration on Light Absorption and Dispersion by Impurity Centres in the Case of Weak Electron–Phonon Coupling
Vol. 18, No. 2, p. 853–862 (1966).
See also 6, p. 58; *20.1*, p. *346*.

G. D. Guseinov, M. Z. Ismailov, and A. G. Talybov
On the New Semiconducting Compound $HgTlS_2$
Vol. 18, No. 2, p. 929–935 (1966).
See also *5*, p. *50*; 8, p. 80; 14.3, p. 234; 16, p. 278; 22.8, p. 548.

R. J. Fleming
Potential Energy Curve for F-Centre Electrons in Calcium Fuoride
Vol. 18, No. 2, p. K139–K142 (1966).
See also *10.2*, p. *138*; 22.5.3, p. 532.

E. D. Aluker and I. P. Mezina
The Temperature Dependence of Radioluminescence Yield in Alkali Halide Crystal Phosphors
Vol. 19, No. 1, p. 35–40 (1967).
See also *20.3*, p. *365*; 22.5.2, p. 520.

V. L. Vinetskii and G. A. Kholodar
On the Electric Conductivity of Semiconductors Caused by the Ionization of Thermal Lattice Defects
Vol. 19, No. 1, p. 41–49 (1967).
See also 10, p. 97; *14.3*, p. *234*; 22.6, p. 536.

K. W. Böer and C. A. Kennedy
*Vacuum Heat Treatment of CdS Single Crystals
Vol. 19, No. 1, p. 203–210 (1967).
See also 14.3, p. 234; 16, p. 278; 22.4.1, p. 499.

Yu. A. Logatchov
*Model Potential Approximation for Z_1-Centres in Ionic Crystals
Vol. 19, No. 1, p. K35–K38 (1967).
See also 10.2, p. 138; 22.5.2, p. 520.

A. Glodeanu
*Helium-Like Impurities in Semiconductors
Vol. **19**, No. 1, p. K43–K46 (1967).
See also 22.1.1, p. 461; 22.1.2, p. 470; 22.2.1, p. 483.

G. Döhler
On the Field Emission of Minority Carriers in Photoconductors
Vol. **19**, No. 2, p. 555–564 (1967).
See also *14.4.1*, p. *257*; 16, p. 278; 22.1.3, p. 477; 22.4.1, p. 499.

R. D. Levine and A. T. Amos
On the Theory of Surface and Impurity States
Vol. **19**, No. 2, p. 587–596 (1967).
See also *13.3*, p. *197*.

P. T. Landsberg
The Condition for Negative Absorption in Semiconductors
Vol. **19**, No. 2, p. 777–780 (1967).
See also *20.2*, p. *360*; 22, p. 443.

R. Böttcher, W. Windsch, and W. Lüdke
EPR Investigations on Mn^{2+}-Doped LiF Single Crystals
Vol **20**, No. 1, p. 121–126 (1967).
See also 9, p. 88; *19*, p. *328*; 22.5.2, p. 521.

P. Devaux and M. Schott
*Thermally Stimulated Currents without Optical Excitation. Application to Copper Phtalocyanine
Vol. **20**, No. 1, p. 301–309 (1967).
See also 14.4.1, p. 257; 22.9, p. 558.

W. Dreybrodt and D. Silber
Electron Spin Resonance of Tl^{++} Centres in KCl Crystals
Vol. **20**, No. 1, p. 337–346 (1967).
See also *19*, p. *328*; 22.5.2, p. 521.

Z. Gołacki and T. Figielski
Extrinsic Photoconductivity in Ge Caused by Dislocations
Vol. **20**, No. 1, p. K1–K3 (1967).
See also *16*, p. *279*; 22.1.1, p. 461.

A. F. Lubchenko and S. I. Dudkin
Effect of Concentration on Light Absorption and Dispersion by Impurity Centres in the Case of Weak Elektron–Phonon Coupling (II)
Vol. **20**, No. 2, p. 427–432 (1967).
See also 6, p. 60; *20.1*, p. *347*.

C. Hamann
On the Electric and Thermoelectric Properties of Copper Phthalocyanine Single Crystals
Vol. **20**, No. 2, p. 481–491 (1967).
See also *14.3*, p. *236*; 14.4.1, p. 258; 15, p. 271; 22.9, p. 559.

G. Blasse and A. Bril
Some Observations on the Cr^{3+} Fluorescence in the Huntite Structure
Vol. **20**, No. 2, p. 551–556 (1967).
See also 20.1, p. 347; *20.3*, p. *365*; 22.8, p. 549.

M. Höhne and M. Stasiw
Te_2^{3-} and Se_2^{3-} Centres in AgBr. (I): The *g*-, Superhyperfine, and Hyperfine Tensors
Vol. **20**, No. 2, p. 657–666 (1967).
See also 10.2, p. 139; *19*, p. *328*; 22.5.1, p. 515.

M. Höhne and M. Stasiw
Te_2^{3-} and Se_2^{3-} Centres in AgBr. (II): Electronic Structure
Vol. **20**, No. 2, p. 667–673 (1967).
See also 10.2, p. 139; *19*, p. *328*; 22.5.1, p. 515.

P. Feltham
F-Centres in Alkaline Earth Fluorides
Vol. **20**, No. 2, p. 675–680 (1967).
See also *10.2*, p. *139*; 22.5.3, p. 533.

M. Jaroš and J. Zítková
Remark on the Theory of the ESR Spectrum of the d^1-Configuration in Rutile (TiO_2)
Vol. **20**, No. 2, p. K99–K101 (1967).
See also *19*, p. *328*; 22.6, p. 537.

W. Haubenreisser and U. Lindner
*Some Symmetry Aspects in the Theory of Localized Magnetic States in Metals
Vol. **20**, No. 2, p. K131–K134 (1967).
See also 18.1, p. 296; 21, p. 378.

W. Dreybrodt
Temperature Dependence of Hyperfine Structure Tensors of Diatomic Halogen Centres in Alkali Halide Crystals
Vol. **21**, No. 1, p. 99–112 (1967).
See also *19*, p. *329*; 22.5.2, p. 522.

W. Schröter
Electric Properties of Dislocations in Germanium
Vol. **21**, No. 1, p. 211–224 (1967).
See also 10, p. 99; *14.3*, p. *236*; 22.1.1, p. 461.

N. F. Mott and R. S. Allgaier
*Localized States in Disordered Lattices
Vol. **21**, No. 1, p. 343–356 (1967).
See also 2, p. 23; 21, p. 378; 22, p. 445.

P. Bräunlich
Light Effects on Space-Charge Controlled Currents in CdS
Vol. **21**, No. 1, p. 383–391 (1967).
See also *14.4.1*, p. *258*; 16, p. 279; 20.3, p. 365; 22.4.1, p. 499.

G. B. Abdullaev, Z. A. Iskender-Zade, E. A. Dzhafarova, and V. E. Chelnokov
Study of Electron Lifetime in p-Si
Vol. **21**, No. 1, p. 423–429 (1967).
See also *14.3*, p. *236*; 14.4.1, p. 258; 22.1.2, p. 471.

V. F. Zolotaryov, D. G. Semak, and D. V. Chepur
*Thermally Stimulated Currents under the Condition of Persistant Internal Polarization
Vol. **21**, No. 1, p. 437–442 (1967).
See also 14.4.1, p. 258; 22, p. 445.

A. A. Kaminskii, Kh. S. Bagdasarov, and L. M. Belyev
Analysis of the Optical Spectra and Stimulated Radiation of $Y_3Al_5O_{12}$–Nd^{3+} Crystals
Vol. **21**, No. 1, p. K23–K29 (1967).
See also *20.1*, p. *348*; 20.2, p. 361; *20.3*, p. *365*; 22.8, p. 550.

J. Trylski
*Energy of the Donor Ground State in CdS, CdSe, and ZnO
Vol. **21**, No. 1, p. K53–K55 (1967).
See also 22.4, p. 495; 22.4.1, p. 499; 22.4.2, p. 506.

L. Müller and Y. Mihai
Absorption of Amorphous Se Thin Layers Doped with Bi and Zn
Vol. **21**, No. 1, p. K65–K68 (1967).
See also 2, p. 23; *20.1*, p. *348*; 22.1.3, p. 477.

G. Getov, P. Simidtchieva, M. Nikiforova, and R. Andreytchin
Effect of Some Elements on the Optical Absorption Edge of Vitreous As_2S_3
Vol. **21**, No. 1, p. K87–K89 (1967).
See also 2, p. 23; *20.1*, p. *348*; 22, p. 445.

V. K. Bashenov, S. P. Fedotov, and V. A. Presnov
*On Deep Acceptors in Iron-Doped Gallium Arsenide
Vol. **21**, No. 1, p. K91–K93 (1967).
See also 19, p. 329; 22.2.1, p. 483.

L. Gouskov, G. Lecoy, and C. Llinares
Carrier Recombination in Nickel Doped Germanium from Lifetime and Noise Measurements
Vol. **21**, No. 2, p. 619–626 (1967).
See also *14.3*, p. *236*; *16*, p. *279*; 22.1.1, p. 462.

M. Averous and G. Bougnot
Effect of Donor Impurity Concentration on the Piezoresistance of n-Type GaSb
Vol. **21**, No. 2, p. 665–670 (1967).
See also 13.1, p. 178; *14.3*, p. *236*; 22.2.3, p. 491.

J. P. Agrawal
Infrared Fluorescence in Manganese Activated Zinc Sulphide Phosphors
Vol. **21**, No. 2, p. 679–682 (1967).
See also 20.1, p. 348; *20.3*, p. *366*; 22.4.1, p. 500.

V. Hizhnyakov and I. Tehver
Theory of Resonant Secondary Radiation due to Impurity Centres in Crystals
Vol. **21**, No. 2, p. 755–768 (1967).
See also 20.1, p. 348; *20.3*, p. *366*; 22, p. 446.

A. F. Lubchenko
*Light Absorption and Dispersion by Impurity Centres at High Concentrations
Vol. **21**, No. 2, p. 785–788 (1967).
See also *20.1*, p. *349*.

M. Damodar-Pai and K. M. van Vliet
Generation–Recombination and Diffusion Noise in Cadmium Sulphide Photoconductors
Vol. **21**, No. 2, p. 819–835 (1967).
See also *16*, p. *279*; 22.4.1, p. 500.

A. ZARĘBA
Photoconductivity in Boron at Low Temperature
Vol. **21**, No. 2, p. K135—K138 (1967).
See also *16*, p. *279*; 22.1, p. 457.

E. V. MATYUSHKIN, L. S. KUKUSHKIN, and V. V. EREMENKO
Peculiarities of the $RbMnF_3:Nd^{3+}$ Crystals Decay Kinetics due to the Migration of Excitation Energy
Vol. **22**, No. 1, p. 65—69 (1967).
See also *20.3*, p. *366*; 22.5, p. 513; 22.8, p. 551.

E. W. CLAFFY
Impurity Absorption Bands in Thermoluminescent LiF
Vol. **22**, No. 1, p. 71—76 (1967).
See also *20.1*, p. *349*; 20.3, p. 366; 22.5.2, p. 523.

H. W. DEN HARTOG and J. ARENDS
F-Centers in SrF_2
Vol. **22**, No. 1, p. 131—140 (1967).
See also *10.2*, p. *139*; 19, p. 329; 22.5.3, p. 533.

J. VIŠČAKAS, J. VAITKUS, and S. STONKUS
Influence of Heat Treatment on the Spectral Dependence of Photoconductivity in CdSe Crystals
Vol. **22**, No. 1, p. K3—K5 (1967).
See also *16*, p. *279*; 22.4.2, p. 506.

L. N. COJOCARU
Effects of Fast Neutrons on Electric Properties of NiO Single Crystals
Vol. **22**, No. 2, p. 361—364 (1967).
See also *11*, p. *149*; 14.3, p. 237; 15, p. 272; 22.6, p. 539.

A. DALL'OLIO, G. DASCOLA, and V. VARACCA
EPR Evidence of Exchange-Coupled Ion Pairs and of Isolated Ions in Some Chlorosubstituted Cupric Acetates
Vol. **22**, No. 2, p. 365—370 (1967).
See also *19*, p. *330*; 22.9, p. 559.

A. ALDEA
Theory of the Hall Effect in Disordered Systems: Impurity Band Conduction (II)
Vol. **22**, No. 2, p. 377—380 (1967).
See also *14.3*, p. *237*; 22, p. 446.

P. BRÄUNLICH
Calculation of Light-Stimulated Luminescence and Current Transients in Phosphors
Vol. **22**, No. 2, p. 391—400 (1967).
See also *16*, p. *280*; *20.3*, p. *366*; 22, p. 447.

S. C. JAIN and G. D. SOOTHA
Electrical and Optical Properties of Titanium-Doped Lithium Fluoride Crystals
Vol. **22**, No. 2, p. 505—516 (1967).
See also *14.4*, p. *253*; *20.1*, p. *349*; 22.5.2, p. 523.

L. ELSTNER and W. KAMPRATH
*Quenched-in Levels in p-Type Silicon
Vol. **22**, No. 2, p. 541—547 (1967).
See also 11, p. 149; 14.3, p. 237; 22.1.2, p. 471.

J. E. CAFFYN, T. L. GOODFELLOW, E. M. OLYMBIOS, and D. J. PERRIN
Dielectric and Optical Absorption Measurements on CaF_2 and CaF_2:Y
Vol. **22**, No. 2, p. 549—557 (1967).
See also 10, p. 101; *14.4*, p. *253*; *20.1*, p. *350*; 22.5.3, p. 533.

R. BOGDANOVIĆ
*Potential of a Vacant Site in the Graphite Lattice
Vol. **22**, No. 2, p. 603—606 (1967).
See also 22.1, p. 457.

M. I. ALIEV, G. I. SAFARALIEV, and S. G. ABDINOVA
About the Role of Minority Carriers in the Impurity Conduction of GaSb
Vol. **22**, No. 2, p. 741—745 (1967).
See also *14.3*, p. *237*; 22.2.3, p. 491.

H. J. ROHDE
Double Injection into CdS Single Crystals
Vol **23**, No. 1, p. 277—288 (1967).
See also *14.4.1*, p. *259*; 22.4.1, p. 500.

C. BENOIT À LA GUILLAUME and M. VOOS
*Optical Measurement of Compensation in Highly Doped Silicon
Vol. **23**, No. 1, p. 295—299 (1967).
See also 20.3, p. 367; 22.1.2, p. 472.

J. H. CRAWFORD, JR., W. A. SIBLEY, and E. SONDER
The Influence of Electron-Trapping Impurity on Defect Creation and Bleaching in KCl Irradiated at 78 °K
Vol. **23**, No. 1, p. 301—306 (1967).
See also *10.2*, p. *140*; 11, p. 150; 22.5.2, p. 523.

P. GÖRLICH, H. KARRAS, G. KÖTITZ, and P. ULLMANN
Thermoluminescence Investigations on Doped Alkaline Earth Fluorides
Vol. **23**, No. 1, p. 313—323 (1967).
See also 20.1, p. 350; *20.3*, p. *367*; 22.5.3, p. 533.

H. BÖTTGER
Study of the Jahn-Teller Effect of Ions with Spin 1
Vol. **23**, No. 1, p. 325—333 (1967).
See also *19*, p. *330*.

S. WANG
*Optical Absorption of an Interstitial Alkali-Earth Cation in an Alkali-Halide Crystal
Vol. **23**, No. 1, p. 387—399 (1967).
See also 10.2, p. 140; 13.2, p. 193; 22.5.2, p. 523.

S. WANG
*Electronic States of the M_A-Centers in KCl
Vol. **23**, No. 1, p. 401—412 (1967).
See also 10.2, p. 140; 22.5.2, p. 523.

P. NOVÁK
*On the Inversion Splitting in Octahedrally Coordinated E_g Electronic States
Vol. **23**, No. 1, p. K45—K47 (1967).

I. P. IPATOVA, A. A. KLOCHICHIN, and A. V. SUBASHIEV
The Influence of Localized and Resonance Modes of Lattice Vibrations on the Radiative Recombination in Semiconductors
Vol. **23**, No. 2, p. 467—473 (1967).
See also *6*, p. *63*; *20.3*, p. *367*; 22.1.2, p. 472.

P. G. Eliseev, A. I. Krasilnikov, M. A. Manko, and I. Z. Pinsker
Band-Filling Model for Injection Luminescence at Higher Temperatures
Vol. **23**, No. 2, p. 587–593 (1967).
See also *20.3*, p. *367*; 22.2.1, p. 484.

H.-P. Hennig
Contribution to the Determination of Gold Centre Cross Sections in Single-Crystal Germanium by Double Injection Experiments
Vol. **23**, No. 2, p. 599–605 (1967).
See also *14.4.1*, p. *260*; 22.1.1, p. 463.

H. W. den Hartog and J. Arends
*Electronic Structure of F-Centers in Alkaline Earth Fluorides
Vol. **23**, No. 2, p. 713–719 (1967).
See also 10.2, p. 140; 19, p. 330; 22.5.3, p. 533.

S. V. Vonsovskii, V. I. Cherepanov, A. N. Men, and A. E. Nikiforov
*Group-Theoretical Classification of Electronic States in Crystals Composed of Pairs and Complexes of Bound Impurity Ions
Vol. **24**, No. 1, p. 51–63 (1967).
See also 3.2, p. 35; 22.6, p. 539.

H. Böttger
Splitting of ESR Lines by Spin–Phonon Interaction for Ions with Spin 1
Vol. **24**, No. 1, p. 65–75 (1967).
See also 6, p. 63; *19*, p. *331*; 22.6, p. 539.

T. O. Ogurtani and R. A. Huggins
Theory of Electric Field Gradient due to Conduction Electron Charge Density Redistribution around Screw Dislocations in Metals
Vol. **24**, No. 1, p. 301–311 (1967).
See also *10.1*, p. *126*; 21.1, p. 401.

H.-E. Gumlich, R. Moser, and E. Neumann
On the Mn–Mn Interaction and the Structure of Optical Spectra of ZnS(Mn)
Vol. **24**, No. 1, p. K13–K16 (1967).
See also *20.3*, p. *367*; 22.4.1, p. 501.

T. A. T. Cowell and J. Woods
*Thermally Stimulated Currents and Infra-Red Luminescence in CdS Crystals
Vol. **24**, No. 1, p. K37–K40 (1967).
See also 10.2, p. 141; 14.3, p. 239; 20.3, p. 367; 22.4.1, p. 501.

T. H. Cheung and S. Wang
*Theoretical Studies on Z_1-Bands in Alkali Halide Crystals
Vol. **24**, No. 2, p. 509–523 (1967).
See also 10.2, p. 141; 22.5.2, p. 524.

M. K. Sheinkman, V. A. Tyagai, O. V. Snitko, I. B. Ermolovich, G. L. Belenky, and V. N. Bondarenko
*Studies on the Nature of CdS Single Crystal Sensitization by Etching
Vol. **24**, No. 2, p. 543–549 (1967).
See also 10, p. 103; 16, p. 281; 20.3, p. 368; 22.4.1, p. 501.

H. D. Koswig, U. Retter und W. Ulrici
*Jahn-Teller-Effekt an Übergangsmetallionen in Silberhalogeniden. (I): $Ti^{3+}(3d^1)$ und $Fe^{2+}(3d^6)$
(Jahn-Teller Effect of Transition-Metal Ions in Silver Halides. (I): $Ti^{3+}(3d^1)$ and $Fe^{2+}(3d^6)$)
Vol. **24**, No. 2, p. 605–614 (1967).
See also 20.1, p. 351; 22.5.1, p. 516.

R. A. Burmeister, Jr. and D. A. Stevenson
Electrical Properties of n-Type CdSe
Vol. **24**, No. 2, p. 683–694 (1967).
See also *14.3*, p. *239*; 22.4.2, p. 507.

H. W. den Hartog
*Fourth Shell Isotropic H.F.S.-Constants of F-Centers in CaF_2
Vol. **24**, No. 2, p. K117–K120 (1967).
See also 10.2, p. 141; 19, p. 331; 22.5.3, p. 533.

G. B. Abdullaev, M. Kh. Alieva, and A. Z. Mamedova
Thermal and Infrared Quenching of Photoconductivity in n-InSe Single Crystals
Vol. **25**, No. 1, p. 75–79 (1968).
See also *16*, p. *281*; 22, p. 449.

P. Krispin
On the Photoimpedance of ZnS Phosphors
Vol. **25**, No. 1, p. 147–154 (1968).
See also *14.4.1*, p. *261*; 16, p. 281; 22.4.1, p. 501.

H. L. Arora, S. Kim, and S. Wang
Optical Absorption of the Impurity Ag and Models for the E-Band in an Alkali Halide
Vol. **25**, No. 1, p. 223–232 (1968).
See also *10.2*, p. *141*; 22.5.2, p. 525.

J. Viščakas, P. Mačkus, and A. Smilga
Space-Charge-Limited Currents and High Electric Field Effect in Vitreous Selenium Films
Vol. **25**, No. 1, p. 331–335 (1968).
See also 2, p. 23; 14.3.1, p. 246; *14.4.1*, p. *261*; 22.1.3, p. 478.

G. Baldini, A. Jean, and G. Spinolo
Lifetime of the Excited Cu^+ in Alkali Halides
Vol. **25**, No. 2, p. 557–561 (1968).
See also 20.1, p. 351; *20.3*, p. *368*; 22.5.2, p. 525.

Yu. A. Logatchov
*Model Potential for Systems with Two Valence Electrons
Vol. **25**, No. 2, p. 763–774 (1968).
See also 20.3, p. 368; 22, p. 450.

H. W. den Hartog
F-Centers in BaF_2
Vol. **25**, No. 2, p. K111–K114 (1968).
See also *10.2*, p. *142*; 19, p. 333; 22.5.3, p. 533.

V. V. Antonov-Romanovskii
*On the Effective Recombination Cross-Section of Charged Centres in Crystal Phosphors
Vol. **26**, No. 1, p. 173–183 (1968).
See also 22, p. 450.

H.-P. Hennig
N- and S-Type Negative Resistances and Relaxation Oscillations in Au-Doped Double Injection Ge Diodes
Vol. **26**, No. 1, p. 225–229 (1968).
See also *14.4.1*, p. *262*; 22.1.1, p. 464.

A. BUROFF and J. MALINOWSKI
Charge Compensation of Trapped Holes in Silver Bromide
Vol. **26**, No. 1, p. 267–276 (1968).
See also 10.2, p. 142; *16*, p. *282*; 22.5.1, p. 516.

C. HAMANN
*On the Trap Distribution in Thin Films of Copper Phthalocyanine
Vol. **26**, No. 1, p. 311–318 (1968).
See also 14.3.1, p. 246; 14.4.1, p. 262; 22.9, p. 560.

P. M. KARAGEORGY-ALKALAEV and A. YU. LEIDERMAN
*Statistics of Inter-Impurity Recombination of Electrons and Holes in Semiconductors
Vol. **26**, No. 2, p. 419–428 (1968).
See also 20.3, p. 368; 22, p. 450.

A. B. KUNZ
*Notes on a Valence Electron Model of Atoms
Vol. **26**, No. 2, p. 447–452 (1968).
See also 22.5.2, p. 526.

E. D. TRIFONOV and A. S. TROSHIN
Nonlinear Polarisation of an Impurity Crystal
Vol. **26**, No. 2, p. 519–525 (1968).
See also *20.1*, p. *353*; *20.3*, p. *368*; 22, p. 451.

G. DÖHLER and K. HACKER
A Possibility for Band Structure Determination: Interband and Donor-Band Transitions of Crystal Electrons in High Electric Fields
Vol. **26**, No. 2, p. 551–564 (1968).
See also *13.1*, p. *185*; 14.4.1, p. 262; 20.1, p. 353; 22, p. 451.

W. B. FOWLER
*One-Band One-Site Model for Crystal Imperfections
Vol. **26**, No. 2, p. 639–644 (1968).

H. BÖTTGER
*Jahn-Teller-Aufspaltung einer ESR-Linie
(Jahn-Teller Splitting of an ESR Line)
Vol. **26**, No. 2, p. 681–692 (1968).
See also 6, p. 65; 19, p. 333; 22.6, p. 541.

A. V. TULUB and K. PATZER
On the Theory of Cooperative Absorption (Emission) of Light in Crystals
Vol. **26**, No. 2, p. 693–700 (1968).
See also *20.1*, p. *353*; *20.3*, p. *369*; 22.5, p. 513.

YU. A. LOGATCHOV
*Two-Electron Correlation in Impurity Centres in Model Potential Approximation
Vol. **26**, No. 2, p. 765–768 (1968).
See also 20.3, p. 369; 22.5.2, p. 526.

R. A. SHATAS and J. D. STETTLER
*Mollwo-Ivey Relation in Aggregate Defect Centers
Vol. **26**, No. 2, p. K155–K157 (1968).
See also 10.2, p. 142; 22.5.2, p. 526; 22.6, p. 541.

R. LINK and H. LUTSCH
Hall Effect Measurements in Pulse-Heated Tellurium Single Crystals
Vol. **26**, No. 2, p. K159–K162 (1968).
See also *14.3*, p. *241*; 22.1.3, p. 479.

I. K. Andronik, E. K. Arushanov, O. V. Emelyanenko, and D. N. Nasledov
Negative Magnetoresistance in p-CdSb
Vol. 27, No. 1, p. 45–49 (1968).
See also *14.3*, p. *241*; 22, p. 451; 22.7, p. 546.

G. Baldini, I. Pollini, and G. Spinolo
Optical Properties of α- and β-$TiCl_3$
Vol. 27, No. 1, p. 95–100 (1968).
See also *20.1*, p. *353*; 22.5, p. 514.

W. Fuhs and J. Stuke
Hopping Recombination in Trigonal Selenium Single Crystals
Vol. 27, No. 1, p. 171–184 (1968).
See also *16*, p. *282*; 22.1.3, p. 479.

J. H. P. van Weeren, G. Koopmans, and J. Blok
*The Position of the Dislocation Acceptor Level in n-Type Ge
Vol. 27, No. 1, p. 219–224 (1968).
See also 10, p. 106; 14.3, p. 241; 22.1.1, p. 465.

H. Friedrich and G. Junkg
*Generation-Recombination Statistics in Semiconductors — A Model with Two Centres
Vol. 27, No. 1, p. 237–248 (1968).
See also 16, p. 282; 22, p. 451.

W. Ulrici
Study of Trivalent Transition-Metal Ions in Silver Halides. (I): AgCl:V
Vol. 27, No. 1, p. 333–344 (1968).
See also *10*, p. *107*; 14.4, p. 254; 20.1, p. 354; 22.5.1, p. 516.

V. M. Agranovich and Yu. V. Konobeev
Diffusion of Free Excitons in Molecular Crystals
Vol. 27, No. 1, p. 435–442 (1968).
See also *13.2*, p. *194*; 22.9, p. 560.

R. J. Fleming
Activation Energies and Emission Spectra of the Thermoluminescence of CaF_2 Crystals Gamma-Irradiated at 77 °K
Vol. 27, No. 1, p. K57–K61 (1968).
See also *20.3*, p. *369*; 22.5.3, p. 534.

A. F. Lubchenko
Theory of Urbach's Rule for Impurity Centres
Vol. 27, No. 1, p. K73–K76 (1968).
See also 10, p. 107; *20.1*, p. *354*; 22, p. 452.

D. Schwarz
ESR Study of Cu^{2+} in AgCl Crystals Additionally Doped with Se or S
Vol. 27, No. 1, p. K77–K80 (1968).
See also 10, p. 107; *19*, p. *333*; 22.5.1, p. 516.

W. Ulrici
Study of Trivalent Transition-Metal Ions in Silver Halides. (II): Chromium and Titanium
Vol. 27, No. 2, p. 489–500 (1968).
See also *10*, p. *107*; 14.4, p. 254; 20.1, p. 354; 22.5.1, p. 517.

J. Sak
*On the Theory of Shallow Impurity States
Vol. 27, No. 2, p. 521–527 (1968).
See also 6, p. 66; 13.1, p. 186; 22, p. 452.

J. AUTH
On the Reverse-Biased Capacitance of Step p^+-n Junctions with Traps
Vol. 27, No. 2, p. 653–656 (1968).
See also *14.3.2*, p. *250*; 22.1.2, p. 474.

P. R. HERCZFELD and K. M. VAN VLIET
*Electronic Processes and Fluctuations in Optically Quenched CdS and CdSe Crystals
Vol. 27, No. 2, p. 681–694 (1968).
See also *16*, p. *282*; 22.4.1, p. 503; 22.4.2, p. 508.

K. MAIER and W. GEBHARDT
Stress-Induced Inhomogeneous Effects of the K-Absorption in RbCl
Vol. 27, No. 2, p. 713–721 (1968).
See also *10.2*, p. *143*; 22.5.2, p. 526.

H. K. MÜLLER
Electrical and Optical Properties of Sputtered In_2O_3 Films. (I): Electrical Properties and Intrinsic Absorption
Vol. 27, No. 2, p. 723–731 (1968).
See also *14.3.1*, p. *247*; 20.1, p. 355; 22.6, p. 541.

W. SZYMAŃSKA and J. GINTER
On the Scattering of Electrons by Ionised Impurities in the Case of Strong Degeneracy
Vol. 27, No. 2, p. K103–K106 (1968).
See also *14.3*, p. *242*; 22.4.3, p. 510.

V. I. CHEREPANOV and A. E. NIKIFOROV
Consideration of Polarization and Covalency Corrections to the Quasi-Coulomb Shift of Optical Lines of Ion Pairs in Crystals
Vol. 28, No. 1, p. 65–73 (1968).
See also *10*, p. *108*; 20.3, p. 370; 22.6, p. 542.

K. ŽĎÁNSKÝ
Vibrational Resonance Mode and the Temperature Dependence of Hyperfine Coupling of Mn^{2+} in BaO
Vol. 28, No. 1, p. 181–186 (1968).
See also 10, p. 108; *19*, p. *334*; 22.6, p. 542.

M. ALI, M. FRIDMAN, M. DENAYER, and P. NAGELS
Electrical Conduction in Li-Doped MnO Single Crystals
Vol. 28, No. 1, p. 193–206 (1968).
See also 13.1, p. 187; *14.3*, p. *242*; 15, p. 274; 22.6, p. 542.

B. G. DICK
*Calculation of Properties of Vacancy-Centered Z_1-Center Models in Alkali-Halides
Vol. 28, No. 1, p. 223–232 (1968).
See also *10.2*, p. *143*; 22.5.2, p. 527.

G. WOLFRAM
Photoconductivity of the Z_2-Center in KCl:Sr Crystals
Vol. 28, No. 1, p. K11–K13 (1968).
See also 10.2, p. 143; *16*, p. *283*; 22.5.2, p. 527.

V. V. ZOTOV and V. V. SERDYUK
Effect of Infrared Quenching on the Spectral Distribution of Induced Impurity Photoconductivity of CdS Single Crystals
Vol. 28, No. 1, p. K31–K33 (1968).
See also *16*, p. *283*; 22.4.1, p. 503.

V. L. BONCH-BRUEVICH
On the Problem of Recombination Autooscillations
Vol. **28**, No. 2, p. 687–693 (1968).
See also 13, p. 172; *14.4.1*, p. *263*; 22.1.1, p. 466.

R. C. WHELAN and D. SHAW
*Evidence of Doubly Ionized Native Donor in CdTe
Vol. **29**, No. 1, p. 145–152 (1968).
See also 10, p. 109; 14.3, p. 243; 22.4.3, p. 511.

H. J. GLÄSER and D. GEIST
Electrical Conductivity and Hall Effect of Manganese-Doped Germanium and the Influence of Heat Treatment on the State of Incorporation of Manganese
Vol. **29**, No. 1, p. 167–170 (1968).
See also *14.3*, p. *243*; 19, p. 335; 22.1.1, p. 466.

R. BOYN
*Optical Absorption due to Intrinsic Defects in CdS Single Crystals
Vol. **29**, No. 1, p. 307–321 (1968).
See also 10, p. 110; *20.1*, p. *356*; 22.4.1, p. 504.

V. M. ASNIN, G. L. ERISTAVI, and A. A. ROGACHEV
Effect of Weak Electric Fields on the Absorption Edge in Doped Germanium
Vol. **29**, No. 1, p. 443–446 (1968).
See also 13.2, p. 195; 14.3, p. 243; *20.1*, p. *356*; 22.1.1, p. 466.

D. T. Y. WEI, C. M. PENCHINA, and Y. S. PARK
Observation of Oscillatory Lifetime in CdS
Vol. **29**, No. 1, p. K7–K10 (1968).
See also *16*, p. *284*; 22.4.1, p. 504.

F. HUTH
Hall Effect and Mobility in n-Type GaAs
Vol. **29**, No. 1, p. K35–K37 (1968).
See also *14.3*, p. *243*; 22.2.1, p. 485.

R. J. FLEMING
Evaluation of 1s–2p Optical Transition Energy for F-Centres in CaF_2
Vol. **29**, No. 1, p. K77–K80 (1968).
See also *10.2*, p. *144*; 22.5.3, p. 534.

O. V. BOGDANKEVICH, N. A. BORISOV, I. V. KRUKOVA, and B. M. LAVRUSHIN
Temperature Dependence of Laser Threshold Current Density and Emission Spectra in Electron-Beam Pumped Gallium Arsenide Lasers
Vol. **29**, No. 2, p. 715–723 (1968).
See also *20.2*, p. *362*; 22.2.1, p. 486.

B. TUCK
Effect of Heat Treatments on GaAs Luminescence
Vol. **29**, No. 2, p. 793–798 (1968).
See also *20.3*, p. *370*; 22.2.1, p. 486.

G. E. ARKHANGELSKII, Z. L. MORGENSHTERN, and V. B. NEUSTRUEV
On the Nature of the Colour Centres of Ruby
Vol. **29**, No. 2, p. 831–836 (1968).
See also *10.2*, p. *144*; *20.3*, p. *371*; 22.6, p. 543.

A. MEHRA
Optical Absorption of Mn^{2+}-Doped Alkali Halides
Vol. **29**, No. 2, p. 847–857 (1968).
See also 6, p. 69; *10*, p. *111*; *20.1*, p. *357*; 22.5.2, p. 529.

J. GYULAI
Induced Impurity Breakdown Oscillations and Observation of Traps Lying Higher than the Indirect Band Edge in GaP
Vol. **29**, No. 2, p. K85–K88 (1968).
See also *14.4.1*, p. *264*; 16, p. 284; 22.2.2, p. 488.

D. K. ROY
*On the Determination of Impurity Concentration in the Alloyed Region of a Tunnel Diode
Vol. **29**, No. 2, p. K111–K115 (1968).
See also *14.3.2*, p. *251*; 22.1.1, p. 466.

M. L. REYNOLDS, W. E. HAGSTON, and G. F. J. GARLICK
Absorption and Infra-Red Emission of Nickel and Cobalt Doped Alkali Halides
Vol. **30**, No. 1, p. 97–103 (1968).
See also *20.1*, p. *358*; *20.3*, p. *371*; 22.5.2, p. 530.

M. L. REYNOLDS, W. E. HAGSTON, and G. F. J. GARLICK
*Experimental Observations of the Absorption, Infra-Red Emission, and Excitation Spectra of Cr^{3+} Ions in $ZnWO_4$
Vol. **30**, No. 1, p. 113–117 (1968).
See also 20.1, p. 358; 20.3, p. 371; 22.8, p. 556.

J. INGELS and G. JACOBS
Faraday Rotation of the U_2 Centre in KCl, KBr, and KI
Vol. **30**, No. 1, p. 163–172 (1968).
See also *18*, p. *295*; *20*, p. *342*; 22.5.2, p. 530.

R. L. GILBERT
Empirical Relationships for the Emission Energies and Lifetimes of the F-Center
Vol. **30**, No. 1, p. 173–182 (1968).
See also *10.2*, p. *144*; 20.3, p. 371; 22.5.2, p. 530.

V. HÁDEK
*Electron and Hole Conductivity of Donor-Acceptor Complexes of Oligoanilinic Substances and Their Thermoelectric Properties
Vol. **30**, No. 1, p. 275–282 (1968).
See also *14.3*, p. *244*; 15, p. 275; 22.9, p. 560.

R. BOYN
Polarization of Optical Absorption and Emission Bands due to Donor-Acceptor Associates in Wurtzite Type Crystals
Vol. **30**, No. 1, p. 305–310 (1968).
See also *20.1*, p. *358*; *20.3*, p. *371*; 22.4, p. 496.

W. STÖSSEL and W. ZIMMERMANN
Photoconduction in CdS Crystals Generated by a Modulated Electron Beam
Vol. **30** No. 1, p. 311–318 (1968).
See also *11*, p. *155*; *16*, p. *284*; 22.4.1, p. 505.

V. V. ANTONOV-ROMANOVSKII
*On the Trapping of Free and Release of Localized Carriers in Solids
Vol. **30**, No. 1, p. 341–347 (1968).
See also 22, p. 454.

A. S. DAVYDOV
*The Radiationless Transfer of Energy of Electronic Excitation between Impurity Molecules in Crystals
Vol. **30**, No. 1, p. 357–366 (1968).
See also 6, p. 69; 20.1, p. 358; 20.3, p. 371; 22, p. 454.

V. V. Boldyrev, G. B. London, and V. K. Zhuravlev
Study of Semiconducting Properties of Potassium Permanganate
Vol. **30**, No. 1, p. K13–K15 (1968).
See also *14.3*, p. *244*; *15*, p. *275*; 22.8, p. 556.

D. N. Bose
*Thermoluminescence and Thermally Stimulated Currents in Bismuth Doped ZnS Crystals
Vol. **30**, No. 1, p. K57–K60 (1968).
See also 16, p. 284; 20.3, p. 371; 22.4.1, p. 505.

H. H. Nickle
*Comment on a Valence Electron Model of Atoms
Vol. **30**, No. 1, p. K65–K67 (1968).
See also 22.5.2, p. 530.

G. G. Kovalevskaya and S. V. Slobodchikov
Lifetimes of Current Carriers in InP
Vol. **30**, No. 2, p. 441–447 (1968).
See also *14.3*, p. *244*; 22.2.2, p. 488.

K. Peuker and E. D. Trifonov
Theory of Light Emission by Impurity Centres from Non-Equilibrium Vibrational States
Vol. **30**, No. 2, p. 479–484 (1968).
See also 6, p. 70; *20.3*, p. *371*; 22, p. 455.

M. L. Reynolds, W. E. Hagston, and G. F. J. Garlick
*The Theory of the Absorption and Infrared Emission of Cr^{3+} Ions in $ZnWO_4$
Vol. **30**, No. 2, p. 507–514 (1968).
See also 20.1, p. 359; 20.3, p. 371; 22.8, p. 557.

N. E. Byer and H. S. Sack
Ultrasonic Velocity and Attenuation in Alkali Halides Containing CN^-. (I): Experimental Results
Vol. **30**, No. 2, p. 569–577 (1968).
See also *7*, p. *78*; 10, p. 112; 22.5.2, p. 530.

N. E. Byer and H. S. Sack
Ultrasonic Velocity and Attenuation in Alkali Halides Containing CN^-. (II): Theoretical Interpretation
Vol. **30**, No. 2, p. 579–586 (1968).
See also *7*, p. *78*; 10, p. 112; 22.5.2, p. 531.

G. Döhler
Critical Conditions for Transitions between Stationary and Non-Stationary High-Field Domains in Semi-Insulators
Vol. **30**, No. 2, p. 627–636 (1968).
See also *14.4.1*, p. *265*; 22.4.1, p. 505.

M. L. Reynolds, W. E. Hagston, and G. F. J. Garlick
*The Optical Absorption Spectra of Co and Ni in $ZnWO_4$
Vol. **30**, No. 2, p. 735–739 (1968).
See also *20.1*, p. *359*; 22.8, p. 557.

P. G. BARANOV, R. A. ZHITNIKOV, and N. I. MELNIKOV
Theoretical Study of Properties of Copper and Silver Atoms Stabilized in Alkali Halide Crystals
Vol. **30**, No. 2, p. 859–868 (1968).
See also 3.2, p. 36; 10.2, p. 144; *19*, p. *336*; 22.5.2, p. 531.

B. SELLE and J. MAEGE
Electrical and Optical Properties of Evaporated Cu_2S Films
Vol. **30**, No. 2, p. K153–K155 (1968).
See also *14.3.1*, p. *247*; 15, p. 275; 16, p. 285; *20.1*, p. *359*; 22, p. 455.

14. Electrical Properties of Solids. Transport Phenomena

Original Papers and Short Notes

B. Baranowski and J. Szymaszek
*The Electrical Resistance Anomaly of Nickel Hydride at Low Temperatures
Vol. **20**, No. 1, p. K37—K39 (1967).

H. Stolz
*Second Order Conductivity and Nonlinear Plasma Excitation of an Isotropic Electron Gas
Vol. **21**, No. 1, p. 77—88 (1967).
See also **13**, p. 165; 20, p. 339.

P. S. Zyryanov and V. I. Okulov
*Effect of Current Carrier Scattering on the Complex Dielectric Tensor
Vol. **21**, No. 1, p. 89—98 (1967).
See also 14.4, p. 253.

J. Mertsching and H. W. Streitwolf
*On the Derivation of the Boltzmann Equation in the Presence of a Magnetic Field
Vol. **21**, No. 2, p. K167—K169 (1967).

H. Stolz
*On the τ-Approximation in the Quantum Theory of the Conductivity Tensor
Vol. **23**, No. 2, p. 729—743 (1967).

14.1 Metals. Conductors

Review Articles

D. P. Morgan
Helicon Waves in Solids
Vol. **24**, No. 1, p. 9—36 (1967).
See also 6, p. 56; *13*, p. *162*; 14.3, p. 231; 19, p. 326; *20*, p. *337*; 21, p. 373; 21.1, p. 390; 21.2, p. 424; 21.6, p. 430; 22.1.1, p. 458; 22.2.1, p. 482; 22.2.3, p. 489.

Original Papers and Short Notes

H. Kronmüller und A. Seeger
*Magnetostriktiver und magnetokritalliner elektrischer Restwiderstand in Ferromagnetika
(Magnetostrictive and Magnetocyrstalline Residual Electric Resistivity in Ferromagnetics)
Vol. **16**, No. 1, p. 253—266 (1966).
See also 13, p. 162; *18.2*, p. *299*; 21, p. 373.

P. Peretto, J. L. Oddou, C. Minier-Cassayre, D. Dautreppe, and P. Moser
Investigation and Interpretation of Stage I of Nickel Electron Irradiated at 20 °K
Vol. **16**, No. 1, p. 281—287 (1966).
See also *10.1*, p. *114*; 18.2, p. 299; 21.1, p. 390.

R. Dietz and W. Hellenthal
Recovery Processes in Nickel Thin Films between 20 and 300 °C.
Vol. **16**, No. 2, p. 789—794 (1966).
See also *10.1*, p. *114*; 21.1, p. 391.

K. Isebeck, F. Rau, W. Schilling, K. Sonnenberg, P. Tischer, and H. Wenzl
Stored Energy, Volume, and Resistivity Change in Neutron Irradiated Aluminium
Vol. **17**, No. 1, p. 259—268 (1966).
See also 10.1, p. 115; *11*, p. *146*; 21, p. 374.

P. Vigier, C. Minier-Cassayre, P. Moser, and V. Hivert
Study of Point Defects Produced by Neutron Irradiation in Doped Iron
Vol. **17**, No. 1, p. 317—321 (1966).
See also *10.1*, p. *116*; 11, p. 146; 18.2, p. 302; 21.1.1, p. 412.

R. Pichon, C. Minier-Cassayre, V. Hivert et P. Moser
*Étude du revenu de la résistivité du niobium irradié aux électrons à 20 °K
(Study of the Recovery of Resistivity of Niobium Electron Irradiated at 20 °K)
Vol. **17**, No. 1, p. K33—K35 (1966).
See also 10.1, p. 116; 21, p. 374.

E. Kovács-Csetényi, C. R. Vassel, and I. Kovács
Determination of the Solubility of Si in Dilute Al-Si Alloys by Low-Temperature Resistivity Measurements
Vol. **17**, No. 1, p. K123—K126 (1966).
See also *1.1*, p. *8*; 21, p. 375.

J. L. Oddou, C. Minier-Cassayre, and P. Moser
Effect of Impurities on the Recovery Stages of the Resistivity of Nickel Electron Irradiated at Low Temperatures
Vol. **17**, No. 2, p. 793—798 (1966).
See also *10.1*, p. *117*; *11*, p. *146*; 21.1, p. 393.

Vu dinh Ky
The Nernst Effect in Permalloy Films
Vol. **17**, No. 2, p. K203—K205 (1966).
See also 15, p. 270; *18.2*, p. *303*; 21.1.1, p. 413.

Vu dinh Ky
The Planar Nernst Effect in Permalloy Films
Vol. **17**, No. 2, p. K207—K209 (1966).
See also 15, p. 270; *18.2*, p. *303*; 21.1.1, p. 413.

J. P. Issi, A. Moureau, and A. Luyckx
*Low-Field Galvanomagnetic Coefficients of Polycrystalline Bismuth at 80 °K
Vol. **17**, No. 2, p. K217—K219 (1966).
See also 21.7, p. 438.

R. Lück
*Zur Temperatur- und Feldabhängigkeit der galvanomagnetischen Eigenschaften von Aluminium und Indium
(Temperature and Field Dependence of the Galvanomagnetic Properties of Aluminium and Indium)
Vol. **18**, No. 1, p. 49—57 (1966).
See also 13.1, p. 176; 21, p. 375.

R. Lück
*Die Temperatur- und Feldabhängigkeit der galvanomagnetischen Eigenschaften von Blei und der Einfluß der endlichen Probedicke hierauf
(Temperature and Field Dependence of the Galvanomagnetic Properties of Lead and the Effect of Finite Sample-Thickness on These Dependences)
Vol. **18**, No. 1, p. 59—66 (1966).
See also 13.1, p. 176; 21, p. 375.

K. Isebeck, R. Müller, W. Schilling, and H. Wenzl
Reaction Kinetics of Stage III Recovery in Aluminium after Neutron Irradiation
Vol. **18**, No. 1, p. 427—439 (1966).
See also *10.1*, p. *118*; 11, p. 147; 21, p. 375.

C. Varotto and A. E. Vidoz
*On the Recovery and Ordering Processes of Cold Worked Ni-Fe Alloys
Vol. **18**, No. 1, p. K49—K52 (1966).
See also 10.1, p. 118; 21.1.1, p. 414.

S. Arajs, G. R. Dunmyre, and S. J. Dechter
*Electrical Resistivity and Longitudinal Electrical Magnetoresistivity of Palladium-Rich Palladium-Erbium Alloys at Low Temperatures
Vol. **18**, No. 2, p. 505—509 (1966).
See also 21, p. 376.

H. Yoshinaga
*Measurements of the Ansiotropy of the Dislocation Resistivity in Au, Ag, and Cu
Vol. **18**, No. 2, p. 625—636 (1966).
See also 10.1, p. 119; 21.1, p. 394; 21.6, p. 431.

L. I. van Torne
*Point Defect Electron Scattering in Beryllium
Vol. **18**, No. 2, p. 737—742 (1966).
See also 10.1, p. 119; 21.3, p. 426.

S. M. KLOTSMAN, A. N. TIMOFEEV, and I. SH. TRAKHTENBERG
On the Mechanism of Lattice Electromigration in Metals
Vol. **18**, No. 2, p. 847–852 (1966).
See also *9*, p. *88*; 21.1, p. 394; 21.6, p. 432.

G. S. NIKOLSKII and V. V. EREMENKO
The Effect of a Magnetic Field on the Heat Conductivity of Er Polycrystals
Vol. **18**, No. 2, p. K123–K125 (1966).
See also *8*, p. *80*; 21.4, p. 428.

M. S. R. CHARI
*The Electrical and Thermal Resistivities of Iron
Vol. **19**, No. 1, p. 169–172 (1967).
See also 8, p. 81; 21.1.1, p. 415.

H. P. VAN DE BRAAK and W. J. CASPERS
*Disorder Resistivity and Spin Depolarization in Rare-Earth Alloys
Vol. **19**, No. 1, p. 385–393 (1967).
See also 13, p. 164; 18.2, p. 305; 21.4, p. 428.

L. I. VAN TORNE
Electron Charge Distribution about Zinc Ions in Aluminium
Vol. **19**, No. 2, p. 855–862 (1967).
See also 10.1, p. 121; *13*, p. *165*; 21, p. 377.

B. SPRUŠIL and P. VOSTRÝ
*Resistivity Changes in Air-Quenched Gold Wires
Vol. **19**, No. 2, p. K83–K86 (1967).
See also 21.6, p. 432.

H. BROSS and G. BOHN
*On the Theory of Electrical Resistivity of Polyvalent Metals
Vol. **20**, No. 1, p. 277–283 (1967).
See also 21, p. 377.

R. LALLEMENT and D. A. WIGLEY
Study of the Variation of Length of α-Plutonium by Self-Irradiation at 4 °K
Vol. **21**, No. 1, p. 145–150 (1967).
See also *11*, p. *148*; 12, p. 157; 21.5, p. 429.

A. KÖTHE and F. SCHLÄT
Recovery of Deformed Niobium and Tantalum Degassed in Ultra-High Vacuum
Vol. **21**, No. 1, p. K73–K76 (1967).
See also *10.1*, p. *122*; 21, p. 379.

B. F. ROTHENSTEIN, A. POLICEC, M. LUPULESCU, and C. ANGHEL
The Electromagnetic Aspect of Magnetostrictively Induced Torsional Oscillations (the Procopiu Effect with a Cylindrical Ferromagnetic Specimen Fixed at one End)
Vol. **21**, No. 2, p. 601–606 (1967).
See also *18.2*, p. *307*; 21.1.1, p. 417.

R. LÜCK und K. E. SAEGER
*Über die Messung der Anisotropie der galvanomagnetischen Transversalspannung an Einkristallfolien
(Measurement of the Anisotropy of the Galvanomagnetic Transverse Voltage in Single-Crystal Foils)
Vol. **21**, No. 2, p. 671–677 (1967).
See also 21.1, p. 398.

R. Herrmann
*Oszillationen der Oberflächenimpedanz von Wolfram in schwachen Magnetfeldern (Oscillations of the Surface Impedance of Tungsten in Weak Magnetic Fields)
Vol. **21**, No. 2, p. 703–707 (1967).
See also **13.1**, p. 178; *19*, p. *329*; 21, p. 379.

F. Dworschak, H. Schuster, H. Wollenberger, and J. Wurm
The Influence of the Size Effect on Electrical Resistivity Measurements in Irradiated Metals
Vol. **21**, No. 2, p. 741–745 (1967).
See also *11*, p. *149*; 21, p. 379.

S. I. Masharov
*Electrical Resistivity in Antiferromagnetic Alloys
Vol. **21**, No. 2, p. 747–753 (1967).
See also 18.4, p. 322; 21, p. 379.

Yu. A. Bogod and V. V. Eremenko
*Peculiarities of the Magnetoresistance in Bi in High Transverse and Longitudial Magnetic Fields
Vol. **21**, No. 2, p. 797–803 (1967).
See also 21.7, p. 438.

K. J. Kedves
On Quenching Experiments in Gold
Vol. **21**, No. 2, p. K183–K186 (1967).
See also *10.1*, p. *124*; 21.6, p. 433.

Yu. P. Irkhin, A. N. Voloshinskii, and Sh. Sh. Abelskii
*On the Theory of Spontaneous Hall Effect at Low Temperatures
Vol. **22**, No. 1, p. 309–318 (1967).
See also 13, p. 166; 18.2., p. 308.

H. Bilger and H. Kronmüller
Recovery Spectrum of Cobalt
Vol. **22**, No. 2, p. 683–688 (1967).
See also *10.1*, p. *124*; 21.1, p. 399.

Vu dinh Ky
*Planar Hall and Nernst Effect in Ferrogmanetic Metals
Vol. **22**, No. 2, p. 729–736 (1967).
See also 13, p. 167; *15*, p. *272*; 18.2, p. 308; 21.1, p. 400; 21.1.1, p. 418.

S. Arajs
Magnetic Susceptibility of Face-Centered Cubic Iron–Rhenium Alloys
Vol. **22**, No. 2, p. 737–740 (1967).
See also *18.4*, p. *323*; 21.1.1, p. 418.

H. Oppermann, G. Weise, J. Barthel, and K.-H. Berthel
Molybdenum Single Crystals with Very Small Residual-Resistivity Ratios
Vol. **22**, No. 2, p. K151–K153 (1967).
See also *3.1*, p. *32*; 21, p. 381.

G. Duesing, H. Hemmerich, D. Meissner, and W. Schilling
The Influence of Vacancies and Impurities on the Damage Production in Platinum Electron-Irradiated at 90 °K
Vol. **23**, No. 2, p. 481–488 (1967).
See also 10.1, p. 125; *11*, p. *150*; 21.6, p. 434.

P. S. Ho and A. L. Ruoff
A Quasi-Harmonic Calculation of Lattice Dynamics for Na
Vol. **23**, No. 2, p. 489–499 (1967).
See also *6*, p. *63*; 13.1, p. 182; 21.2, p. 425.

M. L. Swanson
The Effects of Defect Doping on Low Temperature Neutron-Irradiation Damage and Recovery in Aluminium and Platinum
Vol. **23**, No. 2, p. 649–661 (1967).
See also 10.1, p. 125; *11*, p. *150*; 21, p. 381; 21.6, p. 434.

H. H. Neely and D. W. Keefer
Resistivity Studies of Electron Irradiated Iron
Vol. **24**, No. 1, p. 217–220 (1967).
See also *11*, p. *150*; 21.1.1, p. 419.

W. Glaeser, H. Imgrund und H. Wever
*Zur Frage der reproduzierbaren Restwiderstandsmessung an hochreinem Eisen
(On the Problem of Reproducible Measurements of the Residual Resistivity of High-Purity Iron)
Vol. **24**, No. 1, p. K5–K8 (1967).
See also 18.2, p. 309; 21.1.1, p. 420.

B. Sprušil and P. Vostrý
Reply to "On Quenching Experiments in Gold" by F. J. Kedves
Vol. **24**, No. 1, p. K41–K42 (1967).
See also *10.1*, p. *126*; 21.6, p. 434.

Yu. A. Bogod
*On the Resistance of Bi in High Magnetic Fields
Vol. **24**, No. 1, p. K49–K51 (1967).
See also 21.7, p. 438.

R. Brugière and P. Lucasson
On the Stage III Recovery of Electron Irradiated Aluminium
Vol. **24**, No. 1, p. K77–K81 (1967).
See also *10.1*, p. *127*; 11, p. 150; 21, p. 383.

A. Seeger, E. Mann, and K. Clausecker
*Magnetoresistivity of Metals. (I): General Theory
Vol. **24**, No. 2, p. 721–732 (1967).
See also 21, p. 383

J. H. Evans and B. L. Eyre
*The Effect of Tungsten Content on the Residual Resistivity of Zone Refined Molybdenum
Vol. **25**, No. 1, p. K39–K41 (1968).
See also *3.1*, p. *33*; 21, p. 384.

C. Lea and C. H. B. Mee
The Attenuation Lenght of Photoelectrons in Thin Films of Uranium
Vol. **25**, No. 2, p. 613–619 (1968).
See also *17*, p. *288*; 21.5, p. 429.

G. I. Guseva and P. S. Zyryanov
Quantum Theory of Thermogalvanomagnetic Phenomena (TGMP) in Conductors with Equal Concentration of Electrons and Holes
Vol. **25**, No. 2, p. 775–785 (1968).
See also 14.3, p. 240; *15*, p. *273*; 21.7, p. 438; 22.2, p. 481.

R. Kishore
*Electrical Conductivity of Metals
Vol. **26**, No. 1, p. 133–138 (1968).
See also 21.1, p. 403; 21.2, p. 425; 21.6, p. 435.

H. Beyer and R. Sizmann
Recrystallization of Cold-Worked Gold during α-Irradiation
Vol. **26**, No. 1, p. 359–370 (1968).
See also 4, p. 45; 10.1, p. 129; *11*, p. *151*; 21.6, p. 435.

V. Christoph
*On the Theory of the Hall Effect in Ferromagnetics
Vol. **26**, No. 1, p. K17–K20 (1968).
See also 18.2, p. 311.

H. Wagenblast and S. Arajs
*Electrical Resistivity Studies of Iron-Nitrogen Solid Solutions
Vol. **26**, No. 2, p. 409–418 (1968).
See also 21.1.1, p. 420.

Vu dinh Ky
*Planar Hall Effect in Ferromagnetic Films
Vol. **26**, No. 2, p. 565–569 (1968).
See also *18.2*, p. *311*; 21.1, p. 404; 21.1.1, p. 421.

C. W. Chen, L. R. Edwards, and S. Legvold
*Resistance Anomaly in Palladium-Silver Alloys Containing Rare Earth Impurities
Vol. **26**, No. 2, p. 611–619 (1968).
See also 10.1, p. 129; 21, p. 385; 21.6, p. 435.

I. Y. Dekhtyar, E. G. Madatova, P. S. Nisin, and R. G. Fedchenko
Plastic Deformation Effect on the Electronic Properties of Solid Solutions with Localized Magnetic Moments
Vol. **26**, No. 2, p. K151–K154 (1968).
See also 6.1, p. 73; *10.1*, p. *130*; 18.1, p. 297 ; 21.6, p. 436.

H. J. Fischbeck and J. Mertsching
The Theory of Low-Field Oscillations of the Surface Impedance in Metals
Vol. **27**, No. 1, p. 345–357 (1968).
See also 13.1, p. 186; *19*, p. *333*; 21, p. 386.

S. I. Masharov
Effect of Phonon Drag on the Kinetic Properties of Alloys
Vol. **27**, No. 1, p. 455–461 (1968).
See also *6*, p. *65*; 8, p. 84; 13, p. 170; 15, p. 274; 21, p. 386.

J. Tóth
*The Temperature Dependence of Resistivity in Al-Ta Dilute Alloys
Vol. **27**, No. 1, p. K47–K48 (1968).
See also 21, p. 386.

S. Arajs
*An Investigation of the Electrical Resistivity of Nickel-Rich Nickel-Gold Alloys
Vol. **28**, No. 1, p. 171–174 (1968).
See also 21.1, p. 405; 21.6, p. 436.

K. E. Saeger
*Hall Effect and Magnetoresistance in Copper Single Crystals at Low Temperatures
Vol. **28**, No. 2, p. 589–602 (1968).
See also 21.1, p. 406.

H. H. NEELY, D. W. KEEFER, and A. SOSIN
Electron Irradiation and Recovery of Tungsten
Vol. 28, No. 2, p. 675–682 (1968).
See also 10.1, p. 132; *11*, p. *153*; 21, p. 387.

R. KISHORE
*On Transverse Electrical Conductivity of Metals
Vol. 28, No. 2, p. 695–698 (1968).
See also 21, p. 387.

J. POLÁK
Thermal Quasi-Equilibrium of Vacancies, Divacancies, and Impurities in Platinum
Vol. 28, No. 2, p. 773–782 (1968).
See also *10.1*, p. *132*; 21.6, p. 436.

YU. A. BOGOD, V. V. EREMENKO, and L. K. CHUBOVA
*The Influence of Boundary Surface on the Magnetoresistance of Bismuth
Vol. 28, No. 2, p. K155–K157 (1968).
See also 21.7, p. 438.

K. SCHRÖDER and N. P. BAUM
*Absolute Thermoelectric Power and Electrical Resistivity of Chromium-Rich Chromium–Nickel Alloys
Vol. 29, No. 1, p. 107–114 (1968).
See also 13.1, p. 188; *15*, p. *274*; 18.4, p. 324; 21.1, p. 407.

Z. I. ALIZADE and T. G. MAGARRAMOV
*Galvanomagnetic Effect in Ferromagnetic Alloys of Nickel–Tantalum and Nickel–Niobium Systems
Vol. 29, No. 1, p. K19–K22 (1968).
See also *18.2*, p. *314*; 21.1, p. 407.

J. DELAPLACE, J. C. NICOUD, D. SCHUMACHER, and G. VOGL
Low-Temperature Neutron Radiation Damage and Recovery in Beryllium
Vol. 29, No. 2, p. 819–824 (1968).
See also 10.1, p. 134; *11*, p. *154*; 21.3, p. 427.

J. C. JOUSSET
Irradiation Defects in α-Uranium and Matthiessen's Rule
Vol. 29, No. 2, p. K127–K131 (1968).
See also *11*, p. *154*; 21.5, p. 429.

H. R. VYDYANATH, D. H. SASTRY, and K. I. VASU
On the Kinetics of Clustering in a Magnesium-3%Zinc Alloy
Vol. 29, No. 2, p. K137–K140 (1968).
See also *4*, p. *47*; 10.1, p. 134; 21.1, p. 408; 21.3, p. 427.

I. GAAL and L. URAY
Investigation of Non-Conductive Second Phases by Means of the Shape Factor of Electrical Resistance
Vol. 29, No. 2, p. K163–K166 (1968).
See also *4*, p. *48*; 21, p. 388.

S. J. SIVONEN and E. J. SUONINEN
Temperature Dependence of the Magnetic Ordering in Gd-$GdFe_2$ Phase Mixtures
Vol. 29, No. 2, p. K171–K173 (1968).
See also *18.2*, p. *314*; 21.1.1, p. 423; 21.4, p. 428.

J. DELAPLACE, J. HILLAIRET, J. C. NICOUD, D. SCHUMACHER, and G. VOGL
Low Temperature Neutron Radiation Damage and Recovery in Magnesium
Vol. 30, No. 1, p. 119–126 (1968).
See also 10.1, p. 134; *11*, p. *154*; 21.3, p. 427.

D. Becker, F. Dworschak, C. Lehmann, K. T. Rie, H. Schuster, H. Wollenberger, and J. Wurm
Determination of the Average Number of Replacement Collisions in an Electron-Irradiated F.C.C. Crystal (Ni_3Mn)
Vol. **30**, No. 1, p. 219—229 (1968).
See also 10.1, p. 135; *11*, p. *155*; 18.2, p. 315; 21.1, p. 408.

T. Skośkiewicz and B. Baranowski
*Investigation of the Electrical Resistance Anomaly in the Palladium-Hydrogen System
Vol. **30**, No. 1, p. K33—K35 (1968).
See also 10.1, p. 135; 21, p. 388.

R. E. Hummel and J. W. Koger
Effect of Neutron Irradiation on the Martensitic Transformation in β_1-Copper–Zinc
Vol. **30**, No. 2, p. 515—523 (1968).
See also *1.2*, p. *13*; *11*, p. *155*; 21.1, p. 409.

R. Kishore
Green Function Theory of Transport Properties of Anharmonic Metals
Vol. **30**, No. 2, p. 817—826 (1968).
See also *6*, p. *70*; 8, p. 86; *13*, p. *173*; 21, p. 389.

R. Brugière and P. Lucasson
On the Stage III Recovery of Electron Irradiated Copper
Vol. **30**, No. 2, p. K139—K143 (1968).
See also *10.1*, p. *136*; 11, p. 155; 21.1, p. 409.

14.2 Superconductivity. Superconducting Materials and Devices

Original Papers and Short Notes

J. S. RAJPUT and A. K. GUPTA
*On the Specific Heat Anomaly in Superconducting Lead
Vol. 16, No. 1, p. K51—K53 (1966).
See also *8*, p. *79*; 21, p. 373.

J. NIEMIEC and E. TROJNAR
*The Transition Temperature of Superconducting Technetium–Rhenium Alloys
Vol. 17, No. 1, p. K89—K90 (1966).

G. J. VAN GURP
*Superconducting Fluxline Pinning by Twin Boundaries
Vol. 17, No. 1, p. K135—K137 (1966).
See also 4, p. 38.

H. A. ULLMAIER
*AC Measurements on Hard Superconductors
Vol. 17, No. 2, p. 631—643 (1966).

H. KOPPE
*The Current Carrying Capacity of a Hard Superconducting Wire in Zero External Field
Vol. 17, No. 2, p. K229—K232 (1966).

H. A. ULLMAIER and R. H. KERNOHAN
*The Current Carrying Capacity of a Hard Superconducting Wire in Zero External Field
Vol. 17, No. 2, p. K233—K235 (1966).

V. RADHAKRISHNAN
*Tunnelling in a Two-Band Model Superconductor
Vol. 18, No. 1, p. 113—118 (1966).

E. NEMBACH, H. FREYHARDT und P. HAASEN
*Abhängigkeit des Supraleitverhaltens tordierter Niobeinkristalle vom Probendurchmesser
(Dependence of the Superconducting Behaviour of Twisted Niobium Single Crystals on Sample Diameter)
Vol. 18, No. 2, p. 807—812 (1966).

H. TRÄUBLE und U. ESSMANN
*Ein hochauflösendes Verfahren zur Untersuchung magnetischer Struckturen von Supraleitern
(High-Resolution Technique for Studying Magnetic Structures of Superconductors)
Vol. 18, No. 2, p. 813—828 (1966).
See also 4, p. 39.

G. J. VAN GURP
*Temperature Dependence of the Critical Field in Superconducting Vanadium
Vol. 19, No. 1, p. 173—176 (1967).
See also 21.1, p. 395.

R. LABUSCH
*Elastische Konstanten des Flußfadengitters in Supraleitern zweiter Art
(Elastic Constants of the Flux Line Lattice in Type II Superconductors)
Vol. 19, No. 2, p. 715—719 (1967).
See also 12.1, p. 159.

H. Träuble und U. Essmann
*Die Beobachtung magnetischer Strukturen von Supraleitern zweiter Art
(Observation of Magnetic Structures in Type II Superconductors)
Vol. **20**, No. 1, p. 95–111 (1967).
See also 18, p. 291; 21, p. 377.

V. Radhakrishnan
*The Penetration Depth in Two Band Model Superconductors
Vol. **20**, No. 2, p. 783–789 (1967).
See also 21, p. 378.

R. Freud
*The Influence of Joule's Heat upon Characteristic Magnitudes in a Cylinder with Current in the Intermediate State
Vol. **20**, No. 2, p. K151–K153 (1967).

R. A. French
*Type-1 Superconductivity in Molybdenum
Vol. **21**, No. 1, p. K35–K38 (1967).
See also 13.1, p. 178; 21, p. 379.

S. Takács
*Der Flußeintritt in Supraleiter II. Art
(Penetration of Flux into Type II Superconductors)
Vol. **21**, No. 2, p. 709–716 (1967).

E. J. Kramer and Ch. L. Bauer
Variations of Young's Modulus throughout the Mixed State in Niobium
Vol. **22**, No. 1, p. 199–202 (1967).
See also 1.2, p. 11; *10.1*, p. *124*; 18, p. 292, 21, p. 380.

L. Dobrosavljević et C. Dupuis
*Étude d'eutectiques supraconducteurs plomb–étain à lamelles orientées
(Study of Superconductiong Pb–Sn Eutectic with Oriented Lamellas)
Vol. **23**, No. 2, p. 509–520 (1967).
See also 18, p. 293; 21, p. 381.

R. M. Easson and P. Hlawiczka
*Effect of H_{c1} on ac Losses in Niobium
Vol. **23**, No. 2, p. K129–K132 (1967).
See also 21, p. 382.

S. A. Nemnonov and E. Z. Kurmaev
*Band Structure and Superconductivity of A_3B-Type Intermetallic Compounds with β-W Structure
Vol. **24**, No. 1, p. K43–K47 (1967).
See also *13.1*, p. *183*; 21, p. 383.

A. Hubert
On the Theory of Two-Phase Domain Structures in Superconductors and Ferromagnets
Vol. **24**, No. 2, p. 669–682 (1967).
See also *4*, p. *44*; 18.2, p. 309; 21.1, p. 402.

H. Träuble und U. Essmann
*Fehler im Flußliniengitter von Supraleitern zweiter Art
(Defects in the Flux Thread Lattice in Type II Superconductors)
Vol. **25**, No. 1, p. 373–393 (1968).
See also 10.1, p. 128; 21, p. 384.

H. TRÄUBLE und U. ESSMANN
*Der direkte Nachweis von Flußlinienbewegungen in stromdurchflossenen Supraleitern
(Direct Evidence for Motive of Flux Threads during Current Passage in Superconductors)
Vol. **25**, No. 1, p. 395—402 (1968).
See also 10.1, p. 128; 21, p. 384.

R. H. T. YEH and H. MECHETTI
*Superconductive Current in a Superconductor-Normal Metal-Superconductor Junction
Vol. **25**, No. 1, p. K65—K66 (1968).

R. H. T. YEH
*Ultrasonic Attenuation of Anisotropic Superconductor in $ql \ll 1$ Limit
Vol. **25**, No. 2, p. 551—555 (1968).
See also 7, p. 77.

T. CLAESON
*Superconductivity in the Mercury–Tin System
Vol. **25**, No. 2, p. K95—K98 (1968).

R. FREUD, B. MAKIEJ, and A. SIKORA
*The Character of the Transition of Pure Tin to the Intermediate State under the Influence of Electric Current at Temperatures Close to T_c
Vol. **25**, No. 2, p. K127—K130 (1968).

A. A. GALKIN and V. M. SVISTUNOV
*Tunnelling Effect and High Pressures
Vol. **26**, No. 1, p. K55—K57 (1968).

A. SEEGER and H. KRONMÜLLER
*The Micromagnetic Equations of a Superconductor
Vol. **27**, No. 1, p. 371—382 (1968).
See also 10.1, p. 131.

R. M. F. LINFORD and R. G. RHODES
*AC Losses near the Lower Critical Field of Superconducting Niobium
Vol. **28**, No. 1, p. K63—K65 (1968).

R. H. T. YEH and H. MECHETTI
*Phase-Dependent Superconductive Current Density in a Superconductor–Normal Metal–Superconductor Junction at $T = T_{CN}$
Vol. **28**, No. 2, p. 683—686 (1968).

W. WELLER
*Nonequilibrium Thermodynamics of Superconducting Alloys in High Magnetic Fields
Vol. **30**, No. 1, p. 373—378 (1968).

A. A. GALKIN, A. A. DIKII, and V. M. SVISTUNOV
*Influence of High Pressure on the Energy Gap of Tin
Vol. **30**, No. 2, p. K107—K110 (1968).

14.3 Semiconductors

Review Articles

A. S. BORSHCHEVSKII, N. A. GORYUNOVA, F. P. KESAMANLY, and D. N. NASLEDOV
*Semiconducting $A^{II}B^{IV}C_2^{V}$ Compounds
Vol. **21**, No. 1, p. 9–55 (1967).
See also 3.1, p. 29; 3.2, p. 34; 5, p. 49; 8, p. 79; 10, p. 93; 13.1, p. 174; 15, p. 270; 16, p. 276; 17, p. 286; 18, p. 291; 20.1, p. 344; 22.8, p. 547.

D. P. MORGAN
Helicon Waves in Solids
Vol. **24**, No. 1, p. 9–36 (1967).
See also 6, p. 56; *13*, p. *162*; 14.1, p. 220; 19, p. 326; *20*, p. *337*; 21, p. 373; 21.1, p. 390; 21.2, p. 424; 21.6, p. 430; 22.1.1, p. 458; 22.2.1, p. 482; 22.2.3, p. 489.

G. HARBEKE
On the Band Structure of Anisotropic Crystals
Vol. **27**, No. 1, p. 9–44 (1968).
See also *13.1*, p. *174*; 13.2, p. 191; 20.1, p. 344; 22, p. 439; 22.1, p. 456; 22.1.1, p. 458; 22.1.2, p. 468; 22.1.3, p. 476; 22.4.1, p. 497; 22.4.2, p. 506; 22.4.3, p. 509; 22.5, p. 512; 22.7; p. 545; 22.8, p. 547.

V. L. BONCH-BRUEVICH and E. G. LANDSBERG
*Recombination Mechanisms
Vol. **29**, No. 1, p. 9–43 (1968).
See also 6, p. 56; 13, p. 162; 13.2, p. 191; *13.4*, p. *200*; 20, p. 337; 20.3, p. 363; 22.1.1, p. 458; 22.1.2, p. 468.

Original Papers and Short Notes

E. I. RASHBA, V. A. ROMANOV, I. I. BOIKO, and I. P. ZHADKO
*Electrical Pinch in Elastically Deformed Germanium
Vol. **16**, No. 1, p. 43–56 (1966).
See also 12, p. 156; 14.4.1, p. 255; 22.1.1, p. 459.

E. KAMIENIECKI
*Distribution Function of Hot Carriers
Vol. **16**, No. 1, p. 57–62 (1966).
See also 6, p. 56; *13*, p. *162*; 22.1.1, p. 459; 22.1.2, p. 468.

P. C. EUTHYMIOU and A. L. TANENBAUM
Effect of Neutron and γ-Ray Irradiation on the Carrier Lifetime and Detectivity of InSb
Vol. **16**, No. 1, p. 183–190 (1966).
See also *11*, p. *145*; 22.2.3, p. 489.

V. P. KALASHNIKOV
*Influence of Spatial Inhomogeneities on the Nuclear Polarization due to Direct Current in Semiconductors
Vol. **16**, No. 1, p. K35–K37 (1966).
See also *13*, p. *163*; 22, p. 439.

P. THOMA and G. VAUBEL
Electrical Properties of Anthracene Single Crystals at Very Low Temperatures
Vol. **16**, No. 2, p. 663–674 (1966).
See also 13.4, p. 201; 14.4.1, p. 256; *16*, p. *276*; 22.9, p. 558.

L. GOUSKOV, G. LECOY, C. LLINARES, and M. SAVELLI
Study of Lifetime and Generation–Recombination Noise Controlled by Dislocations in Germanium
Vol. 16, No. 2, p. 721–728 (1966).
See also *13.4*, p. *202*; 16, p. 276; 22.1.1, p. 459.

W. ŻDANOWICZ and A. WOJAKOWSKI
*Semiconducting Properties of CdP_4
Vol. 16, No. 2, p. K129–K131 (1966).
See also 15, p. 270; 20.1, p. 345; 22, p. 440.

B. PÖDÖR
*Electron Mobility in Plastically Deformed Germanium
Vol. 16, No. 2, p. K167–K170 (1966).
See also 10, p. 94; 22.1.1, p. 459.

D. V. EDDOLLS
*Electrical Properties of n-Type Epitaxial Gallium Arsenide
Vol. 17, No. 1, p. 67–76 (1966).
See also 13.4, p. 202; 22.2.1, p. 482.

S. A. ALIEV, F. P. KESAMANLY, T. S. LAGUNOVA, and D. N. NASLEDOV
*Hall Effect and Magnetoresistance of n-InP Crystals at Low Temperatures
Vol. 17, No. 1, p. 105–108 (1966).
See also 13.4, p. 202; 22.2.2, p. 487.

L. ELSTNER
*Die Temperaturabhängigkeit der Defektelektronenbeweglichkeit in Siliziumkristallen
(Temperature Dependence of the Hole Mobility in Silicon Crystals)
Vol. 17, No. 1, p. 139–150 (1966).
See also 6, p. 56; 22.1.2, p. 469.

A. A. ABDURAKHMANOVA and M. I. ALIEV
*Electrical Properties of GaSb-Ga_2Te_3 and GaSb-GaTe Solid Solutions
Vol. 17, No. 1, p. 375–380 (1966).
See also 13.1, p. 175; 15, p. 270; 22.8, p. 547.

G. DORDA and J. VRBA
*Long-Time Effect of External Fields on Surface Silicon Oxide Conductivity
Vol. 17, No. 1, p. K127–K130 (1966).
See also 22.1.2, p. 469; 22.6, p. 535.

A. F. KRAVCHENKO and V. S. SARDARYAN
On the Structure of the Bottom of the Conduction Band in GaAs
Vol. 17, No. 2, p. 479–488 (1966).
See also *13.1*, p. *175*; 22.2.1, p. 482.

N. SINGH and M. D. KARKHANAVALA
*Studies on the Electrical Properties of Uranium Oxides. (II): Non-stoichiometric U_3O_8
Vol. 17, No. 2, p. 501–508 (1966).
See also 13, p. 163; 22.6, p. 535.

W. H. MITCHELL, E. H. PUTLEY, and N. SHAW
*Hall Effect and Extrinsic Photoconductivity in n-Type InSb
Vol. 17, No. 2, p. 605–608 (1966).
See also *16*, p. *277*; 22.2.3, p. 490.

K. OZAWA and S. ANZAI
Effect of Pressure on the α-Transition Point of Iron Monosulphide
Vol. 17, No. 2, p. 697–700 (1966).
See also *1.2*, p. *9*; 12, p. 156; 22, p. 441.

J. DEVREESE, R. DE CONINCK, and H. POLLAK
*On the Conduction Mechanism in Uranium Dioxide
Vol. 17, No. 2, p. 825–829 (1966).
See also 13, p. 163; 15, p. 270; 22.6, p. 535.

C. VERIÉ
On the Band Structure of HgTe-CdTe. (I): Electric Measurements
Vol. 17, No. 2, p. 889–901 (1966).
See also *13.1*, p. *176*; 22.4.3, p. 509; 22.8, p. 548.

F. I. ISMAILOV, G. A. AKHUNDOV, and O. R. VERNICH
*Investigation of Electrical Conductivity and Hall Effect in GaSe Single Crystals
Vol. 17, No. 2, p. K237–K240 (1966).
See also 13.4, p. 203; 15, p. 271; 22, p. 441.

G. JUNGK
On the Generation–Recombination Statistics in Semiconductors
Vol. 18, No. 1, p. 297–306 (1966).
See also *13.4*, p. *203*; 16, p. 278; 22.1.2, p. 470.

P. BRÄUNLICH and A. SCHARMANN
Approximate Solution of Schön's Balance Equations for the Thermoluminescence and the Thermally Stimulated Conductivity of Inorganic Photoconducting Crystals
Vol. 18, No. 1, p. 307–316 (1966).
See also *13.4*, p. *203*; 16, p. 278; 20.3, p. 364; 22.4, p. 495.

V. A. KUTASOV and I. A. SMIRNOV
Effect of Excess Antistructural Bi on the Lattice Thermal Conductivity of Bi_2Te_3, Bi_2Te_3-Sb_2Te_3, and Bi_2Te_3-Bi_2S_3 Containing Iodine
Vol. 18, No. 1, p. 479–488 (1966).
See also *8*, p. *80*; *10*, p. *96*; 15, p. 271; 22, p. 441; 22.8, p. 548.

H. FLIETNER and W. GRÄFE
Large-Signal Field Effect for the Determination of the Surface Levels of Silicon
Vol. 18, No. 1, p. K5–K7 (1966).
See also *13.3*, p. *197*; 22.1.2, p. 470.

I. LICEA
*On the Variation of the Electron Temperature in Polar Semiconductors by an Electric Field
Vol. 18, No. 1, p. K9–K12 (1966).
See also 13, p. 164; 22.2.3, p. 490.

G. B. STREET and W. D. GILL
Photoconductivity and Drift Mobilities in Single Crystal Realgar (As_4S_4)
Vol. 18, No. 2, p. 601–607 (1966).
See also 13.4, p. 204; *16*, p. *278*; 22, p. 441.

B. M. ASKEROV and F. M. GASHIMZADE
*Quantum Theory of the Electrical Conductivity of Semiconductors with a Non-Standard Energy Band
Vol. 18, No. 2, p. 667–675 (1966).
See also 6, p. 58; 13.1, p. 176; 22.2, p. 481; 22.2.3, p. 490.

I. P. Molodyan, D. N. Nasledov, S. I. Radautsan, and V. G. Sidorov
The Effective Mass of Electrons in $(InSb)_x \cdot (InTe)_{1-x}$ Crystals
Vol. 18, No. 2, p. 677–682 (1966).
See also *13.1*, p. *176*; 15, p. 271; 22.8, p. 548.

M. Asche, O. G. Sarbej, and V. M. Vasetskii
*Piezowiderstand von p-Germanium
(Piezoresistance of p-Germanium)
Vol. 18, No. 2, p. 749–754 (1966).
See also 10, p. 97; 12, p. 156; 22.1.1, p. 460.

G. D. Guseinov, M. Z. Ismailov, and A. G. Talybov
On the New Semiconducting Compound $HgTlS_2$
Vol. 18, No. 2, p. 929–935 (1966).
See also *5*, p. *50*; 8, p. 80; 13.4, p. 204; 16, p. 278; 22.8, p. 548.

V. V. Paranjape
*Distribution Function of Hot Electrons
Vol. 18, No. 2, p. K103–K104 (1966).
See also 6, p. 58; *13*, p. *164*; 22, p. 442.

E. Kamieniecki
*Remark on Solving Boltzmann Equation for Hot Carriers
Vol. 18, No. 2, p. K105 (1966).
See also 6, p. 58; *13*, p. *164*; 22, p. 442.

G. Richter
*Zur Bestimmung von Transportkoeffizienten für Minoritätsträger in InSb
(Determination of Transport Coefficients for Minority Carriers in InSb)
Vol. 18, No. 2, p. K149–K152 (1966).
See also 13.1, p. 176; 16, p. 278; 22.2.3, p. 490.

G. B. Abdullaev, F. B. Gadiev, G. M. Aliev, S. I. Mekhtieva, and D. Sh. Abdinov
Interaction of Sodium Admixtures and Oxygen in Selenium
Vol. 18, No. 2, p. K153–K156 (1966).
See also *10*, p. *97*; 22.1.3, p. 477.

V. L. Vinetskii and G. A. Kholodar
*On the Electric Conductivity of Semiconductors Caused by the Ionization of Thermal Lattice Defects
Vol. 19, No. 1, p. 41–49 (1967).
See also 10, p. 97; 13.4, p. 204; 22.6, p. 536.

A. K. Walton and G. J. O. Eyoku
*Piezo-Hall Effect in n-Type Germanium
Vol. 19, No. 1, p. 111–116 (1967).
See also 22.1.1, p. 461.

A. M. George and M. D. Karkhanavala
Studies of the Electrical Properties of Uranium Oxides. (III): Thermoelectric Power of α-U_3O_8
Vol. 19, No. 1, p. 117–121 (1967).
See also *15*, p. *271*; 22.6, p. 536.

K. W. Böer and C. A. Kennedy
Vacuum Heat Treatment of CdS Single Crystals
Vol. 19, No. 1, p. 203–210 (1967).
See also *13.4*, p. *204*; 16, p. 278; 22.4.1, p. 499.

J. Kołodziejczak
*On the Scattering Processes in Semiconductors
Vol. **19**, No. 1, p. 231–237 (1967).

A. S. Filipchenko and D. N. Nasledov
*On the Mixed Mechanism of Electron Scattering in InSb Crystals
Vol. **19**, No. 1, p. 435–439 (1967).
See also 15, p. 271; 22.2.3, p. 490.

J. D. Keys and H. P. Dibbs
On the Bonding in Bismuth Telluride
Vol. **19**, No. 1, p. K11–K13 (1967).
See also *3.2*, p. *35*; 22, p. 442.

S. Żukotyński and J. Kołodziejczak
On the Determination of the Non-Parabolicity of Energy Bands
Vol. **19**, No. 1, p. K51–K54 (1967).
See also *13.1*, p. *176*; 15, p. 271; 22, p. 443.

W. Franz and P. Miosga
Distribution Function and Drift Velocity of Warm Electrons in Germanium
Vol. **19**, No. 2, p. 597–602 (1967).
See also *13*, p. *164*; 22.1.1, p. 461.

E. I. Gavrilitsa and S. I. Radautsan
Influence of Ordering on the Properties of Solid Solutions of the System $(HgSe)_{3x}$-$(In_2Se_3)_{1-x}$
Vol. **19**, No. 2, p. 609–611 (1967).
See also *1.2*, p. *10*; 8, p. 81; 22.8, p. 549.

J. Kołodziejczak and E. Kierzek-Pecold
Free Carrier Electro-Magneto-Optical Phenomena in Semiconductors
Vol. **19**, No. 2, p. 623–630 (1967).
See also *18*, p. *291*; *20*, p. *338*; 22, p. 443.

M. Henzler
Conductivity and Field Effect of Pure Cleavage-Surfaces of Silicon
Vol. **19**, No. 2, p. 833–845 (1967).
See also *13.3*, p. *197*; 22.1.2, p. 470.

J. H. P. van Weeren, R. Struikmans, and J. Blok
*Test of the Read Model Concerning the Electron Mobility in Plastically Deformed n-Type Ge
Vol. **19**, No. 2, p. K107–K110 (1967).
See also 10, p. 98; 22.1.1, p. 461.

R. R. Gałązka and L. Sosnowski
Conduction Band Structure of $Cd_{0.1}Hg_{0.9}Te$
Vol. **20**, No. 1, p. 113–120 (1967).
See also *13.1*, p. *177*; 22.4.3, p. 509; 22.8, p. 549.

L. L. Hench and D. A. Jenkins
*AC Conductivity of a Glass Semiconductor
Vol. **20**, No. 1, p. 327–330 (1967).
See also 2, p. 23; 22.6, p. 537; 22.8, p. 549.

G. B. Abdullaev, E. S. Guseinova, and B. G. Tagiev
*Electric Conductivity Studies of p-GaTe Polycrystals and Single Crystals in Strong Electric Fields
Vol. **20**, No. 2, p. 421–425 (1967).
See also *14.4.1*, p. *258*; 22, p. 444.

C. HAMANN
*On the Electric and Thermoelectric Properties of Copper Phthalocyanine Single Crystals
Vol. 20, No. 2, p. 481–491 (1967).
See also 13.4, p. 205; 14.4.1, p. 258; 15, p. 271; 22.9, p. 559.

G. JONES, G. SMITH, and A. R. BEATTIE
The Drifted Maxwellian Distribution Function in InSb
Vol. 20, No. 2, p. K135–K139 (1967).
See also *13*, p. *165*; 22.2.3, p. 490.

W. P. R. GEORGE, C. H. L. GOODMAN, H. F. STERLING, and R. W. WARREN
*A Possible New Group of Semiconducting Compounds
Vol. 21, No. 1, p. 205–210 (1967).
See also 3.1, p. 31; 3.2, p. 35; 21.7, p. 438; 22.7, p. 545.

W. SCHRÖTER
*Die elektrischen Eigenschaften von Versetzungen in Germanium
(Electric Properties of Dislocations in Germanium)
Vol. 21, No. 1, p. 211–224 (1967).
See also 10, p. 99; 13.4, p. 206; 22.1.1, p. 461.

N. SEXER
*Effets galvanomagnétiques et thermomagnétiques classiques dans l'arséniure de cadmium
(Classical Galvanomagnetic and Thermomagnetic Effects in Cadmium Arsenide)
Vol. 21, No. 1, p. 225–234 (1967).
See also 13.1, p. 178; 15, p. 272; 22.7, p. 545.

G. B. ABDULLAEV, Z. A. ISKENDER-ZADE, E. A. DZHAFAROVA, and V. E. CHELNOKOV
*Study of Electron Lifetime in p-Si
Vol. 21, No. 1, p. 423–429 (1967).
See also 13.4, p. 206; 14.4.1, p. 258; 22.1.2, p. 471.

J. BOUGNOT, E. MONTEIL, and C. LLINARES
Diffusion and Solubility in Gallium Antimonide
Vol. 21, No. 1, p. K31–K34 (1967).
See also *9*, p. *89*; 10, p. 100; 22.2.3, p. 491.

G. B. ABDULLAEV, Z. A. ALIYAROVA, and G. A. ASADOV
Preparation of Cu_2Se Single Crystals and Investigation of Their Electrical Properties
Vol. 21, No. 2, p. 461–464 (1967).
See also 3.1, p. 31; 15, p. 272; 22, p. 445.

L. GOUSKOV, G. LECOY, and C. LLINARES
*Carrier Recombination in Nickel Doped Germanium from Lifetime and Noise Measurements
Vol. 21, No. 2, p. 619–626 (1967).
See also 13.4, p. 207; *16*, p. *279*; 22.1.1 p. 462.

M. AVEROUS and G. BOUGNOT
*Effect of Donor Impurity Concentration on the Piezoresistance of n-Type GaSb
Vol. 21, No. 2, p. 665–670 (1967).
See also 13.1, p. 178; 13.4, p. 207; 22.2.3, p. 491.

I. BUNGET and M. ROSENBERG
*P-Type Conduction in Barium Ferrite Single Crystals
Vol. 21, No. 2, p. K131–K133 (1967).
See also 15, p. 272; 18.3, p. 318.

B. M. Askerov and F. M. Gashimzade
Determination of Parameters of Degenerate Semiconductors
Vol. **21**, No. 2, p. K155—K158 (1967).
See also *13.1*, p. *179*; 15, p. 272; 22, p. 446.

V. L. Bonch-Bruevich
On the Theory of Electrical Domains in Hot Electron Semiconductors
Vol. **22**, No. 1, p. 267—278 (1967).
See also 13, p. 166; *14.4.1*, p. *259*; 22, p. 446; 22.1.1, p. 462; 22.2.1, p. 484.

P. Kordoš
Heat Treatment of InSb Crystals with Different Dislocation Densities
Vol. **22**, No. 1, p. K59—K61 (1967).
See also *9*, p. *89*; 10, p. 101; 22.2.3, p. 491.

L. N. Cojocaru
Effects of Fast Neutrons on Electric Properties of NiO Single Crystals
Vol. **22**, No. 2, p. 361—364 (1967).
See also *11*, p. *149*; 13.4, p. 208; 15, p. 272; 22.6, p. 539.

A. Aldea
*Theory of the Hall Effect in Disordered Systems: Impurity Band Conduction (II)
Vol. **22**, No. 2, p. 377—380 (1967).
See also 13.4, p. 208; 22, p. 446.

L. Elstner and W. Kamprath
Quenched-in Levels in p-Type Silicon
Vol. **22**, No. 2, p. 541—547 (1967).
See also 11, p. 149; *13.4*, p. *208*; 22.1.2, p. 471.

A. I. Aliev, G. I. Safaraliev, and S. G. Abdinova
*About the Role of Minority Carriers in the Impurity Conduction of GaSb
Vol. **22**, No. 2, p. 741—745 (1967).
See also 13.4, p. 209; 22.2.3, p. 491.

R. Yu. Khansevarov, T. V. Mashovets, N. A. Vitovskii, and B. Chelustka
On the Nature of Damages in Indium Antimonide Produced by γ- and X-Irradiation
Vol. **22**, No. 2, p. K95—K101 (1967).
See also *11*, p. *149*; 22.2.3, p. 491.

R. Yu. Khansevarov, T. V. Mashovets, and N. A. Vitovskii
Damage Formation Rates in Ge Subject to Thermal Treatment
Vol. **22**, No. 2, p. K103—K107 (1967).
See also 10, p. 101; *11*, p. *150*; 22.1.1, p. 462.

G. D. Guseinov, A. M. Ramazanzade, E. M. Kerimova, and M. Z. Ismailov
*About a Group of Three-Component Compounds being Analogous to Binary Semiconductors of the $A^{III}B^{VI}$Type
Vol. **22**, No. 2, p. K117—K122 (1967).
See also 22.8, p. 551.

K. D. Tovstyuk and P. I. Voronyuk
*Galvanomagnetic Effects in PbTe
Vol. **23**, No. 1, p. 75—91 (1967).
See also 13.1, p. 180; 22, p. 447.

N. I. Akimenko, P. M. Staryk, and K. D. Tovstyuk
Magnetic Properties of PbTe
Vol. **23**, No. 1, p. 93–96 (1967).
See also *13.1*, p. *180*; 18, p. 293; 22, p. 447.

A. G. Samoilovich, I. S. Buda, and I. V. Dakhovskii
Anisotropy of Thermomagnetic Effects in n-Ge
Vol. **23**, No. 1, p. 229–236 (1967).
See also 13, p. 167; *15*, p. *272*; 22.1.1, p. 463.

J. Kołodziejczak
*Effect of Strong Electric Fields on Magneto-Resistivity and Hall Coefficient in Semiconductors
Vol. **23**, No. 1, p. 423–427 (1967).
See also 6, p. 62; 13.1, p. 181; 14.4.1, p. 260; 22, p. 447; 22.2.3, p. 491.

R. R. Gałązka and T. Zarkzewski
Heavy Hole Effective Mass of $Cd_{0.1}Hg_{0.9}Te$
Vol. **23**, No. 1, p. K39–K43 (1967).
See also *13.1*, p. *181*; 15, p. 272; 22.4.3, p. 510; 22.8, p. 552.

J. Kołodziejczak
Free Carrier Effect on the Generation of Higher Harmonics in Semiconductors
Vol. **23**, No. 1, p. K57–K59 (1967).
See also *13*, p. *167*; 13.1, p. 181; *20*, p. *340*; 22, p. 447.

W. Hörstel and G. Kretzschmar
Lattice Defects Forming Acceptors in Tellurium
Vol. **23**, No. 2, p. 639–647 (1967).
See also *10*, p. *102*; 22.1.3, p. 478.

A. Stella, A. D. Brothers, R. H. Hopkins, and D. W. Lynch
• Pressure Coefficient of the Band Gap in Mg_2Si, Mg_2Ge, and Mg_2Sn
Vol. **23**, No. 2, p. 697–702 (1967).
See also *13.1*, p. *182*; 20.1, p. 350; 22.7, p. 545.

V. L. Bonch-Bruevich
On the Problem of Static Domains in Hot Electron Semiconductors
Vol. **23**, No. 2, p. 761–766 (1967).
See also 13, p. 168; *14.4.1*, p. *260*; 22.1.1, p. 463.

H. Mell and J. Stuke
*Magnetoconductivity of Trigonal Selenium Single Crystals
Vol. **24**, No. 1, p. 183–193 (1967).
See also 13.1, p. 182; 22.1.3, p. 478.

K. H. Herrmann und G. Oelgart
*Piezowiderstandseffekt an Tellur. (I): Experimente mit uniaxialem Druck parallel zur *c*-Achse
(Piezoresistance in Tellurium. (I): Experiments with Uniaxial Stress Parallel tu the *c*-Axis)
Vol. **24**, No. 1, p. 289–296 (1967).
See also 13.1, p. 182; 22.1.3, p. 478.

J. Kołodziejczak
Magnetic Resonance in Higher Harmonics Generated by Free Carriers in Semiconductors
Vol. **24**, No. 1, p. 323–329 (1967).
See also 13.1, p. 183; *19*, p. *331*; 22, p. 448.

C. N. Plăviţu
On the Phonon Drag Effect Contribution to Thermomagnetic Phenomena
Vol. 24, No. 1, p. 361–372 (1967).
See also 6, p. 64; 13, p. 168; *15*, p. *273*; 22, p. 448.

V. A. Ivanova, D. Sh. Abdinov, and G. M. Aliev
*Some Structural, Electrical, and Thermal Parameters of VCr_2Se_4, $FeCr_2Se_4$, and $NiCr_2Se_4$ Ternary Compounds
Vol. 24, No. 1, p. K23–K24 (1967).
See also 5, p. 52; 8, p. 83; 15, p. 273; 22.8, p. 552.

T. A. T. Cowell and J. Woods
Thermally Stimulated Currents and Infra-Red Luminescence in CdS Crystals
Vol. 24, No. 1, p. K37–K40 (1967).
See also 10.2, p. 141; *13.4*, p. *210*; 20.3, p. 367; 22.4.1, p. 501.

D. Siebert, E. Ady, and F. Matossi
*Magnetoresistance of High-Ohmic Semiconductors by a Capacitive Method
Vol. 24, No. 1, p. K65–K67 (1967).
See also 22.2.1, p. 484.

R. Baubinas, A. Sakalas, A. Smilga, and J. Viščakas
*p-Type CdSe Single Crystals
Vol. 24, No. 1, p. K91–K93 (1967).
See also 3.1, p. 33; *16*, p. *280*; 22.4.2, p. 507.

B. Pistoulet, J. L. Robert, and M. Marques
Study of GaSb Valence Band by Magnetoresistance and Hall Effect Measurements at Low Temperature
Vol. 24, No. 2, p. 481–486 (1967).
See also *13.1*, p. *183*; 22.2.3, p. 491.

C. Hammar and P. Weissglas
*Hot Electron Polar Scattering in InSb
Vol. 24, No. 2, p. 531–534 (1967).
See also *13*, p. *168*; 22.2.3, p. 492.

R. A. Burmeister, Jr. and D. A. Stevenson
*Electrical Properties of n-Type CdSe
Vol. 24, No. 2, p. 683–694 (1967).
See also 13.4, p. 211; 22.4.2, p. 507.

V. A. Ivanova, D. Sh. Abdinov, and G. M. Aliev
On Some Characteristics of Chromium Selenides
Vol. 24, No. 2, p. K145–K147 (1967).
See also 5, p. 52; *8*, p. *84*; *15*, p. *273*; 22, p. 449.

Ya. N. Nasirov and Ya. S. Feiziev
Effect of Small Substitutions of Tin by Neodymium on Thermoelectric Properties of SnTe
Vol. 24, No. 2, p. K157–K159 (1967).
See also *15*, p. *273*; 22, p. 449.

D. N. Bose
Field-Effect Measurements on Silicon Carbide
Vol. 24, No. 2, p. K165–K166 (1967).
See also *13.3*, p. *198*; 22.3, p. 494.

G. ARLT and P. QUADFLIEG
*Piezoelectricity in III–V Compounds with a Phenomenological Analysis of the Piezoelectric Effect
Vol. **25**, No. 1, p. 323–330 (1968).
See also 22.2.1, p. 485; 22.2.3, p. 492.

S. OBERLÄNDER and S. SIKORSKI
*On the Mathematical Description of Electrical Processes in Semiconductors. (I): Phase Space Analysis in Subspaces of Dimension Lower than Three
Vol. **25**, No. 1, p. 345–357 (1968).
See also 14.3.2, p. 250; 22, p. 449.

I. LICEA
*Electron Temperature in Polar Semiconductors
Vol. **25**, No. 1, p. 461–471 (1968).
See also *13*, p. *168*; 22.2, p. 481; 22.4, p. 495.

N. N. SIROTA and V. B. SHIPILO
*High-Pressure and Temperature Effect on the Electrical Properties of p-Type Gallium Antimonide
Vol. **25**, No. 1, p. K23–K26 (1968).
See also 22.2.3, p. 492.

N. N. SIROTA and V. B. SHIPILO
*High-Pressure Effect on the Electrical Conductivity and Hall Constant of Indium Antimonide
Vol. **25**, No. 1, p. K27–K30 (1968).
See also 22.2.3, p. 492.

N. A. GORYUNOVA, V. M. ORLOV, V. I. SOKOLOVA, G. P. SHPENKOV, and E. V. TSVETKOVA
Preparation and Some Properties of CuP_2 Single Crystals
Vol. **25**, No. 2, p. 513–519 (1968).
See also *3.1*, p. *33*; 5, p. 53; 15, p. 273; 20.1, p. 351; 22, p. 450.

G. I. GUSEVA and P. S. ZYRYANOV
Quantum Theory of Thermogalvanomagnetic Phenomena (TGMP) in Conductors with Equal Concentration of Electrons and Holes
Vol. **25**, No. 2, p. 775–785 (1968).
See also 14.1, p. 224; *15*, p. *273*; 21.7, p. 438; 22.2, p. 481.

F. M. EL-AKKAD
*Temperature Dependence of Mobility in PbS
Vol. **25**, No. 2, p. K115–K117 (1968).
See also 22, p. 450.

G. B. ABDULLAEV, Z. A. ALIYAROVA, E. H. ZAMANOVA, and G. A. ASADOV
*Investigation of the Electric Properties of Cu_2S Single Crystals
Vol. **26**, No. 1, p. 65–68 (1968).
See also 1.2, p. 12; 15, p. 372; 22, p. 450.

E. O. OSMANOV, YU. V. RUD, and M. E. STRYALKOVSKII
High-Temperature Modifications of the Semiconductor Compounds $CdSnAs_2$ and $CdGeAs_2$
Vol. **26**, No. 1, p. 85–90 (1968).
See also *1.2*, p. *12*; 2, p. 24; 8, p. 84; 12, p. 158; 22.7, p. 546; 22.8, p. 553.

I. LICEA
*Hot Electron Mobility in Polar Semiconductors
Vol. **26**, No. 1, p. 115–123 (1968).
See also *13*, p. *169*; 22.2.3, p. 492.

L. ROTHKIRCH
*Deutung der Anisotropie der galvanomagnetischen Effekte im Tellur durch anisotrope Streuung
(Interpretation of the Anisotropy of the Galvanomagnetic Effects in Tellurium by Anisotropic Scattering)
Vol. **26**, No. 2, p. 709—715 (1968).
See also 13.1, p. 185; 22.1.3, p. 479.

D. I. LAINER and G. K. SAMEDOV
*On the Mechanism of Current Carrier Scattering in Some Solid Solutions on the Basis of Lead Telluride
Vol. **26**, No. 2, p. K123—K125 (1968).
See also 15, p. 273; 22, p. 451.

H. B. WILD and C. A. HOGARTH
*Slow Temporal Variations of Electron Mobility in Orthorhombic Sulphur
Vol. **26**, No. 2, p. K135—K137 (1968).
See also 22.1.3, p. 479.

R. LINK and H. LUTSCH
*Hall Effect Measurements in Pulse-Heated Tellurium Single Crystals
Vol. **26**, No. 2, p. K159—K162 (1968).
See also 13.4, p. 212; 22.1.3, p. 479.

I. K. ANDRONIK, E. K. ARUSHANOV, O. V. EMELYANENKO, and D. N. NASLEDOV
*Negative Magnetoresistance in p-CdSb
Vol. **27**, No. 1, p. 45—49 (1968).
See also 13.4, p. 213; 22, p. 451; 22.7, p. 546.

L. I. ANATYCHUK, V. D. ISKRA, O. J. LUSTE, I. M. RARENKO, and L. I. ZARUBIN
*Anisotropy of Electroconductivity in CdSb
Vol. **27**, No. 1, p. 101—107 (1968).
See also 13.1, p. 185; 22, p. 451; 22.7, p. 546.

J. H. P. VAN WEEREN, G. KOOPMANS, and J. BLOK
The Position of the Dislocation Acceptor Level in n-Type Ge
Vol. **27**, No. 1, p. 219—224 (1968).
See also 10, p. 106; *13.4*, p. *213*; 22.1.1, p. 465.

J. H. P. VAN WEEREN, R. STRUIKMANS, G. KOOPMANS, and J. BLOK
*Low-Field Magnetoresistance and Magnetoconductivity in n-Type Ge with Dislocations
Vol. **27**, No. 1, p. 225—235 (1968).
See also 10, p. 107; 22.1.1, p. 465.

M. M. AKSELROD, K. M. DEMCHUK, I. M. TSIDILKOVSKI, E. L. BROYDA, and K. P. RODIONOV
The Effect of Pressure on the Electron Effective Mass in InSb
Vol. **27**, No. 1, p. 249—254 (1968).
See also *13.1*, p. *186*; 15, p. 274; 22.2.3, p. 492.

U. BIRKHOLZ and J. SCHELM
*Mechanism of Electrical Conduction in β-$FeSi_2$
Vol. **27**, No. 1, p. 413—425 (1968).
See also 13, p. 170; 13.1, p. 186; 15, p. 274; 22, p. 452; 22.7, p. 546.

M. I. KLINGER
On the Experimental Observation of Small Polarons in Semiconductors
Vol. **27**, No. 2, p. 479—488 (1968).
See also *13*, p. *170*; 20.1, p. 354; 22.6, p. 541.

W. SZYMAŃSKA and J. GINTER
*On the Scattering of Electrons by Ionised Impurities in the Case of Strong Degeneracy
Vol. 27, No. 2, p. K103–K106 (1968).
See also 13.4, p. 214; 22.4.3, p. 510.

E. W. KREUTZ, H. PAGNIA, and W. WAIDELICH
X-Ray Induced Conductivity Changes in InSb
Vol. 27, No. 2, p. K111–K114 (1968).
See also *11*, p. *153*; 13.3, p. 198; 22.2.3, p. 492.

I. LICEA
*Warm Electron Temperature and Mobility in Polar Semiconductors
Vol. 27, No. 2, p. K137–K141 (1968).
See also *13*, p. *170*; 13.1, p. 186; 22.2.3, p. 492.

I. LICEA
*Hot Electron Temperature and Mobility in Polar Semiconductors
Vol. 27, No. 2, p. K143–K144 (1968).
See also *13*, p. *170*; 14.4.1, p. 263; 22.2.3, p. 492.

P. GROSSE and B. KRAHL-URBAN
*The Contribution of Free Carriers to the Complex Dielectric Constant of Tellurium at 9 GHz
Vol. 27, No. 2, p. K149–K152 (1968).
See also 22.1.3, p. 479.

V. P. KALASHNIKOV and A. M. ZLOBIN
*On the Theory of Thermogalvanomagnetic Phenomena in Semiconductors at High Magnetic Fields
Vol. 28, No. 1, p. 105–109 (1968).
See also 6, p. 66; *8*, p. *85*; 22, p. 452.

M. ALI, M. FRIDMAN, M. DENAYER, and P. NAGELS
*Electrical Conduction in Li-Doped MnO Single Crystals
Vol. 28, No. 1, p. 193–206 (1968).
See also 13.1, p. 187; 13.4, p. 214; 15, p. 274; 22.6, p. 542.

T. O. TUOMI
Effect of Reactor Irradiation on the Conductivity and the Photoconductivity of Hexagonal Selenium Single Crystals
Vol. 28, No. 1, p. 329–334 (1968).
See also *11*, p. *153*; 16, p. 282; 22.1.3, p. 479.

N. A. GORYUNOVA, S. M. RYVKIN, G. P. SHPENIKOV, I. I. TICHINA, and V. G. FEDOTOV
Investigations of Some Properties of Vitreous and Crystalline $CdGeP_2$
Vol. 28, No. 2, p. 489–494 (1968).
See also 2, p. 24; *20.3*, p. *370*; 22.8, p. 554.

J. SCHNAKENBERG
Polaronic Impurity Hopping Conduction
Vol. 28, No. 2, p. 623–633 (1968).
See also *6*, p. *67*; *13*, p. *171*; 22, p. 453.

P. WEISSGLAS and K. BLÖTEKJAER
*Conductivity of Intrinsic InSb in High Electric Fields
Vol. 28, No. 2, p. 711–718 (1968).
See also 14.4.1, p. 263; 22.2.3, p. 493.

B. M. Askerov and F. M. Gashimzade
*The Transverse Magnetic Resistance in Semiconductors with Non-Standard Energy Band
Vol. **28**, No. 2, p. 783–788 (1968).
See also 13.1, p. 187; 22, p. 453.

W. Zimmermann und F. Stöckmann
*Halleffektsmessungen an Tellur in Magnetfeldern bis 150 kG
(Measurements of Hall Effect on Tellurium in Magnetic Fields up to 150 kG)
Vol. **28**, No. 2, p. K101–K104 (1968).
See also 22.1.3, p. 479.

R. C. Whelan and D. Shaw
Evidence of a Doubly Ionized Native Donor in CdTe
Vol. **29**, No. 1, p. 145–152 (1968).
See also 10, p. 109; *13.4*, p. *215*; 22.4.3, p. 511.

H. J. Gläser und D. Geist
*Elektrische Leitfähigkeit und Halleffekt von mangandotiertem Germanium und der Einfluß von Wärmebehandlungen auf den Einbauzustand des Mangans
(Electrical Conductivity and Hall Effect of Manganese-Doped Germanium and the Influence of Heat Treatment on the State of Incorporation of Manganese)
Vol. **29**, No. 1, p. 167–170 (1968).
See also 13.4, p. 215; 19, p. 335; 22.1.1, p. 466.

K. H. Herrmann and J. Puhlmann
*Piezoresistance of Tellurium. (II): Experiments with Pressures Lowering Symmetry
Vol. **29**, No. 1, p. 193–202 (1968).
See also 13.1, p. 188; 22.1.3, p. 480.

A. A. Vaipolin, N. A. Goryunova, L. I. Kleshchinskii, G. V. Loshakova, and E. O. Osmanov
The Structure and Properties of the Semiconducting Compound $ZnSnP_2$
Vol. **29**, No. 1, p. 435–442 (1968).
See also 4, p. 47; *5*, p. *54*; 6, p. 68; 10, p. 110; 22.8, p. 555.

V. M. Asnin, G. L. Eristavi, and A. A. Rogachev
Effect of Weak Electric Fields on the Absorption Edge in Doped Germanium
Vol. **29**, No. 1, p. 443–446 (1968).
See also 13.2, p. 195; 13.4, p. 215; *20.1*, p. *356*; 22.1.1, p. 466.

F. Huth
*Hall Effect and Mobility in n-Type GaAs
Vol. **29**, No. 1, p. K35–K37 (1968).
See also 13.4, p. 215; 22.2.1, p. 485.

F. B. Gadzhiev, Ch. M. Askerov, and G. M. Aliev
*Effect of Heat Treatment and Natrium Admixtures on the Temperature Dependence of Electric Conductivity of Amorphous Selenium
Vol. **29**, No. 1, p. K47–K50 (1968).
See also 2, p. 24; 22.1.3, p. 480.

E. Mohler
*Linear Electro-Optic Effect of Excitons in CuCl
Vol. **29**, No. 1, p. K55–K57 (1968).
See also 13.2, p. 195; *20*, p. *342*; 22.5, p. 514.

W. SCHILZ
Anisotropy of High-Field Helicon Propagation in PbTe
Vol. **29**, No. 2, p. 559—568 (1968).
See also *13*, p. *172*; *20*, p. *342*; 22, p. 454.

M. AVEROUS, G. BOUGNOT, and J. CALAS
Conduction Bands of GaSb-InSb Mixed Crystals with a Low Concentration of InSb
Vol. **29**, No. 2, p. 807—812 (1968).
See also *13.1*, p. *189*; 22.2.3, p. 493; 22.8, p. 556.

E. Z. DZIUBA, D. NICULESCU, and N. NICULESCU
Energy Band Structure of $Zn_xHg_{1-x}Te$ Solid Solutions
Vol. **29**, No. 2, p. 813—818 (1968).
See also *13.1*, p. *189*; 20.1, p. 357; 22.4.3, p. 511; 22.8, p. 556.

L. N. COJOCARU
Effects of Fission Fragments on Electric Properties of the UO_2-SiO_2 System
Vol. **29**, No. 2, p. K119—K122 (1968).
See also *11*, p. *154*; 15, p. 274; 22.6, p. 543; 22.8, p. 556.

S. HAUSSÜHL
*Piezoelektrisches und elektrisches Verhalten von Lithiumjodat
(Piezoelectrical and Electrical Behaviour of Lithiumjodate)
Vol. **29**, No. 2, p. K159—K162 (1968).
See also *12*, p. *158*; 22.8, p. 556.

M. SODHA and S. SHARMA
*Some Transport Coefficients in Cadmium Sulphide
Vol. **30**, No. 1, p. 239—245 (1968).
See also 6, p. 69; 22.4.1, p. 505.

V. HÁDEK
*Electron and Hole Conductivity of Donor–Acceptor Complexes of Oligoanilinic Substances and Their Thermoelectric Properties
Vol. **30**, No. 1, p. 275—282 (1968).
See also *13.4*, p. *216*; 15, p. 275; 22.9, p. 560.

V. V. BOLDYREV, G. B. LONDON, and V. K. ZHURAVLEV
*Study of Semiconducting Properties of Potassium Permanganate
Vol. **30**, No. 1, p. K13—K15 (1968).
See also 13.4, p. 217; *15*, p. *275*; 22.8, p. 556.

G. G. KOVALEVSKAYA and S. V. SLOBODCHIKOV
*Lifetimes of Current Carriers in InP
Vol. **30**, No. 2, p. 441—447 (1968).
See also 13.4, p. 217; 22.2.2, p. 488.

NGUYEN VAN DONG and PHAM NGU TUNG
*Transport Properties of Silver Telluride in the Solid and Liquid States
Vol. **30**, No. 2, p. 557—567 (1968).
See also 2, p. 25; 14.4, p. 254; *15*, p. *275*; 22, p. 455.

H. R. HUFF, S. KAWAJI, and H. C. GATOS
*Observation of Quantum Galvanomagnetic Phenomena in n-Type Indium Antimonide
Vol. **30**, No. 2, p. 613—618 (1968).
See also 22.2.3, p. 493.

P. P. KONOROV, V. N. SHVETSOV, and S. V. SHEGOLIKHINA
Effects of the Non-Equilibrium Depletion in Germanium at the Electrolyte Interface
Vol. **30**, No. 2, p. 845—850 (1968).
See also *14.4.1*, p. *265*; 22.1.1, p. 467.

14.3.1 Semiconducting Films

Original Papers and Short Notes

R. Grigorovici, N. Croitoru, and A. Dévényi
Thermoelectric Power in Amorphous Germanium
Vol. **16**, No. 2, p. K143–K145 (1966).
See also 2, p. 22; *15*, p. *270*; 22.1.1, p. 459.

M. Hagino and T. Takahashi
Photoelectric Properties of Cs-Sb Films Having Various Compositions
Vol. **17**, No. 2, p. 609–618 (1960).
See also 10, p. 95; 16, p. 277; *17*, p. *286*; 22.7, p. 545.

H. Frank
*Beitrag zur Widerstandsmessung von nn^+- und pp^+-Epitaxialschichten nach dem Durchbruchsverfahren
(Contribution to the Measurement of the Resistance of nn^+ and pp^+ Epitaxial Films by the Breakdown Method)
Vol. **18**, No. 1, p. 401–414 (1966).
See also 14.4.1, p. 257; 22.1.1, p. 460; 22.1.2, p. 470 22.2.1, p. 483.

C. Cherki and R. Coelho
On Charge Storage in Anodic Tantal Oxide Layers
Vol. **19**, No. 2, p. K91–K94 (1967).
See also *14.4.1*, p. *257*; 22.6, p. 537.

P. Schnupp
*Kronig-Penney-Type Calculations for Electron Tunneling through Thin Dielectric Films
Vol. **21**, No. 2, p. 567–573 (1967).
See also 13.1, p. 178; 22, p. 446.

K. H. Gundlach and G. Heldmann
*The Effect of the $E(k)$-Relation on Tunneling through Asymmetric Barriers
Vol. **21**, No. 2, p. 575–579 (1967).
See also 13.1, p. 178; 14.3.2, p. 249; 22.6, p. 538.

J. E. Knappett and S. J. T. Owen
Observation of the Growth Process of Gallium Arsenide on Tungsten
Vol. **21**, No. 2, p. K99–K101 (1967).
See also 1.4, p. 18; *3.1*, p. *31*; 22.2.1, p. 484.

B. Tavger
*Transport Phenomena in Semiconducting Thin Films
Vol. **22**, No. 1, p. 31–38 (1967).
See also 13.1, p. 179; 15, p. 272; 22, p. 446.

I. Licea
*Electrical Conduction of Semiconducting Thin Films in Strong Electric Fields
Vol. **22**, No. 1, p. 147–152 (1967).
See also 13, p. 166; 22, p. 446.

M. Stuart
*Electrode-Limited to Bulk-Limited Conduction in Silicon Oxide Films
Vol. **23**, No. 2, p. 595–597 (1967).
See also *14.4.1*, p. *260*; 22.6, p. 539.

R. Grigorovici, N. Croitoru, and A. Dévényi
Thermoelectric Power in Amorphous Silicon
Vol. **23**, No. 2, p. 621–625 (1967).
See also 2, p. 23; 13.1, p. 182; *15*, p. *273*; 22.1.2, p. 472.

A. G. Revesz
*Noncrystalline Structure and Electronic Conduction of Silicon Dioxide Films
Vol. **24**, No. 1, p. 115–126 (1967).
See also *1.4*, p. *19*; 2, p. 23; 14.3.2, p. 249; 22.6, p. 539.

H. Pagnia
Slow Photoconduction in InSb Films
Vol. **24**, No. 1, p. K97–K100 (1967).
See also *16*, p. *281*; 22.2.3, p. 491.

J. Viščakas, P. Mačkus, and A. Smilga
Space-Charge-Limited Currents and High Electric Field Effect in Vitreous Selenium Films
Vol. **25**, No. 1, p. 331–335 (1968).
See also 2, p. 23; 13.4, p. 211; *14.4.1*, p. *261*; 22.1.3, p. 478.

D. M. Hughes and G. Carter
*The Effects of Oxygen Adsorption and Low Energy Ion Bombardment on the Electrical Properties of Cadmium Sulphide Thin Films
Vol. **25**, No. 1, p. 449–453 (1968).
See also 13.3, p. 198; 22.4.1, p. 501.

P. Schnupp
*A Model for Tunneling through a Non-Crystalline Thin Dielectric Film
Vol. **25**, No. 1, p. 455–460 (1968).
See also 13.1, p. 183; 22, p. 449.

H. P. Geserich
Optical and Electrical Measurements on Thallium(III)Oxide Films
Vol. **25**, No. 2, p. 741–751 (1968).
See also *13.1*, p. *184*; 20.1, p. 352; 22.6, p. 540.

C. Hamann
On the Trap Distribution in Thin Films of Copper Phthalocyanine
Vol. **26**, No. 1, p. 311–318 (1968).
See also *13.4*, p. *212*; 14.4.1, p. 262; 22.9, p. 560.

I. Spînulescu-Carnaru and I. Drăghici
*On Electrical Properties of PbTe Thin Films
Vol. **26**, No. 1, p. K39–K42 (1968).
See also 3.1, p. 33; 22, p. 450.

K. Hirschberg and K. Deutscher
Determination of the Emission Depth of Photoelectrons from Cs–Sb Films
Vol. **26**, No. 2, p. 527–535 (1968).
See also *17*, p. *288*; 20.1, p. 353; 22.7, p. 546.

K. Deutscher and K. Hirschberg
Thickness Dependence of the Quantum Yield of Cesium–Antimony Films
Vol. **27**, No. 1, p. 145–150 (1968).
See also *17*, p. *289*; 20.1, p. 354; 22.7, p. 546.

H. K. MÜLLER
*Electrical and Optical Properties of Sputtered In_2O_3 Films. (I): Electrical Properties and Intrinsic Absorption
Vol. **27**, No. 2, p. 723–731 (1968).
See also 13.4, p. 214; 20.1, p. 355; 22.6, p. 541.

H. K. MÜLLER
Electrical and Optical Properties of Sputtered In_2O_3 Films. (II): Optical Properties in the Near Infrared
Vol. **27**, No. 2, p. 733–740 (1968).
See also *20.1*, p. *355*; 22.6, p. 541.

J. T. CALOW, S. J. T. OWEN, and P. W. WEBB
*The Growth and Electrical Characteristics of Epitaxial Layers of Zinc Selenide on p-Type Germanium
Vol. **28**, No. 1, p. 295–303 (1968).
See also *3.1*, p. *33*; 14.3.2, p. 251; 22.4.2, p. 508.

J. ANTULA
Negative Space-Charge Effect on Schottky Current in Thin Insulating Films
Vol. **28**, No. 1, p. 395–400 (1968).
See also *14.4.1*, p. *263*; 22.6, p. 542.

I. P. ZHADKO and V. A. ROMANOV
On the Origin of High Photovoltages in Semiconductors
Vol. **28**, No. 2, p. 797–800 (1968).
See also *16*, p. *283*; 22, p. 453.

H. BERGER, W. KAHLE, and G. JÄNICHE
*Thickness Dependence of Conductivity due to the Polycrystalline Structure in Evaporated CdS Thin Films
Vol. **28**, No. 2, p. K97–K100 (1968).
See also 1.4, p. 21; 22.4.1, p. 504.

M. ALTWEIN, H. FINKENRATH, Č. KOŇÁK, J. STUKE, and G. ZIMMERER
The Electronic Structure of CdO. (II): Spectral Distribution of Optical Constants
Vol. **29**, No. 1, p. 203–209 (1968).
See also 13.1, p. 188; *20.1*, p. *356*; 22.4, p. 496.

K. H. GUNDLACH
Note on the Effect of the Band Structure on Tunnelling in Metal–Semiconductor–Metal Junctions
Vol. **30**, No. 2, p. K135–K138 (1968).
See also 13.1, p. 190; *14.3.2*, p. *251*; 22, p. 455.

B. SELLE und J. MAEGE
*Elektrische und optische Eigenschaften von Cu_2S-Aufdampfschichten
(Electrical and Optical Properties of Evaporated Cu_2S Films)
Vol. **30**, No. 2, p. K153–K155 (1968).
See also 13.4, p. 218; 15, p. 275; 16, p. 285; *20.1*, p. *359*; 22, p. 455.

14.3.2 Semiconducting Devices. Junctions (Contact Problems See 14.4.1)

Review Articles

J. R. Dale
*Alloyed Semiconductor Heterojunctions
Vol. **16**, No. 2, p. 351–387 (1966).
See also 3.1, p. 29; 5, p. 49; 13.4, p. 200; 16, p. 276; 22.1.1, p. 458; 22.1.2, p. 468; 22.2.1, p. 482; 22.2.2, p. 487; 22.2.3, p. 489.

M. H. Pilkuhn
The Injection Laser
Vol. **25**, No. 1, p. 9–62 (1968).
See also 13.4, p. 200; *14.4.1*, p. *255*; *20.2*, p. *360*; 20.3, p. 363; 22.2.1, p. 482; 22.2.2, p. 487; 22.2.3, p. 489.

Original Papers and Short Notes

G. B. Abdullaev, M. R. Akhundov, and G. A. Akhundov
*Rectifying and Photoconductive Properties of p-GaSe Single Crystals
Vol. **16**, No. 1, p. 209–214 (1966).
See also 16, p. 276; 22, p. 439.

A. S. Antonov
*On the Mechanism of Lithium Ion Drift in the Electrical Field of a p-n Junction in Silicon
Vol. **16**, No. 2, p. 761–770 (1966).
See also *14.4*, p. *252*; 22.1.2, p. 469.

J. Kodeš and K. Šobra
*Pre-Breakdown Region of Selenium p-n Junctions
Vol. **17**, No. 1, p. K19–K20 (1966).
See also 22.1.3, p. 476.

F. Forlani, N. Minnaja, and E. Pagiola
*Theory of the Influence of Surface States on the Impedance of a Semiconductor–Insulator Interface
Vol. **17**, No. 2, p. 657–672 (1966).
See also *13.3*, p. *196*; 22.1.2, p. 469.

T. L. Tansley
*Forward Bias Current-Voltage Characteristics for a Heterojunction in which Tunnelling Dominates
Vol. **18**, No. 1, p. 105–112 (1966).
See also 13, p. 164; 22.2.1, p. 483.

M. Jung
*Pseudo-abrupte legierte p–n-Übergänge in GaP
(Pseudo-Abrupt Alloyed p-n Junctions in GaP)
Vol. **18**, No. 2, p. 743–748 (1966).
See also 13.4, p. 204; 22.2.2, p. 488.

A. G. Revesz
Thermal Oxidation of Silicon: Growth Mechanism and Interface Properties
Vol. **19**, No. 1, p. 193–202 (1967).
See also *1.3*, p. *14*; 13.3, p. 197; 22.1.2, p. 470.

M. P. Mikhailova, D. N. Nasledov, and S. V. Slobodchikov
*The Effect of a Magnetic Field upon Illuminated InAs p-n Junctions
Vol. **19**, No. 1, p. 429—434 (1967).
See also 16, p. 278; 22.2.3, p. 490.

M. J. Małachowski
*Current-Voltage-Temperature Dependence of Symmetric Tunnel Junctions
Vol. **19**, No. 1, p. K27—K30 (1967).

H. Sigmund and K. Berchtold
*Electrical and Photovoltaic Properties of PbS-Si Heterodiodes
Vol. **20**, No. 1, p. 255—259 (1967).
See also 16, p. 278; 22, p. 444; 22.1.2, p. 470.

P. S. Serebrennikov
*On the Theory of Tunneling
Vol. **21**, No. 1, p. 431—436 (1967).
See also 22, p. 445.

A. N. Georgobiani and V. I. Steblin
Electroluminescence of ZnS-Cu_2S Heterojunctions
Vol. **21**, No. 1, p. K45—K48 (1967).
See also *20.3*, p. *365*; 22, p. 445; 22.4.1, p. 499.

A. N. Georgobiani and V. I. Steblin
*Optical and Electrical Characteristics of p–n Junctions in Zinc Sulfide
Vol. **21**, No. 1, p. K49—K51 (1967).
See also 16, p. 279; 20.3, p. 365; 22.4.1, p. 499.

K. H. Gundlach and G. Heldmann
The Effect of the $E(k)$-Relation on Tunneling through Asymmetric Barriers
Vol. **21**, No. 2, p. 575—579 (1967).
See also 13.1, p. 178; *14.3.1*, p. *245*; 22.6, p. 538.

W. Franz
*Duration of the Tunnelling Single Process
Vol. **22**, No. 2, p. K139—K140 (1967).
See also 22, p. 447.

T. L. Tansley
*Spectral Response of p-n Heterojunctions
Vol. **23**, No. 1, p. 241—252 (1967).
See also 14.4.1, p. 259; 16, p. 280; 20.1, p. 350; 22.2.1, p. 484; 22.8, p. 552.

J. Antula
*Tunnel and Schottky Current in Dielectric Thin Films Considering Film Thickness Fluctuations
Vol. **24**, No. 1, p. 89—94 (1967).
See also 14.4.1, p. 260; 22, p. 448.

J. Antula
A Method of Measuring Effective Electron Mass in Thin Insulating Films
Vol. **24**, No. 1, p. 95—98 (1967).
See also *13.1*, p. *182*; 14.4.1, p. 266; 22.6, p. 539.

A. G. Revesz
Noncrystalline Structure and Electronic Conduction of Silicon Dioxide Films
Vol. **24**, No. 1, p. 115—126 (1967).
See also *1.4*, p. *19*; 2, p. 23; *14.3.1*, p. *246*; 22.6, p. 539.

R. GRIGOROVICI, N. CROITORU, L. VESCAN, and M. MARINA
*Hole Injection in Junctions between Amorphous Ge Layers and n-Type Ge Single Crystals
Vol. **24**, No. 1, p. K17—K21 (1967).
See also *14.4.1*, p. *261*; 22.1.1, p. 463.

R. K. PUROHIT
*GaP-GaAs n-p Heterojunctions
Vol. **24**, No. 1, p. K57—K60 (1967).
See also 22.2.1, p. 484; 22.2.2, p. 488.

M. ROUZEYRE
*Contribution à l'étude des phénomènes de transport dans les structures aluminium-alumine-métal
(Contribution to the Study of Transport Phenomena in Aluminium–Alumina–Metal Structures)
Vol. **24**, No. 2, p. 399—411 (1967).
See also 16, p. 281; 21, p. 383; 22.6, p. 540.

T. L. TANSLEY
*Forward Current Injection Modulation of Photocurrent in p-n Heterojunctions
Vol. **24**, No. 2, p. 615—622 (1967).
See also 14.4.1, p. 261; 16, p. 281; 22.2.1, p. 485; 22.2.2, p. 488.

W. LUDWIG and B. KORNEFFEL
*Photoemission Studies on Thin Metal–Insulator–Metal Sandwiches
Vol. **24**, No. 2, p. K137—K140 (1967).
See also *17*, p. 288; 21, p. 383; 22.6, p. 540.

B. SELLE, W. LUDWIG, and R. MACH
*Photovoltaic Response of Cu_2S-CdS p-n Heterojunctions
Vol. **24**, No. 2, p. K149—K152 (1967).
See also *16*, p. *281*; 22, p. 449; 22.4.1, p. 501.

S. OBERLÄNDER and S. SIKORSKI
On the Mathematical Description of Electrical Processes in Semiconductors. (I): Phase Space Analysis in Subspaces of Dimension Lower than Three
Vol. **25**, No. 1, p. 345—357 (1968).
See also *14.3*, p. *240*; 22, p. 449.

N. A. VLASENKO and A. N. GERGELL
On the Mechanism of DC Electroluminescence in pCu_xS-nZnS:Mn, Cu, Cl Film Structures
Vol. **26**, No. 1, p. K77—K81 (1968).
See also *20.3*, p. *368*; 22, p. 450; 22.4.1, p. 502.

L. ASK
*Effect of Mechanical Stress on the I-U Characteristic of a GaP p–n Junction
Vol. **26**, No. 2, p. K143—K145 (1968).
See also 2, p. 24; 12, p. 158; 22.2.2, p. 488.

E. IGRAS and T. WARMIŃSKI
*On Electron Mirror Microscopic Observations of the Details of p-n Junction Regions in Si
Vol. **27**, No. 1, p. 69—76 (1968).
See also *1.3*, p. *15*; 22.1.2, p. 473.

J. AUTH
*On the Reverse-Biased Capacitance of Step p^+-n Junctions with Traps
Vol. **27**, No. 2, p. 653—656 (1968).
See also 13.4, p. 214; 22.1.2, p. 474.

J. Hatz
Some Effects of Material Inhomogeneities on the Near-Field Pattern of GaAs Diode Lasers
Vol. **28**, No. 1, p. 233–245 (1968).
See also 10, p. 109; *20.2*, p. *362*; 22.2.1, p. 485.

J. T. Calow, S. J. T. Owen, and P. W. Webb
The Growth and Electrical Characteristics of Epitaxial Layers of Zinc Selenide on p-Type Germanium
Vol. **28**, No. 1, p. 295–303 (1968).
See also *3.1*, p. *33*; *14.3.1*, p. *247*; 22.4.2, p. 508.

D. K. Roy
*On the Separation of the Phonon-Assisted Current Components in Germanium Tunnel Diodes
Vol. **28**, No. 1, p. K75–K79 (1968).
See also 6, p. 67; 22.1.1, p. 465.

Yu. A. Goldberg, M. E. Levinshtein, and D. N. Nasledov
*Magnetic Field Influence on the Low-Frequency Oscillations of Gunn Diodes
Vol. **29**, No. 2, p. K103–K105 (1968).
See also *14.4.1*, p. *264*; 22.2.1, p. 486.

D. K. Roy
*On the Determination of Impurity Concentration in the Alloyed Region of a Tunnel Diode
Vol. **29**, No. 2, p. K111–K115 (1968).
See also *13.4*, p. *216*; 22.1.1, p. 466.

G. Lecoy and L. Gouskov
*Noise in Thin Films Metal-Oxide-Metal Al-Al_2O_3-Al
Vol. **30**, No. 1, p. 9–17 (1968).
See also 22.6, p. 543.

H. Wienhold
*Zu Relaxationseffekten bei epitaktischen Si-MOS-Strukturen
(On Relaxation Effects of Epitaxial Si-MOS Structures)
Vol. **30**, No. 1, p. K17–K21 (1968).
See also 13.3, p. 199; 22.1.2, p. 474.

W. Ludwig, W. Gericke, and H. Treptow
*Electron Tunnelling in MIS-Structures of the Type Al-Al_2O_3-ZnTe
Vol. **30**, No. 1, p. K61–K64 (1968).
See also 13.3, p. 199; 14.4.1, p. 264; 22.4.3, p. 511.

A. A. Gutkin and T. Janicki
*The Effect of Electric Field in a Space Charge Layer on the Photoresponse Edge of Si p-n Junctions
Vol. **30**, No. 2, p. 827–832 (1968).
See also *14.4.1*, p. *265*; 16, p. 285; 22.1.2, p. 475.

K. H. Gundlach
*Note on the Effect of the Band Structure on Tunnelling in Metal–Semiconductor–Metal Junctions
Vol. **30**, No. 2, p. K135 –K138 (1968).
See also 13.1, p. 190; 14.3.1, p. 247; 22, p. 455.

14.4 Dielectrics

Review Articles

P. SÜPITZ and J. TELTOW
Transport of Matter in Simple Ionic Crystals (Cubic Halides)
Vol. **23**, No. 1, p. 9–56 (1967).
See also *9*, p. *87*; *10*, p. *93*; 15, p. 270; 22.5.1, p. 515; 22.5.2, p. 518; 22.5.3, p. 532.

Original Papers and Short Notes

C. N. PLĂVIŢU
*Some Kinetic Phenomena in Impure Ionic Semiconductors
Vol. **16**, No. 1, p. 69–78 (1966).
See also 8, p. 79; 13, p. 162; 15, p. 270; 22, p. 439.

V. V. RATNAM
On the Electrical and Magnetic Properties of Coloured and Uncoloured Calcium Fluoride Crystals
Vol. **16**, No. 2, p. 549–557 (1966).
See also *10.2*, p. *137*; 13.4, p. 201; 18.1, p. 296; 22.5.3, p. 532.

A. S. ANTONOV
*On the Mechanism of Lithium Ion Drift in the Electrical Field of a p-n Junction in Silicon
Vol. **16**, No. 2, p. 761–770 (1966).
See also *14.3.2*, p. *248*; 22.1.2, p. 469.

E. KRÖNER und D. KESSEL
*Zur nichtlokalen klassischen Theorie der Dielektrika
(On the Non-Local Classical Theory of Dielectrics)
Vol. **17**, No. 1, p. 211–220 (1966).
See also 22, p. 440.

J. H. BEAUMONT and P. W. M. JACOBS
*Effect of Gaseous Impurities on the Structure-Sensitive Conductance of Potassium Chloride Single Crystals
Vol. **17**, No. 1, p. K45–K48 (1966).
See also 10, p. 95; 22.5.2, p. 519.

C. RAMASASTRY and C. V. RAMAIAH
*Dielectric Constant of Sodium Chlorate Crystals at Microwave Frequencies
Vol. **19**, No. 1, p. K15–K17 (1967).
See also 22.8, p. 549.

T. GESZTI
On the Theory of Thermally Activated Processes
Vol. **20**, No. 1, p. 165–177 (1967).
See also 9, p. 88; *10*, p. *98*; 12, p. 157.

P. ASADI
*On the Electrical Conductivity of Fluorite Single Crystals Doped with Uranium Dioxide
Vol. **20**, No. 1, p. K59–K61 (1967).
See also 10, p. 99; 22.5.3, p. 533.

P. ASADI
On the Existence of Piezoelectric Textures in Single Crystals of Antozonite–Fluorite
Vol. **20**, No. 2, p. K73–K75 (1967).
See also *4*, p. *41*; 22.8, p. 550.

I. Bunget and M. Rosenberg
Influence of Changes of the Magnetocrystalline Anisotropy on the Electrical Resistivity of $Ba_3Co_2Fe_{24}O_{41}$
Vol. **20**, No. 2, p. K163–K165 (1967).
See also *18.3*, p. *317*.

P. S. Zyryanov and V. I. Okulov
Effect of Current Carrier Scattering on the Complex Dielectric Tensor
Vol. **21**, No. 1, p. 89–98 (1967).
See also *14*, p. *219*.

J. Hańderek
*Electric Conductivity and Dielectric Absorption Phenomena in Lead Zirconate Ceramics
Vol. **21**, No. 1, p. 323–329 (1967).
See also *14.4.2*, p. *267*; 22.8, p. 550.

A. C. Papadakis
*On the Measurement of Carrier Mobilities in Insulators Using Transient Space-Charge-Perturbed Currents
Vol. **21**, No. 1, p. K77–K80 (1967).
See also 14.4.1, p. 258; 22, p. 445.

V. P. Kalashnikov
Theory of Spin-Lattice Relaxation of Conduction Electrons of Ionic Semiconductors at Extremely High Magnetic Fields
Vol. **21**, No. 2, p. 775–784 (1967).
See also *6*, p. *61*; 13, p. 166; 18, p. 292; 22, p. 446.

K. Lal
*Dielectric Properties of Interstitial Impurity Cations in Alkali Halide Crystals
Vol. **22**, No. 1, p. 141–146 (1967).
See also 10, p. 101; 22.5.2, p. 523.

S. C. Jain and G. D. Sootha
*Electrical and Optical Properties of Titanium-Doped Lithium Fluoride Crystals
Vol. **22**, No. 2, p. 505–516 (1967).
See also 13.4, p. 208; *20.1*, p. *349*; 22.5.2, p. 523.

J. E. Caffyn, T. L. Goodfellow, E. M. Olymbios, and D. J. Perrin
*Dielectric and Optical Absorption Measurements on CaF_2 and CaF_2:Y
Vol. **22**, No. 2, p. 549–557 (1967).
See also 10, p. 101; 13.4, p. 209; *20.1*, p. *350*; 22.5.3, p. 533.

A. S. Davydov and V. A. Onishchuk
*The Dielectric Permeability of Molecular Crystals
Vol. **24**, No. 1, p. 373–384 (1967).
See also 22, p. 448.

F. Fröhlich and G. Hensel
*Low Temperature Conductivity of Ca-Doped KCl Crystals Influenced by Segregation Processes
Vol. **24**, No. 2, p. 535–541 (1967).
See also 10, p. 103; 22.5.2, p. 524.

F. Haberey and H. P. J. Wijn
*Effect of Temperature on the Dielectric Relaxation in Polycrystalline Ferrites
Vol. **26**, No. 1, p. 231–240 (1968).
See also 18.3, p. 319.

J. Arends and H. Nijboer
*The Ionic Conductivity of CsCl Doped with K^+, Rb^+, and Br^-
Vol. **26**, No. 2, p. 537–542 (1968).
See also 10, p. 105; 22.5.2, p. 526.

W. Ulrici
Study of Trivalent Transition-Metal Ions in Silver Halides. (I): AgCl:V
Vol. **27**, No. 1, p. 333–344 (1968).
See also *10*, p. *107*; 13.4, p. 213; 20.1, p. 354; 22.5.1, p. 516.

W. Ulrici
Study of Trivalent Transition-Metal Ions in Silver Halides. (II): Chromium and Titanium
Vol. **27**, No. 2, p. 489–500 (1968).
See also *10*, p. *107*; 13.4, p. 213; 20.1, p. 354; 22.5.1, p. 517.

Yu. E. Perlin and Sh. N. Gifeisman
The Theory of Shallow F-Centres
Vol. **28**, No. 1, p. 45–60 (1968).
See also 6, p. 66; *10.2*, p. *143*; 13, p. 171; 18, p. 294; 22.2.1, p. 485; 22.2.2, p. 488; 22.2.3, p. 493; 22.4, p. 495; 22.4.1, p. 503; 22.4.2, p. 508; 22.4.3, p. 511; 22.5.1, p. 517; 22.5.2, p. 527.

V. K. Jain and P. K. C. Pillai
*Characteristics of Plastic Electrets
Vol. **28**, No. 1, p. 415–419 (1968).
See also 2, p. 24; 22.9, p. 560.

Š. Kupča
Hardness of NaCl Single Crystals Prepared with Various Concentrations of $BaCl_2$ in the Melt
Vol. **29**, No. 1, p. K15–K17 (1968).
See also *10*, p. *110*; 22.5.2, p. 529.

H. Weijma and B. R. Knottnerus
*Ionic Conductivity of Pure and Ba^{2+}-Doped CsCl
Vol. **29**, No. 1, p. K43–K45 (1968).
See also 1.2, p. 12; 10, p. 110; 22.5.2, p. 529.

B. Chalupa, R. Michalec, V. Petržílka, J. Tichý, and J. Zelenka
Diffraction of Neutrons on a Vibrating Quartz Crystal
Vol. **29**, No. 1, p. K51–K54 (1968).
See also *5*, p. *54*; 22.6, p. 543.

S. Haussühl and H. Hesse
Quadratic Electrooptical Effect (Kerr Effect) in Alkali Halides of NaCl Type
Vol. **30**, No. 1, p. 209–214 (1968).
See also *20*, p. *342*; 22.5.2, p. 530.

Nguyen van Dong and Pham Ngu Tung
Transport Properties of Silver Telluride in the Solid and Liquid States
Vol. **30**, No. 2, p. 557–567 (1968).
See also 2, p. 25; *14.3*, p. *244*; *15*, p. *275*; 22, p. 455.

G. Hensel and J. Ziller
*Precipitation Influenced Low Temperature Conductivity of Ca Doped NaCl Crystals
Vol. **30**, No. 2, p. K165–K168 (1968).
See also 10, p. 113; 22.5.2, p. 531.

14.4.1 High Field Phenomena, Space Charge Effects, Inhomogeneities, Injected Carriers (Electroluminescence See 20.3; Junctions See 14.3.2)

Review Articles

L. Eckertová
Field Cathodes with Thin Insulating Films
Vol. **18**, No. 1, p. 3–40 (1966).
See also 13, p. 162; 13.3, p. 196; 13.4, p. 200; *17*, p. *286*; 22.6, p. 535.

M. H. Pilkuhn
*The Injection Laser
Vol. **25**, No. 1, p. 9–62 (1968).
See also 13.4, p. 200; 14.3.2, p. 248; *20.2*, p. *360*; 20. 3, p. 363; 22.2.1, p. 482; 22.2.2, p. 487; 22.2.3, p. 489.

Original Papers and Short Notes

E. I. Rashba, V. A. Romanov, I. I. Boiko, and I. P. Zhadko
Electrical Pinch in Elastically Deformed Germanium
Vol. **16**, No. 1, p. 43–56 (1966).
See also 12, p. 156; *14.3*, p. *231*; 22.1.1, p. 459.

S. G. Kalashnikov and V. L. Bonch-Bruevich
*On the Velocity of Space Charge Waves (Electrical Domains) in Semiconductors
Vol. **16**, No. 1, p. 197–203 (1966).
See also 22.1.1, p. 459.

G. B. Abdullaev, E. S. Guseinova, and B. G. Tagiev
*Electrical Conductivity of p-GaSe Single Crystals in Strong Electric Fields
Vol. **16**, No. 1, p. 205–208 (1966).
See also 13.4, p. 201; 22, p. 439.

W. E. Wilhelm
*A Simple Asymptotic Formula for the Motion of Simultaneous Electric Field and Conductivity Domains in Semiconductors
Vol. **16**, No. 1, p. K59–K62 (1966).
See also 22, p. 439.

H. Lemke
*Zur Zeitabhängigkeit raumladungsbegrenzter Injektionsströme in Halbleitern
(On the Time Dependence of Space-Charge Limited Injection Currents in Semiconductors)
Vol. **16**, No. 2, p. 413–426 (1966).
See also 13.4, p. 201; 22, p. 440.

H. Lemke
*Zeitabhängigkeit raumladungsbegrenzter Injektionsströme in Fe-dotiertem p-Silizium
(Time Dependence of Space-Charge Limited Injection Currents in Fe-Doped p-Silicon)
Vol. **16**, No. 2, p. 427–438 (1966).
See also 13.4, p. 201; 22.1.2, p. 469.

R. Granger
*Distribution électronique anisotrope dans les semiconducteurs purs au voisinage de l'avalanche
(Anisotropic Electron Distribution in Pure Semiconductors near Breakdown)
Vol. **16**, No. 2, p. 599–612 (1966).
See also 6, p. 56; 13, p. 163; 22, p. 440; 22.2.3, p. 489.

P. THOMA and G. VAUBEL
Electrical Properties on Anthracence Single Crystals at Very Low Temperatures
Vol. **16**, No. 2, p. 663—674 (1966).
See also 13.4, p. 201; 14.3, p. 231; *16*, p. *276*, 22.9, p. 558.

H. J. ROHDE
*Einfluß flacher Haftstellen auf die Strom-Spannungs-Charakteristik bei Doppelinjektion
(Effect of Shallow Traps on the Current-Voltage Characteristics in the Case of Double Injection)
Vol. **16**, No. 2, p. K179—K182 (1966).
See also 13.4, p. 202; 22.4.1, p. 498.

K. W. BÖER and P. L. QUINN
*Inhomogeneous Field Distribution in Homogeneous Semiconductors Having an N-Shaped Negative Differential Conductivity
Vol. **17**, No. 1, p. 307—316 (1966).
See also 22, p. 440.

R. FISCHER, F. STÖCKMANN, and J. STUKE
*Raumladungsbegrenzte Ströme in inhomogen belichteten CdS-Einkristallen
(Space-Charge Limited Currents in Inhomogeneously Illuminated CdS Single Crystals)
Vol. **17**, No. 1, p. 335—341 (1966).
See also 16, p. 277; 22.4.1, p. 498.

J. STUKE und G. WEISER
*Franz-Keldysh-Effekt an trigonalem und amorphem Selen
(Franz-Keldysh Effect in Trigonal and Amorphous Selenium)
Vol. **17**, No. 1, p. 343—352 (1966).
See also 13.1, p. 175; *20* p. *337*; 22.1.3, p. 476.

P. VOSS
*Surface Pinned Layer-Like Field Inhomogeneities in CdS
Vol. **17**, No. 1, p. K21—K24 (1966).
See also 22.4.1, p. 498.

P. BAUDUIN et F. BUCHY
*Oscillations de courant dans le sulfure de cadmium photoconducteur
(Current Oscillations in Photoconducting Cadmium Sulphide)
Vol. **17**, No. 2, p. 517—530 (1966).
See also *16*, p. *277*; 22.4.1, p. 498.

K. KIROV
Pulse Method for Studying Trapping Centres in CdS Type Crystals
Vol. **17**, No. 2, p. 581—592 (1966).
See also *13.4*, p. *202*; 16, p. 277; 22.4.1, p. 498.

G. B. ABDULLAEV, E. S. GUSEINOVA, and B. G. TAGIEV
*Electrical Conductivity of n-InSe Single Crystals in Strong Electric Fields
Vol. **17**, No. 2, p. 593—596 (1966).
See also 13.1, p. 175; 22, p. 440.

J. KOŁODZIEJCAK and H. STRAMSKA
On the Theory of Free Carrier Electro-Magneto-Optical Phenomena
Vol. **17**, No. 2, p. 701—707 (1966).
See also 13.1, p. 176; *18*, p. *291*; *20*, p. *337*; 22, p. 441.

V. DIENYS and J. POZHELA
*Anisotropy of the Conductivity of n-Type Germanium at High Electric Fields
Vol. **17**, No. 2, p. 769—776 (1966).
See also 13, p. 163; 22.1.1, p. 460.

E.-H. Weber
Relationship between Space Charge, Surface Conductivity, and Band Bending for Photoconductors under Uniform Optical Excitation
Vol. **17**, No. 2, p. 843–854 (1966).
See also 13.3, p. 197; 13.4, p. 203; *16*, p. *277*; 22.4, p. 495.

H.-J. Dittfeld und H. Streller
*Zur Abhängigkeit der maximalen Sperrspannung eines Metall–Halbleiter-Spitzenkontaktes von der Beweglichkeit der Majoritätsladungsträger
(Dependence of the Maximum Reverse Voltage of a Metal–Semiconductor Point Contact on the Mobility of Majority Carriers)
Vol. **17**, No. 2, p. K245–K248 (1966).
See also 22.1.1, p. 460; 22.1.2, p. 469.

H. Frank
Contribution to the Measurement of the Resistance of nn^+ and pp^+ Epitaxial Films by the Breakdown Method
Vol. **18**, No. 1, p. 401–414 (1966).
See also *14.3.1*, p. *245*; 22.1.1, p. 460; 22.1.2, p. 470; 22.2.1, p. 483.

J. Kołodziejczak
On the Theory of Galvanomagnetic Phenomena in Strong Electric Fields
Vol. **19**, No. 1, p. 373–384 (1967).
See also 13.1, p. 176; *15*, p. *271*; 22, p. 442.

J. Kołodziejczak and E. Kierzek-Pecold
*Free Carrier Electro-Optical Kerr Effect in Semiconductors
Vol. **19**, No. 1, p. K55–K58 (1967).
See also 13.1, p. 176; *20*, p. *338*; 22, p. 443.

G. Döhler
*On the Field Emission of Minority Carriers in Photoconductors
Vol. **19**, No. 2, p. 555–564 (1967).
See also 13.4, p. 205; 16, p. 278; 22.1.3, p. 477; 22.4.1, p. 499.

R. Enderlein and R. Keiper
On the Theory of Electro-Absorption and Electro-Reflectance in Semiconductors
Vol. **19**, No. 2, p. 673–681 (1967).
See also 13.1, p. 177; *20*, p. *338*; 22, p. 443.

C. Cherki and R. Coelho
*On Charge Storage in Anodic Tantal Oxide Layers
Vol. **19**, No. 2, p. K91–K94 (1967).
See also 14.3.1, p. 245; 22.6, p. 537.

R. Enderlein
The Influence of Collisions on the Franz-Keldysh Effect
Vol. **20**, No. 1, p. 295–299 (1967).
See also 13.1, p. 177; *20*, p. *338*; 22, p. 444.

P. Devaux and M. Schott
Thermally Stimulated Currents without Optical Excitation. Application to Copper Phtalocyanine
Vol. **20**, No. 1, p. 301–309 (1967).
See also *13.4*, p. *205*; 22.9, p. 558.

N. A. Vlasenko and S. A. Zynio
Role of Carrier Polarization in the Mechanism of the Electroluminescence of ZnS–Mn Films
Vol. **20**, No. 1, p. 311–320 (1967).
See also *20.3*, p. *365*; 22.4.1, p. 499.

P. Asadi
*On the Existence of a Spontaneous Polarization in Fluorite Single Crystals Doped with Lime
Vol. **20**, No. 1, p. K55–K58 (1967).
See also 10, p. 98; 22.5.3, p. 532.

G. B. Abdullaev, E. S. Guseinova, and B. G. Tagiev
*Electric Conductivity Studies of p-GaTe Polycrystals and Single Crystals in Strong Electric Fields
Vol. **20**, No. 2, p. 421–425 (1967).
See also *14.3*, p. *235*; 22, p. 444.

C. Hamann
On the Electric and Thermoelectric Properties of Copper Phthalocyanine Single Crystals
Vol. **20**, No. 2, p. 481–491 (1967).
See also 13.4, p. 205; *14.3*, p. *236*; 15, p. 271; 22.9, p. 559.

C. Hervouet
*Magnetic Field Dependence of Current Oscillations in Piezoelectric Semiconductors
Vol. **21**, No. 1, p. 117–125 (1967).
See also 22.2.1, p. 483.

P. Bräunlich
*Light Effects on Space-Charge Controlled Currents in CdS
Vol. **21**, No. 1, p. 383–391 (1967).
See also 13.4, p. 206; 16, p. 279; 20.3, p. 365; 22.4.1, p. 499.

G. B. Abdullaev, Z. A. Iskender-Zade, E. A. Dzhafarova, and V. E. Chelnokov
Study of Electron Lifetime in p-Si
Vol. **21**, No. 1, p. 423–429 (1967).
See also 13.4, p. 206; *14.3*, p. *236*; 22.1.2, p. 471.

V. F. Zolotaryov, D. G. Semak, and D. V. Chepur
Thermally Stimulated Currents under the Condition of Persistant Internal Polarization
Vol. **21**, No. 1, p. 437–442 (1967).
See also *13.4*, p. *206*; 22, p. 445.

A. C. Papadakis
On the Measurement of Carrier Mobilities in Insulators Using Transient Space-Charge Perturbed Currents
Vol. **21**, No. 1, p. K77–K80 (1967).
See also *14.4*, p. *253*; 22, p. 445.

J.-P. Ropé
*Étude expérimentale de la dynamique des domaines de haut champ électrique dans le couplage électron-phonon
(Experimental Study of the Dynamics of High-Field Domains in Electron-Phonon Coupling)
Vol. **21**, No. 2, p. 517–524 (1967).
See also 7, p. 76; 22.2.1, p. 484.

A. Bonnot
*Relaxation des domaines de haut champ et émission de lumière associées aux oscillations de courant dans GaAs
(Relaxation of High-Field Domains and Light Emission Associated with Current Oscillations in GaAs)
Vol. **21**, No. 2, p. 525–531 (1967).
See also 7, p. 76; *20.3*, p. *366*; 22.2.1, p. 484.

G. LAUTZ and M. SCHULZ
Helical Density-Waves and Oscillistor Effect in the Electron-Hole Plasma of n-Germanium
Vol. **21**, No. 2, p. K139—K143 (1967).
See also *13*, p. *166*; *20*, p. *339*; 22.1.1, p. 462.

K. W. BÖER and G. A. DUSSEL
*Slow-Moving Field Domains in CdS in the Range of Negative Differential Conductivity
Vol. **21**, No. 2, p. K145—K149 (1967).
See also 22.4.1, p. 500.

R. SITTIG
Stark Effect Measurements of the OH^- Absorption Band in KCl
Vol. **21**, No. 2, p. K175—K177 (1967).
See also 10.2, p. 139; *20.1*, p. *349*; 22.5.2, p. 523.

H. N. SPECTOR
Optical Absorption in Electric and Magnetic Fields
Vol. **22**, No. 1, p. 185—193 (1967).
See also 13.1, p. 179; 18, p. 292; *20.1*, p. *349*; 22, p. 446.

E. GUTSCHE and H. LANGE
*Electroreflectance of CdS and CdSe Single Crystals at the Fundamental Absorption Edge
Vol. **22**, No. 1, p. 229—233 (1967).
See also 13.1, p. 179; 13.2, p. 192; *20*, p. *339*; 20.1, p. 349; 22.4.1, p. 500; 22.4.2, p. 506.

V. L. BONCH-BRUEVICH
*On the Theory of Electrical Domains in Hot Electron Semiconductors
Vol. **22**, No. 1, p. 267—278 (1967).
See also 13, p. 166; 14.3, p. 237; 22, p. 446; 22.1.1, p. 462; 22.2.1, p. 484.

J. PASTRŇÁK
*A Thermodynamic Investigation of the Conditions for the Generation of Oscillations of an Electric Current Flowing through a Non-Ohmic Medium
Vol. **22**, No. 2, p. 407—413 (1967).
See also 22, p. 447.

R. ENDERLEIN and R. KEIPER
On the Dielectric Function of a Semiconductor in an External Electric Field
Vol. **23**, No. 1, p. 127—136 (1967).
See also 13.1, p. 180; *20*, p. *339*; 22, p. 447.

T. L. TANSLEY
Spectral Response of p-n Heterojunctions
Vol. **23**, No. 1, p. 241—252 (1967).
See also *14.3.2*, p. *249*; 16, p. 280; 20.1, p. 350; 22.2.1, p. 484; 22.8, p. 552.

H. J. ROHDE
*Double Injection into CdS Single Crystals
Vol. **23**, No. 1, p. 277—288 (1967).
See also 13.4, p. 209; 22.4.1, p. 500.

B. HILCZER, B. BŁASZCZYK, and S. GODERSKA
*Effect of Storage Temperature on the Time Dependence of Charge Density in Carnauba Wax and Methyl Polymethacrylate Electrets
Vol. **23**, No. 1, p. 419—422 (1967).
See also 22.9, p. 559.

J. Kołodziejczak
Effect of Strong Electric Fields on Magneto-Resistivity and Hall Coefficient in Semiconductors
Vol. **23**, No. 1, p. 423—427 (1967).
See also 6, p. 62; 13.1, p. 181; *14.3*, p. *238*; 22, p. 447; 22.2.3, p. 491.

S. G. Kalashnikov, A. I. Morosov, and V. V. Proklov
*Some Data on Acousto-Electrical Domains in Photoconductive Cadmium Sulphide
Vol. **23**, No. 1, p. K9—K12 (1967).
See also *7*, p. *77*; 22.4.1, p. 501.

W. Güth und H. Bleher
*Lawinendurchbruch in Germanium bei optischer Band–Band-Anregung (Avalanche Breakdown in Germanium in the Case of Optical Band–Band Excitation)
Vol. **23**, No. 1, p. K13—K16 (1967).
See also 13, p. 167; 16, p. 280; 22.1.1, p. 463.

D. E. Aspnes
Note on Collision Broadening of Franz-Keldysh Effect
Vol. **23**, No. 1, p. K79—K81 (1967).
See also 13.1, p. 181; *20*, p. *340*; 22, p. 447.

M. Stuart
*Electrode-Limited to Bulk-Limited Conduction in Silicon Oxide Films
Vol. **23**, No. 2, p. 595—597 (1967).
See also *14.3.1*, p. *245*; 22.6, p. 539.

H.-P. Hennig
*Contribution to the Determination of Gold Centre Cross Sections in Single-Crystal Germanium by Double Injection Experiments
Vol. **23**, No. 2, p. 599—605 (1967).
See also 13.4, p. 210; 22.1.1, p. 463.

V. L. Bonch-Bruevich
*On the Problem of Static Domains in Hot Electron Semiconductors
Vol. **23**, No. 2, p. 761—766 (1967).
See also 13, p. 168; 14.3, p. 238; 22.1.1, p. 463.

J. Antula
Tunnel and Schottky Current in Dielectric Thin Films Considering Film Thickness Fluctuations
Vol. **24**, No. 1, p. 89—94 (1967).
See also *14.3.2*, p. *249*; 22, p. 448.

J. Antula
A Method of Measuring Effective Electron Mass in Thin Insulating Films
Vol. **24**, No. 1, p. 95—98 (1967).
See also *13.1*, p. *182*; 14.3.2, p. 249; 22.6, p. 539.

H. Lemke and G. O. Müller
*Transient Behaviour of Space-Charge-Limited Currents in p-Type Silicon
Vol. **24**, No. 1, p. 127—133 (1967).
See also 22.1.2, p. 472.

G. Döhler
*On Slow High Field Domains in Homogeneous Photoconductors
Vol. **24**, No. 1, p. 331—340 (1967).
See also 16, p. 280; 22.1.3, p. 478.

R. Grigorovici, N. Croitoru, L. Vescan, and M. Marina
*Hole Injection in Junctions between Amorphous Ge Layers and n-Type Ge Single Crystals
Vol. **24**, No. 1, p. K17–K21 (1967).
See also *14.3.2*, p. *250*; 22.1.1, p. 463.

B. Autin
*Transient Acousto-Electric Phenomena in Semiconducting Piezo-Electric Crystals
Vol. **24**, No. 2, p. 501–508 (1967).
See also *7*, p. *77*; 22, p. 448.

M. S. Kagan, S. G. Kalashnikov, and N. G. Zhdanova
*Stationary State of a Semiconductor with Negative Bulk Differential Conductivity
Vol. **24**, No. 2, p. 551–559 (1967).
See also 22.1.1, p. 464.

T. L. Tansley
Forward Current Injection Modulation of Photocurrent in p–n Heterojunctions
Vol. **24**, No. 2, p. 615–622 (1967).
See also *14.3.2*, p. *250*; 16, p. 281; 22.2.1, p. 485; 22.2.2, p. 488.

K.-E. Kröll
*Slowly Propagating High-Field Domains in Semiconductors with Negative Differential Conductivity
Vol. **24**, No. 2, p. 707–714 (1967).
See also 22, p. 449.

H. L. Hartnagel
*Some Ideas on Electroacoustic Domains
Vol. **24**, No. 2, p. K133–K135 (1967).
See also *7*, p. *77*; 22, p. 449.

W. Ludwig and G. Voigt
*Voltage-Controlled Negative Differential Resistance in Zinc Indium Sulfide Crystals
Vol. **24**, No. 2, p. K161–K164 (1967).
See also 22.8, p. 552.

P. Krispin
*Zur Photoimpedanz von ZnS-Phosphoren
(On the Photoimpedance of ZnS Phosphors)
Vol. **25**, No. 1, p. 147–154 (1968).
See also 13.4, p. 211; 16, p. 281; 22.4.1, p. 501.

J. Viščakas, P. Mačkus, and A. Smilga
*Space-Charge-Limited Currents and High Electric Field Effect in Vitreous Selenium Films
Vol. **25**, No. 1, p. 331–335 (1968).
See also 2, p. 23; 13.4, p. 211; 14.3.1, p. 246; 22.1.3, p. 478.

M. Schulz
Helical Density Waves and the Oscillistor Effect in the Electron Hole Plasma of the n-Type Ge (Theory)
Vol. **25**, No. 2, p. 521–530 (1968).
See also *13*, p. *169*; *20*, p. *340*; 22.1.1, p. 464.

G. Krenzke, H. Lemke, and G. O. Müller
*Properties of Gold Contacts on Clean n-Silicon Surfaces
Vol. **25**, No. 2, p. K131–K134 (1968).
See also 13.3, p. 198; 22.1.2, p. 473.

A. A. GRINBERG and A. A. KASTALSKII
*The Ultrasound Generation under the Gunn Effect in a Piezoelectric Material
Vol. **26**, No. 1, p. 219—224 (1968).
See also *7*, p. *77*; 22.2.1, p. 485.

H.-P. HENNIG
*N- and S-Type Negative Resistances and Relaxation Oscillations in Au-Doped Double Injection Ge Diodes
Vol. **26**, No. 1, p. 225—229 (1968).
See also 13.4, p. 211; 22.1.1, p. 464.

C. HAMANN
On the Trap Distribution in Thin Films of Copper Phthalocyanine
Vol. **26**, No. 1, p. 311—318 (1968).
See also *13.4*, p. *212*; 14.3.1, p. 246; 22.9, p. 560.

H. REHME
*Elektronenmikroskopische Untersuchung zum Mechanismus von Bariumtitanat-Kaltleiterkeramik
(Electron-Microscopic Study of the Mechanism of Barium Titanate Ceramics)
Vol. **26**, No. 1, p. K1—K3 (1968).
See also 4, p. 45; 14.4.2, p. 268.

R. ENDERLEIN
*Excitonic Effects in the Electro-Absorption of Semiconductors
Vol. **26**, No. 2, p. 509—517 (1968).
See also 13.1, p. 185; 13.2, p. 193; *20*, p. *340*; 20.1, p. 353; 22, p. 451.

G. DÖHLER and K. HACKER
A Possibility for Band Structure Determination: Intraband and Donor-Band Transitions of Crystal Electrons in High Electric Fields
Vol. **26**, No. 2, p. 551—564 (1968).
See also *13.1*, p. *185*; 13.4, p. 212; 20.1, p. 353; 22, p. 451.

M. BRAUER
*On the Behaviour of a Cylindrical Electron–Hole Plasma in a Magnetic Field with Transverse Component
Vol. **27**, No. 1, p. 151—159 (1968).
See also 18, p. 294; 22.1.1, p. 465.

G. G. ROBERTS
*Electron Injection into a p-Type Semiconductor
Vol. **27**, No. 1, p. 209—218 (1968).
See also 22.1.3, p. 479.

M. PORSCH
Green's Function Calculation of Electro-Optical Effects in Semiconductors. (I): General Formalism and Interband Transitions
Vol. **27**, No. 1, p. 359—370 (1968).
See also 13.1, p. 186; *20*, p. *341*; 22, p. 451.

J. K. POZHELA and K. K. REPSHAS
Thermoelectric Force of Hot Carriers
Vol. **27**, No. 2, p. 757—762 (1968).
See also 13, p. 170; *15*, p. *274*; 22.1.1, p. 465.

R. KOEPP and K. MARINOVA
*High Field Domains in CdS Crystals Moving in the Anode–Cathode Direction
Vol. **27**, No. 2, p. K117—K120 (1968).
See also 16, p. 282; 22.4.1, p. 503.

I. Licea
Hot Electron Temperature and Mobility in Polar Semiconductors
Vol. **27**, No. 2, p. K143—K144 (1968).
See also *13*, p. *170*; *14.3*, p. *242*; 22.2.3, p. 492.

H.-J. Hänsch
Study of the Electroluminescence Mechanism of CdS Single Crystals
Vol. **28**, No. 1, p. 255—268 (1968).
See also 13.2, p. 194; *20.3*, p. *370*; 22.4.1, p. 503.

K. W. Böer and P. Voss
*Stationary Anode-Adjacent High-Field Domains in Cadmium Sulfide
Vol. **28**, No. 1, p. 355—364 (1968).
See also 16, p. 282; 22.4.1, p. 503.

A. B. Fazakas and A. Friedman
*On the Double Injection Current Noise in Solids
Vol. **28**, No. 1, p. 385—394 (1968).
See also 22, p. 452.

J. Antula
*Negative Space-Charge Effect on Schottky Current in Thin Insulating Films
Vol. **28**, No. 1, p. 395—400 (1968).
See also 14.3.1, p. 247; 22.6, p. 542.

H. Heinrich, G. Bauer, and D. Kasperkovitz
*Spatial Dependence of Hot Carrier Oscillations in n-Type Germanium
Vol. **28**, No. 1, p. K51—K53 (1968).
See also 13, p. 171; 22.1.1, p. 465.

R. Enderlein
*On a Certain Type of Instability of Electrical Domains
Vol. **28**, No. 2, p. 519—527 (1968).
See also 22, p. 453.

V. L. Bonch-Bruevich
*On the Problem of Recombination Autooscillations
Vol. **28**, No. 2, p. 687—693 (1968).
See also 13, p. 172; 13.4, p. 215; 22.1.1, p. 466.

A. B. Fazakas and P. H. Handel
Stability of Surface-Barriers in Semiconductors
Vol. **28**, No. 2, p. 699—709 (1968).
See also *13.3*, p. *199*; 22, p. 453.

P. Weissglas and K. Blötekjaer
Conductivity of Intrinsic InSb in High Electric Fields
Vol. **28**, No. 2, p. 711—718 (1968).
See also *14.3*, p. *242*; 22.2.3, p. 493.

Z. S. Gribnikov, G. I. Lomova, and V. A. Romanov
*Injection of Current Carriers in Anisotropic Semiconductor Plates and the Magnetodiode Effect
Vol. **28**, No. 2, p. 815—825 (1968).
See also 22.1.1, p. 466.

M. E. Levinstein and M. S. Shur
*Behaviour of the High-Field Domains below the Voltage of the Nucleation Threshold
Vol. **28**, No. 2, p. 827—834 (1968).
See also 22, p. 453.

S. Ballarò, A. Balzarotti, and V. Grasso
*Electroreflectance at Oblique Incidence in Ge
Vol. 28, No. 2, p. K109–K112 (1968).
See also 13.1, p. 188; *20*, p. *341*; 20.1, p. 356; 22.1.1, p. 466.

S. N. Shestatskii and V. V. Sobolev
*Electroreflectance Spectrum of Cuprous Oxide Crystals
Vol. 28, No. 2, p. K131–K133 (1968).
See also 13.1, p. 188; 13.2, p. 194; *20*, p. *342*; 20.1, p. 356; 22.6, p. 542.

P. H. Handel
*Instabilities and Turbulence in Semiconductors
Vol. 29, No. 1, p. 299–306 (1968).
See also 22, p. 454.

J. Gyulai
*Induced Impurity Breakdown Oscillations and Observation of Traps Lying Higher than the Indirect Band Edge in GaP
Vol. 29, No. 2, p. K85–K88 (1968).
See also 13.4, p. 216; 16, p. 284; 22.2.2, p. 488.

Yu. A. Goldberg, M. E. Levinshtein, and D. N. Nasledov
*Magnetic Field Influence on the Low-Frequency Oscillations of Gunn Diodes
Vol. 29, No. 2, p. K103–K105 (1968).
See also *14.3.2*, p. *251*; 22.2.1, p. 486.

H.-P. Hennig
*On the Conduction Mechanism in the Instable Range of a Double Injection Diode
Vol. 29, No. 2, p. K167–K169 (1968).
See also 22.1.1, p. 466.

K. W. Böer and P. Voss
*Transition between Stationary and Moving High-Field Domains in CdS in a Range of N-Shaped Negative Differential Conductivity due to Field-Quenching
Vol. 30, No. 1, p. 291–304 (1968).
See also 16, p. 284; 20, p. 342; 22.4.1, p. 505.

R. Koepp and A. Urbelis
*Low-Frequency Photocurrent Oscillations in CdS Crystals with Thermally Induced Negative Differential Conductivity
Vol. 30, No. 1, p. K5–K8 (1968).
See also *16*, p. *284*; 22.4.1, p. 505.

W. Ludwig, W. Gericke, and H. Treptow
Electron Tunnelling in MIS-Structures of the Type Al-Al_2O_3-ZnTe
Vol. 30, No. 1, p. K61–K64 (1968).
See also 13.3, p. 199; *14.3.2*, p. *251*; 22.4.3, p. 511.

M. Ya. Bakirov and G. I. Ibaeev
*Charge Accumulation Effect in Selenium Crystals under Applied Electric Field
Vol. 30, No. 1, p. K73–K75 (1968).
See also 22.1.3, p. 480.

D. Boccon-Gibod
*Modulation of Near Infrared Radiation by Acoustoelectric Domains in n-Type GaAs at Room Temperature
Vol. 30, No. 2, p. 525–531 (1968).
See also 7, p. 78; *20.1*, p. *359*; 22.2.1, p. 486.

G. Döhler
*Critical Conditions for Transitions between Stationary and Non-Stationary High-Field Domains in Semi-Insulators
Vol. **30**, No. 2, p. 627–636 (1968).
See also 13.4, p. 217; 22.4.1, p. 505.

A. T. Davidson and A. D. Yoffe
*Hopping Electrical Conduction and Thermal Breakdown in Natural and Synthetic Mica
Vol. **30**, No. 2, p. 741–754 (1968).
See also 3, p. 28; 10, p. 113; 20.1, p. 359; 22, p. 455.

R. K. L. Gay and H. L. Hartnagel
*The Spatial and Time Variation of Electric Fields in Semiconducting CdS
Vol. **30**, No. 2, p. 755–758 (1968).
See also 7, p. 78; 22.4.1, p. 505.

A. A. Gutkin and T. Janicki
*The Effect of Electric Field in a Space Charge Layer on the Photoresponse Edge of Si p-n Junctions
Vol. **30**, No. 2, p. 827–832 (1968).
See also *14.3.2*, p. *251*; 16, p. 285; 22.1.2, p. 475.

P. P. Konorov, V. N. Shvetsov, and S. V. Shegolikhina
*Effects of the Non-Equilibrium Depletion in Germanium at the Electrolyte Interface
Vol. **30**, No. 2, p. 845–850 (1968).
See also 14.3, p. 244; 22.1.1, p. 467.

14.4.2 Ferroelectric Materials and Phenomena

Original Papers and Short Notes

P. GUYON et J. LAJZEROWICZ
*Mesures du temps de retournement ferroélectrique de KH_2PO_4
(Measurement of the Ferroelectric Switching Time of KH_2PO_4)
Vol. **16**, No. 2, p. 525—530 (1966).

H. VOGT and H. HAPP
Infrared Spectrum of $NaNO_2$ in the Ferro- and Paraelectric Phase between 30 and 150 μm
Vol. **16**, No. 2, p. 711—719 (1966).
See also *20.1*, p. *344*.

M. BALARIN and M. SCHENK
On the Defect-Induced Shift of Transition Points between Crystalline Phases
Vol. **17**, No. 1, p. 91—99 (1966).
See also *1.2*, p. *9*; 10, p. 94; 11, p. 146.

E. V. CHISLER and M. S. SHUR
The Raman Spectrum of $NaNO_2$ in the Ferroelectric Phase
Vol. **17**, No. 1, p. 163—171 (1966).
See also 5, p. 49; *6*, p. *56*; *20.1*, p. *345*.

E. V. CHISLER and M. S. SHUR
Raman Spectrum and Phase Transition in the Ferroelectric Crystal $NaNO_2$
Vol. **17**, No. 1, p. 173—176 (1966).
See also 1.2, p. 9; *6*, p. *57*; *20.1*, p. *345*.

G. VÖLKEL and W. WINDSCH
EPR Studies of Copper-Doped Rochelle-Salt Single Crystals
Vol. **17**, No. 1, p. K75—K78 (1966).
See also 10, p. 95; *19*, p. *326*.

S. WARTEWIG and W. WINDSCH
EPR Studies of Chromium Doped Triglycine Sulphate Single Crystals
Vol. **17**, No. 2, p. K221—K224 (1966).
See also *19*, p. *327*.

O. HAUSER and M. SCHENK
Radiation-Induced Phase Transitions of Some Perovskite-Type Substances and Their Thermodynamic Treatment
Vol. **18**, No. 2, p. 547—555 (1966).
See also *1.2*, p. *10*; 8, p. 80; 11, p. 147.

H. REHME
Electron-Microscopic Detection of Barrier Layers in Barium Titanate Ceramics
Vol. **18**, No. 2, p. K101—K102 (1966).
See also *4*, p. *39*.

B. JIMÉNEZ, J. MENDIOLA, and E. MAURER
*Orientation Dependence of the Ferroelectricity of T.G.S.
Vol. **19**, No. 2, p. 805—810 (1967).

G. REMAUT, R. GEVERS, and S. AMELINCKX
Wavy Fringes at Domain Boundaries in Barium Titanate Observed in the Electron Microscope
Vol. **20**, No. 2, p. 613—621 (1967).
See also *4*, p. *41*; 22.8, p. 549.

K. Žďánský, H. Arend, and F. Kubec
Electron Spin Resonance of Co^{2+} Ions in Barium Titanate
Vol. **20**, No, 2, p. 653–656 (1967).
See also *19*, p. *328*.

J. Kobayashi
*Optical Observation of Domain Motion in Ferroelectric $BaTiO_3$
Vol. **21**, No. 1, p. 151–158 (1967).
See also 4, p. 42.

J. Haňderek
*Electric Conductivity and Dielectric Absorption Phenomena in Lead Zirconate Ceramics
Vol. **21**, No. 1, p. 323–329 (1967).
See also *14.4*, p. *253*; 22.8, p. 550.

N. Kristoffel and P. Konsin
Pseudo-Jahn-Teller Effect and Second Order Phase Transitions in Crystals
Vol. **21**, No. 1, p. K39–K43 (1967).
See also *1.2*, p. *10*; 6, p. 61; *13*, p. *166*.

L. A. Shuvalov and N. R. Ivanov
*The Second Ferroelectric Phase Transition in $NaH_3(SeO_3)_2$ Crystals and Their Physical Properties
Vol. **22**, No. 1, p. 279–287 (1967).
See also 1.2, p. 11.

R. V. Damle
*Lead Zirconate by Coprecipitation
Vol. **22**, No. 1, p. K63–K65 (1967).
See also 22.8, p. 551.

R. Vu Huy Dat and C. Baumberger
*Ferroelectricity of Barium Titanate Thin Films of less than One Micron Thickness
Vol. **22**, No. 2, p. K67–K70 (1967).

A. Le Traon, F. Le Traon et S. Le Montagner
*Influence de l'organisation en domaines du titanate de baryum sur sa température de transition
(Influence of Domain Organization in Barium Titanate on the Transition Temperature)
Vol. **23**, No. 2, p. 709–712 (1967).

J. Stankowski, S. Waplak, B. Sczaniecki, and A. Dezor
Temperature Anomaly of EPR Spectra of Paramagnetic Ions in Ferroelectrics
Vol. **23**, No. 2, p. K159–K162 (1967).
See also *19*, p. *331*.

H. E. Müser und J. Pottharst
*Zum dielektrischen Verhalten von Seignettesalz im Bereich der Dezimeter- und Zentimeterwellen
(Dielectric Behaviour of Rochelle Salt in the Centimetre and Decimetre Wavelength Region)
Vol. **24**, No. 1, p. 109–113 (1967).

H. G. Reik and D. Heese
The Small Polaron Problem with an Application to Optical and DC Data of Reduced Barium Titanate
Vol. **24**, No. 1, p. 281–287 (1967).
See also 6, p. 64; *13*, p. *168*; 20.1, p. 351.

F. Kosek and H. Arend
*On High Temperature Conductivity of $BaTiO_3$
Vol. **24**, No. 1, p. K69–K71 (1967).

J. Stankowski
EPR Study of Cu^{2+} Ion in Triglycine Fluoroberyllate
Vol. **24**, No. 2, p. 451—456 (1967).
See also 10, p. 103; *19*, p. *331*.

R. E. Nettleton
Statistical Perturbation Theory of Order-Disorder Ferroelectrics: Zeroth Approximation
Vol. **24**, No. 2, p. 561—570 (1967).
See also *13*, p. *168*.

L. Merten
Orientational Dispersion of Optical Lattice Vibrations of Tetragonal $BaTiO_3$ at Room Temperature
Vol. **25**, No. 1, p. 125—130 (1968).
See also *6*, p. *64*.

W. Cochran and A. Zia
Structure and Dynamics of Perovskite-Type Crystals
Vol. **25**, No. 1, p. 273—283 (1968).
See also *6*, p. *64*.

G. Völkel
Effect of an External Electric Field on the EPR Spectra of Cu^{2+}-Doped Rochelle-Salt Single Crystals
Vol. **25**, No. 1. p. K35—K38 (1968).
See also *19*, p. *332*.

J. Kobayashi, H. Schmid, and E. Ascher
*Optical Study on the Ferroelectric Orthorhombic Phase of Fe-I-Boracite
Vol. **26**, No. 1, p. 277—283 (1968).
See also 4, p. 45; 20, p. 340.

H. Rehme
Electron-Microscopic Study of the Mechanism of Barium Titanate Ceramics
Vol. **26**, No. 1, p. K1—K3 (1968).
See also 4, p. 45; *14.4.1*, p. *262*.

B. J. Wicks and M. H. Lewis
*Direct Observations of Ferroelectric Domains in Lithium Niobate
Vol. **26**, No. 2, p. 571—576 (1968).
See also *4*, p. *45*; 10, p. 105.

R. Vu Huy Dat
*Dielectric Behaviour of Ferroelectric Monocrystalline Antimony Sulfo-Iodide with Electric Field Frequency, up to Microwave Range
Vol. **27**, No. 1, p. K41—K42 (1968).

B. A. Strukov, M. Amin, and V. A. Kopchik
Comparative Investigation of the Specific Heat of KH_2PO_4 (KDP) and KD_2PO_4 (DKDP) Single Crystals
Vol. **27**, No. 2, p. 741—749 (1968).
See also 1.2, p. 12; *8*, p. *85*.

O. G. Vendik
*On the Nonlocal Connection between Polarization and Electric Field in Ferroelectrics
Vol. **27**, No. 2, p. K99—K102 (1968).

A. P. ZHUKOV, I. S. REZ, V. I. PAKHOMOV, and G. K. SEMIN
Temperature Dependences of the Nuclear Quadrupole Resonance Spectra of As^{75} in KH_2AsO_4, RbH_2AsO_4, CsH_2AsO_4, $NH_4H_2AsO_4$, and of Their Deuterated Analogues
Vol. **27**, No. 2, p. K129—K132 (1968).
See also *19*, p. *334*.

H. KONWENT
On the Theory of Hydrogen-Bonded Ferroelectrics
Vol. **28**, No. 1, p. 39—44 (1968).

A. VAN DEN BEUKEL
Specific Heat of Ice near the "Ferroelectric" Transition Temperature
Vol. **28**, No. 2, p. 565—568 (1968).
See also *8*, p. *85*.

N. KRISTOFFEL and P. KONSIN
Displacive Vibronic Phase Transitions in Narrow-Gap Semiconductors
Vol. **28**, No. 2, p. 731—739 (1968).
See also *1.2*, p. *12*; *6*, p. *67*; 22, p. 453.

G. VÖLKEL and W. WINDSCH
Electric Shifts in the Paramagnetic Resonance of Ferroelectric Cu^{2+}-Doped Rochelle Salt
Vol. **28**, No. 2, p. K163—K165 (1968).
See also *19*, p. *335*.

W. WINDSCH and E. HARTMANN
*Switching Effect in Vanadyl-Doped Ferroelectric Triglycine Sulphate Observed by Electron Paramagnetic Resonance
Vol. **29**, No. 1, p. K11—K13 (1968).
See also 19, p. 335.

V. S. CHINCHOLKAR and H.-G. UNRUH
*Surface Layers of Triglycine Sulfate Single Crystals
Vol. **29**, No. 2, p. 669—673 (1968).

G. CHANUSSOT and H. AREND
*Thermocurrents in Paraelectric $BaTiO_3$
Vol. **29**, No. 2, p. K149—K151 (1968).
See also 10, p. 111.

H. VOGT and H. HAPP
The Temperature Dependence of Absorption of Sodium Nitrite in the Far Infrared
Vol. **30**, No. 1, p. 67—72 (1968).
See also *6*, p. *69*; *20.1*, p. *358*.

H. SCHULZE and G. VÖLKEL
Solid State Effect in Cu^{2+}-Doped Single Crystals of Rochelle Salt
Vol. **30**, No. 1, p. K9—K11 (1968).
See also *19*, p. *336*.

R. HSIANG-TAO YEH
*Possible Proximity Effect in Ferroelectric Films
Vol. **30**, No. 1, p. K45—K47 (1968).
See also 1.2, p. 13.

W. COCHRAN
Neutron Scattering and Dielectric Properties of Perovskite-Type Crystals
Vol. **30**, No. 2, p. K157—K160 (1968).
See also 1.2, p. 13; *6*, p. *70*.

15. Thermoelectric and Thermomagnetic Properties of Solids

Review Articles

A. S. Borshchevskii, N. A. Goryunova, F. P. Kesamanly, and D. N. Nasledov
Semiconducting $A^{II}B^{IV}C_2^{V}$ Compounds
Vol. **21**, No. 1, p. 9–55 (1967).
See also 3.1, p. 29; 3.2, p. 34; 5, p. 49; 8, p. 79; 10, p. 93; 13.1, p. 174; *14.3*, p. *231*; 16, p. 276; 17, p. 286; 18, p. 291; 20.1, p. 344; 22.8, p. 547.

P. Süptitz and J. Teltow
Transport of Matter in Simple Ionic Crystals (Cubic Halides)
Vol. **23**, No. 1, p. 9–56 (1967).
See also *9*, p. *87*; *10*, p. *93*; 14.4, p. 252; 22.5.1, p. 515; 22.5.2, p. 518; 22.5.3, p. 532.

Original Papers and Short Notes

C. N. Plăviţu
Some Kinetic Phenomena in Impure Ionic Semiconductors
Vol. **16**, No. 1, p. 69–78 (1966).
See also 8, p. 79; 13, p. 162; *14.4*, p. *252*; 22, p. 439.

A. G. Samoilovich, M. V. Nitsovich, and V. M. Nitsovich
*On the Theory of Anisotropic Thermoelectric Power in Semiconductors
Vol. **16**, No. 2, p. 459–465 (1966).
See also 6, p. 56; 13.1, p. 174; 22, p. 440.

W. Żdanowicz and A. Wojakowski
Semiconducting Properties of CdP_4
Vol. **16**, No. 2, p. K129–K131 (1966).
See also *14.3*, p. *232*; 20.1, p. 345; 22, p. 440.

R. Grigorovici, N. Croitoru, and A. Dévényi
*Thermoelectric Power in Amorphous Germanium
Vol. **16**, No. 2, p. K143–K145 (1966).
See also 2, p. 22; 14.3.1, p. 245; 22.1.1, p. 459.

A. A. Abdurakhmanova and M. I. Aliev
Electrical Properties of GaSb-Ga_2Te_3 and GaSb-GaTe Solid Solutions
Vol. **17**, No. 1, p. 375–380 (1966).
See also 13.1, p. 175; *14.3*, p. *232*; 22.8, p. 547.

I. A. Akimov, A. M. Meshkov, and A. N. Terenin
*On Thermal E.M.F. Measurements in Photo-Semiconducting Dyes
Vol. **17**, No. 1, p. K95–K98 (1966).
See also 13.4, p. 202; 16, p. 277; 22, p. 440.

J. Devreese, R. De Coninck, and H. Pollak
On the Conduction Mechanism in Uranium Dioxide
Vol. **17**, No. 2, p. 825–829 (1966).
See also 13, p. 163; *14.3*, p. *233*; 22.6, p. 535.

Vu dinh Ky
The Nernst Effect in Permalloy Films
Vol. **17**, No. 2, p. K203–K205 (1966).
See also 14.1, p. 221; *18.2*, p. *303*; 21.1.1, p. 413.

Vu dinh Ky
The Planar Nernst Effect in Permalloy Films
Vol. **17**, No. 2, p. K207–K209 (1966).
See also 14.1, p. 221; *18.2*, p. *303*; 21.1.1, p. 413.

F. I. ISMAILOV, G. A. AKHUNDOV, and O. R. VERNICH
Investigation of Electrical Conductivity and Hall Effect in GaSe Single Crystals
Vol. **17**, No. 2, p. K237—K240 (1966).
See also 13.4, p. 203; *14.3*, p. *233*; 22, p. 441.

V. A. KUTASOV and I. A. SMIRNOV
Effect of Excess Antistructural Bi on the Lattice Thermal Conductivity of Bi_2Te_3, Bi_2Te_3-Sb_2Te_3, and Bi_2Te_3-Bi_2S_3 Containing Iodine
Vol. **18**, No. 1, p. 479—488 (1966).
See also *8*, p. *80*; *10*, p. *96*; 14.3, p. 233; 22, p. 441; 22.8, p. 548.

G. B. ABDULLAEV, A. U. MALSAGOV, and V. M. GLAZOV
*Thermoelectric Power of $A^IB^{III}C_2^{VI}$ Type Compounds in the Solid and Liquid State
Vol. **18**, No. 1, p. K29—K31 (1966).
See also 2, p. 22; 22.8, p. 548.

I. P. MOLODYAN, D. N. NASLEDOV, S. I. RADAUTSAN, and V. G. SIDOROV
The Effective Mass of Electrons in $(InSb)_x \cdot (InTe)_{1-x}$ Crystals
Vol. **18**, No. 2, p. 677—682 (1966).
See also *13.1*, p. *176*; 14.3, p. 234; 22.8, p. 548.

A. M. GEORGE and M. D. KARKHANAVALA
*Studies of the Electrical Properties of Uranium Oxides. (III): Thermoelectric Power of α-U_3O_8
Vol. **19**, No. 1, p. 117—121 (1967).
See also 14.3, p. 234; 22.6, p. 536.

J. KOŁODZIEJCZAK
*On the Theory of Galvanomagnetic Phenomena in Strong Electric Fields
Vol. **19**, No. 1, p. 373—384 (1967).
See also 13.1, p. 176; 14.4.1, p. 257; 22, p. 442.

A. S. FILIPCHENKO and D. N. NASLEDOV
On the Mixed Mechanism of Electron Scattering in InSb Crystals
Vol. **19**, No. 1, p. 435—439 (1967).
See also *14.3*, p. *235*; 22.2.3, p. 490.

S. ŻUKOTYŃSKI and J. KOŁODZIEJCZAK
On the Determination of the Non-Parabolicity of Energy Bands
Vol. **19**, No. 1, p. K51—K54 (1967).
See also *13.1*, p. *176*; 14.3, p. 235; 22, p. 443.

G. COSTE
*Oscillatory Thermomagnetic Effects in Lead Telluride
Vol. **20**, No. 1, p. 361—364 (1967).
See also 13.1, p. 177; 22, p. 444.

C. HAMANN
On the Electric and Thermoelectric Properties of Copper Phthalocyanine Single Crystals
Vol. **20**, No. 2, p. 481—491 (1967).
See also 13.4, p. 205; *14.3*, p. *236*; 14.4.1, p. 258; 22.9, p. 559.

I. I. PINCHUK
*Thermomagnetic Effects in Anisotropic Media
Vol. **20**, No. 2, p. 537—543 (1967).
See also 3, p. 27; 21.7, p. 438.

KH. I. AMIRKHANOV, R. I. BASHIROV, and M. M. GADZHIALIEV
*Quantum Oscillations of the Nernst-Ettingshausen Effect in Indium Arsenide
Vol. **20**, No. 2, p. K119–K122 (1967).
See also 22.2.1, p. 483.

N. SEXER
Classical Galvanomagnetic and Thermomagnetic Effects in Cadmium Arsenide
Vol. **21**, No. 1, p. 225–234 (1967).
See also 13.1, p. 178; *14.3*, p. *236*; 22.7, p. 545.

A. A. ADAMENKO and I. J. DEKHTYAR
*Effect of Plastic Deformation on the Absolute Thermopower of α-Brass
Vol. **21**, No. 1, p. K81–K84 (1967).
See also 10.1, p. 122; 21.1, p. 398.

G. B. ABDULLAEV, Z. A. ALIYAROVA, and G. A. ASADOV
Preparation of Cu_2Se Single Crystals and Investigation of Their Electrical Properties
Vol. **21**, No. 2, p. 461–464 (1967).
See also 3.1, p. 31; *14.3*, p. *236*; 22, p. 445.

I. BUNGET and M. ROSENBERG
P-Type Conduction in Barium Ferrite Single Crystals
Vol. **21**, No. 2, p. K131–K133 (1967).
See also *14.3*, p. *236*; 18.3, p. 318.

B. M. ASKEROV and F. M. GASHIMZADE
Determination of Parameters of Degenerate Semiconductors
Vol. **21**, No. 2, p. K155–K158 (1967).
See also *13.1*, p. *179*; 14.3, p. 237; 22, p. 446.

B. TAVGER
Transport Phenomena in Semiconducting Thin Films
Vol. **22**, No. 1, p. 31–38 (1967).
See also 13.1, p. 179; *14.3.1*, p. *245*; 22, p. 446.

L. N. COJOCARU
Effects of Fast Neutrons on Electric Properties of NiO Single Crystals
Vol. **22**, No. 2, p. 361–364 (1967).
See also *11*, p. *149*; 13.4, p. 208; 14.3, p. 237; 22.6, p. 539.

VU DINH KY
*Planar Hall and Nernst Effect in Ferromagnetic Metals
Vol. **22**, No. 2, p. 729–736 (1967).
See also 13, p. 167; *14.1*, p. *223*; 18.2, p. 308; 21.1, p. 400; 21.1.1, p. 418

W. SZYMAŃSKA
Conduction Band Structure of HgTe from Thermomagnetic Measurements
Vol. **23**, No. 1, p. 69–73 (1967).
See also *13.1*, p. *180*; 22.4.3, p. 510.

A. G. SAMOILOVICH, I. S. BUDA, and I. V. DAKHOVSKII
*Anisotropy of Thermomagnetic Effects in n-Ge
Vol. **23**, No. 1, p. 229–236 (1967).
See also 13, p. 167; 14.3, p. 238; 22.1.1, p. 463.

R. R. GAŁĄZKA and T. ZAKRZEWSKI
Heavy Hole Effective Mass of $Cd_{0.1}Hg_{0.9}Te$
Vol. **23**, No. 1, p. K39–K43 (1967).
See also *13.1*, p. *181*; 14.3, p. 238; 22.4.3, p. 510; 22.8, p. 552.

A. R. Adams, F. Baumann, and J. Stuke
Thermal Conductivity of Selenium and Tellurium Single Crystals and Phonon Drag of Tellurium
Vol. **23**, No. 1, p. K99—K104 (1967).
See also *8*, p. *83*; 13, p. 167; 22.1.3, p. 477.

R. Grigorovici, N. Croitoru, and A. Dévényi
*Thermoelectric Power in Amorphous Silicon
Vol. **23**, No. 2, p. 621—625 (1967).
See also 2, p. 23; 13.1, p. 182; 14.3.1, p. 246; 22.1.2, p. 472.

C. N. Plăviţu
*On the Phonon Drag Effect Contribution to Thermomagnetic Phenomena
Vol. **24**, No. 1, p. 361—372 (1967).
See also 6, p. 64; 13, p. 168; 14.3, p. 239; 22, p. 448.

V. A. Ivanova, D. Sh. Abdinov, and G. M. Aliev
Some Structural, Electrical, and Thermal Parameters of VCr_2Se_4, $FeCr_2Se_4$, and $NiCr_2Se_4$ Ternary Compounds
Vol. **24**, No. 1, p. K23—K24 (1967).
See also 5, p. 52; 8, p. 83; *14.3*, p. *239*; 22.8, p. 552.

V. A. Ivanova, D. Sh. Abdinov, and G. M. Aliev
*On Some Characteristics of Chromium Selenides
Vol. **24**, No. 2, p. K145—K147 (1967).
See also 5, p. 52; *8*, p. *84*; 14.3, p. 239; 22, p. 449.

Ya. N. Nasirov and Ya. S. Feiziev
*Effect of Small Substitutions of Tin by Neodymium on Thermoelectric Properties of SnTe
Vol. **24**, No. 2, p. K157—K159 (1967).
See also 14.3, p. 239; 22, p. 449.

N. A. Goryunova, V. M. Orlov, V. I. Sokolova, G. P. Shpenkov, and E. V. Tsvetkova
Preparation and Some Properties of CuP_2 Single Crystals
Vol. **25**, No. 2, p. 513—519 (1968).
See also *3.1*, p. *33*; 5, p. 53; 14.3, p. 240; 20.1, p. 351; 22, p. 450.

F. L. Weichman, R. K. Lomnes, and S. Żukotyński
*Photothermoelectric Effects in CdS
Vol. **25**, No. 2, p. 583—591 (1968).
See also *16*, p. *281*; 22.4.1, p. 501.

G. I. Guseva and P. S. Zyryanov
*Quantum Theory of Thermogalvanomagnetic Phenomena (TGMP) in Conductors with Equal Concentration of Electrons and Holes
Vol. **25**, No. 2, p. 775—785 (1968).
See also 14.1, p. 224; 14.3, p. 240; 21.7, p. 438; 22.2, p. 481.

G. B. Abdullaev, Z. A. Aliyarova, E. H. Zamanova, and G. A. Asadov
Investigation of the Electric Properties of Cu_2S Single Crystals
Vol. **26**, No. 1, p. 65—68 (1968).
See also 1.2, p. 12; *14.3*, p. *240*; 22, p. 450.

D. I. Lainer and G. K. Samedov
On the Mechanism of Current Carrier Scattering in Some Solid Solutions on the Basis of Lead Telluride
Vol. **26**, No. 2, p. K123—K125 (1968).
See also *14.3*, p. *241*; 22, p. 451.

L. GEORGESCU
*Nonlinear Thermodynamics of Thermoelectric Phenomena in Ionic Semiconductors
Vol. **27**, No. 1, p. 125—130 (1968).
See also 22, p. 451.

M. M. AKSELROD, K. M. DEMCHUK, I. M. TSIDILKOVSKI, E. L. BROYDA, and K. P. RODIONOV
The Effect of Pressure on the Electron Effective Mass in InSb
Vol. **27**, No. 1, p. 249—254 (1968).
See also *13.1*, p. *186*; 14.3, p. 241; 22.2.3, p. 492.

U. BIRKHOLZ and J. SCHELM
Mechanism of Electrical Conduction in β-$FeSi_2$
Vol. **27**, No. 1, p. 413—425 (1968).
See also 13, p. 170; 13.1, p. 186; *14.3*, p. *241*; 22, p. 452; 22.7, p. 546.

S. I. MASHAROV
Effect of Phonon Drag on the Kinetic Properties of Alloys
Vol. **27**, No. 1, p. 455—461 (1968).
See also *6*, p. *65*; 8, p. 84; 13, p.170; 14.1, p. 225; 21, p. 386.

U. BIRKHOLZ, W. HEERING, and F. STÖCKMANN
*Thermoelectric Power of Illuminated CdS
Vol. **27**, No. 1, p. K69—K72 (1968).
See also 16, p. 282; 22.4.1, p. 502.

J. K. POZHELA and K. K. REPSHAS
*Thermoelectric Force of Hot Carriers
Vol. **27**, No. 2, p. 757—762 (1968).
See also 13, p. 170; 14.4.1, p. 262; 22.1.1, p. 465.

I. IVANOV-OMSKII, F. P. KESAMANLY, B. T. KOLOMIETS, A. SH. MEKHTIEV, and V. A. SKRIPKIN
Density of States Effective Mass of Holes in HgTe
Vol. **27**, No. 2, p. K169—K172 (1968).
See also *13.1*, p. *186*; 22.4.3, p. 511.

M. ALI, M. FRIDMAN, M. DENAYER, and P. NAGELS
Electrical Conduction in Li-Doped MnO Single Crystals
Vol. **28**, No. 1, p. 193—206 (1968).
See also 13.1, p. 187; 13.4, p. 214; *14.3*, p. *242*; 22.6, p. 542.

K. SCHRÖDER and N. P. BAUM
*Absolute Thermoelectric Power and Electrical Resistivity of Chromium-Rich Chromium–Nickel Alloys
Vol. **29**, No. 1, p. 107—114 (1968).
See also 13.1, p. 188; *14.1*, p. *226*; 18.4, p. 324; 21.1, p. 407.

J. E. LEWIS, H. RODOT, and P. HAEN
*The Low-Temperature Thermoelectric Power and Thermal Conductivity of GeTe and of Some GeTe-MnTe Alloys
Vol. **29**, No. 2, p. 743—754 (1968).
See also 6, p. 68; *8*, p. *86*; 13, p. 173; 22, p. 454; 22.8, p. 555.

L. N. COJOCARU
Effects of Fission Fragments on Electric Properties of the UO_2-SiO_2 System
Vol. **29**, No. 2, p. K119—K122 (1968).
See also *11*, p. *154*; 14.3, p. 244; 22.6, p. 543; 22.8, p. 556.

E. P. BERDNIKOV, V. A. DROZDOV, and L. A. MOZGOVAYA
On the Relation between Microhardness and Point Defects of Crystals
Vol. **29**, No. 2, p. K145—K147 (1968).
See also *10*, p. *111*; 12, p. 158; 22.1.2, p. 474.

V. Hádek
Electron and Hole Conductivity of Donor-Acceptor Complexes of Oligoanilinic Substances and Their Thermoelectric Properties
Vol. **30**, No. 1, p. 275—282 (1968).
See also *13.4*, p. *216*; *14.3*, p. 244; 22.9, p. 560.

V. V. Boldyrev, G. B. London, and V. K. Zhuravlev
*Study of Semiconducting Properties of Potassium Permanganate
Vol. **30**, No. 1, p. K13—K15 (1968).
See also 13.4, p. 217; *14.3*, p. *244*; 22.8, p. 556.

Nguyen van Dong and Pham Ngu Tung
*Transport Properties of Silver Telluride in the Solid and Liquid States
Vol. **30**, No. 2, p. 557—567 (1968).
See also 2, p. 25; *14.3*, p. *244*; 14.4, p. 254; 22, p. 455.

B. Selle and J. Maege
Electrical and Optical Properties of Evaporated Cu_2S Films
Vol. **30**, No. 2, p. K153—K155 (1968).
See also 13.4, p. 218; *14.3.1*, p. *247*; 16, p. 285; *20.1*, p. *359*; 22, p. 455.

16. Photoconductivity. Photovoltaic Effects

Review Articles

J. R. Dale
Alloyed Semiconductor Heterojunctions
Vol. **16**, No. 2, p. 351–387 (1966).
See also 3.1, p. 29; 5, p. 49; 13.4, p. 200; *14.3.2*, p. *248*; 22.1.1, p. 458; 22.1.2, p. 468; 22.2.1, p. 482; 22.2.2, p. 487; 22.2.3, p. 489.

A. S. Borshchevskii, N. A. Goryunova, F. P. Kesamanly, and D. N. Nasledov
Semiconducting $A^{II}B^{IV}C_2^V$ Compounds
Vol. **21**, No. 1, p. 9–55 (1967).
See also 3.1, p. 29; 3.2, p. 34; 5, p. 49; 8, p. 79; 10. p. 93; 13.1, p. 174; *14.3*, p. *231*; 15, p. 270; 17, p. 286; 18, p. 291; 20.1, p. 344; 22.8, p. 547.

Original Papers and Short Notes

G. B. Abdullaev, M. R. Akhundov, and G. A. Akhundov
Rectifying and Photoconductive Properties of p-GaSe Single Crystals
Vol. **16**, No. 1, p. 209–214 (1966).
See also *14.3.2*, p. *248*; 22, p. 439.

K. W. Böer and R. Schubert
*The Influence of Oxygen in the Ultrahigh Vacuum Range on Electrical Properties of CdS
Vol. **16**, No. 1, p. K5–K9 (1966).
See also 13.3, p. 196; 22.4.1, p. 497.

P. Thoma and G. Vaubel
*Electrical Properties of Anthracene Single Crystals at Very Low Temperatures
Vol. **16**, No. 2, p. 663–674 (1966).
See also 13.4, p. 201; 14.3, p. 231; 14.4.1, p. 256; 22.9, p. 558.

K. W. Böer, J. W. Feitknecht, and D. G. Kannenberg
Properties of Recrystallized Evaporated CdS Layers
Vol. **16**, No. 2, p. 697–709 (1966).
See also *1.4*, p. *16*; 13.4, p. 201; 22.4.1, p. 497.

L. Gouskov, G. Lecoy, C. Llinares, and M. Savelli
Study of Lifetime and Generation–Recombination Noise Controlled by Dislocations in Germanium
Vol. **16**, No. 2, p. 721–728 (1966).
See also *13.4*, p. *202*; 14.3, p. 232; 22.1.1, p. 459.

V. A. Zuev and V. G. Litovchenko
*Theory of the Photovoltaic Effect in Semiconductors
Vol. **16**, No. 2, p. 751–760 (1966).
See also 13.3, p. 196; 22.1.1, p. 459; 22.1.2, p. 469.

A. J. Nadolny, J. W. Ostrowski, and M. Z. Przegalińka-Mieszkowska
*Kinetics of Photoconductivity in Boron
Vol. **16**, No. 2, p. K133–K137 (1966).
See also 22.1, p. 456.

R. Schubert
*Gas Desorption from Virginal CdS Crystals
Vol. **16**, No. 2, p. K157–K159 (1966).
See also 13.3, p. 196; 22.4.1, p. 497.

R. Fischer, F. Stöckmann, and J. Stuke
Space-Charge Limited Currents in Inhomogeneously Illuminated CdS Single Crystals
Vol. 17, No. 1, p. 335–341 (1966).
See also *14.4.1*, p. *256*; 22.4.1, p. 498.

R. B. Laibowitz and H. S. Sack
*Luminescence and Photoconductivity in Silver Chloride with Laser Excitation
Vol. 17, No. 1, p. 353–360 (1966).
See also 6, p. 57; 13.4, p. 202; *20.3*, p. *364*; 22.5.1, p. 515.

V. A. Romanov, I. P. Zhadko, and I. I. Boiko
*Photoconductivity of Anisotropically Deformed Intrinsic Germanium
Vol. 17, No. 1, p. 389–396 (1966).
See also 12, p. 156; 22.1.1, p. 459.

E. Gutsche, F. Spiegelberg, and J. Voigt
*Transient Photoconductivity Quenching in CdS Single Crystals due to Surface Trapping
Vol. 17, No. 1, p. K11–K14 (1966).
See also 13.3, p. 196; 22.4.1, p. 498.

J. Gyulai, V. K. Subashiev, and G. A. Chalikyan
*On the Mechanism of Photoconductivity in High-Resistivity GaP
Vol. 17, No. 1, p. K49–K51 (1966).
See also 13.1, p. 175; 22.2.2, p. 487.

I. A. Akimov, A. M. Meshnikov, and A. N. Terenin
On Thermal E.M.F. Measurements in Photo-Semiconducting Dyes
Vol. 17, No. 1, p. K95–K98 (1966).
See also 13.4, p. 202; *15*, p. *270*; 22, p. 440.

P. Bauduin et F. Buchy
*Osciallations de courant dans le sulfure de cadmium photoconducteur
(Current Oscillations in Photoconducting Cadmium Sulphide)
Vol. 17, No. 2, p. 517–530 (1966).
See also *14.4.1*, p. *256*; 22.4.1, p. 498.

K. Kirov
Pulse Method for Studying Trapping Centres in CdS Type Crystals
Vol. 17, No. 2, p. 581–592 (1966).
See also *13.4*, p. *202*; 14.4.1, p. 256; 22.4.1, p. 498.

W. H. Mitchell, E. H. Putley, and N. Shaw
*Hall Effect and Extrinsic Photoconductivity in n-Type InSb
Vol. 17, No. 2, p. 605–608 (1966).
See also *14.3*, p. *232*; 22.2.3, p. 490.

M. Hagino and T. Takahashi
Photoelectric Properties of Cs-Sb Films Having Various Compositons
Vol. 17, No. 2, p. 609–618 (1966).
See also 10, p. 95; 14.3.1, p. 245; *17*, p. *286*; 22.7, p. 545.

E.-H. Weber
*Zusammenhang zwischen Raumladung, Oberflächenleitfähigkeit und Bandverbiegung für Photoleiter bei homogener optischer Anregung
(Relationship between Space Charge, Surface Conductivity, and Band Bending for Photoconductors under Uniform Optical Excitation)
Vol. 17, No. 2, p. 843–854 (1966).
See also 13.3, p. 197; 13.4, p. 203; 14.4.1, p. 257; 22.4, p. 495.

V. D. Egorov
*Zum linearen Anklingen der Photoleitfähigkeit in Halbleitern
(Linear Photoconductive Rise in Semiconductors)
Vol. **18**, No. 1, p. 145–150 (1966).
See also 13.3, p. 197; 22.1.1, p. 460.

G. Jungk
On the Generation-Recombination Statistics in Semiconductors
Vol. **18**, No. 1, p. 297–306 (1966).
See also *13.4*, p. *203*; 14.3, p. 233; 22.1.2, p. 470.

P. Bräunlich and A. Scharmann
Approximate Solution of Schön's Balance Equations for the Thermoluminescence and the Thermally Stimulated Conductivity of Inorganic Photoconducting Crystals
Vol. **18** No. 1, p. 307–316 (1966).
See also *13.4*, p. *203*; 14.3, p. 233; 20.3, p. 364; 22.4, p. 495.

B. P. Kozyrev and T. Boţilă
*Some Photoelectrical Properties of KRS-5 Single Crystals
Vol. **18**, No. 1, p. K57–K61 (1966).
See also 13.4, p. 203; 22.5, p. 512; 22.8, p. 548.

G. B. Street and W. D. Gill
*Photoconductivitity and Drift Mobilities in Single Crystal Realgar (As_4S_4)
Vol. **18**, No. 2, p. 601–607 (1966).
See also 13.4, p. 204; 14.3, p. 233; 22, p. 441.

G. D. Guseinov, M. Z. Ismailov, and A. G. Talybov
On the New Semiconducting Compound $HgTlS_2$
Vol. **18**, No. 2, p. 929–935 (1966).
See also *5*, p. *50*; 8, p. 80; 13.4, p. 204; 14.3, p. 234; 22.8, p. 548.

G. Richter
Determination of Transport Coefficients for Minority Carriers in InSb
Vol. **18**, No. 2, p. K149–K152 (1966).
See also 13.1, p. 176; *14.3*, p. *234*; 22.2.3, p. 490.

K. W. Böer and C. A. Kennedy
Vacuum Heat Treatment of CdS Single Crystals
Vol. **19**, No. 1, p. 203–210 (1967).
See also *13.4*, p. *204*; 14.3, p. 234; 22.4.1, p. 499.

M. P. Mikhailova, D. N. Nasledov, and S. V. Slobodchikov
The Effect of a Magnetic Field upon Illuminated InAs p-n Junctions
Vol. **19**, No. 1, p. 429–434 (1967).
See also *14.3.2*, p. *249*; 22.2.3, p. 490.

G. Döhler
On the Field Emission of Minority Carriers in Photoconductors
Vol. **19**, No. 2, p. 555–564 (1967).
See also 13.4, p. 205; *14.4.1*, p. *257*; 22.1.3, p. 477; 22.4.1, p. 499.

A. R. Beattie and G. Smith
*Recombination in Semiconductors by a Light Hole Auger Transition
Vol. **19**, No. 2, p. 577–586 (1967).
See also *13.1*, p. *177*; 22.2.1, p. 483; 22.2.3, p. 490.

H. Sigmund and K. Berchtold
Electrical and Photovoltaic Properties of PbS-Si Heterodiodes
Vol. **20**, No. 1, p. 255–259 (1967).
See also *14.3.2*, p. *249*; 22, p. 444; 22.1.2, p. 470.

Z. Goŀacki and T. Figielski
*Extrinsic Photoconductivity in Ge Caused by Dislocations
Vol. **20**, No. 1, p. K1–K3 (1967).
See also 13.4, p. 205; 22.1.1, p. 461.

J. Spyridelis, J. Stoimenos, and N. Economou
Optical and Photoconductive Phenomena in Cuprous Oxide
Vol. **20**, No. 2, p. 623–628 (1967).
See also 13.2, p. 192; *20.1*, p. *347*; 22.6, p. 537.

I. A. Akimov, V. M. Bentsa, F. I. Vilesov, and A. N. Terenin
External Photoeffect from Sensitizing Dyes Adsorbed on Semiconductors
Vol. **20**, No. 2, p. 771–776 (1967).
See also 13.3, p. 197; *17*, p. *287*; 22.4, p. 495; 22.5, p. 512; 22.5.1, p. 515.

J. W. Ostrowski and A. Pajączkowska
*Spectral Response of Photosensitivity (SRPS) in Boron
Vol. **20**, No. 2, p. K159–K162 (1967).
See also 22.1, p. 457.

P. Bräunlich
Light Effects on Space-Charge Controlled Currents in CdS
Vol. **21**, No. 1, p. 383–391 (1967).
See also 13.4, p. 206; *14.4.1*, p. *258*; 20.3, p. 365; 22.4.1, p. 499.

A. N. Georgobiani and V. I. Steblin
Optical and Electrical Characteristics of p–n Junctions in Zinc Sulfide
Vol. **21**, No. 1, p. K49–K51 (1967).
See also *14.3.2*, p. *249*; 20.3, p. 365; 22.4.1, p. 499.

L. Gouskov, G. Lecoy, and C. Llinares
*Carrier Recombination in Nickel Doped Germanium from Lifetime and Noise Measurements
Vol. **21**, No. 2, p. 619–626 (1967).
See also 13.4, p. 207; *14.3*, p. *236*; 22.1.1, p. 462.

M. Damodar-Pai and K. M. van Vliet
*Generation–Recombination and Diffusion Noise in Cadmium Sulphide Photoconductors
Vol. **21**, No. 2, p. 819–835 (1967).
See also 13.4, p. 207; 22.4.1, p. 500.

A. Zaręba
*Photoconductivity in Boron at Low Temperature
Vol. **21**, No. 2, p. K135–K138 (1967).
See also 13.4, p. 208; 22.1, p. 457.

W. Degrieck and G. Jacobs
*Luminescence and Photoconductivity of the B-Centre in Additively Coloured KCl:Ag Crystals
Vol. **22**, No. 1, p. 177–184 (1967).
See also 10.2, p. 139; *20.3*, p. *366*; 22.5.2, p. 523.

J. Viščakas, J. Vaitkus, and S. Stonkus
*Influence of Heat Treatment on the Spectral Dependence of Photoconductivity in CdSe Crystals
Vol. **22**, No. 1, p. K3–K5 (1967).
See also 13.4, p. 208; 22.4.2, p. 506.

P. Bräunlich
*Calculation of Light-Stimulated Luminescence and Current Transients in Phosphors
Vol. 22, No. 2, p. 391–400 (1967).
See also 13.4, p. 208; *20.3*, p. *366*; 22, p. 447.

I. Dima and G. Vasiliu
*Photoconductivity of ZnSe Thin Films
Vol. 22, No. 2, p. K79–K82 (1967).
See also 22.4.2, p. 507.

V. A. Drozdov and L. A. Mozgovaya
Infrared Quenching of the Photomechanical Effect in Cadmium Sulphide
Vol. 22, No. 2, p. K109–K112 (1967).
See also *12*, p. *157*; 22.4.1, p. 500.

B. Autin, O. Cahen, E. Dieulesaint, D. Gouvernelle, J. Tavernier, and P. Bauduin
*Two-Step Current Saturation in CdS Crystals
Vol. 22, No. 2, p. K135–K137 (1967).
See also 22.4.1, p. 500.

T. L. Tansley
Spectral Response of p-n Heterojunctions
Vol. 23, No. 1, p. 241–252 (1967).
See also *14.3.2*, p. *249*; 14.4.1, p. 259; 20.1, p. 350; 22.2.1, p. 484; 22.8, p. 552.

W. Güth and H. Bleher
Avalanche Breakdown in Germanium in the Case of Optical Band–Band Excitation
Vol. 23, No. 1, p. K13–K16 (1967).
See also 13, p. 167; *14.4.1*, p. *260*; 22.1.1, p. 463.

F. R. Kessler and E. Sutter
*Absorption and Photoconductivity of a Series of Selenium-Tellurium Mixed Crystals
Vol. 23, No. 1, p. K25–K33 (1967).
See also 13.1, p. 181; *20.1*, p. *350*; 22.1.3, p. 477.

E. Nagy and G. Petö
Ordering in the Alloy Cu_3Au-V
Vol. 23, No. 1, p. K49–K52 (1967).
See also *1.2*, p. *11*; 21.1, p. 400.

R. Grigorovici, N. Croitoru, and A. Dévényi
*Photoconductivity in Amorphous Germanium
Vol. 23, No. 2, p. 627–631 (1967).
See also 2, p. 23; 22.1.1, p. 463.

G. G. Kovalevskaya, D. N. Nasledov, and S. V. Slobodchikov
*Photoconductivity Oscillations in InP
Vol. 23, No. 2, p. 755–760 (1967).
See also *6*, p. *63*; 13.1, p. 182; 22.2.2, p. 488.

G. Döhler
On Slow High Field Domains in Homogenous Photoconductors
Vol. 24, No. 1, p. 331–340 (1967).
See also *14.4.1*, p. *260*; 22.1.3, p. 478.

R. Baubinas, A. Sakalas, A. Smilga, and J. Viščakas
*p-Type CdSe Single Crystals
Vol. 24, No. 1, p. K91–K93 (1967).
See also 3.1, p. 33; *14.3*, p. *239*; 22.4.2, p. 507.

H. Pagnia
*Slow Photoconduction in InSb Films
Vol. **24**, No. 1, p. K97—K100 (1967).
See also 14.3.1, p. 246; 22.2.3, p. 491.

M. Rouzeyre
Contribution to the Study of Transport Phenomena in Aluminium-Alumina-Metal Structures
Vol. **24**, No. 2, p. 399—411 (1967).
See also *14.3.2*, p. *250*; 21, p. 383; 22.6, p. 540.

M. K. Sheinkman, V. A. Tyagai, O. V. Snitko, I. B. Ermolovich, G. L. Belenky, and V. N. Bondarenko
Studies on the Nature of CdS Single Crystal Sensitization by Etching
Vol. **24**, No. 2, p. 543—549 (1967).
See also 10, p. 103; *13.4*, p. *210*; 20.3, p. 368; 22.4.1, p. 501.

T. L. Tansley
Forward Current Injection Modulation of Photocurrent in p-n Heterojunctions
Vol. **24**, No. 2, p. 615—622 (1967).
See also *14.3.2*, p. *250*; 14.4.1, p. 261; 22.2.1, p. 485; 22.2.2, p. 488.

V. A. Besfamilnaya, V. F. Elesin, I. A. Kurova, N. N. Ormont, and V. V. Ostroborodova
*On the Nature of Oscillations in Impurity Photoconductivity Spectra of Ge
Vol. **24**, No. 2, p. 757—762 (1967).
See also 6, p. 64; 13, p. 168; 22.1.1, p. 464.

B. Selle, W. Ludwig, and R. Mach
*Photovoltaic Response of Cu_2S-CdS p-n Heterojunctions
Vol. **24**, No. 2, p. K149—K152 (1967).
See also *14.3.2*, p. *250*; 22, p. 449; 22.4.1, p. 501.

G. B. Abdullaev, M. Kh. Alieva, and A. Z. Mamedova
*Thermal and Infrared Quenching of Photoconductivity in n-InSe Single Crystals
Vol. **25**, No. 1, p. 75—79 (1968).
See also 13.4, p. 211; 22, p. 449.

P. Krispin
On the Photoimpedance of ZnS Phosphors
Vol. **25**, No. 1, p. 147—154 (1968).
See also 13.4, p. 211; *14.4.1*, p. *261*; 22.4.1, p. 501.

G. Jungk and H. Menniger
Influence of Energy Band Structure on Photoeffects in Silicon
Vol. **25**, No. 1, p. K59—K63 (1968).
See also *13.1*, p. *184*; 22.1.2, p. 473.

F. L. Weichman, R. K. Lomnes, and S. Żukotyński
*Photothermoelectric Effects in CdS
Vol. **25**, No. 2, p. 583—591 (1968).
See also *15*, p. *273*; 22.4.1, p. 501.

L. Pelsers and G. Jacobs
*Photoconductivity of CsBr Single Crystals Containing F- and M-Centres
Vol. **25**, No. 2, p. 621—628 (1968).
See also 10.2, p. 142; 22.5.2, p. 525.

H. Friedrich, L. A. Sysoev, and K. Thiessen
*Minority Carriers in Zinc Sulphide
Vol. **26**, No. 1, p. 107—114 (1968).
See also 22.4.1, p. 502.

A. Buroff and J. Malinowski
*Charge Compensation of Trapped Holes in Silver Bromide
Vol. **26**, No. 1, p. 267–276 (1968).
See also 10.2, p. 142; 13.4, p. 212; 22.5.1, p. 516.

W. Picht and W. Ruppel
*Photocurrent, Luminescence, and Absorption Saturation for Electron Beam Excitation in CdS
Vol. **26**, No. 2, p. K163–K165 (1968).
See also 20.1, p. 353; 20.3, p. 369; 22.4.1, p. 502.

W. Fuhs and J. Stuke
*Hopping Recombination in Trigonal Selenium Single Crystals
Vol. **27**, No. 1, p. 171–184 (1968).
See also 13.4, p. 213; 22.1.3, p. 479.

P. Devaux and M. Schott
*Field Effect on Photocarriers in Copper Phthalocyanine Single Crystals
Vol. **27**, No. 1, p. 201–208 (1968).
See also 13.2, p. 194; *13.3*, p. *198*; 22.9, p. 560.

H. Friedrich and G. Jungk
Generation-Recombination Statistics in Semiconductors — A Model with Two Centres
Vol. **27**, No. 1, p. 237–248 (1968).
See also *13.4*, p. *213*; 22, p. 451.

U. Birkholz, W. Heering, and F. Stöckmann
Thermoelectric Power of Illuminated CdS
Vol. **27**, No. 1, p. K69–K72 (1968).
See also *15*, p. *274*; 22.4.1, p. 502.

P. R. Herczfeld, K. M. van Vliet, and M. D. Pai
*Experiments on Fluctuations in Optically Quenched CdS and CdSe Crystals
Vol. **27**, No. 2, p. 671–680 (1968).
See also 22.4.1, p. 503; 22.4.2, p. 508.

P. R. Herczfeld and K. M. van Vliet
*Electronic Processes and Fluctuations in Optically Quenched CdS and CdSe Crystals
Vol. **27**, No. 2, p. 681–694 (1968).
See also *13.4*, p. *214*; 22.4.1, p. 503; 22.4.2, p. 508.

R. Koepp and K. Marinova
High Field Domains in CdS Crystals Moving in the Anode–Cathode Direction
Vol. **27**, No. 2, p. K117–K120 (1968).
See also *14.4.1*, p. *262*; 22.4.1, p. 503.

P. Byszewski, J. Holzman, and J. Kołodziejczak
*Oscillatory Magneto-Photoconductivity Effect in Germanium
Vol. **27**, No. 2, p. K125–K127 (1968).
See also 22.1.1, p. 465.

T. O. Tuomi
Effect of Reactor Irradiation on the Conductivity and the Photoconductivity of Hexagonal Selenium Single Crystals
Vol. **28**, No. 1, p. 329–334 (1968).
See also *11*, p. *153*; 14.3, p. 242; 22.1.3, p. 479.

K. W. Böer and P. Voss
Stationary Anode-Adjacent High-Field Domains in Cadmium Sulfide
Vol. **28**, No. 1, p. 355–364 (1968).
See also *14.4.1*, p. *263*; 22.4.1, p. 503.

G. Wolfram
*Photoconductivity of the Z_2-Center in KCl:Sr Crystals
Vol. 28, No. 1, p. K11—K13 (1968).
See also 10.2, p. 143; 13.4, p. 214; 22.5.2, p. 527.

V. V. Zotov and V. V. Serdyuk
*Effect of Infrared Quenching on the Spectral Distribution of Induced Impurity Photoconductivity of CdS Single Crystals
Vol. 28, No. 1, p. K31—K33 (1968).
See also 13.4, p. 214; 22.4.1, p. 503.

R. V. Grigoriev, B. V. Novikov, and A. E. Cherednichenko
*Influence of Oxygen Adsorption on the Fine Structure of the Spectral Distribution of Photoconductivity in CdS Crystals
Vol. 28, No. 1, p. K85—K88 (1968).
See also 13.2, p. 194; 13.3, p. 198; 22.4.1, p. 504.

H. Pagnia
*Surface Photoconduction in p-InSb Single Crystals
Vol. 28, No. 1, p. K89—K92 (1968).
See also 13.3, p. 198; 22.2.3, p. 493.

H. G. Grimmeiss and G. Olofsson
*Storage Effects in Photoconducting GaP-Cu
Vol. 28, No. 2, p. 547—554 (1968).
See also 22.2.2, p. 488.

E.-H. Weber
*Surface Photoconductivity of CdS Influenced by Chemisorption and Desorption of Oxygen
Vol. 28, No. 2, p. 649—662 (1968).
See also 13.3, p. 199; 22.4.1, p. 504.

V. Dolocan
*Role of Gradual Capture in the Photoelectromagnetic and Photoconductive Effects
Vol. 28, No. 2, p. 719—723 (1968).
See also 22, p. 453.

M. Georgiev, E. Zagorska, and J. Malinowski
Temperature Dependence of the Diffusion of Photoexcited Holes in AgBr
Vol. 28, No. 2, p. 725—730 (1968).
See also *10.2*, p. *143*; 22.5.1, p. 517.

I. P. Zhadko and V. A. Romanov
*On the Origin of High Photovoltages in Semiconductors
Vol. 28, No. 2, p. 797—800 (1968).
See also 14.3.1, p. 247; 22, p. 453.

M. Vaněček, J. Toušek, and E. Klier
*Study of the Kinetics of X-Ray Photoconductivity in CdSe and CdTe and Determination of the Electron-Hole Generation Energy
Vol. 28, No. 2, p. K147—K149 (1968).
See also 22.4.2, p. 508; 22.4.3, p. 511.

P. B. Pickar and H. D. Tiller
Optical Energy Gap in TlSe
Vol. 29, No. 1, p. 153—158 (1968).
See also 5, p. 54; 13.1, p. 188; *20.1*, p. *356*; 22, p. 453.

D. T. Y. WEI, C. M. PENCHINA, and Y. S. PARK
*Observation of Oscillatory Lifetime in CdS
Vol. **29**, No. 1, p. K7—K10 (1968).
See also 13.4, p. 215; 22.4.1, p. 504.

D. BONNET, H. GERMER, and H. RABENHORST
*Photoconduction in $Cd(S_xSe_{1-x})$ Films with Graded Composition
Vol. **29**, No. 1, p. K65—K67 (1968).
See also 22.4.1, p. 504; 22.4.2, p. 508; 22.8, p. 555.

H. PFLEIDERER
*PEM Effect with Short-Circuited Dember Voltage
Vol. **29**, No. 2, p. 597—604 (1968).
See also 22.2.3, p. 493.

J. GYULAI
Induced Impurity Breakdown Oscillations and Observation of Traps Lying Higher than the Indirect Band Edge in GaP
Vol. **29**, No. 2, p. K85—K88 (1968).
See also 13.4, p. 216; *14.4.1*, p. *264*; 22.2.2, p. 488.

E. FORTIN
*Photocurrent Amplification by a Magnetic Field
Vol. **29**, No. 2, p. K153—K155 (1968).
See also 22.2.1, p. 486.

K. W. BÖER and P. VOSS
Transition between Stationary and Moving High-Field Domains in CdS in a Range of N-Shaped Negative Differential Conductivity due to Field-Quenching
Vol. **30**, No. 1, p. 291—304 (1968).
See also *14.4.1*, p. *264*; 20, p. 342; 22.4.1, p. 505.

W. STÖSSEL and W. ZIMMERMANN
*Photoconduction in CdS Crystals Generated by a Modulated Electron Beam
Vol. **30**, No. 1, p. 311—318 (1968).
See also *11*, p. *155*; 13.4, p. 216; 22.4.1, p. 505.

R. KOEPP and A. URBELIS
*Low-Frequency Photocurrent Oscillations in CdS Crystals with Thermally Induced Negative Differential Conductivity
Vol. **30**, No. 1, p. K5—K8 (1968).
See also *14.4.1*, p. *264*; 22.4.1, p. 505.

D. N. BOSE
Thermoluminescence and Thermally Stimulated Currents in Bismuth Doped ZnS Crystals
Vol. **30**, No. 1, p. K57—K60 (1968).
See also *13.4*, p. *217*; 20.3, p. 371; 22.4.1, p. 505.

D. V. CHEPUR, D. M. BERCHA, I. D. TURYANITSA, and V. YU. SLIVKA
Peculiarities of the Energy Spectrum and Edge Absorption in the Chain Compounds $A^{V}B^{VI}C^{VII}$
Vol. **30**, No. 2, p. 461—468 (1968).
See also *13.1*, p. *189*; 20.1, p. 358; 22.8, p. 557.

A. A. GUTKIN and T. JANICKI
The Effect of Electric Field in a Space Charge Layer on the Photoresponse Edge of Si p–n Junctions
Vol. **30**, No. 2, p. 827–832 (1968).
See also *14.3.2*, p. *251*; *14.4.1*, p. *265*; 22.1.2, p. 475.

E. E. GODIK
*The Variation of the Photoelectron Mobility with Photon Energy in Germanium and Silicon
Vol. **30**, No. 2, p. K127–K129 (1968).
See also 22.1.1, p. 467; 22.1.2, p. 475.

B. SELLE and J. MAEGE
Electrical and Optical Properties of Evaporated Cu_2S Films
Vol. **30**, No. 2, p. K153–K155 (1968).
See also **13.4**, p. 218; *14.3.1*, p. *247*; 15, p. 275; *20.1*, p. *359*; 22, p. 455.

17. Emission of Electrons and Ions from Solids

Review Articles

I. Eckertova
*Feldkathoden mit dünnen Isolatorschichten
(Field Cathodes with Thin Insulating Films)
Vol. 18, No. 1, p. 3–40 (1966).
See also 13, p. 162; 13.3, p. 196; 13.4, p. 200; 14.4.1, p. 255; 22.6, p. 535.

A. S. Borshchevskii, N. A. Goryunova, F. P. Kesamanly, and D. N. Nasledov
Semiconducting $A^{II}B^{IV}C_2^{V}$ Compounds
Vol. 21, No. 1, p. 9–55 (1967).
See also 3.1, p. 29; 3.2, p. 34; 5, p. 49; 8, p. 79; 10, p. 93; 13.1, p. 174; *14.3*, p.*231*; 15, p. 270; 16, p. 276; 18, p. 291; 20.1, p. 344; 22.8, p. 547.

F. R. Brotzen
*Emission of Exoelectrons from Metallic Materials
Vol. 22, No. 1, p. 9–30 (1967).
See also 13.3, p. 196; 21, p. 373.

Original Papers and Short Notes

M. Dubejko and S. Olszewski
*Quantum-Statistical Calculations on the Work Function of Metals
Vol. 16, No. 2, p. 399–404 (1966).
See also 13, p. 163; 21, p. 373.

M. Attardo and J. M. Galligan
A Field Ion Microscope Study of Neutron Irradiated Tungsten
Vol. 16, No. 2, p. 449–457 (1966).
See also 4, p. 37; *11*, p. *145*; 21, p. 374.

S. G. Christov
*General Theory of Electron Emission from Metals
Vol. 17, No. 1, p. 11–26 (1966).
See also 21, p. 374.

H. Hora and R. Kantlehner
*The Anomalous Vectorial Photoemission from a Cesium–Antimony Layer
Vol. 17, No. 1, p. 59–66 (1966).
See also 13.1, p. 175; 22.7, p. 545.

M. Hagino and T. Takahashi
*Photoelectric Properties of Cs-Sb Films Having Various Compositions
Vol. 17, No. 2, p. 609–618 (1966).
See also 10, p. 95; 14.3.1, p. 245; 16, p. 277; 22.7, p. 545.

V. E. Yurasova, V. M. Buchanov, and M. Golo
*Directional Emission of Charged Particles from a Single Crystal under Ion Bombardment
Vol. 17, No. 2, p. K187–K190 (1966).
See also 21.1, p. 393.

Hj. Matzke
*Diffusion of Tritium from Lithium Fluoride
Vol. 18, No. 1, p. 317–324 (1966).
See also 11, p. 147; 22.5.2, p. 519.

P. PETRESCU and V. GHEORDĂNESCU
*Anisotropy of the Photoemission from Some Coloured Alkali Halides
Vol. **18**, No. 2, p. 511–515 (1966).
See also 10.2, p. 138; 22.5.2, p. 520.

H. MAYER und J. HÖLZL
*Experimentelle Bestimmung der maximalen Austrittstiefen monoenergetischer Sekundärelektronen
(Experimental Determination of the Maximum Emissions-Depths of Monoenergetic Secondary Electrons)
Vol. **18**, No. 2, p. 779–785 (1966).
See also 13, p. 164; 21.2, p. 424.

I. N. EVDOKIMOV, E. S. MASHKOVA, V. A. MOLCHANOV, and D. D. ODINTSOV
*Dependence of the Ion–Electron Emission Coefficient on the Angle of Incidence
Vol. **19**, No. 1, p. 407–415 (1967).
See also 21, p. 377; 22.1, p. 457.

H. NEUMANN
*Photosensitive Field Electron Emission from In_2S_3
Vol. **20**, No. 1, p. K33–K35 (1967).
See also 22, p. 444.

I. A. AKIMOV, V. M. BENTSA, F. I. VILESOV, and A. N. TERENIN
*External Photoeffect from Sensitizing Dyes Adsorbed on Semiconductors
Vol. **20**, No. 2, p. 771–776 (1967).
See also 13.3, p. 197; 16, p. 279; 22.4, p. 495; 22.5, p. 512; 22.5.1, p. 515.

S. G. CHRISTOV
*Unified Theory of Thermionic and Field Emission from Semiconductors
Vol. **21**, No. 1, p. 159–173 (1967).
See also 13.3, p. 197; 22, p. 444.

M. KARRAS and E. SUONINEN
*Energy Distribution of Photoelectrons Created by Al K_α X-Rays in Alkali Halide Targets
Vol. **21**, No. 2, p. 487–493 (1967).
See also 22.5.2, p. 522.

I. L. SOKOLSKAYA, V. G. IVANOV, and G. N. FURSEY
Study of Barium Adsorption on Germanium by Field-Emission Microscopy
Vol. **21**, No. 2, p. 789–795 (1967).
See also 1.3, p. 15; *13.3*, p. *197*; 22.1.1, p. 462.

G. N. FURSEY, I. L. SOKOLSKAYA, and V. G. IVANOV
*Field Emission from p-Type Germanium
Vol. **22**, No. 1, p. 39–46 (1967).
See also 22.1.1, p. 462.

P. PETRESCU
L-Bands Detected in Photoemission Spectra of NaCl, LiCl, and RbBr
Vol. **23**, No. 2, p. K105–K107 (1967).
See also *10.2*, p. *140*; 22.5.2, p. 524.

L. ERNST
*About the Fowler-Nordheim Plots of Germanium Field Emitters
Vol. **24**, No. 1, p. 177–181 (1967).
See also 13.3, p. 198; 22.1.1, p. 463.

J. Hölzl und B. Heimann
*Experimentelle Ergebnisse zur Temperaturabhängigkeit der Sekundärelektronen-Emission an polykristallinem Nickel im Intervall $100 \leqq T \leqq 600$ °C
(Experimental Results on the Temperature Dependence of Secondary Electron Emission from Polycrystalline Nickel in the Range $100 \leqq T \leqq 600$ °C)
Vol. **24**, No. 2, p. 651—657 (1967).
See also 18.1, p. 297; 18.2, p. 309; 21.1, p. 402.

W. Ludwig and B. Korneffel
*Photoemission Studies on Thin Metal–Insulator–Metal Sandwiches
Vol. **24**, No. 2, p. K137—K140 (1967).
See also *14.3.2*, p. *250*; 21, p. 383; 22.6, p. 540.

William R. Savage
Field Electron Microscope Study of Niobium Surfaces
Vol. **25**, No. 1, p. 131—138 (1968).
See also *1.3*, p. *15*; 21, p. 383.

C. Lea and C. H. B. Mee
*The Attenuation Length of Photoelectrons in Thin Films of Uranium
Vol. **25**, No. 2, p. 613—619 (1968).
See also 14.1, p. 224; 21.5, p. 429.

R. Gauthier, D. Bois, P. Pinard et F. Davoine
*Production d'ions et d'électrons par interaction avec la matière d'un faisceau de lumière cohérente
(Ion and Electron Production by Interaction of a Coherent Light Beam with Substance)
Vol. **25**, No. 2, p. 691—696 (1968).
See also 21.1, p. 403; 21.1.1, p. 420; 22.1.1, p. 464; 22.2.1, p. 485; 22.4.1, p. 502; 22.5.2, p. 525; 22.6, p. 540.

A. Meessen
*Vectorial Selective Surface Effects in the Optical Absorption and Photoelectric Emission of Thin Metal Films
Vol. **26**, No. 1, p. 125—132 (1968).
See also 20.1, p. 352; 21, p. 385.

R. H. Tredgold, R. H. Williams, G. Kavalyauskene, and R. Keezer
*Photoemission Studies on Amorphous Layers and Single Crystals of Selenium
Vol. **26**, No. 1, p. K5—K8 (1968).
See also 2, p. 24; 22.1.3, p. 479.

K. Hirschberg und K. Deutscher
*Die Bestimmung der Austrittstiefe von Photoelektronen in Caesium–Antimon-Schichten
(Determination of the Emission Depth of Photoelectrons from Cs–Sb Films)
Vol. **26**, No. 2, p. 527—535 (1968).
See also 14.3.1, p. 246; 20.1, p. 353; 22.7, p. 546.

G. Vaubel and H. Baessler
Determination of the Energy of Conduction States in Anthracene Crystals by Photoemission of Electrons from Sodium
Vol. **26**, No. 2, p. 599—606 (1968).
See also *13.1*, p. *185*; 22.9, p. 560.

H. BAESSLER, N. RIEHL, and G. VAUBEL
Calculation of the Width of the Second Conduction Band in Anthracene Crystals from Photoemission Data
Vol. **26**, No. 2, p. 607–610 (1968).
See also *13.1*, p. *185*; 22.9, p. 560.

K. MEYER und E. LINKE
*Triboinduzierte Elektronenemission und Lumineszenz an Alkalihalogeniden (Triboinduced Electron-Emission und Luminescence from Alkali Halides)
Vol. **26**, No. 2, p. K89–K92 (1968).
See also *20.3*, p. *369*; 22.5.2, p. 526.

K. DEUTSCHER and K. HIRSCHBERG
*Thickness Dependence of the Quantum Yield of Cesium-Antimony Films
Vol. **27**, No. 1, p. 145–150 (1968).
See also 14.3.1, p. 246; 20.1, p. 354; 22.7, p. 546.

H. HORA and G. B. KABIERSCH
*Combined Infrared Photoemission from Cs_3Sb
Vol. **27**, No. 2, p. 593–600 (1968).
See also 22.7, p. 546.

G. APPELT
*Fine Structure Measurements in the Energy Angular Distribution of Secondary Electrons from a (110) Face of Copper
Vol. **27**, No. 2, p. 657–669 (1968).
See also 13.1, p. 186; 20.1, p. 354; 21.1, p. 405.

P. GÖRLICH and M. LUDWIG
*On the Photostimulation of the Exoelectron Emission in the Infrared Spectral Region
Vol. **28**, No. 2, p. K117–K119 (1968).
See also 21, p. 387.

P. PETRESCU
*On the L-Bands and the First Exciton Bands in the Photoemission Spectra of Alkali Iodides
Vol. **29**, No. 1, p. 333–340 (1968).
See also 10.2, p. 143; 13.2, p. 195; 22.5.2, p. 528.

G. M. DE'MUNARI, F. GIUSIANO, and G. MAMBRIANI
A Phase Transition and the Photoemissive Quantum Yield at Low Temperatures in Cs_3Sb Films
Vol. **29**, No. 1, p. 341–356 (1968).
See also *1.2*, p. *12*; 3, p. 27; 13.1, p. 188; 22.7, p. 546.

P. PETRESCU
*On the Photoemission Spectrum of LiF:Mg
Vol. **29**, No. 1, p. K1–K2 (1968).
See also 10.2, p. 144; 22.5.2, p. 528.

P. PETRESCU
*Evidence for L-Bands in Photoemission Spectra of KF and KCl
Vol. **29**, No. 1, p. K3–K5 (1968).
See also 10.2, p. 144; 22.5.2, p. 529.

C. Simoi, I. Hrianca, and P. Crăciun
*Exoemission of Electrons without Photostimulation
Vol. **29**, No. 2, p. 761–766 (1968).
See also 13, p. 173; 21, p. 388.

J. Hofmann and W. Steinmann
Plasma Resonance in the Photoemission of Silver
Vol. **30**, No. 1, p. K53–K56 (1968).
See also *13*, p. *173*; *20*, p. *342*; 21.6, p. 437.

18. Magnetic Properties of Solids

Review Articles

A. S. BORSHCHEVSKII, N. A. GORYUNOVA, F. P. KESAMANLY, and D. N. NASLEDOV
Semiconducting $A^{II}B^{IV}C_2^V$ Compounds
Vol. **21**, No. 1, p. 9—55 (1967).
See also 3.1, p. 29; 3.2, p. 34; 5, p. 49; 8, p. 79; 10, p. 93; 13.1, p. 174; *14.3*, p. *231*; 15, p. 270; 16, p. 276; 17, p. 286; 20.1, p. 344; 22.8, p. 547.

M. BALKANSKI and E. AMZALLAG
Band Parameters Determination from Faraday Rotation Measurements
Vol. **30**, No. 2, p. 407—439 (1968).
See also *13.1*, p. *174*; 20, p. 337; 22.1.1, p. 458; 22.1.2, p. 468; 22.2.1, p. 482; 22.2.2, p. 487; 22.2.3, p. 489; 22.4, p. 495; 22.4.1, p. 497; 22.4.2, p. 506; 22.4.3, p. 509.

Original Papers and Short Notes

M. ZVÁRA, F. ŽALOUDEK, and V. PROSSER
*Interband Faraday Effect in CdTe Single Crystals
Vol. **16**, No. 1, p. K21—K24 (1966).
See also 13, p. 162; *20*, p. *337*; 22.4.3, p. 509.

TH. KLUPSCH and W. HAUBENREISSER
*A New Nonlinear Mechanism for Establishing the Saturation Susceptibility during Spin Wave Parallel Pumping
Vol. **16**, No. 2, p. K147—K151 (1966).
See also *13*, p. *163*.

H. J. FISCHBECK
On Pippard's Theory of Magnetic Breakdown in Metals
Vol. **17**, No. 1, p. 197—210 (1966).
See also *13.1*, p. *175*; 21, p. 374.

J. KOŁODZIEJCZAK and H. STRAMSKA
*On the Theory of Free Carrier Electro-Magneto-Optical Phenomena
Vol. **17**, No. 2, p. 701—707 (1966).
See also 13.1, p. 176; 14.4.1, p. 256; *20*, p. *337*; 22, p. 441.

J. KOŁODZIEJCZAK and E. KIERZEK-PECOLD
*Free Carrier Electro-Magneto-Optical Phenomena in Semiconductors
Vol. **19**, No. 2, p. 623—630 (1967).
See also 14.3, p. 235; *20*, p. *338*; 22, p. 443.

J. VILLAIN
*Oscillations collectives de spins et ordre à courte distance
(Collective Spin Oscillations and Short-Range Order)
Vol. **19**, No. 2, p. 767—776 (1967).
See also *13*, p. *165*.

R. E. NEWNHAM and R. P. SANTORO
Magnetic and Optical Properties of Dioptase
Vol. **19**, No. 2, p. K87—K90 (1967).
See also *13*, p. *165*; *20*, p. *338*; 22.8, p. 549.

H. TRÄUBLE and U. ESSMANN
Observation of Magnetic Structures in Type II Superconductors
Vol. **20**, No. 1, p. 95—111 (1967).
See also *14.2*, p. *229*; 21, p. 377.

L. Cser, J. Ostanevich, and L. Pál
Mössbauer Effect in Iron-Aluminium Alloys (I)
Vol. **20**, No. 2, p. 581–589 (1967).
See also *6.1*, p. *71*; 21.1.1, p. 416.

L. Cser, J. Ostanevich, and L. Pál
Mössbauer Effect in Iron-Aluminium Alloys (II)
Vol. **20**, No. 2, p. 591–596 (1967).
See also *6.1*, p. *72*; 21.1.1, p. 416.

T. N. Casselman and H. N. Spector
Interaction of Optical Phonons with Spin Waves in Ionic Crystals
Vol. **20**, No. 2, p. 605–612 (1967).
See also *6*, p. *60*; 13, p. 165; *19*, p. *328*; 22, p. 444.

J. Blinowski and M. Grynberg
*On the Direct Interband Magnetoabsorption in Uniaxially Stressed Germanium
Vol. **20**, No. 2, p. K107–K111 (1967).
See also 10, p. 99; *20*, p. *339*; 22.1.1, p. 461.

V. P. Kalashnikov
Theory of Spin-Lattice Relaxation of Conduction Electrons of Ionic Semiconductors at Extremely High Magnetic Fields
Vol. **21**, No. 2, p. 775–784 (1967).
See also *6*, p. *61*; 13, p. 166; 14.4, p. 253; 22, p. 446.

Ž. Dimitrijević, S. Kraśnicki, H. Rżany, J. Todorović, A. Wanic, H. Curien, and A. Milojević
*Anistropy of Magnon Dispersion Relation in Haematite
Vol. **21**, No. 2, p. K163–K166 (1967).
See also *13*, p. *166*; 22.6, p. 538.

E. Majerníková
*Faraday Effect and Spin-Orbit Interaction of the Lowest Exciton States in Molecular Crystals
Vol. **22**, No. 1, p. 113–122 (1967).
See also 13.2, p. 192; *20*, p. *339*; 22.9, p. 559.

H. N. Spector
Optical Absorption in Electric and Magnetic Fields
Vol. **22**, No. 1, p. 185–193 (1967).
See also 13.1, p. 179; 14.4.1, p. 259; *20.1*, p. *349*; 22, p. 446.

E. J. Kramer and Ch. L. Bauer
Variations of Young's Modulus throughout the Mixed State in Niobium
Vol. **22**, No. 1, p. 199–202 (1967).
See also 1.2, p. 11; *10.1*, p. *124*; 14.2, p. 229; 21, p. 380.

H. J. Fischbeck
Theory of Weakly Bound Bloch Electrons in a Magnetic Field. (I): Nearly Free Approximation
Vol. **22**, No. 1, p. 235–244 (1967).
See also *13.1*, p. *179*.

H. J. Fischbeck
*Theory of Weakly Bound Bloch Electrons in a Magnetic Field. (II): Asymptotic Approximation
Vol. **22**, No. 2, p. 649–658 (1967).
See also *13.1*, p. *180*.

A. I. KOSINSKAYA and S. I. URITSKY
On the Classical Theory of the Optical Properties of Free Carriers in Semiconductors
Vol. **23**, No. 1, p. 57–68 (1967).
See also 13.1, p. 180; *20*, p. *339*; 22.1.1, p. 463.

N. I. AKIMENKO, P. M. STARYK, and K. D. TOVSTYUK
Magnetic Properties of PbTe
Vol. **23**, No. 1, p. 93–96 (1967).
See also *13.1*, p. *180*; 14.3, p. 238; 22, p. 447.

J. R. PEVERLEY and P. H. E. MEIJER
*Entropy and Susceptibility of a Pure Dipole-Dipole Substance
Vol. **23**, No. 1, p. 353–360 (1967).
See also *8*, p. *83*; 22.8, p. 552.

P. BYSZEWSKI, B. KALIŃSKA, and J. KOŁODZIEJCZAK
*Interband Faraday Rotation and Ellipticity in Germanium and Gallium Arsenide
Vol. **23**, No. 1, p. K53–K56 (1967).
See also 13, p. 167; *20*, p. *340*; 22.1.1, p. 463; 22.2.1, p. 484.

L. DOBROSAVLJEVIĆ and C. DUPUIS
Study of Superconducting Pb-Sn Eutectic with Oriented Lamellas
Vol. **23**, No. 2, p. 509–520 (1967).
See also *14.2*, p. *229*; 21, p. 381.

M. D. STAFLEU and A. R. DE VROOMEN
De Haas-van Alphen Effect in White Tin
Vol. **23**, No. 2, p. 675–682 (1967).
See also *13.1*, p. *182*; 21, p. 382.

B. I. KULIEV and G. M. NEDLIN
*The Types of Magnetic Ordering in Crystals Having the Structure of $RbNiF_3$
Vol. **24**, No. 1, p. K25–K28 (1967).
See also 1.2, p. 11; 3.2, p. 35; 22.5, p. 513; 22.8, p. 552.

U. K. MISHRA and A. K. WALTON
Infra-Red Faraday Effect in p-Type InAs and Si
Vol. **24**, No. 1, p. K87–K90 (1967).
See also *13.1*, p. *183*; 20, p. 340; 22.1.2, p. 472; 22.2.1, p. 485.

L. ŚNIADOWER and R. KOWALCZYK
*Magnetoplasma Reflection in Mercury Telluride
Vol. **25**, No. 1, p. 233–240 (1968).
See also 13.1, p. 183; *20*, p. *340*; 20.1, p. 351; 22.4.3, p. 510.

J. SÓLYOM and A. ZAWADOWSKI
On the Specific Heat of Dilute Magnetic Alloys
Vol. **25**, No. 1, p. 473–483 (1968).
See also *8*, p. *84*.

E. E. ANDERSON
*Concerning the Uniqueness of a Molecular Field Solution for Two Magnetic Sublattices
Vol. **25**, No. 1, p. K31–K34 (1968).
See also 5, p. 53.

P. GROSSE and K. WINZER
Magnetoabsorption Measurements on Tellurium
Vol. **26**, No. 1, p. 139–150 (1968).
See also *13.1*, p. *184*; *20.1*, p. *352*; 22.1.3, p. 478.

SH. SH. BASHKIROV and G. YA. SELYUTIN
Mössbauer Spectrum Shift Depending on the Degree of Spin-System Ordering
Vol. **26**, No. 1, p. 253–259 (1968).
See also *6.1*, p. *73*; 13, p. 169.

P. DE CHATEL and I. SZABÓ
Spin Density Oscillations in a Polarized Free Electron Gas
Vol. **26**, No. 1, p. 319–325 (1968).
See also *13*, p. *169*.

W. THIELEMANN
Effect of Doping on the Optical Absorption of Circularly Polarized Waves in n-Type GaAs in a Magnetic Field
Vol. **26**, No. 1, p. K29–K31 (1968).
See also 13.1, p. 184; *20.1*, p. *352*; 22.2.1, p. 485.

H. OVERHOF and U. RÖSSLER
Magnetic Space Groups and Their Irreducible Representations
Vol. **26**, No. 2, p. 461–468 (1968).
See also *3*, p. *27*; 13.1, p. 185.

M. BRAUER
On the Behaviour of a Cylindrical Electron-Hole Plasma in a Magnetic Field with Transverse Component
Vol. **27**, No. 1, p. 151–159 (1968).
See also *14.4.1*, p. *262*; 22.1.1, p. 465.

H. STRAMSKA, Z. BACHAN, P. BYSZEWSKI, and J. KOŁODZIEJCZAK
*Interband Faraday Rotation and Ellipticity Observed at the Absorption Edge in Silicon
Vol. **27**, No. 1, p. K25–K28 (1968).
See also 13.1, p. 186; *20*, p. *341*; 22.1.2, p. 473.

M. ZVÁRA
Interband Faraday Rotation in Gallium Arsenide
Vol. **27**, No. 2, p. K157–K160 (1968).
See also 13, p. 170; *20*, p. *341*; 22.2.1, p. 485.

YU. E. PERLIN and SH. N. GIFEISMAN
The Theory of Shallow F-Centres
Vol. **28**, No. 1, p. 45–60 (1968).
See also 6, p. 66; *10.2*, p. *143*; 13, p. 171; 14.4, p. 254; 22.2.1, p. 485; 22.2.2, p. 488; 22.2.3, p. 493; 22.4, p. 495; 22.4.1, p. 503; 22.4.2, p. 508; 22.4.3, p. 511; 22.5.1, p. 517; 22.5.2, p. 527.

H. LE GALL
Quantum-Mechanical Treatment of the Transient Processes with Three-Magnon Non-Linear Transitions under Perpendicular Pumping
Vol. **28**, No. 2, p. 495–505 (1968).
See also 13, p. 171; *19*, p. *334*.

L. ŚNIADOWER
*Temperature Studies of Magnetoplasma Reflection in Mercury Telluride
Vol. **28**, No. 2, p. K125–K130 (1968).
See also 13.1, p. 188; *20.1*, p. *356*; 22.4.3, p. 511.

B. GARBEN und H. SELIGER
*Faraday-Effekt an amorphem Selen
(Faraday Effect on Amorphous Selenium)
Vol. **29**, No. 1, p. K27–K29 (1968).
See also 13, p. 172; *20*, p. *342*; 22.1.3, p. 480.

J. Kołodziejczak
*Nonlinear Magnetooptical Phenomena in Semiconductors (I)
Vol. **29**, No. 2, p. 645–652 (1968).
See also 13.1, p. 189; *20*, p. *342*; 22, p. 454.

J. Ingels and G. Jacobs
*Faraday Rotation of the U_2 Centre in KCl, KBr, and KI
Vol. **30**, No. 1, p. 163–172 (1968).
See also 13.4, p. 216; *20*, p. *342*; 22.5.2, p. 530.

H. J. Fischbeck
General Theory of the Bloch Electron in a Magnetic Field
Vol. **30**, No. 2, p. 779–789 (1968).
See also *13.1*, p. *190*.

18.1 Paramagnetic Properties

Original Papers and Short Notes

W. Brodkorb
Theory of the Ferromagnetism of Magnetically Anisotropic Films with Cubic Crystal Structure. (I): Calculation of Magnetization and Curie Temperature with the Green Function Method
Vol. **16**, No. 1, p. 225–236 (1966).
See also *18.2*, p. *299*.

K. H. J. Buschow and J. F. Fast
Magnetic and Structural Characteristics of Some Equiatomic Rare-Earth Germanides
Vol. **16**, No. 2, p. 467–473 (1966).
See also *3*, p. *26*; 5, p. 49; 18.2, p. 300; 18.3, p. 316; 18.4, p. 321; 22.7, p. 545.

V. V. Ratnam
On the Electrical and Magnetic Properties of Coloured and Uncoloured Calcium Fluoride Crystals
Vol. **16**, No. 2, p. 549–557 (1966).
See also *10.2*, p. *137*; 13.4, p. 201; 14.4, p. 252; 22.5.3, p. 532.

W. Brodkorb and W. Haubenreisser
Theory of the Ferromagnetism of Magnetically Anisotropic Films with Cubic Crystal Structure. (II): Calculation of the Curie Temperature for an Fe-(100)-Film
Vol. **16**, No. 2, p. 577–583 (1966).
See also *18.2*, p. *300*; 21.1.1, p. 411.

W. Haubenreisser and U. Lindner
Some Symmetry Aspects in the Theory of Localized Magnetic States in Metals
Vol. **20**, No. 2, p. K131–K134 (1967).
See also *13.4*, p. *206*; 21, p. 378.

U. Gonser and R. W. Grant
Determination of Spin Directions and Electric Field Gradient Axes in Vivianite by Polarized Recoil-Free γ-Rays
Vol. **21**, No. 1, p. 331–342 (1967).
See also *6.1*, p. *72*; 13, p. 165; 18.4, p. 322; 22.8, p. 550.

K. H. J. Buschow and J. F. Fast
Crystal Structure and Magnetic Properties of Some Rare Earth Germanides
Vol. **21**, No. 2, p. 593–600 (1967).
See also *5*, p. *51*; 18.3, p. 318; 18.4, p. 322; 22.7, p. 545.

K. Tompa, F. Tóth, and G. Grüner
Susceptibility of MnO Measured by the NMR Method
Vol. **22**, No. 1, p. K11–K13 (1967).
See also 18.4, p. 322; *19*, p. *329*; 22.6, p. 538.

H.-G. Kahle, H. Kalbfleisch, and E. Stein
*Paramagnetic Relaxation and Observation of a Phonon Bottleneck in Terbium Ethyl Sulfate
Vol. **22**, No. 2, p. 537–540 (1967).
See also *6*, p. *62*; 19, p. 330; 22.8, p. 551.

F. F. Y. Wang and M. Kestigian
Magnetic Susceptibilities of Sodium Neodymium Tungstate and Sodium Terbium Tungstate
Vol. **23**, No. 1, p. 289–294 (1967).
See also *18.4*, p. *323*; 22.8, p. 552.

G. ALBRECHT and G. WEBER
Hyperfine Fields Induced by Spin-Orbit-Coupling in Binary Intermetallic Compounds
Vol. **23**, No. 1, p. K17–K20 (1967).
See also *19*, p. *330*; 21.1, p. 400.

YU. B. PADERNO and S. POKRZYWNICKI
*Magnetic Properties of Some Heavy Rare Earth Tetraborides
Vol. **24**, No. 1, p. K11–K12 (1967).
See also 22, p. 448.

YU. B. PADERNO, S. POKRZYWNICKI, and B. STALIŃSKI
*Magnetic Properties of Some Rare Earth Hexaborides
Vol. **24**, No. 1, p. K73–K76 (1967).
See also 22, p. 448.

J. HÖLZL and B. HEIMANN
Experimental Results on the Temperature Dependence of Secondary Electron Emission from Polycrystalline Nickel in the Range $100 \leqq T \leqq 600$ °C
Vol. **24**, No. 2, p. 651–657 (1967).
See also *17*, p. *288*; 18.2, p. 309; 21.1, p. 402.

B. HEINRICH, J. ZÍTKOVÁ, and J. KACZÉR
*The Temperature Dependence of the Paramagnetic Susceptibility of Ni-I Boracite
Vol. **26**, No. 2, p. 443–446 (1968).
See also 18.4, p. 324; 22.8, p. 553.

I. Y. DEKHTYAR, E. G. MADATOVA, P. S. NISIN, and R. G. FEDCHENKO
Plastic Deformation Effect on the Electronic Properties of Solid Solutions with Localized Magnetic Moments
Vol. **26**, No. 2, p. K151–K154 (1968).
See also 6.1, p. 73; *10.1*, p. *130*; 14.1, p. 225; 21.6, p. 436.

V. K. SAXENA and D. L. BHATTACHARYA
*Effect of Electron-Lattice Interactions on the Spin-Paramagnetism of Metals
Vol. **27**, No. 1, p. 427–433 (1968).
See also 13, p. 170; 21, p. 386; 21.2, p. 425.

N. N. SIROTA and E. A. VASILEV
*Magnetic Properties and Structure of $MnAs_xSb_{1-x}$ Solid Solutions
Vol. **28**, No. 2, p. K175–K177 (1968).
See also 18.2, p. 313; 22.8, p. 555.

K. PFEUFFER
*Determination of the Electron Spin-Lattice Relaxation Time T_1 in the Direct Process
Vol. **29**, No. 1, p. 171–177 (1968).
See also 6, p. 67; 19, p. 335; 22.8, p. 555.

A. M. VAN DIEPEN, H. W. DE WIJN, and K. H. J. BUSCHOW
Nuclear Magnetic Resonance and Susceptibility of Equiatomic Rare-Earth–Aluminium Compounds
Vol. **29**, No. 1, p. 189–192 (1968).
See also 13, p. 172; *19*, p. *335*; 21, p. 387; 21.4, p. 428.

R. G. HOWELL and D. J. NEWMAN
High Resolution Study of Crystal Field Lines in the Far Infra-Red
Vol. **29**, No. 2, p. 697–706 (1968).
See also 6, p. 68; *20.1*, p. *357*; 22.8, p. 555.

S. P. TANEJA and S. P. PURI
*Crystal Field Calculation of Principal Magnetic Susceptibilities and Anisotropy Variation with Temperature of Anhydrous Neodymium Chloride
Vol. **30**, No. 1, p. 193–197 (1968).
See also **3**, p. 28; 22.5, p. 514.

F. J. SCHEDEWIE
*Paramagnetic Relaxation in Neodymium-Ethylsulfate: Phonon Re-Absorption and Spectral Spin Diffusion
Vol. **30**, No. 1, p. 251–259 (1968).
See also 6, p. 69; 13, p. 173; 19, p. 336; 22.8, p. 556.

V. ANSORGE and A. MEŇOVSKÝ
*Elementary Magnetic Properties of Some U-Ga Compounds
Vol. **30**, No. 1, p. K31–K32 (1968).
See also *18.2*, p. *315*; 22, p. 454.

J. CZERWONKO
On the Thermodynamics and Paramagnetic Properties of Dispersed Metals
Vol. **30**, No. 2, p. 723–734 (1968).
See also *13.1*, p. *190*; 21, p. 389; 21.1, p. 409; 21.6, p. 437.

H. W. DE WIJN, K. H. J. BUSCHOW, and A. M. VAN DIEPEN
Experimental Evidence for Interband Mixing in Rare-Earth Intermetallic Compounds
Vol. **30**, No. 2, p. 759–766 (1968).
See also 13, p. 173; 13.1, p. 190; *19*, p. *336*; 21, p. 389; 21.4, p. 428.

18.2 Ferromagnetic Properties

Review Articles

A. AHARONI
*Magnetization Curling
Vol. **16**, No. 1, p. 3–42 (1966).
See also 4, p. 37; 21.1, p. 390; 21.1.1, p. 410.

Original Papers and Short Notes

W. BRODKORB
*Zur Theorie des Ferromagnetismus magnetisch anisotroper Schichten mit kubischer Kristallstruktur. (I): Berechnung der Magnetisierung und der Curietemperatur mit der Methode der Greenschen Funktionen
(Theory of the Ferromagnetism of Magnetically Anisotropic Films with Cubic Crystal Structure. (I): Calculation of Magnetization and Curie Temperature with the Green Function Method)
Vol. **16**, No. 1, p. 225–236 (1966).
See also 18.1, p. 296.

R. STRAUBEL und D. ROBASCHIK
*Zur Berechnung der Wechselwirkung einer Ansammlung von ausgerichteten magnetischen Einbereichsteilchen
(Calculation of the Interaction in an Aggregate of Aligned Magnetic Single-Domain Particles)
Vol. **16**, No. 1, p. 237–246 (1966).
See also 4, p. 37; 18.3, p. 316.

H. KRONMÜLLER und A. SEEGER
*Magnetostriktiver und magnetokristalliner elektrischer Restwiderstand in Ferromagnetika
(Magnetostrictive and Magnetocrystalline Residual Electric Resistivity in Ferromagnetics)
Vol. **16**, No. 1, p. 253–266 (1966).
See also 13, p. 162; *14.1*, p. *220*; 21, p. 373.

P. PERETTO, J. L. ODDOU, C. MINIER-CASSAYRE, D. DAUTREPPE, and P. MOSER
Investigation and Interpretation of Stage I of Nickel Electron Irradiated at 20 °K
Vol. **16**, No. 1, p. 281–287 (1966).
See also *10.1*, p. *114*; 14.1, p. 220; 21.1, p. 390.

P. ESCUDIER, F. BIRAGNET, J. DEVENYI, and A. YELON
*Structural and Magnetic Properties of Epitaxial Permalloy Films
Vol. **16**, No. 1, p. 295–300 (1966).
See also *1.4*, p. *16*; 21.1.1, p. 410.

D. PACCARD, C. SCHLENKER, O. MASSENET, R. MONTMORY, and A. YELON
*A New Property of Ferromagnetic–Antiferromagnetic Coupling
Vol. **16**, No. 1, p. 301–311 (1966).
See also *18.4*, p. *321*; 21.1, p. 390; 21.1.1, p. 410.

H. LACHOWICZ
*Ramp Field Switching in Thin Magnetic Films
Vol. **16**, No. 1, p. 313–320 (1966).
See also 4, p. 37; 21.1.1, p. 410.

D. J. CRAIK and M. J. WOOD
*Thermally Activated Domain Wall Movement
Vol. **16**, No. 1, p. 321–328 (1966).
See also 4, p. 37; 21.1.1, p. 410.

A. CORCIOVEI and G. COSTACHE
*Influence of the Perpendicular and Parallel Anisotropy on the Ferromagnetic Properties of Thin Films
Vol. **16**, No. 1, p. 329–336 (1966).
See also 1.4, p. 16.

B. HEINRICH and Z. FRAIT
*Temperature Dependence of the FMR Linewidth of Iron Single-Crystal Platelets
Vol. **16**, No. 1, p. K11–K14 (1966).
See also *19*, p. *326*; 21.1.1, p. 410.

H. BÄURICH
*Berechnung der Energie, Magnetisierungsverteilung und Ausdehnung einer Kreuzblochlinie
(Calculation of the Energy, Magnetization Distribution, and Size of a Cross Bloch-Line)
Vol. **16**, No. 1, p. K39–K43 (1966).
See also 21.1.1, p. 410.

H. SPREEN und CH. SCHWINK
*Vollständige Lösung des magnetoelastischen Problems von Bereichstrukturen mit zyklischer Magnetisierung
(Complete Solution of the Magnetoelastic Problem for Domain Structures with Cyclic Magnetization)
Vol. **16**, No. 1, p. K45–K49 (1966).
See also 4, p. 37; 21.1, p. 390.

K. H. J. BUSCHOW and J. F. FAST
Magnetic and Structural Characteristics of Some Equiatomic Rare-Earth Germanides
Vol. **16**, No. 2, p. 467–473 (1966).
See also *3*, p. *26*; 5, p. 49; 18.1, p. 296; 18.3, p. 316; 18.4, p. 321; 22.7, p. 545.

H. A. WRIEDT and S. ARAJS
*Ferromagnetic Curie Temperatures of Some Iron-Zinc Solid Solutions
Vol. **16**, No. 2, p. 475–478 (1966).
See also 21.1.1, p. 410.

F. BIRAGNET, J. DEVENYI, G. CLERC, O. MASSENET, R. MONTMORY, and A. YELON
*Interactions between Domain Walls in Coupled Films
Vol. **16**, No. 2, p. 569–576 (1966).
See also 4, p. 37; 21.1.1, p. 411.

W. BRODKORB und W. HAUBENREISSER
*Zur Theorie des Ferromagnetismus magnetisch anisotroper Schichten mit kubischer Kristallstruktur. (II): Berechnung der Curietemperatur für eine Eisenschicht mit (100)-Oberflächenorientierung
(Theory of the Ferromagnetism of Magnetically Anisotropic Films with Cubic Crystal Structure. (II): Calculation of the Curie Temperature for an Fe-(100)-Film)
Vol. **16**, No. 2, p. 577–583 (1966).
See also 18.1, p. 296; 21.1.1, p. 411.

J. A. CUNDALL and A. P. KING
*Comparison of Magnetic Measurements and Theoretical Predictions for Nickel-Iron Films over a Wide Composition Range
Vol. **16**, No. 2, p. 613–619 (1966).
See also 4, p. 38; 21.1.1, p. 411.

V. A. Buravikhin, V. G. Kazakov, V. S. Khristosenko und P. I. Krukover
*Der Einfluß einachsiger elastischer Spannungen auf die Ummagnetisierungsprozesse dünner Permalloyschichten
(Effect of Uniaxial Elastic Stress on the Remagnetization Processes inThin Permalloy Films)
Vol. **16**, No. 2, p. 651—656 (1966).
See also 12.1, p. 159; 21.1.1, p. 411.

V. A. Buravikhin und V. I. Popov
*Über die Struktur und die Polarität von Bereichsgrenzen in ferromagnetischen Schichten
(Structure and Polarity of Domain Boundaries in Ferromagnetic Films)
Vol. **16**, No. 2, p. 657—662 (1966).
See also 4, p. 38; 12.1, p. 159; 21.1.1, p. 411.

S. K. Sidorov and A. V. Doroshenko
*On the Magnetic Structure of Some Alloys of Transition Metals
Vol. **16**, No. 2, p. 737—744 (1966).
See also *18.4*, p. *321*; 21.1, p. 391; 21.1.1, p. 411.

A. A. Glazer, A. P. Potapov, R. I. Tagirov, and Ya. S. Shur
*On Particularities of Magnetic Properties of Thin Films with Exchange Anisotropy
Vol. **16**, No. 2, p. 745—750 (1966).
See also 4, p. 38; 21.1.1, p. 411.

E. Klugmann
*Magnetic After-Effect of the Reversible Susceptibility of Cobalt
Vol. **16**, No. 2, p. K117—K119 (1966).
See also 21.1, p. 391.

O. S. Kolotov, T. N. Letova, V. A. Pogozhev und R. V. Telesnin
*Über die Ummagnetisierung dünner ferromagnetischer Schichten längs der Achse leichter Magnetisierung
(Remagnetization of Thin Ferromagnetic Layers along the Easy Axis)
Vol. **16**, No. 2, p. K121—K123 (1966).
See also 21.1.1, p. 411.

H. Pascard
*Effets à forte puissance dans les films minces ferromagnétiques, avec champ polarisant oblique
(High-Power Effects in Thin Ferromagnetic Films with Oblique Polarization Field)
Vol. **16**, No. 2, p. K153—K156 (1966).
See also 21.1.1, p. 411.

J. W. Moroń und B. Pilch
*Weitere Untersuchungen über Relaxationserscheinungen in Fe-Si-C(N)-Legierungen
(Further Studies of Relaxation Phenomena in Fe-Si-C(N) Alloys)
Vol. **16**, No. 2, p. K171—K174 (1966).
See also 21.1.1, p. 411.

D. A. Thompson and H. Chang
*Nanosecond Microscopic Measurement of Magnetic Film Switching by Kerr Magneto-Optic Apparatus
Vol. **17**, No. 1, p. 83—90 (1966).
See also 21.1.1, p. 411.

P. Weber and W. Ruske
*Temperature Dependence of Oblique-Incidence Anisotropy in Permalloy Films
Vol. **17**, No. 1, p. 185—190 (1966).
See also 1.4, p. 16; 21.1.1, p. 412.

W. ANDRÄ and K. STEENBECK
*Magnetic Anisotropy Induced in Thin Films at Low Temperatures
Vol. **17**, No. 1, p. 191–195 (1966).
See also 10.1, p. 115; 21.1, p. 392.

L. V. KIRENSKII, V. G. PYNKO, G. P. PYNKO, R. V. SUKHANOVA, N. I. SIVKOV, G. I. RUSOV und I. S. EDELMAN
*Magnetische Eigenschaften von Einkristallschichten aus Eisen, Nickel, Kobalt und deren Legierungen
(Magnetic Properties of Single-Crystal Films of Iron, Nickel, Cobalt, and Their Alloys)
Vol. **17**, No. 1, p. 249–257 (1966).
See also 4, p. 38; 21.1, p. 392; 21.1.1, p. 412.

W. METZDORF and H. E. WIEHL
*Negative Oblique-Incidence Anisotropy in Magnetostriction-Free Permalloy Films
Vol. **17**, No. 1, p. 285–294 (1966).
See also 1.4, p. 16; 21.1.1, p. 412.

P. VIGIER, C. MINIER-CASSAYRE, P. MOSER, and V. HIVERT
Study of Point Defects Produced by Neutron Irradition in Doped Iron
Vol. **17**, No. 1, p. 317–321 (1966).
See also *10.1*, p. *116*; 11, p. 146; 14.1, p. 220; 21.1.1, p. 412.

G. KNEER and W. ZINN
*On the Origin of M-Induced Anisotropy in Nickel and Permalloy Films
Vol. **17**, No. 1, p. 323–334 (1966).
See also 21.1, p. 392; 21.1.1, p. 412.

V. V. KOBELEV
*Ferromagnetic Films in Nonuniform Fields
Vol. **17**, No. 1, p. 381–388 (1966).
See also 21.1.1, p. 412.

Z. KAMBERSKÁ and V. KAMBERSKÝ
*Oblique-Incidence Anisotropy in Cobalt Films
Vol. **17**, No. 1, p. 411–415 (1966).
See also 1.4, p. 17; 21.1, p. 392.

I. B. PUCHALSKA and A. SUKIENNICKI
*Temperature Dependence of Oblique-Incidence Anisotropy Observed by Lorentz Microscopy
Vol. **17**, No. 1, p. K1 (1966).
See also 1.4, p. 17; 21.1.1, p. 412.

P. VIGIER, V. HIVERT, A. BOURRET, and P. MOSER
Precipitation and Re-Solution of Carbon in Iron
Vol. **17**, No. 1, p. K37–K39 (1966).
See also *10.1*, p. *116*; 21.1.1, p. 412.

W. D. DOYLE
*Epitaxial Influence of the Substrates on the Anisotropy in Thin Electrodesposited Ni–Fe Cylindrical Films
Vol. **17**, No. 1, p. K67–K69 (1966).
See also 1.4, p. 17; 21.1.1, p. 412.

O. HENKEL
*Wechselwirkungseffekte von Fe-Einbereichsteilchen in eingefrorenen Suspensionen
(Interaction Effects of Single-Domain Fe-Particles in Frozen-in Suspensions)
Vol. **17**, No. 1, p. K99–K103 (1966).
See also 18.3, p. 316; 21.1.1, p. 413.

H. HOFFMANN
*Grundsätzliche Betrachtungen einer nichtlinearen Theorie der Anfangssuszeptibilität einachsiger ferromagnetischer Schichten
(Principal Considerations about a Non-Linear Theory of the Initial Susceptibility of Uniaxial Ferromagnetic Films)
Vol. **17**, No. 1, p. K105–K107 (1966).
See also 1.4, p. 17.

A. DOBRZAŃSKA, J. W. MOROŃ und J. RASEK
*Magnetische Untersuchungen über Relaxationserscheinungen in α-Fe-C-Legierungen
(Magnetic Studies of Relaxation Phenomena in α-Fe-C Alloys)
Vol. **17**, No. 1, p. K109–K112 (1966).
See also 21.1.1, p. 413.

L. V. KIRENSKII, T. P. ISOTOVA, N. M. SALANSKII und N. S. CHISTYAKOV
*Magnetische Mehrfachschichten
(Magnetic Multiple Layers)
Vol. **17**, No. 2, p. 489–499 (1966).
See also 21.1.1, p. 413.

L. S. PALATNIK, M. YA. FUKS, L. I. LUKASHENKO, A. G. RAVLIK und A. A. KOZMA
*Struktur und magnetische Eigenschaften von kondensierten ferromagnetischen Schichten
(Structure and Magnetic Properties of Condensed Ferromagnetic Films)
Vol. **17**, No. 2, p. 543–554 (1966).
See also *1.4*, p. *17*; 3.1, p. 30; 10.1, p. 116; 21.1, p. 392; 21.1.1, p. 413.

A. G. LESNIK, G. I. LEVIN, and V. M. NEDOSTUP
*The Importance of Vacancies and Interstitial Atoms in the Development of Induced Anisotropy
Vol. **17**, No. 2, p. 745–751 (1966).
See also 3.1, p. 30; 10.1, p. 116; 21.1.1, p. 413.

A. G. LESNIK and L. M. SANDLER
*Effect of Tensile Stress on the Magnetic Properties of Permalloy Films
Vol. **17**, No. 2, p. 753–760 (1966).
See also 10.1, p. 117; 19, p. 327; 21.1.1, p. 413.

B. STALIŃSKI, Z. BIEGAŃSKI, and R. TROĆ
Low-Temperature Heat Capacity and Thermodynamical Functions of Ferromagnetic Uranium Phosphide U_3P_4
Vol. **17**, No. 2, p. 837–841 (1966).
See also 1.2, p. 9; *8*, p. *80*; 22, p. 441.

VU DINH KY
*The Nernst Effect in Permalloy Films
Vol. **17**, No. 2, p. K203–K205 (1966).
See also 14.1, p. 221; 15, p. 270; 21.1.1, p. 413.

VU DINH KY
*The Planar Nernst Effect in Permalloy Films
Vol. **17**, No. 2, p. K207–K209 (1966).
See also 14.1, p. 221; 15, p. 270; 21.1.1, p. 413.

H. BILGER
*Temperatur- und Verformungsabhängigkeit der Koerzitivfeldstärke von Eisen-Einkristallen (II)
(Temperature and Deformation Dependence of the Coercive Field Strength in Iron Single Crystals (II))
Vol. **18**, No. 1, p. 207–219 (1966).
See also 10.1, p. 117; 21.1.1, p. 414.

B. F. Rothenstein, C. Anghel, and M. Lupulescu
*Influence of Hydrogen on the Magnetomechanical Damping in Nickel
Vol. **18**, No. 1, p. K13—K16 (1966).
See also *10.1*, p. *118*; 21.1, p. 394.

E. Praveczki
*Decoupling of Green's Functions in the Theory of Ferromagnetism
Vol. **18**, No. 1, p. K33—K35 (1966).

G. A. Smolenskii, V. P. Zhuze, V. E. Adamyan, and G. M. Loginov
*Magnetic Properties of Ce, Pr, and Nd Monochalcogenides at 4.2 to 1300 °K
Vol. **18**, No. 2, p. 873—880 (1966).
See also *18.4*, p. *321*; 22, p. 442.

O. Henkel
*Winkelabhängigkeit der magnetischen Eigenschaften bei normalerweise positiv wechselwirkenden Teilchenketten
(Angular Dependence of Magnetic Properties for Particle Chains with Normally Positive Interaction)
Vol. **18**, No. 2, p. K113—K116 (1966).
See also 21.1.1, p. 414.

K. Steenbeck
*Zur Magnetisierungsänderung dünner Nickelschichten durch elektrostatische Aufladung
(Change in the Magnetization of Thin Nickel Layers due to Electrostatic Charging)
Vol. **18**, No. 2, p. K131—K134 (1966).
See also 21.1, p. 394.

H. Bilger
On the Recovery of Plastically Deformed Iron Single Crystals above Room Temperature
Vol. **18**, No. 2, p. K135—K138 (1966).
See also *10.1*, p. *120*; 21.1.1, p. 414.

I. V. Ioffe
On the Theory of the Lattice Oscillations of Dielectric Crystals in an External Electric Field
Vol. **19**, No. 1, p. 51—56 (1967).
See also *6*, p. *58*; 18.4, p. 321; 19, p. 327; 22, p. 442.

E. Köster
*Verformungs- und Temperaturabhängigkeit der Magnetisierungsprozesse in Nickeleinkristallen. (I): Blochwandverschiebungen in ⟨110⟩- und ⟨111⟩-Nickeleinkristallen
(Deformation and Temperature Dependence of Magnetization Processes in Nickel Single Crystals. (I): Bloch Wall Displacements in ⟨110⟩ and ⟨111⟩ Nickel Single Crystals)
Vol. **19**, No. 1, p. 153—168 (1967).
See also 10.1, p. 120; 21.1, p. 395.

Ch. Schwink und O. Grüter
*Untersuchungen der Bereichstruktur von Nickeleinkristallen. (II): Messungen und Beobachtungen an prismatischen Kristallen mit ⟨110⟩ als Achse
(Studies of the Domain Structure of Nickel Single Crystals. (II): Measurements and Observations on Prismatic Crystals with an ⟨110⟩ Axis)
Vol. **19**, No. 1, p. 217—229 (1967).
See also 4, p. 40; 21.1, p. 395.

H. P. van de Braak and W. J. Caspers
Disorder Resistivity and Spin Depolarization in Rare-Earth Alloys
Vol. **19**, No. 1, p. 385–393 (1967).
See also 13, p. 164; *14.1*, p. *222*; 21.4, p. 428.

D. D. Mishin and R. M. Gretchishkin
*Electron Microscopic Examination of High-Coercive Cobalt–Platinum Alloys
Vol. **19**, No. 1, p. K1–K3 (1967).
See also 4, p. 40; 21.1, p. 395; 21.6, p. 432.

B. F. Rothenstein, A. Policec, C. Anghel, and M. Lupulescu
*Contributions to the Mechanism of the Procopiu and Inverse Wiedemann Effects
Vol. **19**, No. 2, p. 613–622 (1967).

E. Köster
*Verformungs- und Temperaturabhängigkeit der Magnetisierungsprozesse in Nickeleinkristallen. (II): Blochwandverschiebungen und Drehungen der spontanen Magnetisierung in ⟨001⟩-Nickeleinkristallen
(Deformation and Temperature Dependence of the Magnetization Processes in Nickel Single Crystals. (II): Bloch Wall Displacements and Rotational Processes of the Spontaneous Magnetization in ⟨001⟩ Nickel Single Crystals)
Vol. **19**, No. 2, p. 655–671 (1967).
See also 10.1, p. 121; 21.1, p. 396.

K.-H. Pfeffer
*Wechselwirkung zwischen Versetzungen und ebenen Blochwänden mit starrem Magnetisierungsverlauf
(Interaction between Dislocations and Plane Bloch-Walls with a Rigid Magnetization Curve)
Vol. **19**, No. 2, p. 735–750 (1967).
See also *10.1*, p. *121*; 21.1, p. 396; 21.1.1, p. 415.

G. A. Jones
Preferred Orientation and Domain Structure of Iron Films Deposited on (100) Copper Substrates
Vol. **19**, No. 2, p. 811–821 (1967).
See also *1.4*, p. *18*; 21.1.1, p. 415.

R. H. Wade
*Electron Diffraction from a Magnetic Phase Grating
Vol. **19**, No. 2, p. 847–854 (1967).
See also 4, p. 40; 21.1, p. 396; 21.1.1, p. 415.

R. Straubel
Anhysteretic Magnetization of a Collective of Magnetic Single-Domain Particles
Vol. **20**, No. 1, p. 267–275 (1967).

K.-H. Pfeffer
*Mikromagnetische Behandlung der Wechselwirkung zwischen Versetzungen und ebenen Blochwänden. (I): Allgemeine Theorie
(Micromagnetic Treatment of the Interaction between Dislocations and Plane Bloch-Walls. (I): General Theory)
Vol. **20**, No. 1, p. 395–411 (1967).
See also 10.1, p. 121; 21.1, p. 396; 21.1.1, p. 416.

D. D. Mishin
*On the Theory of Magnetic Hardness of Platinum Cobalt
Vol. **20**, No. 1, p. K9–K10 (1967).
See also 21.1, p. 396; 21.6, p. 433.

L. Zsoldos
*Lattice Parameter Change of FeRh Alloys due to Antiferromagnetic–Ferromagnetic Transformation
Vol. **20**, No. 1, p. K25—K28 (1967).
See also 1.2, p. 10; 5, p. 51; *18.4*, p. *322*; 21.1.1, p. 416.

H. Markert
Decrease of Number and Size of Barkhausen Jumps during Ultrasonic Irradiation
Vol. **20**, No. 1, p. K67—K70 (1967).
See also *7*, p. *76*; 18.3, p. 317; 21.1.1, p. 416.

J. Przystawa and W. Suski
*Magnetic Ordering in Uranium Compounds with Crystal Structures of Anti-Cu_2Sb and PbFCl Types. Molecular Field Approximation
Vol. **20**, No. 2, p. 451—459 (1967).
See also 3.2, p. 35; *18.4*, p. *322*; 22, p. 444.

N. van Dang
*Effets de la désaimantation thermique sur les cycles d'hystérésis
(Effect of Thermal Demagnetization on the Hysteresis Cycles)
Vol. **20**, No. 2, p. 557—565 (1967).
See also 21.1.1, p. 416.

B. Roessler
*Observations of Ferromagnetic Domains by Anomalous Transmission of X-Rays. The Effects of Variation in the Reflecting Plane
Vol. **20**, No. 2, p. 713—723 (1967).
See also 4, p. 41; 21.1.1, p. 416.

V. A. Bokov, N. A. Grigoryan, and M. F. Bryzhina
X-Ray Diffraction and Magnetic Studies of Solid Solutions $Bi_{1-x}Ca_xMnO_3$
Vol. **20**, No. 2, p. 745—754 (1967).
See also *1*, p. *7*; 1.2, p. 10; 18.3, p. 317; 18.4, p. 322; 22.8, p. 549.

G. A. Jones
*Structural and Magnetic Properties of Epitaxial Films of Iron
Vol. **20**, No. 2, p. K95—K97 (1967).
See also *1.4*, p. *18*; 21.1.1, p. 416.

D. D. Mishin
*On the Theory of Susceptibility and Coercive Force of Magnetic Materials
Vol. **20**, No. 2, p. K123—K126 (1967).
See also 21.1.1, p. 416.

O. Henkel
*On the Temperature Dependence of Interaction Effects in Ferromagnetic Fine Particle Assemblies
Vol. **20**, No. 2, p. K127—K130 (1967).
See also 21.1.1, p. 417.

I. B. Puchalska, A. Sukiennicki, Z. Petykiewicz, and T. Tymosz
The Influence of Annealing on the Structure of Polycrystalline Permalloy Thin Films
Vol. **21**, No. 1, p. 183—189 (1967).
See also *1.4*, p. *18*; 21.1.1, p. 417.

S. Arajs and G. R. Dunmyre
*A Note on the Consistency of Values of the Spontaneous or Saturation Magnetization of Polycrystalline Iron and Nickel at 298 °K
Vol. **21**, No. 1, p. 191—195 (1967).
See also 21.1, p. 397; 21.1.1, p. 417.

J. Steinert
*Induced Uniaxial Magnetic Anisotropy of Fe-Al Alloys at Low Concentrations
Vol. **21**, No. 1, p. K13–K15 (1967).
See also 10.1, p. 122; 21, p. 379; 21.1.1, p. 417.

I. B. Puchalska and A. Sukiennicki
*On the Temperature Dependence of the Domain Structure
Vol. **21**, No. 1, p. K57–K58 (1967).
See also 4, p. 42; 21.1.1, p. 417.

B. F. Rothenstein, A. Policec, M. Lupulescu, and C. Anghel
*The Electromagnetic Aspect of Magnetostrictively Induced Torsional Oscillations (the Procopiu Effect with a Cylindrical Ferromagnetic Specimen Fixed at one End)
Vol. **21**, No. 2, p. 601–606 (1967).
See also 14.1, p. 222; 21.1.1, p. 417.

M. Polcarová and J. Kaczér
*X-Ray Diffraction Contrast on Ferromagnetic Domain Walls in Fe-Si Single Crystals
Vol. **21**, No. 2, p. 635–642 (1967).
See also 4, p. 42; 21.1.1, p. 417.

K.-H. Pfeffer
*Mikromagnetische Behandlung der Wechselwirkung zwischen Versetzungen und ebenen Blochwänden. (II): Anwendungen
(Micromagnetic Treatment of the Interaction between Dislocations and Plane Bloch-Walls. (II): Applications)
Vol. **21** , No. 2, p. 837–856 (1967).
See also 10.1, p. 123; 21.1, p. 398.

K.-H. Pfeffer
*Zur Theorie der Koerzitivfeldstärke und Anfangssuszeptibilität
(On the Theory of Coercive Field Strength and Initial Susceptibility)
Vol. **21**, No. 2, p. 857–872 (1967).
See also 10.1, p. 123; 21.1, p. 398; 21.1.1, p. 418

H. Mühe
*Temperature Dependence of Magnetocrystalline Anisotropy of an Fe-Al Alloy near the Concentration Fe_3Al
Vol. **21**, No. 2, p. K115–K119 (1967).
See also 21, p. 380; 21.1.1, p. 418.

Vu dinh Ky
*Susceptibility of Ferromagnetic Films in Paraprocess
Vol. **21**, No. 2, p. K159–K162 (1967).
See also 21.1, p. 398; 21.1.1, p. 418.

A. R. Ferchmin
*Some Remarks on the Paper "Zur Theorie des Ferromagnetismus dünner Schichten" by H. Hemschik
Vol. **21**, No. 2, p. K179–K182 (1967).
See also *13.1*, p. *179*.

M.-P. Stoll
*The Use of the Magneto-Optical Kerr Effect for the Determination of Magnetization Curves
Vol. **22**, No. 1, p. 163–176 (1967).
See also 20, p. 339; 21.1.1, p. 418.

Yu. P. Irkhin, A. N. Voloshinskii, and Sh. Sh. Abelskii
On the Theory of Spontaneous Hall Effect at Low Temperatures
Vol. **22**, No. 1, p. 309–318 (1967).
See also 13, p. 166; *14.1*, p. *223*.

G. P. Huffman, P. R. Errington, and R. M. Fisher
Mössbauer Study of the Fe-Mn Carbides $(Fe_{1-x}Mn_x)_3C$ and $(Fe_{1.1}Mn_{3.9})C_2$
Vol. **22**, No. 2, p. 473–481 (1967).
See also *6.1*, p. *72*; 22.8, p. 551.

L. Gerward
*Magnetic After-Effect in Deformed Cobalt Single Crystals
Vol. **22**, No. 2, p. 659–666 (1967).
See also 10.1, p. 124; 21.1, p. 399.

A. Hubert
*Der Einfluß der Magnetostriktion auf die magnetische Bereichsstruktur einachsiger Kristalle, insbesondere des Kobalts
(Effect of Magnetostriction on the Magnetic Domain Structure of Uniaxial Crystals, in Particular of Cobalt)
Vol. **22**, No. 2, p. 709–727 (1967).
See also 21.1, p. 399.

Vu dinh Ky
Planar Hall and Nernst Effect in Ferromagnetic Metals
Vol. **22**, No. 2, p. 729–736 (1967).
See also 13, p. 167; *14.1*, p. *223*; *15*, p. *272*; 21.1, p. 400; 21.1.1, p. 418.

A. Bourret et M. Kleman
*Rubans d'aimantation dans les lames minces de fer pur
(Stripe Domains in Thin Films of Pure Iron)
Vol. **23**, No. 1, p. 207–221 (1967).
See also 21.1.1, p. 418.

A. E. Lord, Jr.
Coupling of Magnons and Phonons at High Frequencies
Vol. **23**, No. 2, p. 521–526 (1967).
See also 6, p. 63; *13*, p. *167*.

H. Bäurich
*Berechnung der Energie, Magnetisierungsverteilung und Ausdehnung einer Kreuzblochlinie
(Calculation of the Energy, Magnetization Distribution, and Length of a Cross Bloch-Line)
Vol. **23**, No. 2, p. K137–K138 (1967).

O. Henkel
*Positive and Negative Interaction of Fine Particle Assemblies in Correlation with Other Magnetic Properties
Vol. **23**, No. 2, p. K151–K153 (1967).
See also 21.1.1, p. 419.

E. F. Makarov, V. A. Povitskii, E. B. Granovskii, and A. A. Fridman
Study of Alnico 8 by Mössbauer Spectroscopy
Vol. **24**, No. 1, p. 45–50 (1967).
See also *6.1*, p. *73*; 21.1.1, p. 419.

F. J. Krings
High-Temperature Ferromagnetic After-Effect in Iron-Silicon Alloys during Plastic Deformation
Vol. 24, No. 1, p. 163—176 (1967).
See also *10.1*, p. *126*; 21.1.1, p. 419.

J. Przystawa
*The Néel Temperatures and Stability of Magnetic Phases in the Disturbed Body-Centered Tetragonal Lattice. Uranium Compounds
Vol. 24, No 1, p. 313—322 (1967).
See also *18.4*, p. *323*; 22, p. 448; 22.8, p. 552.

V. V. Kelarev, S. K. Sidorov, V. V. Klyushin, and R. Z. Abdulov
*Neutron-Diffraction Study of Antiferro-Ferromagnetic Transition in a System of Ordered $Fe(Pd_xPt_{1-x})_3$ Alloys
Vol. 24, No. 1, p. 385—390 (1967).
See also *18.4*, p. *323*; 21.1.1, p. 419.

W. Glaeser, H. Imgrund, and H. Wever
On the Problem of Reproducible Measurements of the Residual Resistivity of High-Purity Iron
Vol. 24, No. 1, p. K5—K8 (1967).
See also *14.1*, p. *224*; 21.1.1, p. 420.

H. Spreen
*Untersuchungen der Bereichstruktur von Nickeleinkristallen. (III): Die Abschlußstruktur in zylindrischen Kristallen mit ⟨110⟩ als Achse
(Study of the Domain Structure in Nickel Single Crystals. (III): The Closure Domain Structure of Cylindrical Crystals with an ⟨110⟩ Axis)
Vol. 24, No. 2, p. 413—429 (1967).
See also 4, p. 44; 21.1, p. 401.

H. Spreen
*Untersuchungen der Bereichstruktur von Nickeleinkristallen. (IV): Die Grundstruktur in zylindrischen Kristallen mit ⟨100⟩ als Achse
(Study of the Domain Structure in Nickel Single Crystals. (IV): The Basic Structure of Cylindrical Crystals with an ⟨100⟩ Axis)
Vol. 24, No. 2, p. 431—441 (1967).
See also 4, p. 44; 21.1, p. 401.

J. Hölzl and B. Heimann
Experimental Results on the Temperature Dependence of Secondary Electron Emission from Polycrystalline Nickel in the Range $100 \leqq T \leqq 600$ °C
Vol. 24, No. 2, p. 651—657 (1967).
See also *17*, p. *288*; 18.1, p. 297; 21.1, p. 402.

A. Hubert
On the Theory of Two-Phase Domain Structures in Superconductors and Ferromagnets
Vol. 24, No. 2, p. 669—682 (1967).
See also *4*, p. *44*; 14.2, p. 229; 21.1, p. 402.

K. H. J. Buschow, J. F. Fast, A. M. van Diepen, and H. W. de Wijn
*Magnetic Dilution of Rare-Earth Aluminium Cubic Laves Phases RAl_2
Vol. 24, No. 2, p. 715—720 (1967).
See also 1.2, p. 11; 21, p. 383.

H. P. van de Braak and W. J. Caspers
*Point Defects in Ferromagnets
Vol. **24**, No. 2, p. 733–741 (1967).
See also *10.1*, p. *127*.

O. Henkel
*Interaction Effects in Alnico Alloys
Vol. **24**, No. 2, p. K153–K156 (1967).
See also 21.1.1, p. 420.

F. F. Y. Wang and M. Kestigian
Magnetic Properties of Al-Substituted Yttrium Orthoferrite
Vol. **25**, No. 1, p. 119–124 (1968).
See also *18.3*, p. *319*; 18.4, p. 323.

H. A. Brown
*The Critical and High-Temperature Behavior of a Heisenberg Ferromagnet
Vol. **25**, No. 1, p. K11–K14 (1968).

A. Sukiennicki and L. Wojtczak
*Magnetic Domain Structure in Uniaxial Thin Films
Vol. **25**, No. 1, p. K19–K21 (1968).
See also 21.1, p. 402; 21.1.1, p. 420.

A. Holz
*Theoretische Untersuchung der Keimbildung in Ferromagnetika
(Theoretical Study of Nucleation in Ferromagnetics)
Vol. **25**, No. 2, p. 567–582 (1968).
See also 4, p. 44; 21.1, p. 403; 21.1.1, p. 420.

M. Kuriyama and G. M. McManus
X-Ray Interference Fringes and Domain Arrangements in Fe + 3 wt% Si Single Crystals
Vol. **25**, No. 2, p. 667–677 (1968).
See also *4*, p. *45*; 21.1.1, p. 420.

G. Vertogen
Ground State of a System with Ferromagnetic s–f Interaction
Vol. **25**, No. 2, p. 729–740 (1968).
See also *13*, p. *169*.

A. I. Mitsek and N. P. Kolmakova
*On the Theory of Weak Ferromagnetics
Vol. **25**, No. 2, p. 787–799 (1968).
See also *18.4*, p. *323*; 19, p. 332.

L. Valenta, W. Haubenreisser, and W. Brodkorb
*A Contribution to the Problem of the Spatial Distribution of the Spontaneous Magnetization in Thin Films
Vol. **26**, No. 1, p. 191–196 (1968).
See also 6.1, p. 73.

R. Gemperle and A. Gemperle
*Dependence of the Ferromagnetic Domain Width on the Thickness of Cobalt Foils
Vol. **26**, No. 1, p. 207–217 (1968).
See also 1.4, p. 20; 21.1, p. 404.

M. Sorohan
*The Influence of Ion Bombardment upon the Induced Magnetic Anisotropy in Thin Nickel Films
Vol. **26**, No. 1, p. K9–K11 (1968).
See also *11*, p. *151*; 21.1, p. 404.

V. Christoph
On the Theory of the Hall Effect in Ferromagnetics
Vol. **26**, No. 1, p. K17–K20 (1968).
See also *14.1*, p. *225*.

J. Villain
*Critical Exponents in the Dynamics of an Isotropic Magnetic System near Its Transition Temperature
Vol. **26**, No. 2, p. 501–507 (1968).
See also *18.4*, p. *324*.

Vu dinh Ky
*Planar Hall Effect in Ferromagnetic Films
Vol. **26**, No. 2, p. 565–569 (1968).
See also *14.1*, p. *225*; 21.1, p. 404; 21.1.1, p. 421.

A. E. Lord, Jr.
Sound Wave Velocity Change due to the Magnon-Phonon Interaction
Vol. **26**, No. 2, p. 717–724 (1968).
See also 6, p. 65; *7*, p. *77*; 13, p. 169; 22.8, p. 553.

G. F. Bolling, A. Arrott, and R. H. Richman
*A Study of Curie Temperatures and the Effect of Carbon in Face-Centered Cubic Iron-Nickel Alloys
Vol. **26**, No. 2, p. 743–750 (1968).
See also 10.1, p. 130; 21.1.1, p. 421.

A. Holz
*Die Magnetisierungsverteilung an der Stirnfläche zylindrischer Ferromagnetika mit Anwendungen auf die Keimbildung
(Magnetization Distribution at the Basal Faces of Cylindrical Ferromagnetics with Application to Nucleation)
Vol. **26**, No. 2, p. 751–764 (1968).
See also 4, p. 46; 21.1.1, p. 420.

A. R. Ferchmin
*Remarks to the Theory of Ferromagnetism of Thin Films
Vol. **26**, No. 2, p. K107–K108 (1968).
See also *13.1*, p. *185*.

K.-H. Müller und M. Wolf
*Magnetisierung und Spinwellensteifigkeit im Heisenbergmodell für Fe-Al-Legierungen
(Magnetization and Stiffness of Spin Waves in the Heisenberg Model for Fe-Al Alloys)
Vol. **26**, No. 2, p. K109–K111 (1968).
See also *13*, p. *169*; 21.1.1, p. 421.

E. Feldtkeller
*Magnetic Domain Wall Dynamics
Vol. **27**, No. 1, p. 161–170 (1968).
See also 4, p. 46; 21.1, p. 404; 21.1.1, p. 421.

U. Gradmann and J. Müller
*Flat Ferromagnetic, Epitaxial 48Ni/52Fe(111) Films of Few Atomic Layers
Vol. **27**, No. 1, p. 313–324 (1968).
See also *1.4*, p. *20*; 21.1, p. 405; 21.1.1, p. 421.

B. F. Rothenstein, A. Policec, and I. Artzner
*Contributions to the Kinetics of Charging Nickel with Hydrogen
Vol. **27**, No. 1, p. K5–K6 (1968).
See also 21.1, p. 405.

H. A. BROWN and W. F. PARKS
*The Magnetic Dilution Problem
Vol. 27, No. 1, p. K17–K20 (1968).

J. KINEL, L. KOZŁOWSKI, and J. W. MOROŃ
*Diffusional Relaxation Effects in Hydrogen Charged Silicon Iron
Vol. 27, No. 1, p. K37–K39 (1968).
See also 10.1, p. 131; *12.1*, p. *160*; 21.1.1, p. 421.

H. BURKHARD and J. JAUMANN
*The Determination of the "Gyroelectric" and "Gyromagnetic" Constants of Thick, Evaporated Layers of Iron by Means of Measurements in the Reflected Light only
Vol. 27, No. 1, p. K43–K45 (1968).
See also *20*, p. *341*; 20.1, p. 354; 21.1.1, p. 421.

D. D. MISHIN
The Effect of Dislocations on the Properties of Silicon Iron in Alternating Magnetic Fields
Vol. 27, No. 1, p. K49–K50 (1968).
See also *10.1*, p. *131*; 21.1.1, p. 421.

H.-D. DIETZE und K. SCHRÖDER
*Magnetisierung in der Umgebung unmagnetischer Einschlüsse in Ferromagnetika. (I): Mikromagnetische Grundlagen
(Magnetization in the Environment of Non-Magnetic Inclusions in Ferromagnetics. (I): Micromagnetic Basis)
Vol. 27, No. 2, p. 601–610 (1968).

K. SCHRÖDER und H.-D. DIETZE
*Magnetisierung in der Umgebung unmagnetischer Einschlüsse in Ferromagnetika. (II): Kugelförmige Hohlräume in Ferromagnetika ohne Wände
(Magnetization in the Environment of Non-Magnetic Inclusions in Ferromagnetics. (II): Spherical Cavities in Ferromagnetics without Walls)
Vol. 27, No. 2, p. 611–622 (1968).

B. S. TOŠIĆ and R. B. ŽAKULA
*Kinematic and Dynamic Interaction between Elementary Excitations in Ferromagnetics with Spin $S = 1/2$
Vol. 27, No. 2, p. 623–629 (1968).
See also *13*, p. *170*.

A. I. MITSEK, N. P. KOLMAKOVA, V. A. FEIGIN, and P. F. GAIDANSKII
*Spin-Wave Distribution Function and Thermodynamics of the Heisenberg Ferromagnetic
Vol. 28, No. 1, p. 83–96 (1968).
See also 8, p. 85; 13, p. 171.

R. P. FERRIER and I. B. PUCHALSKA
*360° Walls and Strong Stripe Domains in Permalloy Films
Vol. 28, No. 1, p. 335–347 (1968).
See also 1.4, p. 20; 21.1.1, p. 422.

W. BITTERLICH and R. J. JELITTO
Computational Evidence for the Transition Point of a Square Ising Lattice with First and Second Neigbour Interactions
Vol. 28, No. 1, p. 365–372 (1968).
See also *1.2*, p. *12*; 18.4, p. 324.

D. J. Craik and M. J. Wood
*Thermally Activated Domain Wall Movement in Nickel–Iron Films
Vol. **28**, No. 1, p. 407–414 (1968).
See also 1.4, p. 20; 21.1, p. 405; 21.1.1, p. 422.

M. Sorohan
*Oscillating Hysteresis in Thin Nickel Films
Vol. **28**, No. 1, p. K59–K61 (1968).
See also 21.1, p. 406.

E. Klugmann
*Magnetic After-Effect in Cobalt
Vol. **28**, No. 1, p. K67–K70 (1968).
See also 21.1, p. 406.

C. Schlenker
*Couplage ferro–antiferromagnetique et traînage magnétique dans des couches minces multiples Co-CoO et Ni-NiO
(Ferro-Antiferromagnetic Coupling and Magnetic After-Effect in Multiple Co-CoO and Ni-NiO Films)
Vol. **28**, No. 2, p. 507–517 (1968).
See also *18.4*, p. *324*; 21.1, p. 406; 22.6, p. 542.

D. I. Lalović, B. S. Tošić, and R. B. Žakula
*On the Problem of Kinematic Interaction in the Quantum Theory of Magnetism
Vol. **28**, No. 2, p. 635–642 (1968).
See also *13*, p. *171*.

A. Holz und H. Kronmüller
*Der Beitrag der Versetzungen zur Anisotropie des magnetischen Einmündungsgesetzes in Nickel
(Contribution of Dislocations to the Anisotropy of the Magnetic Saturation Curve in Nickel)
Vol. **28**, No. 2, p. 755–771 (1968).
See also 10.1, p. 132; 21.1, p. 406.

V. A. Bokov, N. A. Grigoryan, M. F. Bryzhina, and V. V. Tikhonov
*Effect of Lattice Distortions on the Magnetic Behaviour of Perovskite-Type Manganites
Vol. **28**, No. 2, p. 835–847 (1968).
See also 5, p. 54; *18.4*, p. *324*; 22.8, p. 554.

E. Praveczki
*On the Use of Bose Operators for Ferromagnets at General Temperatures
Vol. **28**, No. 2, p. K105–K107 (1968).
See also 13, p. 172.

S. Völlmar
*A Model System for Hard-Magnetic Particle Ensembles
Vol. **28**, No. 2, p. K171–K173 (1968).

N. N. Sirota and E. A. Vasilev
Magnetic Properties and Structure of $MnAs_xSb_{1-x}$ Solid Solutions
Vol. **28**, No. 2, p. K175–K177 (1968).
See also *18.1*, p. *297*; 22.8, p. 555.

M. A. Krivoglaz and L. B. Kvashnina
The Correlation Function $S_{\cdot z}(\omega)$ for Weakly Coupled Impurity Atom Spin in a Ferromagnetic
Vol. **29**, No. 1, p. 53–59 (1968).
See also *13*, p. *172*; 19, p. 335.

M. A. Krivoglaz and L. B. Kvashnina
On the Spin Broadening of Mössbauer Lines of Impurities in Ferromagnetics and of Atoms in Antiferromagnetics
Vol. **29**, No. 1, p. 61–73 (1968).
See also *6.1*, p. *74*; 13, p. 172; 18.4, p. 324.

H. Gengnagel and U. Hofmann
*Temperature Dependence of the Magnetocrystalline Energy Constants K_1, K_2, and K_3 of Iron
Vol. **29**, No. 1, p. 91–97 (1968).
See also 21.1.1, p. 422.

W. Brückner, R. Perthel, K. Kleinstück, and G. E. R. Schulze
*Magnetic Properties of $ZrFe_2$ and $TiFe_2$ within Their Homogeneity Range
Vol. **29**, No. 1, p. 211–216 (1968).
See also *18.4*, p. *325*; 21.1.1, p. 422.

F. Walz
*Untersuchung der magnetischen Nachwirkung in Nickel nach Neutronenbestrahlung und plastischer Verformung
(Investigation of the Magnetic After-Effect in Nickel after Neutron Irradiation and Plastic Deformation)
Vol. **29**, No. 1, p. 245–258 (1968).
See also 10.1, p. 133; 11, p. 153; 21.1, p. 407.

A. Sukiennicki
*On an New Type of Stripe Domains
Vol. **29**, No. 1, p. 417–421 (1968).
See also 4, p. 47; 21.1.1, p. 422.

Z. I. Alizade and T. G. Magarramov
*Galvanomagnetic Effect in Ferromagnetic Alloys of Nickel-Tantalum and Nickel-Niobium Systems
Vol. **29**, No. 1, p. K19–K22 (1968).
See also *14.1*, p. *226*; 21.1, p. 407.

J. Kinel, J. W. Moroń, and J. Przybyła
*A Diffusional Magnetic Viscosity Effect in Iron-Carbon Martensite
Vol. **29**, No. 1, p. K39–K41 (1968).
See also 21.1.1, p. 422.

D. I. Bower, E. Claridge, and I. S. T. Tsong
Low-Temperature Elastic Constants and Specific Heats of F.C.C. Nickel-Iron Alloys
Vol. **29**, No. 2, p. 617–625 (1968).
See also 6, p. 68; 8, p. 86; *12.1*, p. *160*; 13, p. 172; 21.1, p. 407; 21.1.1, p. 422.

J. P. Guigay and R. H. Wade
Mainly on the Fresnel Mode in Lorentz Microscopy
Vol. **29**, No. 2, p. 799–805 (1968).
See also *4*, p. *47*.

B. F. Rothenstein and I. Artzner
*On the Internal Barkhausen Effect
Vol. **29**, No. 2, p. K117–K118 (1968).
See also 21.1, p. 408; 21.1.1, p. 423.

S. J. Sivonen and E. J. Suoninen
*Temperature Dependence of the Magnetic Ordering in Gd-$GdFe_2$ Phase Mixtures
Vol. **29**, No. 2, p. K171–K173 (1968).
See also 14.1, p. 226; 21.1.1, p. 423; 21.4, p. 428.

J. L. Leveque, T. Anagnostopoulos, H. Bilger, and P. Moser
Point Defects in Iron–Carbon Solid Solution Studied by Magnetic After-Effect Measurements
Vol. **29**, No. 2, p. K179–K181 (1968).
See also *10.1*, p. *134*; *11*, p. *154*; 21.1.1, p. 423.

D. D. Mishin
*Dislocation Theory of Susceptibility and Coercive Force of Silicon Iron
Vol. **30**, No. 1, p. 31–35 (1968).
See also *10.1*, p. *134*; 21.1.1, p. 423.

D. Becker, F. Dworschak, C. Lehmann, K. T. Rie, H. Schuster, H. Wollenberger, and J. Wurm
Determination of the Average Number of Replacement Collisions in an Electron-Irradiated F.C.C. Crystal (Ni_3Mn)
Vol. **30**, No. 1, p. 219–229 (1968).
See also 10.1, p. 135; *11*, p. *155*; 14.1, p. 227; 21.1, p. 408.

S. Völlmar
*Calculations Concerning a Model for Hard-Magnetic Particle Ensembles
Vol. **30**, No. 1, p. K27–K30 (1968).
See also 21.1.1, p. 423.

V. Ansorge and A. Meňovský
*Elementary Magnetic Properties of Some U-Ga Compounds
Vol. **30**, No. 1, p. K31–K32 (1968).
See also *18.1*, p. *298*; 22, p. 454.

E. Köster
*Magnetization and Magnetostriction of Hexagonal Cobalt Single Crystals with Internal Stresses
Vol. **30**, No. 2, p. 455–460 (1968).
See also 10.1, p. 135; 21.1, p. 409.

Ch. Schwink and O. Schärpf
*Electron-Optic Investigation of the Magnetic Stray Field above Bloch Walls in Cylindric Nickel Crystals
Vol. **30**, No. 2, p. 637–644 (1968).
See also 4, p. 48; 21.1, p. 409.

B. F. Rothenstein, J. Artzner, and T. Wächter
*The Electromagnetic Aspect of Magnetostrictively Induced Longitudinal Oscillations (the Procopiu Effect with a Cylindrical Ferromagnetic Sample Free at one End)
Vol. **30**, No. 2, p. K91–K92 (1968).
See also 21.1, p. 409.

Th. Ricker
*Domain Walls in Thin Ferromagnetic Films Investigated by Low-Angle Electron Deflection
Vol. **30**, No. 2, p. K93–K95 (1968).
See also 4, p. 48; 21.1, p. 409; 21.1.1, p. 423.

A. Seeger and H. Kronmüller
*Reversible Magnetization Curve and Magnetostriction of Uniaxial Ferromagnetic Crystals with Internal Stresses
Vol. **30**, No. 2, p. K101–K106 (1968).
See also 10.1, p. 136; 21.1, p. 409.

J. Mathon and E. P. Wohlfarth
*Contributions to the Invar Problem
Vol. **30**, No. 2, p. K131–K133 (1968).
See also 21.1, p. 409; 21.1.1, p. 423.

18.3 Ferrimagnetic Properties. Ferrites

Original Papers and Short Notes

E. Wieser, W. Meisel, and K. Kleinstück
Mössbauer Studies of Mn and Mg–Mn Ferrites
Vol. **16**, No. 1, p. 127–135 (1966).
See also *6.1*, p. *71*.

R. Straubel and D. Robaschik
Calculation of the Interaction in an Aggregate of Aligned Magnetic Single-Domain Particles
Vol. **16**, No. 1, p. 237–246 (1966).
See also 4, p. 37; *18.2*, p. *299*.

K. H. J. Buschow and J. F. Fast
Magnetic and Structural Characteristics of Some Equiatomic Rare-Earth Germanides
Vol. **16**, No. 2, p. 467–473 (1966).
See also *3*, p. *26*; 5, p. 49; 18.1, p. 296; 18.2, p. 300; 18.4, p. 321; 22.7, p. 545.

E. Steinbeiss
*Untersuchungen über Ummagnetisierungsprozesse in Rechteckferriten
(Investigations of Remagnetization Processes in Ferrites with Rectangular Hysteresis Loop)
Vol. **16**, No. 2, p. 499–506 (1966).

R. Gerber, L. Michalowsky, K. Motzke und E. Steinbeiss
*Über die induzierte Anisotropieenergie manganhaltiger Ferrite
(Induced Anisotropy-Energy of Ferrites Containing Manganese)
Vol. **16**, No. 2, p. 795–801 (1966).

R. Perthel, G. Elbinger und W. Keilig
*Über die Anisotropie des magnetischen Momentes und die Kristallanisotropie bei Kobaltferrit
(Anisotropy of Magnetic Moment and Crystal Anisotropy in Cobalt Ferrite)
Vol. **17**, No. 1, p. 151–154 (1966).

S. S. Starobinets, B. M. Lebed, and A. G. Gurevich
Parametric Magneto-Acoustic Resonance in Ca–Bi–V Garnet
Vol. **17**, No. 1, p. K7–K9 (1966).
See also *19*, p. *326*.

B. Lemaire and J. le Gall
*Spin-Wave Instability in Parallel Pumping with Long Duration Pulses and Steady State
Vol. **17**, No. 1, p. K41–K43 (1966).
See also *13*, p. *163*.

O. Henkel
Interaction Effects of Single-Domain Fe-Particles in Frozen-in Suspensions
Vol. **17**, No. 1, p. K99–K103 (1966).
See also *18.2*, p. *302*; 21.1.1, p. 413.

A. Bragiński and Z. Siermiński
*Magnetic Ordering and Relaxation in Iron-Deficient Mn-Ferrites
Vol. **17**, No. 2, p. K211–K215 (1966).

E. Steinbeiss
*Wandbeweglichkeitsmessungen an Ferriten mit rechteckförmiger Hystereseschleife (Measurements of the Wall Mobility in Ferrites Having a Rectangular Hysteresis Loop)
Vol. **18**, No. 1, p. 241—249 (1966).

R. Krishnan
*Influence of Tetrahedral Ni^{2+} Ions on the Magnetic Properties of YIG
Vol. **18**, No. 1, p. K53—K56 (1966).
See also 19, p. 327.

U. Heinecke
*Über ein Modell zur Berechnung der Entmagnetisierungskurven von Hartferriten (A Model for Calculating Demagnetization Curves of Hard Ferrites)
Vol. **18**, No. 2, p. 569—578 (1966).

L. Jahn
*Crystal Anisotropy and Saturation Magnetization of Hard Ferrites with a Composition near the M-Compound Prepared by Means of the Verneuil Method
Vol. **19**, No. 2, p. K75—K77 (1967).

H. Markert
Decrease of Number and Size of Barkhausen Jumps during Ultrasonic Irradiation
Vol. **20**, No. 1, p. K67—K70 (1967).
See also *7*, p. *76*; 18.2, p. 306; 21.1.1, p. 416.

E. Jäger and R. Perthel
Some Remarks to the Crystal Field Theory of Spinel Lattices
Vol. **20**, No. 2, p. 433—441 (1967).
See also 3.2, p. 35; *13*, p. *165*.

V. A. Bokov, N. A. Grigoryan, and M. F. Bryzhina
X-Ray Diffraction and Magnetic Studies of Solid Solutions $Bi_{1-x}Ca_xMnO_3$
Vol. **20**, No. 2, p. 745—754 (1967).
See also *1*, p. *7*; 1.2, p. 10; 18.2, p. 306; 18.4, p. 322; 22.8, p. 549.

I. Bunget and M. Rosenberg
*Influence of Changes of the Magnetocrystalline Anisotropy on the Electrical Resistivity of $Ba_3Co_2Fe_{24}O_{41}$
Vol. **20**, No. 2, p. K163—K165 (1967).
See also 14.4, p. 253.

M. Rosenberg, C. Tănăsoiu, and V. Florescu
*The Influence of the Magnetic Field on the Domain Structure of Ferrimagnetic Oxides with High Anisotropy Field
Vol. **21**, No. 1, p. 197—204 (1967).
See also 4, p. 42; 22.6, p. 537.

G. Elbinger
*On the Magnetocrystalline Anisotropy of Ni-Co Ferrites
Vol. **21**, No. 1, p. 303—307 (1967).
See also 21.1, p. 397.

R. Krishnan
Ionic Redistribution and Anisotropy in Heat Treated NiO-Doped YIG Single Crystals
Vol. **21**, No. 1, p. K5—K7 (1967).
See also *10*, p. *100*; 19, p. 329.

K. H. J. Buschow and J. F. Fast
Crystal Structure and Magnetic Properties of Some Rare Earth Germanides
Vol. **21**, No. 2, p. 593–600 (1967).
See also *5*, p. *51*; 18.1, p. 296; 18.4, p. 322; 22.7, p. 545.

I. Bunget and M. Rosenberg
P-Type Conduction in Barium Ferrite Single Crystals
Vol. **21**, No. 2, p. K131–K133 (1967).
See also *14.3*, p. *236*; 15, p. 272.

Ž. Dimitrijević, S. Krašnicki, J. Todorović, and A. Wanic
*Neutron Investigation of Magnons in Magnetite
Vol. **22**, No. 1, p. K55–K57 (1967).

I. Dézsi, L. Keszthelyi, D. Kulgawczuk, B. Molnár, and N. A. Eissa
Mössbauer Study of β- and δ-FeOOH and Their Disintegration Products
Vol. **22**, No. 2, p. 617–629 (1967).
See also 1.2, p. 11; *6.1*, p. *72*; 18.4, p. 323; 22.8, p. 551.

W. Simonet et M. Paulus
*Sur l'origine et les conséquences de la coexistence des parois de domaines fixes et mobiles dans un cristal à anisotropie induite
(On the Origin and the Consequences of the Coexistence of Fixed and Mobile Domain Walls in a Crystal with Induced Anisotropy)
Vol. **22**, No. 2, p. K87–K90 (1967).

V. Petrescu und Gh. Maxim
*Die magnetische Diffusionsnachwirkung bei Mg-Fe-Ferriten
(The Magnetic Diffusion After-Effect of Mg-Fe Ferrites)
Vol. **22**, No. 2, p. K113–K116 (1967).

A. Bragiński
*High-Temperature After-Effect in Magnetite
Vol. **22**, No. 2, p. K127–K130 (1967).

F. Voigt
The Orientation Dependence of the Spin Wave Damping in Ferromagnetic Insulators with Spinel-Structure
Vol. **23**, No. 2, p. 607–612 (1967).
See also 13, p. 168; *19*, p. *330*.

N. Rezlescu
*Zum Procopiu-Effekt bei Ferriten
(On the Procopiu Effect in Ferrites)
Vol. **23**, No. 2, p. K117–K119 (1967).

Gh. Maxim und V. Petrescu
*Die Magnetische Diffusionsnachwirkung der Ni-Mg-Fe-Ferrite
(The Magnetic Diffusion After-Effect of Ni-Mg-Fe Ferrites)
Vol. **23**, No. 2, p. K125–K128 (1967).

A. Sawaoka, N. Kawai, and S. Kikuchi
*Change of the Magnetic Anisotropy Constant K_1 of Manganese Ferrite and Manganese Zinc Ferrite under Hydrostatic Pressure
Vol. **24**, No. 1, p. K83–K85 (1967).

J. M. Trooster and A. Dymanus
Mössbauer Effect in $Ga_{2-x}Fe_xO_3$ and Related Compounds
Vol. **24**, No. 2, p. 487–499 (1967).
See also *6.1*, p. *73*; 22.6, p. 540; 22.8, p. 552.

A. Marais and T. Merceron
*Orientation Superstructures and Their Effects in Spinel Ferrites
Vol. **24**, No. 2, p. 635–649 (1967).

F. F. Y. Wang and M. Kestigian
*Magnetic Properties of Al-Substituted Yttrium Orthoferrite
Vol. **25**, No. 1, p. 119–124 (1968).
See also 18.2, p. 310; 18.4, p. 323.

E. Jäger
Jahn-Teller Effect at Lattice Sites of Symmetry D_{3d}
Vol. **25**, No. 1, p. K43–K46 (1968).
See also 5, p. 53; *13*, p. *169*.

E. Wieser, V. A. Povitskii, E. F. Makarov, and K. Kleinstück
Mössbauer Investigations of Mg, Mn, and Mg-Mn Ferrites in External Magnetic Fields
Vol. **25**, No. 2, p. 607–611 (1968).
See also *6.1*, p. *73*.

Z. Smetana
The Influence of Mn^{3+} Ion Clustering on Ferromagnetic Resonance Linewidth of Manganese-Rich Mn-Fe Spinels
Vol. **25**, No. 2, p. K87–K89 (1968).
See also 10, p. 104; *19*, p. *332*.

F. Haberey and H. P. J. Wijn
Effect of Temperature on the Dielectric Relaxation in Polycrystalline Ferrites
Vol. **26**, No. 1, p. 231–240 (1968).
See also *14.4*, p. *253*.

M. Manzel and D. Linzen
*Magnetoelastic Interaction in BVC-Garnet
Vol. **26**, No. 1, p. K43–K45 (1968).

R. Krishnan
Preparation and Resonance Properties of CuO Doped YIG Single Crystals
Vol. **26**, No. 1, p. K47–K50 (1968).
See also *3.1*, p. *33*; *19*, p. *333*.

D. S. Kulgawczuk, Z. Obuszko, and A. Szytuła
*Susceptibility and Magnetization of β- and δ-FeOOH
Vol. **26**, No. 1, p. K83–K85 (1968).
See also 18.4, p. 324; 22.8, p. 553.

A. Szytuła, A. Burewicz, Ž. Dimitrijević, S. Kraśnicki, H. Rżany, J. Todorović, A. Wanic, and W. Wolski
Neutron Diffraction Studies of α-FeOOH
Vol. **26**, No. 2, p. 429–434 (1968).
See also *4*, p. *45*; 18.4, p. 324; 22.8, p. 553.

J. Macháčková
Ferromagnetic Resonance Measurements of Magnesium-Manganese Ferrite Films
Vol. **26**, No. 2, p. 435–442 (1968).
See also *19*, p. *333*.

L. Cser, I. Dézsi, I. Gladkih, L. Keszthelyi, D. Kulgawczuk, N. A. Eissa, and E. Sterk
Mössbauer Study of Hyperfine Fields in Mn-Zn Ferrites
Vol. **27**, No. 1, p. 131–138 (1968).
See also *6.1*, p. *73*.

C. Michalk
*Die magnetische Diffusionsnachwirkung in Ni–Fe-Ferriten mit geringen Co-Zusätzen (Magnetic After-Effect due to Diffusion in Ni–Fe Ferrites with Small Admixtures of Co)
Vol. **27**, No. 1, p. K51—K55 (1968).

S. A. Mironov, K. V. Shevlyagin, A. G. Gurevich, and B. M. Lebed
*Excitation of Magnetostatic, Magnetoelastic, and Acoustic Waves in Ca–Bi–V–Fe Garnet
Vol. **27**, No. 2, p. K85—K87 (1968).
See also 7, p. 78.

W. Haubenreisser
*Nonlinear Two-Magnon-One-Phonon Confluence Process in BVC Garnet
Vol. **27**, No. 2, p. K145—K147 (1968).
See also 6, p. 66; 13, p. 170.

I. Bunget
Note on Mössbauer Effect in $MnFe_2O_4$
Vol. **28**, No. 1, p. K39—K42 (1968).
See also *6.1*, p. *74*.

H. Gamari-Seale and P. Papamantellos
A Neutron Diffraction Study on Gallium-Substituted Magnetite
Vol. **29**, No. 1, p. 323—331 (1968).
See also *5*, p. *54*.

V. A. M. Brabers and P. Dekker
Infrared Spectra and Cation Distributions of Manganese Ferrites
Vol. **29**, No. 1, p. K73—K76 (1968).
See also *20.1*, p. *357*.

K. H. J. Buschow, J. F. Fast, and A. S. van der Goot
*Magnetic Properties of Some Co-Rich Erbium Cobalt Intermetallic Compounds
Vol. **29**, No. 2, p. 825—829 (1968).
See also 21.1, p. 407; 21.4, p. 428.

R. V. Pisarev, I. G. Siny, N. N. Nesterova, G. A. Smolensky, and P. P. Syrnikov
*Faraday Rotation and Magnetic Dichroism in $RbNi(Co)F_3$
Vol. **30**, No. 1, p. 367—372 (1968).
See also 13, p. 173; *20*, p. *342*; 22.5, p. 514; 22.8, p. 556.

Gh. Maxim und V. Petrescu
*Die bei Zimmertemperatur induzierte Anisotropie in Ni–Fe-Ferriten mit Co-Zusätzen (Room Temperature Induced Anisotropy in Ni–Fe Ferrites with Co Additions)
Vol. **30**, No. 1, p. K49—K51 (1968).

U. Heinecke
*A Contribution to the Calculation of Magnetization Curves and Minor Loops
Vol. **30**, No. 2, p. 551—556 (1968).

M. P. Petrov, S. A. Kizhaev, and G. A. Smolenskii
Ferromagnetic Resonance in $Ga_{0.85}Fe_{1.15}O_3$
Vol. **30**, No. 2, p. 871—874 (1968).
See also *19*, p. *336*.

V. S. Mandel, E. M. Smokotin, G. A. Petrakovskii, and B. M. Lebed
*Magneto-Elastic Coupling Coefficients of Ca–Bi–V–I-Garnet
Vol. **30**, No. 2, p. K111—K113 (1968).
See also 19, p. 336.

18.4 Antiferromagnetic Properties

Original Papers and Short Notes

D. Paccard, C. Schlenker, O. Massenet, R. Montmory, and A. Yelon
*A New Property of Ferromagnetic–Antiferromagnetic Coupling
Vol. **16**, No. 1, p. 301–311 (1966).
See also *18.2*, p. *299*; 21.1, p. 390; 21.1.1, p. 410.

R. E. Newnham, R. P. Santoro, P. F. Seal, and G. R. Stallings
*Antiferromagnetism in $Mn_3B_2O_6$, $Co_3B_2O_6$, and $Ni_3B_2O_6$
Vol. **16**, No. 1, p. K17–K19 (1966).
See also 22.8, p. 547.

Z. Šroubek
On the Theory of the Superexchange Interaction in Ionic Crystals: $KNiF_3$
Vol. **16**, No. 2, p. 405–412 (1966).
See also *3.2*, p. *34*; 13, p. 163; 22.5, p. 512; 22.8, p. 547.

K. H. J. Buschow and J. F. Fast
Magnetic and Structural Characteristics of Some Equiatomic Rare-Earth Germanides
Vol. **16**, No. 2, p. 467–473 (1966).
See also *3*, p. *26*; 5, p. 49; 18.1, p. 296; 18.2, p. 300; 18.3, p. 316; 22.7, p. 545.

S. K. Sidorov and A. V. Doroshenko
*On the Magnetic Structure of Some Alloys of Transition Metals
Vol. **16**, No. 2, p. 737–744 (1966).
See also *18.2*, p. *301*; 21.1, p. 391; 21.1.1, p. 411.

J. Przystawa
*On the Magnetization Directions in the Heisenberg Antiferromagnet — The Random-Phase Green's Function Method
Vol. **17**, No. 1, p. 155–162 (1966).

F. van der Woude
Mössbauer Effect in α-Fe_2O_3
Vol. **17**, No. 1, p. 417–432 (1966).
See also *6.1*, p. *71*; 22.6, p. 535.

A. Szytuła, A. Burewicz, K. Dýrek, A. Hrynkiewicz, D. Kulgawczuk, Z. Obuszko, H. Rżany, and A. Wanic
*Susceptibility Measurements of the Antiferromagnetic A-Goethite
Vol. **17**, No. 2, p. K195–K197 (1966).
See also 6.1, p. 71; 22, p. 441.

V. V. Eremenko, E. V. Matyushkin, and S. V. Petrov
Study of Energy Transfer from 3d to 4f Electrons in Antiferromagnetic $MnF_2:Eu^{3+}$ Crystals
Vol. **18**, No. 2, p. 683–686 (1966).
See also *13.4*, p. *204*; *20.3*, p. *364*; 22.5.3, p. 532.

G. A. Smolenskii, V. P. Zhuze, V. E. Adamyan, and G. M. Loginov
*Magnetic Properties of Ce, Pr, and Nd Monochalcogenides at 4.2 to 1300 °K
Vol. **18**, No. 2, p. 873–880 (1966).
See also *18.2*, p. *304*; 22, p. 442.

I. V. Ioffe
On the Theory of the Lattice Oscillations of Dielectric Crystals in an External Electric Field
Vol. **19**, No. 1, p. 51–56 (1967).
See also *6*, p. *58*; 18.2, p. 304; 19, p. 327; 22, p. 442.

R. De Batist, R. Gevers, and M. Verschueren
Magnon Contribution to the Low-Temperature Specific Heat of UO_2
Vol. **19**, No. 1, p. 77–88 (1967).
See also *8*, p. *80*; 22.6, p. 536.

L. Zsoldos
*Lattice Parameter Change of FeRh Alloys due to Antiferromagnetic–Ferromagnetic Transformation
Vol. **20**, No. 1, p. K25–K28 (1967).
See also 1.2, p. 10; 5, p. 51; *18.2*, p. *306*; 22.1.1, p. 416.

J. Przystawa and W. Suski
*Magnetic Ordering in Uranium Compounds with Crystal Structures of Anti-Cu_2Sb and PbFCl Types. Molecular Field Approximation
Vol. **20**, No. 2, p. 451–459 (1967).
See also 3.2, p. 35; *18.2*, p. *306*; 22, p. 444.

V. A. Bokov, N. A. Grigoryan, and M. F. Bryzhina
X-Ray Diffraction and Magnetic Studies of Solid Solutions $Bi_{1-x}Ca_xMnO_3$
Vol. **20**, No. 2, p. 745–754 (1967).
See also *1*, p. *7*; 1.2, p. 10; 18.2, p. 306; 18.3, p. 317; 22.8, p. 549.

V. M. Judin and A. B. Sherman
*Magnetic Properties of $NaNiF_3$
Vol. **20**, No. 2, p. 759–766 (1967).
See also 22.5, p. 512; 22.8, p. 550.

U. Gonser and R. W. Grant
Determination of Spin Directions and Electric Field Gradient Axes in Vivianite by Polarized Reçoil-Free γ-Rays
Vol. **21**, No. 1, p. 331–342 (1967).
See also *6.1*, p. *72*; 13, p. 165; 18.1, p. 296; 22.8, p. 550.

A. Burewicz, D. Kulgawczuk, and A. Szytuła
Internal Magnetic Field in α-FeOOD
Vol. **21**, No. 1, p. K85–K86 (1967).
See also *6.1*, p. *72*; 22.8, p. 550.

K. H. J. Buschow and J. F. Fast
Crystal Structure and Magnetic Properties of Some Rare Earth Germanides
Vol. **21**, No. 2, p. 593–600 (1967).
See also *5*, p. *51*; 18.1, p. 296; 18.3, p. 318; 22.7, p. 545.

S. I. Masharov
Electrical Resistivity in Antiferromagnetic Alloys
Vol. **21**, No. 2, p. 747–753 (1967).
See also *14.1*, p. *223*; 21, p. 379.

I. Dézsi, Gy. Erlaki, and L. Keszthelyi
Mössbauer Effect Studies on the $x(Rh_2O_3)1-x(Fe_2O_3)$ System
Vol. **21**, No. 2, p. K121–K125 (1967).
See also *6.1*, p. *72*; 22.6, p. 538; 22.8, p. 551.

U. Bertelsen, J. M. Knudsen, and H. Krogh
Mössbauer Effect in FeF_3
Vol. **22**, No. 1, p. 59–64 (1967).
See also *6.1*, p. *72*; 22.5, p. 513.

K. Tompa, F. Tóth, and G. Grüner
Susceptibility of MnO Measured by the NMR Method
Vol. **22**, No. 1, p. K11–K13 (1967).
See also 18.1, p. 296; *19*, p. *329*; 22.6, p. 538.

J. Leciejewicz, R. Troć, A. Murasik, and A. Zygmunt
*Neutron-Diffraction Study of Antiferromagnetism in USb_2 and UBi_2
Vol. **22**, No. 2, p. 517—526 (1967).
See also 5, p. 52; 22, p. 447.

I. Dézsi, L. Keszthelyi, D. Kulgawczuk, B. Molnár, and N. A. Eissa
Mössbauer Study of β- and δ-FeOOH and Their Disintegration Products
Vol. **22**, No. 2, p. 617—629 (1967).
See also 1.2, p. 11; *6.1*, p. *72*; 18.3, p. 318; 22.8, p. 551.

S. Arajs
*Magnetic Susceptibility of Face-Centered Cubic Iron–Rhenium Alloys
Vol. **22**, No. 2, p. 737—740 (1967).
See also 14.1, p. 223; 21.1.1, p. 418.

F. F. Y. Wang and M. Kestigian
*Magnetic Susceptibilities of Sodium Neodymium Tungstate and Sodium Terbium Tungstate
Vol. **23**, No. 1, p. 289—294 (1967).
See also 18.1, p. 296; 22.8, p. 552.

W. Brodkorb
Green's Function Method in the Theory of Antiferromagnetic Thin Films with Body Centred Cubic Crystal Structure
Vol. **23**, No. 1, p. K35—K38 (1967).

R. Troć, A. Murasik, A. Zygmunt, and J. Leciejewicz
*The Magnetic Ordering in Uranium Monoarsenide
Vol. **23**, No. 2, p. K123—K124 (1967).
See also 22, p. 448.

J. Przystawa
*The Néel Temperatures and Stability of Magnetic Phases in the Disturbed Body-Centered Tetragonal Lattice. Uranium Compounds
Vol. **24**, No. 1, p. 313—322 (1967).
See also *18.2*, p. *309*; 22, p. 448; 22.8, p. 552.

V. V. Kelarev, S. K. Sidorov, V. V. Klyushin, and R. Z. Abdulov
*Neutron-Diffraction Study of Antiferro-Ferromagnetic Transition in a System of Ordered $Fe(Pd_xPt_{1-x})_3$ Alloys
Vol. **24**, No. 1, p. 385—390 (1967).
See also *18.2*, p. *309*; 21.1.1, p. 419.

A. J. Springthorpe
*The Rhombohedral Distortion in NiO
Vol. **24**, No. 1, p. K3—K4 (1967).
See also 4, p. 43; 22.6, p. 539.

F. F. Y. Wang and M. Kerstigian
Magnetic Properties of Al-Substituted Yttrium Orthoferrite
Vol. **25**, No. 1, p. 119—124 (1968).
See also 18.2, p. 310; *18.3*, p. *319*.

A. I. Mitsek and N. P. Kolmakova
*On the Theory of Weak Ferromagnetics
Vol. **25**, No. 2, p. 787—799 (1968).
See also *18.2*, p. *310*; 19, p. 332.

R. Fatehally, N. P. Sastry, and R. Nagarajan
Exchange and Spin-Orbit Coupling in $KFeF_3$
Vol. **26**, No. 1, p. 91—96 (1968).
See also *13*, p. *169*; 22.5, p. 513; 22.8, p. 553.

F. van der Woude
On the Interpretation of Broadened and Shifted Lines in Mössbauer Spectra
Vol. **26**, No. 1, p. K37—K38 (1968).
See also *6.1*, p. *73*; 22.5, p. 513.

D. S. Kulgawczuk, Z. Obuszko, and A. Szytuła
Susceptibility and Magnetization of β- and δ-FeOOH
Vol. **26**, No. 1, p. K83—K85 (1968).
See also *18.3*, p. *319*; 22.8, p. 553.

A. Szytuła, A. Burewicz, Ž. Dimitrijević, S. Kraśnicki, H. Rżany, J. Todorović, A. Wanic, and W. Wolski
Neutron Diffraction Studies of α-FeOOH
Vol. **26**, No. 2, p. 429—434 (1968).
See also *4*, p. *45*; 18.3, p. 319; 22.8, p. 553.

B. Heinrich, J. Zítková, and J. Kaczér
The Temperature Dependence of the Paramagnetic Susceptibility of Ni-I Boracite
Vol. **26**, No. 2, p. 443—446 (1968).
See also *18.1*, p. *297*; 22.8, p. 553.

J. Villain
*Critical Exponents in the Dynamics of an Isotropic Magnetic System near Its Transition Temperature
Vol. **26**, No. 2, p. 501—507 (1968).
See also *18.2*, p. *311*.

W. Bitterlich and R. J. Jelitto
Computational Evidence for the Transition Point of a Square Ising Lattice with First and Second Neighbour Interactions
Vol. **28**, No. 1, p. 365—372 (1968).
See also *1.2*, p. *12*; 18.2, p. 312.

C. Schlenker
*Couplage ferro-antiferromagnétique et traînage magnétique dans des couches minces multiples Co-CoO et Ni-NiO
(Ferro-Antiferromagnetic Coupling and Magnetic After-Effect in Multiple Co-CoO and Ni-NiO Films)
Vol. **28**, No. 2, p. 507—517 (1968).
See also *18.2*, p. *313*; 21.1, p. 406; 22.6, p. 542.

V. A. Bokov, N. A. Grigoryan, M. F. Bryzhina, and V. V. Tikhonov
*Effect of Lattice Distortions on the Magnetic Behaviour of Perovskite-Type Manganites
Vol. **28**, No. 2, p. 835—847 (1968).
See also 5, p. 54; *18.2*, p. *313*; 22.8, p. 554.

M. A. Krivoglaz and L. B. Kvashnina
On the Spin Broadening of Mössbauer Lines of Impurities in Ferromagnetics and of Atoms in Antiferromagnetics
Vol. **29**, No. 1, p. 61—73 (1968).
See also *6.1*, p. *74*; 13, p. 172; 18.2, p. 314.

K. Schröder and N. P. Baum
Absolute Thermoelectric Power and Electrical Resistivity of Chromium-Rich Chromium-Nickel Alloys
Vol. **29**, No. 1, p. 107—114 (1968).
See also 13.1, p. 188; *14.1*, p. *226*; *15*, p. *274*; 21.1, p. 407.

W. Brückner, R. Perthel, K. Kleinstück, and G. E. R. Schulze
*Magnetic Properties of $ZrFe_2$ and $TiFe_2$ within Their Homogeneity Range
Vol. **29**, No. 1, p. 211–216 (1968).
See also *18.2*, p. *314*; 21.1.1, p. 422.

V. Beckmann, W. Bruckner, W. Fuchs, G. Ritter, and H. Wegener
*The Measurement of the Anisotropy Constant of Antiferromagnetic $FePO_4$ by Means of the Mössbauer Effect
Vol. **29**, No. 2, p. 781–791 (1968).
See also 6.1, p. 74; 22.8, p. 555.

R. C. Thiel and C. B. van den Berg
Temperature Dependence of Hyperfine Interactions in Near-Stoichiometric FeS. (I): Experiment
Vol. **29**, No. 2, p. 837–846 (1968).
See also 1.2, p. 13; *6.1*, p. *74*; 22, p. 454.

Yu. G. Litvinenko, V. V. Eremenko, and T. I. Garber
*Antiferromagnetic Ordering Effect on the Light Absorption Spectrum by Crystalline Oxygen
Vol. **30**, No. 1, p. 49–60 (1968).
See also 13.1, p. 189; 20.1, p. 358; 23, p. 561.

A. Murasik, W. Suski, R. Troć, and J. Leciejewicz
*The Magnetic Properties of Uranium Oxyselenide
Vol. **30**, No. 1, p. 61–66 (1968).
See also 22.8, p. 556.

J. Leciejewicz, A. Murasik, and R. Troć
*The Antiferromagnetic Ordering in Uranium Monoarsenide and Monoantimonide
Vol. **30**, No. 1, p. 157–162 (1968).
See also 4, p. 48; 22, p. 454.

J. Przystawa
*Magnetic Symmetry of Uranium Compounds of UX_2 Type
Vol. **30**, No. 2, p. K115–K117 (1968).
See also 3, p. 28; 22, p. 455.

19. Magnetic Resonance in Solids

Review Articles

D. P. Morgan
Helicon Waves in Solids
Vol. **24**, No. 1, p. 9–36 (1967).
See also 6, p. 56; *13*, p. *162*; 14.1, p. 220; 14.3, p. 231; *20*, p. *337*; 21, p. 373; 21.1, p. 390; 21.2, p. 424; 21.6, p. 430; 22.1.1, p. 458; 22.2.1, p. 482; 22.2.3, p. 489.

J. Mertsching
Theory of Electromagnetic Waves in Metals and Their Interaction with Ultrasonic Waves (II)
Vol. **26**, No. 1, p. 9–64 (1968).
See also *7*, p. *76*; 13, p. 162; 20, p. 337; 21, p. 373.

A. Seeger and K. P. Chik
Diffusion Mechanisms and Point Defects in Silicon and Germanium
Vol. **29**, No. 2, p. 455–542 (1968).
See also *9*, p. *87*; *10*, p. *93*; 11, p. 145; 13.4, p. 200; 22.1.1, p. 458; 22.1.2, p. 468.

Original Papers and Short Notes

W. Dreybrodt and D. Silber
*EPR Measurements on Diatomic Halogen Centers in Alkali Halides
Vol. **16**, No. 1, p. 215–223 (1966).
See also 10.2, p. 137; 13.4, p. 201; 22.5.2, p. 518.

B. Heinrich and Z. Frait
*Temperature Dependence of the FMR Linewidth of Iron Single-Crystal Platelets
Vol. **16**, No. 1, p. K11–K14 (1966).
See also *18.2*, p. *300*; 21.1.1, p. 410.

G. B. Abdullaev, N. I. Ibragimov, Sh. V. Mamedov, and T. Ch. Zhuvarly
*ESR Study of the Behaviour of Oxygen in Selenium
Vol. **16**, No. 2, p. K113–K116 (1966).
See also 10, p. 94; 22.1.3, p. 476.

S. S. Starobinets, B. M. Lebed, and A. G. Gurevich
*Parametric Magneto-Acoustic Resonance in Ca-Bi-V Garnet
Vol. **17**, No. 1, p. K7–K9 (1966).
See also 18.3, p. 316.

G. Völkel und W. Windsch
*EPR-Untersuchungen an kupferdotierten Seignettesalz-Einkristallen
(EPR Studies of Copper-Doped Rochelle-Salt Single Crystals)
Vol. **17**, No. 1, p. K75–K78 (1966).
See also 10, p. 95; 14.4.2, p. 266.

M. Höhne und M. Stasiw
*ESR in Se- und Te-dotierten AgBr-Einkristallen
(ESR in Se- and Te-Doped AgBr Single Crystals)
Vol. **17**, No. 1, p. K119–K121 (1966).
See also 22.5.1, p. 515.

B. A. Greenberg
*The Effect of Dislocation Dipoles on the Shape of the Nuclear Magnetic Resonance Line
Vol. **17**, No. 2, p. 673–681 (1966).
See also 10, p. 95; 22, p. 441.

G. Schoffa
*Die Elektronenspinresonanz von strahlenerzeugten Radikalen in Einkristallen des 2,6-Butylhydroxytoluols
(Electron Spin Resonance of Radiation-Produced Radicals in 2,6-Butylhydroxytoluene Single Crystals)
Vol. **17**, No. 2, p. 683–689 (1966).
See also 11, p. 146; 22.9, p. 558.

A. G. Lesnik and L. M. Sandler
Effect of Tensile Stress on the Magnetic Properties of Permalloy Films
Vol. **17**, No. 2, p. 753–760 (1966).
See also 10.1, p. 117; *18.2*, p. *303*; 21.1.1, p. 413.

S. Wartewig und W. Windsch
*EPR-Untersuchungen an chromdotierten Triglyzinsulfat-Einkristallen
(EPR Studies of Chromium Doped Triglycine Sulphate Single Crystals)
Vol. **17**, No. 2, p. K221–K224 (1966).
See also 14.4.2, p. 266.

A. Räuber and J. Schneider
*Localized $^2S_1/_2$-State Centres in ZnS
Vol. **18**, No. 1, p. 125–132 (1966).
See also 13.4, p. 203; 22.4.1, p. 498.

K. Tompa
*Nuclear Magnetic Resonance of ^{63}Cu and ^{65}Cu in Copper
Vol. **18**, No. 1, p. 391–400 (1966).
See also 21.1, p. 393.

R. Krishnan
Influence of Tetrahedral Ni^{2+} Ions on the Magnetic Properties of YIG
Vol. **18**, No. 1, p. K53–K56 (1966).
See also *18.3*, p. *317*.

D. Schmid
*Der *g*-Faktor von F-Zentren in Alkalihalogenid-Kristallen
(The *g*-Factor of F-Centres in Alkali Halide Crystals)
Vol. **18**, No. 2, p. 653–666 (1966).
See also 10.2, p. 138; 22.5.2, p. 520.

K. K. Shvarts, A. Ya. Vitol, A. V. Podin, D. O. Kalnin, and Yu. A. Ekmanis
Radiation Effects in Pile-Irradiated LiF Crystals
Vol. **18**, No. 2, p. 897–909 (1966).
See also 10.2, p. 138 ; *11*, p. *147*; 20.1, p. 346; 22.5.2, p. 520.

I. V. Ioffe
On the Theory of the Lattice Oscillations of Dielectric Crystals in an External Electric Field
Vol. **19**, No. 1, p. 51–56 (1967).
See also *6*, p. *58*; 18.2, p. 304; 18.4, p. 321; 22, p. 442.

W. van der Lugt and S. B. van der Molen
*Nuclear Magnetic Resonance in Liquid Gallium Alloys
Vol. **19**, No. 1, p. 327–331 (1967).
See also 2, p. 22; 21, p. 377.

F. BORSA, R. G. BARNES, and R. A. REESE
*Nuclear Magnetic Resonance and Mössbauer Effect Study of Sn^{119} in Rare Earth–Tin Intermetallic Compounds
Vol. 19, No. 1, p. 359–371 (1967).
See also *6.1*, p. *71*; 22.7, p. 545.

R. BÖTTCHER, W. WINDSCH, and W. LÜDKE
*EPR Investigations on Mn^{2+}-Doped LiF Single Crystals
Vol. 20, No. 1, p. 121–126 (1967).
See also 9, p. 88; 13.4, p. 205; 22.5.2, p. 521.

J. R. PILBROW and J. M. SPAETH
*ESR Studies of Cu^{2+} in NH_4Cl Single Crystals between 4.2 and 453 °K. (I): Spin-Hamiltonian Parameters and Lattice Constant for NH_4Cl
Vol. 20, No. 1, p. 225–235 (1967).
See also 22, p. 444; 22.5, p. 512.

J. R. PILBROW and J. M. SPAETH
*ESR Studies of Cu^{2+} in NH_4Cl Single Crystals between 4.2 and 453 °K. (II): Theoretical Analysis: Vibrational Admixtures, Spin Polarization, and $\langle r^{-3} \rangle$ for Cu^{2+} Ions
Vol. 20, No. 1, p. 237–248 (1967).
See also 22, p. 444; 22.5, p. 512.

W. DREYBRODT and D. SILBER
*Electron Spin Resonance of Tl^{++} Centres in KCl Crystals
Vol. 20, No. 1, p. 337–346 (1967).
See also 13.4, p. 205; 22.5.2, p. 521.

T. N. CASSELMAN and H. N. SPECTOR
*Interaction of Optical Phonons with Spin Waves in Ionic Crystals
Vol. 20, No. 2, p. 605–612 (1967).
See also *6*, p. *60*; 13, p. 165; 18, p. 292; 22, p. 444.

K. ŽDÁNSKÝ, H. AREND, and F. KUBEC
*Electron Spin Resonance of Co^{2+} Ions in Barium Titanate
Vol. 20, No. 2, p. 653–656 (1967).
See also 14.4.2, p. 267.

M. HÖHNE and M. STASIW
*Te_2^{3-} and Se_2^{3-} Centres in AgBr. (I): The *g*-, Superhyperfine, and Hyperfine Tensors
Vol. 20, No. 2, p. 657–666 (1967).
See also 10.2, p. 139; 13.4, p. 206; 22.5.1, p. 515.

M. HÖHNE and M. STASIW
*Te_2^{3-} and Se_2^{3-} Centres in AgBr. (II): Electronic Structure
Vol. 20, No. 2, p. 667–673 (1967).
See also 10.2, p. 139; 13.4, p. 206; 22.5.1, p. 515.

M. JAROŠ and J. ZÍTKOVÁ
*Remark on the Theory of the ESR Spectrum of the d^1-Configuration in Rutile (TiO_2)
Vol. 20, No. 2, p. K99–K101 (1967).
See also 13.4, p. 206; 22.6, p. 537.

B. A. GREENBERG
*Effect of Segregation at Dislocations on the Shape of the NMR Line
Vol. 20, No. 2, p. K103–K106 (1967).
See also 4, p. 42; 10, p. 99.

W. Dreybrodt
*Temperature Dependence of Hyperfine Structure Tensors of Diatomic Halogen Centres in Alkali Halide Crystals
Vol. **21**, No. 1, p. 99–112 (1967).
See also 13.4, p. 206; 22.5.2, p. 522.

A. Weiss and K. Zohner
*Nuclear Quadrupole Interaction of ^{23}Na and ^{11}B and the Crystal Structure of $NaBF_4$
Vol. **21**, No. 1, p. 257–270 (1967).
See also 3.2, p. 35; 5, p. 51; 22.5, p. 513; 22.8, p. 550.

R. Krishnan
Ionic Redistribution and Anisotropy in Heat Treated NiO-Doped YIG Single Crystals
Vol. **21**, No. 1, p. K5–K7 (1967).
See also *10*, p. *100*; 18.3, p. 317.

V. K. Bashenov, S. P. Fedotov, and V. A. Presnov
On Deep Acceptors in Iron-Doped Gallium Arsenide
Vol. **21**, No. 1, p. K91–K93 (1967).
See also *13.4*, p. *207*; 22.2.1, p. 483.

R. Herrmann
*Oszillationen der Oberflächenimpedanz von Wolfram in schwachen Magnetfeldern (Oscillations of the Surface Impedance of Tungsten in Weak Magnetic Fields)
Vol. **21**, No. 2, p. 703–707 (1967).
See also 13.1, p. 178; *14.1*, p. *223*; 21, p. 379.

D. Demco and F. Constantinescu
*Dispersion Relations between the Relaxation Times and the Shift of the Resonance Frequency
Vol. **21**, No. 2, p. K151–K154 (1967).

A. Ya. Blank and E. A. Kaner
*Waves with Discrete Spectrum in Metals in the Vicinity of the Cyclotron Resonance Frequency
Vol. **22**, No. 1, p. 47–58 (1967).
See also 13.1, p. 179; 21, p. 380.

H. W. den Hartog and J. Arends
F-Centers in SrF_2
Vol. **22**, No. 1, p. 131–140 (1967).
See also *10.2*, p. *139*; 13.4, p. 208; 22.5.3, p. 533.

E. K. Kaner and V. L. Falko
*Magnetoacoustic Resonance Effects in Metals in Inclined Magnetic Fields
Vol. **22**, No. 1, p. 319–331 (1967).
See also 7, p. 77; 13.1, p. 179; 21, p. 380.

K. Tompa, F. Tóth, and G. Grüner
*Susceptibility of MnO Measured by the NMR Method
Vol. **22**, No. 1, p. K11–K13 (1967).
See also 18.1, p. 296; 18.4, p. 322; 22.6, p. 538.

H. Puszkarski
Surface Spin Waves in Hexagonal Cobalt Thin Films
Vol. **22**, No. 2, p. 355–359 (1967).
See also *13*, p. *166*; 13.3, p. 197; 21.1, p. 399.

A. Dall'Olio, G. Dascola, and V. Varacca
*EPR Evidence of Exchange-Coupled Ion Pairs and of Isolated Ions in Some Chloro-substituted Cupric Acetates
Vol. **22**, No. 2, p. 365—370 (1967).
See also 13.4, p. 208; 22.9, p. 559.

J. O. Henningsen
*Cyclotron Resonance in Zinc
Vol. **22**, No. 2, p. 441—452 (1967).
See also *13.1*, p. *180*; 21.1, p. 399.

H.-G. Kahle, H. Kalbfleisch, and E. Stein
Paramagnetic Relaxation and Observation of a Phonon Bottleneck in Terbium Etyl Sulfate
Vol. **22**, No. 2, p. 537—540 (1967).
See also *6*, p. *62*; *18.1*, p. *296*; 22.8, p. 551.

O. P. Revokatov and U. Eichhoff
*NMR Spectrum and Line Shape of Two Interacting Two-Spin Systems in Crystals
Vol. **22**, No. 2, p. K91—K94 (1967).

J. Arends and J. F. Verwey
*ESR on UV Irradiated Lead Halides at 80 °K
Vol. **23**, No. 1, p. 137—145 (1967).
See also 4, p. 43; 10.2, p. 140; 22.5, p. 513.

H. Böttger
*Untersuchung des Jahn-Teller-Effektes an Ionen mit Spin 1
(Study of the Jahn-Teller Effect of Ions with Spin 1)
Vol. **23**, No. 1, p. 325—333 (1967).
See also 13.4, p. 209.

G. Albrecht and G. Weber
*Hyperfine Fields Induced by Spin-Orbit-Coupling in Binary Intermetallic Compounds
Vol. **23**, No. 1, p. K17—K20 (1967).
See also 18.1, p. 297; 21.1, p. 400.

W. van der Lugt and J. S. Knol
*The Knight Shift in Potassium from 23 to 123 °C
Vol. **23**, No. 1, p. K83—K86 (1967).
See also 21.2, p. 424.

B. Heinrich, D. Fraitová, and V. Kamberský
*The Influence of s-d Exchange on Relaxation of Magnons in Metals
Vol. **23**, No. 2, p. 501—507 (1967).
See also *13*, p. *167*; 21.1.1, p. 419.

F. Voigt
*The Orientation Dependence of the Spin Wave Damping in Ferromagnetic Insulators with Spinel-Structure
Vol. **23**, No. 2, p. 607—612 (1967).
See also 13, p. 168; 18.3, p. 318.

D. Haarer, D. Schmid und H. C. Wolf
*Elektronenspin-Resonanz von Triplett-Exzitonen in Anthrazen
(Electron Spin Resonance of Triplet Excitons in Anthracene)
Vol. **23**, No. 2, p. 633—638 (1967).
See also *13.2*, p. *193*; 22.9, p. 559.

H. W. den Hartog and J. Arends
Electronic Structure of F-Centers in Alkaline EarthFluorides
Vol. **23**, No. 2, p. 713—719 (1967).
See also 10.2, p. 140; *13.4*, p. *210*; 22.5.3, p. 533.

J. Loos
*The Nonlinear Excitation of Strongly Coupled Spin Waves
Vol. **23**, No. 2, p. 721–728 (1967).
See also 13, p. 168.

J. Stankowski, S. Waplak, B. Sczaniecki, and A. Dezor
*Temperature Anomaly of EPR Spectra of Paramagnetic Ions in Ferroelectrics
Vol. **23**, No. 2, p. K159–K162 (1967).
See also 14.4.2, p. 267.

H. Böttger
*Aufspaltung der ESR-Linie durch Spin-Phonon-Wechselwirkung an Ionen mit Spin 1
(Splitting of ESR Lines by Spin–Phonon Interaction for Ions with Spin 1)
Vol. **24**, No. 1, p. 65–75 (1967).
See also 6, p. 63; 13.4, p. 210; 22.6, p. 539.

R. Dupree and B. W. Holland
*The Range of g-Factors and the Breakdown of Motional Narrowing in Conduction Electron Spin Resonance
Vol. **24**, No. 1, p. 275–279 (1967).
See also 21, p. 382; 21.1, p. 401; 21.2, p. 425.

J. Kołodziejczak
*Magnetic Resonance in Higher Harmonics Generated by Free Carriers in Semiconductors
Vol. **24**, No. 1, p. 323–329 (1967).
See also 13.1, p. 183; 14.3, p. 238; 22, p. 448.

I. V. Ioffe and A. L. Kazakov
*On the Theory of Ferroácoustic Resonance
Vol. **24**, No. 1, p. K9–K10 (1967).
See also 7, p. 77.

H. Götz, H. Schulze, and K. Werner
*Overhauser Effect in Semiconducting Crystalline TCNQ-Complexes
Vol. **24**, No. 1, p. K95–K96 (1967).
See also 22.9, p. 560.

J. Stankowski
*EPR Study of Cu^{2+} Ion in Triglycine Fluoroberyllate
Vol. **24**, No. 2, p. 451–456 (1967).
See also 10, p. 103; 14.4.2, p. 268.

R. Dupree, C. T. Forwood, and M. J. A. Smith
*Conduction Electron Spin Resonance in Small Particles of Gold
Vol. **24**, No. 2, p. 525–530 (1967).
See also 21.6, p. 435.

M. Höhne and M. Stasiw
*ESR Studies of Different Se_2^{3-} Centres in Silver Halides
Vol. **24**, No. 2, p. 591–604 (1967).
See also 10, p. 103; 22.5.1, p. 516.

H. W. den Hartog
Fourth Shell Isotropic H.F.S.-Constants of F-Centers in CaF_2
Vol. **24**, No. 2, p. K117–K120 (1967).
See also 10.2, p. 141; *13.4*, p. *211*; 22.5.3, p. 533.

R. Herrmann
Cyclotron Resonance Study of the Fermi Surface of Molybdenum
Vol. **25**, No. 1, p. 427–435 (1968).
See also *13.1*, p. *183*; 21, p. 384.

L. K. KEYS
*ESR of Ti_2O_3
Vol. **25**, No. 1, p. K5—K6 (1968).
See also 22.6, p. 540.

G. VÖLKEL
*Die Beeinflussung der EPR-Spektren von Cu^{2+}-dotierten Seignettesalz-Einkristallen durch ein äußeres elektrisches Feld
(Effect of an External Electric Field on the EPR Spectra of Cu^{2+}-Doped Rochelle-Salt Single Crystals)
Vol. **25**, No. 1, p. K35—K38 (1968).
See also 14.4.2, p. 268.

K. N. SHRIVASTAVA
*Temperature Dependence of Paramagnetic Resonance Transition Probabilities: Mn^{2+} in MgO
Vol. **25**, No. 1, p. K47—K50 (1968).
See also 10, p. 104; 22.6, p. 540.

V. K. BASHENOV, V. A. PRESNOV, and S. P. FEDOTOV
*Some Remarks on the Antishielding Effect in III–V Compounds
Vol. **25**, No. 1, p. K51—K54 (1968).
See also 22.2, p. 481.

M. HÖHNE and M. STASIW
*Detection of Ag^{2+} in Doped AgCl Crystals by ESR
Vol. **25**, No. 1, p. K55—K57 (1968).
See also 10.2, p. 141; 22.5.1, p. 516.

Y. KAZUMATA
Atomic ^3H Center in Neutron-Irradiated LiF
Vol. **25**, No. 2, p. 563—566 (1968).
See also *10.2*, p. *142*; 11, p. 151; 22.5.2, p. 525.

R. HERRMANN and KARIN HERRMANN
Cyclotron Resonance in Tellurium
Vol. **25**, No. 2, p. 655—659 (1968).
See also *13.1*, p. *184*; 22.1.3, p. 478.

R. HERRMANN
Fermi Surface of Tungsten and Molybdenum
Vol. **25**, No. 2, p. 661—666 (1968).
See also *13.1*, p. *184*; 21, p. 384.

A. I. MITSEK and N. P. KOLMAKOVA
On the Theory of Weak Ferromagnetics
Vol. **25**, No. 2, p. 787—799 (1968).
See also *18.2*, p. *310*; *18.4*, p. *323*.

P. FELTHAM and R. GHOSH
On Colour Centres in Plastically Deformed Alkaline Earth Fluorides
Vol. **25**, No. 2, p. K71—K73 (1968).
See also 10, p. 104; *10.2*, p. *142*; 22.5.3, p. 533.

Z. SMETANA
*The Influence of Mn^{3+} Ion Clustering on Ferromagnetic Resonance Linewidth of Manganese-Rich Mn–Fe Spinels
Vol. **25**, No. 2, p. K87—K89 (1968).
See also 10, p. 104; 18.3, p. 319.

R. G. Bessent and P. Feltham
Formation of F-Centres in Alkaline Earth Oxides by Plastic Deformation
Vol. **25**, No. 2, p. K107—K110 (1968).
See also 10, p. 105; *10.2*, p. *142*; 22.6, p. 541.

H. W. den Hartog
F-Centers in BaF_2
Vol. **25**, No. 2, p. K111—K114 (1968).
See also *10.2*, p. *142*; 13.4, p. 211; 22.5.3, p. 533.

R. A. Zhitnikov, V. B. Koltsov, and N. I. Melnikov
*Paramagnetic Resonance of Copper Atoms Stabilized in Alkali-Halide Crystals
Vol. **26**, No. 1, p. 371—380 (1968).
See also 10, p. 105; 22.5.2, p. 526.

J. Dziesiaty and R. Böttcher
*Electron Paramagnetic Resonance of Ti^{2+} Ions in Cadmium Sulfide Single Crystals
Vol. **26**, No. 1, p. K21—K24 (1968).
See also 22.4.1, p. 502.

R. Krishnan
*Preparation and Resonance Properties of CuO Doped YIG Single Crystals
Vol. **26**, No. 1, p. K47—K50 (1968).
See also *3.1*, p. *33*; 18.3, p. 319.

J. Macháčková
*Ferromagnetic Resonance Measurements of Magnesium-Manganese Ferrite Films
Vol. **26**, No. 2, p. 435—442 (1968).
See also 18.3, p. 319.

H. Böttger
Jahn-Teller Splitting of an ESR Line
Vol. **26**, No. 2, p. 681—692 (1968).
See also 6, p. 65; *13.4*, p. *212*; 22.6, p. 541.

Y. Kazumata
*Self-Trapped Holes in Neutron Irradiated CaF_2 and BaF_2 Crystals
Vol. **26**, No. 2, p. K119—K122 (1968).
See also 10.2, p. 142; 11, p. 152; 22.5.3, p. 534.

H. J. Fischbeck and J. Mertsching
*The Theory of Low-Field Oscillations of the Surface Impedance in Metals
Vol. **27**, No. 1, p. 345—357 (1968).
See also 13.1, p. 186; 14.1, p. 225; 21, p. 386.

M. Jaroš
On the Covalency in Transition-Metal Impurity Covalent Complexes
Vol. **27**, No. 1, p. K21—K24 (1968).
See also *3.2*, p. *36*; 22.4.3, p. 510.

D. Schwarz
*ESR Study of Cu^{2+} in AgCl Crystals Additionally Doped with Se or S
Vol. **27**, No. 1, p. K77—K80 (1968).
See also 10, p. 107; 13.4, p. 213; 22.5.1, p. 516.

S. D. McLaughlan and H. W. Evans
Production of Colloidal Calcium by Electron Irradiation of CaF_2 Crystals
Vol. **27**, No. 2, p. 695—700 (1968).
See also 10.2, p. 142; *11*, p. *152*; 20.1, p. 354; 22.5.3, p. 534.

A. P. Zhukov, I. S. Rez, V. I. Pakhomov, and G. K. Semin
*Temperature Dependences of the Nuclear Quadrupole Resonance Spectra of As^{75} in KH_2AsO_4, RbH_2AsO_4, CsH_2AsO_4, $NH_4H_2AsO_4$, and of Their Deuterated Analogues
Vol. 27, No. 2, p. K129—K132 (1968).
See also 14.4.2, p. 269.

C. Marti, R. Romestain, and R. Visocekas
*Forbidden Transitions ($\Delta M = 2$) Induced by an Axial Crystal Field in Electronic Paramagnetic Resonance
Vol. 28, No. 1, p. 97—103 (1968).
See also 13, p. 171; 22.4.1, p. 503.

B. W. Holland
*Pseudopotentials and the Knight Shift
Vol. 28, No. 1, p. 121—125 (1968).
See also 13, p. 171; 21, p. 386; 21.2, p. 425.

L. Kratěna, K. Ždánský, and V. Šik
*Paramagnetic Resonance Spectrum of Gadolinium in $CdWO_4$ Single Crystals
Vol. 28, No. 1, p. 175—179 (1968).
See also 22.8, p. 554.

K. Ždánský
*Vibrational Resonance Mode and the Temperature Dependence of Hyperfine Coupling of Mn^{2+} in BaO
Vol. 28, No. 1, p. 181—186 (1968).
See also 10, p. 108; 13.4, p. 214; 22.6, p. 542.

M. Owner-Petersen and M. R. Samuelsen
*Cyclotron Resonance of Holes in Si
Vol. 28, No. 1, p. 211—222 (1968).
See also 13.1, p. 187; 22.1.2, p. 474.

M. Höhne and M. Stasiw
*ESR Detection of Self-Trapped Holes in AgCl
Vol. 28, No. 1, p. 247—253 (1968).
See also 10.2, p. 143; 20.1, p. 355; 22.5.1, p. 517.

Yu. A. Sherstkov, V. I. Nepsha, V. A. Vagenin, A. E. Nikiforov, and A. I. Krotkii
*Effect of Applied Electric Field on EPR Spectra of Crystal Impurity Centres
Vol. 28, No. 1, p. 269—275 (1968).
See also 10, p. 109; 13, p. 171; 22.6, p. 542.

K. E. Mann and L. V. Holroyd
*The EPR Spectra of Gd^{3+} in Single Crystals of BaO
Vol. 28, No. 1, p. K27—K29 (1968).
See also 22.6, p. 542.

H. le Gall
*Quantum-Mechanical Treatment of the Transient Processes with Three-Magnon Non-Linear Transitions under Perpendicular Pumping
Vol. 28, No. 2, p. 495—505 (1968).
See also 13, p. 171; 18, p. 294.

G. K. Born, R. J. Grasser, and A. O. Scharmann
EPR and Thermoluminescence Measurements on $CaWO_4$ Crystals
Vol. 28, No. 2, p. 583—588 (1968).
See also 10.2, p. 143; *11*, p. *153*; 20.3, p. 370; 22.8, p. 554.

G. Völkel and W. Windsch
*Electric Shifts in the Paramagnetic Resonance of Ferroelectric Cu^{2+}-Doped Rochelle Salt
Vol. 28, No. 2, p. K163—K165 (1968).
See also 14.4.2, p. 269.

M. A. Krivoglaz and L. B. Kvashnina
The Correlation Function $S_{zz}(\omega)$ for Weakly Coupled Impurity Atom Spin in a Ferromagnetic
Vol. 29, No. 1, p. 53—59 (1968).
See also *13*, p. *172*; 18.2, p. 313.

H. J. Gläser and D. Geist
Electrical Conductivity and Hall Effect of Manganese-Doped Germanium and the Influence of Heat Treatment on the State of Incorporation of Manganese
Vol. 29, No. 1, p. 167—170 (1968).
See also 13.4, p. 215; *14.3*, p. *243*; 22.1.1, p. 466.

K. Pfeuffer
Determination of the Electron Spin-Lattice Relaxation Time T_1 in the Direct Process
Vol. 29, No. 1, p. 171—177 (1968).
See also 6, p. 67; *18.1*, p. *297*; 22.8, p. 555.

A. M. van Diepen, H. W. de Wijn, and K. H. J. Buschow
*Nuclear Magnetic Resonance and Susceptibility of Equiatomic Rare-Earth–Aluminium Compounds
Vol. 29, No. 1, p. 189—192 (1968).
See also 13, p. 172; 18.1, p. 297; 21, p. 387; 21.4, p. 428.

W. Windsch and E. Hartmann
Switching Effect in Vanadyl-Doped Ferroelectric Triglycine Sulphate Observed by Electron Paramagnetic Resonance
Vol. 29, No. 1, p. K11—K13 (1968).
See also *14.4.2*, p. *269*.

R. Kersten
*Hyperfine and Quadrupole Interactions of the F-Centre in KCl as Measured by ENDOR
Vol. 29, No. 2, p. 575—586 (1968).
See also 10.2, p. 144; 22.5.2, p. 529.

K. N. Shrivastava
Phonon-Induced Hyperfine Coupling of V^{2+} Ion in Cubic Crystalline Fields
Vol. 29, No. 2, p. 737—742 (1968).
See also *6*, p. *68*; 22.6, p. 543.

I. S. McLintock and J. C. Orr
*Determination of Integrated Absorption in ESR Measurements
Vol. 29, No. 2, p. K157—K158 (1968).

A. J. Watts and J. E. Cousins
*Temperature Dependent Spin-Lattice Relaxation Times for Conduction Electrons in Lithium Particles in n-Irradiated Lithium Fluoride
Vol. 30, No. 1, p. 105—112 (1968).
See also 21.2, p. 425; 22.5.2, p. 530.

O. Kanert
Calculation of Quadrupole Broadening from (111)⟨110⟩ Dislocations in Cubic Cristals
Vol. 30, No. 1, p. 127—134 (1968).
See also *10.1*, p. *134*; 21, p. 388; 21.1, p. 408.

F. J. SCHEDEWIE
Paramagnetic Relaxation in Neodymium-Ethylsulfate: Phonon Re-Absorption and Spectral Spin Diffusion
Vol. **30**, No. 1, p. 251–259 (1968).
See also 6, p. 69; 13, p. 173; *18.1*, p. *298*; 22.8, p. 556.

H. SCHULZE and G. VÖLKEL
*Solid State Effect in Cu^{2+}-Doped Single Crystals of Rochelle Salt
Vol. **30**, No. 1, p. K9–K11 (1968).
See also 14.4.2, p. 269.

H. IMAGAWA
*ESR Studies of Cupric Ion in Various Oxide Glasses
Vol. **30**, No. 2, p. 469–478 (1968).
See also 2, p. 25; 3.2, p. 36; 22.8, p. 557.

H. W. DE WIJN, K. H. J. BUSCHOW, and A. M. VAN DIEPEN
*Experimental Evidence for Interband Mixing in Rare-Earth Intermetallic Compounds
Vol. **30**, No. 2, p. 759–766 (1968).
See also 13, p. 173; 13.1, p. 190; 18.1, p. 298; 21, p. 389; 21.4, p. 428.

P. G. BARANOV, R. A. ZHITNIKOV, and N. I. MELNIKOV
*Paramagnetic Resonance of Silver Atoms Stabilized in Alkali Halide Crystals
Vol. **30**, No. 2, p. 851–858 (1968).
See also 10.2, p. 144; 22.5.2, p. 531.

P. G. BARANOV, R. A. ZHITNIKOV, and N. I. MELNIKOV
*Theoretical Study of Properties of Copper and Silver Atoms Stabilized in Alkali Halide Crystals
Vol. **30**, No. 2, p. 859–866 (1968).
See also 3.2, p. 36; 10.2, p. 144; 13.4, p. 218; 22.5.2, p. 531.

M. P. PETROV, S. A. KIZHAEV, and G. A. SMOLENSKII
*Ferromagnetic Resonance in $Ga_{0.85}Fe_{1.15}O_3$
Vol. **30**, No. 2, p. 871–874 (1968).
See also 18.3, p. 320.

V. S. MANDEL, E. M. SMOKOTIN, G. A. PETRAKOVSKII, and B. M. LEBED
Magneto-Elastic Coupling Coefficients of Ca-Bi-V-I-Garnet
Vol. **30**, No. 2, p. K111–K113 (1968).
See also *18.3*, p. *320*.

H. BÖTTGER
Zur Spin-Phonon-Wechselwirkung im äußeren elektrischen Feld
(On the Spin-Phonon Interaction in External Electric Field)
Vol. **30**, No. 2, p. K119–K123 (1968).
See also 6, p. 70; 13, p. 173; 22.6, p. 544.

20. Optical Properties of Solids

Review Articles

D. P. MORGAN
*Helicon Waves in Solids
Vol. **24**, No. 1, p. 9–36 (1967).
See also 6, p. 56; *13*, p. *162*; 14.1, p. 220; 14.3, p. 231; 19, p. 326; 21, p. 373; 21.1, p. 390; 21.2, p. 424; 21.6, p. 430; 22.1.1, p. 458; 22.2.1, p. 482; 22.2.3, p. 489.

J. MERTSCHING
Theory of Electromagnetic Waves in Metals and Their Interaction with Ultrasonic Waves (II)
Vol. **26**, No. 1, p. 9–64 (1968).
See also *7*, p. *76*; 13, p. 162; 19, p. 326; 21, p. 373.

W. STEINMANN
*Optical Plasma Resonances in Solids
Vol. **28**, No. 2, p. 437–462 (1968).
See also *13*, p. *162*; 13.3, p. 196; 20.1, p. 344; 21, p. 373; 21.2, p. 424; 21.6, p. 430.

V. L. BONCH-BRUEVICH and E. G. LANDSBERG
Recombination Mechanisms
Vol. **29**, No. 1, p. 9–43 (1968).
See also 6, p. 56; 13, p. 162; 13.2, p. 191; *13.4*, p. *200*; *14.3*, p. *231*; 20.3, p. 363; 22.1.1, p. 458; 22.1.2, p. 468.

M. BALKANSKI and E. AMZALLAG
Band Parameters Determination from Faraday Rotation Masurements
Vol. **30**, No. 2, p. 407–439 (1968).
See also *13.1*, p. *174*; 18, p. 291; 22.1.1, p. 458; 22.1.2, p. 468; 22.2.1, p. 482; 22.2.2, p. 487; 22.2.3, p. 489; 22.4, p. 495; 22.4.1, p. 497; 22.4.2, p. 506; 22.4.3, p. 509.

Original Papers and Short Notes

W. L. MCCUBBIN
Symmetry Orbitals and UV Selection Rules for the Polymethylene Chain
Vol. **16**, No. 1, p. 289–293 (1966).
See also *3*, p. *26*; *13.1*, p. *174*; 22.9, p. 558.

M. ZVÁRA, F. ŽALOUDEK, and V. PROSSER
*Interband Faraday Effect in CdTe Single Crystals
Vol. **16**, No. 1, p. K21–K24 (1966).
See also 13, p. 162; *18*, p. *291*; 22.4.3, p. 509.

J. STUKE und G. WEISER
*Franz-Keldysh-Effekt an trigonalem und amorphem Selen
(Franz-Keldysh Effect in Trigonal and Amorphous Selenium)
Vol. **17**, No. 1, p. 343–352 (1966).
See also 13.1, p. 175; *14.4.1*, p. *256*; 22.1.3, p. 476.

J. KOŁODZIEJCZAK and H. STRAMSKA
*On the Theory of Free Carrier Electro-Magneto-Optical Phenomena
Vol. **17**, No. 2, p. 701–707 (1966).
See also 13.1, p. 176; 14.4.1, p. 256; *18*, p. *291*; 22, p. 441.

J. MERTSCHING
*Theorie des Dielektrizitätstensors in Relaxationszeit-Approximation
(Theory of the Dielectric Tensor in the Relaxation Time Approximation)
Vol. **17**, No. 2, p. 799–806 (1966).
See also *13*, p. *163*.

K. GODWOD, R. KOWALCZYK, and M. LEFELD-SOSNOWSKA
The Influence of the Primary X-Ray Beam Divergence on the X-Ray Wave Field Absorption Coefficients Measured by Means of the Three-Crystal X-Ray Spectrometer
Vol. **17**, No. 2, p. 869—874 (1966).
See also *4*, p. *38*; 22.1.2, p. 469.

G. HAURET and M. GIRARD-NOTTIN
Study of Light Scattering in Potassium Chloride Crystals Doped with Calcium
Vol. **17**, No. 2, p. 881—888 (1966).
See also *10*, p. *95*; 22.5.2, p. 519.

H. FLIETNER und N. KEMPE
*Untersuchung der Schwingungseigenschaften des Oszillistors
(Study of the Oscillatory Properties of the Oscillistor)
Vol. **18**, No. 1, p. 415—426 (1966).
See also *13*, p. *164*; 22.1.1, p. 460.

J. KOŁODZIEJCZAK and E. KIERZEK-PECOLD
*Free Carrier Electro-Optical Kerr Effect in Semiconductors
Vol. **19**, No. 1, p. K55—K58 (1967).
See also 13.1, p. 176; *14.4.1*, p. *257*; 22, p. 443.

J. KOŁODZIEJCZAK and E. KIERZEK-PECOLD
*Free Carrier Electro-Magneto-Optical Phenomena in Semiconductors
Vol. **19**, No. 2, p. 623—630 (1967).
See also 14.3, p. 235; *18*, p. *291*; 22, p. 443.

R. ENDERLEIN and R. KEIPER
*On the Theory of Electro-Absorption and Electro-Reflectance in Semiconductors
Vol. **19**, No. 2, p. 673—681 (1967).
See also 13.1, p. 177; 14.4.1, p. 257; 22, p. 443.

R. FISCHER
*Zur Theorie der Erzeugung der zweiten Harmonischen an der Oberfläche eines unmagnetischen, quadratischen, optisch einachsigen Kristalls
(On the Theory of the Generation of the Second Harmonics at the Surface of a Non-Magnetic, Quadratic, Optically Uniaxial Crystal)
Vol. **19**, No. 2, p. 757—765 (1967).

R. E. NEWNHAM and R. P. SANTORO
*Magnetic and Optical Properties of Dioptase
Vol. **19**, No. 2, p. K87—K90 (1967).
See also *13*, p. *165*; 18, p. 291; 22.8, p. 549.

R. FISCHER
*Ein Beispiel für die Existenz eines „Brewster-Winkels“ in der nichtlinearen Optik anisotroper Kristalle
(Example for the Existence of a "Brewster Angle" in the Non-Linear Optics of Anisotropic Crystals)
Vol. **19**, No. 2, p. K103—K105 (1967).

R. ENDERLEIN
*The Influence of Collisions on the Franz-Keldysh Effect
Vol. **20**, No. 1, p. 295—299 (1967).
See also 13.1, p. 177; 14.4.1, p. 257; 22, p. 444.

S. Y. CHUANG and B. G. HOGG
Positron Annihilation in Alnico and $(La_{0.7}Pb_{0.3})MnO_3$
Vol. **20**, No. 1, p. 331—335 (1967).
See also 11, p. 148; *13*, p. *165*; 21.1.1, p. 415; 22.8, p. 549.

J. Blinowski and M. Grynberg
*On the Direct Interband Magnetoabsorption in Uniaxially Stressed Germanium
Vol. **20**, No. 2, p. K107—K111 (1967).
See also 10, p. 99; *18*, *292*; 22.1.1, p. 461.

H. Stolz
Second Order Conductivity and Nonlinear Plasma Excitation of an Isotropic Electron Gas
Vol. **21**, No. 1, p. 77—88 (1967).
See also 13, p. 165; *14*, p. *219*.

G. Lautz and M. Schulz
*Schraubenförmige Dichtewellen und Oszillistor-Effekt im Elektron-Loch-Plasma des n-Germaniums
(Helical Density-Waves and Oscillistor Effect in the Electron-Hole Plasma of n-Germanium)
Vol. **21**, No. 2, p. K139—K143 (1967).
See also *13*, p. *166*; 14.4.1, p. 259; 22.1.1, p. 462.

E. Majerníková
*Faraday Effect and Spin-Orbit Interaction of the Lowest Exciton States in Molecular Crystals
Vol. **22**, No. 1, p. 113—122 (1967).
See also 13.2, p. 192; *18*, p. *292*; 22.9, p. 559.

M.-P. Stoll
The Use of the Magneto-Optical Kerr Effect for the Determination of Magnetization Curves
Vol. **22**, No. 1, p. 163—176 (1967).
See also *18.2*, p. *307*; 21.1.1, p. 418.

F. Brouers
*Plasmon Satellites of Soft X-Ray Spectra
Vol. **22**, No. 1, p. 213—221 (1967).
See also *13*, p. *166*; 21, p. 380; 21.2, p. 424.

E. Gutsche and H. Lange
*Electroreflectance of CdS and CdSe Single Crystals at the Fundamental Absorption Edge
Vol. **22**, No. 1, p. 229—233 (1967).
See also 13.1, p. 179; 13.2, p. 192; *14.4.1*, p. *259*; 20.1, p. 349; 22.4.1, p. 500; 22.4.2, p. 506.

E. A. Kaner and V. G. Skobov
*On the Possibility of Existence of Quantum Electromagnetic Waves in Metals
Vol. **22**, No. 1, p. 333—340 (1967).
See also *13*, p. *166*; 13.1, p. 179; 21, p. 380.

A. Otto
Theory of Plasmon Excitation in Thin Films by Electrons of Non-Normal Incidence
Vol. **22**, No. 2, p. 401—406 (1967).
See also *13*, p. *166*; 13.3, p. 197.

A. I. Kosinskaya and S. I. Uritsky
*On the Classical Theory of the Optical Properties of Free Carriers in Semiconductors
Vol. **23**, No. 1, p. 57—68 (1967).
See also 13.1, p. 180; 18, p. 293; 22.1.1, p. 463.

R. Enderlein and R. Keiper
*On the Dielectric Function of a Semiconductor in an External Electric Field
Vol. **23**, No. 1, p. 127—136 (1967).
See also 13.1, p. 180; 14.4.1, p. 259; 22, p. 447.

J. B. Swan, A. Otto, and H. Fellenzer
*Observed Retardation Effects on the Energy of the ω_+-Surface Plasmons in Thin Aluminium Foils
Vol. 23, No. 1, p. 171—176 (1967).
See also *13*, p. *167*; 13.3, p. 198; 21, p. 381.

P. Byszewski, B. Kalińska, and J. Kołodziejczak
*Interband Faraday Rotation and Ellipticity in Germanium and Gallium Arsenide
Vol. 23, No. 1, p. K53—K56 (1967).
See also 13, p. 167; *18*, p. *293*; 22.1.1, p. 463; 22.2.1, p. 484.

J. Kołodziejczak
*Free Carrier Effect on the Generation of Higher Harmonics in Semiconductors
Vol. 23, No. 1, p. K57—K59 (1967).
See also *13*, p. *167*; 13.1, p. 181; 14.3, p. 238; 22, p. 447.

D. E. Aspnes
*Note on Collision Broadening of Franz-Keldysh Effect
Vol. 23, No. 1, p. K79—K81 (1967).
See also 13.1, p. 181; 14.4.1, p. 260; 22, p. 447.

W. Schülke and U. Berg
Measurement of the Compton Band on Beryllium Single Crystals
Vol. 23, No. 1, p. K87—K91 (1967).
See also *13.1*, p. *181*; 21.3, p. 426.

U. K. Mishra and A. K. Walton
Infra-Red Faraday Effect in p-Type InAs and Si
Vol. 24, No. 1, p. K87—K90 (1967).
See also *13.1*, p. *183*; 18, p. 293; 22.1.2, p. 472; 22.2.1, p. 485.

J. Geiger
Inelastic Electron Scattering in Thin Films at Oblique Incidence
Vol. 24, No. 2, p. 457—460 (1967).
See also *13*, p. *168*; 21, p. 383.

L. Śniadower and R. Kowalczyk
*Magnetoplasma Reflection in Mercury Telluride
Vol. 25, No. 1, p. 233—240 (1968).
See also 13.1, p. 183; *18*, p. *293*; 20.1, p. 351; 22.4.3, p. 510.

M. Schulz
*Helical Density Waves and the Oscillistor Effect in the Electron Hole Plasma of the n-Type Ge (Theory)
Vol. 25, No. 2, p. 521—530 (1968).
See also *13*, p. *169*; 14.4.1, p. 261; 22.1.1, p. 464.

V. V. Nemoshkalenko and M. A. Mindlina
*On the Form of the $K\beta_{2,5}$-Zinc Band
Vol. 25, No. 2, p. K83—K86 (1968).
See also 13.1, p. 184; 21.1, p. 403.

J. Kobayashi, H. Schmid, and E. Ascher
Optical Study on the Ferroelectric Orthorhombic Phase of Fe-I-Boracite
Vol. 26, No. 1, p. 277—283 (1968).
See also 4, p. 45; *14.4.2*, p. *268*.

R. Enderlein
*Excitonic Effects in the Electro-Absorption of Semiconductors
Vol. 26, No. 2, p. 509—517 (1968).
See also 13.1, p. 185; 13.2, p. 193; *14.4.1*, p. *262*; 20.1, p. 353; 22, p. 451.

A. Otto
*Eine neue Methode der Anregung nichtstrahlender Oberflächenplasmaschwingungen (A New Method for Exciting Non-Radiative Surface Plasma Oscillations)
Vol. 26, No. 2, p. K99—K101 (1968).
See also *13*, p. *169*; 21.6, p. 436.

M. Porsch
*Green's Function Calculation of Electro-Optical Effects in Semiconductors. (I): General Formalism and Interband Transitions
Vol. 27, No. 1, p. 359—370 (1968).
See also 13.1, p. 186; 14.4.1, p. 262; 22, p. 451.

H. Stramska, Z. Bachan, P. Byszewski, and J. Kołodziejczak
*Interband Faraday Rotation and Ellipticity Observed at the Absorption Edge in Silicon
Vol. 27, No. 1, p. K25—K28 (1968).
See also 13.1, p. 186; *18*, p. *294*; 22.1.2, p. 473.

H. Burkhard and J. Jaumann
*The Determination of the "Gyroelectric" and "Gyromagnetic" Constants of Thick, Evaporated Layers of Iron by Means of Measurements in the Reflected Light only
Vol. 27, No. 1, p. K43—K45 (1968).
See also *18.2*, p. *312*; 20.1, p. 354; 21.1.1, p. 421.

O. Brümmer und G. Dräger
*Untersuchungen zur Deutung der kantenfernen Feinstruktur der Röntgen-K-Absorptionsspektren
(On the Interpretation of the Fine Structure of X-Ray K Absorption Spectra Far from the Edge)
Vol. 27, No. 2, p. 513—516 (1968).
See also 2, p. 24; 22.1.1, p. 465; 22.6, p. 541; 22.8, p. 554.

A. Graja
*Production of the Second Harmonic of Light in Ammonium Pentaborate and Other Powdered Piezoelectric Crystals
Vol. 27, No. 2, p. K93—K97 (1968).
See also *20.2*, p. *362*; 22.8, p. 554.

Ya. O. Dovhyj
*Connection between the Circular Dichroism and Rotatory Dispersion of Gyrotropic Crystals and Its Experimental Verification in the Region of the Excitation of Circular Excitons
Vol. 27, No. 2, p. K121—K124 (1968).
See also 13.2, p. 194; 22, p. 452.

M. Zvára
*Interband Faraday Rotation in Gallium Arsenide
Vol. 27, No. 2, p. K157—K160 (1968).
See also 13, p. 170; 18, p. 294; 22.2.1, p. 485.

V. V. Nemoshkalenko and V. V. Gorskii
*Investigation of the Valence Electron Distribution in Mg-Al Alloys
Vol. 28, No. 1, p. K15—K17 (1968).
See also 13.1, p. 187; 21, p. 387; 21.3, p. 427.

S. Ballarò, A. Balzarotti, and V. Grasso
*Electroreflectance at Oblique Incidence in Ge
Vol. 28, No. 2, p. K109—K112 (1968).
See also 13.1, p. 188; *14.4.1*, p. *264*; 20.1, p. 356; 22.1.1, p. 466.

S. N. Shestatskii and V. V. Sobolev
*Electroreflectance Spectrum of Cuprous Oxide Crystals
Vol. **28**, No. 2, p. K131–K133 (1968).
See also 13.1, p. 188; 13.2, p. 194; *14.4.1*, p. *264*; 20.1, p. 356; 22.6, p. 542.

B. Garben und H. Seliger
*Faraday-Effekt an amorphem Selen
(Faraday Effect on Amorphous Selenium)
Vol. **29**, No. 1, p. K27–K29 (1968).
See also 13, p. 172; *18*, p. *294*; 22.1.3, p. 480.

E. Mohler
*Linear Electro-Optic Effect of Excitons in CuCl
Vol. **29**, No. 1, p. K55–K57 (1968).
See also 13.2, p. 195; *14.3*, p. *243*; 22.5, p. 514.

W. Schilz
*Anisotropy of High-Field Helicon Propagation in PbTe
Vol. **29**, No. 2, p. 559–568 (1968).
See also *13*, p. *172*; 14.3, p. 244; 22, p. 454.

J. Kołodziejczak
*Nonlinear Magnetooptical Phenomena in Semiconductors (I)
Vol. **29**, No. 2, p. 645–652 (1968).
See also 13.1, p. 189; *18*, p. *295*; 22, p. 454.

J. Ingels and G. Jacobs
*Faraday Rotation of the U_2 Centre in KCl, KBr, and KI
Vol. **30**, No. 1, p. 163–172 (1968).
See also 13.4, p. 216; *18*, p. *295*; 22.5.2, p. 530.

S. Haussühl und H. Hesse
*Quadratischer elektrooptischer Effekt (Kerr-Effekt) in Alkalihalogeniden vom NaCl-Typ
(Quadratic Electrooptical Effect (Kerr Effect) in Alkali Halides of NaCl Type)
Vol. **30**, No. 1, p. 209–214 (1968).
See also 14.4, p. 254; 22.5.2, p. 530.

L. I. Datsenko, M. Ya. Skorokhod, and A. S. Vasilkovskii
The Effect of Dislocations on the Intensity Jumps at the Interferential Transmission of X-Rays near the K-Edge of Absorption in Ge
Vol. **30**, No. 1, p. 231–237 (1968).
See also *10*, p. *112*; 22.1.1,p. 466.

K. W. Böer and P. Voss
Transition between Stationary and Moving High-Field Domains in CdS in a Range of N-Shaped Negative Differential Conductivity due to Field-Quenching
Vol. **30**, No. 1, p. 291–304 (1968).
See also *14.4.1*, p. *264*; 16, p. 284; 22.4.1, p. 505.

R. V. Pisarev, I. G. Siny, N. N. Nesterova, G. A. Smolensky, and P. P. Syrnikov
*Faraday Rotation and Magnetic Dichroism in $RbNi(Co)F_3$
Vol. **30**, No. 1, p. 367–372 (1968).
See also 13, p. 173; *18.3*, p. *320*; 22.5, p. 514; 22.8, p. 556.

J. Hofmann and W. Steinmann
*Plasma Resonance in the Photoemission of Silver
Vol. **30**, No. 1, p. K53–K56 (1968).
See also *13*, p. *173*; 17, p. 290; 21.6, p. 437.

L. MERTEN
*Polariton Dispersion in Biaxial and Uniaxial Crystals
Vol. **30**, No. 2, p. 449–454 (1968).
See also *6*, p. *70*; 22.6, p. 544.

V. V. NEMOSHKALENKO, M. A. MINDLINA, and B. P. MAMKO
*X-Ray Spectra and Electronic Structure of Elements at the End of the First Long Period from ^{29}Cu to ^{34}Se
Vol. **30**, No. 2, p. 703–712 (1968).
See also 13.1, p. 189; 21.1, p. 409; 22.1.1, p. 467; 22.1.3, p. 480.

L. A. ALMASOV and I. M. DYKMAN
*Plasma Oscillations in Semiconductors with Non-Standard Dispersion Law in the Presence of Electric and Magnetic Fields
Vol. **30**, No. 2, p. 767–778 (1968).
See also *13*, p. *173*; 22.1.1, p. 467.

20.1 Spectra, Optical Constants (Colour Centres See 10.2; X-Ray Spectra See 20)

Review Articles

Y. P. Varshni
Band-to-Band Radiative Recombination in Groups IV, VI, and III–V Semiconductors (I)
Vol. **19**, No. 2, p. 459–514 (1967).
See also 13.1, p. 174; 13.2, p. 191; 13.4, p. 200; 20.2, p. 360; *20.3*, p. *363*; 22.1, p. 456; 22.1.1, p. 458; 22.1.2, p. 468; 22.1.3, p. 476; 22.2.1, p. 482; 22.2.2, p. 487; 22.2.3, p. 489.

Y. P. Varshni
Band-to-Band Radiative Recombination in Groups IV, VI, and III–V Semiconductors (II)
Vol. **20**, No. 1, p. 9–36 (1967).
See also 13.1, p. 174; 13.2, p. 191; 13.4, p. 200; 20.2, p. 360; *20.3*, p. *363*; 22.1, p. 456; 22.1.1, p. 458; 22.1.2, p. 468; 22.1.3, p. 476; 22.2.1, p. 482; 22.2.2, p. 487; 22.2.3, p. 489;

A. S. Borshchevskii, N. A. Goryunova, F. P. Kesamanly, and D. N. Nasledov
Semiconducting $A^{II}B^{IV}C_2^{V}$ Compounds
Vol. **21**, No. 1, p. 9–55 (1967).
See also 3.1, p. 29; 3.2, p. 34; 5, p. 49; 8, p. 79; 10, p. 93; 13.1, p. 174; *14.3*, p. *231*; 15, p. 270; 16, p. 276; 17, p. 286; 18, p. 291; 22.8, p. 547.

G. Harbeke
On the Band Structure of Anisotropic Crystals
Vol. **27**, No. 1, p. 9–44 (1968).
See also *13.1*, p. *174*; 13.2, p. 191; 14.3, p. 231; 22, p. 439; 22.1, p. 456; 22.1.1, p. 458; 22.1.2, p. 468; 22.1.3, p. 476; 22.4.1, p. 497; 22.4.2, p. 506; 22.4.3, p. 509; 22.5, p. 512; 22.7, p. 545; 22.8, p. 547.

W. Steinmann
Optical Plasma Resonances in Solids
Vol. **28**, No. 2, p. 437–462 (1968).
See also *13*, p. *162*; 13.3, p. 196; *20*, p. *337*; 21, p. 373; 21.2, p. 424; 21.6, p. 430

Original Papers and Short Notes

B. Velický and J. Sak
Excitonic Effects in the Interband Absorption of Semiconductors
Vol. **16**, No. 1, p. 147–157 (1966).
See also *13.1*, p. *174*; *13.2*, p. *191*; 22, p. 439.

G. A. Akhundov and T. G. Kerimova
*Infrared Absorption of $A^{III}B^{VI}$ Single Crystals
Vol. **16**, No. 1, p. K15–K16 (1966).
See also 22, p. 439.

G. B. Abdullaev, S. I. Mekhtieva, G. M. Aliev, D. Sh. Abdinov, and T. G. Kerimova
*Determination of the Oxygen Content in Selenium
Vol. **16**, No. 1, p. K31–K34 (1966).
See also 2, p. 22; 10, p. 94; 22.1.3, p. 476.

H. Vogt und H. Happ
*Das Ultrarotspektrum von $NaNO_2$ zwischen 30 und 150 μm in der ferro- und paraelektrischen Phase
(Infrared Spectrum of $NaNO_2$ in the Ferro- and Paraelectric Phase between 30 and 150 μm)
Vol. **16**, No. 2, p. 711–719 (1966).
See also 14.4.2, p. 266.

M. Musa
Some Results Concerning the Co^{++} Site Symmetry in Alkali Halide Crystals
Vol. **16**, No. 2, p. 771–776 (1966).
See also *10*, p. *94*; 22.5.2, p. 518.

W. Żanowicz and A. Wojakowski
Semiconducting Properties of CdP_4
Vol. **16**, No. 2, p. K129–K131 (1966).
See also *14.3*, p. *232*; 15, p. 270; 22, p. 440.

M. V. Kurik and N. I. Vitrikhovskii
*Optical Properties and Energy Structure of CdS-CdTe Crystals
Vol. **16**, No. 2, p. K139–K142 (1966).
See also *13.1*, p. *175*; 22.4.1, p. 497; 22.4.3, p. 509; 22.8, p. 547.

G. A. Bogomolova, A. A. Kaminskii, and V. A. Timofeeva
*Nd^{3+} Optical Centres in $Y_3Al_5O_{12}$ Crystals
Vol. **16**, No. 2, p. K165–K166 (1966).
See also 10, p. 94; 20.2, p. 360; 22.8, p. 547.

E. V. Chisler and M. S. Shur
*The Raman Spectrum of $NaNO_2$ in the Ferroelectric Phase
Vol. **17**, No. 1, p. 163–171 (1966).
See also 5, p. 49, *6*, p. *56*; 14.4.2, p. 266.

E. V. Chisler and M. S. Shur
*Raman Spectrum and Phase Transition in the Ferroelectric Crystal $NaNO_2$
Vol. **17**, No. 1, p. 173–176 (1966).
See also 1.2, p. 9; *6*, p. *57*; 14.4.2, p. 266.

H. D. Koswig
Dependence of the Inner-Electronic Repulsion Parameters of the $3d^3$ and $3d^7$ Configuration on the Ligands in Octahedral Crystal Fields
Vol. **17**, No. 1, p. K57–K61 (1966).
See also *13.4*, p. *202*; 22, p. 440.

K. Owen and R. L. Anderson
*Determination of Optical Absorption Using Heterocrystals
Vol. **17**, No. 1, p. K79–K81 (1966).
See also 22.1.1, p. 459.

H. D. Koswig
*„Optisches Bleichen" Ni^{2+}-dotierter AgCl-Kristalle
("Optical Bleaching" of Ni^{2+}-Doped AgCl Crystals)
Vol. **17**, No. 1, p. K83–K87 (1966).
See also 10.2, p. 137; 22.5.1, p. 515.

A. S. Filipchenko, G. G. Kulikov, D. N. Naslëdov, and Yu. I. Ukhanov
Effective Mass of Holes in InSb
Vol. **17**, No. 1, p. K113–K114 (1966).
See also *13.1*, p. *175*; 22.2.3, p. 490.

N. A. Gasanova, G. A. Akhundov, and M. A. Nizametdinova
*Polarization Effects in the Ultraviolet Reflection from GaS and GaSe Single Crystals
Vol. **17**, No. 1, p. K115–K117 (1966).
See also 22, p. 440.

N. A. Gasanova, G. A. Akhundov, and M. A. Nizametdinova
*Optical Reflection and Absorption of GaS_xSe_{1-x} Single Crystals
Vol. **17**, No. 1, p. K131–K134 (1966).
See also 22.8, p. 547.

U. RÖSSLER and M. LIETZ
Band Structure of Cubic and Hexagonal ZnS (APW Method)
Vol. **17**, No. 2, p. 597–604 (1966).
See also 5, p. 50; *13.1*, p. *175*; 22.4.1, p. 498.

T. KAMIKAWA, Y. KAZUMATA, and K. OZAWA
New Infrared Absorption in Neutron-Irradiated Lithium Fluoride
Vol. **17**, No. 2, p. 645–655 (1966).
See also *10.2*, p. *137*; 11, p. 146; 22.5.2, p. 519.

H. BÖTTGER
*Temperaturabhängigkeit der optischen Absorption von Übergangselementionen in Silberchlorid
(Temperature Dependence of the Optical Absorption of Transition Element Ions in Silver Chloride)
Vol. **17**, No. 2, p. 855–868 (1966).
See also 13.4, p. 203; 22.5.1, p. 515.

W. DREYBRODT and K. FUSSGAENGER
*Interpretation of the Temperature Dependence of Ag^+ Dipole Strengths in Alkali Halides
Vol. **18**, No. 1, p. 133–144 (1966).
See also 13.4, p. 203; 22.5.2, p. 519.

H. C. BOLTON, L. A. BURSILL, A. C. MCLAREN, and R. G. TURNER
*On the Origin of the Colour of Labradorite
Vol. **18**, No. 1, p. 221–230 (1966).
See also 4, p. 38; 22.8, p. 548.

S. A. MOSKALENKO and P. I. KHADSHI
Infrared Absorption by Excitons due to Photoionization and Intraband Lattice Scattering
Vol. **18**, No. 1, p. 379–390 (1966).
See also 6, p. 57; *13.2*, p. *191*; 22.1.1, p. 460; 22.2.2, p. 487; 22.2.3, p. 490; 22.6, p. 536.

U. GERHARDT and E. MOHLER
Piezooptical Experiments on the Excitons in KBr and KI
Vol. **18**, No. 1, p. K45–K48 (1966).
See also *13.2*, p. *192*; 22.5.2, p. 519.

D. SCHWARZ
*Die Ausläuferabsorption chalkogendotierter AgBr-Kristalle bei variabler Fremdkationenkonzentration
(Extrinsic Absorption of Chalogen Doped AgBr Crystals in the Case of Variable Cation-Impurity Concentration)
Vol. **18**, No. 2, p. 787–794 (1966).
See also 10, p. 97; 22.5.1, p. 515.

A. F. LUBCHENKO and S. I. DUDKIN
*Effect of Concentration on Light Absorption and Dispersion by Impurity Centres in the Case of Weak Electron-Phonon Coupling
Vol. **18**, No. 2, p. 853–862 (1966).
See also 6, p. 58, 13.4, p. 204.

K. K. SHVARTS, A. YA. VITOL, A. V. PODIN, D. O. KALNIN, and YU. A. EKMANIS
Radiation Effects in Pile-Irradiated LiF Crystals
Vol. **18**, No. 2, p. 897–909 (1966).
See also 10.2, p. 138; *11*, p. *147*; 19, p. 327; 22.5.2, p. 520.

V. L. Broude, E. I. Rashba, and E. F. Sheka
A New Approach to the Vibronic Spectra of Molecular Crystals
Vol. **19**, No. 1, p. 395–406 (1967).
See also *6*, p. *58*; 22.9, p. 558.

B. Velický
*Polaron Effect on the Optical Absorption Edge of Semiconductors
Vol. **19**, No. 1, p. K39–K42 (1967).
See also 6, p. 59; 13, p. 164; 13.1, p. 176; 22, p. 442.

M. A. Nizametdinova
*The Reflection Spectrum of GaSe and GaS Single Crystals near the Fundamental Absorption Edge
Vol. **19**, No. 2, p. K111–112 (1967).
See also 13.1, p. 177; 22, p. 443.

A. S. Davydov and E. N. Myasnikov
*Absorption and Dispersion of Light by Molecular Excitons
Vol. **20**, No. 1, p. 153–163 (1967).
See also *13.2*, p. *192*; 22, p. 443.

L. Roskovcová, J. Pastrňák, and R. Babušková
*The Dispersion of the Refractive Index and the Birefringence of AlN
Vol. **20**, No. 1, p. K29–K32 (1967).
See also 22.2, p. 481.

Z. Bodó and I. Hevesi
*Optical Absorption near the Absorption Edge in V_2O_5 Single Crystals
Vol. **20**, No. 1, p. K45–K49 (1967).
See also 13.1, p. 177; 22.6, p. 537.

A. F. Lubchenko and S. I. Dudkin
*Effect of Concentration on Light Absorption and Dispersion by Impurity Centres in the Case of Weak Elektron-Phonon Coupling (II)
Vol. **20**, No. 2, p. 427–432 (1967).
See also 6, p. 60; 13.4, p. 205.

L. Żdanowicz
*Some Optical Properties of Thin Evaporated Cd_3As_2 Films
Vol. **20**, No. 2, p. 473–480 (1967).
See also 22.7, p. 545.

G. Blasse and A. Bril
Some Observations on the Cr^{3+} Fluorescence in the Huntite Structure
Vol. **20**, No. 2, p. 551–556 (1967).
See also 13.4, p. 205; *20.3*, p. *365*; 22.8, p. 549.

J. Spyridelis, J. Stoimenos, and N. Economou
*Optical and Photoconductive Phenomena in Cuprous Oxide
Vol. **20**, No. 2, p. 623–628 (1967).
See also 13.2, p. 192; 16, p. 279; 22.6, p. 537.

V. M. Asnin and A. A. Rogachev
*Exciton Absorption in Doped Germanium
Vol. **20**, No. 2, p. 755–757 (1967).
See also *13.2*, p. *192*, 22.1.1, p. 461.

W. Henrion
On an Exciton Structure in the Reflection Spectrum of Trigonal Selenium Single Crystals
Vol. **20**, No. 2, p. K145–K149 (1967).
See also 13.1, p. 178; *13.2*, p. *192*; 22.1.3, p. 477.

H. BILZ, R. ZEYHER, and R. K. WEHNER
Anharmonic Sideband Effects of the U-Center Local Mode
Vol. **20**, No. 2, p. K167—K171 (1967).
See also *6*, p. *60*; 10.2, p. 139; 22.5.2, p. 522.

J. S. ZIOMEK and P. B. PICKAR
*Optical Studies of β-SiC
Vol. **21**, No. 1, p. 271—278 (1967).
See also 5, p. 51; 6, p. 61; 13.1, p. 178; 22.3, p. 494.

A. A. KAMINSKII, YU. K. VORONKO, and V. V. OSIKO
Mixed Systems on the Basis of Fluorides as New Laser Materials for Quantum Electronics. The Optical and Emission Parameters
Vol. **21**, No. 1, p. K17—K21 (1967).
See also *20.2*, p. *361*; 20.3, p. 365; 22.5, p. 513; 22.8, p. 550.

A. A. KAMINSKII, KH. S. BAGDASAROV, and L. M. BELYEV
*Analysis of the Optical Spectra and Stimulated Radiation of $Y_3Al_5O_{12}$-Nd^{3+} Crystals
Vol. **21**, No. 1, p. K23—K29 (1967).
See also 13.4, p. 207; 20.2, p. 361; *20.3*, p. *365*; 22.8, p. 550.

Z. A. DEMIDENKO and V. S. MASHKEVICH
*Indirect Optical Transitions at the Centre of the Brillouin Zone
Vol. **21**, No. 1, p. K59—K63 (1967).
See also 6, p. 61; 22, p. 445.

L. MÜLLER and Y. MIHAI
*Absorption of Amorphous Se Thin Layers Doped with Bi and Zn
Vol. **21**, No. 1, p. K65—K68 (1967).
See also 2, p. 23; 13.4, p. 207; 22.1.3, p. 477.

G. GETOV, P. SIMIDTCHIEVA, M. NIKIFOROVA, and R. ANDREYTCHIN
*Effect of Some Elements on the Optical Absorption Edge of Vitreous As_2S_3
Vol. **21**, No. 1, p. K87—K89 (1967).
See also 2, p. 23; 13.4, p. 207; 22, p. 445.

M. HULIN
*Selection Rules for Two-Phonon Absorption Processes
Vol. **21**, No. 2, p. 607—618 (1967).
See also *6*, p. *61*.

J. P. AGRAWAL
Infrared Fluorescence in Manganese Activated Zinc Sulphide Phosphors
Vol. **21**, No. 2, p. 679—682 (1967).
See also 13.4, p. 207; *20.3*, p. *366*; 22.4.1, p. 500.

V. HIZHNYAKOV and I. TEHVER
Theory of Resonant Secondary Radiation due to Impurity Centres in Crystals
Vol. **21**, No. 2, p. 755—768 (1967).
See also 13.4, p. 207; *20.3*, p. *366*; 22, p. 446.

D. M. BERCHA, A. N. BORETS, I. M. STAKHYRA, and K. D. TOVSTYUK
*The Band Edge and the Energy Spectrum of In_2Se
Vol. **21**, No. 2, p. 769—774 (1967).
See also 13.1, p. 178; 22, p. 446.

A. F. Lubchenko
*Light Absorption and Dispersion by Impurity Centres at High Concentrations
Vol. **21**, No. 2, p. 785–788 (1967).
See also *13.4*, p. *207*.

V. K. Koniukhov, L. A. Kulevskii, and A. M. Prokhorov
*Two-Photon Absorption Spectrum in CdS near the Fundamental Absorption Edge
Vol. **21**, No. 2, p. K107–K110 (1967).
See also 22.4.1, p. 500.

R. Sittig
*Stark Effect Measurements of the OH^- Absorption Band in KCl
Vol. **21**, No. 2, p. K175–K177 (1967).
See also 10.2, p. 139; 14.4.1, p. 259; 22.5.2, p. 523.

E. W. Claffy
*Impurity Absorption Bands in Thermoluminescent LiF
Vol. **22**, No. 1, p. 71–76 (1967).
See also 13.4, p. 208; 20.3, p. 366; 22.5.2, p. 523.

H. N. Spector
*Optical Absorption in Electric and Magnetic Fields
Vol. **22**, No. 1, p. 185–193 (1967).
See also 13.1, p. 179; 14.4.1, p. 259; 18, p. 292; 22, p. 446.

E. Gutsche and H. Lange
Electroreflectance of CdS and CdSe Single Crystals at the Fundamental Absorption Edge
Vol. **22**, No. 1, p. 229–233 (1967).
See also 13.1, p. 179; 13.2, p. 192; *14.4.1*, p. *259*; *20*, p. *339*; 22.4.1, p. 500; 22.4.2, p. 506.

G. E. Arkhangelskii, Z. L. Morgenshtern, and V. B. Neustruev
Colour Centres in Ruby Crystals
Vol. **22**, No. 1, p. 289–295 (1967).
See also *10.2*, p. *140*; 20.2, p. 361; 20.3, p. 366; 22.6, p. 538.

W. Henrion
*Reflectivity Measurements on Trigonal Seleniüm Single Crystals in the Spectral Region between 1.6 and 6.0 eV
Vol. **22**, No. 1, p. K33–K37 (1967).
See also 13.1, p. 179; 13.2, p. 192; 22.1.3, p. 477.

H. L. Jetter und H. C. Wolf
*Zur Frage der Davydov-Aufspaltung des ersten Singulett-Anregungszustandes von Anthracen-Kristallen
(On the Davydov Splitting of the First Excited Singulet-State of Anthracene Crystals)
Vol. **22**, No. 1, p. K39–K43 (1967).
See also 12, p. 157; 13.2, p. 193; 22.9, p. 559.

E. Mohler, J. Stuke, and G. Zimmerer
*Some New Features of the Reflectance Spectrum of Trigonal Selenium Single Crystals
Vol. **22**, No. 1, p. K49–K53 (1967).
See also 13.1, p. 179; 22.1.3, p. 477.

S. C. Jain and G. D. Sootha
*Electrical and Optical Properties of Titanium-Doped Lithium Fluoride Crystals
Vol. **22**, No. 2, p. 505–516 (1967).
See also 13.4, p. 208; *14.4*, p. *253*; 22.5.2, p. 523.

J. E. Caffyn, T. L. Goodfellow, E. M. Olymbios, and D. J. Perrin
*Dielectric and Optical Absorption Measurements on CaF_2 and CaF_2:Y
Vol. **22**, No. 2, p. 549—557 (1967).
See also 10, p. 101; 13.4, p. 209; *14.4*, p. *253*; 22.5.3, p. 533.

E. Kierzek-Pecold, J. Kołodziejczak, and I. Pracka
*Optical Constants of β-Rhombohedral Boron in the Region 1.2 te 6.2 eV
Vol. **22**, No. 2, p. K147—K150 (1967).
See also 22.1, p. 457.

T. L. Tansley
Spectral Response of. p-n Heterojunctions
Vol. **23**, No. 1, p. 241—252 (1967).
See also *14.3.2*, p. *249*; 14.4.1, p. 259; 16, p. 280; 22.2.1, p. 484; 22.8, p. 552.

P. Görlich, H. Karras, G. Kötitz, and P. Ullmann
Thermoluminescence Investigations on Doped Alkaline Earth Fluorides
Vol. **23**, No. 1, p. 313—323 (1967).
See also 13.4, p. 209; *20.3*, p. *367*; 22.5.3, p. 533.

R. Ludeke and W. Paul
*Growth and Optical Properties of Wurtzite and Sphalerite CdSe Epitaxial Thin Films
Vol. **23**, No. 1, p. 413—418 (1967).
See also 1.4, p. 19; *3.1*, p. *32*; 13.1, p. 181; 22.4.2, p. 507.

J.-Y. Roncin and K. Moorjani
*On the Absorption Spectrum of Gaseous and Solid Xenon
Vol. **23**, No. 1, p. K1—K3 (1967).
See also 13.2, p. 193; 23, p. 561.

A. A. Kaminskii, Y. E. Lapsker, and B. P. Sobolev
Induced Emission of $NaCaCeF_6:Nd^{3+}$ at Room Temperature
Vol. **23**, No. 1, p. K5—K7 (1967).
See also *20.2*, p. *361*; 20.3, p. 367; 22.5, p. 513; 22.8, p. 552.

F. R. Kessler and E. Sutter
*Absorption and Photoconductivity of a Series of Selenium-Tellurium Mixed Crystals
Vol. **23**, No. 1, p. K25—K33 (1967).
See also 13.1, p. 181; *16*, p. *280*; 22.1.3, p. 477.

H. Lange and W. Henrion
*Thermoreflectance of CdS and Se Single Crystals
Vol. **23**, No. 1, p. K67—K70 (1967).
See also 13.1, p. 181; 22.1.3, p. 477; 22.4.1, p. 501.

V. K. Subashiev and A. A. Kukharskii
*The Reflection Coefficient of Optically Inhomogeneous Solids
Vol. **23**, No. 2, p. 447—452 (1967).
See also 22.1.2, p. 472.

R. G. J. Grisar, K. P. Reiners, K. F. Renk, and L. Genzel
*Impurity-Induced Far-Infrared Absorption in the Phonon Gap Regions of KI and KBr
Vol. **23**, No. 2, p. 613—620 (1967).
See also *6*, p. *63*; 22.5.2, p. 524.

A. Stella, A. D. Brothers, R. H. Hopkins, and D. W. Lynch
Pressure Coefficient of the Band Gap in Mg_2Si, Mg_2Ge, and Mg_2Sn
Vol. **23**, No. 2, p. 697—702 (1967).
See also *13.1*, p. *182*; 14.3, p. 238; 22.7, p. 545.

R. Geick, U. Schröder, and J. Stuke
Lattice Vibrational Properties of Trigonal Selenium
Vol. **24**, No. 1, p. 99–108 (1967).
See also *6*, p. *63*; 22.1.3, p. 478.

D. Bäuerle and B. Fritz
*Infrared Vibrational Absorption by U-Centers in NaI
Vol. **24**, No. 1, p. 207–215 (1967).
See also *6*, p. *63*; 10.2, p. 141; 22.5.2, p. 524.

H. G. Reik and D. Heese
The Small Polaron Problem with an Application to Optical and DC Data of Reduced Barium Titanate
Vol. **24**, No. 1, p. 281–287 (1967).
See also 6, p. 64; *13*, p. *168*; 14.4.2, p. 267.

P. Grosse and M. Selders
The p-Band of Tellurium at Very Low Temperatures
Vol. **24**, No. 1, p. K33–K36 (1967).
See also *13.1*, p. *183*; 22.1.3, p. 478.

H. D. Koswig, U. Retter, and W. Ulrici
Jahn-Teller Effect of Transition-Metal Ions in Silver Halides. (I): $Ti^{3+}(3d^1)$ and $Fe^{2+}(3d^6)$
Vol. **24**, No. 2, p. 605–614 (1967).
See also *13.4*, p. *210*; 22.5.1, p. 516.

J. Auleytner, E. F. Gross, A. A. Kralina, and B. S. Razbirin
Investigation of the Real Structure of CdS Crystals in Connection with the Shift of Exciton Lines
Vol. **24**, No. 2, p. K107–K110 (1967).
See also *10*, p. *103*; *13.2*, p. *193*; 22.4.1, p. 501.

L. Ghiţă and E. Ţenescu
*On the Exciton Spectrum of PbI_2 on Different Substrates
Vol. **24**, No. 2, p. K175–K178 (1967).
See also 13.2, p. 193; 22.5, p. 513.

L. Śniadower and R. Kowalczyk
Magnetoplasma Reflection in Mercury Telluride
Vol. **25**, No. 1, p. 233–240 (1968).
See also 13.1, p. 183; *18*, p. *293*; *20*, p. *340*; 22.4.3, p. 510.

B. L. Evans and P. A. Young
Delocalized Excitons in Thin Anisotropic Crystals
Vol. **25**, No. 1, p. 417–425 (1968).
See also *13.2*, p. *193*; 22, p. 449.

N. A. Goryunova, V. M. Orlov, V. I. Sokolova, G. P. Shpenkov, and E. V. Tsvetkova
Preparation and Some Properties of CuP_2 Single Crystals
Vol. **25**, No. 2, p. 513–519 (1968).
See also *3.1*, p. *33*; 5, p. 53; 14.3, p. 240; 15, p. 273; 22, p. 450.

G. Baldini, A. Jean, and G. Spinolo
Lifetime of the Excited Cu^+ in Alkali Halides
Vol. **25**, No. 2, p. 557–561 (1968).
See also 13.4, p. 211; *20.3*, p. *368*; 22.5.2, p. 525.

H. P. Geserich
Optical and Electrical Measurements on Thallium(III)Oxide Films
Vol. **25**, No. 2, p. 741—751 (1968).
See also *13.1*, p. *184*; 14.3.1, p. 246; 22.6, p. 540.

M. V. Dmitruk, A. A. Kaminskii, V. V. Osiko, and T. A. Tevosyan
Induced Emission of Hexagonal LaF_3-SrF_2-Nd^{3+} Crystals at Room-Temperature
Vol. **25**, No. 2, p. K75—K78 (1968).
See also *20.2*, p. *361*; 20.3, p. 368; 22.5, p. 513; 22.8, p. 552.

V. Vyšín and V. Janků
*Note on the Interpretation of the Rotatory Dispersion of Potassium Dithionate
Vol. **25**, No. 2, p. K103—K104 (1968).
See also 22.8, p. 553.

Yu. K. Voronko and V. A. Sychugov
The Stimulated Emission of Er^{3+} Ions in CaF_2 at $\lambda_1 = 8456$ Å and $\lambda_2 = 8548$ Å
Vol. **25**, No. 2, p. K119—K121 (1968).
See also *20.2*, p. *361*; 20.3, p. 368; 22.5.3, p. 534.

P. G. Harper and J. A. Hilder
Exciton Spectra in Thin Crystals
Vol. **26**, No. 1, p. 69—76 (1968).
See also *13.2*, p. *193*; 22, p. 450.

S. Yamada, Y. Kawasaki, and O. Nishida
*Free Carrier Absorption in n-Type CdTe Crystals
Vol. **26**, No. 1, p. 77—84 (1968).
See also 22.4.3, p. 510.

A. Meessen
Vectorial Selective Surface Effects in the Optical Absorption and Photoelectric Emission of Thin Metal Films
Vol. **26**, No. 1, p. 125—132 (1968).
See also *17*, p. *288*; 21, p. 385.

P. Grosse und K. Winzer
*Magnetoabsorptionsmessungen an Tellur
(Magnetoabsorption Measurements on Tellurium)
Vol. **26**, No. 1, p. 139—150 (1968).
See also *13.1*, p. *184*; 18, p. 293; 22.1.3, p. 478.

A. Hadni, J. Claudel et P. Strimer
*Spectres d'absorption et de réflexion, dans l'infrarouge lointain, de ZnSe, ZnTe et CdSe à basse température
(Far Infrared Absorption and Reflection Spectra of ZnSe, ZnTe, and CdSe at Low Temperature)
Vol. **26**, No. 1, p. 241—252 (1968).
See also 6, p. 65; 22.4.2, p. 507; 22.4.3, p. 510.

T. A. Davis and K. Vedam
*Variation of the Refractive Indices of Calcite, with Pressure to 7 kbar
Vol. **26**, No. 1, p. 285—290 (1968).
See also 22, p. 450.

W. Thielemann
*Effect of Doping on the Optical Absorption of Circularly Polarized Waves in n-Type GaAs in a Magnetic Field
Vol. **26**, No. 1, p. K29—K31 (1968).
See also 13.1, p. 184; 18, p. 294; 22.2.1, p. 485.

G. A. Sikharulidze and Yu. I. Ukhanov
*Absorption of Polarized Light in $CdSnAs_2$ Crystals
Vol. **26**, No. 1, p. K33–K36 (1968).
See also 13.1, p. 184; 22.7, p. 546; 22.8, p. 553.

R. Enderlein
Excitonic Effects in the Electro-Absorption of Semiconductors
Vol. **26**, No. 2, p. 509–517 (1968).
See also 13.1, p. 185; 13.2, p. 193; *14.4.1*, p. *262*; *20*, p. *340*; 22, p. 451.

E. D. Trifonov and A. S. Troshin
*Nonlinear Polarisation of an Impurity Crystal
Vol. **26**, No. 2, p. 519–525 (1968).
See also 13.4, p. 212; *20.3*, p. *368*; 22, p. 451.

K. Hirschberg and K. Deutscher
Determination of the Emission Depth of Photoelectrons from Cs–Sb Films
Vol. **26**, No. 2, p. 527–535 (1968).
See also 14.3.1, p. 246; *17*, p. *288*; 22.7, p. 546.

G. Döhler and K. Hacker
A Possibility for Band Structure Determination: Intraband and Donor-Band Transitions of Crystal Electrons in High Electric Fields
Vol. **26**, No. 2, p. 551–564 (1968).
See also *13.1*, p. *185*; 13.4, p. 212; 14.4.1, p. 262; 22, p. 451.

J. Pastrňák and L. Roskovcová
*Optical Absorption Edge of AlN Single Crystals
Vol. **26**, No. 2, p. 591–597 (1968).
See also 13.1, p. 185; 22.2, p. 481.

A. V. Tulub und K. Patzer
*Zur Theorie der kooperativen Absorption (Emission) von Licht in Kristallen
(On the Theory of Cooperative Absorption (Emission) of Light in Crystals)
Vol. **26**, No. 2, p. 693–700 (1968).
See also 13.4, p. 212; *20.3*, p. *369*; 22.5, p. 513.

W. Picht and W. Ruppel
Photocurrent, Luminescence, and Absorption Saturation for Electron Beam Excitation in CdS
Vol. **26**, No. 2, p. K163–K165 (1968).
See also *16*, p. *282*; 20.3, p. 369; 22.4.1, p. 502.

A. S. Davydov
*Theory of Urbach's Rule
Vol. **27**, No. 1, p. 51–56 (1968).
See also 22, p. 451.

E. Schmidt
*Optical Properties of GeSi Alloys in the Energy Region from 1 to 13 eV
Vol. **27**, No. 1, p. 57–67 (1968).
See also 13.1, p. 185; 22.1.1, p. 464; 22.1.2, p. 473.

G. Baldini, I. Pollini, and G. Spinolo
*Optical Properties of α- and β-$TiCl_3$
Vol. **27**, No. 1, p. 95–100 (1968).
See also 13.4, p. 213; 22.5, p. 514.

K. DEUTSCHER and K. HIRSCHBERG
Thickness Dependence of the Quantum Yield of Cesium-Antimony Films
Vol. 27, No. 1, p. 145—150 (1968).
See also 14.3.1, p. 246; *17*, p. *289*; 22.7, p. 546.

M. GRYNBERG
Influence of Uniaxial Stress on the Optical Properties of CdSe
Vol. 27, No. 1, p. 255—261 (1968).
See also *13.1*, p. *186*; 20.3, p. 369; 22.4.2, p. 507.

W. ULRICI
Study of Trivalent Transition-Metal Ions in Silver Halides. (I): AgCl:V
Vol. 27, No. 1, p. 333—344 (1968).
See also *10*, p. *107*; 13.4, p. 213; 14.4, p. 254; 22.5.1, p. 516.

V. N. BOGOMOLOV and D. N. MIRLIN
*Optical Absorption by Polarons in Rutile (TiO_2) Single Crystals
Vol. 27, No. 1, p. 443—453 (1968).
See also 13, p. 170; 22.6, p. 541.

H. BURKHARD and J. JAUMANN
The Determination of the "Gyroelectric" and "Gyromagnetic" Constants of Thick, Evaporated Layers of Iron by Means of Measurements in the Reflected Light only
Vol. 27, No. 1, p. K43—K45 (1968).
See also *18.2*, p. *312*; *20*, p. *341*; 21.1.1, p. 421.

A. F. LUBCHENKO
*Theory of Urbach's Rule for Impurity Centres
Vol. 27, No. 1, p. K73—K76 (1968).
See also 10, p. 107; 13.4, p. 213; 22, p. 452.

M. YA. VALAKH, N. I. VITRIKHOVSKII, and M. P. LISITSA
The Electron Effective Mass in CdS_xSe_{1-x} Crystals
Vol. 27, No. 1, p. K81—K83 (1968).
See also *13.1*, p. *186*; 22.4.1, p. 502; 22.4.2, p. 507; 22.8, p. 554.

M. I. KLINGER
On the Experimental Observation of Small Polarons in Semiconductors
Vol. 27, No. 2, p. 479—488 (1968).
See also *13*, p. *170*; 14.3, p. 241; 22.6, p. 541.

W. ULRICI
Study of Trivalent Transition-Metal Ions in Silver Halides. (II): Chromium and Titanium
Vol. 27, No. 2, p. 489—500 (1968).
See also *10*, p. *107*; 13.4, p. 213; 14.4, p. 254; 22.5.1, p. 517.

G. APPELT
Fine Structure Measurements in the Energy Angular Distribution of Secondary Electrons from a (110) Face of Copper
Vol. 27, No. 2, p. 657—669 (1968).
See also 13.1, p. 186; *17*, p. *289*; 21.1, p. 405.

S. D. MCLAUGHLAN and H. W. EVANS
Production of Colloidal Calcium by Electron Irradiation of CaF_2 Crystals
Vol. 27, No. 2, p. 695—700 (1968).
See also 10.2, p. 142; *11*, p. *152*; 19, p. 333; 22.5.3, p. 534.

H. K. Müller
Electrical and Optical Properties of Sputtered In_2O_3 Films. (I): Electrical Properties and Intrinsic Absorption
Vol. 27, No. 2, p. 723–731 (1968).
See also 13.4, p. 214; *14.3.1*, p. *247*; 22.6, p. 541.

H. K. Müller
*Electrical and Optical Properties of Sputtered In_2O_3 Films. (II): Optical Properties in the Near Infrared
Vol. 27, No. 2, p. 733–740 (1968).
See also 14.3.1, p. 247; 22.6, p. 541.

V. Fedoseev and V. Hizhnyakov
On the Theory of Low-Temperature Exciton Absorption Spectra
Vol. 27, No. 2, p. 751–755 (1968).
See also 6, p. 66; *13.2*, p. *194*; 22, p. 452.

E. Kierzek-Pecold, J. Kołodziejczak, and I. Pracka
*Anisotropy in the Reflection Coefficient of β-Rhombohedral Boron in the Region 1 to 6 eV
Vol. 27, No. 2, p. K107–K109 (1968).
See also 22.1, p. 457.

J. S. Tiwari, A. Mehra, and K. G. Srivastava
*Electronic Absorption Spectrum of K_2CoF_4
Vol. 28, No. 1, p. 61–63 (1968).
See also 5, p. 53; 13, p. 171; 22.5, p. 514; 22.8, p. 554.

L. Merten
Directional Dispersion of Extraordinary Optical Phonons of α-Quartz
Vol. 28, No. 1, p. 111–119 (1968).
See also *6*, p. *66*; 22.6, p. 542.

M. Höhne and M. Stasiw
ESR Detection of Self-Trapped Holes in AgCl
Vol. 28, No. 1, p. 247–253 (1968).
See also 10.2, p. 143; *19*, p. *334*; 22.5.1, p. 517.

M. V. Andriyashik, M. Yu. Sakhnovskii, V. B. Timofeev, and A. S. Yakimova
*Optical Transitions in the Spectra of the Fundamental Absorption and Reflection of InSe Single Crystals
Vol. 28, No. 1, p. 277–285 (1968).
See also 13.1, p. 187; 13.2, p. 194; 22, p. 542.

A. Farwer und F. Matossi†
*Innere optische Reflexionen an kubischem ZnS mit eindimensionaler Lagefehlordnung
(Internal Optical Reflection in Cubic ZnS with One-Dimensional Stacking Faults)
Vol. 28, No. 1, p. 373–383 (1968).
See also 4, p. 47; 10, p. 109; 22.4.1, p. 503.

B. F. Bilenkii, A. G. Mikolaichuk und D. M. Freik
*Struktur und optische Eigenschaften von epitaxialen SnTe-, SnSe- und SnS-Schichten
(Structure and Optical Properties of Epitaxial SnTe, SnSe, and SnS Films)
Vol. 28, No. 1, p. K5–K7 (1968).
See also *1.4*, p. *21*; 22, p. 452.

C. MUNTONI and A. SERPI
On an Allotropic Transformation of ZnS after Pressure Treatment
Vol. 28, No. 1, p. K43–K45 (1968).
See also *1.2*, p. *12*; 13.1, p. 187; 22.4.1, p. 503.

I. F. CHANG, S. S. MITRA, J. N. PLENDL, and L. C. MANSUR
Long-Wavelength Longitudinal Phonons of Multi-Mode Crystals
Vol. 28, No. 2, p. 663–673 (1968).
See also *6*, p. *67*; 22.4.1, p. 504; 22.4.2, p. 508; 22.6, p. 542; 22.8, p. 554.

S. BALLARÒ, A. BALZAROTTI, and V. GRASSO
Electroreflectance at Oblique Incidence in Ge
Vol. 28, No. 2, p. K109–K112 (1968).
See also 13.1, p. 188; *14.4.1*, p. *264*; *20*, p. *341*; 22.1.1, p. 466.

L. ŚNIADOWER
*Temperature Studies of Magnetoplasma Reflection in Mercury Telluride
Vol. 28, No. 2, p. K125–K130 (1968).
See also 13.1, p. 188; *18*, p. *294*; 22.4.3, p. 511.

S. N. SHESTATSKII and V. V. SOBOLEV
Electroreflectance Spectrum of Cuprous Oxide Crystals
Vol. 28, No. 2, p. K131–K133 (1968).
See also 13.1, p. 188; 13.2, p. 194; *14.4.1*, p. *264*; *20*, p. *342*; 22.6, p. 542.

M. R. TUBBS
*The Variation of Exciton Peak Position with Lattice Parameter in Metal Iodide Crystals
Vol. 28, No. 2, p. K135–K137 (1968).
See also 13.2, p. 195; 22.5, p. 514; 22.5.2, p. 528; 22.5.3, p. 534.

YU. B. ROSENFELD, B. G. VEKHTER, and B. S. TSUKERBLAT
*Optical Bands in Electron-Vibrational Systems with Degeneracy
Vol. 28, No. 2, p. K179–K182 (1968).
See also 6, p. 67.

W. KRESS, H. BORIK, and R. K. WEHNER
Infrared Lattice Absorption of Silicon and Germanium
Vol. 29, No. 1, p. 133–143 (1968).
See also *6*, p. *67*; 22.1.1, p. 466; 22.1.2, p. 474.

P. B. PICKAR and H. D. TILLER
*Optical Energy Gap in TlSe
Vol. 29, No. 1, p. 153–158 (1968).
See also 5, p. 54; 13.1, p. 188; 16, p. 283; 22, p. 453.

M. ALTWEIN, H. FINKENRATH, Č. KOŇÁK, J. STUKE, and G. ZIMMERER
*The Electronic Structure of CdO. (II): Spectral Distribution of Optical Constants
Vol. 29, No. 1, p. 203–209 (1968).
See also 13.1, p. 188; 14.3.1, p. 247; 22.4, p. 496.

R. BOYN
*Optical Absorption due to Intrinsic Defects in CdS Single Crystals
Vol. 29, No. 1, p. 307–321 (1968).
See also 10, p. 110; *13.4*, p. *215*; 22.4.1, p. 504.

V. M. ASNIN, G. L. ERISTAVI, and A. A. ROGACHEV
*Effect of Weak Electric Fields on the Absorption Edge in Doped Germanium
Vol. 29, No. 1, p. 443–446 (1968).
See also 13.2, p. 195; 13.4, p. 215; 14.3, p. 243; 22.1.1, p. 466.

P. Reimers and W. Ruppel
The Preparation of CdS-CdSe Graded Single Crystals
Vol. **29**, No. 1, p. K31–K33 (1968).
See also *3.1*, p. *33*; 5, p. 54; 22.4.1, p. 504; 22.4.2, p. 508; 22.8, p. 555.

K. Platzöder
*Temperature Effects on the Vacuum Ultraviolet Reflectance of α-Quartz
Vol. **29**, No. 1, p. K63–K64 (1968).
See also 13.1, p. 189; 13.2, p. 195; 22.6, p. 543.

V. A. M. Brabers and P. Dekker
*Infrared Spectra and Cation Distributions of Manganese Ferrites
Vol. **29**, No. 1, p. K73–K76 (1968).
See also 18.3, p. 320.

R. A. Burley
An Infrared Spectrophotometric Study of Vitreous Selenium Doped with Selenium Dioxide
Vol. **29**, No. 2, p. 551–558 (1968).
See also 2, p. 24; *10*, p. *110*; 22.1.3, p. 480.

A. R. Ruffa
The Valence Bond Approximation in Crystals — Application to an Analysis of the Ultraviolet Spectrum in Quartz
Vol. **29**, No. 2, p. 605–616 (1968).
See also 3.2, p. 36; *13.1*, p. *189*; 22.6, p. 543.

D. Bäuerle and B. Fritz
*Vibrational Spectra of U_1-Centres in KI
Vol. **29**, No. 2, p. 639–643 (1968).
See also *6*, p. *68*; 10.2, p. 144; 22.5.2, p. 529.

R. G. Howell and D. J. Newman
*High Resolution Study of Crystal Field Lines in the Far Infra-Red
Vol. **29**, No. 2, p. 697–706 (1968).
See also 6, p. 68; 18.1, p. 297; 22.8, p. 555.

J. Dillinger, Č. Konák, V. Prosser, J. Sak, and M. Zvára
Phonon-Assisted Exciton Transitions in $A^{II}B^{VI}$ Semiconductors
Vol. **29**, No. 2, p. 707–714 (1968).
See also 6, p. 68; *13.2*, p. *195*; 22.4.1, p. 505; 22.4.2, p. 508; 22.4.3, p. 511.

E. Z. Dziuba, D. Niculescu, and N. Niculescu
Energy Band Structure of $Zn_xHg_{1-x}Te$ Solid Solutions
Vol. **29**, No. 2, p. 813–818 (1968).
See also *13.1*, p. *189*; 14.3, p. 244; 22.4.3, p. 511; 22.8, p. 556.

A. Mehra
*Optical Absorption of Mn^{2+}-Doped Alkali Halides
Vol. **29**, No. 2, p. 847–857 (1968).
See also 6, p. 69; *10*, p. *111*; 13.4, p. 215; 22.5.2, p. 529.

E. Kierzek-Pecold, J. Kołodziejczak, and I. Pracka
*Optical Constants of Crystalline CaB_6
Vol. **29**, No. 2, p. K183–K185 (1968).
See also 13.1, p. 189; 22, p. 454.

Yu. G. Litvinenko, V. V. Eremenko, and T. I. Garber
Antiferromagnetic Ordering Effect on the Light Absorption Spectrum by Crystalline Oxygen
Vol. **30**, No. 1, p. 49–60 (1968).
See also 13.1, p. 189; *18.4*, p. *325*; 23, p. 561.

H. Vogt and H. Happ
*The Temperature Dependence of Absorption of Sodium Nitrite in the Far Infrared
Vol. **30**, No. 1, p. 67–72 (1968).
See also *6*, p. *69*; 14.4.2, p. 269.

K. H. Timmesfeld
*Anharmonic Broadening of Resonant Modes
Vol. **30**, No. 1, p. 73–85 (1968).
See also 3.2, p. 36; *6*, p. *69*; 22.5.2, p. 530.

M. L. Reynolds, W. E. Hagston, and G. F. J. Garlick
*Absorption and Infra-Red Emission of Nickel and Cobalt Doped Alkali Halides
Vol. **30**, No. 1, p. 97–103 (1968).
See also 13.4, p. 216; *20.3*, p. *371*; 22.5.2, p. 530.

M. L. Reynolds, W. E. Hagston, and G. F. J. Garlick
Experimental Observations of the Absorption, Infra-Red Emission, and Excitation Spectra of Cr^{3+} Ions in $ZnWO_4$
Vol. **30**, No. 1, p. 113–117 (1968).
See also *13.4*, p. *216*; 20.3, p. 371; 22.8, p. 556.

R. Boyn
*Polarization of Optical Absorption and Emission Bands due to Donor–Acceptor Associates in Wurtzite Type Crystals
Vol. **30**, No. 1, p. 305–310 (1968).
See also 13.4, p. 216; *20.3*, p. *371*; 22.4, p. 496.

V. V. Sobolev, S. D. Shutov, Yu. V. Popov, and S. N. Shestatskii
*Reflectivity Spectra of the Rhombohedral Crystals Bi_2Te_3, Bi_2Se_3, and Sb_2Te_3 over the Range from 0.7 to 12.5 eV
Vol. **30**, No. 1, p. 349–355 (1968).
See also 13.1, p. 189; 22, p. 454.

A. S. Davydov
The Radiationless Transfer of Energy of Electronic Excitation between Impurity Molecules in Crystals
Vol. **30**, No. 1, p. 357–366 (1968).
See also 6, p. 69; *13.4*, p. *216*; 20.3, p. 371; 22, p. 454.

U. Birkholz, H. Finkenrath, J. Naegele, and N. Uhle
*Infrared Reflectivity of Semiconducting $FeSi_2$
Vol. **30**, No. 1, p. K81–K84 (1968).
See also 22, p. 455; 22.7, p. 546.

D. V. Chepur, D. M. Bercha, I. D. Turyanitsa, and V. Yu. Slivka
Peculiarities of the Energy Spectrum and Edge Absorption in the Chain Compounds $A^{V}B^{VI}C^{VII}$
Vol. **30**, No. 2, p. 461–468 (1968).
See also *13.1*, p. *189*; 16, p. 284; 22.8, p. 557.

E. F. Gross, B. S. Razbirin, V. P. Fedorov, and Yu. P. Naumov
Bound Exciton Complexes in CdSe Crystals
Vol. **30**, No. 2, p. 485–494 (1968).
See also *13.2*, p. *195*; 20.3, p. 371; 22.4.2, p. 508.

D. Strauch
*Shell Model Treatment of the Vibrations of Crystals Containing Substitutional Defects. (II): On the Theory of Defect-Induced One-Phonon Absorption
Vol. **30**, No. 2, p. 495–505 (1968).
See also *6*, p. *70*; 10, p. 112; 22.1, p. 457; 22.1.1, p. 466; 22.1.2, p. 475.

M. L. Reynolds, W. E. Hagston, and G. F. J. Garlick
The Theory of the Absorption and Infrared Emission of Cr^{3+} Ions in $ZnWO_4$
Vol. **30**, No. 2, p. 507–514 (1968).
See also *13.4*, p. *217*; 20.3, p. 371; 22.8, p. 557.

D. Boccon-Gibod
*Modulation of Near Infrared Radiation by Acoustoelectric Domains in n-Type GaAs at Room Temperature
Vol. **30**, No. 2, p. 525–531 (1968).
See also 7, p. 78; *14.4.1*, p. *264*; 22.2.1, p. 486.

G. Baldini and M. Guzzi
Zero-Phonon Transitions of Eu^{++}- and Sm^{++}-Doped Alkali Halides
Vol. **30**, No. 2, p. 601–612 (1968).
See also *6*, p. *70*; 10, p. 112; 20.3, p. 372; 22.5.2, p. 531.

K. Dos, A. Haug, and P. Rohner
Consideration of the Exchange Interaction in theTwo-Body Picture of Excitons and Its Influence on the Exciton Spectra
Vol. **30**, No. 2, p. 619–625 (1968).
See also *13.2*, p. *195*; 22.4.1, p. 505.

J. Voigt and F. Spiegelberg
On the Dynamical Behaviour of Excitons in CdS Single Crystals. (I): Correlation between Intrinsic Excitonic Absorption and Emission
Vol. **30**, No. 2, p. 659–670 (1968).
See also 6, p. 70; *13.2*, p. *195*; 20.3, p. 372; 22.4.1, p. 505.

M. L. Reynolds, W. E. Hagston, and G. F. J. Garlick
*The Optical Absorption Spectra of Co and Ni in $ZnWO_4$
Vol. **30**, No. 2, p. 735–739 (1968).
See also *13.4*, p. *217*; 22.8, p. 557.

A. T. Davidson and A. D. Yoffe
Hopping Electrical Conduction and Thermal Breakdown in Natural and Synthetic Mica
Vol. **30**, No. 2, p. 741–754 (1968).
See also 3, p. 28; 10, p. 113; *14.4.1*, p. *265*; 22, p. 455.

E. M. Averbakh, V. V. Sobolev, N. N. Syrbu, and Ya. A. Ugai
*The Optical Reflectivity Spectra of CdSb and ZnSb Single Crystals
Vol. **30**, No. 2, p. K145–K147 (1968).
See also 13.1, p. 190; 22.7, p. 546.

E. Cruceanu, St. Slădaru, and T. Boţilă
*Optical Energy Gap of Tl_2Te_3
Vol. **30**, No. 2, p. K149–K151 (1968).
See also 13.1, p. 190; 22, p. 455.

B. Selle und J. Maege
*Elektrische und optische Eigenschaften von Cu_2S-Aufdampfschichten
(Electrical and Optical Properties of Evaporated Cu_2S Films)
Vol. **30**, No. 2, p. K153–K155 (1968).
See also 13.4, p. 218; *14.3.1*, p. *247*; 15, p. 275; 16, p. 285; 22, p. 455.

20.2 Lasers

Review Articles

Y. P. VARSHNI
Band-to-Band Radiative Recombination in Groups IV, VI, and III-V Semiconductors (I)
Vol. **19**, No. 2, p. 459–514 (1967).
See also 13.1, p. 174; 13.2, p. 191; 13.4, p. 200; 20.1, p. 344; *20.3*, p. *363*; 22.1, p. 456; 22.1.1, p. 458; 22.1.2, p. 468; 22.1.3, p. 476; 22.2.1, p. 482; 22.2.2, p. 487; 22.2.3, p. 489.

Y. P. VARSHNI
Band-to-Band Radiative Recombination in Groups IV, VI, and III-V Semiconductors (II)
Vol. **20**, No. 1, p. 9–36 (1967).
See also 13.1, p. 174; 13.2, p. 191; 13.4, p. 200; 20.1, p. 344; *20.3*, p. *363*; 22.1, p. 456; 22.1.1, p. 458; 22.1.2, p. 468; 22.1.3, p. 476; 22.2.1, p. 482; 22.2.2, p. 487; 22.2.3, p. 489.

M. H. PILKUHN
*The Injection Laser
Vol. **25**, No. 1, p. 9–62 (1968).
See also 13.4, p. 200; 14.3.2, p. 248; *14.4.1*, p. *255*; 20.3, p. 363; 22.2.1, p. 482; 22.2.2, p. 487; 22.2.3, p. 489.

Original Papers and Short Notes

L. I. VAN TORNE
*Pumping Induced Imperfections in Glass:Nd^{3+} Lasers
Vol. **16**, No. 1, p. 171–182 (1966).
See also 2, p. 22; 4, p. 37; 22.6, p. 535.

G. A. BOGOMOLOVA, A. A. KAMINSKII, and V. A. TIMOFEEVA
Nd^{3+} Optical Centres in $Y_3Al_5O_{12}$ Crystals
Vol. **16**, No. 2, p. K165–K166 (1966).
See also 10, p. 94; *20.1*, p. *345*; 22.8, p. 547.

C. BENOIT À LA GUILLAUME et J.-M. DEBEVER
*Effet laser dans le CdS par bombardement électronique
(Laser Effect in CdS by Electron Bombardment)
Vol. **17**, No. 2, p. 875–879 (1966).
See also 22.4.1, p. 498.

O. V. BOGDANKEVICH, M. M. ZVEREV, A. I. KRASILNIKOV, and A. N. PECHENOV
*Laser Emission in Electron-Beam Excited ZnSe
Vol. **19**, No. 1, p. K5–K6 (1967).
See also 22.4.2, p. 506.

S. RAAB, H. BACHERT, and A. KEIPER
*Spectral Investigation of Emission Inhomogeneities in GaAs Injection Lasers
Vol. **19**, No. 1, p. K59–K62 (1967).
See also 22.2.1, p. 483.

P. T. LANDSBERG
*The Condition for Negative Absorption in Semiconductors
Vol. **19**, No. 2, p. 777–780 (1967).
See also 13.4, p. 205; 22, p. 443.

A. A. Kaminskii
*The Nature of "Aging" of CaF_2-YF_3-Nd^{3+} Crystals (Type I) under Stimulated Emission Conditions
Vol. **20**, No. 1, p. K51—K54 (1967).
See also 22.5, p. 512; 22.5.3, p. 532; 22.8, p. 549.

V. D. Egorov
Luminescence of CdS Excited by Electrons of Various Energies
Vol. **20**, No. 2, p. 705—711 (1967).
See also 13.2, p. 192; *20.3*, p. *365*; 22.4.1, p. 499.

A. A. Kaminskii, Yu. K. Voronko, and V. V. Osiko
*Mixed Systems on the Basis of Fluorides as New Laser Materials for Quantum Electronics. The Optical and Emission Parameters
Vol. **21**, No. 1, p. K17—K21 (1967).
See also 20.1, p. 348; 20.3, p. 365; 22.5, p. 513; 22.8, p. 550.

A. A. Kaminskii, Kh. S. Bagdasarov, and L. M. Belyev
Analysis of the Optical Spectra and Stimulated Radiation of $Y_3Al_5O_{12}$-Nd^{3+} Crystals
Vol. **21**, No. 1, p. K23—K29 (1967).
See also 13.4, p. 207; *20.1*, p. *348*, *20.3*, p. *365*; 22.8, p. 550.

H. Pink
*The Determination of Chromium Distribution in Laser-Rubies with Neutron Activation
Vol. **21**, No. 2, p. K111—K113 (1967).
See also *10*, p. *101*; 22.6, p. 538.

G. E. Arkhangelskii, Z. L. Morgenshtern, and V. B. Neustruev
Colour Centres in Ruby Crystals
Vol. **22**, No. 1, p. 289—295 (1967).
See also *10.2*, p. *140*; 20.1, p. 349; 20.3, p. 366; 22.6, p. 538.

A. A. Kaminskii, Y. E. Lapsker, and B. P. Sobolev
*Induced Emission of $NaCaCeF_6$:Nd^{3+} at Room Temperature
Vol. **23**, No. 1, p. K5—K7 (1967).
See also 20.1, p. 350; 20.3, p. 367; 22.5, p. 513; 22.8, p. 552.

V. D. Egorov, G. O. Müller, and H. H. Weber
Cathodoluminescence Yields in Highly Excited CdS
Vol. **23**, No. 2, p. 703—708 (1967).
See also 13.2, p. 193; *20.3*, p. *367*; 22.4.1, p. 501.

M. V. Dmitruk, A. A. Kaminskii, V. V. Osiko, and T. A. Tevosyan
*Induced Emission of Hexagonal LaF_3-SrF_2-Nd^{3+} Crystals at Room-Temperature
Vol. **25**, No. 2, p. K75—K78 (1968).
See also 20.1, p. 352; 20.3, p. 368; 22.5, p. 513; 22.8, p. 552.

Yu. K. Voronko and V. A. Sychugov
*The Stimulated Emission of Er^{3+} Ions in CaF_2 at $\lambda_1 = 8456$ Å and $\lambda_2 = 8548$ Å
Vol. **25**, No. 2, p. K119—K121 (1968).
See also 20.1, p. 352; 20.3, p. 368; 22.5.3, p. 534.

L. N. Kurbatov, E. F. Gross, V. E. Mashenko, B. S. Razbirin, N. N. Mochalkin, and A. D. Britov
*Coherent Emission of Free Excitons in CdSe Crystals
Vol. **26**, No. 2, p. K139—K142 (1968).
See also 6, p. 65; *13.2*, p. *194*; 22.4.2, p. 507.

Kh. S. Bagdasarov, A. A. Kaminskii, V. S. Krylov, and V. I. Popov
*Room-Temperature Induced Emission of Tetragonal YVO_4 Crystals Containing Nd^{3+}
Vol. **27**, No. 1, p. K1—K3 (1968).
See also 22.8, p. 553.

A. Graja
*Production of the Second Harmonic of Light in Ammonium Pentaborate and Other Powdered Piezoelectric Crystals
Vol. **27**, No. 2, p. K93—K97 (1968).
See also *20*, p. *341*; 22.8, p. 554.

J. Hatz
*Some Effects of Material Inhomogeneities on the Near-Field Pattern of GaAs Diode Lasers
Vol. **28**, No. 1, p. 233—245 (1968).
See also 10, p. 109; 14.3.2, p. 251; 22.2.1, p. 485.

O. V. Bogdankevich, N. A. Borisov, I. V. Krukova, and B. M. Lavrushin
*Temperature Dependence of Laser Threshold Current Density and Emission Spectra in Electron-Beam Pumped Gallium Arsenide Lasers
Vol. **29**, No. 2, p. 715—723 (1968).
See also 13.4, p. 215; 22.2.1, p. 486.

H. Bachert and S. Raab
*The Influence of External Optical Coupling on the Threshold Current Density of GaAs Injection Lasers
Vol. **29**, No. 2, p. K175—K178 (1968).
See also 22.2.1, p. 486.

A. A. Kaminskii and L. Li
*The Photoreduction $Nd^{3+} \rightarrow Nd^{2+}$ in SrF_2 (Type I) Crystals under Stimulated Emission Conditions ("Aging")
Vol. **30**, No. 1, p. K77—K79 (1968).
See also 22.5.3, p. 534.

20.3 Luminescence of Solids (X-Ray Spectra See 20)

Review Articles

Y. P. Varshni
*Band-to-Band Radiative Recombination in Groups IV, VI, and III–V Semiconductors (I)
Vol. **19**, No. 2, p. 459–514 (1967).
See also 13.1, p. 174; 13.2, p. 191; 13.4, p. 200; 20.1, p. 344; 20.2, p. 360; 22.1 p. 456; 22.1.1, p. 458; 22.1.2, p. 468; 22.1.3, p. 476; 22.2.1, p. 482; 22.2.2, p. 487; 22.2.3, p. 489.

Y. P. Varshni
*Band-to-Band Radiative Recombination in Groups IV, VI, and III–V Semiconductors (II)
Vol. **20**, No. 1, p. 9–36 (1967).
See also 13.1, p. 174; 13.2, p. 191; 13.4, p. 200; 20.1, p. 344; 20.2, p. 360; 22.1, p. 456; 22.1.1, p. 458; 22.1.2, p. 468; 22.1.3, p. 476; 22.2.1, p. 482; 22.2.2, p. 487; 22.2.3, p. 489.

M. H. Pilkuhn
The Injection Laser
Vol. **25**, No. 1, p. 9–62 (1968).
See also 13.4, p. 200; 14.3.2, p. 248, *14.4.1*, p. *255*; *20.2*, p. *360*; 22.2.1, p. 482; 22.2.2, p. 487; 22.2.3, p. 489.

F. Williams
Donor-Acceptor Pairs in Semiconductors
Vol. **25**, No. 2, p. 493–512 (1968).
See also 10, p. 93; *13.4*, p. *200*; 22.1, p. 456; 22.1.1, p. 458; 22.1.2, p. 468; 22.2.2, p. 487; 22.4.1, p. 497.

V. L. Bonch-Bruevich and E. G. Landsberg
Recombination Mechanisms
Vol. **29**, No. 1, p. 9–43 (1968).
See also 6, p. 56; 13, p. 162; 13.2, p. 191; *13.4*, p. *200*; *14.3*, p. *231*; 20, p. 337; 22.1.1, p. 458; 22.1.2, p. 468.

Original Papers and Short Notes

N. Riehl and P. Thoma
*Delayed Fluorescence and Thermoluminescence of Anthracene Single Crystals at Very Low Temperatures
Vol. **16**, No. 1, p. 159–170 (1966).
See also 13.4, p. 201; 22.9, p. 558.

G. Gallus and H. C. Wolf
Direct Measurement of the Diffusion Length of Singulet Excitons in Solid Phenanthrene
Vol. **16**, No. 1, p. 277–280 (1966).
See also *13.2*, p. *191*; 22.9, p. 558.

M. Bancie-Grillot et P. Bourtayre
*Émission photoluminescente visible du gadolinium inclus dans le réseau cristallin du sulfure de zinc
(Visible Photoluminescence Emission of Gadolinium Incorporated in the Crystal Lattice of Zinc Sulphide)
Vol. **16**, No. 2, p. 517–523 (1966).
See also 13.4, p. 201; 22.4.1, p. 497.

V. V. Ratnam
*On the Colour Centres and X-Ray Luminescence of Calcium Fluoride Crystals
Vol. **16**, No. 2, p. 559—567 (1966).
See also *10.2*, p. *137*; 22.5.3, p. 532.

M. L. Mukherjee and H. N. Bose
Nonradiative Destruction of F-Centres in X-Irradiated KCl Doped with Anionic Impurities
Vol. **16**, No. 2, p. 591—598 (1966).
See also *10.2*, p. *137*; 22.5.2, p. 518.

R. B. Laibowitz and H. S. Sack
*Luminescence and Photoconductivity in Silver Chloride with Laser Excitation
Vol. **17**, No. 1, p. 353—360 (1966).
See also 6, p. 57; **13.4**, p. 202; *16*, p. *277*; 22.5.1, p. 515.

H.-E. Gumlich, R. Moser und E. Neumann
*Die Rekombinationsverteilung in der instationären Phase der Elektrolumineszenz von ZnS-Phosphoren mit Eisengruppenelementen
(Recombination Distribution in the Non-Stationary Phase of the Electroluminescence of ZnS Phosphors Containing Iron-Group Elements)
Vol. **17**, No. 2, p. 783—792 (1966).
See also 13.4, p. 203; 22.4.1, p. 498.

O. Goede and E. Gutsche
*Edge Emission in CdS as Internal Radiative Transition in Donor–Acceptor Associates
Vol. **17**, No. 2, p. 911—923 (1966).
See also 6, p. 57; 13.4, p. 203; 22.4.1, p. 498.

G. A. Akhundov, I. G. Aksyanov, and A. G. Bagirov
*Electroluminescence of GaSe Single Crystals Excited by Square Pulses
Vol. **17**, No. 2, p. K225—K227 (1966).
See also 22, p. 441.

P. Bräunlich and A. Scharmann
Approximate Solution of Schön's Balance Equations for the Thermoluminescence and the Thermally Stimulated Conductivity of Inorganic Photoconducting Crystals
Vol. **18**, No. 1, p. 307—316 (1966).
See also *13.4*, p. *203*; 14.3, p. 233; 16, p. 278; 22.4, p. 495.

K. F. Lider, B. V. Novikov, and S. A. Permogorov
*Application of Bound-Exciton Spectra in the Study of Radiation Damage in Crystals
Vol. **18**, No. 1, p. K1—K4 (1966).
See also 11, p. 147; *13.2*, p. *191*; 22.4.1, p. 499.

B. S. Razbirin, I. Filiński, and B. Wojtowicz-Natanson
Hot Excitons in Cadmium Selenide
Vol. **18**, No. 1, p. K37—K40 (1966).
See also *13.2*, p. *192*; 22.4.2, p. 506.

B. Tuck
*Photoluminescence of $GaAs_{0.7}P_{0.3}$
Vol. **18**, No. 2, p. 541—545 (1966).
See also 13.4, p. 204; 22.2.1, p. 483; 22.2.2, p. 488; 22.8, p. 548.

V. V. Eremenko, E. V. Matyushkin, and S. V. Petrov
*Study of Energy Transfer from 3d to 4f Electrons in Antiferromagnetic MnF_2:Eu^{3+} Crystals
Vol. **18**, No. 2, p. 683—686 (1966).
See also *13.4*, p. *204*; 18.4, p. 321; 22.5.3, p. 532.

E. D. Aluker and I. P. Mezina
*The Temperature Dependence of Radioluminescence Yield in Alkali Halide Crystal Phosphors
Vol. **19**, No. 1, p. 35–40 (1967).
See also 13.4, p. 204; 22.5.2, p. 520.

V. V. Antonov-Romanovskii
*On the Kinetics of the Crystal Phosphor Luminescence
Vol. **19**, No. 1, p. 417–424 (1967).
See also 22, p. 442.

N. A. Vlasenko and S. A. Zynio
*Role of Carrier Polarization in the Mechanism of the Electroluminescence of ZnS-Mn Films
Vol. **20**, No. 1, p. 311–320 (1967).
See also 14.4.1, p. 257; 22.4.1, p. 499.

F. T. Goldstein
F-Center Formation by Fundamental Absorption in KI
Vol. **20**, No. 1, p. 379–393 (1967).
See also *10.2*, p. *139*; 22.5.2, p. 521.

G. Blasse and A. Bril
Some Observations on the Cr^{3+} Fluorescence in the Huntite Structure
Vol. **20**, No. 2, p. 551–556 (1967).
See also 13.4, p. 205; 20.1, p. 347; 22.8, p. 549.

V. D. Egorov
*Luminescence of CdS Excited by Electrons of Various Energies
Vol. **20**, No. 2, p. 705–711 (1967).
See also 13.2, p. 192; 20.2, p. 361; 22.4.1, p. 499.

F. J. Bryant and E. Webster
*Threshold Energy for Atomic Displacement in Cadmium Telluride
Vol. **21**, No. 1, p. 315–321 (1967).
See also *11*, p. *148*; 22.4.3, p. 509.

P. Bräunlich
Light Effects on Space-Charge Controlled Currents in CdS
Vol. **21**, No. 1, p. 383–391 (1967).
See also 13.4, p. 206; *14.4.1*, p. *258*; 16, p. 279; 22.4.1, p. 499.

A. A. Kaminskii, Yu. K. Voronko, and V. V. Osiko
Mixed Systems on the Basis of Fluorides as New Laser Materials for Quantum Electronics. The Optical and Emission Parameters
Vol. **21**, No. 1, p. K17–K21 (1967).
See also 20.1, p. 348; *20.2*, p. *361*; 22.5, p. 513; 22.8, p. 550.

A. A. Kaminskii, Kh. S. Bagdasarov, and L. M. Belyev
*Analysis of the Optical Spectra and Stimulated Radiation of $Y_3Al_5O_{12}$-Nd^{3+} Crystals
Vol. **21**, No. 1, p. K23–K29 (1967).
See also 13.4, p. 207; *20.1*, p. *348*; 20.2, p. 361; 22.8, p. 550.

A. N. Georgobiani and V. I. Steblin
*Electroluminescence of ZnS-Cu_2S Heterojunctions
Vol. **21**, No. 1, p. K45–K48 (1967).
See also 14.3.2, p. 249; 22, p. 445; 22.4.1, p. 499.

A. N. Georgobiani and V. I. Steblin
Optical and Electrical Characteristics of p-n Junctions in Zinc Sulfide
Vol. **21**, No. 1, p. K49–K51 (1967).
See also *14.3.2*, p. *249*; 16, p. 279; 22.4.1, p. 499.

A. Bonnot
*Relaxation des domaines de haut champ et émission de lumière associées aux oscillations de courant dans GaAs
(Relaxation of High-Field Domains and Light Emission Associated with Current Oscillations in GaAs)
Vol. **21**, No. 2, p. 525–531 (1967).
See also 7, p. 76; *14.4.1*, p. *258*; 22.2.1, p. 484.

J. P. Agrawal
*Infrared Fluorescence in Manganese Acticated Zinc Sulphide Phosphors
Vol. **21**, No. 2, p. 679–682 (1967).
See also 13.4, p. 207; 20.1, p. 348; 22.4.1, p. 500.

V. Hizhnyakov and I. Tehver
*Theory of Resonant Secondary Radiation due to Impurity Centres in Crystals
Vol. **21**, No. 2, p. 755–768 (1967).
See also 13.4, p. 207; 20.1, p. 348; 22, p. 446.

H. Kühnert
*Polarization of F-Center Luminescence in KCl due to Electric Fields
Vol. **21**, No. 2, p. K171–K173 (1967).
See also 10.2, p. 139; 22.5.2, p. 522.

E. V. Matyushkin, L. S. Kukushkin, and V. V. Eremenko
*Peculiarities of the $RbMnF_3:Nd^{3+}$ Crystals Decay Kinetics due to the Migration of Excitation Energy
Vol. **22**, No. 1, p. 65 –69 (1967).
See also 13.4, p. 208; 22.5, p. 513; 22.8, p. 551.

E. W. Claffy
Impurity Absorption Bands in Thermoluminescent LiF
Vol. **22**, No. 1, p. 71–76 (1967).
See also 13.4, p. 208; *20.1*, p. *349*; 22.5.2, p. 523.

W. Degrieck and G. Jacobs
*Luminescence and Photoconductivity of the B-Centre in Additively Coloured KCl: Ag Crystals
Vol. **22**, No. 1, p. 177–184 (1967).
See also 10.2, p. 139; *16*, p. *279*; 22.5.2, p. 523.

G. E. Arkhangelskii, Z. L. Morgenshtern, and V. B. Neustruev
Colour Centres in Ruby Crystals
Vol. **22**, No. 1, p. 289–295 (1967).
See also *10.2*, p. *140*; 20.1, p. 349; 20.2, p. 361; 22.6, p. 538.

K. Jeges
*Elektrolumineszenz von SnO_2-Einkristallen
(Electroluminescence of SnO_2 Single Crystals)
Vol. **22**, No. 1, p. K7–K9 (1967).
See also 22.6, p. 538.

P. Bräunlich
*Calculation of Light-Stimulated Luminescence and Current Transients in Phosphors
Vol. **22**, No. 2, p. 391–400 (1967).
See also 13.4, p. 208; *16*, p. *280*; 22, p. 447.

D. Obrikat, K. Meyer, and F. Polly
*Triboluminescence Spectra of Doped Zinc Sulphides
Vol. **22**, No. 2, p. K123–K126 (1967).
See also 22.4.1, p. 500.

J. B. GRUN, A. MYSYROWICZ, F. RAGA, A. BIVAS, R. LEVY, and S. NIKITINE
*Photon-Assisted Emission in CdSe
Vol. **22**, No. 2, p. K155—K156 (1967).
See also 6, p. 62; 13.2, p. 193; 22.4.2, p. 507.

K. VACEK
*Electroluminescence of AgCl Crystals Excited by Short Field Pulses
Vol. **23**, No. 1, p. 105—111 (1967).
See also 22.5.1, p. 516.

H. P. MÜLLER, P. THOMA, and G. VAUBEL
*The Phosphorescence of Anthracene Single Crystals and Its Spectrum
Vol. **23**, No. 1, p. 253—262 (1967).
See also 13, p. 167; 22.9, p. 559.

C. BENOIT À LA GUILLAUME and M. VOOS
Optical Measurement of Compensation in Highly Doped Silicon
Vol. **23**, No. 1, p. 295—299 (1967).
See also *13.4*, p. *209*; 22.1.2, p. 472.

P. GÖRLICH, H. KARRAS, G. KÖTITZ, and P. ULLMANN
*Thermoluminescence Investigations on Doped Alkaline Earth Fluorides
Vol. **23**, No. 1, p. 313—323 (1967).
See also 13.4, p. 209; 20.1, p. 350; 22.5.3, p. 533.

A. A. KAMINSKII, Y. E. LAPSKER, and B. P. SOBOLEV
Induced Emission of $NaCaCeF_6$:Nd^{3+} at Room Temperature
Vol. **23**, No. 1, p. K5—K7 (1967).
See also 20.1, p. 350; *20.2*, p. *361*; 22.5, p. 513; 22.8, p. 552.

S. IKONOPISOV, L. AVRAMOV, and S. VELEVA
*On the Electroluminescence of Indium Oxide Layers in Contact with an Electrolyte
Vol. **23**, No. 1, p. K61—K65 (1967).
See also 22.6, p. 539.

I. P. IPATOVA, A. A. KLOCHICHIN, and A. V. SUBASHIEV
*The Influence of Localized and Resonance Modes of Lattice Vibrations on the Radiative Recombination in Semiconductors
Vol. **23**, No. 2, p. 467—473 (1967).
See also *6*, p. *63*; 13.4, p. 209; 22.1.2, p. 472.

P. G. ELISEEV, A. I. KRASILNIKOV, M. A. MANKO, and I. Z. PINSKER
*Band-Filling Model for Injection Luminescence at Higher Temperatures
Vol. **23**, No. 2, p. 587—593 (1967).
See also 13.4, p. 210; 22.2.1, p. 484.

V. D. EGOROV, G. O. MÜLLER, and H. H. WEBER
*Cathodoluminescence Yields in Highly Excited CdS
Vol. **23**, No. 2, p. 703—708 (1967).
See also 13.2, p. 193; 20.2, p. 361; 22.4.1, p. 501.

H.-E. GUMLICH, R. MOSER, and E. NEUMANN
*On the Mn–Mn Interaction and the Structure of Optical Spectra of ZnS(Mn)
Vol. **24**, No. 1, p. K13—K16 (1967).
See also 13.4, p. 210; 22.4.1, p. 501.

T. A. T. COWELL and J. WOODS
Thermally Stimulated Currents and Infra-Red Luminescence in CdS Crystals
Vol. **24**, No. 1, p. K37—K40 (1967).
See also 10.2, p. 141; *13.4*, p. *210*; 14.3, p. 239; 22.4.1, p. 501.

M. K. Sheinkman, V. A. Tyagai, O. V. Snitko, I. B. Ermolovich, G. L. Belenky, and V. N. Bondarenko
Studies on the Nature of CdS Single Crystal Sensitization by Etching
Vol. **24**, No. 2, p. 543–549 (1967).
See also 10, p. 103; *13.4*, p. *210*; 16, p. 281; 22.4.1, p. 501.

P. Görlich, H. Karras, G. Kötitz, and R. Rauch
*Phonon-Assisted Colour Centre Fluorescence of Additively Coloured Alkali Earth Fluoride Crystals
Vol. **25**, No. 1, p. K15–K18 (1968).
See also 6, p. 65; *10.2*, p. *141*; 22.5.3, p. 533.

G. Baldini, A. Jean, and G. Spinolo
*Lifetime of the Excited Cu^+ in Alkali Halides
Vol. **25**, No. 2, p. 557–561 (1968).
See also 13.4, p. 211; 20.1, p. 351; 22.5.2, p. 525.

U. Kümmel und G. Zollfrank
*Zur Phänomenologie der Leuchtverteilungen in elektrolumineszierenden, hexagonalen ZnS-Kristallen
(Phenomenological Study of Brightness Distributions in Electroluminescent Hexagonal ZnS Crystals)
Vol. **25**, No. 2, p. 711–720 (1968).
See also 10, p. 104; 22.4.1, p. 502.

Yu. A. Logatchov
Model Potential for Systems with Two Valence Electrons
Vol. **25**, No. 2, p. 763–774 (1968).
See also *13.4*, p. *211*; 22, p. 450.

M. V. Dmitruk, A. A. Kaminskii, V. V. Osiko, and T. A. Tevosyan
Induced Emission of Hexagonal LaF_3-SrF_2-Nd^{3+} Crystals at Room-Temperature
Vol. **25**, No. 2, p. K75–K78 (1968).
See also 20.1, p. 352, *20.2*, p. *361*; 22.5, p. 513; 22.8, p. 552.

Yu. K. Voronko and V. A. Sychugov
The Stimulated Emission of Er^{3+} Ions in CaF_2 at $\lambda_1 = 8456$ Å and $\lambda_2 = 8548$ Å
Vol. **25**, No. 2, p. K119–K121 (1968).
See also 20.1, p. 352; *20.2*, p. *361*; 22.5.3, p. 534.

V. F. Pissarenko and P. I. Bykovskii
*Luminescence of Sm^{2+} in $CaSO_4$
Vol. **26**, No. 1, p. K13–K15 (1968).
See also 22.8, p. 553.

N. A. Vlasenko and A. N. Gergell
*On the Mechanism of DC Electroluminescence in pCu_xS-nZnS:Mn, Cu, Cl Film Structures
Vol. **26**, No. 1, p. K77–K81 (1968).
See also 14.3.2, p. 250; 22, p. 450; 22.4.1, p. 502.

P. M. Karageorgy-Alkalaev and A. Yu. Leiderman
Statistics of Inter-Impurity Recombination of Electrons and Holes in Semiconductors
Vol. **26**, No. 2, p. 419–428 (1968).
See also *13.4*, p. *212*; 22, p. 450.

E. D. Trifonov and A. S. Troshin
*Nonlinear Polarisation of an Impurity Crystal
Vol. **26**, No. 2, p. 519–525 (1968).
See also 13.4, p. 212; *20.1*, p. *353*; 22, p. 451.

A. KAWSKI und A. J. HORODECKI
*Die Elektrolumineszenzeigenschaften von ZnS-Cu in Abhängigkeit von der Art des Koaktivators
(Coactivator Dependence of the Electroluminescent Properties of ZnS-Cu)
Vol. **26**, No. 2, p. 543—549 (1968).
See also 22.4.1, p. 502.

A. V. TULUB und K. PATZER
*Zur Theorie der kooperativen Absorption (Emission) von Licht in Kristallen
(On the Theory of Cooperative Absorption (Emission) of Light in Crystals)
Vol. **26**, No. 2, p. 693—700 (1968).
See also 13.4, p. 212; *20.1*, p. *353*; 22.5, p. 513.

YU. A. LOGATCHOV
Two-Electron Correlation in Impurity Centres in Model Potential Approximation
Vol. **26**, No. 2, p. 765—768 (1968).
See also *13.4*, p. *212*; 22.5.2, p. 526.

K. MEYER und E. LINKE
*Triboinduzierte Elektronenemission und Lumineszenz an Alkalihalogeniden
(Triboinduced Electron-Emission and Luminescence from Alkali Halides)
Vol. **26**, No. 2, p. K89—K92 (1968).
See also *17*, p. *289*; 22.5.2, p. 526.

W. PICHT and W. RUPPEL
Photocurrent, Luminescence, and Absorption Saturation for Electron Beam Excitation in CdS
Vo. **26**, No. 2, p. K163—K165 (1968).
See also *16*, p. *282*; 20.1, p. 353; 22.4.1, p. 502.

P. GÖRLICH, H. KARRAS, G. KÖTITZ, and R. RAUCH
*Polarized Luminescence of X-Irradiated CaF_2:Y and SrF_2:Y Crystals and the Structure of Their Luminescence Centres
Vol. **27**, No. 1, p. 109—115 (1968).
See also 10, p. 106; 22.5.3, p. 534.

M. GRYNBERG
Influence of Uniaxial Stress on the Optical Properties of CdSe
Vol. **27**, No. 1, p. 255—261 (1968).
See also *13.1*, p. *186*; 20.1, p. 354; 22.4.2, p. 507.

B. T. KOLOMIETS, T. N. MAMONTOVA, and V. V. NEGRESKUL
*On the Recombination Radiation of Vitreous Semiconductors
Vol. **27**, No. 1, p. K15—K16 (1968).
See also *2*, p. *24*, 22.8, p. 553.

K. VACEK, A. MYSYROWICZ, R. LEVY, A. BIVAS, and S. NIKITINE
*Luminescence of AgCl Excited by Ultraviolet Laser
Vol. **27**, No. 1, p. K29—K31 (1968).
See also 22.5.1, p. 516.

R. J. FLEMING
*Activation Energies and Emission Spectra of the Thermoluminescence of CaF_2 Crystals Gamma-Irradiated at 77 °K
Vol. **27**, No. 1, p. K57—K61 (1968).
See also 13.4, p. 213; 22.5.3, p. 534.

B. K. DANIELS and D. B. MEADOWCROFT
*Twist Boundaries and Electroluminescence
Vol. **27**, No. 2, p. 535—539 (1968).
See also *10*, p. *107*; 22.4.1, p. 502.

G. F. ALFREY and D. B. MEADOWCROFT
*The Influence of Structure on the Electroluminescence of Zinc Sulphide
Vol. **27**, No. 2, p. 541–544 (1968).
See also 4, p. 46; 10, p. 107; 22.4.1, p. 502.

D. HAHN and J. MIMKES
*On the Frequency Dependence of Electroluminescence of ZnS:Cu Phosphors
Vol. **27**, No. 2, p. K133–K135 (1968).
See also 22.4.1, p. 503.

V. I. CHEREPANOV and A. E. NIKIFOROV
Consideration of Polarization and Covalency Corrections to the Quasi-Coulomb Shift of Optical Lines of Ion Pairs in Crystals
Vol. **28**, No. 1, p. 65–73 (1968).
See also *10*, p. *108*; 13.4, p. 214; 22.6, p. 542.

H.-J. HÄNSCH
*Untersuchungen zum Mechanismus der Elektrolumineszenz von CdS-Einkristallen (Study of the Electroluminescence Mechanism of CdS Single Crystals)
Vol. **28**, No. 1, p. 255–268 (1968).
See also 13.2, p. 194; 14.4.1, p. 263; 22.4.1, p. 503.

L. K. THOMAS
*The Normal Spectral Emissivity of Tungsten-Molybdenum Alloys
Vol. **28**, No. 1, p. 401–406 (1968).
See also 13.1, p. 187; 21, p. 386.

N. A. GORYUNOVA, S. M. RYVKIN, G. P. SHPENIKOV, I. I. TICHINA, and V. G. FEDOTOV
*Investigations of Some Properties of Vitreous and Crystalline $CdGeP_2$
Vol. **28**, No. 2, p. 489–494 (1968).
See also 2, p. 24; 14.3, p. 242; 22.8, p. 554.

G. K. BORN, R. J. GRASSER, and A. O. SCHARMANN
EPR and Thermoluminescence Measurements on $CaWO_4$ Crystals
Vol. **28**, No. 2, p. 583–588 (1968).
See also 10.2, p. 143; *11*, p. *153*; 19, p. 334; 22.8, p. 554.

A. CINGOLANI, A. RIZZO, and A. LEVIALDI
*AC Modulated Green Edge Emission in CdS
Vol. **28**, No. 2, p. K151–K154 (1968).
See also 22.4.1, p. 504.

W. W. HOLLOWAY, JR. and M. KESTIGIAN
*Temperature Dependence of the Broad-Band Fluorescence of $RbMnF_3$
Vol. **28**, No. 2, p. K159–K162 (1968).
See also 22.5, p. 514; 22.8, p. 554.

O. GOEDE
*Investigation of the Emission of CdS Single Crystals in the 2.0 eV Region
Vol. **28**, No. 2, p. K167–K170 (1968).
See also 10, p. 109; 22.4.1, p. 504.

M. P. FONTANA, H. BLUME, and W. J. VAN SCIVER
Properties of Exciton States in NaI. (I): The Intrinsic Photoluminescence
Vol. **29**, No. 1, p. 159–166 (1968).
See also *13.2*, p. *195*; 22.5.2, p. 528.

B. TUCK
*Effect of Heat Treatments on GaAs Luminescence
Vol. **29**, No. 2, p. 793–798 (1968).
See also 13.4, p. 215; 22.2.1, p. 486.

G. E. Arkhangelskii, Z. L. Morgenshtern, and V. B. Neustruev
*On the Nature of the Colour Centres of Ruby
Vol. **29**, No. 2, p. 831—836 (1968).
See also *10.2*, p. *144*; 13.4, p. 215; 22.6, p. 543.

M. L. Reynolds, W. E. Hagston, and G. F. J. Garlick
*Absorption and Infra-Red Emission of Nickel and Cobalt Doped Alkali Halides
Vol. **30**, No. 1, p. 97—103 (1968).
See also 13.4, p. 216; *20.1*, p. *358*; 22.5.2, p. 530.

M. L. Reynolds, W. E. Hagston, and G. F. J. Garlick
Experimental Observations of the Absorption, Infra-Red Emission, and Excitation Spectra of Cr^{3+} Ions in $ZnWO_4$
Vol. **30**, No. 1, p. 113—117 (1968).
See also *13.4*, p. *216*; 20.1, p. 358; 22.8, p. 556.

R. L. Gilbert
Empirical Relationships for the Emission Energies and Lifetimes of the F-Center
Vol. **30**, No. 1, p.173—182 (1968).
See also *10.2*, p. *144*; 13.4, p. 216; 22.5.2, p. 530.

R. Boyn
*Polarization of Optical Absorption and Emission Bands due to Donor–Acceptor Associates in Wurtzite Type Crystals
Vol. **30**, No. 1, p. 305—310 (1968).
See also 13.4, p. 216; *20.1*, p. *358*; 22.4, p. 496.

E. L. Frankevich and B. M. Rumyantsev
*Charge-Transfer Excitons and Delayed Fluorescence of Anthracene
Vol. **30**, No. 1, p. 329—340 (1968).
See also 13.2, p. 195; 22.9, p. 560.

A. S. Davydov
The Radiationless Transfer of Energy of Electronic Excitation between Impurity Molecules in Crystals
Vol. **30**, No. 1, p. 357—366 (1968).
See also 6, p. 69; *13.4*, p. *216*; 20.1, p. 358; 22, p. 454.

D. N. Bose
Thermoluminescence and Thermally Stimulated Currents in Bismuth Doped ZnS Crystals
Vol. **30**, No. 1, p. K57—K60 (1968).
See also *13.4*, p. *217*; 16, p. 284; 22.4.1, p. 505.

K. Peuker and E. D. Trifonov
*Theory of Light Emission by Impurity Centres from Non-Equilibrium Vibrational States
Vol. **30**, No. 2, p. 479—484 (1968).
See also 6, p. 70; 13.4, p. 217; 22, p. 455.

E. F. Gross, B. S. Razbirin, V. P. Fedorov, and Yu. P. Naumov
Bound Exciton Complexes in CdSe Crystals
Vol. **30**, No. 2, p. 485—494 (1968).
See also *13.2*, p. *195*; 20.1, p. 358; 22.4.2, p. 508.

M. L. Reynolds, W. E. Hagston, and G. F. J. Garlick
The Theory of the Absorption and Infrared Emission of Cr^{3+} Ions in $ZnWO_4$
Vol. **30**, No. 2, p. 507—514 (1968).
See also *13.4*, p. *217*; 20.1, p. 359; 22.8, p. 557.

G. Baldini and M. Guzzi
Zero-Phonon Transitions of Eu^{++}- and Sm^{++}-Doped Alkali Halides
Vol. **30**. No. 2, p. 601—612 (1968).
See also *6*, p. *70*; 10, p. 112; 20.1, p. 359; 22.5.2, p. 531.

J. Voigt and F. Spiegelberg
On the Dynamical Behaviour of Excitons in CdS Single Crystals. (I): Correlation between Intrinsic Excitonic Absorption and Emission
Vol. **30**, No. 2, p. 659—670 (1968).
See also 6, p. 70; *13.2*, p. *195*; 20.1, p. 359; 22.4.1, p. 505.

G. Richter and M. Schleifstein
*Photoluminescence in the Presence of Crossed Fields
Vol. **30**, No. 2, p. K97—K100 (1968).
See also 22.1.1, p. 467.

A. M. White, E. W. Wiilliams, and P. Porteous
*Infra-Red and Visible Photoluminescence in $In_{1-x}Ga_xP$
Vol. **30**, No. 2, p. K125—K126 (1968).
See also 22.2.2, p. 488; 22.8, p. 557.

K.-J. Berg and K.-P. Meyer
*Thermoluminescence of Z_1-Centers in Ca Doped KCl Crystals after X-Irradiation
Vol. **30**, No. 2, p. K161—K164 (1968).
See also *10.2*, p. *144*; 22.5.2, p. 531.

21. Metals. Alloys

Review Articles

F. R. Brotzen
Emission of Exoelectrons from Metallic Materials
Vol. **22**, No. 1, p. 9–30 (1967).
See also 13.3, p. 196; *17*, p. *286*.

D. P. Morgan
Helicon Waves in Solids
Vol. **24**, No. 1, p. 9–36 (1967).
See also 6, p. 56; *13*, p. *162*; 14.1, p. 220; 14.3, p. 231; 19, p. 326; *20*, p. *337*; 21.1, p. 390; 21.2, p. 424; 21.6, p. 430; 22.1.1, p. 458; 22.2.1, p. 482; 22.2.3, p. 489.

J. Mertsching
Theory of Electromagnetic Waves in Metals and Their Interaction with Ultrasonic Waves (II)
Vol. **26**, No. 1, p. 9–64 (1968).
See also *7*, p. *76*; 13, p. 162; 19, p. 326; 20, p. 337.

K. N. Subramanian
Cyclic Stressing in Crystals of Rock-Salt Structure
Vol. **28**, No. 1, p. 9–37 (1968).
See also *10*, p. *93*; 10.1, p. 114; 21.1, p. 390; 22.5.1, p. 515; 22.5.2, p. 518; 22.6, p. 535.

W. Steinmann
Optical Plasma Resonances in Solids
Vol. **28**, No. 2, p. 437–462 (1968).
See also *13*, p. *162*; 13.3, p. 196; *20*, p. *337*; 20.1, p. 344; 21.2, p. 424; 21.6, p. 430.

Original Papers and Short Notes

H. Kronmüller and A. Seeger
Magnetostrictive and Magnetocrystalline Residual Electric Resistivity in Ferromagnetics
Vol. **16**, No. 1, p. 253–266 (1966).
See also 13, p. 162; *14.1*, p. *220*; *18.2*, p. *299*.

J. Mertsching
On the Determination of the Electron Free Path from Measurements of Ultrasonic Absorption in Metals
Vol. **16**, No. 1, p. 267–275 (1966).
See also *7*, p. *76*; 13, p. 162.

J. S. Rajput and A. K. Gupta
On the Specific Heat Anomaly in Superconducting Lead
Vol. **16**, No. 1, p. K51–K53 (1966).
See also *8*, p. *79*; *14.2*, p. *228*.

M. Dubejko and S. Olszewski
Quantum-Statistical Calculations on the Work Function of Metals
Vol. **16**, No. 2, p. 399–404 (1966).
See also 13, p. 163; *17*, p. *286*.

S. Ceresara, T. Federighi, and F. Pieragostini
Determination of Diffusion Coefficients in Metals by Resistometric Method — Application to the Diffusion of Zn in Al
Vol. **16**, No. 2, p. 439–447 (1966).
See also *9*, p. *87*.

M. Attardo and J. M. Galligan
A Field Ion Microscope Study of Neutron Irradiated Tungsten
Vol. **16**, No. 2, p. 449–457 (1966).
See also **4**, p. 37; *11*, p. *145*; 17, p. 286.

M. M. Shukla and B. Dayal
Lattice Vibrations of Aluminium on the Basis of Krebs's Model
Vol. **16**, No. 2, p. 513–516 (1966).
See also *6*, p. *56*; 8, p. 79.

Chr. Nanev
Electron-Microscopic Study of the Growth Morphology of Zinc and Cadmium Single Crystals
Vol. **16**, No. 2, p. 777–787 (1966).
See also *1.3*, p. *14*; 3.1, p. 29; 21.1, p. 391.

D. A. Blackburn
Concentration of Defects near Interfaces in a Metal in the Presence of Electric Potential and Temperature Gradients
Vol. **17**, No. 1, p. 3–9 (1966).
See also *10.1*, p. *115*.

S. G. Christov
General Theory of Electron Emission from Metals
Vol. **17**, No. 1, p. 11–26 (1966).
See also *17*, p. *286*.

H. J. Fischbeck
On Pippard's Theory of Magnetic Breakdown in Metals
Vol. **17**, No. 1, p. 197–210 (1966).
See also *13.1*, p. *175*; 18, p. 291.

K. Isebeck, F. Rau, W. Schilling, K. Sonnenberg, P. Tischer, and H. Wenzl
Stored Energy, Volume, and Resistivity Change in Neutron Irradiated Aluminium
Vol. **17**, No. 1, p. 259–268 (1966).
See also 10.1, p. 115; *11*, p. *146*; 14.1, p. 220.

R. von Jan
The Interpretation of Recovery Stages in Irradiated Metals
Vol. **17**, No. 1, p. 361–373 (1966).
See also *10.1*, p. *116*; 11, p. 146; 21.1, p. 392; 21.6, p. 430.

R. Pichon, C. Minier-Cassayre, V. Hivert, and P. Moser
Study of the Recovery of Resistivity of Niobium Electron Irradiated at 20°K
Vol. **17**, No. 1, p. K33–K35 (1966).
See also 10.1, p. 116; *14.1*, p. *220*.

J. Bara, H. U. Hrynkiewicz, A. Z. Hrynkiewicz, M. Karapandźić, and T. Matlak
Magnetic Hyperfine Structure of the ^{57}Fe Impurity Nucleus in Non-Magnetic Materials
Vol. **17**, No. 1, p. K53–K55 (1966).
See also *6.1*, p. *71*.

P. Kratochvíl and B. Smola
The Effect of Vacuum Annealing on the Strain Behaviour of Cadmium Single Crystals
Vol. **17**, No. 1, p. K91–K94 (1966).
See also *10.1*, p. *116*.

E. Kovács-Csetényi, C. R. Vassel, and I. Kovács
Determination of the Solubility of Si in Dilute Al-Si Alloys by Low-Temperature Resistivity Measurements
Vol. **17**, No. 1, p. K123—K126 (1966).
See also *1.1*, p. *8*; **14.1**, p. 220.

A. Seeger and G. Wobser
Parallel Dislocation Arrays in Anisotropic Media
Vol. **17**, No. 2, p. 709—724 (1966).
See also *10*, p. *95*; *10.1*, p. *116*; 21.1, p. 392; 21.3, p. 426; 21.6, p. 431; 22.1.1, p. 460; 22.1.2, p. 469.

A. I. Gubanov
The LCAO Method for an Arbitrary Mixture of Atoms
Vol. **17**, No. 2, p. 807—814 (1966).
See also *2*, p. *22*; *13*, p. *163*; 22, p. 441.

M. F. Rose
Pressure Dependence of the Second Order Elastic Constants
Vol. **17**, No. 2, p. K199—K201 (1966).
See also *12.1*, p. *159*; 21.1, p. 393; 21.6, p. 431.

R. Lück
Temperature and Field Dependence of the Galvanomagnetic Properties of Aluminium and Indium
Vol. **18**, No. 1, p. 49—57 (1966).
See also 13.1, p. 176; *14.1*, p. *221*.

R. Lück
Temperature and Field Dependence of the Galvanomagnetic Properties of Lead and the Effect of Finite Sample-Thickness on These Dependences
Vol. **18**, No. 1, p. 59—66 (1966).
See also 13.1, p. 176; *14.1*, p. *221*.

D. Bergner and W. Lange
Diffusion of Sn and Zn in Sn and in a Homogeneous Sn-Zn Alloy
Vol. **18**, No. 1, p. 67—74 (1966).
See also *9*, p. *87*.

D. Bergner and W. Lange
Volume- and Grain-Boundary Diffusion of Sn and Zn in Heterogeneous Sn-Zn Alloys
Vol. **18**, No. 1, p. 75—84 (1966).
See also *9*, p. *87*.

G. B. Gibbs
Solvent Diffusivity in a Dilute B.C.C. Binary Alloy (Weak Perturbation Case)
Vol. **18**, No. 1, p. 85—88 (1966).
See also *9*, p. *87*.

T. R. Duncan and D. Kuhlmann-Wilsdorf
The Stress Fields of Dislocation Tilt Boundaries in Anisotropic Cubic and Diamond-Type Crystals
Vol. **18**, No. 1, p. 231—239 (1966).
See also *10*, p. *96*; *10.1*, p. *117*; 21.1, p. 393; 21.5, p. 429; 21.6, p. 431; 22.1, p. 456; 22.1.1, p. 460; 22.1.2, p. 470.

K. Isebeck, R. Müller, W. Schilling, and H. Wenzl
Reaction Kinetics of Stage III Recovery in Aluminium after Neutron Irradiation
Vol. **18**, No. 1, p. 427—439 (1966).
See also *10.1*, p. *118*; 11, p. 147; 14.1, p. 221.

K. M. Kunz, A. K. Green, and E. Bauer
On the Formation of Single Crystal Films of F.C.C. Metals on Alkali Halide Cleavage Planes in Ultrahigh Vacuum
Vol. **18**, No. 1, p. 441–457 (1966).
See also *1.4*, p. *17*; *3.1*, p. *30*; 21.6, p. 431.

W. Frank
Critical Shear Stress of Cubic Crystals Containing Defects of Tetragonal Symmetry. (III): Electron- und Neutron-Irradiated Aluminium and Electron-Irradiated Copper
Vol. **18**, No. 1, p. 459–464 (1966).
See also *10.1*, p. *118*; 11, p. 147; 21.1, p. 394.

K. C. A. Blasdale, R. King, and K. E. Puttick
The Microstructure of Tensile Kinks in Cadmium Crystals
Vol. **18**, No. 2, p. 491–503 (1966).
See also 4, p. 39; *10.1*, p. *118*.

S. Arajs, G. R. Dunmyre, and S. J. Dechter
Electrical Resistivity and Longitudinal Electrical Magnetoresistivity of Palladium-Rich Palladium–Erbium Alloys at Low Temperatures
Vol. **18**, No. 2, p. 505–509 (1966).
See also *14.1*, p. *221*.

G. C. T. Liu and J. C. M. Li
Energy of Hexagonal Dislocation Loops
Vol. **18**, No. 2, p. 527–539 (1966).
See also *10*, p. *96*; *10.1*, p. *118*; 22, p. 441; 22.1.2, p. 470.

V. Vítek
Thermally Activated Motion of Screw Dislocations in B.C.C. Metals
Vol. **18**, No. 2, p. 687–701 (1966).
See also *10.1*, p. *119*.

B. Harris
Solution Hardening in Niobium
Vol. **18**, No. 2, p. 715–730 (1966).
See also *10.1*, p. *119*.

R. I. Anishchenko
Calculation of the Mean Inner Potential of a Crystal in the Statistical Theory
Vol. **18**, No. 2, p. 923–928 (1966).
See also *3.2*, p. *34*; 13, p. 164; 21.1, p. 394; 21.1.1, p. 414; 21.2, p. 424; 21.3, p. 426; 21.4, p. 428; 21.6, p. 432; 21.7, p. 438; 22.1.1, p. 460.

R. H. Wade and J. Silcox
Small Angle Electron Scattering from Vacuum Condensed Metallic Films. (I): Theory
Vol. **19**, No. 1, p. 57–62 (1967).
See also *1.4*, p. *17*; 4, p. 39.

G. R. Barsch and Z. P. Chang
Adiabatic, Isothermal, and Intermediate Pressure Derivatives of the Elastic Constants for Cubic Symmetry. (II): Numerical Results for 25 Materials
Vol. **19**, No. 1, p. 139–151 (1967).
See also *12*, p. *156*; 22, p. 442.

A. K. Head
Unstable Dislocations in Anisotropic Crystals
Vol. **19**. No. 1, p. 185–192 (1967).
See also *10.1*, p. *120*; 21.1, p. 395; 21.1.1, p. 415.

W. van der Lugt and S. B. van der Molen
Nuclear Magnetic Resonance in Liquid Gallium Alloys
Vol. **19**, No. 1, p. 327–331 (1967).
See also 2, p. 22; *19*, p. *327*.

F. Guiu
Temperature and Strain Rate Dependence of the Flow Stress in Molybdenum
Vol. **19**, No. 1, p. 339–351 (1967).
See also *10.1*, p. *121*.

I. N. Evdokimov, E. S. Mashkova, V. A. Molchanov, and D. D. Odintsov
Dependence of the Ion-Electron Emission Coefficient on the Angle of Incidence
Vol. **19**, No. 1, p. 407–415 (1967).
See also *17*, p. *287*; 22.1, p. 457.

P. Lukáč
Stress Relaxation in Cd Single Crystals
Vol. **19**, No. 1, p. K47–K50 (1967).
See also *10.1*, p. *121*.

P. Charsley and J. A. M. Salter
Further Comments on the Analysis of Measurements of the Lattice Thermal Conductivity
Vol. **19**, No. 1, p. K63–K65 (1967).
See also *8*, p. *81*; 21.1, p. 396.

L. I. van Torne
Electron Charge Distribution about Zinc Ions in Aluminium
Vol. **19**, No. 2, p. 855–862 (1967).
See also 10.1, p. 121; *13*, p. *165*; 14.1, p. 222.

K. Wetzig
Investigations of Surfaces of the Binary System Ta-Fe by Means of an Emission Electron Microscope
Vol. **19**, No. 2, p. K71–K73 (1967).
See also *1.3*, p. *14*; 4, p. 40; 21.1.1, p. 415.

H. Träuble and U. Essmann
Observation of Magnetic Structures in Type II Superconductors
Vol. **20**, No. 1, p. 95–111 (1967).
See also *14.2*, p. *229*; 18, p. 291.

H. Bross and G. Bohn
On the Theory of Electrical Resistivity of Polyvalent Metals
Vol. **20**, No. 1, p. 277–283 (1967).
See also *14.1*, p. *222*.

R. P. Gupta
Lattice Specific Heats of Thallium and Yttrium
Vol. **20**, No. 1, p. 291–294 (1967).
See also 6, p. 59; *8*, p. *81*.

N. A. Pangarov
Twinning Processes in the Electrocrystallization of Face-Centred Cubic Metals
Vol. **20**, No. 1, p. 371–377 (1967).
See also *3.1*, p. *31*; 4, p. 41.

A. Czachor and A. Pindor
Dependence of Calculated Elastic Constants on the c/a Ratio of Hexagonal Close-Packed Metals
Vol. **20**, No. 1, p. K17–K20 (1967).
See also 5, p. 50; *12.1*, p. *159*; 21.1, p. 396; 21.3, p. 426.

S. Milošević and L. Dobrosavljević
Evalution of the One-Electron Potential for Al Crystals
Vol. **20**, No. 1, p. K63—K66 (1967).
See also *13.1*, p. *177*.

R. F. Peart
Effect of Pressure on the Diffusion of Fe in Ti and Ti + 10% Fe
Vol. **20**, No. 2, p. 545—550 (1967).
See also *9*, p. *88*; 12.1, p. 160; 21.1.1, p. 416.

D. Vaughan and J. M. Silcock
The Orientation and Shape of θ Precipitates Formed in an Al–Cu Alloy
Vol. **20**, No. 2, p. 725—736 (1967).
See also *4*, p. *41*.

V. P. Naberezhnykh and A. A. Maryakhin
Radio-Frequency Size Effects in Cadmium in Magnetic Fields Inclined to the Sample Surface
Vol. **20**, No. 2, p. 737—743 (1967).
See also *13.1*, p. *178*.

V. Radhakrishnan
The Penetration Depth in Two Band Model Superconductors
Vol. **20**, No. 2, p. 783—789 (1967).
See also *14.2*, p. *229*.

H.-J. Ullrich
Precision Lattice Parameter Measurements by Interferences from Lattice Sources (Kossel Lines) and Divergent Beam X-Ray Diffraction (Pseudo-Kossel Lines) in Back Reflection
Vol. **20**, No. 2, p. K113—K117 (1967).
See also *5*, p. *51*; 22, p. 444.

W. Haubenreisser and U. Lindner
Some Symmetry Aspects in the Theory of Localized Magnetic States in Metals
Vol. **20**, No. 2, p. K131—K134 (1967).
See also *13.4*, p. *206*; 18.1, p. 296.

H. Goretzki
Neutron Diffraction Studies on Titanium-Carbon and Zirconium-Carbon Alloys
Vol. **20**, No. 2, p. K141—K143 (1967).
See also *5*, p. *51*.

M. F. Rose
Higher Order Elastic Constants and the Equation of State of F.C.C. Metals
Vol. **21**, No. 1, p. 235—238 (1967).
See also 6, p. 60; 8, p. 82; *12.1*, p. *160*; 21.1, p. 397; 21.6, p. 433.

S. S. Sheinin
The Effect of Systematic Reflections on the (110) Extinction Distance in Molybdenum
Vol. **21**, No. 1, p. 247—256 (1967).
See also *5*, p. *51*.

N. F. Mott and R. S. Allgaier
Localized States in Disordered Lattices
Vol. **21**, No. 1, p. 343—356 (1967).
See also 2, p. 23; *13.4*, p. *206*; 22, p. 445.

J. A. M. Salter and P. Charsley
The Effect of Grain Size on the Lattice Thermal Conductivity of Copper Aluminium Alloys
Vol. **21**, No. 1, p. 357–368 (1967).
See also 4, p. 42; 6, p. 61; *8*, p. *82*; 21.1, p. 397.

J. Steinert
Induced Uniaxial Magnetic Anisotropy of Fe–Al Alloys at Low Concentrations
Vol. **21**, No. 1, p. K13–K15 (1967).
See also 10.1, p. 122; *18.2*, p. *307*; 21.1.1, p. 417.

R. A. French
Type-1 Superconductivity in Molybdenum
Vol. **21**, No. 1, p. K35–K38 (1967).
See also 13.1, p. 178; *14.2*, p. *229*.

A. Köthe and F. Schlät
Recovery of Deformed Niobium and Tantalum Degassed in Ultra-High Vacuum
Vol. **21**, No. 1, p. K73–K76 (1967).
See also *10.1*, p. *122*; 14.1, p. 222.

W. A. Jesser and D. Kuhlmann-Wilsdorf
The Geometry and Energy of a Twist Boundary between Crystals with Unequal Lattice Parameters
Vol. **21**, No. 2, p. 533–544 (1967).
See also *3.1*, p. *31*; 10, p. 100; 10.1, p. 123; 22, p. 445.

J. A. Sigler and D. Kuhlmann-Wilsdorf
Calculations on the Mechanical Energy of Vacancy Condensation Loops, Stacking Fault Tetrahedra, and Voids
Vol. **21**, No. 2, p. 545–556 (1967).
See also *10.1*, p. *123*; 21.1, p. 398; 21.6, p. 433.

K. C. A. Blasdale
Arrays of Dislocation Dipoles in Cadmium
Vol. **21**, No. 2, p. 649–660 (1967).
See also *10.1*, p. *123*.

R. Herrmann
Oscillations of the Surface Impedance of Tungsten in Weak Magnetic Fields
Vol. **21**, No. 2, p. 703–707 (1967).
See also 13.1, p. 178; *14.1*, p. *223*; *19*, p. *329*.

P. Humble
A Comparison of Electron Microscope Images of Dislocations Based on Isotropic and Anisotropic Elasticity Theory
Vol. **21**, No. 2, p. 733–740 (1967).
See also *10.1*, p. *123*; 21.1, p. 398; 21.1.1, p. 417; 21.6, p. 433.

F. Dworschak, H. Schuster, H. Wollenberger, and J. Wurm
The Influence of the Size Effect on Electrical Resistivity Measurements in Irradiated Metals
Vol. **21**, No. 2, p. 741–745 (1967).
See also *11*, p. *149*; 14.1, p. 223.

S. I. Masharov
Electrical Resistivity in Antiferromagnetic Alloys
Vol. **21**, No. 2, p. 747–753 (1967).
See also *14.1*, p. *223*; 18.4, p. 322.

H. Mühe
Temperature Dependence of Magnetocrystalline Anisotropy of an Fe-Al Alloy near the Concentration Fe_3Al
Vol. **21**, No. 2, p. K115–K119 (1967).
See also *18.2*, p. *307*; 21.1.1, p. 418.

A. Ya. Blank and E. A. Kaner
Waves with Discrete Spectrum in Metals in the Vicinity of the Cyclotron Resonance Frequency
Vol. **22**, No. 1, p. 47–58 (1967).
See also 13.1, p. 179; *19*, p. *329*.

M. Paunov
Field Electronmicroscopic Study of the Nucleation of Mercury
Vol. **22**, No. 1, p. 103–112 (1967).
See also *3.1*, p. *31*.

E. J. Kramer and Ch. L. Bauer
Variations of Young's Modulus throughout the Mixed State in Niobium
Vol. **22**, No. 1, p. 199–202 (1967).
See also 1.2, p. 11; *10.1*, p. *124*; 14.2, p. 229; 18, p. 292.

J. L. Routbort and H. S. Sack
Low Temperature Internal Friction of Aluminium at 1 Hz
Vol. **22**, No. 1, p. 203–212 (1967).
See also *10.1*, p. *124*.

F. Brouers
Plasmon Satellites of Soft X-Ray Spectra
Vol. **22**, No. 1, p. 213–221 (1967).
See also *13*, p. *166*; *20*, p. *339*; 21.2, p. 424.

E. A. Kaner and V. L. Falko
Magnetoacoustic Resonance Effects in Metals in Inclined Magnetic Fields
Vol. **22**, No. 1, p. 319–331 (1967).
See also 7, p. 77; 13.1, p. 179; *19*, p. *329*.

E. A. Kaner and V. G. Skobov
On the Possibility of Existence of Quantum Electromagnetic Waves in Metals
Vol. **22**, No. 1, p. 333–340 (1967).
See also *13*, p. *166*; 13.1, p. 179; *20*, p. *339*.

J. Demny
Stacking Faults in Tungsten
Vol. **22**, No. 1, p. K1–K2 (1967).
See also *10.1*, p. *124*.

G. Ya. Selyutin
On the Shift and Intensity of the Mössbauer Line in Metals
Vol. **22**, No. 1, p. K23–K26 (1967).
See also *6.1*, p. *72*.

K. Dettmann, G. Leibfried, and K. Schroeder
Spontaneous Recombination of Frenkel Pairs in Neutron-Irradiated Solids
Vol. **22**, No. 2, p. 433–440 (1967).
See also *11*, p. *149*; 21.1, p. 399; 21.6, p. 434.

V. Vítek
The Effect of the Type of Splitting of Screw Dislocations on the Slip Geometry in B.C.C. Metals
Vol. **22**, No. 2, p. 453–462 (1967).
See also *10.1*, p. *124*; 21.1.1, p. 418.

M. O. Speidel
Interaction of Dislocations with Coherent Particles and Stress Corrosion Cracking of High-Strength Aluminium Alloys
Vol. **22**, No. 2, p. K71–K74 (1967).
See also *10.1*, p. *124*.

H. Oppermann, G. Weise, J. Barthel, and K.-H. Berthel
Molybdenum Single Crystals with Very Small Residual-Resistivity Ratios
Vol. **22**, No. 2, p. K151–K153 (1967).
See also *3.1*, p. *32*; 14.1, p. 223.

A. van den Beukel
On the Relation between Self-Diffusion, Melting, and Vacancy-Impurity Binding in Dilute Alloys
Vol. **23**, No. 1, p. 165–170 (1967).
See also *9*, p. *90*; *10.1*, p. *125*.

J. B. Swan, A. Otto, and H. Fellenzer
Observed Retardation Effects on the Energy of the ω_+-Surface Plasmons in Thin Aluminium Foils
Vol. **23**, No. 1, p. 171–176 (1967).
See also *13*, p. *167*; 13.3, p. 198; *20*, p. *340*.

D. A. Blackburn
Non-Equilibrium Vacancy Concentrations in Metals Subject to Thermal Gradients and Electric Fields
Vol. **23**, No. 1, p. 177–183 (1967).
See also *10.1*, p. *125*; 21.6, p. 434.

R. F. Peart and J. Askill
The Mechanism of Diffusion in B.C.C. Transition Metals
Vol. **23**, No. 1, p. 263–275 (1967).
See also *9*, p. *90*; 10, p. 102; 21.1, p. 400; 21.1.1, p. 419.

S. Pal and P. K. Sharma
On Lindemann's Melting Criterion
Vol. **23**, No. 1, p. 361–364 (1967).
See also *6*, p. *62*; 21.1, p. 400; 21.1.1, p. 419; 21.2, p. 424; 21.6, p. 434.

J. Askill
Tracer Diffusion of Tungsten in Molybdenum
Vol. **23**, No. 1, p. K21–K23 (1967).
See also *9*, p. *90*.

L. Dobrosavljević and C. Dupuis
Study of Superconducting Pb-Sn Eutectic with Oriented Lamellas
Vol. **23**, No. 2, p. 509–520 (1967).
See also *14.2*, p. *229*; 18, p. 293.

M. L. Swanson
The Effects of Defect Doping on Low Temperature Neutron-Irradiation Damage and Recovery in Aluminium and Platinum
Vol. **23**, No. 2, p. 649–661 (1967).
See also 10.1, p. 125; *11*, p. *150*; 14.1, p. 224; 21.6, p. 434.

CHR. NANEV and D. IWANOV
Growth of Zinc and Cadmium Whiskers from the Vapour Phase on a Single Crystal Substrate of the Same Material
Vol. **23**, No. 2, p. 663–673 (1967).
See also *3.1*, p. *32*; 21.1, p. 400.

M. D. STAFLEU and A. R. DE VROOMEN
De Haas-van Alphen Effect in White Tin
Vol. **23**, No. 2, p. 675–682 (1967).
See also *13.1*, p. *182*; 18, p. 293.

M. D. STAFLEU and A. R. DE VROOMEN
Fermi Surface and Pseudopotential Coefficients in White Tin
Vol. **23**, No. 2, p. 683–696 (1967).
See also *13.1*, p. *182*.

R. M. EASSON and P. HLAWICZKA
Effect of H_{cl} on ac Losses in Niobium
Vol. **23**, No. 2, p. K129–K132 (1967).
See also *14.2*, p. *229*.

A. A. MARYAKHIN and I. V. SVECHKAREV
On the Pseudopotential Coefficients and Spin–Orbit Coupling in Metallic Cadmium
Vol. **23**, No. 2, p. K133–K136 (1967).
See also *13.1*, p. *182*.

F. KROUPA
A Note on Work Hardening in B.C.C. Metals
Vol. **23**, No. 2, p. K143–K145 (1967).
See also *10.1*, p. *126*.

L. WOJTCZAK
The Melting Point of Thin Films
Vol. **23**, No. 2, p. K163–K166 (1967).
See also 6, p. 63; *8*, p. *83*; 21.1, p. 400; 21.6, p. 434.

R. DUPREE and B. W. HOLLAND
The Range of g-Factors and the Breakdown of Motional Narrowing in Conduction Electron Spin Resonance
Vol. **24**, No. 1, p. 275–279 (1967).
See also *19*, p. *331*; 21.1, p. 401; 21.2, p. 425.

B. K. JONES
Measurement of the Magnitude of the Electronic Attenuation of Sound in Metals
Vol. **24**, No. 1, p. 297–300 (1967).
See also *7*, p. *77*; 13.1, p. 182; 21.1, p. 401; 21.2, p. 425.

D. HEITKAMP
On the Thermodynamics of Thermal Diffusion in Substitutional Alloys
Vol. **24**, No. 1, p. 341–348 (1967).
See also *9*, p. *90*; 10.1, p. 126.

S. A. Nemnonov and E. Z. Kurmaev
Band Structure and Superconductivity of A_3B-Type Intermetallic Compounds with β-W Structure
Vol. **24**, No. 1, p. K43—K47 (1967).
See also *13.1*, p. *183*; *14.2*, p. *229*.

R. Brugière and P. Lucasson
On the Stage III Recovery of Electron Irradiated Aluminium
Vol. **24**, No. 1, p. K77—K81 (1967).
See also *10.1*, p. *127*; 11, p. 150; 14.1, p. 224.

M. Rouzeyre
Contribution to the Study of Transport Phenomena in Aluminium-Alumina-Metal Structures
Vol. **24**, No. 2, p. 399—411 (1967).
See also *14.3.2*, p. *250*; 16, p. 281; 22.6, p. 540.

J. Geiger
Inelastic Electron Scattering in Thin Films at Oblique Incidence
Vol. **24**, No. 2, p. 457—460 (1967).
See also *13*, p. *168*; 20, p. 340.

K. H. J. Buschow, J. F. Fast, A. M. van Diepen, and H. W. de Wijn
Magnetic Dilution of Rare-Earth Aluminium Cubic Laves Phases RAl_2
Vol. **24**, No. 2, p. 715—720 (1967).
See also 1.2, p. 11; *18.2*, p. *309*.

A. Seeger, E. Mann, and K. Clausecker
Magnetoresistivity of Metals. (I): General Theory
Vol. **24**, No. 2, p. 721—732 (1967).
See also *14.1*, p. *224*.

W. Ludwig and B. Korneffel
Photoemission Studies on Thin Metal–Insulator–Metal Sandwiches
Vol. **24**, No. 2, p. K137—K140 (1967).
See also *14.3.2*, p. *250*; *17*, p. *288*; 22.6, p. 540.

J. Perez and P. Gobin
Room Temperature Peak of Internal Friction of Aluminium
Vol. **24**, No. 2, p. K167—K172 (1967).
See also *10.1*, p. *127*.

M. F. Rose and R. T. Ramsey
Higher Order Elastic Constants in H.C.P. Crystals
Vol. **25**, No. 1, p. 103—108 (1968).
See also 3.2, p. 36; *12.1*, p. *160*; 21.3, p. 426.

H. Dübler and H. Wever
Thermo- and Electrotransport in β-Titanium and β-Zirconium
Vol. **25**, No. 1, p. 109—118 (1968).
See also *9*, p. *91*; 10.1, p. 127.

William R. Savage
Field Electron Microscope Study of Niobium Surfaces
Vol. **25**, No. 1, p. 131—138 (1968).
See also *1.3*, p. *15*; 17, p. 288.

T. E. Volin and R. W. Balluffi
Annealing Kinetics of Voids and the Self-Diffusion Coefficient in Aluminium
Vol. **25**, No. 1, p. 163–173 (1968).
See also **4**, p. **44**; *9*, p. *91*.

F. Guiu
Deformation of Molybdenum Single Crystals at Slow Rates of Strain
Vol. **25**, No. 1, p. 189–202 (1968).
See also *10.1*, p. *127*.

F. Guiu
The Stress Dependence of the Dislocation Velocity in Molybdenum
Vol. **25**, No. 1, p. 203–207 (1968).
See also *10.1*, p. *128*.

H. Träuble and U. Essmann
Defects in the Flux Thread Lattice in Type II Superconductors
Vol. **25**, No. 1, p. 373–393 (1968).
See also 10.1, p. 128; *14.2*, p. *229*.

H. Träuble and U. Essmann
Direct Evidence for Motive of Flux Threads during Current Passage in Superconductors
Vol. **25**, No. 1, p. 395–402 (1968).
See also 10.1, p. 128; *14.2*, p. *230*.

R. Herrmann
Cyclotron Resonance Study of the Fermi Surface of Molybdenum
Vol. **25**, No. 1, p. 427–435 (1968).
See also *13.1*, p. *183*; 19, p. 331.

J. C. Williams and M. J. Blackburn
The Identification of a Non-Basal Slip Vector in Titanium and Titanium-Aluminum Alloys
Vol. **25**, No. 1, p. K1–K3 (1968).
See also *10.1*, p. *128*.

J. H. Evans and B. L. Eyre
The Effect of Tungsten Content on the Residual Resistivity of Zone Refined Molybdenum
Vol. **25**, No. 1, p. K39–K41 (1968).
See also *3.1*, p. *33*; *14.1*, p. *224*.

P. Charsley, J. A. M. Salter, and A. D. W. Leaver
An Experimental Investigation of Dislocation-Phonon Scattering in Some Copper Alloys
Vol. **25**, No. 2, p. 531–540 (1968).
See also *6*, p. *65*; 8, p. 84; *10.1*, p. *128*; 21.1, p. 402.

R. Herrmann
Fermi Surface of Tungsten and Molybdenum
Vol. **25**, No. 2, p. 661–666 (1968).
See also *13.1*, p. *184*; 19, p. 332.

F. C. Greisen
A Systematic Numerical Error in APW Calculations
Vol. **25**, No. 2, p. 753–756 (1968).
See also *13.1*, p. *154*; 21.1, p. 403.

A. S. Sastri, M. J. Marcinkowski, and D. Koskimaki
Nature of the Ni-Ti Martensite Transformation
Vol. **25**, No. 2, p. K67–K69 (1968).
See also 1.1, p. 8; *1.2*, p. *11*; 21.1, p. 403.

M. Luukkala
On the Hysteresis of Fractional Phonons
Vol. **25**, No. 2, p. K99–K101 (1968).
See also *6*, p. *65*.

J. T. Moore, T. R. Duncan, and D. Kuhlmann-Wilsdorf
Corrected Data for the Shear Stresses of Infinite Tilt Boundaries Composed of {111}, 1/2 ⟨1$\bar{1}$0⟩ Dislocations in Elastically Anisotropic F.C.C. Metals
Vol. **25**, No. 2, p. K105–K106 (1968).
See also *10*, p. *105*; *10.1*, p. *129*; 21.1, p. 403; 21.5, p. 429; 21.6, p. 435; 22.1, p. 457; 22.1.1, p. 464; 22.1.2, p. 473.

B. N. Oshcherin
On Enthalpy Calculation of Vacancy Formation in Inorganic Substances
Vol. **25**, No. 2, p. K123–K125 (1968).
See also *10*, p. *105*; *10.1*, p. *129*; 22, p. 450.

A. Meessen
Vectorial Selective Surface Effects in the Optical Absorption and Photoelectric Emission of Thin Metal Films
Vo. **26**, No. 1, p. 125–132 (1968).
See also *17*, p. *288*; 20.1, p. 352.

M. E. de Morton
Atomic Relaxations in a Face-Centred-Tetragonal Alloy
Vol. **26**, No. 1, p. K73–K75 (1968).
See also *10.1*, p. *129*; 12.1, p. 160.

L. Kaun, A. Luft, J. Richter, and D. Schulze
Slip Line Pattern and Active Slip Systems of Tungsten and Molybdenum Single Crystals Weakly Deformed in Tension at Room Temperature
Vol. **26**, No. 2, p. 485–499 (1968).
See also *10.1*, p. *129*.

C. W. Chen, L. R. Edwards, and S. Legvold
Resistance Anomaly in Palladium-Silver Alloys Containing Rare Earth Impurities
Vol. **26**, No. 2, p. 611–619 (1968).
See also 10.1, p. 129; *14.1*, p. *225*; 21.6, p. 435.

Y. Haven
Relation between Relaxation Time and Jump Frequencies for Zener Relaxation in Disordered F.C.C. Alloys
Vol. **26**, No. 2, p. 653–660 (1968).
See also *10.1*, p. *130*; *12.1*, p. *160*.

J. Biersack and W. Dietz
Motion of Markers and Bubbles in Solids by Self-Diffusion in a Temperature Gradient
Vol. **27**, No. 1, p. 139–144 (1968).
See also *9*, p. *91*; 22, p. 451.

T. C. Tisone, J. O. Brittain, and M. Meshii
Stacking Faults in a Cu-15 at% Al Alloy. (I): The Short Range Order and Temperature Dependence of the Stacking Fault Energy
Vol. **27**, No. 1, p. 185–194 (1968).
See also *10.1*, p. *130*; 21.1, p. 404.

H. J. Fischbeck and J. Mertsching
The Theory of Low-Field Oscillations of the Surface Impedance in Metals
Vol. 27, No. 1, p. 345–357 (1968).
See also 13.1, p. 186; 14.1, p. 225; *19*, p. *333*.

V. K. Saxena and D. L. Bhattacharya
Effect of Electron-Lattice Interactions on the Spin-Paramagnetism of Metals
Vol. 27, No. 1, p. 427–433 (1968).
See also 13, p. 170; *18.1*, p. *297*; 21.2, p. 425.

S. I. Masharov
Effect of Phonon Drag on the Kinetic Properties of Alloys
Vol. 27, No. 1, p. 455–461 (1968).
See also *6*, p. *65*; 8, p. 84; 13, p. 170; 14.1, p. 225; 15, p. 274.

K. Wetzig
Some Investigations Concerning the Position and the Melting Point of TaFe Lamellas in the System TaFe-$TaFe_2$
Vol. 27, No. 1, p. K7–K9 (1968).
See also 1.2, p. 12; *4*, p. *46*; 21.1.1, p. 421.

J. Tóth
The Temperature Dependence of Resistivity in Al-Ta Dilute Alloys
Vol. 27, No. 1, p. K47–K48 (1968).
See also *14.1*, p. *225*.

S. Ceresara
A Step Annealing Procedure for the Determination of Diffusion Coefficients in Metals by the Resistometric Method — Application to the Diffusion of Cu in Al
Vol. 27, No. 2, p. 517–520 (1968).
See also *9*, p. *91*.

J. L. Brimhall and B. Mastel
Defect Clusters in Neutron Irradiated Rhenium
Vol. 27, No. 2, p. K89–K92 (1968).
See also *10.1*, p. *131*; 11, p. 152.

B. W. Holland
Pseudopotentials and the Knight Shift
Vol. 28, No. 1, p. 121–125 (1968).
See also 13, p. 171; *19*, p. *334*; 21.2, p. 425.

S. S. Sheinin and C. D. Cann
Effect of Absorption on Multi-Beam Intensity Profiles of Wedge Crystals
Vol. 28, No. 1, p. 187–192 (1968).
See also *5*, p. *53*; 22.6, p. 542.

E. Ruedl, P. Delavignette, and S. Amelinckx
Electron Diffraction and Electron Microscopic Study of Long- and Short-Range Order in Ni_4Mo and of the Substructure Resulting from Ordering
Vol. 28, No. 1, p. 305–328 (1968).
See also *1.1*, p. *8*; 4, p. 46; 21.1, p. 405.

L. K. Thomas
The Normal Spectral Emissivity of Tungsten-Molybdenum Alloys
Vol. 28, No. 1, p. 401–406 (1968).
See also 13.1, p. 187; *20.3*, p. *370*.

J. Van der Planken and A. Deruyttere
Twinning and Fracture along {10$\bar{1}$1} Planes in a Magnesium–Tin Alloy
Vol. 28, No. 1, p. K9–K10 (1968)
See also *10.1*, p. *132*; 21.3, p. 427.

V. V. Nemoshkalenko and V. V. Gorskii
Investigation of the Valence Electron Distribution in Mg–Al Alloys
Vol. 28, No. 1, p. K15–K17 (1968).
See also 13.1, p. 187; *20*, p. *341*; 21.3, p. 427.

P. Regnier and J. M. Dupouy
Dissociation and Glide of Dislocations on a Prismatic Plane of Hexagonal Metals
Vol. 28, No. 1, p. K55–K57 (1968).
See also *10.1*, p. *132*

T. Paakkari and P. Suortti
Absolute Atomic Scattering Factors of Tungsten
Vol. 28, No. 2, p. 539–545 (1968).
See also *5*, p. *54*.

B. Pegel
Stacking Faults on {110} Planes in the B.C.C. Lattice
Vol. 28, No. 2, p. 603–609 (1968).
See also *10.1*, p. *132*; 21.1, p. 406; 21.1.1, p. 422.

A. K. Singh and P. K. Sharma
Thermal Expansion of Cubic Metals in Krebs's Model
Vol. 28, No. 2, p. 643–648 (1968).
See also *6*, p. *67*; *8*, p. *85*; 21.1, p. 406; 21.2, p. 425; 21.6, p. 436.

H. H. Neely, D. W. Keefer, and A. Sosin
Electron Irradiation and Recovery of Tungsten
Vol. 28, No. 2, p. 675–682 (1968).
See also 10.1, p. 132; *11*, p. *153*; 14.1, p. 226.

R. Kishore
On Transverse Electrical Conductivity of Metals
Vol. 28, No. 2, p. 695–698 (1968).
See also *14.1*, p. *226*.

P. Görlich and M. Ludwig
On the Photostimulation of the Exoelectron Emission in the Infrared Spectral Region
Vol. 28, No. 2, p. K117–K119 (1968).
See also *17*, p. *289*.

A. M. van Diepen, H. W. de Wijn, and K. H. J. Buschow
Nuclear Magnetic Resonance and Susceptibility of Equiatomic Rare-Earth–Aluminium Compounds
Vol. 29, No. 1, p. 189–192 (1968).
See also 13, p. 172; 18.1, p. 297; *19*, p. *335*; 21.4, p. 428.

B. Harris
Solution-Hardening and Deformation of Niobium–Tungsten Alloys
Vol. 29, No. 1, p. 383–390 (1968).
See also *10.1*, p. *133*.

V. P. Naberezhnykh
Radio-Frequency Size Effect on Electrons in Boundary Cross-Sections Having Extremum Displacement during a Cyclotron Period
Vol. 29, No. 1, p. K23–K26 (1968).
See also *13.1*, p. *188*.

D. Veselý
The Study of Deformation of Thin Foils of Mo under the Electron Microscope
Vol. **29**, No. 2, p. 675–683 (1968).
See also *10.1*, p. *133*.

D. Veselý
The Study of Slip Bands on the Surface of Mo Single Crystals
Vol. **29**, No. 2, p. 685–696 (1968).
See also *10.1*, p. *133*.

C. Simoi, I. Hrianca, and P. Crăciun
Exoemission of Electrons without Photostimulation
Vol. **29**, No. 2, p. 761–766 (1968).
See also 13, p. 173; *17*, p. *290*.

I. Gaal and L. Uray
Investigation of Non-Conductive Second Phases by Means of the Shape Factor of Electrical Resistance
Vol. **29**, No. 2, p. K163–K166 (1968).
See also *4*, p. *48*; 14.1, p. 226.

T. J. Turner, C. M. Dozier, and G. P. Williams, Jr.
Activation Energies for Zener Relaxation in α-AgCd
Vol. **30**, No. 1, p. 87–91 (1968).
See also 9, p. 92; *10.1*, p. *134*; *12.1*, p. *161*; 21.6, p. 437.

A. B. Gardner, R. L. Sanders, and L. M. Slifkin
Self-Diffusion in α-Silver-Cadmium
Vol. **30**, No. 1, p. 93–96 (1968).
See also *9*, p. *92*; 10.1, p. 134; 21.6, p. 437.

O. Kanert
Calculation of Quadrupole Broadening from (111) $\langle 110 \rangle$ Dislocations in Cubic Crystals
Vol. **30**, No. 1, p. 127–134 (1968).
See also *10.1*, p. *134*; 19, p. 335; 21.1, p. 408.

P. Feltham
A Stochastic Model of Creep
Vol. **30**, No. 1, p. 135–146 (1968).
See also *10*, p. *111*; *10.1*, p. *134*; 22.5.2, p. 530.

J. Gemperlová
Force on Dislocations in Anisotropic Bicrystals and Half-Spaces
Vol. **30**, No. 1, p. 261–274 (1968).
See also *10.1*, p. *135*; 21.1, p. 408; 21.1.1, p. 423; 21.2, p. 425.

M. C. Naik, M. S. Anand, A. R. Paul, and R. P. Agarwala
Impurity Diffusion in Face-Centred Cubic Metals
Vol. **30**, No. 1, p. 319–328 (1968).
See also *9*, p. *92*; 21.1, p. 408; 21.6, p. 437.

T. Skośkiewicz and B. Baranowski
Investigation of the Electrical Resistance Anomaly in the Palladium-Hydrogen System
Vol. **30**, No. 1, p. K33–K35 (1968).
See also 10.1, p. 135; *14.1*, p. *227*.

L. E. Tanner
The Ordering of Ni_3V
Vol. **30**, No. 2, p. 685–701 (1968).
See also *1.2*, p. *13*; 3.1, p. 33; 4, p. 48.

Y. Quéré
Dechannelling Cylinder of Dislocations
Vol. **30**, No. 2, p. 713–722 (1968).
See also *10.1*, p. *135*; *11*, p. *155*.

J. Czerwonko
On the Thermodynamics and Paramagnetic Properties of Dispersed Metals
Vol. **30**, No. 2, p. 723–734 (1968).
See also *13.1*, p. *190*; 18.1, p. 298; 21.1, p. 409; 21.6, p. 437.

H. W. de Wijn, K. H. J. Buschow, and A. M. van Diepen
Experimental Evidence for Interband Mixing in Rare-Earth Intermetallic Compounds
Vol. **30**, No. 2, p. 759–766 (1968).
See also 13, p. 173; 13.1, p. 190; 18.1, p. 298; *19*, p. *336*; 21.4, p. 428.

R. Kishore
Green Function Theory of Transport Properties of Anharmonic Metals
Vol. **30**, No. 2, p. 817–826 (1968).
See also *6*, p. *70*; 8, p. 86; *13*, p. *173*; 14.1, p. 227.

P. Gassenmeier and M. Wilkens
X-Ray Investigation of the Transition between Stages I and II of the Work Hardening Curve of Cubic Face Centered Copper-Gallium Single Crystals
Vol. **30**, No. 2, p. 833–843 (1968).
See also *10.1*, p. *136*; 21.1, p. 409.

21.1 Transition Metals (Group Sc to Zn) and Their Alloys

Review Articles

A. Aharoni
Magnetization Curling
Vol. **16**, No. 1, p. 3—42 (1966).
See also 4, p. 37; *18.2*, p. *299*; 21.1.1, p. 410.

L. E. Murr
Calibration and Use of an Electron Microscope for Precision Micromeasurements in Thin Film Materials
Vol. **19**, No. 1, p. 7—34 (1967).
See also *1.4*, p. *16*; 4, p. 37; 10, p. 93; 21.1.1, p. 410.

D. P. Morgan
Helicon Waves in Solids
Vol. **24**, No. 1, p. 9—36 (1967).
See also 6, p. 56; *13*, p. *162*; 14.1, p. 220; 14.3, p. 231; 19, p. 326; *20*, p. *337*; 21, p. 373; 21.2, p. 424; 21.6, p. 430; 22.1.1, p. 458; 22.2.1, p. 482; 22.2.3, p. 489

K. N. Subramanian
Cyclic Stressing in Crystals of Rock-Salt Structure
Vol. **28**, No. 1, p. 9—37 (1968).
See also *10*, p. *93*; 10.1, p. 114; 21, p. 373; 22.5.1, p. 515; 22.5.2, p. 518; 22.6, p. 535.

Original Papers and Short Notes

J. Melngailis
A Model for the Change in Elastic Modulus of a Solid Caused by Point Defects
Vol. **16**, No. 1, p. 247—252 (1966).
See also *10.1*, p. *114*; 12.1, p. 159.

P. Peretto, J. L. Oddou, C. Minier-Cassayre, D. Dautreppe, and P. Moser
Investigation and Interpretation of Stage I of Nickel Electron Irradiated at 20 °K
Vol. **16**, No. 1, p. 281—287 (1966).
See also *10.1*, p. *114*; 14.1, p. 220; 18.2, p. 299.

D. Paccard, C. Schlenker, O. Massenet, R. Montmory, and A. Yelon
A New Property of Ferromagnetic–Antiferromagnetic Coupling
Vol. **16**, No. 1, p. 301—311 (1966).
See also *18.2*, p. *299*; *18.4*, p. *321*; 21.1.1, p. 410.

H. Spreen and Ch. Schwink
Complete Solution of the Magnetoelastic Problem for Domain Structures with Cyclic Magnetization
Vol. **16**, No. 1, p. K45—K49 (1966).
See also 4, p. 37; *18.2*, p. *300*.

J. Askill
Environmental Effects on the Diffusion of Ta^{182} in B.C.C. Titanium
Vol. **16**, No. 1, p. K63—K65 (1966).
See also *9*, p. *87*.

Chr. Lehmann and P. Sigmund
On the Mechanism of Sputtering
Vol. **16**, No. 2, p. 507—511 (1966).
See also 1.3, p. 14; *11*, p. *145*.

V. Gerold and H. Haberkorn
On the Critical Resolved Shear Stress of Solid Solutions Containing Coherent Precipitates
Vol. **16**, No. 2, p. 675–684 (1966).
See also 4, p. 38; *10.1*, p. *114*.

N. K. Archipova, S. M. Klotsman, A. N. Timofeev, and I. Sh. Trakhtenberg
Effect of a DC Field on the Lattice Diffusion of Silver-110 in Copper and Gold
Vol. **16**, No. 2, p. 729–736 (1966).
See also *9*, p. *87*; 21.6, p. 430.

S. K. Sidorov and A. V. Doroshenko
On the Magnetic Structure of Some Alloys of Transition Metals
Vol. **16**, No. 2, p. 737–744 (1966).
See also *18.2*, p. *301*; *18.4*, p. *321*; 21.1.1, p. 411.

Chr. Nanev
Electron-Microscopic Study of the Growth Morphology of Zinc and Cadmium Single Crystals
Vol. **16**, No. 2, p. 777–787 (1966).
See also *1.3*, p. *14*; 3.1, p. 29; 21, p. 374.

R. Dietz and W. Hellenthal
Recovery Processes in Nickel Thin Films between 20 and 300 °C
Vol. **16**, No. 2, p. 789–794 (1966).
See also *10.1*, p. *114*; 14.1, p. 220.

M. Rühle and M. Wilkens
Nature of the Electron-Microscopically Observable Defects in Neutron-Irradiated Copper
Vol. **16**, No. 2, p. K105–K108 (1966).
See also *10.1*, p. *115*; 11, p. 145.

H. Diepers and J. Diehl
Detection and Formation of Interstitial Agglomerates in Ion-Irradiated Copper Foils
Vol. **16**, No. 2, p. K109–K112 (1966).
See also *10.1*, p. *115*; 11, p. 145.

E. Klugmann
Magnetic After-Effect of the Reversible Susceptibility of Cobalt
Vol. **16**, No. 2, p. K117–K119 (1966).
See also *18.2*, p. *301*.

V. K. Nikulin
Cohesive Energy of Noble Metals
Vol. **16**, No. 2, p. K125–K127 (1966).
See also *3.2*, p. *34*; 5, p. 49; 21.6, p. 430.

T. J. Koppenaal and R. J. Arsenault
Some Notes on the Strengthening Mechanism in Neutron Irradiated Copper Single Crystals
Vol. **17**, No. 1, p. 27–33 (1966).
See also *10.1*, p. *115*; 11, p. 145.

Ch. Schwink and H. Neuhäuser
On the Strengthening Mechanism in Neutron Irradiated Copper Single Crystals. (I): On Experiments Concerning Strain Rate Changes
Vol. **17**, No. 1, p. 35–42 (1966).
See also *10.1*, p. *115*; 11, p. 145.

J. Diehl and G. P. Seidel
On the Strengthening Mechanism in Neutron Irradiated Copper Single Crystals. (II): Analysis of the Thermally Activated Processes
Vol. **17**, No. 1, p. 43–48 (1966).
See also *10.1*, p. *115*; 11, p. 146.

W. Andrä and K. Steenbeck
Magnetic Anisotropy Induced in Thin Films at Low Temperatures
Vol. **17**, No. 1, p. 191–195 (1966).
See also 10.1, p. 115; *18.2*, p. *302*.

L. V. Kirenskii, V. G. Pynko, N. I. Sivkov, G. P. Pynko, R. V. Sukhanova, and M. A. Ovsyannikov
Growth of Single-Crystal Films of Iron, Nickel, Cobalt, and Their Alloys
Vol. **17**, No. 1, p. 243–248 (1966).
See also 1.4, p. 16; *3.1*, p. *29*; 21.1.1, p. 412.

L. V. Kirenskii, V. G. Pynko, G. P. Pynko, R. V. Sukhanova, N. I. Sivkov, G. I. Rusov, and I. S. Edelman
Magnetic Properties of Single-Crystal Films of Iron, Nickel, Cobalt, and Their Alloys
Vol. **17**, No. 1, p. 249–257 (1966).
See also 4, p. 38; *18.2*, p. *302*; 21.1.1, p. 412.

D. Ast and J. Diehl
Study of the Decay of Vacancy Agglomerates in Neutron-Irradiated Nickel by Measurements of Critical Shear Stress
Vol. **17**, No. 1, p. 269–284 (1966).
See also *10.1*, p. *115*; 11, p. 146.

G. Kneer and W. Zinn
On the Origin of M-Induced Anisotropy in Nickel and Permalloy Films
Vol. **17**, No. 1, p. 323–334 (1966).
See also *18.2*, p. *302*; 21.1.1, p. 412.

R. von Jan
The Interpretation of Recovery Stages in Irradiated Metals
Vol. **17**, No. 1, p. 361–373 (1966).
See also *10.1*, p. *116*; 11, p. 146; 21, p. 374; 21.6, p. 430.

Z. Kamberská and V. Kamberský
Oblique-Incidence Anisotropy in Cobalt Films
Vol. **17**, No. 1, p. 411–415 (1966).
See also 1.4, p. 17; *18.2*, p. *302*.

L. S. Palatnik, M. Ya. Fuks, L. I. Lukashenko, A. G. Ravlik, and A. A. Kozma
Structure and Magnetic Properties of Condensed Ferromagnetic Films
Vol. **17**, No. 2, p. 543–554 (1966).
See also *1.4*, p. *17*; 3.1, p. 30; 10.1, p. 116; *18.2*, p. *303*; 21.1.1, p. 413.

A. Seeger and G. Wobser
Parallel Dislocation Arrays in Anisotropic Media
Vol. **17**, No. 2, p. 709–724 (1966).
See also *10*, p. *95*; *10.1*, p. *116*; 21, p. 375; 21.3, p. 426; 21.6, p. 431; 22.1.1, p. 460; 22.1.2, p. 469.

U. Essmann
Electron-Microscopic Study of the Dislocation Arrangement in Deformed Copper Single Crystals. (III): Determination of the Dislocation Density
Vol. **17**, No. 2, p. 725–737 (1966).
See also *10.1*, p. *116*.

R. Gleiss
Calculation of the Debye-Waller Factor of Nickel
Vol. **17**, No. 2, p. 761—767 (1966).
See also *6*, p. *57*; 8, p. 79.

J. L. Oddou, C. Minier-Cassayre, and P. Moser
Effect of Impurities on the Recovery Stages of the Resistivity of Nickel Electron Irradiated at Low Temperatures
Vol. **17**, No. 2, p. 793—798 (1966).
See also *10.1*, p. *117*; *11*, p. *146*; 14.1, p. 220.

A. I. Gubanov and V. K. Nikulin
The Theory of the Cohesion of Noble Metals
Vol. **17**, No. 2, p. 815—823 (1966).
See also *3.2*, p. *34*; 5, p. 50; 12.1, p. 159; 21.6, p. 431.

V. E. Yurasova, V. M. Buchanov, and M. Golo
Directional Emission of Charged Particles from a Single Crystal under Ion Bombardment
Vol. **17**, No. 2, p. K187—K190 (1966).
See also *17*, p. *286*.

M. F. Rose
Pressure Dependence of the Second Order Elastic Constants
Vol. **17**, No. 2, p. K199—K201 (1966).
See also *12.1*, p. *159*; 21, p. 375; 21.6, p. 431.

H. Gärtner and E. Schelzke
Structure of Thin Cobalt Layers Prepared by Cathodic Sputtering
Vol. **17**, No. 2, p. K241—K244 (1966).
See also 1.2, p. 9; *1.4*, p. *17*.

R. P. Gupta, P. K. Sharma, and S. Pal
Temperature Variation of the Grüneisen Parameter of Copper
Vol. **18**, No. 1, p. 119—124 (1966).
See also 6, p. 57; *8*, p. *80*.

K. L. Merkle
Radiation-Induced Point Defect Clusters in Copper and Gold. (I): Clusters Produced in Energetic Displacement Cascades
Vol. **18**, No. 1, p. 173—188 (1966).
See also 10.1, p. 117; *11*, p. *146*; 21.6, p. 431.

A. Seeger and G. Wobser
Stacking Fault Dipoles in Face-Centered Cubic Metals
Vol. **18**, No. 1, p. 189—206 (1966).
See also *10.1*, p. *117*; 21.6, p. 431.

T. R. Duncan and D. Kuhlmann-Wilsdorf
The Stress Fields of Dislocation Tilt Boundaries in Anisotropic Cubic and Diamond-Type Crystals
Vol. **18**, No. 1, p. 231—239 (1966).
See also *10*, p. *96*; *10.1*, p. *117*; 21, p. 375; 21.5, p. 429; 21.6, p. 431; 22.1, p. 456; 22.1.1, p. 460; 22.1.2, p. 470.

K. Tompa
Nuclear Magnetic Resonance of ^{63}Cu and ^{65}Cu in Copper
Vol. **18**, No. 1, p. 391—400 (1966).
See also *19*, p. *327*.

W. FRANK
Critical Shear Stress of Cubic Crystals Containing Defects of Tetragonal Symmetry. (III): Electron- und Neutron-Irradiated Aluminium and Electron-Irradiated Copper
Vol. 18, No. 1, p. 459–464 (1966).
See also *10.1*, p. *118*; 11, p. 147; 21, p. 376.

B. F. ROTHENSTEIN, C. ANGHEL, and M. LUPULESCU
Influence of Hydrogen on the Magnetomechanical Damping in Nickel
Vol. 18, No. 1, p. K13–K16 (1966).
See also *10.1*, p. *118*; *18.2*, p. *304*.

CH. SCHWINK
On the Concept of "Active Slip-Volume" and Its Meaning
Vol. 18, No. 2, p. 557–567 (1966).
See also *10.1*, p. *118*.

D. H. HOWLING
Identification and Morphology of Oxide Particles in Internally Oxidized Cu-Mn Alloys
Vol. 18, No. 2, p. 579–591 (1966).
See also *1.1*, p. *8*; *4*, p. *39*.

F. HÄUSSERMANN and M. WILKENS
Determination of the Stacking Fault Energy of Face-Centered Cubic Metals from the Electron-Microscopic Diffraction Contrast of Stacking Fault Dipoles
Vol. 18, No. 2, p. 609–624 (1966).
See also *10.1*, p. *119*; 21.6, p. 431.

H. YOSHINAGA
Measurements of the Anisotropy of the Dislocation Resistivity in Au, Ag, and Cu
Vol. 18, No. 2, p. 625–636 (1966).
See also 10.1, p. 119; *14.1*, p. *221*; 21.6, p. 431.

G. SCHMID
Creep Experiments on High-Purity Copper Single Crystals
Vol. 18, No. 2, p. 829–845 (1966).
See also *10.1*, p. *119*.

S. M. KLOTSMAN, A. N. TIMOFEEV, and I. SH. TRAKHTENBERG
On the Mechanism of Lattice Electromigration in Metals
Vol. 18, No. 2, p. 847–852 (1966).
See also *9*, p. *88*; 14.1, p. 222; 21.6, p. 432.

R. I. ANISHCHENKO
Calculation of the Mean Inner Potential of a Crystal in the Statistical Theory
Vol. 18, No. 2, p. 923–928 (1966).
See also *3.2*, p. *34*; 13, p. 164; 21, p. 376; 21.1.1, p. 414; 21.2, p. 424; 21.3, p. 426; 21.4, p. 428; 21.6, p. 432; 21.7, p. 438; 22.1.1, p. 460.

C. P. KEMPTER
Vegard's "Law"
Vol. 18, No. 2, p. K117–K118 (1966).
See also *5*, p. *50*; 21.6, p. 432.

K. STEENBECK
Change in the Magnetization of Thin Nickel Layers due to Electrostatic Charging
Vol. 18, No. 2, p. K131–K134 (1966).
See also *18.2*, p. *304*.

R. H. Wade and J. Silcox
Small Angle Electron Scattering from Vacuum Condensed Metallic Films. (II): Experimental Results
Vol. **19**, No. 1, p. 63–76 (1967).
See also *1.4*, p. *18*; 4, p. 40; 21.6, p. 432.

E. Köster
Deformation and Temperature Dependence of Magnetization Processes in Nickel Single Crystals. (I): Bloch Wall Displacements in ⟨110⟩ and ⟨111⟩ Nickel Single Crystals
Vol. **19**, No. 1, p. 153–168 (1967).
See also 10.1, p. 120; *18.2*, p. *304*.

G. J. van Gurp
Temperature Dependence of the Critical Field in Superconducting Vanadium
Vol. **19**, No. 1, p. 173–176 (1967).
See also *14.2*, p. *228*.

A. van den Beukel, P. C. J. Coremans, and M. M. A. Vrijhoef
On the Kinetics of Short-Range Ordering in AuAg (50,50) and CuAl (85,15)
Vol. **19**, No. 1, p. 177–183 (1967).
See also *1.1*, p. *8*; 10.1, p. 120; 21.6, p. 432.

A. K. Head
Unstable Dislocations in Anisotropic Crystals
Vol. **19**, No. 1, p. 185–192 (1967).
See also *10.1*, p. *120*; 21, p. 376; 21.1.1, p. 415.

Ch. Schwink and O. Grüter
Studies of the Domain Structure of Nickel Single Crystals. (II): Measurements and Observations on Prismatic Crystals with an ⟨110⟩ Axis
Vol. **19**, No. 1, p. 217–229 (1967).
See also 4, p. 40; *18.2*, p. *304*.

H. Mughrabi and A. Seeger
The Study of Defects in Quenched Nickel
Vol. **19**, No. 1, p. 251–262 (1967).
See also *10.1*, p. *120*.

M. Rühle
Electron-Microscopic Studies of Small Defect-Agglomerates in Irradiated Metals. (I): Theory of Contrast and Experimental Methods for Determining the Defect Type
Vol. **19**, No. 1, p. 263–278 (1967).
See also *10.1*, p. *120*; 11, p. 147.

M. Rühle
Electron-Microscopic Studies of Small Defect-Agglomerates in Irradiated Metals. (II): Investigations on Neutron- and Ion-Irradiated Cu and Neutron-Irradiated Ni
Vol. **19**, No. 1, p. 279–295 (1967).
See also *10.1*, p. *120*; 11, p. 147.

E. S. Mashkova, V. A. Molchanov, and V. Soshka
Influence of the Thermal Vibrations of the Crystal Lattice upon the Energy Distribution of Scattered Ions
Vol. **19**, No. 1, p. 425–428 (1967).
See also *6*, p. *58*.

D. D. Mishin and R. M. Gretchishkin
Electron Microscopic Examination of High-Coercive Cobalt–Platinum Alloys
Vol. **19**, No. 1, p. K1–K3 (1967).
See also 4, p. 40; *18.2*, p. *305*; 21.6, p. 432.

P. Charsley and J. A. M. Salter
Further Comments on the Analysis of Measurements of the Lattice Thermal Conductivity
Vol. **19**, No. 1, p. K63–K65 (1967).
See also *8*, p. *81*; 21, p. 377.

J. Man, M. Holzmann, and B. Vlach
Microstrain Region and Transition to Macrostrain in 99.9% Polycrystalline Copper
Vol. **19**, No. 2, p. 543–553 (1967).
See also *10.1*, p. *121*.

E. Köster
Deformation and Temperature Dependence of the Magnetization Processes in Nickel Single Crystals. (II): Bloch Wall Displacements and Rotational Processes of the Spontaneous Magnetization in ⟨001⟩ Nickel Single Crystals
Vol. **19**, No. 2, p. 655–671 (1967).
See also 10.1, p. 121; *18.2*, p. *305*.

M. M. Shukla and B. Dayal
The Lattice Vibrations of Nickel in Krebs's Model
Vol. **19**, No. 2, p. 729–733 (1967).
See also *6*, p. *59*.

K.-H. Pfeffer
Interaction between Dislocations and Plane Bloch-Walls with a Rigid Magnetization Curve
Vol. **19**, No. 2, p. 735–750 (1967).
See also *10.1*, p. *121*; *18.2*, p. *305*; 21.1.1, p. 415.

O. Vöhringer and E. Macherauch
The Yield Point of α-Copper-Tin Alloys
Vol. **19**, No. 2, p. 793–803 (1967).
See also *10.1*, p. *121*.

R. H. Wade
Electron Diffraction from a Magnetic Phase Grating
Vol. **19**, No. 2, p. 847–854 (1967).
See also 4, p. 40; *18.2*, p. *305*; 21.1.1, p. 415.

K.-H. Pfeffer
Micromagnetic Treatment of the Interaction between Dislocations and Plane Bloch-Walls. (I): General Theory
Vol. **20**, No. 1, p. 395–411 (1967).
See also 10.1, p. 121; *18.2*, p. *305*; 21.1.1, p. 416.

D. D. Mishin
On the Theory of Magnetic Hardness of Platinum Cobalt
Vol. **20**, No. 1, p. K9–K10 (1967).
See also *18.2*, p. *305*; 21.6, p. 433.

A. Czachor and A. Pindor
Dependence of Calculated Elastic Constants on the c/a Ratio of Hexagonal Close-Packed Metals
Vol. **20**, No. 1, p. K17–K20 (1967).
See also 5, p. 50; *12.1*, p. *159*; 21, p. 377; 21.3, p. 426.

E. P. Lautenschlager, T. C. Tisone, and J. O. Brittain
Electron Transmission Microscopy of NiAl
Vol. **20**, No. 2, p. 443–450 (1967).
See also *4*, p. *41*; 10.1, p. 122.

A. K. Head, M. H. Loretto, and P. Humble
The Influence of Large Elastic Anisotropy on the Determination of Burgers Vectors of Dislocations in β-Brass by Electron Microscopy
Vol. **20**, No. 2, p. 505–519 (1967).
See also *10.1*, p. *122*; 12.1, p. 159.

A. K. Head, M. H. Loretto, and P. Humble
The Identification of Burgers Vectors and Unstable Directions of Dislocations in β-Brass
Vol. **20**, No. 2, p. 521–536 (1967).
See also *10.1*, p. *122*; 12.1, p. 159.

E. A. Metzbower
Noncentral Force Model for Hexagonal Close-Packed Crystal Lattices
Vol. **20**, No. 2, p. 681–692 (1967).
See also 3.2, p. 35; *6*, p. *60*; 8, p. 81; 12.1, p. 160; 21.3, p. 426.

N. A. Chernoplekov, G. Kh. Panova, M. G. Zemlyanov, B. N. Samoilov, and V. I. Kutaitsev
Investigation of the Quasi-Local Level in the Vibrational Spectrum of the Ti Lattice with Heavy Impurity Atoms
Vol. **20**, No. 2, p. 767–770 (1967).
See also *6*, p. *60*; 8, p. 81.

F. J. Worzala
Hardening in Irradiated Copper Crystals
Vol. **20**, No. 2, p. K81–K83 (1967).
See also *10.1*, p. *122*; 11, p. 148.

S. Arajs and G. R. Dunmyre
A Note on the Consistency of Values of the Spontaneous or Saturation Magnetization of Polycrystalline Iron and Nickel at 298 °K
Vol. **21**, No. 1, p. 191–195 (1967).
See also *18.2*, p. *306*; 21.1.1, p. 417.

M. F. Rose
Higher Order Elastic Constants and the Equation of State of F.C.C. Metals
Vol. **21**, No. 1, p. 235–238 (1967).
See also 6, p. 60; 8, p. 82; *12.1*, p. *160*; 21, p. 378; 21.6, p. 433.

G. Elbinger
On the Magnetocrystalline Anisotropy of Ni–Co Ferrites
Vol. **21**, No. 1, p. 303–307 (1967).
See also *18.3*, p. *317*.

J. A. M. Salter and P. Charsley
The Effect of Grain Size on the Lattice Thermal Conductivity of Copper Aluminium Alloys
Vol. **21**, No. 1, p. 357–368 (1967).
See also 4, p. 42; 6, p. 61; *8*, p. *82*; 21, p. 379.

L. M. Clarebrough, D. Michell, and A. P. Smith
A Source of Dislocations in Copper Crystals
Vol. **21**, No. 1, p. 369–373 (1967).
See also *10.1*, p. *122*.

J. Levasseur and J. Philibert
Determination of Intrinsic Diffusion Coefficients by Measuring the Kirkendall Effect
Vol. **21**, No. 1, p. K1–K4 (1967).
See also *9*, p. *89*; 21.1.1, p. 417.

A. A. Adamenko and I. J. Dekhtyar
Effect of Plastic Deformation on the Absolute Thermopower of α-Brass
Vol. **21**, No. 1, p. K81–K84 (1967).
See also 10.1, p. 122; *15*, p. *272*.

V. I. Syutkina and E. S. Yakovleva
The Mechanism of Deformation of the Ordered CuAu Alloy
Vol. **21**, No. 2, p. 465–480 (1967).
See also 1.1, p. 8; 4, p. 42; *10.1*, p. *123*; 21.6, p. 433.

J. A. Sigler and D. Kuhlmann-Wilsdorf
Calculations on the Mechanical Energy of Vacancy Condensation Loops, Stacking Fault Tetrahedra, and Voids
Vol. **21**, No. 2, p. 545–556 (1967).
See also *10.1*, p. *123*; 21, p. 379; 21.6, p. 433.

R. Lück and K. E. Saeger
Measurement of the Anisotropy of the Galvanomagnetic Transverse Voltage in Single-Crystal Foils
Vol. **21**, No. 2, p. 671–677 (1967).
See also *14.1*, p. *222*.

P. Lukáš and M. Klesnil
Fatigue Hardening of Cu–Zn Alloys
Vol. **21**, No. 2, p. 717–731 (1967).
See also *10.1*, p. *123*.

P. Humble
A Comparison of Electron Microscope Images of Dislocations Based on Isotropic and Anisotropic Elasticity Theory
Vol. **21**, No. 2, p. 733–740 (1967).
See also *10.1*, p. *123*; 21, p. 379; 21.1.1, p. 417; 21.6, p. 433.

K.-H. Pfeffer
Micromagnetic Treatment of the Interaction between Dislocations and Plane Bloch-Walls. (II): Applications
Vol. **21**, No. 2, p. 837–856 (1967).
See also 10.1, p. 123; *18.2*, p. *307*.

K.-H. Pfeffer
On the Theory of Coercive Field Strength and Initial Susceptibility
Vol. **21**, No. 2, p. 857–872 (1967).
See also 10.1; p. 123; *18.2*, p. *307*; 21.1.1, p. 418.

A. Clement, N. Clement, and P. Coulomb
Intrinsic and Extrinsic Defect Pairs in a Nonoxidizable Steel and a Copper–Silicon Alloy
Vol. **21**, No. 2, p. K97–K98 (1967)
See also *10.1*, p. *123*; 21.1.1, p. 418.

Vu dinh Ky
Susceptibility of Ferromagnetic Films in Paraprocess
Vol. **21**, No. 2, p. K159–K162 (1967).
See also *18.2*, p. *307*; 21.1.1, p. 418.

K. R. Evans and W. F. Flanagan
An Analysis of the Cottrell-Stokes Law
Vol. **22**, No. 1, p. 195–198 (1967).
See also *10.1*, p. *124*.

H. Puszkarski
Surface Spin Waves in Hexagonal Cobalt Thin Films
Vol. **22**, No. 2, p. 355–359 (1967).
See also *13*, p. *166*; 13.3, p. 197; 19, p. 329.

K. Dettmann, G. Leibfried, and K. Schroeder
Spontaneous Recombination of Frenkel Pairs for Electron Irradiation
Vol. **22**, No. 2, p. 423–432 (1967).
See also *11*, p. *149*.

K. Dettmann, G. Leibfried, and K. Schroeder
Spontaneous Recombination of Frenkel Pairs in Neutron-Irradiated Solids
Vol. **22**, No. 2, p. 433–440 (1967).
See also *11*, p. *149*; 21, p. 380; 21.6, p. 434.

J. O. Henningsen
Cyclotron Resonance in Zinc
Vol. **22**, No. 2, p. 441–452 (1967).
See also *13.1*, p. *180*; *19*, p. *330*.

K. Otsuka and C. M. Wayman
Epitaxial Growth of Vaccuum-Evaporated Co on NaCl. (I): Conditions Leading to Epitaxy
Vol. **22**, No. 2, p. 559–578 (1967).
See also 1.4, p. 19; *3.1*, p. *32*.

K. Otsuka and C. M. Wayman
Epitaxial Growth of Vacuum-Evaporated Co on NaCl. (II): Analysis of Diffraction Patterns
Vol. **22**, No. 2, p. 579–592 (1967).
See also 1.4, p. 19; *3.1*, p. *32*.

K. R. Evans, D. J. Bailey, and W. F. Flanagan
An Interpretation of the Dislocation Velocity-Stress Exponent
Vol. **22**, No. 2, p. 607–616 (1967).
See also *10.1*, p. *124*; 21.1.1, p. 418.

L. Gerward
Magnetic After-Effect in Deformed Cobalt Single Crystals
Vol. **22**, No. 2, p. 659–666 (1967).
See also 10.1, p. 124; *18.2*, p. *308*.

H. Bilger and H. Kronmüller
Recovery Spectrum of Cobalt
Vol. **22**, No. 2, p. 683–688 (1967).
See also *10.1*, p. *124*; 14.1, p. 223.

M. Wilkens, M. Rühle, and F. Häussermann
On the Nature of the Long-Range Dislocation Contrast in Electron Transmission Micrographs
Vol. **22**, No. 2, p. 689–708 (1967).
See also *10.1*, p. *124*.

A. Hubert
Effect of Magnetostriction on the Magnetic Domain Structure of Uniaxial Crystals, in Particular of Cobalt
Vol. **22**, No. 2, p. 709–727 (1967).
See also *18.2*, p. *308*.

Vu dinh Ky
Planar Hall and Nernst Effect in Ferromagnetic Metals
Vol. **22**, No. 2, p. 729–736 (1967).
See also 13, p. 167; *14.1*, p. *223*; *15*, p. *272*; 18.2, p. 308; 21.1.1, p. 418.

J. Pelleg
A Note on Nickel Diffusion in Copper
Vol. **22**, No. 2, p. K83–K85 (1967).
See also *9*, p. *90*.

R. F. Peart and J. Askill
The Mechanism of Diffusion in B.C.C. Transition Metals
Vol. **23**, No. 1, p. 263–275 (1967).
See also *9*, p. *90*; 10, p. 102; 21, p. 381; 21.1.1, p. 419.

S. Pal and P. K. Sharma
On Lindemann's Melting Criterion
Vol. **23**, No. 1, p. 361–364 (1967).
See also *6*, p. *62*; 21, p. 381; 21.1.1, p. 419; 21.2, p. 424; 21.6, p. 434.

G. Albrecht and G. Weber
Hyperfine Fields Induced by Spin-Orbit-Coupling in Binary Intermetallic Compounds
Vol. **23**, No. 1, p. K17–K20 (1967).
See also 18.1, p. 297; *19*, p. *330*.

E. Nagy and G. Petö
Ordering in the Alloy Cu_3Au-V
Vol. **23**, No. 1, p. K49–K52 (1967).
See also *1.2*, p. *11*; 16, p. 280.

R. D. Gretz
Nucleation of Zinc Crystals in Multilayer Adsorption
Vol. **23**, No. 2, p. 453–460 (1967).
See also *3.1*, p. *32*.

Chr. Nanev and D. Iwanov
Growth of Zinc and Cadmium Whiskers from the Vapour Phase on a Single Crystal Substrate of the Same Material
Vol. **23**, No. 2, p. 663–673 (1967).
See also *3.1*, p. *32*; 21, p. 382.

A. G. Khachaturyan and M. P. Usikov
A New Type of X-Ray and Electron Diffuse Scattering and Corresponding Strain Contrast on Electron Microscopic Images
Vol. **23**, No. 2, p. 745–754 (1967).
See also *4*, p. *43*; 22.6, p. 539.

J. L. Davidson and J. M. Galligan
Kinetics of Vacancy Annihilation in Quenched Copper
Vol. **23**, No. 2, p. K139–K141 (1967).
See also *10.1*, p. *126*.

L. Wojtczak
The Melting Point of Thin Films
Vol. **23**, No. 2, p. K163–K166 (1967).
See also 6, p. 63; *8*, p. *83*; 21, p. 382; 21.6, p. 434.

P. Paufler and G. E. R. Schulze
Plastic Deformation of the Intermetallic Compound $MgZn_2$
Vol. **24**, No. 1, p. 77–87 (1967).
See also *10.1*, p. *126*; 21.3, p. 426.

A. Fourdeux, R. Gevers, and S. Amelinckx
Electron Microscopic Contrast Effects at the Interface between Two Different Polytypes. (I): Interface Cubic-Hexagonal
Vol. **24**, No. 1, p. 195–206 (1967).
See also 4, p. 43; *10*, p. *103*; *10.1*, p. *126*.

R. Zeyfang
Lattice Thermal Resistivity due to Dislocations in Alloys at Low Temperatures
Vol. **24**, No. 1, p. 221–234 (1967).
See also 6, p. 63; *8*, p. *83*; 10.1, p. 126.

H. Diepers
Electron-Microscopic Study of Defect Agglomerates in Ion-Irradiated Copper Foils. (I): Analysis of Agglomerate Type and Configuration
Vol. **24**, No. 1, p. 235–244 (1967).
See also 10.1, p. 126; *11*, p. *150*.

B. Roessler and P. A. Flinn
A Theory of X-Ray Diffraction from an Ordered Alloy Containing Antiphase Domains
Vol. **24**, No. 1, p. 263–274 (1967).
See also 1.1, p. 8; *4*, p. *43*; 21.6, p. 434.

R. Dupree and B. W. Holland
The Range of g-Factors and the Breakdown of Motional Narrowing in Conduction Electron Spin Resonance
Vol. **24**, No. 1, p. 275–279 (1967).
See also *19*, p. *331*; 21, p. 382; 21.2, p. 425.

B. K. Jones
Measurement of the Magnitude of the Electronic Attenuation of Sound in Metals
Vol. **24**, No. 1, p. 297–300 (1967).
See also *7*, p. *77*; 13.1, p. 182; 21, p. 382; 21.2, p. 425.

T. O. Ogurtani and R. A. Huggins
Theory of Electric Field Gradient due to Conduction Electron Charge Density Redistribution around Screw Dislocations in Metals
Vol. **24**, No. 1, p. 301–311 (1967).
See also *10.1*, p. *126*; 13.4, p. 210.

H. Spreen
Study of the Domain Structure in Nickel Single Crystals. (III): The Closure Domain Structure of Cylindrical Crystals with an ⟨110⟩ Axis
Vol. **24**, No. 2, p. 413–429 (1967).
See also 4, p. 44; *18.2*, p. *309*.

H. Spreen
Study of the Domain Structure in Nickel Single Crystals. (IV): The Basic Structure of Cylindrical Crystals with an ⟨100⟩ Axis
Vol. **24**, No. 2, p. 431–441 (1967).
See also 4, p. 44; *18.2*, p. *309*.

A. S. Nowick and G. J. Dienes
On the Inherent Curvature of the Arrhenius Plot in Diffusion Experiments
Vol. **24**, No. 2, p. 461–467 (1967).
See also *9*, p. *90*; 21.6, p. 435.

H. Diepers
Electron-Microscopic Study of Defect Agglomerates in Ion-Irradiated Copper Foils. (II): Formation of Agglomerates
Vol. **24**, No. 2, p. 623–633 (1967).
See also 4, p. 44; 10.1, p. 127; *11*, p. *150*.

J. Hölzl and B. Heimann
Experimental Results on the Temperature Dependence of Secondary Electron Emission from Polycrystalline Nickel in the Range $100 \leqq T \leqq 600\,°C$
Vol. **24**, No. 2, p. 651–657 (1967).
See also *17*, p. *288*; 18.1, p. 297; 18.2, p. 309.

A. Hubert
On the Theory of Two-Phase Domain Structures in Superconductors and Ferromagnets
Vol. **24**, No. 2, p. 669–682 (1967).
See also *4*, p. *44*; 14.2, p. 229; 18.2, p. 309.

L. Delaey
Anomalous Spots in Electron Diffraction Patterns of Lamellar Mixture of Two Closed-Packed Phases
Vol. **24**, No. 2, p. K103–K105 (1967).
See also *4*, p. *44*.

A. Bourret and D. Dautreppe
Type of Small Dislocation Loops in Neutron Irradiated Copper
Vol. **24**, No. 2, p. K173–K174 (1967).
See also 10.1, p. 127; *11*, p. *151*.

T. O. Baldwin
The Temperature Dependence of the Borrmann Effect in Copper
Vol. **25**, No. 1, p. 71–74 (1968).
See also 5, p. 52; *6*, p. *64*.

H. P. Bonzel and N. A. Gjostein
Surface Self-Diffusion Measurements on Copper
Vol. **25**, No. 1, p. 209–222 (1968).
See also *9*, p. *91*.

T. Leffers
Computer Simulation of the Plastic Deformation in Face-Centred Cubic Polycrystals and the Rolling Texture Derived
Vol. **25**, No. 1, p. 337–344 (1968).
See also *10.1*, p. *128*.

E. A. Metzbower
Noncentral Force Model for Hexagonal Close-Packed Crystal Lattices (II)
Vol. **25**, No. 1, p. 403–416 (1968).
See also *3.2*, p. *36*; 6, p. 64; 8, p. 84; 12.1, p. 160.

A. Sukiennicki and L. Wojtczak
Magnetic Domain Structure in Uniaxial Thin Films
Vol. **25**, No. 1, p. K19–K21 (1968).
See also *18.2*, p. *310*; 21.1.1, p. 420.

P. Charsley, J. A. M. Salter, and A. D. W. Leaver
An Experimental Investigation of Dislocation-Phonon Scattering in Some Copper Alloys
Vol. **25**, No. 2, p. 531–540 (1968).
See also *6*, p. *65*; 8, p. 84; *10.1*, p. *128*; 21, p. 384.

A. Holz
Theoretical Study of Nucleation in Ferromagnetics
Vol. **25**, No. 2, p. 567–582 (1968).
See also **4**, p. 44; *18.2*, p. *310*; 21.1.1, p. 420.

H. Neuhäuser
Study of the Active Crystal-Length in the Yield Range of Neutron-Irradiated Copper Single Crystals. (I): Slip Band Formation
Vol. **25**, No. 2, p. 593–606 (1968).
See also *10.1*, p. *128*; **11**, p. 151.

R. Gauthier, D. Bois, P. Pinard, and F. Davoine
Ion and Electron Production by Interaction of a Coherent Light Beam with Substance
Vol. **25**, No. 2, p. 691–696 (1968).
See also *17*, p. *288*; 21.1.1, p. 420; 22.1.1, p. 464; 22.2.1, p. 485; 22.4.1, p. 502; 22.5.2, p. 525; 22.6, p. 540.

L. Delaey
Electron Microscopic Investigation of Lamellar Mixtures of Two Close-Packed Phases
Vol. **25**, No. 2, p. 697–710 (1968).
See also 1.1, p. 8; *4*, p. *45*.

F. C. Greisen
A Systematic Numerical Error in APW Calculations
Vol. **25**, No. 2, p. 753–756 (1968).
See also *13.1*, p. *184*; 21, p. 384.

A. S. Sastri, M. J. Marcinkowski, and D. Koskimaki
Nature of the Ni–Ti Martensite Transformation
Vol. **25**, No. 2, p. K67–K69 (1968).
See also 1.1, p. 8; *1.2*, p. *11*; 21, p. 385.

V. V. Nemoshkalenko and M. A. Mindlina
On the Form of the $K\beta_{2,5}$-Zinc Band
Vol. **25**, No. 2, p. K83–K86 (1968).
See also 13.1, p. 184; *20*, p. *340*.

A. Schweighofer and F. W. Marsh
Deformation of Zinc Single Crystals (99.999% Purity) at High Strain Rates
Vol. **25**, No. 2, p. K91–K94 (1968).
See also *10.1*, p. *128*.

J. T. Moore, T. R. Duncan, and D. Kuhlmann-Wilsdorf
Corrected Data for the Shear Stresses of Infinite Tilt Boundaries Composed of {111}, 1/2 $\langle 1\bar{1}0 \rangle$ Dislocations in Elastically Anisotropic F.C.C. Metals
Vol. **25**, No. 2, p. K105–K106 (1968).
See also *10*, p. *105*; *10.1*, p. *129*; 21, p. 385; 21.5, p. 429; 21.6, p. 435; 22.1, p. 457; 22.1.1, p. 464; 22.1.2, p. 473.

R. Kishore
Electrical Conductivity of Metals
Vol. **26**, No. 1, p. 133–138 (1968).
See also *14.1*, p. *225*; 21.2, p. 425; 21.6, p. 435.

W. Frank
Thermally Activated Dislocation Motion in a Solid Containing a Multiple Spectrum of Dislocation Obstacles
Vol. **26**, No. 1, p. 197–206 (1968).
See also *10.1*, p. *129*.

R. GEMPERLE and A. GEMPERLE
Dependence of the Ferromagnetic Domain Width on the Thickness of Cobalt Foils
Vol. **26**, No. 1, p. 207—217 (1968).
See also 1.4, p. 20; *18.2*, p. *310*.

K. DIFFERT
Dislocation Corners in Face Centered Cubic Crystals. (I): Geometry of Dislocation Corners
Vol. **26**, No. 1, p. 327—344 (1968).
See also *10.1*, p. *129*.

J. L. DAVIDSON and J. M. GALLIGAN
A Study of Quench Hardening in Copper
Vol. **26**, No. 1, p. 345—358 (1968).
See also *10.1*, p. *129*.

M. SOROHAN
The Influence of Ion Bombardment upon the Induced Magnetic Anisotropy in Thin Nickel Films
Vol. **26**, No. 1, p. K9—K11 (1968).
See also *11*, p. *151*; *18.2*, p. *310*.

VU DINH KY
Planar Hall Effect in Ferromagnetic Films
Vol. **26**, No. 2, p. 565—569 (1968).
See also *14.1*, p. *225*; *18.2*, p. *311*; 21.1.1, p. 421.

M. RÜHLE
Radiation Hardening of Neutron-Irradiated Copper Single Crystals. (I): Experimental Results
Vol. **26**, No. 2, p. 661—670 (1968).
See also *10.1*, p. *130*; *11*, p. *151*; 21.6, p. 435.

W. FRANK, M. RÜHLE, and M. SAXLOVÁ
Radiation Hardening of Neutron-Irradiated Copper Single Crystals. (II): Interpretation
Vol. **26**, No. 2, p. 671—680 (1968).
See also *10.1*, p. *130*; *11*, p. *151*.

L. K. IVES and A. W. RUFF, JR.
Studies of the Effect of Annealing on Extended Dislocation Nodes in Silver-Tin Alloys
Vol. **27**, No. 1, p. 117—123 (1968).
See also *10.1*, p. *130*; 21.6, p. 436.

E. FELDTKELLER
Magnetic Domain Wall Dynamics
Vol. **27**, No. 1, p. 161—170 (1968).
See also 4, p. 46; *18.2*, p. *311*; 21.1.1, p. 421.

T. C. TISONE, J. O. BRITTAIN, and M. MESHII
Stacking Faults in a Cu-15 at% Al Alloy. (I): The Short Range Order and Temperature Dependence of the Stacking Fault Energy
Vol. **27**, No. 1, p. 185—194 (1968).
See also *10.1*, p. *130*; 21, p. 385.

H. NEUHÄUSER
Study of the Active Crystal-Length in the Yield Range of Neutron-Irradiated Copper Single Crystals. (II): The Active Crystal-Length
Vol. **27**, No. 1, p. 281—290 (1968).
See also *10.1*, p. *130*; 11, p. 152.

U. Gradmann and J. Müller
Flat Ferromagnetic, Epitaxial 48 Ni/52 Fe (111) Films of Few Atomic Layers
Vol. **27**, No. 1, p. 313—324 (1968).
See also *1.4*, p. *20*; *18.2*, p. *311*; 21.1.1, p. 421.

B. F. Rothenstein, A. Policec, and I. Artzner
Contributions to the Kinetics of Charging Nickel with Hydrogen
Vol. **27**, No. 1, p. K5—K6 (1968).
See also *18.2*, p. *311*.

F. P. Bullen and S. McK. Cousland
The Temperature Dependence of the Flow Stress of Copper Single Crystals
Vol. **27**, No. 2, p. 501—512 (1968).
See also *10.1*, p. *131*.

P. Lukáš, M. Klesnil, and J. Krejčí
Dislocations and Persistent Slip Bands in Copper Single Crystals Fatigued at Low Stress Amplitude
Vol. **27**, No. 2, p. 545—558 (1968).
See also 1.3, p. 15; *10.1*, p. *131*.

G. Appelt
Fine Structure Measurements in the Energy Angular Distribution of Secondary Electrons from a (110) Face of Copper
Vol. **27**, No. 2, p. 657—669 (1968).
See also 13.1, p. 186; *17*, p. *289*; 20.1, p. 354.

M. Victoria and A. E. Vidoz
Tensile Behavior upon Ordering of Ni_3Fe Single Crystals
Vol. **28**, No. 1, p. 131—144 (1968).
See also 4, p. 46; *10.1*, p. *131*; 21.1.1, p. 422.

A. E. Vidoz
On the Work Hardening of Superstructures
Vol. **28**, No. 1, p. 145—156 (1968).
See also 4, p. 46; *10.1*, p. *131*; 21.1.1, p. 422; 21.6, p. 436.

S. Arajs
An Investigation of the Electrical Resistivity of Nickel-Rich Nickel-Gold Alloys
Vol. **28**, No. 1, p. 171—174 (1968).
See also *14.1*, p. *225*; 21.6, p. 436.

E. Ruedl, P. Delavignette, and S. Amelinckx
Electron Diffraction and Electron Microscopic Study of Long- and Short-Range Order in Ni_4Mo and of the Substructure Resulting from Ordering
Vol. **28**, No. 1, p. 305—328 (1968).
See also *1.1*, p. *8*; 4, p. 46; 21, p. 386.

D. J. Craik and M. J. Wood
Thermally Activated Domain Wall Movement in Nickel–Iron Films
Vol. **28**, No. 1, p. 407—414 (1968).
See also 1.4, p. 20; *18.2*, p. *313*; 21.1.1, p. 422.

P. Ehrhart, W. Rupp, and R. Sizmann
Stopping Power of 0.5 to 3.5 MeV α-Particles in Cu-Au Alloys
Vol. **28**, No. 1, p. K35—K37 (1968).
See also 1.4, p. 21; *11*, p. *153*; 21.6, p. 436.

M. Sorohan
Oscillating Hysteresis in Thin Nickel Films
Vol. 28, No. 1, p. K59—K61 (1968).
See also *18.2*, p. *313*.

E. Klugmann
Magnetic After-Effect in Cobalt
Vol. 28, No. 1, p. K67—K70 (1968).
See also *18.2*, p. *313*.

C. Schlenker
Ferro–Antiferromagnetic Coupling and Magnetic After-Effect in Multiple Co-CoO and Ni-NiO Films
Vol. 28, No. 2, p. 507—517 (1968).
See also *18.2*, p. *313*; *18.4*, p. *324*; 22.6, p. 542.

H. Bakker
A Curvature in the ln D versus $1/T$ Plot for Self-Diffusion in Nickel at Temperatures from 980 to 1400 °C
Vol. 28, No. 2, p. 569—576 (1968).
See also *9*, p. *92*.

K. E. Saeger
Hall Effect and Magnetoresistance in Copper Single Crystals at Low Temperatures
Vol. 28, No. 2, p. 589—602 (1968).
See also *14.1*, p. *225*.

B. Pegel
Stacking Faults on {110} Planes in the B.C.C. Lattice
Vol. 28, No. 2, p. 603—609 (1968).
See also *10.1*, p. *132*; 21, p. 387; 21.1.1, p. 422.

A. K. Singh and P. K. Sharma
Thermal Expansion of Cubic Metals in Krebs's Model
Vol. 28, No. 2, p. 643—648 (1968).
See also *6*, p. *67*; *8*, p. *85*; 21, p. 387; 21.2, p. 425; 21.6, p. 436.

A. Holz and H. Kronmüller
Contribution of Dislocations to the Anisotropy of the Magnetic Saturation Curve in Nickel
Vol. 28, No. 2, p. 755—771 (1968).
See also 10.1, p. 132; *18.2*, p. *313*.

V. K. Nikulin and M. B. Trzhaskovskaya
The Theory of Cohesion of Noble Metals (II)
Vol. 28, No. 2, p. 801—806 (1968).
See also *3.2*, p. *36*; 13.1, p. 187; 21.6, p. 437.

F. Dworschak, H. Schuster, H. Wollenberger, and J. Wurm
Analysis of Point Defect States in Copper. (I): The Influence of the Point Defect State on the Irradiation Damage Rate at 80 °K
Vol. 29, No. 1, p. 75—80 (1968).
See also *10.1*, p. *132*; 11, p. 153.

F. Dworschak, H. Schuster, H. Wollenberger, and J. Wurm
Analysis of Point Defect States in Copper. (II): The Effects of Stage I and Stage III Annealing on Electron-Irradiation-Induced Frenkel Pairs.
Vol. 29, No. 1, p. 81—89 (1968).
See also *10.1*, p. *133*; 11, p. 153.

P. Humble and C. T. Forwood
A Dynamical Treatment of the Stress-Induced Dissociation of Triangular Frank Dislocation Loops in F.C.C. Metals
Vol. **29**, No. 1, p. 99–106 (1968).
See also *10.1*, p. *133*; 21.6, p. 437.

K. Schröder and N. P. Baum
Absolute Thermoelectric Power and Electrical Resistivity of Chromium-Rich Chromium–Nickel Alloys
Vol. **29**, No. 1, p. 107–114 (1968).
See also 13.1, p. 188; *14.1*, p. *226*; *15*, p. *274*; 18.4, p. 324.

F. Walz
Investigation of the Magnetic After-Effect in Nickel after Neutron Irradiation and Plastic Deformation
Vol. **29**, No. 1, p. 245–258 (1968).
See also 10.1, p. 133; 11, p. 153; *18.2*, p. *314*.

E. Taglauer
Small Angle Scattering of Subthermal Neutrons from Deformed Polycrystalline Copper
Vol. **29**, No. 1, p. 259–268 (1968).
See also 4, p. 47; *10.1*, p. *133*.

A. Bourret and D. Dautreppe
Study of Agglomerated Defects in Irradiated Pure Nickel by Electron Microscopy
Vol. **29**, No. 1, p. 283–298 (1968).
See also 4, p. 47; *10.1*, p. *133*; 11, p. 154.

R. T. Pascoe and C. W. A. Newey
Deformation Modes of the Intermediate Phase NiAl
Vol. **29**, No. 1, p. 357–366 (1968).
See also *10.1*, p. *133*.

Z. I. Alizade and T. G. Magarramov
Galvanomagnetic Effect in Ferromagnetic Alloys of Nickel–Tantalum and Nickel–Niobium Systems
Vol. **29**, No. 1, p. K19–K22 (1968).
See also *14.1*, p. *226*; *18.2*, p. *314*.

F. F. Lavrentev, O. P. Salita, and V. L. Vladimirova
Dislocation Mobility in the $\{11\bar{2}2\}$ $\langle 11\bar{2}3\rangle$ Slip System of Zinc Single Crystals
Vol. **29**, No. 2, p. 569–574 (1968).
See also *10.1*, p. *133*.

D. I. Bower, E. Claridge, and I. S. T. Tsong
Low-Temperature Elastic Constants and Specific Heats of F.C.C. Nickel–Iron Alloys
Vol. **29**, No. 2, p. 617–625 (1968).
See also 6, p. 68; 8, p. 86; *12.1*, p. *160*; 13, p. 172; 18.2, p. 314; 21.1.1, p. 322.

K. H. J. Buschow, J. F. Fast, and A. S. van der Goot
Magnetic Properties of Some Co-Rich Erbium Cobalt Intermetallic Compounds
Vol. **29**, No. 2, p. 825–829 (1968).
See also *18.3*, p. *320*; 21.4, p. 428.

E. E. Galloni and R. Zimmerman
Double Diffraction in Epitaxially Grown Nickel Polycrystalline Films
Vol. **29**, No. 2, p. K91–K94 (1968).
See also *1.4*, p. *21*.

B. F. ROTHENSTEIN and I. ARTZNER
On the Internal Barkhausen Effect
Vol. **29**, No. 2, p. K117—K118 (1968).
See also *18.2*, p. *314*; 21.1.1, p. 423.

G. ROTH and V. NAUNDORF
Measurement of the Activation Energies of Dislocation Pinning in Copper after Electron Irradiation at 78 °K
Vol. **29**, No. 2, p. K123—K125 (1968).
See also 10.1, p. 134; *11*, p. *154*; 12.1, p. 161.

H. R. VYDYANATH, D. H. SASTRY, and K. I. VASU
On the Kinetics of Clustering in a Magnesium-3%-Zinc Alloy
Vol. **29**, No. 2, p. K137—K140 (1968).
See also *4*, p. *47*; 10.1, p. 134; 14.1, p. 226; 21.3, p. 427.

O. KANERT
Calculation of Quadrupole Broadening from (111)⟨110⟩ Dislocations in Cubic Crystals
Vol. **30**, No. 1, p. 127—134 (1968).
See also *10.1*, p. *134*; 19, p. 335; 21, p. 388.

D. BECKER, F. DWORSCHAK, C. LEHMANN, K. T. RIE, H. SCHUSTER, H. WOLLENBERGER, and J. WURM
Determination of the Average Number of Replacement Collisions in an Electron-Irradiated F.C.C. Crystal (Ni_3Mn)
Vol. **30**, No. 1, p. 219—229 (1968).
See also 10.1, p. 135; *11*, p. *155*; 14.1, p. 227; 18.2, p. 315.

J. GEMPERLOVÁ
Force on Dislocations in Anisotropic Bicrystals and Half-Spaces
Vol. **30**, No. 1, p. 261—274 (1968).
See also *10.1*, p. *135*; 21, p. 388; 21.1.1, p. 423; 21.2, p. 425.

A. ANDERMAN and W. G. GEHMAN
Xenon-Vacancy Defect Clusters in Copper
Vol. **30**, No. 1, p. 283—289 (1968).
See also 3.2, p. 36; *10.1*, p. *135*.

M. C. NAIK, M. S. ANAND, A. R. PAUL, and R. P. AGARWALA
Impurity Diffusion in Face-Centred Cubic Metals
Vol. **30**, No. 1, p. 319—328 (1968).
See also *9*, p. *92*; 21, p. 388; 21.6, p. 437.

M. A. KUMAKHOV
Formation of Radiation Defects by Channeling Particles
Vol. **30**, No. 1, p. 379—385 (1968).
See also 3.2, p. 36; *11*, p. *155*; 22.1.2, p. 474.

CH. SCHWINK and H. TRAUB
The Active Slip Volume and the Temperature-Dependent Processes in Stage I of Single Crystals of Cu Alloys
Vol. **30**, No. 1, p. 387—397 (1968).
See also *10.1*, p. *135*.

F. SECCO D'ARAGONA, R. C. RAU, and R. L. LADD
Domain Boundaries in Vanadium
Vol. **30**, No. 1, p. K1—K3 (1968).
See also *4*, p. *48*; 10.1, p. 135.

E. Köster
Magnetization and Magnetostriction of Hexagonal Cobalt Single Crystals with Internal Stresses
Vol. **30**, No. 2, p. 455–460 (1968).
See also 10.1, p. 135; *18.2*, p. *315*.

R. E. Hummel and J. W. Koger
Effect of Neutron Irradiation on the Martensitic Transformation in β_1-Copper-Zinc
Vol. **30**, No. 2, p. 515–523 (1968).
See also *1.2*, p. *13*; *11*, p. *155*; 14.1, p. 227.

Ch. Schwink and O. Schärpf
Electron-Optic Investigation of the Magnetic Stray Field above Bloch Walls in Cylyndric Nickel Crystals
Vol. **30**, No. 2, p. 637–644 (1968).
See also 4, p. 48; *18.2*, p. *315*.

V. V. Nemoshkalenko, M. A. Mindlina, and B. P. Mamko
X-Ray Spectra and Electronic Structure of Elements at the End of the First Long Period from ^{29}Cu to ^{34}Se
Vol. **30**, No. 2, p. 703–712 (1968).
See also 13.1, p. 189; *20*, p. *343*; 22.1.1, p. 467; 22.1.3, p. 480.

J. Czerwonko
On the Thermodynamics and Paramagnetic Properties of Dispersed Metals
Vol. **30**, No. 2, p. 723–734 (1968).
See also *13.1*, p. *190*; 18.1, p. 298; 21, p. 389; 21.6, p. 437.

P. Gassenmeier and M. Wilkens
X-Ray Investigation of the Transition between Stages I and II of the Work Hardening Curve of Cubic Face Centered Copper-Gallium Single Crystals
Vol. **30**, No. 2, p. 833–843 (1968).
See also *10.1*, p. *136*; 21, p. 389.

B. F. Rothenstein, J. Artzner, and T. Wächter
The Electromagnetic Aspect of Magnetostrictively Induced Longitudinal Oscillations (the Procopiu Effect with a Cylindrical Ferromagnetic Sample Free at one End)
Vol. **30**, No. 2, p. K91–K92 (1968).
See also *18.2*, p. *315*.

Th. Ricker
Domain Walls in Thin Ferromagnetic Films Investigated by Low-Angle Electron Deflection
Vol. **30**, No. 2, p. K93–K95 (1968).
See also 4, p. 48; *18.2*, p. *315*; 21.1.1, p. 423.

A. Seeger and H. Kronmüller
Reversible Magnetization Curve and Magnetostriction of Uniaxial Ferromagnetic Crystals with Internal Stresses
Vol. **30**, No. 2, p. K101–K106 (1968).
See also 10.1, p. 136; *18.2*, p. *315*.

J. Mathon and E. P. Wohlfarth
Contributions to the Invar Problem
Vol. **30**, No. 2, p. K131–K133 (1968).
See also *18.2*, p. *315*; 21.1.1, p. 423.

R. Brugière and P. Lucasson
On the Stage III Recovery of Electron Irradiated Copper
Vol. **30**, No. 2, p. K139–K143 (1968).
See also *10.1*, p. *136*; 11, p. 155; 14.1, p. 227.

21.1.1 Fe and Fe Alloys

Review Articles

A. Aharoni
Magnetization Curling
Vol. **16**, No. 1, p. 3–42 (1966).
See also 4, p. 37; *18.2*, p. *299*; 21.1, p. 390.

L. E. Murr
Calibration and Use of an Electron Microscope for Precision Micromeasurements in Thin Film Materials
Vol. **19**, No. 1, p. 7–34 (1967).
See also *1.4*, p. *16*; 4, p. 37; 10, p. 93; 21.1, p. 390.

Original Papers and Short Notes

B. Šesták and S. Kadečková
Etch Pits at Dislocations in High-Purity Iron Single Crystals
Vol. **16**, No. 1, p. 89–94 (1966).
See also *10.1*, p. *114*.

P. Escudier, F. Biragnet, J. Devenyi, and A. Yelon
Structural and Magnetic Properties of Epitaxial Permalloy Films
Vol. **16**, No. 1, p. 295–300 (1966).
See also *1.4*, p. *16*; *18.2*, p. *299*.

D. Paccard, C. Schlenker, O. Massenet, R. Montmory, and A. Yelon
A New Property of Ferromagnetic–Antiferromagnetic Coupling
Vol. **16**, No. 1, p. 301–311 (1966).
See also *18.2*, p. *299*; *18.4*, p. *321*; 21.1, p. 390.

H. Lachowicz
Ramp Field Switching in Thin Magnetic Films
Vol. **16**, No. 1, p. 313–320 (1966).
See also 4, p. 37; *18.2*, p. *299*.

D. J. Craik and M. J. Wood
Thermally Activated Domain Wall Movement
Vol. **16**, No. 1, p. 321–328 (1966).
See also 4, p. 37; *18.2*, p. *300*.

B. Heinrich and Z. Frait
Temperature Dependence of the FMR Linewidth of Iron Single-Crystal Platelets
Vol. **16**, No. 1, p. K11–K14 (1966).
See also *18.2*, p. *300*; *19*, p. *326*.

H. Bäurich
Calculation of the Energy, Magnetization Distribution, and Size of a Cross Bloch-Line
Vol. **16**, No. 1, p. K39–K43 (1966).
See also *18.2*, p. *300*.

H. A. Wriedt and S. Arajs
Ferromagnetic Curie Temperatures of Some Iron–Zinc Solid Solutions
Vol. **16**, No. 2, p. 475–478 (1966).
See also *18.2*, p. *300*.

F. Biragnet, J. Devenyi, G. Clerc, O. Massenet, R. Montmory, and A. Yelon
Interactions between Domain Walls in Coupled Films
Vol. **16**, No. 2, p. 569–576 (1966).
See also 4, p. 37; *18.2*, p. *300.*

W. Brodkorb and W. Haubenreisser
Theory of the Ferromagnetism of Magnetically Anisotropic Films with Cubic Crystal Structure. (II): Calculation of the Curie Temperature for an Fe-(100)-Film
Vol. **16**, No. 2, p. 577–583 (1966).
See also 18.1, p. 296; *18.2*, p. *300.*

J. A. Cundall and A. P. King
Comparison of Magnetic Measurements and Theoretical Predictions for Nickel-Iron Films over a Wide Composition Range
Vol. **16**, No. 2, p. 613–619 (1966).
See also 4, p. 38; *18.2*, p. *300.*

V. A. Buravikhin, V. G. Kazakov, V. S. Khristosenko, and P. I. Krukover.
Effect of Uniaxial Elastic Stress on the Remagnetization Processes in Thin Permalloy Films
Vol. **16**, No. 2, p. 651–656 (1966).
See also 12.1, p. 159; *18.2*, p. *301.*

V. A. Buravikhin and V. I. Popov
Structure and Polarity of Domain Boundaries in Ferromagnetic Films
Vol. **16**, No. 2, p. 657–662 (1966).
See also 4, p. 38; 12.1, p. 159; *18.2*, p. *301.*

S. K. Sidorov and A. V. Doroshenko
On the Magnetic Structure of Some Alloys of Transition Metals
Vol. **16**, No. 2, p. 737–744 (1966).
See also *18.2*, p. *301*; *18.4*, p. *321*; 21.1, p. 391.

A. A. Glazer, A. P. Potapov, R. I. Tagirov, and Ya. S. Shur
On Particularities of Magnetic Properties of Thin Films with Exchange Anisotropy
Vol. **16**, No. 2, p. 745–750 (1966).
See also 4, p. 38; *18.2*, p. *301.*

O. S. Kolotov, T. N. Letova, V. A. Pogozhev, and R. V. Telesnin
Remagnetization of Thin Ferromagnetic Layers along the Easy Axis
Vol. **16**, No. 2, p. K121–K123 (1966).
See also *18.2*, p. *301.*

H. Pascard
High-Power Effects in Thin Ferromagnetic Films with Oblique Polarization Field
Vol. **16**, No. 2, p. K153–K156 (1966).
See also *18.2*, p. *301.*

J. W. Moroń and B. Pilch
Further Studies of Relaxation Phenomena in Fe-Si-C(N) Alloys
Vol. **16**, No. 2, p. K171–K174 (1966).
See also *18.2*, p. *301.*

D. A. Thompson and H. Chang
Nanosecond Microscopic Measurement of Magnetic Film Switching by Kerr Magneto-Optic Apparatus
Vol. **17**, No. 1, p. 83–90 (1966).
See also *18.2*, p. *301.*

A. J. Morton and C. M. Wayman
On the Feasibility of a "Composite" Martensite Model for the $(225)_f$ Austenite–Martensite Transformation
Vol. **17**, No. 1, p. 101–103 (1966).
See also *1.2*, p. *9*; 3, p. 26.

P. Weber and W. Ruske
Temperature Dependence of Oblique-Incidence Anisotropy in Permalloy Films
Vol. **17**, No. 1, p. 185–190 (1966).
See also 1.4, p. 16; *18.2*, p. *301*.

L. V. Kirenskii, V. G. Pynko, N. I. Sivkov, G. P. Pynko, R. V. Sukhanova, and M. A. Ovsyannikov
Growth of Single-Crytal Films of Iron, Nickel, Cobalt, and Their Alloys
Vol. **17**, No. 1, p. 243–248 (1966).
See also 1.4, p. 16; *3.1*, p. *29*; 21.1, p. 392.

L. V. Kirenskii, V. G. Pynko, G. P. Pynko, R. V. Sukhanova, N. I. Sivkov, G. I. Rusov, and I. S. Edelman
Magnetic Properties of Single-Crystal Films of Iron, Nickel, Cobalt, and Their Alloys
Vol. **17**, No. 1, p. 249–257 (1966).
See also 4, p. 38; *18.2*, p. *302*; 21.1, p. 392.

W. Metzdorf and H. E. Wiehl
Negative Oblique-Incidence Anisotropy in Magnetostriction-Free Permalloy Films
Vol. **17**, No. 1, p. 285–294 (1966).
See also 1.4, p. 16; *18.2*, p. *302*.

P. Vigier, C. Minier-Cassayre, P. Moser, and V. Hivert
Study of Point Defects Produced by Neutron Irradiation in Doped Iron
Vol. **17**, No. 1, p. 317–321 (1966).
See also *10.1*, p. *116*; 11, p. 146; 14.1, p. 220; 18.2, p. 302.

G. Kneer and W. Zinn
On the Origin of M-Induced Anisotropy in Nickel and Permalloy Films
Vol. **17**, No. 1, p. 323–334 (1966).
See also *18.2*, p. *302*; 21.1, p. 392.

V. V. Kobelev
Ferromagnetic Films in Nonuniform Fields
Vol. **17**, No. 1, p. 381–388 (1966).
See also *18.2*, p. *302*.

I. B. Puchalska and A. Sukiennicki
Temperature Dependence of Oblique-Incidence Anisotropy Observed by Lorentz Microscopy
Vol. **17**, No. 1, p. K1 (1966).
See also 1.4, p. 17; *18.2*, p. *302*.

P. Vigier, V. Hivert, A. Bourret, and P. Moser
Precipitation and Re-Solution of Carbon in Iron
Vol. **17**, No. 1, p. K37–K39 (1966).
See also *10.1*, p. *116*; 18.2, p. 302.

W. D. Doyle
Epitaxial Influence of the Substrates on the Anisotropy in Thin Electrodeposited Ni-Fe Cylindrical Films
Vol. **17**, No. 1, p. K67–K69 (1966).
See also 1.4, p. 17; *18.2*, p. *302*.

O. Henkel
Interaction Effects of Single-Domain Fe-Particles in Frozen-in Suspensions
Vol. **17**, No. 1, p. K99—K103 (1966).
See also *18.2*, p. *302*; 18.3, p. 316.

A. Dobrzańska, J. W. Moroń, and J. Rasek
Magnetic Studies of Relaxation Phenomena in α-Fe-C Alloys
Vol. **17**, No. 1, p. K109—K112 (1966).
See also *18.2*, p. *303*.

L. V. Kirenskii, T. P. Isotova, N. M. Salanskii, and N. S. Chistyakov
Magnetic Multiple Layers
Vol. **17**, No. 2, p. 489—499 (1966).
See also *18.2*, p. *303*.

L. S. Palatnik, M. Ya. Fuks, L. I. Lukashenko, A. G. Ravlik, and A. A. Kozma
Structure and Magnetic Properties of Condensed Ferromagnetic Films
Vol. **17**, No. 2, p. 543—554 (1966).
See also *1.4*, p. *17*; 3.1, p. 30; 10.1, p. 116; *18.2*, p. *303*; 21.1, p. 392.

A. G. Lesnik, G. I. Levin, and V. M. Nedostup
The Importance of Vacancies and Interstitial Atoms in the Development of Induced Anisotropy
Vol. **17**, No. 2, p. 745—751 (1966).
See also 3.1, p. 30; 10.1, p. 116; *18.2*, p. *303*.

A. G. Lesnik and L. M. Sandler
Effect of Tensile Stress on the Magnetic Properties of Permalloy Films
Vol. **17**, No. 2, p. 753—760 (1966).
See also 10.1, p. 117; *18.2*, p. *303*; 19, p. 327.

L. Valenta and A. Sukiennicki
Order-Disorder Phenomena at the Surface of Solids
Vol. **17**, No. 2, p. 903—909 (1966).
See also *1.3*, p. *14*; *3.2*, p. *34*.

Vu dinh Ky
The Nernst Effect in Permalloy Films
Vol. **17**, No. 2, p. K203—K205 (1966).
See also 14.1, p. 221; 15, p. 270; *18.2*, p. *303*.

Vu dinh Ky
The Planar Nernst Effect in Permalloy Films
Vol. **17**, No. 2, p. K207—K209 (1966).
See also 14.1, p. 221; 15, p. 270; *18.2*, p. *303*.

S. B. McRickard
Twinning Deformation in Iron at 4 °K
Vol. **18**, No. 1, p. 89—97 (1966).
See also *10.1*, p. *117*.

R. Chang and L. J. Graham
Edge Dislocation Core Structure and the Peierls Barrier in Body-Centered Cubic Iron
Vol. **18**, No. 1, p. 99—103 (1966).
See also *10.1*, p. *117*.

J. van Landuyt, R. Gevers, and S. Amelinckx
On the Determination of the Nature of Stacking Faults in F. C. C. Metals from the Bright Field Image
Vol. **18**, No. 1, p. 167—172 (1966).
See also *10.1*, p. *117*.

H. Bilger
Temperature and Deformation Dependence of the Coercive Field Strength in Iron Single Crystals (II)
Vol. **18**, No. 1, p. 207–219 (1966).
See also 10.1, p. 117; *18.2*, p. *303*.

J. van Landuyt, R. Gevers, and S. Amelinckx
The Fine Structure of Spots in Electron Diffraction Resulting from the Presence of Planar Interfaces and Dislocations. (II): Observations on Crystals Containing Stacking Faults
Vol. **18**, No. 1, p. 363–378 (1966).
See also 4, p. 39; *10*, p. *96*; *10.1*, p. *118*; 22.6, p. 536.

C. Varotto and A. E. Vidoz
On the Recovery and Ordering Processes of Cold Worked Ni–Fe Alloys
Vol. **18**, No. 1, p. K49–K52 (1966).
See also 10.1, p. 118; *14.1*, p. *221*.

H. Bilger
Electron-Microscopic Surface and Transmission Studies of Iron Single Crystals and the Effect of Orientation on Their Strengthening Behaviour
Vol. **18**, No. 2, p. 637–651 (1966).
See also *10.1*, p. *119*.

V. Vítek and F. Kroupa
Dislocation Theory of Slip Geometry and Temperature Dependence of Flow Stress in B.C.C. Metals
Vol. **18**, No. 2, p. 703–713 (1966).
See also *10.1*, p. *119*.

V. F. Moiseev and V. I. Trefilov
Change of the Deformation Mechanism (Slip $\rightleftarrows$ Twinning) in Polycrystalline α-Iron
Vol. **18**, No. 2, p. 881–895 (1966).
See also *10.1*, p. *119*.

R. I. Anishchenko
Calculation of the Mean Inner Potential of a Crystal in the Statistical Theory
Vol. **18**, No. 2, p. 923–928 (1966).
See also *3.2*, p. *34*; 13, p. 164; 21, p. 376; 21.1, p. 394; 21.2, p. 424; 21.3, p. 426; 21.4, p. 428; 21.6, p. 432; 21.7, p. 438; 22.1.1, p. 460.

O. Henkel
Angular Dependence of Magnetic Properties for Particle Chains with Normally Positive Interaction
Vol. **18**, No. 2, p. K113–K116 (1966).
See also *18.2*, p. *304*.

H. Bilger
On the Recovery of Plastically Deformed Iron Single Crystals above Room Temperature
Vol. **18**, No. 2, p. K135–K138 (1966).
See also *10.1*, p. *120*; 18.2, p. 304.

Ir. Boesono, G. J. Ernst, M. C. Lemmens, M. J. van Langen, and G. de Vries
A Zener-Damping Peak in Fe–4 Wt% Si
Vol. **19**, No. 1, p. 107–110 (1967).
See also 10.1, p. 120; *12.1*, p. *159*.

M. S. R. Chari
The Electrical and Thermal Resistivities of Iron
Vol. **19**, No. 1, p. 169—172 (1967).
See also 8, p. 81; *14.1*, p. *222*.

A. K. Head
Unstable Dislocations in Anisotropic Crystals
Vol. **19**, No. 1, p. 185—192 (1967).
See also *10.1*, p. *120*; 21, p. 376; 21.1, p. 395.

W. Frank
Critical Shear Stress of Cubic Crystals Containing Defects of Tetragonal Symmetry. (IV): The Lower Field Stress of α-Iron
Vol. **19**, No. 1, p. 239—250 (1967).
See also *10.1*, p. *120*.

A. Gemperle and J. Kočík
Transmission Electron Microscopic Observation of Lattice Deformation in Antiphase Boundaries of Ordered Fe-Si Alloys
Vol. **19**, No. 1, p. 333—338 (1967).
See also 4, p. 40, *5*, p. *50*.

K.-H. Pfeffer
Interaction between Dislocations and Plane Bloch-Walls with a Rigid Magnetization Curve
Vol. **19**, No. 2, p. 735—750 (1967).
See also *10.1*, p. *121*; *18.2*, p. *305*; 21.1, p. 396.

G. A. Jones
Preferred Orientation and Domain Structure of Iron Films Deposited on (100) Copper Substrates
Vol. **19**, No. 2, p. 811—821 (1967).
See also *1.4*, p. *18*; 18.2, p. 305.

R. H. Wade
Electron Diffraction from a Magnetic Phase Grating
Vol. **19**, No. 2, p. 847—854 (1967).
See also 4, p. 40; *18.2*, p. *305*; 21.1, p. 396.

K. Wetzig
Investigations of Surfaces of the Binary System Ta-Fe by Means of an Emission Electron Microscope
Vol. **19**, No. 2, p. K71—K73 (1967).
See also *1.3*, p. *14*; 4, p. 40; 21, p. 377.

P. Guyot
Camel-Humps Peierls Hills in Iron
Vol. **19**, No. 2, p. K95—K98 (1967).
See also *10.1*, p. *121*.

M. Bevis and N. Swindells
The Determination of the Orientation of Micro-Crystals Using a Back-Reflection Kossel Technique and an Electron Probe Microanalyser
Vol. **20**, No. 1, p. 197—212 (1967).
See also 3, p. 27; *4*, p. *41*.

S. Y. Chuang and B. G. Hogg
Positron Annihilation in Alnico and $(La_{0.7}Pb_{0.3})MnO_3$
Vol. **20**, No. 1, p. 331—335 (1967).
See also 11, p. 148; *13*, p. *165*; 20, p. 338; 22.8, p. 549.

K.-H. PFEFFER
Micromagnetic Treatment of the Interaction between Dislocations and Plane Bloch-Walls. (I): General Theory
Vol. **20**, No. 1, p. 395–411 (1967).
See also 10.1, p. 121; *18.2*, p. *305*; 21.1, p. 396.

L. ZSOLDOS
Lattice Parameter Change of FeRh Alloys due to Antiferromagnetic–Ferromagnetic Transformation
Vol. **20**, No. 1, p. K25–K28 (1967).
See also 1.2, p. 10; 5, p. 51; *18.2*, p. *306*; *18.4*, p. *322*.

H. MARKERT
Decrease of Number and Size of Barkhausen Jumps during Ultrasonic Irradiation
Vol. **20**, No. 1, p. K67–K70 (1967).
See also *7*, p. *76*; 18.2, p. 306; 18.3, p. 317.

R. F. PEART
Effect of Pressure on the Diffusion of Fe in Ti and Ti +10% Fe
Vol. **20**, No. 2, p. 545–550 (1967).
See also *9*, p. *88*; 12.1, p. 160; 21, p. 378.

N. VAN DANG
Effect of Thermal Demagnetization on the Hysteresis Cycles
Vol. **20**, No. 2, p. 557–565 (1967).
See also *18.2*, p. *306*.

L. CSER, J. OSTANEVICH, and L. PÁL
Mössbauer Effect in Iron–Aluminium Alloys (I)
Vol. **20**, No. 2, p. 581–589 (1967).
See also *6.1*, p. *71*; 18, p. 292.

L. CSER, J. OSTANEVICH, and L. PÁL
Mössbauer Effect in Iron–Aluminium Alloys (II)
Vol. **20**, No. 2, p. 591–596 (1967).
See also *6.1*, p. *72*; 18, p. 292.

B. ROESSLER
Observations of Ferromagnetic Domains by Anomalous Transmission of X-Rays. The Effects of Variation in the Reflecting Plane
Vol. **20**, No. 2, p. 713–723 (1967).
See also 4, p. 41; *18.2*, p. *306*.

S. LIBOVICKÝ
Antiphase Boundaries in Silicon Iron Single Crystals Revealed by Etching
Vol. **20**, No. 2, p. K85–K87 (1967).
See also *4*, p. *42*.

G. A. JONES
Structural and Magnetic Properties of Epitaxial Films of Iron
Vol. **20**, No. 2, p. K95–K97 (1967).
See also *1.4*, p. *18*; *18.2*, p. *306*.

D. D. MISHIN
On the Theory of Susceptiblity and Coercive Force of Magnetic Materials
Vol. **20**, No. 2, p. K123–K126 (1967).
See also *18.2*, p. *306*.

O. Henkel
On the Temperature Dependence of Interaction Effects in Ferromagnetic Fine Particle Assemblies
Vol. **20**, No. 2, p. K127—K130 (1967).
See also *18.2*, p. *306*.

I. B. Puchalska, A. Sukiennicki, Z. Petykiewicz, and T. Tymosz
The Influence of Annealing on the Structure of Polycrystalline Permalloy Thin Films
Vol. **21**, No. 1, p. 183—189 (1967).
See also *1.4*, p. *18*; 18.2, p. 306.

S. Arajs and G. R. Dunmyre
A Note on the Consistency of Values of the Spontaneous or Saturation Magnetization of Polycrystalline Iron and Nickel at 298 °K
Vol. **21**, No. 1, p. 191—195 (1967).
See also *18.2*, p. *306*; 21.1, p. 397.

E. D. Hondros
Herring-Nabarro Creep Studies on the Self-Diffusion of Iron in the Presence of Solutes
Vol. **21**, No. 1, p. 375—382 (1967).
See also *9*, p. *89*; 10.1, p. 122.

R. Gevers, J. van Landuyt, and S. Amelinckx
Two Beam Kinematical Theory for the Diffraction of Electrons by Crystals with Stacking Faults
Vol. **21**, No. 1, p. 393—411 (1967).
See also *10*, p. *99*; *10.1*, p. *122*; 22.6, p. 538.

J. Levasseur and J. Philibert
Determination of Intrinsic Diffusion Coefficients by Measuring the Kirkendall Effect
Vol. **21**, No. 1, p. K1—K4 (1967).
See also *9*, p. *89*; 21.1, p. 397.

J. Steinert
Induced Uniaxial Magnetic Anisotropy of Fe-Al Alloys at Low Concentrations
Vol. **21**, No. 1, p. K13—K15 (1967).
See also 10.1, p. 122; *18.2*, p. *307*; 21, p. 379.

I. B. Puchalska and A. Sukiennicki
On the Temperature Dependence of the Domain Structure
Vol. **21**, No. 1, p. K57—K58 (1967).
See also 4, p. 42; *18.2*, p. *307*.

B. F. Rothenstein, A. Policec, M. Lupulescu, and C. Anghel
The Electromagnetic Aspect of Magnetostrictively Induced Torsional Oscillations (the Procopiu Effect with a Cylindrical Ferromagnetic Specimen Fixed at one End)
Vol. **21**, No. 2, p. 601—606 (1967).
See also 14.1, p. 222; *18.2*, p. *307*.

M. Polcarová and J. Kaczér
X-Ray Diffraction Contrast on Ferromagnetic Domain Walls in Fe-Si Single Crystals
Vol. **21**, No. 2, p. 635—642 (1967).
See also 4, p. 42; *18.2*, p. *307*.

P. Humble
A Comparison of Electron Microscope Images of Dislocations Based on Isotropic and Anisotropic Elasticity Theory
Vol. **21**, No. 2, p. 733—740 (1967).
See also *10.1*, p. *123*; 21, p. 379; 21.1, p. 398; 21.6, p. 433

K.-H. Pfeffer
On the Theory of Coercive Field Strength and Initial Susceptibility
Vol. 21, No. 2, p. 857—872 (1967).
See also 10.1, p. 123; *18.2*, p. *307*; 21.1, p. 398.

A. Clement, N. Clement, and P. Coulomb
Intrinsic and Extrinsic Defect Pairs in a Nonoxidizable Steel and a Copper-Silicon Alloy
Vol. 21, No. 2, p. K97—K98 (1967).
See also *10.1*, p. *123*; 21.1, p. 398.

H. Mühe
Temperature Dependence of Magnetocrystalline Anisotropy of an Fe-Al Alloy near the Concentration Fe_3Al
Vol. 21, No. 2, p. K115—K119 (1967).
See also *18.2*, p. *307*; 21, p. 380.

Vu dinh Ky
Susceptibility of Ferromagnetic Films in Paraprocess
Vol. 21, No. 2, p. K159—K162 (1967).
See also *18.2*, p. *307*; 21.1, p. 398.

M.-P. Stoll
The Use of the Magneto-Optical Kerr Effect for the Determination of Magnetization Curves
Vol. 22, No. 1, p. 163—176 (1967).
See also *18.2*, p. *307*; 20, p. 339.

V. Vítek
The Effect of the Type of Splitting of Screw Dislocations on the Slip Geometry in B.C.C. Metals
Vol. 22, No. 2, p. 453—462 (1967).
See also *10.1*, p. *124*; 21, p. 381.

K. R. Evans, D. J. Bailey, and W. F. Flanagan
An Interpretation of the Dislocation Velocity-Stress Exponent
Vol. 22, No. 2, p. 607—616 (1967).
See also *10.1*, p. *124*; 21.1, p. 399.

Vu dinh Ky
Planar Hall and Nernst Effect in Ferromagnetic Metals
Vol. 22, No. 2, p. 729—736 (1967).
See also 13, p. 167; *14.1*, p. *223*; *15*, p. *272*; 18.2, p. 308; 21.1, p. 400.

S. Arajs
Magnetic Susceptibility of Face-Centered Cubic Iron-Rhenium Alloys
Vol. 22, No. 2, p. 737—740 (1967).
See also 14.1, p. 223; *18.4*, p. *323*.

V. G. Kurdyumov, L. M. Morgulis, and L. M. Utevskii
Slip Band Structure of Deformed Silicon Iron Single Crystals
Vol. 22, No. 2, p. K75—K78 (1967).
See also *10.1*, p. *125*.

A. Bourret and M. Kleman
Stripe Domains in Thin Films of Pure Iron
Vol. 23, No. 1, p. 207—221 (1967).
See also *18.2*, p. *308*.

R. F. Peart and J. Askill
The Mechanism of Diffusion in B.C.C. Transition Metals
Vol. 23, No. 1, p. 263—275 (1967).
See also *9*, p. *90*; 10, p. 102; 21, p. 381; 21.1, p. 400.

S. Pal and P. K. Sharma
On Lindemann's Melting Criterion
Vol. 23, No. 1, p. 361—364 (1967).
See also *6*, p. *62* ; 21, p. 381; 21.1, p. 400. 21.2, p. 424; 21.6, p. 434.

W. Brückner, K. Kleinstück, and G. E. R. Schulze
Atomic Arrangement in the Homogeneity Range of the Laves Phases $ZrFe_2$ and $TiFe_2$
Vol. 23, No. 2, p. 475—480 (1967).
See also *5*, p. *52*.

B. Heinrich, D. Fraitová, and V. Kamberský
The Influence of s-d Exchange on Relaxation of Magnons in Metals
Vol. 23, No. 2, p. 501—507 (1967).
See also *13*, p. *167*; *19*, p. *330*.

R. Gevers, J. van Landuyt, and S. Amelinckx
Fine Structure of Weak Kinematical Electron Diffraction Beams for a Foil Containing Stacking Fault or Anti-Phase Boundaries — Two-Beam Case
Vol. 23, No. 2, p. 549—562 (1967).
See also *4*, p. *43*; *10.1*, p. *125*.

O. Henkel
Positive and Negative Interaction of Fine Particle Assemblies in Correlation with Other Magnetic Properties
Vol. 23, No. 2, p. K151—K153 (1967).
See also *18.2*, p. *308*.

E. F. Makarov, V. A. Povitskii, E. B. Granovskii, and A. A. Fridman
Study of Alnico 8 by Mössbauer Spectroscopy
Vol. 24, No. 1, p. 45—50 (1967).
See also *6.1*, p. *73*; 18.2, p. 308.

M. J. Marcinkowski and H. J. Leamy
Analysis of Dislocation Loops in Superlattices
Vol. 24, No. 1, p. 149—162 (1967).
See also *10*, p. *103*; *10.1*, p. *126*.

F. J. Krings
High-Temperature Ferromagnetic After-Effect in Iron–Silicon Alloys during Plastic Deformation
Vol. 24. No. 1, p. 163—176 (1967).
See also *10.1*, p. *126*; 18.2, p. 309.

H. H. Neely and D. W. Keefer
Resistivity Studies of Electron Irradiated Iron
Vol. 24, No. 1, p. 217—220 (1967).
See also *11*, p. *150*; 14.1, p. 224.

V. V. Kelarev, S. K. Sidorov, V. V. Klyushin, and R. Z. Abdulov
Neutron-Diffraction Study of Antiferro-Ferromagnetic Transition in a System of Ordered $Fe(Pd_xPt_{1-x})_3$ Alloys
Vol. 24, No. 1, p. 385—390 (1967).
See also *18.2*, p. *309*; *18.4*, p. *323*.

W. Glaeser, H. Imgrund, and H. Wever
On the Problem of Reproducible Measurements of the Residual Resistivity of High-Purity Iron
Vol. **24**, No. 1, p. K5—K8 (1967).
See also *14.1*, p. *224*; 18.2, p. 309.

O. Henkel
Interaction Effects in Alnico Alloys
Vol. **24**, No. 2, p. K153—K156 (1967).
See also *18.2*, p. *310*.

G. P. Seidel
Effect of Neutron Irradiation on the Plastic Deformation of Iron Single Crystals with Different Carbon Concentrations. (I): The Field Stress at the Deformation Onset
Vol. **25**, No. 1, p. 175—188 (1968).
See also *10.1*, p. *127*; *11*, p. *151*.

A. Sukiennicki and L. Wojtczak
Magnetic Domain Structure in Uniaxial Thin Films
Vol. **25**, No. 1, p. K19—K21 (1968).
See also *18.2*, p. *310*; 21.1, p. 402.

A. Holz
Theoretical Study of Nucleation in Ferromagnetics
Vol. **25**, No. 2, p. 567—582 (1968).
See also 4, p. 44; *18.2*, p. *310*; 21.1, p. 403.

M. Kuriyama and G. M. McManus
X-Ray Interference Fringes and Domain Arrangements in Fe + 3wt% Si Single Crystals
Vol. **25**, No. 2, p. 667—677 (1968).
See also *4*, p. *45*; 18.2, p. 310.

R. Gauthier, D. Bois, P. Pinard, and F. Davoine
Ion and Electron Production by Interaction of a Coherent Light Beam with Substance
Vol. **25**, No. 2, p. 691—696 (1968).
See also *17*, p. *288*; 21.1, p. 403; 22.1.1, p. 464; 22.2.1, p. 485; 22.4.1, p. 502; 22.5.2, p. 525; 22.6, p. 540.

H.-J. Bunge
The Orientation Distribution Function of the Crystallites in Cold-Rolled and Annealed Low-Carbon Steel Sheets
Vol. **26**, No. 1, p. 167—172 (1968).
See also 1.1, p. 8; *4*, p. *45*.

P. Rowlands and M. Bevis
A Rapid Method for Orientating Crystals from Kossel Patterns
Vol. **26**, No. 1, p. K25—K28 (1968).
See also 3, p. 27; *4*, p. *45*.

S. J. Lewis and P. A. Flinn
Disappearance of Quadrupole Splitting of the Mössbauer Resonance in Iron–Carbon Austenite at Elevated Temperatures
Vol. **26**, No. 1, p. K51—K54 (1968).
See also *6.1*, p. *73*.

H. Wagenblast and S. Arajs
Electrical Resistivity Studies of Iron–Nitrogen Solid Solutions
Vol. **26**, No. 2, p. 409—418 (1968).
See also *14.1*, p. *225*.

Vu dinh Ky
Planar Hall Effect in Ferromagnetic Films
Vol. **26**, No. 2, p. 565—569 (1968).
See also *14.1*, p. *225*; *18.2*, p. *311*; 21.1, p. 404.

G. F. Bolling, A. Arrott, and R. H. Richman
A Study of Curie Temperatures and the Effect of Carbon in Face-Centered Cubic Iron–Nickel Alloys
Vol. **26**, No. 2, p. 743—750 (1968).
See also 10.1, p. 130; *18.2*, p. *311*.

A. Holz
Magnetization Distribution at the Basal Faces of Cylindrical Ferromagnetics with Application to Nucleation
Vol. **26**, No. 2, p. 751—764 (1968).
See also 4, p. 46; *18.2*, p. *311*.

K.-H. Müller and M. Wolf
Magnetization and Stiffness of Spin Waves in the Heisenberg Model for Fe–Al Alloys
Vol. **26**, No. 2, p. K109—K111 (1968).
See also *13*, p. *169*; *18.2*, p. *311*.

E. Feldtkeller
Magnetic Domain Wall Dynamics
Vol. **27**, No. 1, p. 161—170 (1968).
See also 4, p. 46; *18.2*, p. *311*; 21.1, p. 404.

U. Gradmann and J. Müller
Flat Ferromagnetic, Epitaxial 48Ni/52Fe(111) Films of Few Atomic Layers
Vol. **27**, No. 1, p. 313—324 (1968).
See also *1.4*, p. *20*; *18.2*, p. *311*; 21.1, p. 405.

K. Wetzig
Some Investigations Concerning the Position and the Melting Point of TaFe Lamellas in the System TaFe–$TaFe_2$
Vol. **27**, No. 1, p. K7—K9 (1968).
See also 1.2, p. 12; *4*, p. *46*; 21, p. 386.

J. Kinel, L. Kozłowski, and J. W. Moroń
Diffusional Relaxation Effects in Hydrogen Charged Silicon Iron
Vol. **27**, No. 1, p. K37—K39 (1968).
See also 10.1, p. 131; *12.1*, p. *160*; *18.2*, p. *312*.

H. Burkhard and J. Jaumann
The Determination of the "Gyroelectric" and "Gyromagnetic" Constants of Thick, Evaporated Layers of Iron by Means of Measurements in the Reflected Light only
Vol. **27**, No. 1, p. K43—K45 (1968).
See also *18.2*, p. *312*; *20*, p. *341*; 20.1, p. 354.

D. D. Mishin
The Effect of Dislocations on the Properties of Silicon Iron in Alternating Magnetic Fields
Vol. **27**, No. 1, p. K49—K50 (1968).
See also *10.1*, p. *131*; 18.2, p. 312.

M. Wuttig, J. T. Stanley, and H. K. Birnbaum
Interstitial Solute Trapping in Irradiated and Quenched Iron
Vol. **27**, No. 2, p. 701—712 (1968).
See also *10.1*, p. *131*.

M. Victoria and A. E. Vidoz
Tensile Behaviour upon Ordering of Ni_3Fe Single Crystals
Vol. **28**, No. 1, p. 131–144 (1968).
See also 4, p. 46; *10.1*, p. *131*; 21.1, p. 405.

A. E. Vidoz
On the Work Hardening of Superstructures
Vol. **28**, No. 1, p. 145–156 (1968).
See also 4, p. 46; *10.1*, p. *131*; 21.1, p. 405; 21.6, p. 436.

R. P. Ferrier and I. B. Puchalska
360° Walls and Strong Stripe Domains in Permalloy Films
Vol. **28**, No. 1, p. 335–347 (1968).
See also 1.4, p. 20; *18.2*, p. *312*.

D. J. Craik and M. J. Wood
Thermally Activated Domain Wall Movement in Nickel–Iron Films
Vol. **28**, No. 1, p. 407–414 (1968).
See also 1.4, p. 20; *18.2*, p. *313*; 21.1, p. 405.

R. Conte, P. Groh, and B. Escaig
Study of Plastic Flow of Iron Whiskers between 300 and 4 °K
Vol. **28**, No. 2, p. 475–488 (1968).
See also *10.1*, p. *132*.

B. Pegel
Stacking Faults on {110} Planes in the B.C.C. Lattice
Vol. **28**, No. 2, p. 603–609 (1968).
See also *10.1*, p. *132*; 21, p. 387; 21.1, p. 406.

V. V. Nemoshkalenko, O. N. Rasumov, and V. V. Gorskii
Investigation of the Mössbauer Effect in Some Fe–Al Alloys
Vol. **29**, No. 1, p. 45–48 (1968).
See also *6.1*, p. *74*; 13, p. 172.

H. Gengnagel and U. Hofmann
Temperature Dependence of the Magnetocrystalline Energy Constants K_1, K_2, and K_3 of Iron
Vol. **29**, No. 1, p. 91–97 (1968).
See also *18.2*, p. *314*.

W. Brückner, R. Perthel, K. Kleinstück, and G. E. R. Schulze
Magnetic Properties of $ZrFe_2$ and $TiFe_2$ within Their Homogeneity Range
Vol. **29**, No. 1, p. 211–216 (1968).
See also *18.2*, p. *314*; *18.4*, p. *325*.

A. Sukiennicki
On a New Type of Stripe Domains
Vol. **29**, No. 1, p. 417–421 (1968).
See also 4, p. 47; *18.2*, p. *314*.

J. Kinel, J. W. Moroń, and J. Przybyła
A Diffussional Magnetic Viscosity Effect in Iron-Carbon Martensite
Vol. **29**, No. 1, p. K39–K41 (1968).
See also *18.2*, p. *314*.

D. I. Bower, E. Claridge, and I. S. T. Tsong
Low-Temperature Elastic Constants and Specific Heats of F.C.C. Nickel-Iron Alloys
Vol. **29**, No. 2, p. 617–625 (1968).
See also 6, p. 68; 8, p. 86; *12.1*, p. *160*; 13, p. 172; 18.2, p. 314; 21.1, p. 407.

B. F. Rothenstein and I. Artzner
On the Internal Barkhausen Effect
Vol. **29**, No. 2, p. K117–K118 (1968).
See also *18.2*, p. *314*; 21.1, p. 408.

S. J. Sivonen and E. J. Suoninen
Temperature Dependence of the Magnetic Ordering in Gd-$GdFe_2$ Phase Mixtures
Vol. **29**, No. 2, p. K171–K173 (1968).
See also 14.1, p. 226; *18.2*, p. *314*; 21.4, p. 428.

J. L. Leveque, T. Anagnostopoulos, H. Bilger, and P. Moser
Point Defects in Iron–Carbon Solid Solution Studied by Magnetic After-Effect Measurements
Vol. **29**, No. 2, p. K179–K181 (1968).
See also *10.1*, p. *134*; *11*, p. *154*; 18.2, p. 315.

D. D. Mishin
Dislocation Theory of Susceptibility and Coercive Force of Silicon Iron
Vol. **30**, No. 1, p. 31–35 (1968).
See also *10.1*, p. *134*; *18.2*, p. *315*.

J. Gemperlová
Force on Dislocations in Anisotropic Bicrystals and Half-Spaces
Vol. **30**, No. 1, p. 261–274 (1968).
See also *10.1*, p. *135*; 21, p. 388; 21.1, p. 408; 21.2, p. 425.

S. Völlmar
Calculations Concerning a Model for Hard-Magnetic Particle Ensembles
Vol. **30**, No. 1, p. K27–K30 (1968).
See also *18.2*, p. *315*.

A. Gemperle
Crystallography of Antiphase Boundaries in Fe–Si Alloys
Vol. **30**, No. 2, p. 541–550 (1968).
See also 3, p. 28; *4*, p. *48*.

R. de Ridder, J. van Landuyt, R. Gevers, and S. Amelinckx
The Fine Structure of Spots in Electron Diffraction Resulting from the Presence of Planar Interfaces and Dislocations. (IV): Wedge Crystals
Vol. **30**, No. 2, p. 797–815 (1968).
See also *4*, p. *48*; 10, p. 113; 10.1, p. 136; 22.6, p. 544.

Th. Ricker
Domain Walls in Thin Ferromagnetic Films Investigated by Low-Angle Electron Deflection
Vol. **30**, No. 2, p. K93–K95 (1968).
See also 4, p. 48; *18.2*, p. *315*; 21.1, p. 409.

J. Mathon and E. P. Wohlfarth
Contributions to the Invar Problem
Vol. **30**, No. 2, p. K131–K133 (1968).
See also *18.2*, p. *315*; 21.1, p. 409.

21.2 Alkali Metals

Review Articles

D. P. Morgan
Helicon Waves in Solids
Vol. **24**, No. 1, p. 9–36 (1967).
See also 6, p. 56; *13*, p. *162*; 14.1, p. 220; 14.3, p. 231; 19, p. 326; *20*, p. *337*; 21. p. 373; 21.1, p. 390; 21.6, p. 430; 22.1.1, p. 458; 22.2.1, p. 482; 22.2.3, p. 489.

W. Steinmann
Optical Plasma Resonances in Solids
Vol. **28**, No. 2, p. 437–462 (1968).
See also *13*, p. *162*; 13.3, p. 196; *20*, p. *337*; 20.1, p. 344; 21, p. 373; 21.6, p. 430.

Original Papers and Short Notes

H. Mayer and J. Hölzl
Experimental Determination of the Maximum Emission-Depths of Monoenergetic Secondary Electrons
Vol. **18**, No. 2, p. 779–785 (1966).
See also 13, p. 164; *17*, p. *287*.

R. I. Anishchenko
Calculation of the Mean Inner Potential of a Crystal in the Statistical Theory
Vol. **18**, No. 2, p. 923–928 (1966).
See also *3.2*, p. *34*; 13, p. 164; 21, p. 376; 21.1, p. 394; 21.1.1, p. 414; 21.3, p. 426; 21.4, p. 428; 21.6, p. 432; 21.7, p. 438; 22.1.1, p. 460.

G. Björkman and G. Grimvall
Polarization Vectors for Lattice Vibrations in Sodium
Vol. **19**, No. 2, p. 863–865 (1967).
See also *6*, p. *59*.

A. Lodding
Thermotransport and the Associated Isotope Effect in Solids
Vol. **22**, No. 1, p. 157–161 (1967).
See also *9*, p. *89*; 22.8, p. 551.

F. Brouers
Plasmon Satellites oft Soft X-Ray Spectra
Vol. **22**, No. 1, p. 213–221 (1967).
See also *13*, p. *166*; *20*, p. *339*; 21, p. 380.

M. Matyáš and L. Švob
Remarks on Weiser's Theory of Interstitial Diffusion in the Diamond Lattice
Vol. **22**, No. 2, p. K141–K145 (1967).
See also *9*, p. *90*.

S. Pal and P. K. Sharma
On Lindemann's Melting Criterion
Vol. **23**, No. 1, p. 361–364 (1967).
See also *6*, p. *62*; 21, p. 381; 21.1, p. 400; 21.1.1, p. 419; 21.6, p. 434.

W. van der Lugt and J. S. Knol
The Knight Shift in Potassium from 23 to 123 °C
Vol. **23**, No. 1, p. K83–K86 (1967).
See also *19*, p. *330*.

P. S. Ho and A. L. Ruoff
A Quasi-Harmonic Calculation of Lattice Dynamics for Na
Vol. **23**, No. 2, p. 489–499 (1967).
See also *6*, p. *63*; 13.1, p. 182; 14.1, p. 224.

I. M. Bernstein and J. C. M. Li
Thermally Activated Deformation of Potassium
Vol. **23**, No. 2, p. 539–548 (1967).
See also *10.1*, p. *125*.

R. Dupree and B. W. Holland
The Range of *g*-Factors and the Breakdown of Motional Narrowing in Conduction Electron Spin Resonance
Vol. **24**, No. 1, p. 275–279 (1967).
See also *19*, p. *331*; 21, p. 382; 21.1, p. 401.

B. K. Jones
Measurement of the Magnitude of the Electronic Attenuation of Sound in Metals
Vol. **24**, No. 1, p. 297–300 (1967).
See also *7*, p. *77*; 13.1, p. 182; 21, p. 382; 21.1, p. 401.

R. Kishore
Electrical Conductivity of Metals
Vol. **26**, No. 1, p. 133–138 (1968).
See also *14.1*, p. *225*; 21.1, p. 403; 21.6, p. 435.

V. K. Saxena and D. L. Bhattacharya
Effect of Electron-Lattice Interactions on the Spin-Paramagnetism of Metals
Vol. **27**, No. 1, p. 427–433 (1968).
See also 13, p. 170; *18.1*, p. *297*; 21, p. 386.

B. W. Holland
Pseudopotentials and the Knight Shift
Vol. **28**, No. 1, p. 121–125 (1968).
See also 13, p. 171; *19*, p. *334*; 21, p. 386.

A. K. Singh and P. K. Sharma
Thermal Expansion of Cubic Metals in Krebs's Model
Vol. **28**, No. 2, p. 643–648 (1968).
See also *6*, p. *67*; *8*, p. *85*; 21, p. 387; 21.1, p. 406; 21.6, p. 436.

A. J. Watts and J. E. Cousins
Temperature Dependent Spin-Lattice Relaxation Times for Conduction Electrons in Lithium Particles in n-Irradiated Lithium Fluoride
Vol. **30**, No. 1, p. 105–112 (1968).
See also *19*, p. *335*; 22.5.2, p. 530.

J. Gemperlová
Force on Dislocations in Anisotropic Bicrystals and Half-Spaces
Vol. **30**, No. 1, p. 261–274 (1968).
See also *10.1*, p. *135*; 21, p. 388; 21.1, p. 408; 21.1.1, p. 423.

21.3 Alkali-Earth Metals

Original Papers and Short Notes

D. Rühlicke
Slip Mechanism of Plastically Deformed Magnesium Single Crystals above Room Temperature
Vol. **16**, No. 1, p. K1—K4 (1966).
See also *10.1*, p. *114*.

A. Seeger and G. Wobser
Parallel Dislocation Arrays in Anisotropic Media
Vol. **17**, No. 2, p. 709—724 (1966).
See also *10*, p. *95*; *10.1*, p. *116*; 21, p. 375; 21.1, p. 392; 21.6, p. 431; 22.1.1, p. 460; 22.1.2, p. 469.

L. I. van Torne
Point Defect Electron Scattering in Beryllium
Vol. **18**, No. 2, p. 737—742 (1966).
See also 10.1, p. 119; *14.1*, p. *221*.

R. I. Anishchenko
Calculation of the Mean Inner Potential of a Crystal in the Statistical Theory
Vol. **18**, No. 2, p. 923—928 (1966).
See also *3.2*, p. *34*; 13, p. 164; 21, p. 376; 21.1, p. 394; 21.1.1, p. 414; 21.2, p. 424; 21.4, p. 428; 21.6, p. 432; 21.7, p. 438; 22.1.1, p. 460.

A. Czachor and A. Pindor
Dependence of Calculated Elastic Constants on the c/a Ratio for Hexagonal Close-Packed Metals
Vol. **20**, No. 1, p. K17—K20 (1967).
See also 5, p. 50; *12.1*, p. *159*; 21, p. 377; 21.1, p. 396.

E. A. Metzbower
Noncentral Force Model for Hexagonal Close-Packed Crystal Lattices
Vol. **20**, No. 2, p. 681—692 (1967).
See also 3.2, p. 35; *6*, p. *60*; 8, p. 81; 12.1, p. 160; 21.1, p. 397.

W. Schülke and U. Berg
Measurement of the Compton Band on Beryllium Single Crystals
Vol. **23**, No. 1, p. K87—K91 (1967).
See also *13.1*, p. *181*; 20, p. 340.

P. Regnier and J. M. Dupouy
Prismatic Glide of Beryllium Monocrystals
Vol. **23**, No. 2, p. K109—K111 (1967).
See also *10.1*, p. *125*.

P. Paufler and G. E. R. Schulze
Plastic Deformation of the Intermetallic Compound $MgZn_2$
Vol. **24**, No. 1, p. 77—87 (1967).
See also *10.1*, p. *126*; 21.1, p. 401.

M. F. Rose and R. T. Ramsey
Higher Order Elastic Constants in H.C.P. Crystals
Vol. **25**, No. 1, p. 103—108 (1968).
See also 3.2, p. 36; *12.1*, p. *160*; 21, p. 383.

J. Van der Planken and A. Deruyttere
Twinning and Fracture along {10$\bar{1}$1} Planes in a Magnesium-Tin Alloy
Vol. **28**, No. 1, p. K9–K10 (1968).
See also *10.1*, p. *132*; 21, p. 387.

V. V. Nemoshkalenko and V. V. Gorskii
Investigation of the Valence Electron Distribution in Mg-Al Alloys
Vol. **28**, No. 1, p. K15–K17 (1968).
See also 13.1, p. 187; *20*, p. *341*; 21, p. 387.

A. Czachor
On the Lattice Dynamics of Hexagonal Structure Metals
Vol. **29**, No. 1, p. 423–434 (1968).
See also *6*, p. *68*.

J. Delaplace, J. C. Nicoud, D. Schumacher, and G. Vogl
Low-Temperature Neutron Radiation Damage and Recovery in Beryllium
Vol. **29**, No. 2, p. 819–824 (1968).
See also 10.1, p. 134; *11*, p. *154*; 14.1, p. 226.

H. R. Vydyanath, D. H. Sastry, and K. I. Vasu
On the Kinetics of Clustering in a Magnesium-3% Zinc Alloy
Vol. **29**, No. 2, p. K137–K140 (1968).
See also *4*, p. *47*; 10.1, p. 134; 14.1, p. 226; 21.1, p. 408.

J. Delaplace, J. Hillairet, J. C. Nicoud, D. Schumacher, and G. Vogl
Low Temperature Neutron Radiation Damage and Recovery in Magnesium
Vol. **30**, No. 1, p. 119–126 (1968).
See also 10.1, p. 134; *11*, p. *154*; 14.1, p. 226.

21.4 Rare-Earth Metals

Original Papers and Short Notes

R. I. ANISHCHENKO
Calculation of the Mean Inner Potential of a Crystal in the Statistical Theory
Vol. **18**, No. 2, p. 923–928 (1966).
See also *3.2*, p. *34*; **13**, p. 164; 21, p. 376; 21.1, p. 394; 21.1.1, p. 414; 21.2, p. 424; 21.3, p. 426; 21.6, p. 432; 21.7, p. 438; 22.1.1, p. 460.

G. S. NIKOLSKII and V. V. EREMENKO
The Effect of a Magnetic Field on the Heat Conductivity of Er Polycrystals
Vol. **18**, No. 2, p. K123–K125 (1966).
See also *8*, p. *80*; 14.1, p. 222.

H. P. VAN DE BRAAK and W. J. CASPERS
Disorder Resistivity and Spin Depolarization in Rare-Earth Alloys
Vol. **19**, No. 1, p. 385–393 (1967).
See also 13, p. 164; *14.1*, p. *222*; 18.2, p. 305.

V. A. FINKEL, Y. N. SMIRNOV, and V. V. VOROBJOV
On the Problem of the Thermal Expansion of Gadolinium
Vol. **23**, No. 1, p. K97–K98 (1967).
See also *8*, p. *83*.

A. M. VAN DIEPEN, H. W. DE WIJN, and K. H. J. BUSCHOW
Nuclear Magnetic Resonance and Susceptibility of Equiatomic Rare-Earth–Aluminium Compounds
Vol. **29**, No. 1, p. 189–192 (1968).
See also 13, p. 172; 18.1, p. 297; *19*, p. *335*; 21, p. 387.

K. H. J. BUSCHOW, J. F. FAST, and A. S. VAN DER GOOT
Magnetic Properties of Some Co-Rich Erbium Cobalt Intermetallic Compounds
Vol. **29**, No. 2, p. 825–829 (1968).
See also *18.3*, p. *320*; 21.1, p. 407.

S. J. SIVONEN and E. J. SUONINEN
Temperature Dependence of the Magnetic Ordering in Gd-$GdFe_2$ Phase Mixtures
Vol. **29**, No. 2, p. K171–K173 (1968).
See also 14.1, p. 226; *18.2*, p. *314*; 21.1.1, p. 423.

H. W. DE WIJN, K. H. J. BUSCHOW, and A. M. VAN DIEPEN
Experimental Evidence for Interband Mixing in Rare-Earth Intermetallic Compounds
Vol. **30**, No. 2, p. 759–766 (1968).
See also 13, p. 173; 13.1, p. 190; 18.1, p. 298; *19*, p. *336*; 21, p. 389.

21.5 Actinides

Original Papers and Short Notes

T. R. Duncan and D. Kuhlmann-Wilsdorf
The Stress Fields of Dislocation Tilt Boundaries in Anisotropic Cubic and Diamond-Type Crystals
Vol. **18**, No. 1, p. 231–239 (1966).
See also *10*, p. *96*; *10.1*, p. *117*; 21, p. 375; 21.1, p. 393; 21.6, p. 431; 22.1, p. 456; 22.1.1, p. 460; 22.1.2, p. 470.

R. Lallement and D. A. Wigley
Study of the Variation of Length of α-Plutonium by Self-Irradiation at 4 °K
Vol. **21**, No. 1, p. 145–150 (1967).
See also *11*, p. *148*; 12, p. 157; 14.1, p. 222.

C. Lea and C. H. B. Mee
The Attenuation Length of Photoelectrons in Thin Films of Uranium
Vol. **25**, No. 2, p. 613–619 (1968).
See also 14.1, p. 224; *17*, p. *288*.

J. T. Moore, T. R. Duncan, and D. Kuhlmann-Wilsdorf
Corrected Data for the Shear Stresses of Infinite Tilt Boundaries Composed of {111}, 1/2 ⟨1$\bar{1}$0⟩ Dislocations in Elastically Anisotropic F.C.C. Metals
Vol. **25**, No. 2, p. K105–K106 (1968).
See also *10*, p. *105*; *10.1*, p. *129*; 21, p. 385; 21.1, p. 403; 21.6, p. 435; 22.1, p. 457; 22.1.1, p. 464; 22.1.2, p. 473.

J. C. Jousset
Irradiation Defects in α-Uranium and Matthiessen's Rule
Vol. **29**, No. 2, p. K127–K131 (1968).
See also *11*, p. *154*; 14.1, p. 226.

21.6 Noble Metals

Review Articles

D. P. Morgan
Helicon Waves in Solids
Vol. **24**, No. 1, p. 9–36 (1967).
See also 6, p. 56; *13*, p. *162*; 14.1, p. 220; 14.3, p. 231; 19, p. 326; *20*, p. *337*; 21, p. 373; 21.1, p. 390; 21.2, p. 424; 22.1.1, p. 458; 22.2.1, p. 482; 22.2.3, p. 489.

W. Steinmann
Optical Plasma Resonances in Solids
Vol. **28**, No. 2, p. 437–462 (1968).
See also *13*, p. *162*; 13.3, p. 196; *20*, p. *337*; 20.1, p. 344; 21, p. 373; 21.2, p. 424.

Original Papers and Short Notes

P. C. J. Gallagher
An Absolute Determination of the Extrinsic and Intrinsic Stacking Fault Energies in Ag-In Alloys
Vol. **16**, No. 1, p. 95–115 (1966).
See also *10.1*, p. *114*.

M. J. Stowell and T. J. Law
The Migration of Double Positioning Boundaries during the Growth of Epitaxial (111) Gold Films
Vol. **16**, No. 1, p. 117–125 (1966).
See also *1.4*, p. *16*; *3.1*, p. *29*.

M. J. Stowell and T. J. Law
Interrupted Vapour Deposition of Epitaxial Gold Films Grown Inside an Electron Microscope
Vol. **16**, No. 2, p. 479–485 (1966).
See also *1.4*, p. *16*; 3.1, p. 29.

K.-P. Chik
Electron Diffraction Contrast of Small Stacking Fault Tetrahedra with Applications to Quenched Gold
Vol. **16**, No. 2, p. 685–696 (1966).
See also *10.1*, p. *114*.

N. K. Archipova, S. M. Klotsman, A. N. Timofeev, and I. Sh. Trakhtenberg
Effect of a DC Field on the Lattice Diffusion of Silver-110 in Copper and Gold
Vol. **16**, No. 2, p. 729–736 (1966).
See also *9*, p. *87*; 21.1, p. 391.

V. K. Nikulin
Cohesive Energy of Noble Metals
Vol. **16**, No. 2, p. K125–K127 (1966).
See also *3.2*, p. *34*; 5, p. 49; 21.1, p. 391.

R. von Jan
The Interpretation of Recovery Stages in Irradiated Metals
Vol. **17**, No. 1, p. 361–373 (1966).
See also *10.1*, p. *116*; 11, p. 146; 21, p. 374; 21.1, p. 392.

D. N. Seidman and R. W. Balluffi
On the Efficiency of Dislocation Climb in Gold
Vol. **17**, No. 2, p. 531—541 (1966).
See also *10.1*, p. *116*.

H. Jahrreiss and H.-J. Isken
Epitaxial Behaviour of Thin Evaporated Silver Layers
Vol. **17**, No. 2, p. 619—629 (1966).
See also *1.4*, p. *17*; *3.1*, p. *30*.

A. Seeger and G. Wobser
Parallel Dislocation Arrays in Anisotropic Media
Vol. **17**, No. 2, p. 709—724 (1966).
See also *10*, p. *95*; *10.1*, p. *116*; 21, p. 375; 21.1, p. 392; 21.3, p. 426; 22.1.1, p. 460; 22.1.2, p. 469.

A. I. Gubanov and V. K. Nikulin
The Theory of the Cohesion of Noble Metals
Vol. **17**, No. 2, p. 815—823 (1966).
See also *3.2*, p. *34*; 5, p. 50; 12.1, p. 159; 21.1, p. 393.

M. F. Rose
Pressure Dependence of the Second Order Elastic Constants
Vol. **17**, No. 2, p. K199—K201 (1966).
See also *12.1*, p. *159*; 21, p. 375; 21.1, p. 393.

K. L. Merkle
Radiation-Induced Point Defect Clusters in Copper and Gold.(I): Clusters Produced in Energetic Displacement Cascades
Vol. **18**, No. 1, p. 173—188 (1966).
See also 10.1, p. 117; *11*, p. *146*; 21.1, p. 393.

A. Seeger and G. Wobser
Stacking Fault Dipoles in Face-Centered Cubic Metals
Vol. **18**, No. 1, p. 189—206 (1966).
See also *10.1*, p. *117*; 21.1, p. 393.

T. R. Duncan and D. Kuhlmann-Wilsdorf
The Stress Fields of Dislocation Tilt Boundaries in Anisotropic Cubic and Diamond-Type Crystals
Vol. **18**, No. 1, p. 231—239 (1966).
See also *10*, p. *96*; *10.1*, p. *117*; 21, p. 375; 21.1, p. 393; 21.5, p. 429; 22.1, p. 456; 22.1.1, p. 460; 22.1.2, p. 470.

K. M. Kunz, A. K. Green, and E. Bauer
On the Formation of Single Crystal Films of F.C.C. Metals on Alkali Halide Cleavage Planes in Ultrahigh Vacuum
Vol. **18**, No. 1, p. 441—457 (1966).
See also *1.4*, p. *17*; *3.1*, p. *30*; 21, p. 376.

F. Häussermann and M. Wilkens
Determination of the Stacking Fault Energy of Face-Centered Cubic Metals from the Electron-Microscopic Diffraction Contrast of Stacking Fault Dipoles
Vol. **18**, No. 2, p. 609—624 (1966).
See also *10.1*, p. *119*; 21.1, p. 394.

H. Yoshinaga
Measurements of the Anisotropy of the Dislocation Resistivity in Au, Ag, and Cu
Vol. **18**, No. 2, p. 625—636 (1966).
See also 10.1, p. 119; *14.1*, p. *221*; 21.1, p. 394.

S. M. KLOTSMAN, A. N. TIMOFEEV, and I. SH. TRAKHTENBERG
On the Mechanism of Lattice Electromigration in Metals
Vol. **18**, No. 2, p. 847—852 (1966).
See also *9*, p. *88*; 14.1, p. 222; 21.1, p. 394.

R. I. ANISHCHENKO
Calculation of the Mean Inner Potential of a Crystal in the Statistical Theory
Vol. **18**, No. 2 , p. 923—928 (1966).
See also *3.2*, p. *34*; 13, p. 164; 21, p. 376; 21.1, p. 394; 21.1.1, p. 414; 21.2, p. 424; 21.3, p. 426; 21.4, p. 428; 21.7, p. 438; 22.1.1, p. 460.

W. HEYE and G. WASSERMANN
Mechanical Twinning in Cold-Rolled Silver Single Crystals
Vol. **18**, No. 2, p. K107—K111 (1966).
See also *10.1*, p. *119*.

C. P. KEMPTER
Vegard's "Law"
Vol. **18**, No. 2, p. K117—K118 (1966).
See also *5*, p. *50*; 21.1, p. 394.

R. H. WADE and J. SILCOX
Small Angle Electron Scattering from Vacuum Condensed Metallic Films. (II): Experimental Results
Vol. **19**, No. 1, p. 63—76 (1967).
See also *1.4*, p. *18*; 4, p. 40; 21.1, p. 395.

A. VAN DEN BEUKEL, P. C. J. COREMANS, and M. M. A. VRIJHOEF
On the Kinetics of Short-Range Ordering in AuAg (50,50) and CuAl (85,15)
Vol. **19**, No. 1, p. 177—183 (1967).
See also *1.1*, p. *8*; 10.1, p. 120; 21.1, p. 395.

D. D. MISHIN and R. M. GRETCHISHKIN
Electron Microscopic Examination of High-Coercive Cobalt-Platinum Alloys
Vol. **19**, No. 1, p. K1—K3 (1967).
See also 4, p. 40; *18.2*, p. *305*; 21.1, p. 395.

B. SPRUŠIL and P. VOSTRÝ
Resistivity Changes in Air-Quenched Gold Wires
Vol. **19**, No. 2, p. K83—K86 (1967).
See also *14.1*, p. *222*.

B. YA. PINES, I. P. GREBENNIK, R. I. BAZYURA, and G. V. ZOZULYA
Surface and Volume Diffusion in Thin Films of the System Ag-Se
Vol. **20**, No. 1, p. 213—223 (1967).
See also *9*, p. *88*; 22, p. 443; 22.1.3, p. 477.

J. ŠEDIVÝ and H. ŠICHOVA
X-Ray Measurement of the Characteristic Debye Temperature of Polycrystalline Silver Correcting for Primary Extinction
Vol. **20**, No. 1, p. 249—254 (1967).
See also *8*, p. *81*.

N. A. PANGAROV
Calculation of the Specific Surface Energies of Crystal Planes by Investigating the Orientation of Crystal Nuclei on Inert Substrates
Vol. **20**, No. 1, p. 365—369 (1967).
See also *3.1*, p. *31*; 3.2, p. 35.

D. D. Mishin
On the Theory of Magnetic Hardness of Platinum Cobalt
Vol. **20**, No. 1, p. K9–K10 (1967).
See also *18.2*, p. *305*; 21.1, p. 396.

R. Vanselow
Changes due to Solved Impurities at Pure Platinum Surfaces: Observations with the Field Electron-Microscope
Vol. **21**, No. 1, p. 69–76 (1967).
See also *1.3*, p. *15*; 10.1, p. 122.

M. F. Rose
Higher Order Elastic Constants and the Equation of State of F.C.C. Metals
Vol. **21**, No. 1, p. 235–238 (1967).
See also 6, p. 60; 8, p. 82; *12.1*, p. *160*; 21, p. 378; 21.1, p. 397.

V. I. Syutkina and E. S. Yakovleva
The Mechanism of Deformation of the Ordered CuAu Alloy
Vol. **21**, No. 2, p. 465–480 (1967).
See also 1.1, p. 8; 4, p. 42; *10.1*, p. *123*; 21.1, p. 398.

J. A. Sigler and D. Kuhlmann-Wilsdorf
Calculations on the Mechanical Energy of Vacancy Condensation Loops, Stacking Fault Tetrahedra, and Voids
Vol. **21**, No. 2, p. 545–556 (1967).
See also *10.1*, p. *123*; 21, p. 379; 21.1, p. 398.

J. Polák
Kinetics of Quenched-in Vacancies in Pure Platinum
Vol. **21**, No. 2, p. 581–591 (1967).
See also *10.1*, p. *123*.

P. Humble
A Comparison of Electron Microscope Images of Dislocations Based on Isotropic and Anisotropic Elasticity Theory
Vol. **21**, No. 2, p. 733–740 (1967).
See also *10.1*, p. *123*; 21, p. 379; 21.1, p. 398; 21.1.1, p. 417.

N. S. Natarajan and M. S. R. Chari
The Anomalous Thermal Conductivity of Dilute Ag-Mn Alloys at Helium Temperatures
Vol. **21**, No. 2, p. K127–K130 (1967).
See also *8*, p. *82*.

F. J. Kedves
On Quenching Experiments in Gold
Vol. **21**, No. 2, p. K183–K186 (1967).
See also *10.1*, p. *124*; 14.1, p. 223.

P. Reimers
Change of the Penetration Curves in Diffusion Measurements due to the Evaporation of the Diffusion Material
Vol. **22**, No. 1, p. K27–K31 (1967).
See also *9*, p. *89*.

K. Dettmann, G. Leibfried, and K. Schroeder
Spontaneous Recombination of Frenkel Pairs in Neutron-Irradiated Solids
Vol. 22, No. 2, p. 433–440 (1967).
See also *11*, p. *149*; 21, p. 380; 21.1, p. 399.

B. Persson and H. Norden
Epitaxial Growth of Gold on Cadmium Iodide
Vol. 22, No. 2, p. 631–647 (1967).
See also 1.4, p. 19; *3.1*, p. *32*.

D. A. Blackburn
Non-Equilibrium Vacancy Concentrations in Metals Subject to Thermal Gradients and Electric Fields
Vol. 23, No. 1, p. 177–183 (1967).
See also *10.1*, p. *125*; 21, p. 381.

G. Martin, D. A. Blackburn, and Y. Adda
Self-Diffusion in the Grain Boundaries of Silver Bicrystals under Hydrostatic Pressure
Vol. 23, No. 1, p. 223–228 (1967).
See also *9*, p. *90*; 10.1, p. 125.

S. Pal and P. K. Sharma
On Lindemann's Melting Criterion
Vol. 23, No. 1, p. 361–364 (1967).
See also *6*, p. *62*; 21, p. 381; 21.1, p. 400; 21.1.1, p. 419; 21.2, p. 424.

G. Duesing, H. Hemmerich, D. Meissner, and W. Schilling
The Influence of Vacancies and Impurities on the Damage Production in Platinum Electron-Irradiated at 90 °K
Vol. 23, No. 2, p. 481–488 (1967).
See also 10.1, p. 125; *11*, p. *150*; 14.1, p. 223.

M. L. Swanson
The Effects of Defect Doping on Low Temperature Neutron-Irradiation Damage and Recovery in Aluminium and Platinum
Vol. 23, No. 2, p. 649–661 (1967).
See also 10.1, p. 125; *11*, p. *150*; 14.1, p. 224; 21, p. 381.

L. Wojtczak
The Melting Point of Thin Films
Vol. 23, No. 2, p. K163–K166 (1967).
See also 6, p. 63; *8*, p. *83*; 21, p. 382; 21.1, p. 400.

B. Roessler and P. A. Flinn
A Theory of X-Ray Diffraction from an Ordered Alloy Containing Antiphase Domains
Vol. 24, No. 1, p. 263–274 (1967).
See also 1.1, p. 8; *4*, p. *43*; 21.1, p. 401.

B. Sprušil and P. Vostrý
Reply to "On Quenching Experiments in Gold" by F. J. Kedves
Vol. 24, No. 1, p. K41–K42 (1967).
See also *10.1*, p. *126*; 14.1, p. 224.

A. S. Nowick and G. J. Dienes
On the Inherent Curvature of the Arrhenius Plot in Diffusion Experiments
Vol. **24**, No. 2, p. 461–467 (1967).
See also *9*, p. *90*; 21.1, p. 401.

R. Dupree, C. T. Forwood, and M. J. A. Smith
Conduction Electron Spin Resonance in Small Particles of Gold
Vol. **24**, No. 2, p. 525–530 (1967).
See also *19*, p. *331*.

M. J. Stowell and T. J. Law
The Growth and Defect Structure of Gold Films Formed on Molybdenite in Ultra-High Vacuum
Vol. **25**, No. 1, p. 139–146 (1968).
See also *1.4*, p. *19*; *3.1*, p. *33*; 10.1, p. 127.

D. Schumacher, A. Seeger, and O. Härlin
Vacancies, Divacancies, and Self-Diffusion in Platinum
Vol. **25**, No. 1, p. 359–371 (1968).
See also 9, p. 91; *10.1*, p. *128*.

J. T. Moore, T. R. Duncan, and D. Kuhlmann-Wilsdorf
Corrected Data for the Shear Stresses of Infinite Tilt Boundaries Composed of {111}, 1/2 ⟨1$\bar{1}$0⟩ Dislocations in Elastically Anisotropic F.C.C. Metals
Vol. **25**, No. 2, p. K105–K106 (1968).
See also *10*, p. *105*; *10.1*, p. *129*; 21, p. 385; 21.1, p. 403; 21.5, p. 429; 22.1, p. 457; 22.1.1, p. 464; 22.1.2, p. 473.

R. Kishore
Electrical Conductivity of Metals
Vol. **26**, No. 1, p. 133–138 (1968).
See also *14.1*, p. *225*; 21.1, p. 403; 21.2, p. 425

H. Beyer and R. Sizmann
Recrystallization of Cold-Worked Gold during α-Irradiation
Vol. **26**, No. 1, p. 359–370 (1968).
See also 4, p. 45; 10.1, p. 129; *11*, p. *151*; 14.1, p. 225.

A. Sobaszek
Experimental Comparison of the Grain Boundary Diffusion Models
Vol. **26**, No. 1, p. K59–K63 (1968).
See also *9*, p. *91*.

C. W. Chen, L. R. Edwards, and S. Legvold
Resistance Anomaly in Palladium-Silver Alloys Containing Rare Earth Impurities
Vol. **26**, No. 2, p. 611–619 (1968).
See also 10.1, p. 129; *14.1*, p. *225*; 21, p. 385.

G. P. Williams, Jr. and T. J. Turner
Zener Relaxation in Silver Solid Solutions
Vol. **26**, No. 2, p. 645–651 (1968).
See also *10.1*, p. *130*; *12.1*, p. *160*.

M. Rühle
Radiation Hardening of Neutron-Irradiated Copper Single Crystals. (I): Experimental Results
Vol. **26**, No. 2, p. 661–670 (1968).
See also *10.1*, p. *130*; *11*, p. *151*; 21.1, p. 404.

A. OTTO
A New Method for Exciting Non-Radiative Surface Plasma Oscillations
Vol. **26**, No. 2, p. K99—K101 (1968).
See also *13*, p. *169*; *20*, p. *341*.

I. Y. DEKHTYAR, E. G. MADATOVA, P. S. NISIN, and R. G. FEDCHENKO
Plastic Deformation Effect on the Electronic Properties of Solid Solutions with Localized Magnetic Moments
Vol. **26**, No. 2, p. K151—K154 (1968).
See also 6.1, p. 73; *10.1*, p. *130*; 14.1, p. 225; 18.1, p. 297.

L. K. IVES and A. W. RUFF, JR.
Studies of the Effect of Annealing on Extended Dislocation Nodes in Silver-Tin Alloys
Vol. **27**, No. 1, p. 117—123 (1968).
See also *10.1*, p. *130*; 21.1, p. 404.

T. SCHOBER and R. W. BALLUFFI
Direct Evidence for Dechanneling of 40 kV Gold Ions at Planar Defects in Silver and Gold
Vol. **27**, No. 1, p. 195—200 (1968).
See also *10.1*, p. *130*; *11*, p. *152*.

J. M. GALLIGAN and M. J. ATTARDO
Radiation Hardening in Platinum
Vol. **27**, No. 1, p. 383—390 (1968).
See also 10.1, p. 131; *11*, p. *152*.

S. RADELAAR
The Kinetics of Short-Range Order in Some Ag-Au Alloys
Vol. **27**, No. 1, p. K63—K67 (1968).
See also *1.1*, p. *8*; 9, p. 91; 10.1, p. 131.

A. E. VIDOZ
On the Work Hardening of Superstructures
Vol. **28**, No. 1, p. 145—156 (1968).
See also 4, p. 46; *10.1*, p. *131*; 21.1, p. 405; 21.1.1, p. 422.

S. ARAJS
An Investigation of the Electrical Resistivity of Nickel-Rich Nickel-Gold Alloys
Vol. **28**, No. 1, p. 171—174 (1968).
See also *14.1*, p. *225*; 21.1, p. 405.

P. EHRHART, W. RUPP, and R. SIZMANN
Stopping Power of 0.5 to 3.5 MeV α-Particles in Cu-Au Alloys
Vol. **28**, No. 1, p. K35—K37 (1968).
See also 1.4, p. 21; *11*, p. *153*; 21.1, p. 405.

A. K. SINGH and P. K. SHARMA
Thermal Expansion of Cubic Metals in Krebs's Model
Vol. **28**, No. 2, p. 643—648 (1968).
See also *6*, p. *67*; *8*, p. *85*; 21, p. 387; 21.1, p. 406; 21.2, p. 425.

J. POLÁK
Thermal Quasi-Equilibrium of Vacancies, Divacancies, and Impurities in Platinum
Vol. **28**, No. 2, p. 773—782 (1968).
See also *10.1*, p. *132*; 14.1, p. 226.

V. K. Nikulin and M. B. Trzhaskovskaya
The Theory of Cohesion of Noble Metals (II)
Vol. **28**, No. 2, p. 801–806 (1968).
See also *3.2*, p. *36*; 13.1, p. 187; 21.1, p. 406.

P. Humble and C. T. Forwood
A Dynamical Treatment of the Stress-Induced Dissociation of Triangular Frank Dislocation Loops in F.C.C. Metals
Vol. **29**, No. 1, p. 99–106 (1968).
See also *10.1*, p. *133*; 21.1, p. 407.

A. Seeger and H. Mehrer
Interpretation of Self-Diffusion and Vacancy Properties in Gold
Vol. **29**, No. 1, p. 231–243 (1968).
See also *9*, p. *92*; *10.1*, p. *133*.

K. Ibel and R. Sizmann
Energy Loss of ⟨110⟩ Channelled α-Recoil Atoms in Gold
Vol. **29**, No. 1, p. 403–415 (1968).
See also *11*, p. *154*.

M. S. R. Chari and N. S. Natarajan
The Phonon–Electron Scattering Coefficients in Dilute Silver–Manganese Alloys
Vol. **29**, No. 1, p. K69–K72 (1968).
See also 6, p. 68; *8*, p. *85*; 13, p. 172.

T. J. Turner, C. M. Dozier, and G. P. Williams, Jr.
Activation Energies for Zener Relaxation in α-AgCd
Vol. **30**, No. 1, p. 87–91 (1968).
See also 9, p. 92; *10.1*, p. *134*; *12.1*, p. *161*; 21, p. 388.

A. B. Gardner, R. L. Sanders, and L. M. Slifkin
Self-Diffusion in α-Silver-Cadmium
Vol. **30**, No. 1, p. 93–96 (1968).
See also *9*, p. *92*; 10.1, p. 134; 21, p. 388.

M. C. Naik, M. S. Anand, A. R. Paul, and R. P. Agarwala
Impurity Diffusion in Face-Centred Cubic Metals
Vol. **30**, No. 1, p. 319–328 (1968).
See also *9*, p. *92*; 21, p. 388; 21.1, p. 408.

J. Hofmann and W. Steinmann
Plasma Resonance in the Photoemission of Silver
Vol. **30**, No. 1, p. K53–K56 (1968).
See also *13*, p. *173*; 17, p. 290; *20*, p. *342*.

J. Czerwonko
On the Thermodynamics and Paramagnetic Properties of Dispersed Metals
Vol. **30**, No. 2, p. 723–734 (1968).
See also *13.1*, p. *190*; 18.1, p. 298; 21, p. 389; 21.1, p. 409.

21.7 Semi-Metals (Semiconducting Alloys See 22.7)

Original Papers and Short Notes

M. E. Ertl
Thermal Resistivity due to Mass Fluctuation Scattering in Bismuth-Antimony Alloys
Vol. **17**, No. 1, p. K63—K65 (1966).
See also 6, p. 57; *8*, p. *79*.

J. P. Issi, A. Moureau, and A. Luyckx
Low-Field Galvanomagnetic Coefficients of Polycrystalline Bismuth at 80 °K
Vol. **17**, No. 2, p. K217—K219 (1966).
See also *14.1*, p. *221*.

V. I. Startsev, V. P. Soldatov, and M. M. Brodsky
The Rate of Twin Layer Growth in Bismuth Single Crystals
Vol. **18**, No. 2, p. 863—871 (1966).
See also *4*, p. *39*.

R. I. Anishchenko
Calculation of the Mean Inner Potential of a Crystal in the Statistical Theory
Vol. **18**, No. 2, p. 923—928 (1966).
See also *3.2*, p. *34*; 13, p. 164; 21, p. 376; 21.1, p. 394; 21.1.1, p. 414; 21.2, p. 424; 21.3, p. 426; 21.4, p. 428; 21.6, p. 432; 22.1.1, p. 460.

I. I. Pinchuk
Thermomagnetic Effects in Anisotropic Media
Vol. **20**, No. 2, p. 537—543 (1967).
See also 3, p. 27; *15*, p. *271*.

W. P. R. George, C. H. L. Goodman, H. F. Sterling, and R. W. Warren
A Possible New Group of Semiconducting Compounds
Vol. **21**, No. 1, p. 205—210 (1967).
See also 3.1, p. 31; 3.2, p. 35; *14.3*, p. *236*; 22.7, p. 545.

G. Holtmeyer and J. Kalus
Effect of the Short-Range Order around Magnesium Impurities in Bismuth Melts on the Yield of Nuclear Resonance Fluorescence
Vol. **21**, No. 1, p. 297—302 (1967).
See also 2, p. 23; 6.1, p. 72; *11*, p. *148*.

Yu. A. Bogod and V. V. Eremenko
Peculiarities of the Magnetoresistance in Bi in High Transverse and Longitudinal Magnetic Fields
Vol. **21**, No. 2, p. 797—803 (1967).
See also *14.1*, p. *223*.

Yu. A. Bogod
On the Resistance of Bi in High Magnetic Fields
Vol. **24**, No. 1, p. K49—K51 (1967).
See also *14.1*, p. *224*.

G. I. Guseva and P. S. Zyryanov
Quantum Theory of Thermogalvanomagnetic Phenomena (TGMP) in Conductors with Equal Concentration of Electrons and Holes
Vol **25**, No. 2, p. 775—785 (1968).
See also 14.1, p. 224; 14.3, p. 240; *15*, p. *273*; 22.2, p. 481.

Yu. A. Bogod, V. V. Eremenko, and L. K. Chubova
The Influence of Boundary Surface on the Magnetoresistance of Bismuth
Vol. **28**, No. 2, p. K155—K157 (1968).
See also *14.1*, p. *226*.

22. Semiconductors and Ionic Crystals

Review Articles

G. Harbeke
On the Band Structure of Anisotropic Crystals
Vol. **27**, No. 1, p. 9–44 (1968).
See also *13.1*, p. *174*; 13.2, p. 191; 14.3, p. 231; 20.1, p. 344; 22.1, p. 456; 22.1.1, p. 458; 22.1.2, p. 468; 22.1.3, p. 476; 22.4.1, p. 497; 22.4.2, p. 506; 22.4.3, p. 509; 22.5, p. 512; 22.7, p. 545; 22.8, p. 547.

Original Papers and Short Notes

I. B. Bersuker and B. G. Vekhter
Jahn-Teller Distortions and Inversion Splitting for Transition Metal Complexes and Local Centres in Crystals with Degenerate Electronic T-Terms
Vol. **16**, No. 1, p. 63–68 (1966).
See also 10, p. 93; *13.4*, p. *200*.

C. N. Plăviţu
Some Kinetic Phenomena in Impure Ionic Semiconductors
Vol. **16**, No. 1, p. 69–78 (1966).
See also 8, p. 79; 13, p. 162; *14.4*, p. *252*; 15, p. 270.

B. Velický and J. Sak
Excitonic Effects in the Interband Absorption of Semiconductors
Vol. **16**, No. 1, p. 147–157 (1966).
See also *13.1*, p. *174*; *13.2*, p. *191*; 20.1, p. 344.

G. B. Abdullaev, E. S. Guseinova, and B. G. Tagiev
Electrical Conductivity of p-GaSe Single Crystals in Strong Electric Fields
Vol. **16**, No. 1, p. 205–208 (1966).
See also 13.4, p. 201; *14.4.1*, p. *255*.

G. B. Abdullaev, M. R. Akhundov, and G. A. Akhundov
Rectifying and Photoconductive Properties of p-GaSe Single Crystals
Vol. **16**, No. 1, p. 209–214 (1966).
See also *14.3.2*, p. *248*; 16, p. 276.

A. I. Bobrysheva
The Interaction of Two Excitons in a Crystal
Vol. **16**, No. 1, p. 337–348 (1966).
See also 6, p. 56; *13.2*, p. *191*.

G. A. Akhundov and T. G. Kerimova
Infrared Absorption of $A^{III}B^{VI}$ Single Crystals
Vol. **16**, No. 1, p. K15–K16 (1966).
See also *20.1*, p. *344*.

V. P. Kalashnikov
Influence of Spatial Inhomogeneities on the Nuclear Polarization due to Direct Current in Semiconductors
Vol. **16**, No. 1, p. K35–K37 (1966).
See also *13*, p. *163*; *14.3*, p. *231*.

W. E. Wilhelm
A Simple Asymptotic Formula for the Motion of Simultaneous Electric Field and Conductivity Domains in Semiconductors
Vol. **16**, No. 1, p. K59–K62 (1966).
See also *14.4.1*, p. *255*.

M. TOMÁŠEK
Some Remarks about Electronic Correlation in Crystal Surfaces: The Ideal (100) Surface of a Semi-Infinite Diamond-Like Crystal
Vol. **16**, No. 2, p. 389–397 (1966).
See also *13.3*, p. *196*.

H. LEMKE
On the Time Dependence of Space-Charge Limited Injection Currents in Semiconductors
Vol. **16**, No. 2, p. 413–426 (1966).
See also 13.4, p. 201; *14.4.1*, p. *255*.

A. G. SAMOILOVICH, M. V. NITSOVICH, and V. M. NITSOVICH
On the Theory of Anisotropic Thermoelectric Power in Semiconductors
Vol. **16**, No. 2, p. 459–465 (1966).
See also 6, p. 56; 13.1, p. 174; *15*, p. *270*.

R. GRANGER
Anisotropic Electron Distribution in Pure Semiconductors near Breakdown
Vol. **16**, No. 2, p. 599–612 (1966).
See also 6, p. 56; 13, p. 163; *14.4.1*, p. *255*; 22.2.3, p. 489.

YA. A. IOSILEVSKII
Mössbauer Effect in Some Polyatomic Crystals
Vol. **16**, No. 2, p. 633–649 (1966).
See also *6.1*, p. *71*.

W. ŻDANOWICZ and A. WOJAKOWSKI
Semiconducting Properties of CdP_4
Vol. **16**, No. 2, p. K129–K131 (1966).
See also *14.3*, p. *232*; 15, p. 270; 20.1, p. 345.

E. KRÖNER and D. KESSEL
On the Non-Local Classical Theory of Dielectrics
Vol. **17**, No. 1, p. 211–220 (1966).
See also *14.4*, p. *252*.

K. W. BÖER and P. L. QUINN
Inhomogeneous Field Distribution in Homogeneous Semiconductors Having an N-Shaped Negative Differential Conductivity
Vol. **17**, No. 1, p. 307–316 (1966).
See also *14.4.1*, p. *256*.

H. D. KOSWIG
Dependence of the Inner-Electronic Repulsion Parameters of the $3d^3$ and $3d^7$ Configuration on the Ligands in Octahedral Crystal Fields
Vol. **17**, No. 1, p. K57–K61 (1966).
See also *13.4*, p. *202*; 20.1, p. 345.

I. A. AKIMOV, A. M. MESHKOV, and A. N. TERENIN
On Thermal E.M.F. Measurements in Photo-Semiconducting Dyes
Vol. **17**, No. 1, p. K95–K98 (1966).
See also 13.4, p. 202; *15*, p. *270*; 16, p. 277.

N. A. GASANOVA, G. A. AKHUNDOV, and M. A. NIZAMETDINOVA
Polarization Effects in the Ultraviolett Reflection from GaS and GaSe Single Crystals
Vol. **17**, No. 1, p. K115–K117 (1966).
See also *20.1*, p. *345*.

G. B. ABDULLAEV, E. S. GUSEINOVA, and B. G. TAGIEV
Electrical Conductivity of n-InSe Single Crystals in Strong Electric Fields
Vol. **17**, No. 2, p. 593–596 (1966).
See also 13.1, p. 175; *14.4.1*, p. *256*.

B. A. Greenberg
The Effect of Dislocation Dipoles on the Shape of the Nuclear Magnetic Resonance Line
Vol. **17**, No. 2, p. 673–681 (1966).
See also 10, p. 95; *19*, p. *327*.

K. Ozawa and S. Anzai
Effect of Pressure on the α-Transition Point of Iron Monosulphide
Vol. **17**, No. 2, p. 697–700 (1966).
See also *1.2*, p. *9*; 12, p. 156; 14.3, p. 233.

J. Kołodziejczak and H. Stramska
On the Theory of Free Carrier Electro-Magneto-Optical Phenomena
Vol. **17**, No. 2, p. 701–707 (1966).
See also 13.1, p. 176; 14.4.1, p. 256; *18*, p. *291*; *20*, p. *337*.

A. I. Gubanov
The LCAO Method for an Arbitrary Mixture of Atoms
Vol. **17**, No. 2, p. 807–814 (1966).
See also *2*, p. *22*; *13*, p. *163*; 21, p. 375.

B. Staliński, Z. Biegański, and R. Troć
Low-Temperature Heat Capacity and Thermodynamical Functions of Ferromagnetic Uranium Phosphide U_3P_4
Vol. **17**, No. 2, p. 837–841 (1966).
See also 1.2, p. 9; *8*, p. *80*; 18.2, p. 303.

A. Szytuła, A. Burewicz, K. Dyrek, A. Hrynkiewicz, D. Kulgawczuk, Z. Obuszko, H. Rżany, and A. Wanic
Susceptibility Measurements of the Antiferromagnetic A-Goethite
Vol. **17**, No. 2, p. K195–K197 (1966).
See also 6.1, p. 71; *18.4*, p. *321*.

G. A. Akhundov, I. G. Aksyanov, and A. G. Bagirov
Electroluminescence of GaSe Single Crystals Excited by Square Pulses
Vol. **17**, No. 2, p. K225–K227 (1966).
See also *20.3*, p. *364*.

F. I. Ismailov, G. A. Akhundov, and O. R. Vernich
Investigation of Electrical Conductivity and Hall Effect in GaSe Single Crystals
Vol. **17**, No. 2, p. K237–K240 (1966).
See also 13.4, p. 203; *14.3*, p. *233*; 15, p. 271.

V. A. Kutasov and I. A. Smirnov
Effect of Excess Antistructural Bi on the Lattice Thermal Conductivity of Bi_2Te_3, Bi_2Te_3-Sb_2Te_3, and Bi_2Te_3-Bi_2S_3 Containing Iodine
Vol. **18**, No. 1, p. 479–488 (1966).
See also *8*, p. *80*; *10*, p. *96*; 14.3, p. 233; 15, p. 271; 22.8, p. 548.

G. C. T. Liu and J. C. M. Li
Energy of Hexagonal Dislocation Loops
Vol. **18**, No. 2, p. 527–539 (1966).
See also *10*, p. *96*; *10.1*, p. *118*; 21, p. 376; 22.1.2, p. 470.

G. B. Street and W. D. Gill
Photoconductivity and Drift Mobilities in Single Crystal Realgar (As_4S_4)
Vol. **18**, No. 2, p. 601–607 (1966).
See also 13.4, p. 204; 14.3, p. 233; *16*, p. *278*.

J. H. C. Hogg and W. J. Duffin
Twinning in InS
Vol. **18**, No. 2, p. 755–760 (1966).
See also *4*, p. *39*; 5, p. 50.

G. A. SMOLENSKII, V. P. ZHUZE, V. E. ADAMYAN, and G. M. LOGINOV
Magnetic Properties of Ce, Pr, and Nd Monochalcogenides at 4.2 to 1300°K
Vol. **18**, No. 2, p. 873–880 (1966).
See also *18.2*, p. *304*; *18.4*, p. *321*.

G. D. GUSEINOV and A. I. RASULOV
Heat Conductivity Study of GaSe Monocrystals
Vol. **18**, No. 2, p. 911–922 (1966).
See also 6, p. 58; *8*, p. *80*.

V. V. PARANJAPE
Distribution Function of Hot Electrons
Vol. **18**, No. 2, p. K103–K104 (1966).
See also 6, p. 58; *13*, p. *164*; *14.3*, p. *234*.

E. KAMIENIECKI
Remark on Solving Boltzmann Equation for Hot Carriers
Vol. **18**, No. 2, p. K105 (1966).
See also 6, p. 58; *13*, p. *164*; *14.3*, p. *234*.

G. ALBRECHT and G. WOLF
Electron Specific Heat and λ-Anomaly of $CrH_{0.84}$
Vol. **18**, No. 2, p. K119–K122 (1966).
See also 8, p. 80; *13*, p. *164*.

I. V. IOFFE
On the Theory of the Lattice Oscillations of Dielectric Crystals in an External Electric Field
Vol. **19**, No. 1, p. 51–56 (1967).
See also *6*, p. *58*; 18.2, p. 304; 18.4, p. 321; 19, p. 327.

G. R. BARSCH and Z. P. CHANG
Adiabatic, Isothermal, and Intermediate Pressure Derivatives of the Elastic Constants for Cubic Symmetry. (II): Numerical Results for 25 Materials
Vol. **19**, No. 1, p. 139–151 (1967).
See also *12*, p. *156*; 21, p. 376.

J. KOŁODZIEJCZAK
On the Theory of Galvanomagnetic Phenomena in Strong Electric Fields
Vol. **19**, No. 1, p. 373–384 (1967).
See also 13.1, p. 176; 14.4.1, p. 257; *15*, p. *271*.

V. V. ANTONOV-ROMANOVSKII
On the Kinetics of the Crystal Phosphor Luminescence
Vol. **19**, No. 1, p. 417–424 (1967).
See also *20.3*, p. *365*.

G. D. GUSEINOV, A. I. RASULOV, E. M. KERIMOVA, and M. Z. ISMAILOV
On Heat Conductivity of $A^{III}B^{VI}$-Type Semiconductors
Vol. **19**, No. 1, p. K7–K10 (1967).
See also *8*, p. *81*.

J. D. KEYS and H. P. DIBBS
On the Bonding in Bismuth Telluride
Vol. **19**, No. 1, p. K11–K13 (1967).
See also *3.2*, p. *35*; 14.3, p. 235.

B. VELICKÝ
Polaron Effect on the Optical Absorption Edge of Semiconductors
Vol. **19**, No. 1, p. K39–K42 (1967).
See also 6, p. 59; 13, p. 164; 13.1, p. 176; *20.1*, p. *347*.

S. Żukotyński and J. Kołodziejczak
On the Determination of the Non-Parabolicity of Energy Bands
Vol. **19**, No. 1, p. K51—K54 (1967).
See also *13.1*, p. *176*; 14.3, p. 235; 15, p. 271.

J. Kołodziejczak and E. Kierzek-Pecold
Free Carrier Electro-Optical Kerr Effect in Semiconductors
Vol. **19**, No. 1, p. K55—K58 (1967).
See also 13.1, p. 176; *14.4.1*, p. *257*; *20*, p. *338*.

J. Kołodziejczak and E. Kierzek-Pecold
Free Carrier Electro-Magneto-Optical Phenomena in Semiconductors
Vol. **19**, No. 2, p. 623—630 (1967).
See also 14.3, p. 235; *18*, p. *291*; *20*, p. *338*.

R. Enderlein and R. Keiper
On the Theory of Electro-Absorption and Electro-Reflectance in Semiconductors
Vol. **19**, No. 2, p. 673—681 (1967).
See also 13.1, p. 177; 14.4.1, p. 257; *20*, p. *338*.

P. T. Landsberg
The Condition for Negative Absorption in Semiconductors
Vol. **19**, No. 2, p. 777—780 (1967).
See also 13.4, p. 205; *20.2*, p. *360*.

M. A. Nizametdinova
The Reflection Spectrum of GaSe and GaS Single Crystals near the Fundamental Absorption Edge
Vol. **19**, No. 2, p. K111—K112 (1967).
See also 13.1, p. 177; *20.1*, p. *347*.

V. V. Tarassov
Heat Capacity and Structure of Vitreous Silica and Diamond-Like Lattices (I)
Vol. **20**, No. 1, p. 37—43 (1967).
See also 2, p. 22; *6*, p. *59*; 8, p. 81.

V. V. Tarassov
Heat Capacity and Structure of Vitreous Silica and Diamond-Like Lattices (II)
Vol. **20**, No. 1, p. 45—57 (1967).
See also 2, p. 22; *6*, p. *59*; 8, p. 81.

D. J. D. Thomas
The Effect of Heat Treatment on Silicon Nitride Layers on Silicon
Vol. **20**, No. 1, p. 131—142 (1967).
See also *1.4*, p. *18*.

A. S. Davydov
On the Operator of Exciton–Phonon Interaction
Vol. **20**, No. 1, p. 143—151 (1967).
See also *6*, p. *59*; *13.2*, p. *192*.

A. S. Davydov and E. N. Myasnikov
Absorption and Dispersion of Light by Molecular Excitons
Vol. **20**, No. 1, p. 153—163 (1967).
See also *13.2*, p. *192*; *20.1*, p. *347*.

B. Ya. Pines, I. P. Grebennik, R. I. Bazyura, and G. V. Zozulya
Surface and Volume Diffusion in Thin Films of the System Ag-Se
Vol. **20**, No. 1, p. 213—223 (1967).
See also *9*, p. *88*; 21.6, 432; 22.1.3, p. 477.

J. R. PILBROW and J. M. SPAETH
ESR Studies of Cu^{2+} in NH_4Cl Single Crystals between 4.2 and 453 °K. (I): Spin-Hamiltonian Parameters and Lattice Constant for NH_4Cl
Vol. **20**, No. 1, p. 225–235 (1967).
See also *19*, p. *328*; 22.5, p. 512.

J. R. PILBROW and J. M. SPAETH
ESR Studies of Cu^{2+} in NH_4Cl Single Crystals between 4.2 and 453 °K. (II): Theoretical Analysis: Vibrational Admixtures, Spin Polarization, and $\langle r^{-3} \rangle$ for Cu^{2+} Ions
Vol. **20**, No. 1, p. 237–248 (1967).
See also *19*, p. *328*; 22.5, p. 512.

H. SIGMUND and K. BERCHTOLD
Electrical and Photovoltaic Properties of PbS–Si Heterodiodes
Vol. **20**, No. 1, p. 255–259 (1967).
See also *14.3.2*, p. *249*; 16, p. 278; 22.1.2, p. 470.

R. ENDERLEIN
The Influence of Collisions on the Franz-Keldysh Effect
Vol. **20**, No. 1, p. 295–299 (1967).
See also 13.1, p. 177; 14.4.1, p. 257; *20*, p. *338*.

G. COSTE
Oscillatory Thermomagnetic Effects in Lead Telluride
Vol. **20**, No. 1, p. 361–364 (1967).
See also 13.1, p. 177; *15*, p. *271*.

H. NEUMANN
Photosensitive Field Electron Emission from In_2S_3
Vol. **20**, No. 1, p. K33–K35 (1967).
See also *17*, p. *287*.

G. B. ABDULLAEV, E. S. GUSEINOVA, and B. G. TAGIEV
Electric Conductivity Studies of p-GaTe Polycrystals and Single Crystals in Strong Electric Fields
Vol. **20**, No. 2, p. 421–425 (1967).
See also *14.3*, p. *235*; *14.4.1*, p. *258*.

J. PRZYSTAWA and W. SUSKI
Magnetic Ordering in Uranium Compounds with Crystal Structures of Anti-Cu_2Sb and PbFCl Types. Molecular Field Approximation
Vol. **20**, No. 2, p. 451–459 (1967).
See also 3.2, p. 35; *18.2*, p. *306*; *18.4*, p. *322*.

T. N. CASSELMAN and H. N. SPECTOR
Interaction of Optical Phonons with Spin Waves in Ionic Crystals
Vol. **20**, No. 2, p. 605–612 (1967).
See also *6*, p. *60*; 13, p. 165; 18, p. 292; *19*, p. *328*.

H.-J. ULLRICH
Precision Lattice Parameter Measurements by Interferences from Lattice Sources (Kossel Lines) and Divergent Beam X-Ray Diffraction (Pseudo-Kossel Lines) in Back Reflection
Vol. **20**, No. 2, p. K113–K117 (1967).
See also *5*, p. *51*; 21, p. 378.

S. G. CHRISTOV
Unified Theory of Thermionic and Field Emission from Semiconductors
Vol. **21**, No. 1, p. 159–173 (1967).
See also 13.3, p. 197; *17*, p. *287*.

N. F. Mott and R. S. Allgaier
Localized States in Disordered Lattices
Vol. **21**, No. 1, p. 343–356 (1967).
See also 2, p. 23; *13.4*, p. *206*; 21, p. 378.

P. S. Serebrennikov
On the Theory of Tunneling
Vol. **21**, No. 1, p. 431–436 (1967).
See also *14.3.2*, p. *249*.

V. F. Zolotaryov, D. G. Semak, and D. V. Chepur
Thermally Stimulated Currents under the Condition of Persistant Internal Polarization
Vol. **21**, No. 1, p. 437–442 (1967).
See also *13.4*, p. *206*; 14.4.1, p. 258.

J. Tobisch and W. Hase
Structure Investigations of Two Uranium Nitrides U_2N_{3+x}
Vol. **21**, No. 1, p. K11–K12 (1967).
See also *5*, p. *51*.

A. N. Georgobiani and V. I. Steblin
Electroluminescence of ZnS-Cu_2S Heterojunctions
Vol. **21**, No. 1, p. K45–K48 (1967).
See also 14.3.2, p. 249; *20.3*, p. *365*; 22.4.1, p. 499.

Z. A. Demidenko and V. S. Mashkevich
Indirect Optical Transitions at the Centre of the Brillouin Zone
Vol. **21**, No. 1, p. K59–K63 (1967).
See also 6, p. 61; *20.1*, p. *348*.

A. C. Papadakis
On the Maesurement of Carrier Mobilities in Insulators Using Transient Space-Charge Perturbed Currents
Vol. **21**, No. 1, p. K77–K80 (1967).
See also *14.4*, p. *253*; 14.4.1, p. 258.

G. Getov, P. Simidtchieva, M. Nikiforova, and R. Andreytchin
Effect of Some Elements on the Optical Absorption Edge of Vitreous As_2S_3
Vol. **21**, No. 1, p. K87–K89 (1967).
See also 2, p. 23; 13.4, p. 207; *20.1*, p. *348*.

G. B. Abdullaev, Z. A. Aliyarova, and G. A. Asadov
Preparation of Cu_2Se Single Crystals and Investigation of Their Electrical Properties
Vol. **21**, No. 2, p. 461–464 (1967).
See also 3.1, p. 31; *14.3*, p. *236*; 15, p. 272.

Č. Jech
Ion-Bombardement Enhanced Solubility in Solids
Vol. **21**, No. 2, p. 481–486 (1967).
See also *11*, p. *149*; 22.6, p. 538.

W. A. Jesser and D. Kuhlmann-Wilsdorf
The Geometry and Energy of a Twist Boundary between Crystals with Unequal Lattice Parameters
Vol. **21**, No. 2, p. 533–544 (1967).
See also *3.1*, p. *31*; 10, p. 100; 10.1, p. 123; 21, p. 379.

P. Schnupp
Kronig-Penney-Type Calculations for Electron Tunneling trough Thin Dielectric Films
Vol. **21**, No. 2, p. 567–573 (1967).
See also 13.1, p. 178; *14.3.1*, p. *245*.

V. Hizhnyakov and I. Tehver
Theory of Resonant Secondary Radiation due to Impurity Centres in Crystals
Vol. **21**, No. 2, p. 755–768 (1968).
See also 13.4, p. 207; 20.1, p. 348; *20.3*, p. *366*.

D. M. Bercha, A. N. Borets, I. M. Stakhyra, and K. D. Tovstyuk
The Band Edge and the Energy Spectrum of In_2Se
Vol. **21**, No. 2, p. 769–774 (1967).
See also 13.1, p. 178; *20.1*, p. *348*.

V. P. Kalashnikov
Theory of Spin-Lattice Relaxation of Conduction Electrons of Ionic Semiconductors at Extremely High Magnetic Fields
Vol. **21**, No. 2, p. 775–784 (1967).
See also *6*, p. *61*; 13, p. 166; 14.4, p. 253; 18, p. 292.

B. M. Askerov and F. M. Gashimzade
Determination of Parameters of Degenerate Semiconductors
Vol. **21**, No. 2, p. K155–K158 (1967).
See also *13.1*, p. *179*; 14.3, p. 237; 15, p. 272.

B. Tavger
Transport Phenomena in Semiconducting Thin Films
Vol. **22**, No. 1, p. 31–38 (1967).
See also 13.1, p. 179; *14.3.1*, p. *245*; 15, p. 272.

I. Licea
Electrical Conduction of Semiconducting Thin Films in Strong Electric Fields
Vol. **22**, No. 1, p. 147–152 (1967).
See also 13, p. 166; *14.3.1*, p. *245*.

H. N. Spector
Optical Absorption in Electric and Magnetic Fields
Vol. **22**, No. 1, p. 185–193 (1967).
See also 13.1, p. 179; 14.4.1, p. 259; 18, p. 292; *20.1*, p. *349*.

V. L. Bonch-Bruevich
On the Theory of Electrical Domains in Hot Electron Semiconductors
Vol. **22**, No. 1, p. 267–278 (1967).
See also 13, p. 166; 14.3, p. 237; *14.4.1*, p. *259*; 22.1.1, p. 462; 22.2.1, p. 484.

A. Aldea
Theory of the Hall Effect in Disordered Systems: Impurity Band Conduction (II)
Vol. **22**, No. 2, p. 377–380 (1967).
See also 13.4, p. 208; *14.3*, p. *237*.

V. Frei
On Energy Bands of Crystals with Lowered Point Symmetry
Vol. **22**, No. 2, p. 381–390 (1967).
See also *13.1*, p. *180*.

P. Bräunlich
Calculation of Light-Stimulated Luminescence and Current Transients in Phosphors
Vol. **22**, No. 2, p. 391—400 (1967).
See also 13.4, p. 208; *16*, p. *280*; *20.3*, p. *366*.

J. Pastrňák
A Thermodynamic Investigation of the Conditions for the Generation of Oscillations of an Electric Current Flowing through a Non-Ohmic Medium
Vol. **22**, No. 2, p. 407—413 (1967).
See also *14.4.1*, p. *259*.

J. Leciejewicz, R. Troć, A. Murasik, and A. Zygmunt
Neutron-Diffraction Study of Antiferromagnetism in USb_2 and UBi_2
Vol. **22**, No. 2, p. 517—526 (1967).
See also 5, p. 52; *18.4*, p. *323*.

W. Franz
Duration of the Tunnelling Single Process
Vol. **22**, No. 2, p. K139—K140 (1967).
See also *14.3.2*, p. *249*.

K. D. Tovstyuk and P. I. Voronyuk
Galvanomagnetic Effects in PbTe
Vol. **23**, No. 1, p. 75—91 (1967).
See also 13.1, p. 180; *14.3*, p. *237*.

N. I. Akimenko, P. M. Staryk, and K. D. Tovstyuk
Magnetic Properties of PbTe
Vol. **23**, No. 1, p. 93—96 (1967).
See also *13.1*, p. *180*; 14.3, p. 238; 18, p. 293.

R. Enderlein and R. Keiper
On the Dielectric Function of a Semiconductor in an External Electric Field
Vol. **23**, No. 1, p. 127—136 (1967).
See also 13.1, p. 180; 14.4.1, p. 259; *20*, p. *339*.

M. Porsch and J. Röseler
Recoil Effects in the Polaron Problem
Vol. **23**, No. 1, p. 365—376 (1967).
See also 6, p. 62; *13*, p. *167*; 13.1, p. 181.

J. Kołodziejczak
Effect of Strong Electric Fields on Magneto-Resistivity and Hall Coefficient in Semiconductors
Vol. **23**, No. 1, p. 423—427 (1967).
See also 6, p. 62; 13.1, p. 181; *14.3*, p. *238*; 14.4.1, p. 260; 22.2.3, p. 491.

V. G. Lyapin and K. B. Tolpygo
Partial Account of Interelectron Correlation in the Theory of Band Structure of Crystals with Sphalerite and Diamond Lattices
Vol. **23**, No. 1, p. 429—437 (1967).
See also *13.1*, p. *181*.

J. Kołodziejczak
Free Carrier Effect on the Generation of Higher Harmonics in Semiconductors
Vol. **23**, No. 1, p. K57—K59 (1967).
See also *13*, p. *167*; 13.1, p. 181; 14.3, p. 238; *20*, p. *340*.

D. E. Aspnes
Note on Collision Broadening of Franz-Keldysh Effect
Vol. **23**, No. 1, p. K79—K81 (1967).
See also 13.1, p. 181; 14.4.1, p. 260; *20*, p. *340*.

G. D. Guseinov and A. M. Ramazanzade
Visualization of Dislocations on Basal Planes of GaSe Single Crystals
Vol. **23**, No. 2, p. 461–465 (1967).
See also *10*, p. *102*.

R. Troć, A. Murasik, A. Zygmunt, and J. Leciejewicz
The Magnetic Ordering in Uranium Monoarsenide
Vol. **23**, No. 2, p. K123–K124 (1967).
See also *18.4*, p. *323*.

J. Antula
Tunnel and Schottky Current in Dielectric Thin Films Considering Film Thickness Fluctuations
Vol. **24**, No. 1, p. 89–94 (1967).
See also *14.3.2*, p. *249*; 14.4.1, p. 260.

J. Przystawa
The Néel Temperatures and Stability of Magnetic Phases in the Disturbed Body-Centered Tetragonal Lattice. Uranium Compounds
Vol. **24**, No. 1, p. 313–323 (1967).
See also *18.2*, p. *309*; *18.4*, p. *323*; 22.8, p. 552.

J. Kołodziejczak
Magnetic Resonance in Higher Harmonics Generated by Free Carriers in Semiconductors
Vol. **24**, No. 1, p. 323–329 (1967).
See also 13.1, p. 183; 14.3, p. 238; *19*, p. *331*.

S. Toschev and I. Gutzow
Derivation of Nucleation Kinetics in Solids by Examination of Plane Sections through the Samples
Vol. **24**, No. 1, p. 349–359 (1967).
See also 1.2, p. 11; *3.1*, p. *32*.

C. N. Plăviţu
On the Phonon Drag Effect Contribution to Thermomagnetic Phenomena
Vol. **24**, No. 1, p. 361–372 (1967).
See also 6, p. 64; 13, p. 168; 14.3, p. 239; *15*, p. *273*.

A. S. Davydov and V. A. Onishchuk
The Dielectric Permeability of Molecular Crystals
Vol. **24**, No. 1, p. 373–384 (1967).
See also *14.4*, p. *253*.

Yu. B. Paderno and S. Pokrzywnicki
Magnetic Properties of Some Heavy Rare Earth Tetraborides
Vol. **24**, No. 1, p. K11–K12 (1967).
See also *18.1*, p. *297*.

Yu. B. Paderno, S. Pokrzywnicki and B. Staliński
Magnetic Properties of Some Rare Earth Hexaborides
Vol. **24**, No. 1, p. K73–K76 (1967).
See also *18.1*, p. *297*.

B. Autin
Transient Acousto-Electric Phenomena in Semiconducting Piezo-Electric Crystals
Vol. **24**, No. 2, p. 501–508 (1967).
See also *7*, p. *77*; *14.4.1*, p. *261*.

K.-E. KRÖLL
Slowly Propagating High-Field Domains in Semiconductors with Negative Differential Conductivity
Vol. **24**, No. 2, p. 707—714 (1967).
See also *14.4.1*, p. *261*.

H. L. HARTNAGEL
Some Ideas on Electroacoustic Domains
Vol. **24**, No. 2, p. K133—K135 (1967).
See also *7*, p. *77*; *14.4.1*, p. *261*.

V. A. IVANOVA, D. SH. ABDINOV, and G. M. ALIEV
On Some Characteristics of Chromium Selenides
Vol. **24**, No. 2, p. K145—K147 (1967).
See also 5, p. 52; *8*, p. *84*; 14.3, p. 239; *15*, p. *273*.

B. SELLE, W. LUDWIG, and R. MACH
Photovoltaic Response of Cu_2S-CdS p–n Heterojunctions
Vol. **24**, No. 2, p. K149—K152 (1967).
See also *14.3.2*, p. *250*; *16*, p. *281*; 22.4.1, p. 501.

YA. N. NASIROV and YA. S. FEIZIEV
Effect of Small Substitutions of Tin by Neodymium on Thermoelectric Properties of SnTe
Vol. **24**, No. 2, p. K157—K159 (1967).
See also 14.3, p. 239; *15*, p. *273*.

G. B. ABDULLAEV, M. KH. ALIEVA, and A. Z. MAMEDOVA
Thermal and Infrared Quenching of Photoconductivity in n-InSe Single Crystals
Vol. **25**, No. 1, p. 75—79 (1968).
See also 13.4, p. 211; *16*, p. *281*.

M. V. COLEMAN and D. J. D. THOMAS
The Structure of Amorphous Silicon Nitride Films
Vol. **25**, No. 1, p. 241—252 (1968).
See also *1.4*, p. *20*; 2, p. 23.

A. G. FITZGERALD and G. THOMAS
Electron Microscope Observations of Twinning and Phase Transformations in Indium Sulfide Crystals
Vol. **25**, No. 1, p. 263—271 (1968).
See also *1.2*, p. *11*; *4*, p. *44*.

S. OBERLÄNDER and S. SIKORSKI
On the Mathematical Description of Electrical Processes in Semiconductors. (I): Phase Space Analysis in Subspaces of Dimension Lower than Three
Vol. **25**, No. 1, p. 345—357 (1968).
See also *14.3*, p. *240*; 14.3.2, p. 250.

B. L. EVANS and P. A. YOUNG
Delocalized Excitons in Thin Anisotropic Crystals
Vol. **25**, No. 1, p. 417—425 (1968).
See also *13.2*, p. *193*; 20.1, p. 351.

P. SCHNUPP
A Model for Tunneling through a Non-Crystalline Thin Dielectric Film
Vol. **25**, No. 1, p. 455—460 (1968).
See also 13.1, p. 183; *14.3.1*, p. *246*.

N. A. Goryunova, V. M. Orlov, V. I. Sokolova, G. P. Shpenkov, and E. V. Tsvetkova
Preparation and Some Properties of CuP_2 Single Crystals
Vol. **25**, No. 2, p. 513–519 (1968).
See also *3.1*, p. *33*; 5, p. 53; 14.3, p. 240; 15, p. 273; 20.1, p. 351.

Yu. A. Logatchov
Model Potential for Systems with Two Valence Electrons
Vol. **25**, No. 2, p. 763–774 (1968).
See also *13.4*, p. *211*; 20.3, p. 368.

F. M. El-Akkad
Temperature Dependence of Mobility in PbS
Vol. **25**, No. 2, p. K115–K117 (1968).
See also *14.3*, p. *240*.

B. N. Oshcherin
On Enthalpy Calculation of Vacancy Formation in Inorganic Substances
Vol. **25**, No. 2, p. K123–K125 (1968).
See also *10*, p. *105*; *10.1*, p. *129*; 21, p. 385.

G. B. Abdullaev, Z. A. Aliyarova, E. H. Zamanova, and G. A. Asadov
Investigation of the Electric Properties of Cu_2S Single Crystals
Vol. **26**, No. 1, p. 65–68 (1968).
See also 1.2, p. 12; *14.3*, p. *240*; 15, p. 273.

P. G. Harper and J. A. Hilder
Exciton Spectra in Thin Crystals
Vol. **26**, No. 1, p. 69–76 (1968).
See also *13.2*, p. *193*; 20.1, p. 352.

V. V. Antonov-Romanovskii
On the Effective Recombination Cross-Section of Charged Centres in Crystal Phosphors
Vol. **26**, No. 1, p. 173–183 (1968).
See also *13.4*, p. *211*.

T. A. Davis and K. Vedam
Variation of the Refractive Indices of Calcite, with Pressure to 7 kbar
Vol. **26**, No. 1, p. 285–290 (1968).
See also *20.1*, p. *352*.

I. Spînulescu-Carnaru and I. Drăghici
On Electrical Properties of PbTe Thin Films
Vol. **26**, No. 1, p. K39–K42 (1968).
See also 3.1, p. 33; *14.3.1*, p. *246*.

N. A. Vlasenko and A. N. Gergell
On the Mechanism of DC Electroluminescence in pCu_xS-nZnS:Mn, Cu, Cl Film Structures
Vol. **26**, No. 1, p. K77–K81 (1968).
See also 14.3.2, p. 250; *20.3*, p. *368*; 22.4.1, p. 502.

P. M. Karageorgy-Alkalaev and A. Yu. Leiderman
Statics of Inter-Impurity Recombination of Electrons and Holes in Semiconductors
Vol. **26**, No. 2, p. 419–428 (1968).
See also *13.4*, p. *212*; 20.3, p. 368.

R. Enderlein
Excitonic Effects in the Electro-Absorption of Semiconductors
Vol. **26**, No. 2, p. 509–517 (1968).
See also 13.1, p. 185; 13.2, p. 193; *14.4.1*, p. *262*; *20*, p. *340*; 20.1, p. 353.

E. D. Trifonov and A. S. Troshin
Nonlinear Polarisation of an Impurity Crystal
Vol. **26**, No. 2, p. 519–525 (1968).
See also 13.4, p. 212; *20.1*, p. *353*; *20.3*, p. *368*.

G. Döhler and K. Hacker
A Possibility for Band Structure Determination: Intraband and Donor-Band Transitions of Crystal Electrons in High Electric Fields
Vol. **26**, No. 2, p. 551–564 (1968).
See also *13.1*, p. *185*; 13.4, p. 212; 14.4.1, p. 262; 20.1, p. 353.

D. I. Lainer and G. K. Samedov
On the Mechanism of Current Carrier Scattering in Some Solid Solutions on the Basis of Lead Telluride
Vol. **26**, No. 2, p. K123–K125 (1968).
See also *14.3*, p. *241*; 15, p. 273.

I. K. Andronik, E. K. Arushanov, O. V. Emelyanenko, and D. N. Nasledov
Negative Magnetoresistance in p-CdSb
Vol. **27**, No. 1, p. 45–49 (1968).
See also 13.4, p. 213; *14.3*, p. *241*; 22.7, p. 546.

A. S. Davydov
Theory of Urbach's Rule
Vol. **27**, No. 1, p. 51–56 (1968).
See also *20.1*, p. *353*.

L. I. Anatychuk, V. D. Iskra, O. J. Luste, I. M. Rarenko, and L. I. Zarubin
Anisotropy of Electroconductivity in CdSb
Vol. **27**, No. 1, p. 101–107 (1968).
See also 13.1, p. 185; *14.3*, p. *241*; 22.7, p. 546.

L. Georgescu
Nonlinear Thermodynamics of Thermoelectric Phenomena in Ionic Semiconductors
Vol. **27**, No. 1, p. 125–130 (1968).
See also *15*, p. *274*.

J. Biersack and W. Diez
Motion of Markers and Bubbles in Solids by Self-Diffusion in a Temperature Gradient
Vol. **27**, No. 1, p. 139–144 (1968).
See also *9*, p. *91*; 21, p. 385.

H. Friedrich and G. Jungk
Generation-Recombination Statistics in Semiconductors — A Model with Two Centres
Vol. **27**, No. 1, p. 237–248 (1968).
See also *13.4*, p. *213*; 16, p. 282.

M. Porsch
Green's Function Calculation of Electro-Optical Effects in Semiconductors. (I): General Formalism and Interband Transitions
Vol. **27**, No. 1, p. 359–370 (1968).
See also 13.1, p. 186; 14.4.1, p. 262; *20*, p. *341*.

U. Birkholz and J. Schelm
Mechanism of Electrical Conduction in β-$FeSi_2$
Vol. **27**, No. 1, p. 413–425 (1968).
See also **13**, p. 170; **13.1**, p. 186; ***14.3***, p. ***241***; **15**, p. 274; **22.7**, p. 546.

A. F. Lubchenko
Theory of Urbach's Rule for Impurity Centres
Vol. **27**, No. 1, p. K73–K76 (1968).
See also 10, p. 107; 13.4, p. 213; *20.1*, p. *354*.

J. Sak
On the Theory of Shallow Impurity States
Vol. **27**, No. 2, p. 521–527 (1968).
See also 6, p. 66; 13.1, p. 186; *13.4*, p. *213*.

Č. Jech
Depth Distribution Measurements Using Bombardment-Enhanced Solubility
Vol. **27**, No. 2, p. 573–577 (1968).
See also 1.4, p. 20; *11*, p. *152*; 22.1.2, p. 473.

V. Fedoseev and V. Hizhnyakov
On the Theory of Low-Temperature Exciton Absorption Spectra
Vol. **27**, No. 2, p. 751–755 (1968).
See also 6, p. 66; *13.2*, p. *194*; 20.1, p. 355.

Ya. O. Dovhyj
Connection between the Circular Dichroism and Rotary Dispersion of Gyrotropic Crystals and Its Experimental Verification in the Region of the Excitation of Circular Excitons
Vol. **27**, No. 2, p. K121–K124 (1968).
See also 13.2, p. 194; *20*, p. *341*.

A. A. Lipnik and M. A. Ruvinskii
Attenuation and Amplification of Hypersound by Mott Excitons in Semiconductors with Piezoelectric and Deformational Electron–Phonon Interactions
Vol. **28**, No. 1, p. 75–81 (1968).
See also 6, p. 66; *7*, p. *78*; 13.1, p. 187; 13.2, p. 194.

V. P. Kalashnikov and A. M. Zlobin
On the Theory of Thermogalvanomagnetic Phenomena in Semiconductors at High Magnetic Fields
Vol. **28**, No. 1, p. 105–109 (1968).
See also 6, p. 66; *8*, p. *85*; *14.3*, p. *242*.

M. V. Andriyashik, M. Yu. Sakhnovskii, V. B. Timofeev, and A. S. Yakimova
Optical Transitions in the Spectra of the Fundamental Absorption and Reflection of InSe Single Crystals
Vol. **28**, No. 1, p. 277–285 (1968).
See also 13.1, p. 187; 13.2, p. 194; *20.1*, p. *355*.

A. B. Fazakas and A. Friedman
On the Double Injection Current Noise in Solids
Vol. **28**, No. 1, p. 385–394 (1968).
See also *14.4.1*, p. *263*.

B. F. Bilenkii, A. G. Mikolaichuk, and D. M. Freik
Structure and Optical Properties of Epitaxial SnTe, SnSe, and SnS Films
Vol. **28**, No. 1, p. K5–K7 (1968).
See also *1.4*, p. *21*; *20.1*, p. *355*.

S. M. Atakishiev, D. Sh. Abdinov, and G. A. Akhundov
Heat Conductivity of Indium Selenide
Vol. 28, No. 1, p. K47–K50 (1968).
See also 6, p. 66; *8*, p. *85*.

R. Enderlein
On a Certain Type of Instability of Electrical Domains
Vol. 28, No. 2, p. 519–527 (1968).
See also *14.4.1*, p. *263*.

J. Schnakenberg
Polaronic Impurity Hopping Conduction
Vol. 28, No. 2, p. 623–633 (1968).
See also *6*, p. *67*; *13*, p. *171*; 14.3, p. 242.

A. B. Fazakas and P. H. Handel
Stability of Surface-Barriers in Semiconductors
Vol. 28, No. 2, p. 699–709 (1968).
See also *13.3*, p. *199*; 14.4.1, p. 263.

V. Dolocan
Role of Gradual Capture in the Photoelectromagnetic and Photoconductive Effects
Vol. 28, No. 2, p. 719 –723 (1968).
See also *16*, p. *283*.

N. Kristoffel and P. Konsin
Displacive Vibronic Phase Transitions in Narrow-Gap Semiconductors
Vol 28, No. 2, p. 731–739 (1968).
See also *1.2*, p. *12*; *6*, p. *67*; 14.4.2, p. 269.

B. M. Askerov and F. M. Gashimzade
The Transverse Magnetic Resistance in Semiconductors with Non-Standard Energy Band
Vol. 28, No. 2, p. 783–788 (1968).
See also 13.1, p. 187; *14.3*, p. *243*.

I. P. Zhadko and V. A. Romanov
On the Origin of High Photovoltages in Semiconductors
Vol 28, No. 2, p. 797–800 (1968).
See also 14.3.1, p. 247; *16*, p. *283*.

T. A. Kudykina, K. B. Tolpygo, and S. M. Zubkova
Exciton Theory in Crystals with Complicated Band Structure with Electron-Phonon Interaction
Vol. 28, No. 2, p. 807–814 (1968).
See also 13.1, p. 188; *13.2*, p. *194*; 22.5.2, p. 528.

M. E. Levinstein and M. S. Shur
Behaviour of the High-Field Domains below the Voltage of the Nucleation Threshold
Vol. 28, No. 2, p. 827–834 (1968).
See also *14.4.1*, p. *263*.

Ya. S. Feiziev, Ya. N. Nasirov, and O. N. Novruzov
Thermal Properties of Tin Sulphide Single Crystals
Vol. 28, No. 2, p. K113–K115 (1968).
See also *8*, p. *85*.

P. B. Pickar and H. D. Tiller
Optical Energy Gap in TlSe
Vol. 29, No. 1, p. 153–158 (1968).
See also 5, p. 54; 13.1, p. 188; 16, p. 283; *20.1*, p. *356*.

P. H. HANDEL
Instabilities and Turbulence in Semiconductors
Vol. **29**, No. 1, p. 299–306 (1968).
See also *14.4.1*, p. *264*.

W. SCHILZ
Anisotropy of High-Field Helicon Propagation in PbTe
Vol. **29**, No. 2, p. 559–568 (1968).
See also *13*, p. *172*; 14.3, p. 244; *20*, p. *342*.

J. KOŁODZIEJCZAK
Nonlinear Magnetooptical Phenomena in Semiconductors (I)
Vol. **29**, No. 2, p. 645–652 (1968).
See also 13.1, p. 189; *18*, p. *295*; *20*, p. *342*.

J. E. LEWIS, H. RODOT, and P. HAEN
The Low-Temperature Thermoelectric Power and Thermal Conductivity of GeTe and of Some GeTe-MnTe Alloys
Vol. **29**, No. 2, p. 743–754 (1968).
See also 6, p. 68; *8*, p. *86*; 13, p. 173; *15*, p. *274*; 22.8, p. 555.

R. C. THIEL and C. B. VAN DEN BERG
Temperature Dependence of Hyperfine Interactions in Near-Stoichiometric FeS. (I): Experiment
Vol. **29**, No. 2, p. 837–846 (1968).
See also 1.2, p. 13; *6.1*, p. *74*; 18.4, p. 325.

E. KIERZEK-PECOLD, J. KOŁODZIEJCZAK, and I. PRACKA
Optical Constants of Crystalline CaB_6
Vol. **29**, No. 2, p. K183–K185 (1968).
See also 13.1, p. 189; *20.1*, p. *357*.

J. LECIEJEWICZ, A. MURASIK, and R. TROĆ
The Antiferromagnetic Ordering in Uranium Monoarsenide and Monoantimonide
Vol. **30**, No. 1, p. 157–162 (1968).
See also 4, p. 48; *18.4*, p. *325*.

V. V. ANTONOV-ROMANOVSKII
On the Trapping of Free and Release of Localized Carriers in Solids
Vol. **30**, No. 1, p. 341–347 (1968).
See also *13.4*, p. *216*.

V. V. SOBOLEV, S. D. SHUTOV, YU. V. POPOV, and S. N. SHESTATSKII
Reflectivity Spectra of the Rhombohedral Crystals Bi_2Te_3, Bi_2Se_3, and Sb_2Te_3 over the Range from 0.7 to 12.5 eV
Vol. **30**, No. 1, p. 349–355 (1968).
See also 13.1, p. 189; *20.1*, p. *358*.

A. S. DAVYDOV
The Radiationless Transfer of Energy of Electronic Excitation between Impurity Molecules in Crystals
Vol. **30**, No. 1, p. 357–366 (1968).
See also 6, p. 69; *13.4*, p. *216*; 20.1, p. 358; 20.3, p. 371.

V. ANSORGE and A. MEŇOVSKÝ
Elementary Magnetic Properties of Some U-Ga Compounds
Vol. **30**, No. 1, p. K31–K32 (1968).
See also *18.1*, p. *298*; *18.2*, p. *315*.

U. Birkholz, H. Finkenrath, J. Naegele, and N. Uhle
Infrared Reflectivity of Semiconducting $FeSi_2$
Vol. **30**, No. 1, p. K81–K84 (1968).
See also *20.1*, p. *358*; 22.7, p. 546.

K. Peuker and E. D. Trifonov
Theory of Light Emission by Impurity Centres from Non-Equilibrium Vibrational States
Vol. **30**, No. 2, p. 479–484 (1968).
See also 6, p. 70; 13.4, p. 217; *20.3*, p. *371*.

Nguyen van Dong and Pham Ngu Tung
Transport Properties of Silver Telluride in the Solid and Liquid States
Vol. **30**, No. 2, p. 557–567 (1968).
See also 2, p. 25; *14.3*, p. *244*; 14.4, p. 254; *15*, p. *275*.

A. T. Davidson and A. D. Yoffe
Hopping Electrical Conduction and Thermal Breakdown in Natural and Synthetic Mica
Vol. **30**, No. 2, p. 741–754 (1968).
See also 3, p. 28; 10, p. 113; *14.4.1*, p. *265*; 20.1, p. 359.

J. Przystawa
Magnetic Symmetry of Uranium Compounds of UX_2 Type
Vol. **30**, No. 2, p. K115–K117 (1968).
See also 3, p. 28; *18.4*, p. *325*.

K. H. Gundlach
Note on the Effect of the Band Structure on Tunnelling in Metal–Semiconductor–Metal Junctions
Vol. **30**, No. 2, p. K135–K138 (1968).
See also 13.1, p. 190; 14.3.1, p. 247; *14.3.2*, p. *251*.

E. Cruceanu, St. Slădaru, and T. Boţilă
Optical Energy Gap of Tl_2Te_3
Vol. **30**, No. 2, p. K149–K151 (1968).
See also 13.1, p. 190; *20.1*, p. *359*.

B. Selle and J. Maege
Electrical and Optical Properties of Evaporated Cu_2S Films
Vol. **30**, No. 2, p. K153–K155 (1968).
See also 13.4, p. 218; *14.3.1*, p. *247*; 15, p. 275; 16, p. 285; *20.1*, p. *359*.

22.1 Elements

Review Articles

Y. P. VARSHNI
Band-to-Band Radiative Recombination in Groups IV, VI, and III-V Semiconductors (I)
Vol. **19**, No. 2, p. 459—514 (1967).
See also 13.1, p. 174; 13.2, p. 191; 13.4, p. 200; 20.1, p. 344; 20.2, p. 360; *20.3*, p. *363*; 22.1.1, p. 458; 22.1.2, p. 468; 22.1.3, p. 476; 22.2.1, p. 482; 22.2.2, p. 487; 22.2.3, p. 489.

Y. P. VARSHNI
Band-to-Band Radiative Recombination in Groups IV, VI, and III-V Semiconductors (II)
Vol. **20**, No. 1, p. 9—36 (1967).
See also 13.1, p. 174; 13.2, p. 191; 13.4, p. 200; 20.1, p. 344; 20.2, p. 360; *20.3*, p. *363*; 22.1.1, p. 458; 22.1.2, p. 468; 22.1.3, p. 476; 22.2.1, p. 482; 22.2.2, p. 487; 22.2.3, p. 489.

F. WILLIAMS
Donor–Acceptor Pairs in Semiconductors
Vol. **25**, No. 2, p. 493—512 (1968).
See also 10, p. 93; *13.4*, p. *200*; 20.3, p. 363; 22.1.1, p. 458; 22.1.2, p. 468; 22.2.2, p. 487; 22.4.1, p. 497.

G. HARBEKE
On the Band Structure of Anisotropic Crystals
Vol. **27**, No. 1, p. 9—44 (1968).
See also *13.1*, p. *174*; 13.2, p. 191; 14.3, p. 231; 20.1, p. 344; 22, p. 439; 22.1.1, p. 458; 22.1.2, p. 468; 22.1.3, p. 476; 22.4.1, p. 497; 22.4.2, p. 506; 22.4.3, p. 509; 22.5, p. 512; 22.7, p. 545; 22.8, p. 547.

Original Papers and Short Notes

A. J. NADOLNY, J. W. OSTROWSKI, and M. Z. PRZEGALIŃSKA-MIESZKOWSKA
Kinetics of Photoconductivity in Boron
Vol. **16**, No. 2, p. K133—K137 (1966).
See also *16*, p. *276*.

P. DELAVIGNETTE, R. TRIVEDI, R. GEVERS, and S. AMELINCKX
Electron Microscopic Images of Dislocation Ribbons
Vol. **17**, No. 1, p. 221—241 (1966).
See also *10*, p. *94*.

P. F. LINDQUIST, M. L. HAMMOND, and R. H. BRAGG
Comments on the Paper "On the Interpretation of Electron Diffraction Patterns from 'Amorphous' Boron"
Vol. **17**, No. 1, p. K25—K31 (1966).
See also 1.2, p. 9; *3*, p. *26*; 4, p. 38.

W. GÜTH and W. HAIST
Electronic States of Dislocations in Semiconductors
Vol. **17**, No. 2, p. 691—696 (1966).
See also 13.1, p. 175; *13.4*, p. *202*.

T. R. DUNCAN and D. KUHLMANN-WILSDORF
The Stress Fields of Dislocation Tilt Boundaries in Anisotropic Cubic and Diamond-Type Crystals
Vol. **18**, No. 1, p. 231—239 (1966).
See also *10*, p. *96*; *10.1*, p. *117*; 21, p. 375; 21.1, p. 393; 21.5, p. 429; 21.6, p. 431; 22.1.1, p. 460; 22.1.2, p. 470.

I. N. Evdokimova, E. S. Mashkova, V. A. Molchanov, and D. D. Odintsov
Dependence of the Ion–Electron Emission Coefficient on the Angle of Incidence
Vol. **19**, No. 1, p. 407–415 (1967).
See also *17*, p. *287*; 21, p. 377.

H. M. Otte, H. A. Lipsitt, P. F. Lindquist, M. L. Hammond, and R. H. Bragg
Further Comments on the Paper "On the Interpretation of Electron Diffraction Patterns from 'Amorphous' Boron"
Vol. **19**, No. 2, p. K99–K101 (1967).
See also 1.2, p. 10; *3*, p. *26*; 4, p. 40.

J. W. Ostrowski and A. Pajączkowska
Spectral Response of Photosensitivity (SRPS) in Boron
Vol. **20**, No. 2, p. K159–K162 (1967).
See also *16*, p. *279*.

A. Zaręba
Photoconductivity in Boron at Low Temperature
Vol. **21**, No. 2, p. K135–K138 (1967).
See also 13.4, p. 208; *16*, p. *279*.

R. Bogdanović
Potential of a Vacant Site in the Graphite Lattice
Vol. **22**, No. 2, p. 603–606 (1967).
See also *13.4*, p. *209*.

B. Mukherjee and A. J. F. Boyle
On the Debye Characteristic Temperature of Diamond
Vol. **22**, No. 2, p. K131–K134 (1967).
See also *6*, p. *62*; 8, p. 83; 12, p. 157.

E. Kierzek-Pecold, J. Kołodziejczak, and I. Pracka
Optical Constants of β-Rhombohedral Boron in the Region 1.2 to 6.2 eV
Vol. **22**, No. 2, p. K147–K150 (1967).
See also *20.1*, p. *350*.

J. T. Moore, T. R. Duncan, and D. Kuhlmann-Wilsdorf
Corrected Data for the Shear Stresses of Infinite Tilt Boundaries Composed of {111} 1/2 ⟨1$\bar{1}$0⟩ Dislocations in Elastically Anisotropic F.C.C. Metals
Vol. **25**, No. 2, p. K105–K106 (1968).
See also *10*, p. *105*; *10.1*, p. *129*; 21, p. 385; 21.1, p. 403; 21.5, p. 429; 21.6, p. 435; 22.1.1, p. 464; 22.1.2, p. 473.

E. Kierzek-Pecold, J. Kołodziejczak, and I. Pracka
Anisotropy in the Reflection Coefficient of β-Rhombohedral Boron in the Region 1 to 6 eV
Vol. **27**, No. 2, p. K107–K109 (1968).
See also *20.1*, p. *355*.

K. Kleinhenz and P. Runow
Preparation of Thin Boron Foils and Electron Microscopic Observation of Their Lattice Faults
Vol. **29**, No. 2, p. 627–637 (1968).
See also 1.4, p. 21; *10*, p. *110*.

D. Strauch
Shell Model Treatment of the Vibrations of Crystals Containing Substitutional Defects. (II): On the Theory of Defect-Induced One-Phonon Absorption
Vol. **30**, No. 2, p. 495–505 (1968).
See also *6*, p. *70*; 10, p. 112; *20.1*, p. *359*; 22.1.1, p. 466; 22.1.2, p. 475.

22.1.1 Germanium

Review Articles

J. R. DALE
Alloyed Semiconductor Heterojunctions
Vol. **16**, No. 2, p. 351—387 (1966).
See also 3.1, p. 29; 5, p. 49; 13.4, p. 200; *14.3.2*, p. *248*; 16, p. 276; 22.1.2, p. 468; 22.2.1, p. 482; 22.2.2, p. 487; 22.2.3, p. 489.

Y. P. VARSHNI
Band-to-Band Radiative Recombination in Groups IV, VI, and III-V Semiconductors (I)
Vol. **19**, No. 2, p. 459—514 (1967).
See also 13.1, p. 174; 13.2, p. 191; 13.4, p. 200; 20.1, 344; p. 20.2, p. 360; *20.3*, p. *363*; 22.1, p. 456; 22.1.2, p. 468; 22.1.3, p. 476; 22.2.1, p. 482; 22.2.2, p. 487; 22.2.3, p. 489.

Y. P. VARSHNI
Band-to-Band Radiative Recombination in Groups IV, VI, and III-V Semiconductors (II)
Vol. **20**, No. 1, p. 9—36 (1967).
See also 13.1, p. 174; 13.2, p. 191; 13.4, p. 200; 20.1, p. 344; 20.2, p. 360; *20.3*, p. *363*; 22.1, p. 456; 22.1.2, p. 468; 22.1.3, p. 476; 22.2.1, p. 482; 22.2.2, p. 487; 22.2.3, p. 489.

D. P. MORGAN
Helicon Waves in Solids
Vol. **24**, No. 1, p. 9—36 (1967).
See also 6, p. 56; *13*, p. *162*; 14.1, p. 220; 14.3, p. 231; 19, p. 326; *20*, p. *337*; 21, p. 373; 21.1, p. 390; 21.2, p. 424; 21.6, p. 430; 22.2.1, p. 482; 22.2.3, p. 489.

F. WILLIAMS
Donor-Acceptor Pairs in Semiconductors
Vol. **25**, No. 2, p. 493—512 (1968).
See also 10, p. 93; *13.4*, p. *200*; 20.3, p. 363; 22.1, p. 456; 22.1.2, p. 468; 22.2.2, p. 487; 22.4.1, p. 497.

G. HARBEKE
On the Band Structure of Anisotropic Crystals
Vol **27**, No. 1, p. 9—44 (1968).
See also *13.1*, p. *174*; 13.2, p. 191; 14.3, p. 231; 20.1, p. 344; 22, p. 439; 22.1, p. 456; 22.1.2, p. 468; 22.1.3, p. 476; 22.4.1, p. 497; 22.4.2, p. 506; 22.4.3, p. 509; 22.5, p. 512; 22.7, p. 545; 22.8, p. 547.

V. L. BONCH-BRUEVICH and E. G. LANDSBERG
Recombination Mechanisms
Vol. **29**, No. 1, p. 9—43 (1968).
See also 6, p. 56; 13, p. 162; 13.2, p. 191; *13.4*, p. *200*; *14.3*, p. *231*; 20, p. 337; 20.3, p. 363; 22.1.2, p. 468.

A. SEEGER and K. P. CHIK
Diffusion Mechanisms and Point Defects in Silicon and Germanium
Vol. **29**, No. 2, p. 455—542 (1968).
See also *9*, p. *87*; *10*, p. *93*; 11, p. 145; 13.4, p. 200; 19, p. 326; 22.1.2, p. 468.

M. BALKANSKI and E. AMZALLAG
Band Parameters Determination from Faraday Rotation Measurements
Vol. **30**, No. 2, p. 407—439 (1968).
See also *13.1*, p. *174*; 18, p. 291; 20, p. 337; 22.1.2, p. 468; 22.2.1, p. 482; 22.2.2, p. 487; 22.2.3, p. 489; 22.4, p. 495; 22.4.1, p. 497; 22.4.2, p. 506; 22.4.3, p. 509.

Original Papers and Short Notes

E. I. Rashba, V. A. Romanov, I. I. Boiko, and I. P. Zhadko
Electrical Pinch in Elastically Deformed Germanium
Vol. **16**, No. 1, p. 43–56 (1966).
See also 12, p. 156; *14.3*, p. *231*; 14.4.1, p. 255.

E. Kamieniecki
Distribution Function of Hot Carriers
Vol. **16**, No. 1, p. 57–62 (1966).
See also 6, p. 56; *13*, p. *162*; *14.3*, p. *231*; 22.1.2, p. 468.

S. G. Kalashnikov and V. L. Bonch-Bruevich
On the Velocity of Space Charge Waves (Electrical Domains) in Semiconductors
Vol. **16**, No. 1, p. 197–203 (1966).
See also *14.4.1*, p. *255*.

L. Gouskov, G. Lecoy, C. Llinares, and M. Savelli
Study of Lifetime and Generation–Recombination Noise Controlled by Dislocations in Germanium
Vol. **16**, No. 2, p. 721–728 (1966).
See also *13.4*, p. *202*; 14.3, p. 232; 16, p. 276.

V. A. Zuev and V. G. Litovchenko
Theory of the Photovoltaic Effect in Semiconductors
Vol. **16**, No. 2, p. 751–760 (1966).
See also 13.3, p. 196; *16*, p. *276*; 22.1.2, p. 469.

R. Grigorovici, N. Croitoru, and A. Dévényi
Thermoelectric Power in Amorphous Germanium
Vol. **16**, No. 2, p. K143–K145 (1966).
See also 2, p. 22; 14.3.1, p. 245; *15*, p. *270*.

B. Pödör
Electron Mobility in Plastically Deformed Germanium
Vol. **16**, No. 2, p. K167–K170 (1966).
See also 10, p. 94; *14.3*, p. *232*.

V. A. Romanov, I. P. Zhadko, and I. I. Boiko
Photoconductivity of Anisotropically Deformed Intrinsic Germanium
Vol. **17**, No. 1, p. 389–396 (1966).
See also 12, p. 156; *16*, p. *277*.

O. N. Efimov and E. K. Kovev
Integral Characteristics of Anomalous Transmission of X-Rays
Vol. **17**, No. 1, p. 397–410 (1966).
See also *5*, p. *49*; 10, p. 94.

J. Pfeifer
Epitaxial Growth of Germanium on Germanium Substrates Cleaned and Heated by Electron Bombardment
Vol. **17**, No. 1, p. K15–K18 (1966).
See also 1.3, p. 14; *3.1*, p. *29*.

K. Owen and R. L. Anderson
Determination of Optical Absorption Using Heterocrystals
Vol. **17**, No. 1, p. K79–K81 (1966).
See als *20.1*, p. *345*.

A. SEEGER and G. WOBSER
Parallel Dislocation Arrays in Anisotropic Media
Vol. 17, No. 2, p. 709–724 (1966).
See also *10*, p. *95*; *10.1*, p. *110*; 21, p. 375; 21.1, p. 392; 21.3, p. 426; 21.6, p. 431; 22.1.2 p. 469.

V. DIENYS and J. POZHELA
Anisotropy of the Conductivity of n-Type Germanium at High Electric Fields
Vol. 17, No. 2, p. 769–776 (1966).
See also 13, p. 163; *14.4.1*, p. *256*.

H.-J. DITTFELD and H. STRELLER
Dependence of the Maximum Reverse Voltage of a Metal–Semiconductor Point Contact on the Mobility of Majority Carriers
Vol. 17, No. 2, p. K245–K248 (1966).
See also *14.4.1*, p. *257*; 22.1.2, p. 469.

V. D. EGOROV
Linear Photoconductive Rise in Semiconductors
Vol. 18, No. 1, p. 145–150 (1966).
See also 13.3, p. 197; *16*, p. *278*.

R. COLELLA and A. MERLINI
A Study of the (222) "Forbidden" Reflection in Germanium and Silicon
Vol. 18, No. 1, p. 157–166 (1966).
See also *5*, p. *50*; 22.1.2, p. 469.

T. R. DUNCAN and D. KUHLMANN-WILSDORF
The Stress Fields of Dislocation Tilt Boundaries in Anisotropic Cubic and Diamond-Type Crystals
Vol. 18, No. 1, p. 231–239 (1966).
See also *10*, p. *96*; *10.1*, p. *117*; 21, p. 375; 21.1, p. 393; 21.5, p. 429; 21.6, p. 431; 22.1, p. 456; 22.1.2, p. 470.

S. A. MOSKALENKO and P. I. KHADSHI
Infrared Absorption by Excitons due to Photoionization and Intraband Lattice Scattering
Vol. 18, No. 1, p. 379–390 (1966).
See also 6, p. 57; *13.2*, p. *191*; 20.1, p. 346; 22.2.2, p. 487; 22.2.3, p. 490; 22.6, p. 536.

H. FRANK
Contribution to the Measurement of the Resistance of nn^+ and pp^+ Epitaxial Films by the Breakdown Method
Vol. 18, No. 1, p. 401–414 (1966).
See also *14.3.1*, p. *245*; 14.4.1, p. 257; 22.1.2, p. 470; 22.2.1, p. 483;

H. FLIETNER and N. KEMPE
Study of the Oscillatory Properties of the Oscillistor
Vol. 18, No. 1, p. 415–426 (1966).
See also *13*, p. *164*; *20*, p. *338*.

M. ASCHE, O. G. SARBEJ, and V. M. VASETSKII
Piezorestistance of p-Germanium
Vol. 18, No. 2, p. 749–754 (1966).
See also 10, p. 97; 12, p. 156; *14.3*, p. *234*.

R. I. ANISHCHENKO
Calculation of the Mean Inner Potential of a Crystal in the Statistical Theory
Vol. 18, No. 2, p. 923–928 (1966).
See also *3.2*, p. *34*; 13, p. 164; 21, p. 376; 21.1, p. 394; 21.1.1, p. 414; 21.2, p. 424; 21.3, p. 426; 21.4, p. 428; 21.6, p. 432; 21.7, p. 438.

L. Ernst
Field-Emission Microscopy of Germanium
Vol. **19**, No. 1, p. 89–94 (1967).
See also *1.3*, p. *14*.

A. K. Walton and G. J. O. Eyoku
Piezo-Hall Effect in n-Type Germanium
Vol. **19**, No. 1, p. 111–116 (1967).
See als *14.3*, p. *234*.

S. Schäfer
Measurement of Dislocation Velocities in Germanium
Vol. **19**, No. 1, p. 297–312 (1967).
See also *10*, p. *97*.

A. Glodeanu
Helium-Like Impurities in Semiconductors
Vol **19**, No. 1, p. K43–K46 (1967).
See also *13.4*, p. *205*; 22.1.2, p. 470; 22.2.1, p. 483.

W. Franz and P. Miosga
Distribution Function and Drift Velocity of Warm Electrons in Germanium
Vol. **19**, No. 2, p. 597–602 (1967).
See also *13*, p. *164*; 14.3, p. 235.

J. H. P. van Weeren, E. Struikmans, and J. Blok
Test of the Read Model Concerning the Electron Mobility in Plastically Deformed n-Type Ge
Vol. **19**, No. 2, p. K107–K110 (1967).
See also 10, p. 98; *14.3*, p. *235*.

Z. Gołacki and T. Figielski
Extrinsic Photoconductivity in Ge Caused by Dislocations
Vol. **20**, No. 1, p. K1–K3 (1967).
See also 13.4, p. 205; *16*, p. *279*.

V. M. Asnin and A. A. Rogachev
Exciton Absorption in Doped Germanium
Vol. **20**, No. 2, p. 755–757 (1967).
See also *13.2*, p. *192*; *20.1*, p. *347*.

J. Auleytner, K. Godwod, J. Litwin, and Z. Wołoszyn
Application of X-Ray Methods to the Study of the Real Structure of Monocrystalline Epitaxial Layers
Vol. **20**, No. 2, p. K77–K79 (1967).
See also *1.4*, p. *18*; 22.2.1, p. 483.

J. Blinowski and M. Grynberg
On the Direct Interband Magnetoabsorption in Uniaxially Stressed Germanium
Vol. **20**, No. 2, p. K107–K111 (1967).
See also 10, p. 99; *18*, p. *292*; *20*, p. *339*.

W. Schröter
Electric Properties of Dislocations in Germanium
Vol. **21**, No. 1, p. 211–224 (1967).
See also 10, p. 99; 13.4, p. 206; *14.3*, p. *236*.

L. Gouskov, G. Lecoy, and C. Llinares
Carrier Recombination in Nickel Doped Germanium from Lifetime and Noise Measurements
Vol. **21**, No. 2, p. 619–626 (1967).
See also 13.4, p. 207; *14.3*, p. *236*; *16*, p. *279*.

D. G. Arasli and M. I. Aliev
Influence of Defects and of the Interaction between Them on Phonon Scattering in Heavily Doped Ge and Si Crystals
Vol. **21**, No. 2, p. 643–648 (1967).
See also 6, p. 61; *8*, p. *82*; 10, p. 100; 22.1.2, p. 471

I. L. Sokolskaya, V. G. Ivanov, and G. N. Fursey
Study of Barium Adsorption on Germanium by Field-Emission Microscopy
Vol. **21**, No. 2, p. 789–795 (1967).
See also 1.3, p. 15; *13.3*, p. *197*; 17, p. 287.

G. Lautz and M. Schulz
Helical Density-Waves and Oscillistor Effect in the Electron-Hole Plasma of n-Germanium
Vol. **21**, No. 2, p. K139–K143 (1967).
See also *13*, p. *166*; 14.4.1, p. 259; *20*, p. *339*.

G. N. Fursey, I. L. Sokolskaya, and V. G. Ivanov
Field Emission from p-Type Germanium
Vol. **22**, No, 1, p. 39–46 (1967).
See also *17*, p. *287*.

N. Paschoff
Growth and Structure of Thin Germanium Crystals Crystallized from Molten Films
Vol. **22**, No. 1, p. 83–91 (1967).
See also *1.4*, p. *18*; *3.1*, p. *31*.

N. Paschoff and J. Heydenreich
Dislocation Arrangement in Thin Germanium Foils Prepared by Crystallization
Vol. **22**, No. 1, p. 93–101 (1967).
See also 1.4, p. 18; *10*, p. *101*.

V. L. Bonch-Bruevich
On the Theory of Electrical Domains in Hot Electron Semiconductors
Vol. **22**, No. 1, p. 267–278 (1967).
See also 13, p. 166; 14.3, p. 237; *14.4.1*, p. *259*; 22, p. 446; 22.2.1, p. 484.

O. N. Efimov
Contribution of Thermal Vibrations to the Anomalous Transmission of X-Rays
Vol. **22**, No. 1, p. 297–308 (1967).
See also *5*, p. *51*; 6, p. 62.

D. J. Morgan and J. A. Galloway
Energy Band Structure of Germanium
Vol. **22**, No. 2, p. 491–497 (1967).
See also *13.1*, p. *180*.

R. Yu. Khansevarov, T. V. Mashovets, and N. A. Vitovskii
Damage Formation Rates in Ge Subject to Thermal Treatment
Vol. **22**, No. 2, p. K103–K107 (1967).
See also 10, p. 101; *11*, p. *150*; 14.3, p. 237.

A. I. Kosinskaya and S. I. Uritsky
On the Classical Theory of the Optical Properties of Free Carriers in Semiconductors
Vol. **23**, No. 1, p. 57–68 (1967).
See also 13.1, p. 180; 18, p. 293; *20*, p. *339*.

D. J. Morgan and J. A. Galloway
Sum Rules for Effective Masses in Energy Band Theory
Vol. **23**, No. 1, p. 97–103 (1967).
See also *13.1*, p. *180*.

A. G. Samoilovich, I. S. Buda, and I. V. Dakhovskii
Anisotropy of Thermomagnetic Effects in n-Ge
Vol. **23**, No. 1, p. 229–236 (1967).
See also 13, p. 167; 14.3, p. 238; *15*, p. *272*.

H. Teichler
Calculation of the Peierls Potential in the Diamond Lattice by the Pseudopotential Method
Vol. **23**, No. 1, p. 341–352 (1967).
See also *10*, p. *102*; 22.1.2, p. 472.

W. Güth and H. Bleher
Avalanche Breakdown in Germanium in the Case of Optical Band–Band Excitation
Vol. **23**, No. 1, p. K13–K16 (1967).
See also 13, p. 167; *14.4.1*, p. *260*; 16, p. 280.

P. Byszewski, B. Kalińska, and J. Kołodziejczak
Interband Faraday Rotation and Ellipticity in Germanium and Gallium Arsenide
Vol. **23**, No. 1, p. K53–K56 (1967).
See also 13, p. 167; *18*, p. *293*; *20*, p. *340*; 22.2.1, p. 484.

H.-P. Hennig
Contribution to the Determination of Gold Centre Cross Sections in Single-Crystal Germanium by Double Injection Experiments
Vol. **23**, No. 2, p. 599–605 (1967).
See also 13.4, p. 210; *14.4.1*, p. *260*.

R. Grigorovici, N. Croitoru, and A. Dévényi
Photoconductivity in Amorphous Germanium
Vol. **23**, No. 2, p. 627–631 (1967).
See also 2, p. 23; *16*, p. *280*.

V. L. Bonch-Bruevich
On the Problem of Static Domains in Hot Electron Semiconductors
Vol. **23**, No. 2, p. 761–766 (1967).
See also 13, p. 168; 14.3, p. 238; *14.4.1*, p. *260*.

L. Ernst
About the Fowler-Nordheim Plots of Germanium Field Emitters
Vol. **24**, No. 1, p. 177–181 (1967).
See also 13.3, p. 198; *17*, p. *287*.

R. Grigorovici, N. Croitoru, L. Vescan, and M. Marina
Hole Injection in Junctions between Amorphous Ge Layers and n-Type Ge Single Crystals
Vol. **24**, No. 1, p. K17–K21 (1967).
See also *14.3.2*, p. *250*; *14.4.1*, p. *261*.

M. S. Kagan, S. G. Kalashnikov, and N. G. Zhdanova
Stationary State of a Semiconductor with Negative Bulk Differential Conductivity
Vol. **24**, No. 2, p. 551–559 (1967).
See also *14.4.1*, p. *261*.

V. A. Besfamilnaya, V. F. Elesin, I. A. Kurova, N. N. Ormont, and V. V. Ostroborodova
On the Nature of Oscillations in Impurity Photoconductivity Spectra of Ge
Vol. **24**, No. 2, p. 757–762 (1967).
See also 6, p. 64; 13, p. 168; *16*, p. *281*.

M. Schulz
Helical Density Waves and the Oscillistor Effect in the Electron Hole Plasma of the n-Type Ge (Theory)
Vol. **25**, No. 2, p. 521–530 (1968).
See also *13*, p. *169*; 14.4.1, p. 261; *20*, p. *340*.

R. Gauthier, D. Bois, P. Pinard, and F. Davoine
Ion and Electron Production by Interaction of a Coherent Light Beam with Substance
Vol. **25**, No. 2, p. 691–696 (1968).
See also *17*, p. *288*; 21.1, p. 403; 21.1.1, p. 420; 22.2.1, p. 485; 22.4.1, p. 502; 22.5.2, p. 525; 22.6, p. 540.

J. G. Broerman
Approximate Symmetry Relations for Double Group $\boldsymbol{k}\cdot\boldsymbol{p}$ Expansions
Vol. **25**, No. 2, p. 757–762 (1968).
See also *13.1*, p. *184*.

J. T. Moore, T. R. Duncan, and D. Kuhlmann-Wilsdorf
Corrected Data for the Shear Stresses of Infinite Tilt Boundaries Composed of {111} 1/2 $\langle 1\bar{1}0\rangle$ Dislocations in Elastically Anisotropic F.C.C. Metals
Vol. **25**, No. 2, p. K105–K106 (1968).
See also *10*, p. *105*; *10.1*, p. *129*; 21, p. 385; 21.1, p. 403; 21.5, p. 429; 21.6, p. 435; 22.1, p. 457; 22.1.2, p. 473.

H.-P. Hennig
N- and S-Type Negative Resistances and Relaxation Oscillations in Au-Doped Double Injection Ge Diodes
Vol. **26**, No. 1, p. 225–229 (1968).
See also 13.4, p. 211; *14.4.1*, p. *262*.

J. Łagowski
On Surface Recombination Changes due to Edge Dislocations
Vol. **26**, No. 1, p. K69–K71 (1968).
See also *10*, p. *105*; *13.3*, p. *198*.

H. Alexander
Electron Microscopic Study of Frozen-in Dislocations. (I): Basic Considerations on the Method and Results at the Deformation Onset of Germanium Crystals
Vol. **26**, No. 2, p. 725–741 (1968).
See also *10*, p. *106*.

E. Schmidt
Optical Properties of GeSi Alloys in the Energy Region from 1 to 13 eV
Vol. **27**, No. 1, p. 57–67 (1968).
See also 13.1, p. 185; *20.1*, p. *353*; 22.1.2, p. 473.

M. Brauer
On the Behaviour of a Cylindrical Electron-Hole Plasma in a Magnetic Field with Transverse Component
Vol. 27, No. 1, p. 151–159 (1968).
See also *14.4.1*, p. *262*; 18, p. 294.

J. H. P. van Weeren, G. Koopmans, and J. Blok
The Position of the Dislocation Acceptor Level in n-Type Ge
Vol. 27, No. 1, p. 219–224 (1968).
See also 10, p. 106; *13.4*, p. *213*; 14.3, p. 241.

J. H. P. van Weeren, R. Struikmans, G. Koopmans, and J. Blok
Low-Field Magnetoresistance and Magnetoconductivity in n-Type Ge with Dislocations
Vol. 27, No. 1, p. 225–235 (1968).
See also 10, p. 107; *14.3*, p. *241*.

D. N. Bose
Effect of Low Energy Electron and UV Irradiations on the Charge Condition of Etched Germanium Surfaces
Vol. 27, No. 1, p. 325–331 (1968).
See also 11, p. 152; *13.3*, p. *198*.

H. Alexander
Electron Microscopic Study of Frozen-in Dislocations. (II): Results Obtained for Stages I and II of the Work-Hardening Curve
Vol. 27, No. 1, p. 391–412 (1968).
See also *10*, p. *107*.

O. Brümmer and G. Dräger
On the Interpretation of the Fine Structure of X-Ray K Absorption Spectra Far from the Edge
Vol. 27, No. 2, p. 513–516 (1968).
See also 2, p. 24; *20*, p. *341*; 22.6, p. 541; 22.8, p. 554.

B. H. Zimmermann, H. Jena, G. Ischenko, H. Kilian, and D. Seyboth
Mössbauer Experiments with Coulomb-Excited ^{73}Ge after Coulomb Recoil Implantation
Vol. 27, No. 2, p. 639–651 (1968).
See also *6.1*, p. *74*.

J. K. Pozhela and K. K. Repshas
Thermoelectric Force of Hot Carriers
Vol. 27, No. 2, p. 757–762 (1968).
See also 13, p. 170; 14.4.1, p. 262; *15*, p. *274*.

P.Byszewski, J. Holzman, and J. Kołodziejczak
Oscillatory Magneto-Photoconductivity Effect in Germanium
Vol. 27, No. 2, p. K125–K127 (1968).
See also *16*, p. *282*.

H. Heinrich, G. Bauer, and D. Kasperkovitz
Spatial Dependence of Hot Carrier Oscillations in n-Type Germanium
Vol. 28, No. 1, p. K51–K53 (1968).
See also 13, p. 171; *14.4.1*, p. *263*.

D. K. Roy
On the Separation of the Phonon-Assisted Current Components in Germanium Tunnel Diodes
Vol. 28, No. 1, p. K75–K79 (1968).
See also 6, p. 67; *14.3.2*, p. *251*.

V. L. BONCH-BRUEVICH
On the Problem of Recombination Autooscillations
Vol. **28**, No. 2, p. 687–693 (1968)
See also 13, p. 172; 13.4, p. 215; *14.4.1*, p. *263*.

Z. S. GRIBNIKOV, G. I. LOMOVA, and V. A. ROMANOV
Injection of Current Carriers in Anisotropic Semiconductor Plates and the Magnetodiode Effect
Vol. **28**, No. 2, p. 815–825 (1968).
See also *14.4.1*, p. *263*.

S. BALLARÒ, A. BALZAROTTI, and V. GRASSO
Electroreflectance at Oblique Incidence in Ge
Vol. **28**, No. 2, p. K109–K112 (1968).
See also 13.1, p. 188; *14.4.1*, p. *264*; *20*, p. *341*; 20.1, p. 356.

W. KRESS, H. BORIK, and R. K. WEHNER
Infrared Lattice Absorption of Silicon and Germanium
Vol. **29**, No. 1, p. 133–143 (1968).
See also *6*, p. *67*; 20.1, p. 356; 22.1.2, p. 474.

H. J. GLÄSER and D. GEIST
Electrical Conductivity and Hall Effect of Manganese-Doped Germanium and the Influence of Heat Treatment on the State of Incorporation of Manganese
Vol. **29**, No. 1, p. 167–170 (1968).
See also 13.4, p. 215; *14.3*, p. *243*; 19, p. 335.

V. M. ASNIN, G. L. ERISTAVI, and A. A. ROGACHEV
Effect of Weak Electric Fields on the Absorption Edge in Doped Germanium
Vol. **29**, No. 1, p. 443–446 (1968).
See also 13.2, p. 195; 13.4, p. 215; 14.3, p. 243; *20.1*, p. *356*.

D. K. ROY
On the Determination of Impurity Concentration in the Alloyed Region of a Tunnel Diode
Vol. **29**, No. 2, p. K111–K115 (1968).
See also *13.4*, p. *216*; *14.3.2*, p. *251*.

H.-P. HENNIG
On the Conduction Mechanism in the Instable Range of a Double Injection Diode
Vol. **29**, No. 2, p. K167–K169 (1968).
See also *14.4.1*, p. *264*.

V. V. TARASSOV and A. F. DEMIDENKO
Heat Capacity and Quasi-Chain Dynamics of Diamond-Like Structures
Vol. **30**, No. 1, p. 147–155 (1968).
See also *6*, p. *69*; *8*, p. *86*; 13.1, p. 189; 22.1.2, p. 474; 22.2.1, p. 486; 22.2.2, p. 488; 22.2.3, p. 493.

L. I. DATSENKO, M. YA. SKOROKHOD, and A. S. VASILKOVSKII
The Effect of Dislocations on the Intensity Jumps at the Interferential Transmission of X-Rays near the K-Edge of Absorption in Ge
Vol. **30**, No. 1, p. 231–237 (1968).
See also *10*, p. *112*; 20, p. 342.

D. STRAUCH
Shell Model Treatment of the Vibrations of Crystals Containing Substitutional Defects. (II): On the Theory of Defect-Induced One-Phonon Absorption
Vol. **30**, No. 2, p. 495–505 (1968).
See also *6*, p. *70*; 10, p. 112; *20.1*, p. *359*; 22.1, p. 457; 22.1.2, p. 475.

V. V. Nemoshkalenko, M. A. Mindlina, and B. P. Mamko
X-Ray Spectra and Electronic Structure of Elements at the End of the First Long Period from ^{29}Cu to ^{34}Se
Vol. **30**, No. 2, p. 703–712 (1968).
See also 13.1, p. 189; *20*, p. *343*; 21.1, p. 409; 22.1.3, p. 480.

L. A. Almasov and I. M. Dykman
Plasma Oscillations in Semiconductors with Non-Standard Dispersion Law in the Presence of Electric and Magnetic Fields
Vol. **30**, No. 2, p. 767–778 (1968).
See also *13*, p. *173*; *20*, p. *343*.

P. Konorov, V. N. Shvetsov, and S. V. Shegolikhina
Effects of the Non-Equilibrium Depletion in Germanium at the Electrolyte Interface
Vol. **30**, No. 2, p. 845–850 (1968).
See also 14.3, p. 244; *14.4.1*, p. *265*.

G. Richter and M. Schleifstein
Photoluminescence in the Presence of Crossed Fields
Vol. **30**, No. 2, p. K97–K100 (1968).
See also *20.3*, p. *372*.

E. E. Godik
The Variation of the Photoelectron Mobility with Photon Energy in Germanium and Silicon
Vol. **30**, No. 2, p. K127–K129 (1968).
See also *16*, p. *285*; 22.1.2, p. 475.

22.1.2 Silicon

Review Articles

J. R. Dale
Alloyed Semiconductor Heterojunctions
Vol. **16**, No. 2, p. 351—387 (1966).
See also 3.1, p. 29; 5, p. 49; 13.4, p. 200; *14.3.2*, p. *248*; 16, p. 276; 22.1.1, p. 458; 22.2.1, p. 482; 22.2.2, p. 487; 22.2.3, p. 489.

Y. P. Varshni
Band-to-Band Radiative Recombination in Groups IV, VI, and III-V Semiconductors (I)
Vol. **19**, No. 2, p. 459—514 (1967).
See also 13.1, p. 174; 13.2, p. 191; 13.4, p. 200; 20.1, p. 344; 20.2, p. 360; *20.3*, p. *363*; 22.1, p. 456; 22.1.1, p. 458; 22.1.3, p. 476; 22.2.1, p. 482; 22.2.2, p. 487; 22.2.3, p. 489.

Y. P. Varshni
Band-to-Band Radiative Recombination in Groups IV, VI, and III-V Semiconductors (II)
Vol. **20**, No. 1, p. 9—36 (1967).
See also 13.1, p. 174; 13.2, p. 191; 13.4, p. 200; 20.1, p. 344; 20.2, p. 360; *20.3*, p. *363*; 22.1, p. 456; 22.1.1, p. 458; 22.1.3, p. 476; 22.2.1, p. 482; 22.2.2, p. 487; 22.2.3, p. 489.

F. Williams
Donor-Acceptor Pairs in Semiconductors
Vol. **25**, No. 2, p. 493—512 (1968).
See also 10, p. 93; *13.4*, p. *200*; 20.3, p. 363; 22.1, p. 456; 22.1.1, p. 458; 22.2.2, p. 487; 22.4.1, p. 497.

G. Harbeke
On the Band Structure of Anisotropic Crystals
Vol. **27**, No. 1, p. 9—44 (1968).
See also *13.1*, p. *174*; 13.2, p. 191; 14.3, p. 231; 20.1, p. 344; 22, p. 439; 22.1, p. 456; 22.1.1, p. 458; 22.1.3, p. 476; 22.4.1, p. 497; 22.4.2, p. 506; 22.4.3, p. 509; 22.5, p. 512; 22.7, p. 545; 22.8, p. 547.

V. L. Bonch-Bruevich and E. G. Landsberg
Recombination Mechanisms
Vol. **29**, No. 1, p. 9—43 (1968).
See also 6, p. 56; 13, p. 162; 13.2, p. 191; *13.4*, p. *200*; *14.3*, p. *231*; 20, p. 337; 20.3, p. 363; 22.1.1, p. 458.

A. Seeger and K. P. Chik
Diffusion Mechanisms and Point Defects in Silicon and Germanium
Vol. **29**, No. 2, p. 455—542 (1968).
See also *9*, p. *87*; *10*, p. *93*; 11, p. 145; 13.4, p. 200; 19, p. 326; 22.1.1, p. 458.

M. Balkanski and E. Amzallag
Band Parameters Determination from Faraday Rotation Measurements
Vol. **30**, No. 2, p. 407—439 (1968).
See also *13.1*, p. *174*; 18, p. 291; 20, p. 337; 22.1.1, p. 458; 22.2.1, p. 482; 22.2.2, p. 487; 22.2.3, p. 489; 22.4, p. 495; 22.4.1, p. 497; 22.4.2, p. 506; 22.4.3, p. 509.

Original Papers and Short Notes

E. Kamieniecki
Distribution Function of Hot Carriers
Vol. **16**, No. 1, p. 57—62 (1966).
See also 6, p. 56; *13*, p. *162*; *14.3*, p. *231*; 22.1.1, p. 459.

H. Lemke
Time Dependence of Space-Charge Limited Injection Currents in Fe-Doped p-Silicon
Vol. **16**, No. 2, p. 427–438 (1966).
See also 13.4, p. 201; *14.4.1*, p. *255*.

V. A. Zuev and V. G. Litovchenko
Theory of the Photovoltaic Effect in Semiconductors
Vol. **16**, No. 2, p. 751–760 (1966).
See also 13.3, p. 196; *16*, p. *276*; 22.1.1, p. 459.

A. S. Antonov
On the Mechanism of Lithium Ion Drift in the Electrical Field of a p–n Junction in Silicon
Vol. **16**, No. 2, p. 761–770 (1966).
See also *14.3.2*, p. *248*; *14.4*, p. *252*.

L. Elstner
Temperature Dependence of the Hole Mobility in Silicon Crystals
Vol. **17**, No. 1, p. 139–150 (1966).
See also 6, p. 56; *14.3*, p. *232*.

G. Dorda and J. Vrba
Long-Time Effect of External Fields on Surface Silicon Oxide Conductivity
Vol. **17**, No. 1, p. K127–K130 (1966).
See also *14.3*, p. *232*; 22.6, p. 535.

S. S. Sheinin and C. D. Cann
On the Measurement of Two Beam Values of Extinction Distance
Vol. **17**, No. 1, p. K143–K145 (1966).
See also *10*, p. *95*.

F. Forlani, N. Minnaja, and E. Pagiola
Theory of the Influence of Surface States on the Impedance of a Semiconductor–Insulator Interface
Vol. **17**, No. 2, p. 657–672 (1966).
See also *13.3*, p. *196*; *14.3.2*, p. *248*.

A. Seeger and G. Wobser
Parallel Dislocation Arrays in Anisotropic Media
Vol. **17**, No. 2, p. 709–724 (1966).
See also *10*, p. *95*; *10.1*, p. *116*; 21, p. 375; 21.1, p. 392, 21.3, p. 426; 21.6, p. 431; 22.1.1, p. 460.

K. Godwod, R. Kowalczyk, and M. Lefeld-Sosnowska
The Influence of the Primary X-Ray Beam Divergence on the X-Ray Wave Field Absorption Coefficients Measured by Means of the Three-Crystal X-Ray Spectrometer
Vol. **17**, No. 2, p. 869–874 (1966).
See also *4*, p. *38*; 20, p. 338.

H.-J. Dittfeld and H. Streller
Dependence of the Maximum Reverse Voltage of a Metal–Semiconductor Point Contact on the Mobility of Majority Carriers
Vol. **17**, No. 2, p. K245–K248 (1966).
See also *14.4.1*, p. *257*; 22.1.1, p. 460.

R. Colella and A. Merlini
A Study of the (222) "Forbidden" Reflection in Germanium and Silicon
Vol. **18**, No. 1, p. 157–166 (1966).
See also *5*, p. *50*; 22.1.1, p. 460.

T. R. Duncan and D. Kuhlmann-Wilsdorf
The Stress Fields of Dislocation Tilt Boundaries in Anisotropic Cubic and Diamond Type Crystals
Vol. **18**, No. 1, p. 231–239 (1966).
See also *10*, p. *96*; *10.1*, p. *117*; 21, p. 375; 21.1, p. 393; 21.5, p. 429; 21.6, p. 431; 22.1, p. 456; 22.1.1, p. 460.

G. Jungk
On the Generation–Recombination Statistics in Semiconductors
Vol. **18**, No. 1, p. 297–306 (1966).
See also *13.4*, p. *203*; 14.3, p. 233; 16, p. 278.

H. Frank
Contribution to the Measurement of the Resistance of nn^+ and pp^+ Epitaxial Films by the Breakdown Method
Vol. **18**, No. 1, p. 401–414 (1966).
See also *14.3.1*, p. *245*; 14.4.1, p. 257; 22.1.1, p. 460; 22.2.1, p. 483.

H. Flietner and W. Gräfe
Large-Signal Field Effect for the Determination of the Surface Levels of Silicon
Vol. **18**, No. 1, p. K5–K7 (1966).
See also *13.3*, p. *197*; 14.3, p. 233.

G. C. T. Liu and J. C. M. Li
Energy of Hexagonal Dislocation Loops
Vol. **18**, No. 2, p. 527–539 (1966).
See also *10*, p. *96*; *10.1*, p. *118*; 21, p. 376; 22, p. 441.

A. G. Revesz
Thermal Oxidation of Silicon: Growth Mechanism and Interface Properties
Vol. **19**, No. 1, p. 193–202 (1967).
See also *1.3*, p. *14*; 13.3, p. 197; 14.3.2, p. 248.

A. Glodeanu
Helium-Like Impurities in Semiconductors
Vol. **19**, No. 1, p. K43–K46 (1967).
See also *13.4*, p. *205*; 22.1.1, p. 461; 22.2.1, p. 483.

M. Henzler
Conductivity and Field Effect of Pure Cleavage-Surfaces of Silicon
Vol. **19**, No. 2, p. 833–845 (1967).
See also *13.3*, p. *197*; 14.3, p. 235.

E. Igras and T. Warmiński
Electron Mirror Observation of Lithium Precipitation at Scratch-Induced Dislocations in Silicon
Vol. **19**, No. 2, p. K67–K69 (1967).
See also *1.3*, p. *14*; 4, p. 40; 10, p. 98.

M. Arndt and A. F. Rudolph
Density Changes due to Dislocations in Silicon Single Crystals
Vol. **19**, No. 2, p. K79–K82 (1967).
See also *10*, p. *98*.

H. Sigmund and K. Berchtold
Electrical and Photovoltaic Properties of PbS-Si Heterodiodes
Vol. **20**, No. 1, p. 255–259 (1967).
See also *14.3.2*, p. *249*; 16, p. 278; 22, p. 444.

E. IGRAS and T. WARMIŃSKI
Electron Mirror Microscopic Investigation of Surface Diffusion of Lithium in Silicon
Vol. **20**, No. 1, p. K5–K8 (1967).
See also 1.3, p. 14; *9*, p. *88*.

H. P. BONZEL
Diffusion of Nickel in Silicon
Vol. **20**, No. 2, p. 493–504 (1967).
See also *9*, p. *88*; 10, p. 99.

R. N. GHOSHTAGORE
On the Mechanism of Substitutional Diffusion in Silicon
Vol. **20**, No. 2, p. K89–K94 (1967).
See also *9*, p. *89*; 10, p. 99.

F. BALIBAR and A. AUTHIER
Experimental and Theoretical Study of the Contrast of Dislocation Images
Vol. **21**, No. 1, p. 413–422 (1967).
See also *10*, p. *100*.

G. B. ABDULLAEV, Z. A. ISKENDER-ZADE, E. A. DZHAFAROVA, and V. E. CHELNOKOV
Study of Electron Lifetime in p-Si
Vol. **21**, No. 1, p. 423–429 (1967).
See also 13.4, p. 206; *14.3*, p. *236*; 14.4.1, p. 258.

L. FIERMANS and J. VENNIK
Microprobe Investigations of Copper Precipitates in Silicon Single Crystals
Vol. **21**, No. 2, p. 627–634 (1967).
See also *4*, p. *42*.

D. G. ARASLI and M. I. ALIEV
Influence of Defects and of the Interaction between Them on Phonon Scattering in Heavily Doped Ge and Si Crystals
Vol. **21**, No. 2, p. 643–648 (1967).
See also 6, p. 61; *8*, p. *82*; 10, p. 100; 22.1.1, p. 462.

L. P. KHIZNICHENKO, P. F. KROMER, D. K. KAIPNAZAROV, E. OTENYAZOV, D. YUSUPOVA, and L. G. ZOTOVA
The Properties of the Low-Temperature Internal Friction Peak in Silicon
Vol. **21**, No. 2, p. 805–810 (1967).
See also 10, p. 100; 11, p. 149; *12*, p. *157*.

P. F. KROMER and L. P. KHIZNICHENKO
On the Low-Temperature Internal Friction in Silicon
Vol. **21**, No. 2, p. 811–818 (1967).
See also 10, p. 100; *12*, p. *157*.

L. FIERMANS and J. VENNIK
The Influence of Carbon on Precipitation of Copper in Silicon Single Crystals
Vol. **22**, No. 2, p. 463–471 (1967).
See also *4*, p. *43*.

L. ELSTNER and W. KAMPRATH
Quenched-in Levels in p-Type Silicon
Vol. **22**, No. 2, p. 541–547 (1967).
See also 11, p. 149; *13.4*, p. *208*; 14.3, p. 237.

M. V. COLEMAN and D. J. D. THOMAS
The Structure of Silicon Oxide Films
Vol. **22**, No. 2, p. 593–602 (1967).
See also *1.4*, p. *19*; 22.6, p. 539.

F. EICHHORN, D. SIPPEL, and K. KLEINSTÜCK
Influence of Oxygen Segregations in Silicon Single Crystals on the Halfwidth of the Double-Crystal Rocking Curve of Thermal Neutrons
Vol. **23**, No. 1, p. 237–240 (1967).
See also *4*, p. *43*.

C. BENOIT À LA GUILLAUME and M. VOOS
Optical Measurement of Compensation in Highly Doped Silicon
Vol. **23**, No. 1, p. 295–299 (1967).
See also *13.4*, p. *209*; 20.3, p. 367.

H. TEICHLER
Calculation of the Peierls Potential in the Diamond Lattice by the Pseudopotential Method
Vol. **23**, No. 1, p. 341–352 (1967).
See also *10*, p. *102*; 22.1.1, p. 463.

V. K. SUBASHIEV and A. A. KUKHARSKII
The Reflection Coefficient of Optically Inhomogeneous Solids
Vol. **23**, No. 2, p. 447–452 (1967).
See also *20.1*, p. *350*.

I. P. IPATOVA, A. A. KLOCHICHIN, and A. V. SUBASHIEV
The Influence of Localized and Resonance Modes of Lattice Vibrations on the Radiative Recombination in Semiconductors
Vol. **23**, No. 2, p. 467–473 (1967).
See also *6*, p. *63*; 13.4, p. 209; *20.3*, p. *367*.

R. GRIGOROVICI, N. CROITORU, and A. DÉVÉNYI
Thermoelectric Power in Amorphous Silicon
Vol. **23**, No. 2, p. 621–625 (1967).
See also 2, p. 23; 13.1, p. 182; 14.3.1, p. 246; *15*, p. *273*.

B. W. BATTERMAN and G. HILDEBRANDT
Observation of X-Ray Pendellösung Fringes in Darwin Reflection
Vol. **23**, No. 2, p. K147–K149 (1967).
See also *5*, p. *52*.

H. LEMKE and G. O. MÜLLER
Transient Behaviour of Space-Charge-Limited Currents in p-Type Silicon
Vol. **24**, No. 1, p. 127–133 (1967).
See also *14.4.1*, p. *260*.

J. PUNZEL
Temperature Dependence of the Surface Structure of Ion-Bombarded Si Single Crystals
Vol. **24**, No. 1, p. K1–K2 (1967).
See also *1.3*, p. *15*; 11, p. 150.

U. K. MISHRA and A. K. WALTON
Infra-Red Faraday Effect in p-Type InAs and Si
Vol. **24**, No. 1, p. K87–K90 (1967).
See also *13.1*, p. *183*; 18, p. 293; 20, p. 340; 22.2.1, p. 485.

M. V. COLEMAN and D. J. D. THOMAS
The Structure of Amorphous Silicon Films
Vol. **24**, No. 2, p. K111–K115 (1967).
See also *1.4*, p. *19*; 2, p. 23.

G. Jungk and H. Menniger
Influence of Energy Band Structure on Photoeffects in Silicon
Vol. **25**, No. 1, p. K59—K63 (1968).
See also *13.1*, p. *184*; 16, p. 281.

J. T. Moore, T. R. Duncan, and D. Kuhlmann-Wilsdorf
Corrected Data for the Shear Stresses of Infinite Tilt Boundaries Composed of $\{111\}$, 1/2 $\langle 1\bar{1}0 \rangle$ Dislocations in Elastilly Anisotropic F.C.C. Metals
Vol. **25**, No. 2, p. K105—K106 (1968).
See also *10*, p. *105*; *10.1*, p. *129*; 21, p. 385; 21.1, p. 403; 21.5, p. 429; 21.6, p. 435; 22.1, p. 457; 22.1.1, p. 464.

G. Krenzke, H. Lemke, and G. O. Müller
Properties of Gold Contacts on Clean n-Silicon Surfaces
Vol. **25**, No. 2, p. K131—K134 (1968).
See also 13.3, p. 198; *14.4.1*, p. *261*.

S. Dash and M. L. Joshi
Climb of Extrinsic Faults in Silicion
Vol. **26**, No. 1, p. 159—166 (1968).
See also *10*, p. *105*.

M. Hart and A. D. Milne
Direct Observations of Plane Wave and Spherical Wave Pendellösung Fringes
Vol. **26**, No. 1, p. 185—189 (1968).
See also 3, p. 27; *5*, p. *53*.

J. M. Pankratz and M. L. Rudee
An Application of Stereo Electron Microscopy to Anomalously Wide Images in Neutron Irradiated Silicon
Vol. **26**, No. 2, p. K97—K98 (1968).
See also 4, p. 46; *10*, p. *106*.

E. Schmidt
Optical Properties of GeSi Alloys in the Energy Region from 1 to 13 eV
Vol. **27**, No. 1, p. 57—67 (1968).
See also 13.1, p. 185; *20.1*, p. *353*; 22.1.1, p. 464.

E. Igras and T. Warmiński
On Electron Mirror Microscopic Observations of the Details of p–n Junction Regions in Si
Vol. **27**, No. 1, p. 69—76 (1968).
See also *1.3*, p. *15*; *14.3.2*, p. *250*.

A. Authier
Contrast of a Stacking Fault on X-Ray Topographs
Vol. **27**, No. 1, p. 77—93 (1968).
See also *10*, p. *106*.

H. Stramska, Z. Bachan, P. Byszewski, and J. Kołodziejczak
Interband Faraday Rotation and Ellipticity Observed at the Absorption Edge in Silicon
Vol. **27**, No. 1, p. K25—K28 (1968).
See also 13.1, p. 186; *18*, p. *294*; *20*, p. *341*.

Č. Jech
Depth Distribution Measurements Using Bombardment-Enhanced Solubility
Vol. **27**, No. 2, p. 573—577 (1968).
See also 1.4, p. 20; *11*, p. *152*; 22, p. 452.

J. Auth
On the Reverse-Biased Capacitance of Step p⁺-n Junctions with Traps
Vol. **27**, No. 2, p. 653—656 (1968).
See also 13.4, p. 214; *14.3.2*, p. *250*.

M. Owner-Petersen and M. R. Samuelsen
Cyclotron Resonance of Holes in Si
Vol. **28**, No. 1, p. 211—222 (1968).
See also 13.1, p. 187; *19*, p. *334*.

I. Alstrup and K. Johansen
Empirical Third Neighbour LCAO Energy Bands of Silicon
Vol. **28**, No. 2, p. 555—564 (1968).
See also *13.1*, p. *187*.

E. Zielińska-Rohozińska
Observations of the Effects of X-Ray Polarization on Pendellösung Fringes with the Three-Crystal X-Ray Spectrometer
Vol. **28**, No. 2, p. K139—K142 (1968).
See also *4*, p. *47*; 5, p. 54.

W. Kress, H. Borik, and R. K. Wehner
Infrared Lattice Absorption of Silicon and Germanium
Vol. **29**, No. 1, p. 133—143 (1968).
See also *6*, p. *67*; 20.1, p. 356; 22.1.1, p. 466.

E. S. Meieran and I. A. Blech
Contrast Asymmetries in Lang Topographs of Crystals Strained by Thin Films
Vol. **29**, No. 2, p. 653—667 1968).
See also *4*, p. *47*; *10*, p. *111*; 22.2.1, p. 486.

E. P. Berdnikov, V. A. Drozdov, and L. A. Mozgovaya
On the Relation between Microhardness and Point Defects of Crystals
Vol. **29**, No. 2, p. K145—K147 (1968).
See also *10*, p. *111*; 12, p. 158; 15, p. 274.

V. V. Tarassov and A. F. Demidenko
Heat Capacity and Quasi-Chain Dynamics of Diamond-Like Structures
Vol. **30**, No. 1, p. 147—155 (1968).
See also *6*, p. *69*; *8*, p. *86*; 13.1, p. 189; 22.1.1, p. 466; 22.2.1, p. 486; 22.2.2, p. 488; 22.2.3, p. 493.

M. A. Kumakhov
Formation of Radiation Defects by Channeling Particles
Vol. **30**, No. 1, p. 379—385 (1968).
See also 3.2, p. 36; *11*, p. *155*; 21.1, p. 408.

H. Wienhold
On Relaxation Effects of Epitaxial Si-MOS Structures
Vol. **30**, No. 1, p. K17—K21 (1968).
See also 13.3, p. 199; *14.3.2*, p. *251*.

M. Lefeld-Sosnowska and C. Malgrange
Observation of Oscillations in Rocking Curves of the Laue Reflected and Refracted Beams from Thin Si Single Crystals
Vol. **30**, No. 1, p. K23—K25 (1968).
See also *5*. p. *55*.

D. Strauch

Shell Model Treatment of the Vibrations of Crystals Containing Substitutional Defects. (II): On the Theory of Defect-Induced One-Phonon Absorption
Vol. **30**, No. 2, p. 495–505 (1968).
See also *6*, p. *70*; 10, p. 112; *20.1*, p. *359*; 22.1, p. 457; 22.1.1, p. 466.

C. D. Cann and S. S. Sheinin

The Dependence of Extinction Distance on Deviation from the Bragg Condition in Silicon
Vol. **30**, No. 2, p. 791–796 (1968).
See also *5*, p. *55*.

A. A. Gutkin and T. Janicki

The Effect of Electric Field in a Space Charge Layer on the Photoresponse Edge of Si p-n Junctions
Vol. **30**, No. 2, p. 827–832 (1968).
See also *14.3.2*, p. *251*; *14.4.1*, p. *265*; 16, p. 285.

E. E. Godik

The Variation of the Photoelectron Mobility with Photon Energy in Germanium and Silicon
Vol. **30**, No. 2, p. K127–K129 (1968).
See also *16*, p. *285*; 22.1.1, p. 467.

22.1.3 Group VI Elements

Review Articles

Y. P. Varshni
Band-to-Band Radiative Recombination in Groups IV, VI, and III-V Semiconductors (I)
Vol. **19**, No. 2, p. 459–514 (1967).
See also 13.1, p. 174; 13.2, p. 191; 13.4, p. 200; 20.1, p. 344; 20.2, p. 360; *20.3*, p. *363*; 22.1, p. 456; 22.1.1, p. 458; 22.1.2, p. 468; 22.2.1, p. 482; 22.2.2, p. 487; 22.2.3, p. 489.

Y. P. Varshni
Band-to-Band Radiative Recombination in Groups IV, VI, and III-V Semiconductors (II)
Vol. **20**, No. 1, p. 9–36 (1967).
See also 13.1, p. 174; 13.2, p. 191; 13.4, p. 200; 20.1, p. 344; 20.2, p. 360; *20.3*, p. *363*; 22.1, p. 456; 22.1.1, p. 458; 22.1.2, p. 468; 22.2.1, p. 482; 22.2.2, p. 487; 22.2.3, p. 489.

G. Harbeke
On the Band Structure of Anisotropic Crystals
Vol. **27**, No. 1, p. 9–44 (1968).
See also *13.1*, p. *174*; 13.2, p. 191; 14.3, p. 231; 20.1, p. 344; 22, p. 439; 22.1, p. 456; 22.1.1, p. 458; 22.1.2, p. 468; 22.4.1, p. 497; 22.4.2, p. 506; 22.4.3, p. 509; 22.5, p. 512; 22.7, p. 545; 22.8, p. 547.

Original Papers and Short Notes

G. B. Abdullaev, S. I. Mekhtieva, G. M. Aliev, D. Sh. Abdinov, and T. G. Kerimova
Determination of the Oxygen Content in Selenium
Vol. **16**, No. 1, p. K31–K34 (1966).
See also 2, p. 22; 10, p. 94; *20.1*, p. *344*.

J. Treusch and R. Sandrock
Energy Band Structures of Selenium and Tellurium (Kohn-Rostoker Method)
Vol. **16**, No. 2, p. 487–497 (1966).
See also 5, p. 49; *13.1*, p. *175*.

G. B. Abdullaev, N. I. Ibragimov, Sh. V. Mamedov, and T. Ch. Zhuvarly
ESR Study of the Behaviour of Oxygen in Selenium
Vol. **16**, No. 2, p. K113–K116 (1966).
See also 10, p. 94; *19*, p. *326*.

J. Stuke and G. Weiser
Franz-Keldysh Effect in Trigonal and Amorphous Selenium
Vol. **17**, No. 1, p. 343–352 (1966).
See also 13.1, p. 175; *14.4.1*, p. *256*; *20*, p. *337*.

J. Kodeš and K. Šobra
Pre-Breakdown Region of Selenium p-n Junctions
Vol. **17**, No. 1, p. K19–K20 (1966).
See also *14.3.2*, p. *248*.

H. P. D. Lanyon and E. F. Hockings
The Selenium-Tellurium System
Vol. **17**, No. 2, p. K185–K186 (1966).
See also *3.1*, p. *30*.

G. B. Abdullaev, F. B. Gadiev, G. M. Aliev, S. I. Mekhtieva, and D. Sh. Abdinov
Interaction of Sodium Admixtures and Oxygen in Selenium
Vol. **18**, No. 2, p. K153—K156 (1966).
See also *10*, p. *97*; 14.3, p. 234.

G. Döhler
On the Field Emission of Minority Carriers in Photoconductors
Vol. **19**, No. 2, p. 555—564 (1967).
See also 13.4, p. 205; *14.4.1*, p. *257*; 16, p. 278; 22.4.1, p. 499.

A. Axmann and W. Gissler
Scattering of Slow Neutrons on Polycrystalline Selenium and Tellurium
Vol. **19**, No. 2, p. 721—727 (1967).
See also *6*, p. *59*.

B. Ya. Pines, I. P. Grebennik, R. I. Bazyura, and G. V. Zozulya
Surface and Volume Diffusion in Thin Films of the System Ag-Se
Vol. **20**, No. 1, p. 213—223 (1967).
See also *9*, p. *88*; 21.6, p. 432; 22, p. 443.

S. U. Dzhalilov and K. I. Rzaev
On the Phenomenon of Selenium Vitrification
Vol. **20**, No. 1, p. 261—266 (1967).
See also *2*, p. *22*.

W. Henrion
On an Exciton Structure in the Reflection Spectrum of Trigonal Selenium Single Crystals
Vol. **20**, No. 2, p. K145—K149 (1967).
See also 13.1, p. 178; *13.2*, p. *192*; 20.1, p. 347.

L. Müller and Y. Mihai
Absorption of Amorphous Se Thin Layers Doped with Bi and Zn
Vol. **21**, No. 1, p. K65—K68 (1967).
See also 2, p. 23; 13.4, p. 207; *20.1*, p. *348*.

W. Henrion
Reflectivity Measurements on Trigonal Selenium Single Crystals in the Spectral Region between 1.6 and 6.0 eV
Vol. **22**, No. 1, p. K33—K37 (1967).
See also 13.1, p. 179; 13.2, p. 192; *20.1*, p. *349*.

E. Mohler, J. Stuke, and G. Zimmerer
Some New Features of the Reflectance Spectrum of Trigonal Selenium Single Crystals
Vol. **22**, No. 1, p. K49—K53 (1967).
See also 13.1, p. 179; *20.1*, p. *349*.

F. R. Kessler and E. Sutter
Absorption and Photoconductivity of a Series of Selenium-Tellurium Mixed Crystals
Vol. **23**, No. 1, p. K25—K33 (1967).
See also 13.1, p. 181; *16*, p. *280*; *20.1*, p. *350*; *350*.

H. Lange and W. Henrion
Thermoreflectance of CdS and Se Single Crystals
Vol. **23**, No. 1, p. K67—K70 (1967).
See also 13.1, p. 181; *20.1*, p. *350*; 22.4.1, p. 501.

A. R. Adams, F. Baumann, and J. Stuke
Thermal Conductivity of Selenium and Tellurium Single Crystals and Phonon Drag of Tellurium
Vol. **23**, No. 1, p. K99—K104 (1967).
See also *8*, p. *83*; 13, p. 167; 15, p. 273.

M. Picard and M. Hulin
A Pseudopotential Approach to the Electron Band Structure of Tellurium
Vol. **23**, No. 2, p. 563–570 (1967).
See also *13.1*, p. *182*.

W. Hörstel and G. Kretzschmar
Lattice Defects Forming Acceptors in Tellurium
Vol. **23**, No. 2, p. 639–647 (1967).
See also *10*, p. *102*; 14.3, p. 238.

R. Geick, U. Schröder, and J. Stuke
Lattice Vibrational Properties of Trigonal Selenium
Vol. **24**, No. 1, p. 99–108 (1967).
See also *6*, p. *63*; 20.1, p. 351.

H. Mell and J. Stuke
Magnetoconductivity of Trigonal Selenium Single Crystals
Vol. **24**, No. 1, p. 183–193 (1967).
See also 13.1, p. 182; *14.3*, p. *238*.

K. H. Herrmann and G. Oelgart
Piezoresistance in Tellurium. (I): Experiments with Uniaxial Stress Parallel to the *c*-Axis
Vol. **24**, No. 1, p. 289–296 (1967).
See also 13.1, p. 182; *14.3*, p. *238*.

G. Döhler
On Slow High Field Domains in Homogeneous Photoconductors
Vol. **24**, No. 1, p. 331–340 (1967).
See also *14.4.1*, p. *260*; 16, p. 280.

P. Grosse and M. Selders
The p-Band of Tellurium at Very Low Temperatures
Vol. **24**, No. 1, p. K33–K36 (1967).
See also *13.1*, p. *183*; 20.1, p. 351.

W. Schüz
Forbidden Reflexes in Electron Diffraction from Tellurium Films Evaporated in Vacuum
Vol. **25**, No. 1, p. 253–261 (1968).
See also 1.4, p. 20; *5*, p. *53*.

J. Viščakas, P. Mačkus, and A. Smilga
Space-Charge-Limited Currents and High Electric Field Effect in Vitreous Selenium Films
Vol. **25**, No. 1, p. 331–335 (1968).
See also 2, p. 23; 13.4, p. 211; 14.3.1, p. 246; *14.4.1*, p. *261*.

R. Herrmann and Karin Herrmann
Cyclotron Resonance in Tellurium
Vol. **25**, No. 2, p. 655–659 (1968).
See also *13.1*, p. *184*; 19, p. 332.

P. Grosse and K. Winzer
Magnetoabsorption Measurements on Tellurium
Vol. **26**, No. 1, p. 139–150 (1968).
See also *13.1*, p. *184*; 18, p. 293; *20.1*, p. *352*.

B. Kramer and P. Thomas
Relativistic KKR-Calculations on the Band Structure of Selenium and Tellurium
Vol. **26**, No. 1, p. 151–158 (1968).
See also *13.1*, p. *184*.

R. H. Tredgold, R. H. Williams, G. Kavalyauskene, and R. Keezer
Photoemission Studies on Amorphous Layers and Single Crystals of Selenium
Vol. **26**, No. 1, p. K5–K8 (1968).
See also 2, p. 24; *17*, p. *288*.

L. Rothkirch
Interpretation of the Anisotropy of the Galvanomagnetic Effects in Tellurium by Anisotropic Scattering
Vol. **26**, No. 2, p. 709–715 (1968).
See also 13.1, p. 185; *14.3*, p. *241*.

Kh. M. Khalilov, B. B. Kuliev, and F. A. Gambarov
Effect of Phosphorus Admixtures on Viscoelastic Properties of Selenium
Vol. **26**, No. 2, p. K113–K117 (1968).
See also *2*, p. *24*; 10, p. 106; *12*, p. *158*.

H. B. Wild and C. A. Hogarth
Slow Temporal Variations of Electron Mobility in Orthorhombic Sulphur
Vol. **26**, No. 2, p. K135–K137 (1968).
See also *14.3*, p. *241*.

R. Link and H. Lutsch
Hall Effect Measurements in Pulse-Heated Tellurium Single Crystals
Vol. **26**, No. 2, p. K159–K162 (1968).
See also 13.4, p. 212; *14.3*, p. *241*.

W. Fuhs and J. Stuke
Hopping Recombination in Trigonal Selenium Single Crystals
Vol. **27**, No. 1, p. 171–184 (1968).
See also 13.4, p. 213; *16*, p. *282*.

G. G. Roberts
Electron Injection into a p-Type Semiconductor
Vol. **27**, No. 1, p. 209–218 (1968).
See also *14.4.1*, p. *262*.

P. Grosse and B. Krahl-Urban
The Contribution of Free Carriers to the Complex Dielectric Constant of Tellurium at 9 GHz
Vol. **27**, No. 2, p. K149–K152 (1968).
See also *14.3*, p. *242*.

T. O. Tuomi
Effect of Reactor Irradiation on the Conductivity and the Photoconductivity of Hexagonal Selenium Single Crystals
Vol. **28**, No. 1, p. 329–334 (1968).
See also *11*, p. *153*; 14.3, p. 242; 16, p. 282.

J.-U. Arnold and P. Grosse
The Temperature Dependence of the Atom Coordinates in the Tellurium Lattice
Vol. **28**, No. 1, p. K93–K96 (1968).
See also *3*, p. *27*; 5, p. 54; 8, p. 85.

W. Zimmermann and F. Stöckmann
Measurements of Hall Effect on Tellurium in Magnetic Fields up to 150 kG
Vol. **28**, No. 2, p. K101–K104 (1968).
See also *14.3*, p. *243*.

K. H. HERRMANN and J. PUHLMANN
Piezoresistance of Tellurium. (II): Experiments with Pressures Lowering Symmetry
Vol. **29**, No. 1, p. 193–202 (1968).
See also 13.1, p. 188; *14.3*, p. *243*.

B. GARBEN and H. SELIGER
Faraday Effect on Amorphous Selenium
Vol. **29**, No. 1, p. K27–K29 (1968).
See also 13, p. 172; *18*, p. *294*; *20*, p. *342*.

F. B. GADZHIEV, CH. M. ASKEROV, and G. M. ALIEV
Effect of Heat Treatment and Natrium Admixtures on the Temperature Dependence of Electric Conductivity of Amorphous Selenium
Vol. **29**, No. 1, p. K47–K50 (1968).
See also 2, p. 24; *14.3*, p. *243*.

R. A. BURLEY
An Infrared Spectrophotometric Study of Vitreous Selenium Doped with Selenium Dioxide
Vol. **29**, No. 2, p. 551–558 (1968).
See also 2, p. 24; *10*, p. *110*; 20.1, p. 357.

M. YA. BAKIROV and G. I. IBAEEV
Charge Accumulation Effect in Selenium Crystals under Applied Electric Field
Vol. **30**, No. 1, p. K73–K75 (1968).
See also *14.4.1*, p. *264*.

V. V. NEMOSHKALENKO, M. A. MINDLINA, and B. P. MAMKO
X-Ray Spectra and Electronic Structure of Elements at the End of the First Long Period from ^{29}Cu to ^{34}Se
Vol. **30**, No. 2, p. 703–712 (1968).
See also 13.1, p. 189; *20*, p. *343*; 21.1, p. 409; 22.1.1, p. 467.

22.2 III–V Compounds

Original Papers and Short Notes

B. M. Askerov and F. M. Gashimzade
Quantum Theory of the Electrical Conductivity of Semiconductors with a Non-Standard Energy Band
Vol. **18**, No. 2, p. 667–675 (1966).
See also 6, p. 58; 13.1, p. 176; *14.3*, p. *233*; 22.2.3, p. 490.

L. Roskovcová, J. Pastrňák, and R. Babušková
The Dispersion of the Refractive Index and the Birefringence of AlN
Vol. **20**, No. 1, p. K29–K32 (1967).
See also *20.1*, p. *347*.

I. Licea
Electron Temperature in Polar Semiconductors
Vol. **25**, No. 1, p. 461–471 (1968).
See also *13*, p. *168*; *14.3*, p. *240*; 22.4, p. 495.

V. K. Bashenov, V. A. Presnov, and S. P. Fedotov
Some Remarks on the Antishielding Effect in III-V Compounds
Vol. **25**, No. 1, p. K51–K54 (1968).
See also *19*, p. *332*.

G. I. Guseva and P. S. Zyryanov
Quantum Theory of Thermogalvanomagnetic Phenomena (TGMP) in Conductors with Equal Concentration of Electrons and Holes
Vol. **25**, No. 2, p. 775–785 (1968).
See also 14.1, p. 224; 14.3, p. 240; *15*, p. *273*; 21.7, p. 438.

J. Pastrňák and L. Roskovcová
Optical Absorption Edge of AlN Single Crystals
Vol. **26**, No. 2, p. 591–597 (1968).
See also 13.1, p. 185; *20.1*, p. *353*.

22.2.1 Arsenides

Review Articles

J. R. Dale
Alloyed Semiconductor Heterojunctions
Vol. **16**, No. 2, p. 351–387 (1966).
See also 3.1, p. 29; 5, p. 49; 13.4, 200; *14.3.2*, p. *248*; 16, p. 276; 22.1.1, p. 458; 22.1.2, p. 468; 22.2.2, p. 487; 22.2.3, p. 489.

Y. P. Varshni
Band-to-Band Radiative Recombination in Groups IV, VI, and III-V Semiconductors (I)
Vol. **19**, No. 2, p. 459–514 (1967).
See also 13.1, p. 174; 13.2, p. 191; 13.4, p. 200; 20.1, p. 344; 20.2, p. 360; *20.3*, p. *363*; 22.1, p. 456; 22.1.1, p. 458; 22.1.2, p. 468; 22.1.3, p. 476; 22.2.2, p. 487; 22.2.3, p. 489.

Y. P. Varshni
Band-to-Band Radiative Recombination in Groups IV, VI, and III-V Semiconductors (II)
Vol. **20**, No. 1, p. 9–36 (1967).
See also 13.1, p. 174; 13.2, p. 191; 13.4, p. 200; 20.1, p. 344; 20.2, p. 360; *20.3*, p. *363*; 22.1, p. 456; 22.1.1, p. 458; 22.1.2, p. 468; 22.1.3, p. 476; 22.2.2, p. 487; 22.2.3, p. 489.

D. P. Morgan
Helicon Waves in Solids
Vol. **24**, No. 1, p. 9–36 (1967).
See also 6, p. 56; *13*, p. *162*; 14.1, p. 220; 14.3, p. 231; 19, p. 326; *20*, p. *337*; 21, p. 373; 21.1, p. 390; 21.2, p. 424; 21.6, p. 430; 22.1.1, p. 458; 22.2.3, p. 489.

M. H. Pilkuhn
The Injection Laser
Vol. **25**, No. 1, p. 9–62 (1968).
See also 13.4, p. 200; 14.3.2, p. 248; *14.4.1*, p. *255*; *20.2*, p. *360*; 20.3, p. 363; 22.2.2, p. 487; 22.2.3, p. 489.

M. Balkanski and E. Amzallag
Band Parameters Determination from Faraday Rotation Measurements
Vol. **30**, No. 2, p. 407–439 (1968).
See also *13.1*, p. *174*; 18, p. 291; 20, p. 337; 22.1.1, p. 458; 22.1.2, p. 468; 22.2.2, p. 387; 22.2.3, p. 489; 22.4, p. 495; 22.4.1, p. 497; 22.4.2, p. 506; 22.4.3, p. 509.

Original Papers and Short Notes

J. Kołodziejczak and S. Żukotyński
The Effect of Higher Bands on the Effective Masses in InSb and InAs
Vol. **16**, No. 1, p. K55–K58 (1966).
See also *13.1*, p. *174*; 22.2.3, p. 489.

D. V. Eddolls
Electrical Properties of n-Type Epitaxial Gallium Arsenide
Vol. **17**, No. 1, p. 67–76 (1966).
See also 13.4, p. 202; *14.3*, p. *232*.

A. F. Kravchenko and V. S. Sardaryan
On the Structure of the Bottom of the Conduction Band in GaAs
Vol. **17**, No. 2, p. 479–488 (1966).
See also *13.1*, p. *175*; 14.3, p. 232.

T. L. Tansley
Forward Bias Current-Voltage Characteristics for a Heterojunction in which Tunnelling Dominates
Vol. **18**, No. 1, p. 105–112 (1966).
See also 13, p. 164; *14.3.2*, p. *248*.

H. Frank
Contribution to the Measurement of the Resistance of nn^+ and pp^+ Epitaxial Films by the Breakdown Method
Vol. **18**, No. 1, p. 401–414 (1966).
See also *14.3.1*, p. *245*; 14.4.1, p. 257; 22.1.1, p. 460; 22.1.2, p. 470.

B. Tuck
Photoluminescence of $GaAs_{0.7}P_{0.3}$
Vol. **18**, No. 2, p. 541–545 (1966).
See also 13.4, p. 204; *20.3*, p. *364*; 22.2.2, p. 488; 22.8, p. 548.

A. Glodeanu
Helium-Like Impurities in Semiconductors
Vol. **19**, No. 1, p. K43–K46 (1967).
See also *13.4*, p. *205*; 22.1.1, p. 461; 22.1.2, p. 470.

S. Raab, H. Bachert, and A. Keiper
Spectral Investigation of Emission Inhomogeneities in GaAs Injection Lasers
Vol. **19**, No. 1, p. K59–K62 (1967).
See also *20.2*, p. *360*.

A. R. Beattie and G. Smith
Recombination in Semiconductors by a Light Hole Auger Transition
Vol. **19**, No. 2, p. 577–586 (1967).
See also *13.1*, p. *177*; *16*, p. *278*; 22.2.3, p. 490.

B. I. Boltaks and T. D. Dzhafarov
The Effect of Applied Electric Field on Diffusion of Impurities in Gallium Arsenide
Vol. **19**, No. 2, p. 705–714 (1967).
See also *9*, p. *88*.

J. Auleytner, K. Godwod, J. Litwin, and Z. Wołoszyn
Application of X-Ray Methods to the Study of the Real Structure of Monocrystalline Epitaxial Layers
Vol. **20**, No. 2, p. K77–K79 (1967).
See also *1.4*, p. *18*; 22.1.1, p. 461.

Kh. I. Amirkhanov, R. I. Bashirov, and M. M. Gadzhialiev
Quantum Oscillations of the Nernst-Ettingshausen Effect in Indium Arsenide
Vol. **20**, No. 2, p. K119–K122 (1967).
See also *15*, p. *272*.

C. Hervouet
Magnetic Field Dependence of Current Oscillations in Piezoelectric Semiconductors
Vol. **21**, No. 1, p. 117–125 (1967).
See also *14.4.1*, p. *258*.

V. K. Bashenov, S. P. Fedotov, and V. A. Presnov
On Deep Acceptors in Iron-Doped Gallium Arsenide
Vol. **21**, No. 1, p. K91–K93 (1967).
See also *13.4*, p. *207*; 19, p. 329.

J.-P. Ropé
Experimental Study of the Dynamics of High-Field Domains in Electron–Phonon Coupling
Vol. **21**, No. 2, p. 517–524 (1967).
See also 7, p. 76, *14.4.1*, p. *258*.

A. Bonnot
Relaxation of High-Field Domains and Light Emission Associated with Current Oscillations in GaAs
Vol. **21**, No. 2, p. 525–531 (1967).
See also 7, p. 76; *14.4.1*, p. *258*; *20.3*, p. *366*.

J. E. Knappett and S. J. T. Owen
Observation of the Growth Process of Gallium Arsenide on Tungsten
Vol. **21**, No. 2, p. K99–K101 (1967).
See also 1.4, p. 18; *3.1*, p. *31*; 14.3.1, p. 245.

V. L. Bonch-Bruevich
On the Theory of Electrical Domains in Hot Electron Semiconductors
Vol. **22**, No. 1, p. 267–278 (1967).
See also 13, p. 166; 14.3, p. 237; *14.4.1*, p. *259*; 22, p. 446; 22.1.1, p. 462.

F. E. Roberts
Evaporation of As from GaAs
Vol **22**, No. 2, p. 373–376 (1967).
See also *1*, p. *7*.

T. L. Tansley
Spectral Response of p–n Heterojunctions
Vol. **23**, No. 1, p. 241–252 (1967).
See also *14.3.2*, p. *249*; 14.4.1, p. 259; 16, p. 280; 20.1, p. 350; 22.8, p. 552.

P. Byszewski, B. Kalińska, and J. Kołodziejczak
Interband Faraday Rotation and Ellipticity in Germanium and Gallium Arsenide
Vol. **23**, No. 1, p. K53–K56 (1967).
See also 13, p. 167; *18*, p. *293*; *20*, p. *340*; 22.1.1, p. 463.

P. G. Eliseev, A. I. Krasilnikov, M. A. Manko, and I. Z. Pinsker
Band-Filling Model for Injection Luminescence at Higher Temperatures
Vol. **23**, No. 2, p. 587–593 (1967).
See also 13.4, p. 210; *20.3*, p. *367*.

P. Guetin
Note on the Density-of-State Effective Mass in GaAs
Vol. **24**, No. 1, p. K29–K31 (1967).
See also *13.1*, p. *183*.

R. K. Purohit
GaP–GaAs n–p Heterojunctions
Vol. **24**, No. 1, p. K57–K60 (1967).
See also *14.3.2*, p. *250*; 22.2.2, p. 488.

D. Siebert, E. Ady, and F. Matossi
Magnetoresistance of High-Ohmic Semiconductors by a Capacitive Method
Vol. **24**, No. 1, p. K65–K67 (1967).
See also *14.3*, p. *239*.

U. K. Mishra and A. K. Walton
Infra-Red Faraday Effect in p-Type InAs and Si
Vol. **24**, No. 1, p. K87—K90 (1967).
See also *13.1*, p. *183*; 18, p. 293; 20, p. 340; 22.1.2, p. 472.

T. L. Tansley
Forward Current Injection Modulation of Photocurrent in p-n Heterojunctions
Vol. **24**, No. 2, p. 615—622 (1967).
See also *14.3.2*, p. *250*; 14.4.1, p. 261; 16, p. 281; 22.2.2, p. 488.

G. Arlt and P. Quadflieg
Piezoelectricity in III-V Compounds with a Phenomenological Analysis of the Piezoelectric Effect
Vol. **25**, No. 1, p. 323—330 (1968).
See also *14.3*, p. *240*; 22.2.3, p. 492.

R. Gauthier, D. Bois, P. Pinard, and F. Davoine
Iron and Electron Production by Interaction of a Coherent Light Beam with Substance
Vol. **25**, No. 2, p. 691—696 (1968).
See also *17*, p. *288*; 21.1, p. 403; 21.1.1, p. 420; 22.1.1, p. 464; 22.4.1, p. 502; 22.5.2, p. 525; 22.6, p. 540.

A. A. Grinberg and A. A. Kastalskii
The Ultrasound Generation under the Gunn Effect in a Piezoelectric Material
Vol. **26**, No. 1, p. 219—224 (1968).
See also *7*, p. *77*; *14.4.1*, p. *262*.

W. Thielemann
Effect of Doping on the Optical Absorption of Circularly Polarized Waves in n-Type GaAs in a Magnetic Field
Vol. **26**, No. 1, p. K29—K31 (1968)
See also 13.1, p. 184; 18, p. 294; *20.1*, p. *352*.

D. G. Coates
Improvements in the Orientation Effect Observed by Scanning Electron Microscopy
Vol. **27**, No. 1, p. K11—K13 (1968).
See also *4*, p. *46*; 5, p. 53.

M. Zvára
Interband Faraday Rotation in Gallium Arsenide
Vol. **27**, No. 2, p. K157—K160 (1968).
See also 13, p. 170; 18, p. 294; *20*, p. *341*.

Yu. E. Perlin and Sh. N. Gifeisman
The Theory of Shallow F-Centres
Vol. **28**, No. 1, p. 45—60 (1968).
See also 6, p. 66; *10.2*, p. *143*; 13, p. 171; 14.4, p. 254; 18, p. 294; 22.2.2, p. 488; 22.2.3, p. 493; 22.4, p. 495; 22.4.1, p. 503; 22.4.2, p. 508; 22.4.3, p. 511; 22.5.1, p. 517; 22.5.2, p. 527.

J. Hatz
Some Effects of Material Inhomogeneities on the Near-Field Pattern of GaAs Diode Lasers
Vol. **28**, No. 1, p. 233—245 (1968).
See also 10, p. 109; 14.3.2, p. 251; *20.2*, p. *362*.

F. Huth
Hall Effect and Mobility in n-Type GaAs
Vol. **29**, No. 1, p. K35—K37 (1968).
See also 13.4, p. 215; *14.3*, p. *243*.

E. S. Meieran and I. A. Blech
Contrast Asymmetries in Lang Topographs of Crystals Strained by Thin Films
Vol. **29**, No. 2, p. 653–667 (1968).
See also *4*, p. *47*; *10*, p. *111*; 22.1.2, p. 474.

O. V. Bogdankevich, N. A. Borisov, I. V. Krukova, and B. M. Lavrushin
Temperature Dependence of Laser Threshold Current Density and Emission Spectra in Electron-Beam Pumped Gallium Arsenide Lasers
Vol. **29**, No. 2, p. 715–723 (1968).
See also 13.4, p. 215; *20.2*, p. *362*.

B. Tuck
Effect of Heat Treatments on GaAs Luminescence
Vol. **29**, No. 2, p. 793–798 (1968).
See also 13.4, p. 215; *20.3*, p. *370*.

Yu. A. Goldberg, M. E. Levinshtein, and D. N. Nasledov
Magnetic Field Influence on the Low-Frequency Oscillations of Gunn Diodes
Vol. **29**, No. 2, p. K103–K105 (1968).
See also *14.3.1*, p. *251*; *14.4.1*, p. *264*.

E. Fortin
Photocurrent Amplification by a Magnetic Field
Vol. **29**, No. 2, p. K153–K155 (1968).
See also *16*, p. *284*.

H. Bachert and S. Raab
The Influence of External Optical Coupling on the Threshold Current Density of GaAs Injection Lasers
Vol. **29**, No. 2, p. K175–K178 (1968).
See also *20.2*, p. *362*.

V. V. Tarassov and A. F. Demidenko
Heat Capacity and Quasi-Chain Dynamics of Diamond-Like Structures
Vol. **30**, No. 1, p. 147–155 (1968).
See also *6*, p. *69*; *8*, p. *86*; 13.1, p. 189; 22.1.1, p. 466; 22.1.2, p. 474; 22.2.2, p. 488; 22.2.3, p. 493.

D. Boccon-Gibod
Modulation of Near Infrared Radiation by Acoustoelectric Domains in n-Type GaAs at Room Temperature
Vol. **30**, No. 2, p. 525–531 (1968).
See also 7, p. 78; *14.4.1*, p. *264*; *20.1*, p. *359*.

22.2.2 Phosphides

Review Articles

J. R. Dale
Alloyed Semiconductor Heterojunctions
Vol. **16**, No. 2, p. 351–387 (1966).
See also 3.1, p. 29; 5, p. 49; 13.4, p. 200; *14.3.2*, p. *248*; 16, p. 276; 22.1.1, p. 458; 22.1.2, p. 468; 22.2.1, p. 482; 22.2.3, p. 489.

Y. P. Varshni
Band-to-Band Radiative Recombination in Groups IV, VI and III–V Semiconductors (I)
Vol. **19**, No. 2, p. 459–514 (1967).
See also 13.1, p. 174; 13.2, p. 191; 13.4, p. 200; 20.1, p. 344; 20.2, p. 360; *20.3*, p. *363*; 22.1, p. 456; 22.1.1, p. 458; 22.1.2, p. 468; 22.1.3, p. 476; 22.2.1, p. 482; 22.2.3, p. 489.

Y. P. Varshni
Band-to-Band Radiative Recombination in Groups IV, VI, and III–V Semiconductors (II)
Vol. **20**, No. 1, p. 9–36 (1967).
See also 13.1, p. 174; 13.2, p. 191; 13.4, p. 200; 20.1, p. 344; 20.2, p. 360; *20.3*, p. *363*; 22.1, p. 456; 22.1.1, p. 458; 22.1.2, p. 468; 22.1.3, p. 476; 22.2.1, p. 482; 22.2.3, p. 489.

M. H. Pilkuhn
The Injection Laser
Vol. **25**, No. 1, p. 9–62 (1968).
See also 13.4, p. 200; 14.3.2, p. 248; *14.4.1*, p. *255*; *20.2*, p. *360*; 20.3, p. 363; 22.2.1, p. 482; 22.2.3, p. 489.

F. Williams
Donor–Acceptor Pairs in Semiconductors
Vol. **25**, No. 2, p. 493–512 (1968).
See also 10, p. 93; *13.4*, p. *200*; 20.3, p. 363; 22.1, p. 456; 22.1.1, p. 458; 22.1.2, p. 468; 22.4.1, p. 497.

M. Balkanski and E. Amzallag
Band Parameters Determination from Faraday Rotation Measurements
Vol. **30**, No. 2, p. 407–439 (1968).
See also *13.1*, p. *174*; 18, p. 291; 20, p. 337; 22.1.1, p. 458; 22.1.2, p. 468; 22.2.1, p. 482; 22.2.3, p. 489; 22.4, p. 495; 22.4.1, p. 497; 22.4.2, p. 506; 22.4.3, p. 509.

Original Papers and Short Notes

S. A. Aliev, F. P. Kesamanly, T. S. Lagunova, and D. N. Nasledov
Hall Eeffect and Magnetoresistance of n-InP Crystals at Low Temperatures
Vol. **17**, No. 1, p. 105–108 (1966).
See also 13.4, p. 202; *14.3*, p. *232*.

J. Gyulai, V. K. Subashiev, and G. A. Chalikyan
On the Mechanism of Photoconductivity in High-Resistivity GaP
Vol. **17**, No. 1, p. K49–K51 (1966).
See also 13.1, p. 175; *16*, p. *277*.

S. A. Moskalenko and P. I. Khadshi
Infrared Absorption by Excitons due to Photoionization and Intraband Lattice Scattering
Vol. **18**, No. 1, p. 379–390 (1966).
See also 6, p. 57; *13.2*, p. *191*; 20.1, p. 346; 22.1.1, p. 460; 22.2.3, p. 490; 22.6, p. 536.

B. Tuck
Photoluminescence of $GaAs_{0.7}P_{0.3}$.
Vol. **18**, No. 2, p. 541—545 (1966).
See also 13.4, p. 204; *20.3*, p. *364*; 22.2.1, p. 483; 22.8, p. 548.

M. Jung
Pseudo-Abrupt Alloyed p-n Junctions in GaP
Vol. **18**, No. 2, p. 743—748 (1966).
See also 13.4, p. 204; *14.3.2*, p. *248*.

G. G. Kovalevskaya, D. N. Nasledov, and S. V. Slobodchikov
Photoconductivity Oscillations in InP
Vol. **23**, No. 2, p. 755—760 (1967).
See also *6*, p. *63*; 13.1, p. 182; *16*, p. *280*.

R. K. Purohit
GaP-GaAs n-p Heterojunctions
Vol. **24**, No. 1, p. K57—K60 (1967).
See also *14.3.2*, p. *250*; 22.2.1, p. 484.

T. L. Tansley
Forward Current Injection Modulation of Photocurrent in p-n Heterojunctions
Vol. **24**, No. 2, p. 615—622 (1967).
See also *14.3.2*, p. *250*; 14.4.1, p. 261; 16, p. 281; 22.2.1, p. 485.

L. Ask
Effect of Mechanical Stress on the I-U Characteristic of a GaP p-n Junction
Vol. **26**, No. 2, p. K143—K145 (1968).
See also 2, p. 24; 12, p. 158; *14.3.2*, p. *250*.

Yu. E. Perlin and Sh. N. Gifeisman
The Theory of Shallow F-Centres
Vol. **28**, No. 1, p. 45—60 (1968).
See also 6, p. 66; *10.2*, p. *143*; 13, p. 171; 14.4, p. 254; 18, p. 294; 22.2.1, p. 485; 22.2.3, p. 493; 22.4, p. 495; 22.4.1, p. 503; 22.4.2, p. 508; 22.4.3, p. 511; 22.5.1, p. 517; 22.5.2, p. 527.

H. G. Grimmeiss and G. Olofsson
Storage Effects in Photoconducting GaP-Cu
Vol. **28**, No. 2, p. 547—554 (1968).
See also *16*, p. *283*.

J. Gyulai
Induced Impurity Breakdown Oscillations and Observation of Traps Lying Higher than the Indirect Band Edge in GaP
Vol. **29**, No. 2, p. K85—K88 (1968).
See also 13.4, p. 216; *14.4.1*, p. *264*; 16, p. 284.

V. V. Tarassov and A. F. Demidenko
Heat Capacity and Quasi-Chain Dynamics of Diamond-Like Structures
Vol **30**, No. 1, p. 147—155 (1968).
See also *6*, p. *69*; *8*, p. *86*; 13.1, p. 189; 22.1.1, p. 466; 22.1.2, p. 474; 22.2.1, p. 486; 22.2.3, p. 493.

G. G. Kovalevskaya and S. V. Slobodchikov
Lifetimes of Current Carriers in InP
Vol. **30**, No. 2, p. 441—447 (1968).
See also 13.4, p. 217; *14.3*, p. *244*.

A. M. White, E. W. Williams, and P. Porteous
Infra-Red and Visible Photoluminescence in $In_{1-x}Ga_xP$
Vol. **30**, No. 2, p. K125—K126 (1968).
See also *20.3*, p. *372*; 22.8, p. 557.

22.2.3 Antimonides

Review Articles

J. R. Dale
Alloyed Semiconductor Heterojunctions
Vol. **16**, No. 2, p. 351—387 (1966).
See also 3.1, p. 29; 5, p. 49; 13.4, p. 200; *14.3.2*, p. *248*; 16, p. 276; 22.1.1, p. 458; 22.1.2, p. 468; 22.2.1, p. 482; 22.2.2, p. 487.

Y. P. Varshni
Band-to-Band Radiative Recombination in Groups IV, VI, and III–V Semiconductors (I)
Vol. **19**, No. 2, p. 459—514 (1967).
See also 13.1, p. 174; 13.2, p. 191; 13.4, p. 200; 20.1, p. 344; 20.2, p. 360; *20.3*, p. *363*; 22.1, p. 456; 22.1.1, p. 458; 22.1.2, p. 468; 22.1.3, p. 476; 22.2.1, p. 482; 22.2.2, p. 487.

Y. P. Varshni
Band-to-Band Radiative Recombination in Groups IV, VI, and III–V Semiconductors (II)
Vol. **20**, No. 1, p. 9—36 (1967).
See also 13.1, p. 174; 13.2, p. 191; 13.4, p. 200; 20.1, p. 344; 20.2, p. 360; *20.3*, p. *363*; 22.1 p. 456; 22.1.1, p. 458; 22.1.2, p. 468; 22.1.3, p. 476; 22.2.1, p. 482; 22.2.2, p. 487.

D. P. Morgan
Helicon Waves in Solids
Vol. **24**, No. 1, p. 9—36 (1967).
See also 6, p. 56; *13*, p. *162*; 14.1, p. 220; 14.3, p. 231; 19, p. 326; *20*, p. *337*; 21, p. 373; 21.1, p. 390; 21.2, p. 424; 21.6, p. 430; 22.1.1, p. 458; 22.2.1, p. 482.

M. H. Pilkuhn
The Injection Laser
Vol. **25**, No. 1, p. 9—62 (1968).
See also 13.4, p. 200; 14.3.2, p. 248; *14.4.1*, p. *255*; *20.2*, p. *360*; 20.3, p. 363; 22.2.1, p. 482; 22.2.2, p. 487.

M. Balkanski and E. Amzallag
Band Parameters Determination from Faraday Rotation Measurements
Vol. **30**, No. 2, p. 407—439 (1968).
See also *13.1*, p. *174*; 18, p. 291; 20, p. 337; 22.1.1, p. 458; 22.1.2, p. 468; 22.2.1, p. 482; 22.2.2, p. 487; 22.4, p. 495; 22.4.1, p. 497; 22.4.2, p. 506; 22.4.3, p. 509.

Original Papers and Short Notes

P. C. Euthymiou and A. L. Tanenbaum
Effect of Neutron and γ-Ray Irradiation on the Carrier Lifetime and Detectivity of InSb
Vol. **16**, No. 1, p. 183—190 (1966).
See also *11*, p. *145*; 14.3, p. 231.

J. Kołodziejczak and S. Żukotyński
The Effect of Higher Bands on the Effective Masses in InSb and InAs
Vol. **16**, No. 1, p. K55—K58 (1966).
See also *13.1*, p. *174*; 22.2.1, p. 482.

R. Granger
Anisotropic Electron Distribution in Pure Semiconductors near Breakdown
Vol. **16**, No. 2, p. 599—612 (1966).
See also 6, p. 56; 13, p. 163; *14.4.1*, p. *255*; 22, p. 440.

A. S. Filipchenko, G. G. Kulikov, D. N. Nasledov, and Yu. I. Ukhanov
Effective Mass of Holes in InSb
Vol. **17**, No. 1, p. K113–K114 (1966).
See also *13.1*, p. *175*; 20.1, p. 345.

W. H. Mitchell, E. H. Putley, and N. Shaw
Hall Effect and Extrinsic Photoconductivity in n-Type InSb
Vol. **17**, No. 2, p. 605–608 (1966).
See also *14.3*, p. *232*; *16*, p. *277*.

G. B. Abdullaev, M. I. Aliev, and G. I. Safaraliev
About Phonon Scattering in Doped GaSb Crystals
Vol. **17**, No. 2, p. 777–781 (1966).
See also 6, p. 57; *8*, p. *79*.

S. A. Moskalenko and P. I. Khadshi
Infrared Absorption by Excitons due to Photoionization and Intraband Lattice Scattering
Vol. **18**, No. 1, p. 379–390 (1966).
See also 6, p. 57; *13.2*, p. *191*; 20.1, p. 346; 22.1.1, p. 460; 22.2.2, p. 487; 22.6, p. 536.

I. Licea
On the Variation of the Electron Temperature in Polar Semiconductors by an Electric Field
Vol. **18**, No. 1, p. K9–K12 (1966).
See also 13, p. 164; *14.3*, p. *233*.

B. M. Askerov and F. M. Gashimzade
Quantum Theory of the Electrial Conductivity of Semiconductors with a Non-Standard Energy Band
Vol. **18**, No. 2, p. 667–675 (1966).
See also 6, p. 58; 13.1, p. 176; *14.3*, p. *233*; 22.2, p. 481.

G. Richter
Determination of Transport Coefficients for Minority Carriers in InSb
Vol. **18**, No. 2, p. K149–K152 (1966).
See also 13.1, p. 176; *14.3*, p. *234*; 16, p. 278.

M. P. Mikhailova, D. N. Nasledov, and S. V. Slobodchikov
The Effect of a Magnetic Field upon Illuminated InAs p-n Junctions
Vol. **19**, No. 1, p. 429–434 (1967).
See also *14.3.2*, p. *249*; 16, p. 278.

A. S. Filipchenko and D. N. Nasledov
On the Mixed Mechanism of Electron Scattering in InSb Crystals
Vol. **19**, No. 1, p. 435–439 (1967).
See also *14.3*, p. *235*; 15, p. 271.

A. R. Beattie and G. Smith
Recombination in Semiconductors by a Light Hole Auger Transition
Vol. **19**, No. 2, p. 577–586 (1967).
See also *13.1*, p. *177*; *16*, p. *278*; 22.2.1, p. 483.

G. Jones, G. Smith, and A. R. Beattie
The Drifted Maxwellian Distribution Function in InSb
Vol. **20**, No. 2, p. K135–K139 (1967).
See also *13*, p. *165*; 14.3, p. 236.

O. Budke and J. Krempaský
Measurement of Thermal Parameters of Thin Films and Plane-Parallel Plates Located on a Thermally Conductive Base
Vol. **21**, No. 1, p. 175–182 (1967).
See also *8*, p. *82*; 22.5.2, p. 522.

J. Bougnot, E. Monteil, and C. Llinares
Diffusion and Solubility in Gallium Antimonide
Vol. **21**, No. 1, p. K31–K34 (1967).
See also *9*, p. *89*; 10, p. 100; 14.3, p. 236.

M. Averous and G. Bougnot
Effect of Donor Impurity Concentration on the Piezoresistance of n-Type GaSb
Vol. **21**, No. 2, p. 665–670 (1967).
See also 13.1, p. 178; 13.4, p. 207; *14.3*, p. *236*.

S. A. Zeinalov and M. I. Aliev
Investigation of the Thermal Conducticity of InSb-GaSb Solid Solutions
Vol. **22**, No. 1, p. 153–156 (1967).
See also *8*, p. *83*; 22.8, p. 551.

P. Kordoš
Heat Treatment of InSb Crystals with Different Dislocation Densities
Vol. **22**, No. 1, p. K59–K61 (1967).
See also *9*, p. *89*; 10, p. 101; 14.3, p. 237.

M. I. Aliev, G. I. Safaraliev, and S. G. Abdinova
About the Role of Minority Carriers in the Impurity Conduction of GaSb
Vol. **22**, No. 2, p. 741–745 (1967).
See also 13.4, p. 209; *14.3*, p. *237*.

R. Yu. Khansevarov, T. V. Mashovets, N. A. Vitovskii, and B. Chelustka
On the Nature of Damages in Indium Antimonide Produced by γ-and X-Irradiation
Vol. **22**, No. 2, p. K95–K101 (1967).
See also *11*, p. *149*; 14.3, p. 237.

Ľ. Hrivnák and J. Kovár
Absorption of Ultrasound in Indium Antimonide
Vol. **23**, No. 1, p. 189–193 (1967).
See also *7*, p. *77*; 8, p. 83.

J. Kołodziejczak
Effect of Strong Electric Fields on Magneto-Resistivity and Hall Coefficient in Semiconductors
Vol. **23**, No. 1, p. 423–427 (1967).
See also 6, p. 62; 13.1, p. 181; *14.3*, p. *238*; 14.4.1, p. 260; 22, p. 447.

H. Pagnia
Slow Photoconduction in InSb Films
Vol. **24**, No. 1, p. K97–K100 (1967).
See also 14.3.1, p. 246; *16*, p. *281*.

B. Pistoulet, J. L. Robert, and M. Marqes
Study of GaSb Valence Band by Magnetoresistance and Hall Effect Measurements at Low Temperature
Vol. **24**, No. 2, p. 481–486 (1967).
See also *13.1*, p. *183*; 14.3, p. 239.

C. HAMMAR and P. WEISSGLAS
Hot Electron Polar Scattering in InSb
Vol. 24, No. 2, p. 531–534 (1967).
See also *13*, p. *168*; *14.3*, p. *239*.

M. H. VAN MAAREN
Note on the Diffusion and Solubility of Lithium in Gallium Antimonide
Vol. 24, No. 2, p. K125–K128 (1967).
See also *9*, p. *90*; 10, p. 103.

G. ARLT and P. QUADFLIEG
Piezoelectricity in III-V Compounds with a Phenomenological Analysis of the Piezoelectric Effect
Vol. 25, No. 1, p. 323–330 (1968).
See also *14.3*, p. *240*; 22.2.1, p. 485.

N. N. SIROTA and V. B. SHIPILO
High-Pressure and Temperature Effect on the Electrical Properties of p-Type Gallium Antimonide
Vol. 25, No. 1, p. K23–K26 (1968).
See also *14.3*, p. *240*.

N. N. SIROTA and V. B. SHIPILO
High-Pressure Effect on the Electrical Conductivity and Hall Constant of Indium Antimonide
Vol. 25, No. 1, p. K27–K30 (1968).
See also *14.3*, p. *240*.

I. LICEA
Hot Electron Mobility in Polar Semiconductors
Vol. 26, No. 1, p. 115–123 (1968).
See also *13*, p. *169*; *14.3*, p. *240*.

J. BOUGNOT, L. SZEPESSY, and S. F. DA CUNHA
Diffusion of Cadmium in Gallium Antimonide
Vol. 26, No. 2, p. K127–K129 (1968).
See also *9*, p. *91*.

M. M. AKSELROD, K. M. DEMCHUK, I. M. TSIDILKOVSKI, E. L. BROYDA, and K. P. RODIONOV
The Effect of Pressure on the Electron Effective Mass in InSb
Vol. 27, No. 1, p. 249–254 (1968).
See also *13.1*, p. *186*; 14.3, p. 241; 15, p. 274.

E. W. KREUTZ, H. PAGNIA, and W. WAIDELICH
X-Ray Induced Conductivity Changes in InSb
Vol. 27, No. 2, p. K111–K114 (1968).
See also *11*, p. *153*; 13.3, p. 198; 14.3, p. 242.

I. LICEA
Warm Electron Temperature and Mobility in Polar Semiconductors
Vol. 27, No. 2, p. K137–K141 (1968).
See also *13*, p. *170*; 13.1, p. 186; *14.3*, p. *242*.

I. LICEA
Hot Electron Temperature and Mobility in Polar Semiconductors
Vol. 27, No. 2, p. K143–K144 (1968).
See also *13*, p. *170*; *14.3*, p. *242*; 14.4.1, p. 263.

Yu. E. Perlin and Sh. N. Gifeisman
The Theory of Shallow F-Centres
Vol. **28**, No. 1, p. 45–60 (1968).
See also 6, p. 66; *10.2*, p. *143*; 13, p. 171; 14.4, p. 254; 18, p. 294; 22.2.1, p. 485; 22.2.2, p. 488; 22.4, p. 495; 22.4.1, p. 503; 22.4.2, p. 508; 22.4.3, p. 511; 22.5.1, p. 517; 22.5.2, p. 527.

H. Pagnia
Surface Photoconduction in p-InSb Single Crystals
Vol. **28**, No. 1, p. K89–K92 (1968).
See also 13.3, p. 198; *16*, p. *283*.

P. Weissglas and K. Blötekjaer
Conductivity of Intrisic InSb in High Electric Fields
Vol. **28**, No. 2, p. 711–718 (1968).
See also *14.3*, p. *242*; 14.4.1, p. 263.

H. Pfleiderer
PEM Effect with Short-Circuited Dember Voltage
Vol. **29**, No. 2, p. 597–604 (1968).
See also *16*, p. *284*.

M. Averous, G. Bougnot, and J. Calas
Conduction Bands of GaSb-InSb Mixed Crystals with a Low Concentration of InSb
Vol. **29**, No. 2, p. 807–812 (1968).
See also *13.1*, p. *189*; 14.3, p. 244; 22.8, p. 556.

V. V. Tarassov and A. F. Demidenko
Heat Capacity and Quasi-Chain Dynamics of Diamond-Like Structures
Vol. **30**, No. 1, p. 147–155 (1968).
See also *6*, p. *69*; *8*, p. *86*; 13.1, p. 189; 22.1.1, p. 466; 22.1.2, p. 474; 22.2.1, p. 486; 22.2.2, p. 488.

H. R. Huff, S. Kawaji, and H. C. Gatos
Observation of Quantum Galvanomagnetic Phenomena in n-Type Indium Antimonide
Vol. **30**, No. 2, p. 613–618 (1968).
See also *14.3*, p. *244*.

22.3 Carbides

Review Articles

P. KRISHNA and AJIT RAM VERMA
Crystal-Polymorphism in One Dimension
Vol. **17**, No. 2, p. 437—477 (1966).
See also *3*, p. *26*; *5*, p. *49*; 10, p. 93; 22.4.1, p. 497; 22.5, p. 512.

Original Papers and Short Notes

H. L. BROWN and C. P. KEMPTER
Elastic Properties of Zirconium Carbide
Vol. **18**, No. 1, p. K21—K23 (1966).
See also *12*, p. *156*.

J. S. ZIOMEK and P. B. PICKAR
Optical Studies of β-SiC
Vol. **21**, No. 1, p. 271—278 (1967).
See also 5, p. 51; 6, p. 61; 13.1, p. 178; *20.1*, p. *348*.

J. L. MARTIN, B. JOUFFREY, and P. COSTA
Stacking Faults in a Non-Stoichiometric Face-Centred Cubic TaC
Vol. **22**, No. 2, p. 349—354 (1967).
See also *10*, p. *101*.

D. N. BOSE
Field-Effect Measurements on Silicon Carbide
Vol. **24**, No. 2, p. K165—K166 (1967).
See also *13.3*, p. *198*; 14.3, p. 239.

D. KORDES
Vanadium Monocarbide Containing Ordered Vacancies
Vol. **26**, No. 2, p. K103—K105 (1968).
See also *5*, p. *53*; 10, p. 106.

22.4 II–VI Compounds

Review Articles

M. Balkanski and E. Amzallag
Band Parameters Determination from Faraday Rotation Measurements
Vol. **30**, No. 2, p. 407–439 (1968).
See also *13.1*, p. *174*; 18, p. 291; 20, p. 337; 22.1.1, p. 458; 22.1.2, p. 468; 22.2.1, p. 482; 22.2.2, p. 487; 22.2.3, p. 489; 22.4.1, p. 497; 22.4.2, p. 506; 22.4.3, p. 509.

Original Papers and Short Notes

E.-H. Weber
Relationship between Space Charge, Surface Conductivity, and Band Bending for Photoconductors under Uniform Optical Excitation
Vol. **17**, No. 2, p. 843–854 (1966).
See also 13.3, p. 197; 13.4, p. 203; 14.4.1, p. 257; *16*, p. *277*.

P. Bräunlich and A. Scharmann
Approximate Solution of Schön's Balance Equations for the Thermoluminescence and the Thermally Stimulated Conductivity of Inorganic Photoconducting Crystals
Vol. **18**, No. 1, p. 307–316 (1966).
See also *13.4*, p. *203*; 14.3, p. 233; 16, p. 278; 20.3, p. 364.

E. Gutsche and E. Jahne
Spin–Orbit Splitting of the Valence Band of Wurtzite Type Crystals
Vol. **19**, No. 2, p. 823–832 (1967).
See also *13.1*, p. *177*.

I. A. Akimov, V. M. Bentsa, F. I. Vilesov, and A. N. Terenin
External Photoeffect from Sensitizing Dyes Adsorbed on Semiconductors
Vol. **20**, No. 2, p. 771–776 (1967).
See also 13.3, p. 197; 16, p. 279; *17*, p. *287*; 22.5, p. 512; 22.5.1, p. 515.

E. Jahne and E. Gutsche
Valence Band Structure of Wurtzite Type Crystals
Vol. **21**, No. 1, p. 57–68 (1967).
See also *13.1*, p. *178*; 22.4.1, p. 499; 22.4.2, p. 506.

J. Trylski
Energy of the Donor Ground State in CdS, CdSe, and ZnO
Vol. **21**, No. 1, p. K53–K55 (1967).
See also *13.4*, p. *207*; 22.4.1, p. 499; 22.4.2, p. 506.

I. Licea
Electron Temperature in Polar Semiconductors
Vol. **25**, No. 1, p. 461–471 (1968).
See also *13*, p. *168*; *14.3*, p. *240*; 22.2, p. 481.

Yu. E. Perlin and Sh. N. Gifeisman
The Theory of Shallow F-Centres
Vol. **28**, No. 1, p. 45–60 (1968).
See also 6, p. 66; *10.2*, p. *143*; 13, p. 171; 14.4, p. 254; 18, p. 294; 22.2.1, p. 485; 22.2.2, p. 488; 22.2.3, p. 493; 22.4.1, p. 503; 22.4.2, p. 508; 22.4.3, p. 511; 22.5.1, p. 517; 22.5.2, p. 527.

K. MASCHKE and U. RÖSSLER
The Electronic Structure of CdO. (I): The Energy-Band Structure (APW Method)
Vol. 28, No. 2, p. 577–581 (1968).
See also *13.1*, p. *187*.

M. ALTWEIN, H. FINKENRATH, Č. KOŇÁK, J. STUKE, and G. ZIMMERER
The Electronic Structure of CdO. (II): Spectral Distribution of Optical Constants
Vol. 29, No. 1, p. 203–209 (1968).
See also 13.1, p. 188; 14.3.1, p. 247; *20.1*, p. *356*.

YU. A. OSIPIYAN and I. S. SMIRNOVA
Perfect Dislocations in the Wurtzite Lattice
Vol. 30, No. 1, p. 19–29 (1968).
See also *10*, p. *111*.

R. BOYN
Polarizatation of Optical Absorption and Emission Bands due to Donor-Accecptor Associates in Wurtzite Type Crystals
Vol. 30, No. 1, p. 305–310 (1968).
See also 13.4, p. 216; *20.1*, p. *358*; *20.3*, p. *371*.

22.4.1 Sulphides

Review Articles

P. Krishna and Ajit Ram Verma
Crystal-Polymorphism in One Dimension
Vol. **17**, No. 2, p. 437–477 (1966).
See also *3*, p. *26*; *5*, p. *49*; 10, p. 93; 22.3, p. 494; 22.5, p. 512.

F. Williams
Donor–Acceptor Pairs in Semiconductors
Vol. **25**, No. 2, p. 493–512 (1968).
See also 10, p. 93; *13.4*, p. *200*; 20.3, p. 363; 22.1, p. 456; 22.1.1, p. 458; 22.1.2, p. 468; 22.2.2, p. 487.

G. Harbeke
On the Band Structure of Anisotropic Crystals
Vol. **27**, No. 1, p. 9–44 (1968).
See also *13.1*, p. *174*; 13.2, p. 191; 14.3, p. 231; 20.1, p. 344; 22, p. 439; 22.1, p. 456; 22.1.1, p. 458; 22.1.2, p. 468; 22.1.3, p. 476; 22.4.2, p. 506; 22.4.3, p. 509; 22.5, p. 512; 22.7, p. 545; 22.8, p. 547.

M. Balkanski and E. Amzallag
Band Parameters Determination from Faraday Rotation Measurements
Vol. **30**, No. 2, p. 407–439 (1968).
See also *13.1*, p. *174*; 18, p. 291; 20, p. 337; 22.1.1, p. 458; 22.1.2, p. 468; 22.2.1, p. 482; 22.2.2, p. 487; 22.2.3, p. 489; 22.4, p. 495; 22.4.2, p. 506; 22.4.3, p. 509.

Original Papers and Short Notes

K. W. Böer and R. Schubert
The Influence of Oxygen in the Ultrahigh Vacuum Range on Electrical Properties of CdS
Vol. **16**, No. 1, p. K5–K9 (1966).
See also 13.3, p. 196; *16*, p. *276*.

M. Bancie-Grillot and P. Bourtayre
Visible Photoluminescence Emission of Gadolinium Incorporated in the Crystal Lattice of Zinc Sulphide
Vol. **16**, No. 2, p. 517–523 (1966).
See also 13.4, p. 201; *20.3*, p. *363*.

K. W. Böer, J. W. Feitknecht, and D. G. Kannenberg
Properties of Recrystallized Evaporated CdS Layers
Vol. **16**, No. 2, p. 697–709 (1966).
See also *1.4*, p. *16*; 13.4, p. 201; 16, p. 276.

M. V. Kurik and N. I. Vitrikhovskii
Optical Properties and Energy Structure of CdS–CdTe Crystals
Vol. **16**, No. 2, p. K139–K142 (1966).
See also *13.1*, p. *175*; *20.1*, p. *345*; 22.4.3, p. 509; 22.8, p. 547.

R. Schubert
Gas Desorption from Virginal CdS Crystals
Vol. **16**, No. 2, p. K157–K159 (1966).
See also 13.3, p. 196; *16*, p. *276*.

H. J. Rohde
Effect of Shallow Traps on the Current-Voltage Characteristics in the Case of Double Injection
Vol. **16**, No. 2, p. K179—K182 (1966).
See also 13.4, p. 202; *14.4.1*, p. *256*.

R. Fischer, F. Stöckmann, and J. Stuke
Space-Charge Limited Currents in Inhomogeneously Illuminated CdS Single Crystals
Vol. **17**, No. 1, p. 335—341 (1966).
See also *14.4.1*, p. *256*; 16, p. 277.

E. Gutsche, F. Spiegelberg, and J. Voigt
Transient Photoconductivity Quenching in CdS Single Crystals due to Surface Trapping
Vol. **17**, No. 1, p. K11—K14 (1966).
See also 13.3, p. 196; *16*, p. *277*.

P. Voss
Surface Pinned Layer-Like Field Inhomogeneities in CdS
Vol. **17**, No. 1, p. K21—K24 (1966).
See also *14.4.1*, p. *256*.

P. Bauduin and F. Buchy
Current Oscillations in Photoconducting Cadmium Sulphide
Vol. **17**, No. 2, p. 517—530 (1966).
See also *14.4.1*, p. *256*; *16*, p. *277*.

K. Kirov
Pulse Method for Studying Trapping Centres in CdS Type Crystals
Vol. **17**, No. 2, p. 581—592 (1966).
See also *13.4*, p. *202*; 14.4.1, p. 256; 16, p. 277.

U. Rössler and M. Lietz
Band Structure of Cubic and Hexagonal ZnS (APW Method)
Vol. **17**, No. 2, p. 597—604 (1966).
See also 5, p. 50; *13.1*, p. *175*; 20.1, p. 346.

H.-E. Gumlich, R. Moser, and E. Neumann
Recombination Distribution in the Non-Stationary Phase of the Electroluminescence of ZnS Phosphors Containing Iron-Group Elements
Vol. **17**, No. 2, p. 783—792 (1966).
See also 13.4, p. 203; *20.3*, p. *364*.

C. Benoit à la Guillaume and J.-M. Debever
Laser Effect in CdS by Electron Bombardment
Vol. **17**, No. 2, p. 875—879 (1966).
See also *20.2*, p. *360*.

O. Goede and E. Gutsche
Edge Emission in CdS as Internal Radiative Transition in Donor–Acceptor Associates
Vol. **17**, No. 2, p. 911—923 (1966).
See also 6, p. 57; 13.4, p. 203; *20.3*, p. *364*.

A. Räuber and J. Schneider
Localized $^2S_{1/2}$-State Centres in ZnS
Vol. **18**, No. 1, p. 125—132 (1966).
See also 13.4, p. 203; *19*, p. *327*.

K. F. Lider, B. V. Novikov, and S. A. Permogorov
Application of Bound-Exciton Optical Spectra in the Study of Radiation Damage in Crystals
Vol. **18**, No. 1, p. K1—K4 (1966).
See also 11, p. 147; *13.2*, p. *191*; *20.3*, p. *364*.

K. W. Böer and C. A. Kennedy
Vacuum Heat Treatment of CdS Single Crystals
Vol. **19**, No. 1, p. 203—210 (1967).
See also *13.4*, p. *204*; 14.3, p. 234; 16, p. 278.

G. Döhler
On the Field Emission of Minority Carriers in Photoconductors
Vol. **19**, No. 2, p. 555—564 (1967).
See also 13.4, p. 205; *14.4.1*, p. *257*; 16, p. 278; 22.1.3, p. 477.

N. A. Vlasenko and S. A. Zynio
Role of Carrier Polarization in the Mechanism of the Electroluminescence of ZnS-Mn Films
Vol. **20**, No. 1, p. 311—320 (1967).
See also 14.4.1, p. 257; *20.3*, p. *365*.

V. D. Egorov
Luminescence of CdS Excited by Electrons of Various Energies
Vol. **20**, No. 2, p. 705—711 (1967).
See also 13.2, p. 192; 20.2, p. 361; *20.3*, p. *365*.

G. O. Müller, H. Peibst, E. Schnürer, and H. Thiel
The Influence of High Phonon Flux Densities on the X-Ray Reflectivity of Nearly Ideal CdS Crystals
Vol. **20**, No. 2, p. K173—K175 (1967).
See also *6*, p. *60*; 7, p. 76.

E. Jahne and E. Gutsche
Valence Band Structure of Wurtzite Type Crystals
Vol. **21**, No. 1, p. 57—68 (1967).
See also *13.1*, p. *178*; 22.4, p. 495; 22.4.2, p. 506.

P. Bräunlich
Light Effects on Space-Charge Controlled Currents in CdS
Vol. **21**, No. 1, p. 383—391 (1967).
See also 13.4, p. 206; *14.4.1*, p. *258*; 16, p. 279; 20.3, p. 365.

A. N. Georgobiani and V. I. Steblin
Electroluminescence of ZnS-Cu_2S Heterojunctions
Vol. **21**, No. 1, p. K45—K48 (1967).
See also 14.3.2, p. 249; *20.3*, p. *365*; 22, p. 445.

A. N. Georgobiani and V. I. Steblin
Optical and Electrical Characteristics of p–n Junctions in Zinc Sulfide
Vol. **21**, No. 1, p. K49—K51 (1967).
See also *14.3.2*, p. *249*; 16, p. 279; 20.3, p. 365.

J. Trylski
Energy of the Donor Ground State in CdS, CdSe, and ZnO
Vol. **21**, No. 1, p. K53—K55 (1967).
See also *13.4*, p. *207*; 22.4, p. 495; 22.4.2, p. 506.

F. Kelemen, A. Néda, D. Niculescu, and E. Cruceanu
On the Thermal Conductivity of CdTe and Some $CdTe_{1-x}S_x$ and $CdTe_{1-x}Se_x$ Solid Solutions
Vol. **21**, No. 2, p. 557—565 (1967).
See also *8*, p. *82*; 22.4.2, p. 506; 22.4.3, p. 509; 22.8, p. 550.

J. P. Agrawal
Infrared Fluorescence in Manganese Activated Zinc Sulphide Phosphors
Vol. **21**, No. 2, p. 679—682 (1967).
See also 13.4, p. 207; 20.1, p. 348; *20.3*, p. *366*.

M. Damodar-Pai and K. M. van Vliet
Generation-Recombination and Diffusion Noise in Cadmium Sulphide Photoconductors
Vol. **21**, No. 2, p. 819—835 (1967).
See also 13.4, p. 207; *16*, p. *279*.

V. K. Koniukhov, L. A. Kulevskii, and A. M. Prokhorov
Two-Photon Absorption Spectrum in CdS near the Fundamental Absorption Edge
Vol. **21**, No. 2, p. K107—K110 (1967).
See also *20.1*, p. *349*.

K. W. Böer and G. A. Dussel
Slow Moving Field Domains in CdS in the Range of Negative Differential Conductivity
Vol. **21**, No. 2, p. K145—K149 (1967).
See also *14.4.1*, p. *259*.

E. Gutsche and H. Lange
Electroreflectance of CdS and CdSe Single Crystals at the Fundamental Absorption Edge
Vol. **22**, No. 1, p. 229—233 (1967).
See also 13.1, p. 179; 13.2, p. 192; *14.4.1*, p. *259*; *20*, p. *339*; 20.1, p. 349; 22.4.2, p. 506

V. A. Drozdov and L. A. Mozgovaya
Infrared Quenching of the Photomechanical Effect in Cadmium Sulphide
Vol. **22**, No. 2, p. K109—K112 (1967).
See also *12*, p. *157*; 16, p. 280.

D. Obrikat, K. Meyer, and F. Polly
Triboluminescence Spectra of Doped Zinc Sulphides
Vol. **22**, No. 2, p. K123—K126 (1967).
See also *20.3*, p. *366*.

B. Autin, O. Cahen, E. Dieulesaint, D. Gouvernelle, J. Tavernier, and P. Bauduin
Two-Step Current Saturation in CdS Crystals
Vol. **22**, No. 2, p. K135—K137 (1967).
See also *16*, p. *280*.

H. J. Rohde
Double Injection into CdS Single Crystals
Vol. **23**, No. 1, p. 277—288 (1967).
See also 13.4, p. 209; *14.4.1*, p. *259*.

P. Eckelt
Energy Band Structures of Cubic ZnS, ZnSe, ZnTe, and CdTe (Korringa-Kohn-Rostoker Method)
Vol. **23**, No. 1, p. 307—312 (1967).
See also *13.1*, p. *181*; 22.4.2, p. 507; 22.4.3, p. 510.

S. G. Kalashnikov, A. I. Morosov, and V. V. Proklov
Some Data of Acousto-Electrical Domains in Photoconductive Cadmium Sulphide
Vol. **23**, No. 1, p. K9–K12 (1967).
See also 7, p. 77; *14.4.1*, p. *260*.

H. Lange and W. Henrion
Thermoreflectance of CdS and Se Single Crystals
Vol. **23**, No. 1, p. K67–K70 (1967).
See also 13.1, p. 181; *20.1*, p. *350*; 22.1.3, p. 477.

V. D. Egorov, G. O. Müller, and H. H. Weber
Cathodoluminescence Yields in Highly Excited CdS
Vol. **23**, No. 2, p. 703–708 (1967).
See also 13.2, p. 193; 20.2, p. 361; *20.3*, p. *367*.

H.-E. Gumlich, R. Moser, and E. Neumann
On the Mn–Mn Interaction and the Structure of Optical Spectra of ZnS(Mn)
Vol. **24**, No. 1, p. K13–K16 (1967).
See also 13.4, p. 210; *20.3*, p. *367*.

T. A. T. Cowell and J. Woods
Thermally Stimulated Currents and Infra-Red Luminescence in CdS Crystals
Vol. **24**, No. 1, p. K37–K40 (1967).
See also 10.2, p. 141; *13.4*, p. *210*; 14.3, p. 239; 20.3, p. 367.

M. K. Sheinkman, V. A. Tyagai, O. V. Snitko, I. B. Ermolovich, G. L. Belenky, and V. N. Bondarenko
Studies on the Nature of CdS Single Crystal Sensitization by Etching
Vol. **24**, No. 2, p. 543–549 (1967).
See also 10, p. 103; *13.4*, p. *210*; 16, p. 281; 20.3, p. 368.

J. Auleytner, E. F. Gross, A. A. Kralina, and B. S. Razbirin
Investigation of the Real Structure of CdS Crystals in Connection with the Shift of Exciton Lines
Vol. **24**, No. 2, p. K107–K110 (1967).
See also *10*, p. *103*; *13.2*, p. *193*; 20.1, p. 351.

B. Selle, W. Ludwig, and R. Mach
Photovoltaic Response of Cu_2S-CdS p–n Heterojunctions
Vol. **24**, No. 2, p. K149–K152 (1967).
See also *14.3.2*, p. *250*; *16*, p. *281*; 22, p. 449.

P. Krispin
On the Photoimpedance of ZnS Phosphors
Vol. **25**, No. 1, p. 147–154 (1968).
See also 13.4, p. 211; *14.4.1*, p. *261*; 16, p. 281.

D. M. Hughes and G. Carter
The Effects of Oxygen Adsorption and Low Energy Ion Bombardment on the Electrical Properties of Cadmium Sulphide Thin Films
Vol. **25**, No. 1, p. 449–453 (1968).
See also 13.3, p. 198; *14.3.1*, p. *246*.

F. L. Weichman, R. K. Lomnes, and S. Żukotyński
Photothermoelectric Effects in CdS
Vol. **25**, No. 2, p. 583–591 (1968).
See also *15*, p. *273*; *16*, p. *281*.

R. Gauthier, D. Bois, P. Pinard, and F. Davoine
Ion and Electron Production by Interaction of a Coherent Light Beam with Substance
Vol. **25**, No. 2, p. 691–696 (1968).
See also *17*, p. *288*; 21.1, p. 403; 21.1.1, p. 420; 22.1.1, p. 464; 22.2.1, p. 485; 22.5.2, p. 525; 22.6, p. 540.

U. Kümmel and G. Zollfrank
Phenomenological Study of Brightness Distributions in Electroluminescent Hexagonal ZnS Crystals
Vol. **25**, No. 2, p. 711–720 (1968).
See also 10, p. 104; *20.3*, p. *368*.

H. Friedrich, L. A. Sysoev, and K. Thiessen
Minority Carriers in Zinc Sulphide
Vol. **26**, No. 1, p. 107–114 (1968).
See also *16*, p. *281*.

J. Dziesiaty and R. Böttcher
Electron Paramagnetic Resonance of Ti^{2+} Ions in Cadmium Sulfide Single Crystals
Vol. **26**, No. 1, p. K21–K24 (1968).
See also *19*, p. *333*.

N. A. Vlasenko and A. N. Gergell
On the Mechanism of DC Electroluminescence in pCu_xS-nZnS:Mn, Cu, Cl Film Structures
Vol. **26**, No. 1, p. K77–K81 (1968).
See also 14.3.2, p. 250; *20.3*, p. *368*; 22, p. 450.

A. Kawski and A. J. Horodecki
Coactivator Depencence of the Electroluminescent Properties of ZnS-Cu
Vol. **26**, No. 2, p. 543–549 (1968).
See also *20.3*, p. *369*.

W. Picht and W. Ruppel
Photocurrent, Luminescence, and Absorption Saturation for Electron Beam Excitation in CdS
Vol. **26**, No. 2, p. K163–K165 (1968).
See also *16*, p. *282*; 20.1, p. 353; 20.3, p. 369.

U. Birkholz, W. Heering, and F. Ströckmann
Thermoelectric Power of Illuminated CdS
Vol. **27**, No. 1, p. K69–K72 (1968).
See also *15*, p. *274*; 16, p. 282.

M. Ya. Valakh, N. I. Vitrikhovskii, and M. P. Lisitsa
The Electron Effective Mass in CdS_xSe_{1-x} Crystals
Vol. **27**, No. 1, p. K81–K83 (1968).
See also *13.1*, p. *186*; 20.1, p. 354; 22.4.2, p. 507; 22.8, p. 554.

B. K. Daniels and D. B. Meadowcroft
Twist Boundaries and Electroluminescence
Vol. **27**, No. 2, p. 535–539 (1968).
See also *10*, p. *107*; *20.3*, p. *369*.

G. F. Alfrey and D. B. Meadowcroft
The Influence of Structure on the Electroluminescence of Zinc Sulphide
Vol. **27**, No. 2, p. 541–544 (1968).
See also 4, p. 46; 10, p. 107; *20.3*, p. *370*.

P. R. Herczfeld, K. M. van Vliet, and M. D. Pai
Experiments on Fluctuations in Optically Quenched CdS and CdSe Crystals
Vol. 27, No. 2, p. 671—680 (1968).
See also *16*, p. *282*; 22.4.2, p. 508.

P. R. Herczfeld and K. M. van Vliet
Electronic Processes and Fluctuations in Optically Quenched CdS and CdSe Crystals
Vol. 27, No. 2, p. 681—694 (1968).
See also *13.4*, p. *214*; *16*, p. *282*; 22.4.2, p. 508.

R. Koepp and K. Marinova
High Field Domains in CdS Crystals Moving in the Anode–Cathode Direction
Vol. 27, No. 2, p. K117—K120 (1968).
See also *14.4.1*, p. *262*; 16, p. 282.

D. Hahn and J. Mimkes
On the Frequency Dependence of Electroluminescence of ZnS:Cu Phosphors
Vol. 27, No. 2, p. K133—K135 (1968).
See also *20.3*, p. *370*.

Yu. E. Perlin and Sh. N. Gifeisman
The Theory of Shallow F-Centres
Vol. 28, No. 1, p. 45—60 (1968).
See also 6, p. 66; *10.2*, p. *143*; 13, p. 171; 14.4, p. 254; 18, p. 294; 22.2.1, p. 485; 22.2.2, p. 488; 22.2.3, p. 493; 22.4, p. 495; 22.4.2, p. 508; 22.4.3, p. 511; 22.5.1, p. 517; 22.5.2, p. 527.

C. Marti, R. Romestain, and R. Visocekas
Forbidden Transitions ($\Delta M = 2$) Induced by an Axial Crystal Field in Electronic Paramagnetic Resonance
Vol. 28, No. 1, p. 97—103 (1968).
See also 13, p. 171; *19*, p. *334*.

H.-J. Hänsch
Study of the Electroluminescence Mechanism of CdS Single Crystals
Vol. 28, No. 1, p. 255—268 (1968).
See also 13.2, p. 194; 14.4.1, p. 263; *20.3*, p. *370*.

K. W. Böer and P. Voss
Stationary Anode-Adjacent High-Field Domains in Cadmium Sulfide
Vol. 28, No. 1, p. 355—364 (1968).
See also *14.4.1*, p. *263*; 16, p. 282.

A. Farwer and F. Matossi†
Internal Optical Reflection in Cubic ZnS with One-Dimensional Stacking Faults
Vol. 28, No. 1, p. 373—383 (1968).
See also 4, p. 47; 10, p. 109; *20.1*, p. *355*.

V. V. Zotov and V. V. Serdyuk
Effect of Infrared Quenching on the Spectral Distribution of Induced Impurity Photoconductivity of CdS Single Crystals
Vol. 28, No. 1, p. K31—K33 (1968).
See also 13.4, p. 214; *16*, p. *283*.

C. Muntoni and A. Serpi
On an Allotropic Transformation of ZnS after Pressure Treatment
Vol. 28, No. 1, p. K43—K45 (1968).
See also *1.2*, p. *12*; 13.1, p. 187; 20.1, p. 356.

R. V. Grigoriev, B. V. Novikov, and A. E. Cherednichenko
Influence of Oxygen Adsorption on the Fine Structure of the Spectral Distribution of Photoconductivity in CdS Crystals
Vol. 28, No. 1, p. K85—K88 (1968).
See also 13.2, p. 194; 13.3, p. 198; *16*, p. *283*.

E.-H. Weber
Surface Photoconductivity of CdS Influenced by Chemisorption and Desorption of Oxygen
Vol. 28, No. 2, p. 649—662 (1968).
See also 13.3, p. 199; *16*, p. *283*.

I. F. Chang, S. S. Mitra, J. N. Plendl, and L. C. Mansur
Long-Wavelengh Longitudinal Phonons of Multi-Mode Crystals
Vol. 28, No. 2, p. 663—673 (1968).
See also *6*, p. *67*; 20.1, p. 356; 22.4.2, p. 508; 22.6, p. 542; 22.8, p. 554.

H. Berger, W. Kahle, and G. Jäniche
Thickness Dependence of Conductivity due to the Polycrystalline Structure in Evaporated CdS Thin Films
Vol. 28, No. 2, p. K97—K100 (1968).
See also 1.4, p. 21; *14.3.1*, p. *247*.

A. Cingolani, A. Rizzo, and A. Levialdi
AC Modulated Green Edge Emission in CdS
Vol. 28, No. 2, p. K151—K154 (1968).
See also *20.3*, p. *370*.

O. Goede
Investigation of the Emission of CdS Crystals in the 2.0 eV Region
Vol. 28, No. 2, p. K167—K170 (1968).
See also 10, p. 109; *20.3*, p. *370*.

E. Nebauer
Vapour-Pressure Investigations of Impurity Diffusion and Solubility in A^{II}-B^{VI} Compounds Demonstrated for the System $CdS:Au:S_2$
Vol. 29. No. 1, p. 269—281 (1968).
See also *9*, p. *92*; 10, p. 110.

R. Boyn
Optical Absorption due to Intrinsic Defects in CdS Single Crystals
Vol. 29, No. 1, p. 307—321 (1968).
See also 10, p. 110; *13.4*, p. *215*; *20.1*, p. *356*.

D. T. Y. Wei, C. M. Penchina, and Y. S. Park
Observation of Oscillatory Lifetime in CdS
Vol. 29, No. 1, p. K7—K10 (1968).
See also 13.4, p. 215; *16*, p. *284*.

P. Reimers and W. Ruppel
The Preparation of CdS-CdSe Graded Single Crystals
Vol. 29, No. 1, p. K31—K33 (1968).
See also *3.1*, p. *33*; 5, p. 54; 20.1, p. 357; 22.4.2, p. 508; 22.8, p. 555.

D. Bonnet, H. Germer, and H. Rabenhorst
Photoconduction in $Cd(S_xSe_{1-x})$ Films with Graded Composition
Vol. 29, No. 1, p. K65—K67 (1968).
See also *16*, p. *284*; 22.4.2, p. 508; 22.8, p. 555.

J. Dillinger, Č. Koňák, V. Prosser, J. Sak, and M. Zvára
Phonon-Assisted Exciton Transitions in $A^{II}B^{VI}$ Semiconductors
Vol. **29**, No. 2, p. 707–714 (1968).
See also 6, p. 68; *13.2*, p. *195*; 20.1, p. 357; 22.4.2, p. 508; 22.4.3, p. 511.

P. Flögel, E. Neubaer, H.-J. Ullrich, S. Däbritz, and F. Zimmer
Electron Microprobe and Scanning Electron Microscope Investigations on Plate-Shaped CdS Single Crystals
Vol. **29**, No. 2, p. K89–K90 (1968).
See also 3.1, p. 33; *4*, p. *47*.

M. Sodha and S. Sharma
Some Transport Coefficients in Cadmium Sulphide
Vol. **30**, No. 1, p. 239–245 (1968).
See also 6, p. 69; *14.3*, p. *244*.

K. W. Böer and P. Voss
Transition between Stationary and Moving High-Field Domains in CdS in a Range of N-Shaped Negative Differential Conductivity due to Field-Quenching
Vol. **30**, No. 1, p. 291–304 (1968).
See also *14.4.1*, p. *264*; 16, p. 284; 20, p. 342.

W. Stössel and W. Zimmermann
Photoconduction in CdS Crystals Generated by a Modulated Electron Beam
Vol. **30**, No. 1, p. 311–318 (1968).
See also *11*, p. *155*; 13.4, p. 216; *16*, p. *284*.

R. Koepp and A. Urbelis
Low-Frequency Photocurrent Oscillations in CdS Crystals with Thermally Induced Negative Differential Conductivity
Vol. **30**, No. 1, p. K5–K8 (1968).
See also *14.4.1*, p. *264*; *16*, p. *284*.

D. N. Bose
Thermoluminescence and Thermally Stimulated Currents in Bismuth Doped ZnS Crystals
Vol. **30**, No. 1, p. K57–K60 (1968).
See also *13.4*, p. *217*; 16, p. 284; 20.3, p. 371.

K. Dos, A. Haug, and P. Rohner
Consideration of the Exchange Interaction in the Two-Body Picture of Excitons and Its Influence on the Exciton Spectra
Vol. **30**, No. 2, p. 619–625 (1968).
See also *13.2*, p. *195*; 20.1, p. 359.

G. Döhler
Critical Conditions for Transitions between Stationary and Non-Stationary High-Field Domains in Semi-Insulators
Vol. **30**, No. 2, p. 627–636 (1968).
See also 13.4, p. 217; *14.4.1*, p. *265*.

J. Voigt and F. Spiegelberg
On the Dynamical Behaviour of Excitons in CdS Single Crystals. (I): Correlation between Intrinsic Excitonic Absorption and Emission
Vol. **30**, No. 2, p. 659–670 (1968).
See also 6, p. 70; *13.2*, p. *195*; 20.1, p. 359; 20.3, p. 372.

R. K. L. Gay and H. L. Hartnagel
The Spatial and Time Variation of Electric Fields in Semiconducting CdS
Vol. **30**, No. 2, p. 755–758 (1968).
See also 7, p. 78; *14.4.1*, p. *265*.

22.4.2 Selenides

Review Articles

G. Harbeke
On the Band Structure of Anisotropic Crystals
Vol. **27**, No. 1, p. 9—44 (1968).
See also *13.1*, p. *174*; 13.2, p. 191; 14.3, p. 231; 20.1, p. 344; 22, p. 439; 22.1, p. 456; 22.1.1, p. 458; 22.1.2, p. 468; 22.1.3, p. 476; 22.4.1, p. 497; 22.4.3, p. 509; 22.5, p. 512; 22.7, p. 545; 22.8, p. 547.

M. Balkanski and E. Amzallag
Band Parameters Determination from Faraday Rotation Measurements
Vol. **30**, No. 2, p. 407—439 (1968).
See also *13.1*, p. *174*; 18, p. 291; 20, p. 337; 22.1.1, p. 458; 22.1.2, p. 468; 22.2.1, p. 482; 22.2.2, p. 487; 22.2.3, p. 489; 22.4, p. 495; 22.4.1, p. 497; 22.4.3, p. 509.

Original Papers and Short Notes

B. S. Razbirin, I. Filiński, and B. Wojtowicz-Natanson
Hot Excitons in Cadmium Selenide
Vol. **18**, No. 1, p. K37—K40 (1966).
See also *13.2*, p. *192*; 20.3, p. 364.

O. V. Bogdankevich, M. M. Zverev, A. I. Krasilnikov, and A. N. Pechenov
Laser Emission in Electron-Beam Excited ZnSe
Vol. **19**, No. 1, p. K5—K6 (1967).
See also *20.2*, p. *360*.

E. Jahne and E. Gutsche
Valence Band Structure of Wurtzite Type Crystals
Vol. **21**, No. 1, p. 57—68 (1967).
See also *13.1*, p. *178*; 22.4, p. 495; 22.4.1, p. 499.

J. Trylski
Energy of the Donor Ground State in CdS, CdSe, and ZnO
Vol. **21**, No. 1, p. K53—K55 (1967).
See also *13.4*, p. *207*; 22.4, p. 495; 22.4.1, p. 499.

F. Kelemen, A. Néda, D. Niculescu, and E. Cruceanu
On the Thermal Conductivity of CdTe and Some $CdTe_{1-x}S_x$ and $CdTe_{1-x}Se_x$ Solid Solutions
Vol. **21**, No. 2, p. 557—565 (1967).
See also *8*, p. *82*; 22.4.1, p. 500; 22.4.3, p. 509; 22.8, p. 550.

E. Gutsche and H. Lange
Electroreflectance of CdS and CdSe Single Crystals at the Fundamental Absorption Edge
Vol. **22**, No. 1, p. 229—233 (1967).
See also 13.1, p. 179; 13.2, p. 192; *14.4.1*, p. *259*; *20*, p. *339*; 20.1, p. 349; 22.4.1, p. 500

J. Viščakas, J. Vaitkus, and S. Stonkus
Influence of Heat Treatment on the Spectral Dependence of Photoconductivity in CdSe Crystals
Vol. **22**, No. 1, p. K3—K5 (1967).
See also 13.4, p. 208; *16*, p. *279*.

I. Dima and G. Vasiliu
Photoconductvity of ZnSe Thin Films
Vol. **22**, No. 2, p. K79–K82 (1967).
See also *16*, p. *280*.

J. B. Grun, A. Mysyrowicz, F. Raga, A. Bivas, R. Levy, and S. Nikitine
Photon-Assisted Emission in CdSe
Vol. **22**, No. 2, p. K155–K156 (1967).
See also 6, p. 62; 13.2, p. 193; *20.3*, p. *367*.

P. Eckelt
Energy Band Structures of Cubic ZnS, ZnSe, ZnTe, and CdTe (Korringa-Kohn-Rostoker Method)
Vol. **23**, No. 1, p. 307–312 (1967).
See also *13.1*, p. *181*; 22.4.1, p. 500; 22.4.3, p. 510.

R. Ludeke and W. Paul
Growth and Optical Properties of Wurtzite and Sphalerite CdSe Epitaxial Thin Films
Vol. **23**, No. 1, p. 413–418 (1967).
See also 1.4, p. 19; *3.1*, p. *32*; 13.1, p. 181; *20.1*, p. *350*.

H. P. Singh and B. Dayal
X-Ray Determination of the Thermal Expansion of Zinc Selenide
Vol. **23**, No. 1, p. K93–K95 (1967).
See also 5, p. 52; *8*, p. *83*.

R. Baubinas, A. Sakalas, A. Smilga, and J. Viščakas
p-Type CdSe Single Crystals
Vol. **24**, No. 1, p. K91–K93 (1967).
See also 3.1, p. 33; *14.3*, p. *239*; *16*, p. *280*.

R. A. Burmeister, Jr. and D. A. Stevenson
Electrical Properties of n-Type CdSe
Vol. **24**, No. 2, p. 683–694 (1967).
See also 13.4, p. 211; *14.3*, p. *239*.

A. Hadni, J. Claudel, and P. Strimer
Far Infrared Absorption and Reflection Spectra of ZnSe, ZnTe, and CdSe at Low Temperature
Vol. **26**, No. 1, p. 241–252 (1968).
See also 6, p. 65; *20.1*, p. *352*; 22.4.3, p. 510.

L. N. Kurbatov, E. F. Gross, V. E. Mashenko, B. S. Razbirin, N. N. Mochalkin, and A. D. Britov
Coherent Emission of Free Excitons in CdSe Crystals
Vol. **26**, No. 2, p. K139–K142 (1968).
See also 6, p. 65; *13.2*, p. *194*; *20.2*, p. *361*.

M. Grynberg
Influence of Uniaxial Stress on the Optical Properties of CdSe
Vol. **27**, No. 1, p. 255–261 (1968).
See also *13.1*, p. *186*; 20.1, p. 354; 20.3, p. 369.

M. Ya. Valakh, N. I. Vitrikhovskii, and M. P. Lisitsa
The Electron Effective Mass in CdS_xSe_{1-x}Crystals
Vol. **27**, No. 1, p. K81–K83 (1968).
See also *13.1*, p. *186*; 20.1, p. 354; 22.4.1, p. 502; 22.8, p. 554.

P. R. Herczfeld, K. M. van Vliet, and M. D. Pai
Experiments on Fluctuations in Optically Quenched CdS and CdSe Crystals
Vol. 27, No. 2, p. 671–680 (1968).
See also *16*, p. *282*; 22.4.1, p. 503.

P. R. Herczfeld and K. M. van Vliet
Electronic Processes and Fluctuations in Optically Quenched CdS and CdSe Crystals
Vol. 27, No. 2, p. 681–694 (1968).
See also *13.4*, p. *214*; *16*, p. *282*; 22.4.1, p. 503.

Yu. E. Perlin and Sh. N. Gifeisman
The Theory of Shallow F-Centres
Vol. 28, No. 1, p. 45–60 (1968).
See also 6, p. 66; *10.2*, p. *143*; 13, p. 171; 14.4, p. 254; 18, p. 294; 22.2.1, p. 485; 22.2.2, p. 488; 22.2.3, p. 493; 22.4, p. 495; 22.4.1, p. 503; 22.4.3, p. 511; 22.5.1, p. 517; 22.5.2, p. 527.

J. T. Calow, S. J. T. Owen, and P. W. Webb
The Growth and Electrical Characteristics of Epitaxial Layers of Zinc Selenide on p-Type Germanium
Vol. 28, No. 1, p. 295–303 (1968).
See also *3.1*, p. *33*; *14.3.1*, p. *247*; 14.3.2, p. 251.

I. F. Chang, S. S. Mitra, J. N. Plendl, and L. C. Mansur
Long-Wavelenght Longitudinal Phonons of Multi-Mode Crystals
Vol. 28, No. 2, p. 663–673 (1968).
See also *6*, p. *67*; 20.1, p. 356; 22.4.1, p. 504; 22.6, p. 542; 22.8, p. 554.

M. Vaněček, J. Toušek, and E. Klier
Study of the Kinetics of X-Ray Photoconductivity in CdSe and CdTe and Determination of the Electron-Hole Generation Energy
Vol. 28, No. 2, p. K147–K149 (1968).
See also *16*, p. *283*; 22.4.3, p. 511.

P. Reimers and W. Ruppel
The Preparation of CdS-CdSe Graded Single Crystals
Vol. 29, No. 1, p. K31–K33 (1968).
See also *3.1*, p. *33*; 5, p. 54; 20.1, p. 357; 22.4.1, p. 504; 22.8, p. 555.

D. Bonnet, H. Germer, and H. Rabenhorst
Photoconduction in $Cd(S_xSe_{1-x})$ Films with Graded Composition
Vol. 29, No. 1, p. K65-K67 (1968).
See also *16*, p. *284*; 22.4.1, p. 504; 22.8, p. 555.

J. Dillinger, Č. Koňák, V. Prosser, J. Sak, and M. Zvára
Phonon-Assisted Exciton Transitions in $A^{II}B^{VI}$ Semiconductors
Vol. 29, No. 2, p. 707–714 (1968).
See also 6, p. 68; *13.2*, p. *195*; 20.1, p. 357; 22.4.1, p. 505; 22.4.3, p. 511.

E. F. Gross, B. S. Razbirin, V. P. Fedorov, and Yu. P. Naumov
Bound Exciton Complexes in CdSe Crystals
Vol. 30, No. 2, p. 485–494 (1968).
See also *13.2*, p. *195*; 20.1, p. 358; 20.3, p. 371.

22.4.3 Tellurides

Review Articles

G. Harbeke
On the Band Structure of Anisotropic Crystals
Vol. **27**, No. 1, p. 9–44 (1968).
See also *13.1*, p. *174*; 13.2, p. 191; 14.3, p. 231; 20.1, p. 344; 22, p. 439; 22.1, p. 456; 22.1.1, p. 458; 22.1.2, p. 468; 22.1.3, p. 476; 22.4.1, p. 497; 22.4.2, p. 506; 22.5, p. 512; 22.7, p. 545; 22.8, p. 547.

M. Balkanski and E. Amzallag
Band Parameters Determination from Faraday Rotation Measurements
Vol. **30**, No. 2, p. 407–439 (1968).
See also *13.1*, p. *174*; 18, p. 291; 20, p. 337; 22.1.1, p. 458; 22.1.2, p. 468; 22.2.1, p. 482; 22.2.2, p. 487; 22.2.3, p. 489; 22.4, p. 495; 22.4.1, p. 497; 22.4.2, p. 506.

Original Papers and Short Notes

M. Zvára, F. Žaloudek, and V. Prosser
Interband Faraday Effect in CdTe Single Crystals
Vol. **16**, No. 1, p. K21–K24 (1966).
See also 13, p. 162; *18*, p. *291*; *20*, p. *337*.

M. V. Kurik and N. I. Vitrikhovskii
Optical Properties and Energy Structure of CdS–CdTe Crystals
Vol. **16**, No. 2, p. K139–K142 (1966).
See also *13.1*, p. *175*; *20.1*, p. *345*; 22.4.1, p. 497; 22.8, p. 547.

C. Verié
On the Band Structure of HgTe-CdTe. (I): Electric Measurements
Vol. **17**, No. 2, p. 889–901 (1966).
See also *13.1*, p. *176*; 14.3, p. 233; 22.8, p. 548.

I. Spînulescu-Carnaru
The Crystalline Structure of ZnTe Thin Films
Vol. **18**, No. 2, p. 769–778 (1966).
See also *1.4*, p. *17*; 3.1, p. 30; 5, p. 50.

R. R. Gałązka and L. Sosnowski
Conduction Band Structure of $Cd_{0.1}Hg_{0.9}Te$
Vol. **20**, No. 1, p. 113–120 (1967).
See also *13.1*, p. *177*; 14.3, p. 235; 22.8, p. 549.

F. J. Bryant and E. Webster
Threshold Energy for Atomic Displacement in Cadmium Telluride
Vol. **21**, No. 1, p. 315–321 (1967).
See also *11*, p. *148*; *20.3*, p. *365*.

F. Kelemen, A. Néda, D. Niculescu, and E. Cruceanu
On the Thermal Conductivity of CdTe and Some $CdTe_{1-x}S_x$ and $CdTe_{1-x}Se_x$ Solid Solutions
Vol. **21**, No. 2, p. 557–565 (1967).
See also *8*, p. *82*; 22.4.1, p. 500; 22.4.2, p. 506; 22.8, p. 550.

B. Ray and P. M. Spencer
Phase Diagram of the Alloy System HgTe-CdTe
Vol. **22**, No. 2, p. 371–372 (1967).
See also *1*, p. *7*; *1.2*, p. *11*; 22.8, p. 551.

W. Szymańska
Conduction Band Structure of HgTe from Thermomagnetic Measurements
Vol. **23**, No. 1, p. 69—73 (1967).
See also *13.1*, p. *180*; 15, p. 272.

I. Spînulescu-Carnaru
The Orientation of Crystallites in ZnTe Thin Films
Vol. **23**, No. 1, p. 157—163 (1967).
See also *1.4*, p. *19*; 4, p. 43.

P. Eckelt
Energy Band Structures of Cubic ZnS, ZnSe, ZnTe, and CdTe (Korringa-Kohn-Rostoker Method)
Vol. **23**, No. 1, p. 307—312 (1967).
See also *13.1*, p. *181*; 22.4.1, p. 500; 22.4.2, p. 507.

R. R. Gałązka and T. Zakrzewski
Heavy Hole Effective Mass of $Cd_{0.1}Hg_{0.9}Te$
Vol. **23**, No. 1, p. K39—K43 (1967).
See also *13.1*, p. *181*; 14.3, p. 238; 15, p. 272; 22.8, p. 552.

I. Dima and D. Borşan
The Polymorphism of CdTe Thin Films
Vol. **23**, No. 2, p. K113—K115 (1967).
See also 1.2, p. 11; *1.4*, p. *19*.

L. Śniadower and R. Kowalczyk
Magnetoplasma Reflection in Mercury Telluride
Vol. **25**, No. 1, p. 233—240 (1968).
See also 13.1, p. 183; *18*, p. *293*; *20*, p. *340*; 20.1, p. 351.

F. Bailly
Energies of Formation of Metal Vacancies in II-VI Semiconducting Tellurides (HgTe, CdTe, ZnTe)
Vol. **25**, No. 1, p. 317—322 (1968).
See also *10*, p. *104*.

S. Yamada, Y. Kawasaki, and O. Nishida
Free Carrier Absorption in n-Type CdTe Crystals
Vol. **26**, No. 1, p. 77—84 (1968).
See also *20.1*, p. *352*.

A. Hadni, J. Claudel, and P. Strimer
Far Infrared Absorption and Reflection Spectra of ZnSe, ZnTe and CdSe at Low Temperature
Vol. **26**, No. 1, p. 241—252 (1968).
See also 6, p. 65; *20.1*, p. *352*; 22.4.2, p. 507.

M. Jaroš
On the Covalency in Transition-Metal Impurity Covalent Complexes
Vol. **27**, No. 1, p. K21—K24 (1968).
See also *3.2*, p. *36*; 19, p. 333.

W. Szymańska and J. Ginter
On the Scattering of Electrons by Ionised Impurities in the Case of Strong Degeneracy
Vol. **27**, No. 2, p. K103—K106 (1968).
See also 13.4, p. 214; *14.3*, p. *242*.

I. Ivanov-Omskii, F. P. Kesamanly, B. T. Kolomiets, A. Sh. Mekhtiev, and V. A. Skripkin
Density of States Effective Mass of Holes in HgTe
Vol. **27**, No. 2, p. K169—K172 (1968).
See also *13.1*, p. *186*; 15, p. 274.

Yu. E. Perlin and Sh. N. Gifeisman
The Theory of Shallow F-Centres
Vol. **28**, No. 1, p. 45—60 (1968).
See also 6, p. 66; *10.2*, p. *143*; 13, p. 171; 14.4, p. 254; 18, p. 294; 22.2.1, p. 485; 22.2.2, p. 488; 22.2.3, p. 493; 22.4, p. 495; 22.4.1, p. 503; 22.4.2, p. 508; 22.5.1, p. 517; 22.5.2, p. 527.

F. Kelemen, A. Néda, E. Cruceanu, and D. Niculescu
On the Thermal Conductivity of $Zn_xHg_{1-x}Te$ Solid Solutions
Vol. **28**, No. 1, p. 421—428 (1968).
See also 6, p. 66; *8*, p. *85*; 22.8, p. 554.

L. Śniadower, M. Psoda, and R. R. Gałązka
X-Ray Measurements of Lattice Dilatation in HgTe
Vol. **28**, No. 2, p. K121—K123 (1968).
See also *5*, p. *54*; *8*, p. *85*.

L. Śniadower
Temperature Studies of Magnetoplasma Reflection in Mercury Telluride
Vol. **28**, No. 2, p. K125—K130 (1968).
See also 13.1, p. 188; *18*, p. *294*; *20.1*, p. *356*.

M. Vaněček, J. Toušek, and E. Klier
Study of the Kinetics of X-Ray Photoconductivity in CdSe and CdTe and Determination of the Electron-Hole Generation Energy
Vol. **28**, No. 2, p. K147—K149 (1968).
See also *16*, p. *283*; 22.4.2, p. 508.

R. C. Whelan and D. Shaw
Evidence of a Doubly Ionized Native Donor in CdTe
Vol. **29**, No. 1, p. 145—152 (1968).
See also 10, p. 109; *13.4*, p. *215*; 14.3, p. 243.

J. Dillinger, Č. Koňák, V. Prosser, J. Sak, and M. Zvára
Phonon-Assisted Exciton Transitions in $A^{II}B^{VI}$ Semiconductors
Vol. **29**, No. 2, p. 707—714 (1968).
See also 6, p. 68; *13.2*, p. *195*; 20.1, p. 357; 22.4.1, p. 505; 22.4.2, p. 508.

E. Z. Dziuba, D. Niculescu, and N. Niculescu
Energy Band Structure of $Zn_xHg_{1-x}Te$ Solid Solutions
Vol. **29**, No. 2, p. 813—818 (1968).
See also *13.1*, p. *189*; 14.3, p. 244; 20.1, p. 357; 22.8, p. 556.

W. Ludwig, W. Gericke, and H. Treptow
Electron Tunnelling in MIS-Structures of the Type Al-Al_2O_3-ZnTe
Vol. **30**, No. 1, p. K61—K64 (1968).
See also 13.3, p. 199; *14.3.2*, p. *251*; 14.4.1, p. 264.

22.5 Halides

Review Articles

P. Krishna and Ajit Ram Verma
Crystal-Polymorphism in One Dimension
Vol. **17**, No. 2, p. 437–477 (1966).
See also *3*, p. *26*; *5*, p. *49*; 10, p. 93; 22.3, p. 494; 22.4.1, p. 497.

G. Harbeke
On the Band Structure of Anisotropic Crystals
Vol. **27**, No. 1, p. 9–44 (1968).
See also *13.1*, p. *174*; 13.2, p. 191; 14.3, p. 231; 20.1, p. 344; 22, p. 439; 22.1, p. 456; 22.1.1, p. 458; 22.1.2, p. 468; 22.1.3, p. 476; 22.4.1, p. 497; 22.4.2, p. 506; 22.4.3, p. 509; 22.7, p. 545; 22.8, p. 547.

Original Papers and Short Notes

Z. Šroubek
On the Theory of the Superexchange Interaction in Ionic Crystals: $KNiF_3$
Vol. **16**, No. 2, p. 405–412 (1966).
See also *3.2*, p. *34*; 13, p. 163; 18.4, p. 321; 22.8, p. 547.

K. S. Aleksandrov, L. M. Reshchikova, and B. V. Beznosikov
Behaviour of the Elastic Constants of $KMnF_3$ Single Crystals near the Transition of Puckering
Vol. **18**, No. 1, p. K17–K20 (1966).
See also 1.2, p. 10; *12*, p. *156*; 22.8, p. 548.

B. P. Kozyrev and T. Boţilă
Some Photoelectrical Properties of KRS-5 Single Crystals
Vol. **18**, No. 1, p. K57–K61 (1966).
See also 13.4, p. 203; *16*, p. *278*; 22.8, p. 548.

J. R. Pilbrow and J. M. Spaeth
ESR Studies of Cu^{2+} in NH_4Cl Single Crystals between 4.2 and 453 °K. (I): Spin-Hamiltonian Parameters and Lattice Constant for NH_4Cl
Vol. **20**, No. 1, p. 225–235 (1967).
See also *19*, p. *328*; 22, p. 444.

J. R. Pilbrow and J. M. Spaeth
ESR Studies of Cu^{2+} in NH_4Cl Single Crystals between 4.2 and 453°K. (II): Theoretical Analysis: Vibrational Admixtures, Spin Polarization, and $\langle r^{-3} \rangle$ for Cu^{2+} Ions
Vol. **20**, No. 1, p. 237–248 (1967).
See also *19*, p. *328*; 22, p. 444.

A. A. Kaminskii
The Nature of "Aging" of CaF_2–YF_3–Nd^{3+} Crystals (Type I) under Stimulated Emission Conditions
Vol. **20**, No. 1, p. K51–K54 (1967).
See also *20.2*, p. *361*; 22.5.3, p. 532; 22.8, p. 549.

V. M. Judin and A. B. Sherman
Magnetic Properties of $NaNiF_3$
Vol. **20**, No. 2, p. 759–766 (1967).
See also *18.4*, p. *322*; 22.8, p. 550.

I. A. Akimov, V. M. Bentsa, F. I. Vilesov, and A. N. Terenin
External Photoeffect from Sensitizing Dyes Adsorbed on Semiconductors
Vol. **20**, No. 2, p. 771–776 (1967).
See also 13.3, p. 197; 16, p. 279; *17*, p. *287*; 22.4, p. 495; 22.5.1, p. 515.

A. WEISS and K. ZOHNER
Nuclear Quadrupole Interaction of ^{23}Na and ^{11}B and the Crystal Structure of $NaBF_4$
Vol. **21**, No. 1, p. 257–270 (1967).
See also 3.2, p. 35; 5, p. 51; *19*, p. *329*; 22.8, p. 550.

A. A. KAMINSKII, YU. K. VORONKO, and V. V. OSIKO
Mixed Systems on the Basis of Fluorides as New Laser Materials for Quantum Electronics. The Optical and Emission Parameters
Vol. **21**, No. 1, p. K17–K21 (1967).
See also 20.1, p. 348; *20.2*, p. *361*; 20.3, p. 365; 22.8, p. 550.

U. BERTELSEN, J. M. KNUDSEN, and H. KROGH
Mössbauer Effect in FeF_3
Vol. **22**, No. 1, p. 59–64 (1967).
See also *6.1*, p. *72*; 18.4, p. 322.

E. V. MATYUSHKIN, L. S. KUKUSHKIN, and V. V. EREMENKO
Peculiarities of the $RbMnF_3:Nd^{3+}$ Crystals Decay Kinetics due to the Migration of Excitation Energy
Vol. **22**, No. 1, p. 65–69 (1967).
See also 13.4, p. 208; *20.3*, p. *366*; 22.8, p. 551.

J. ARENDS and J. F. VERWEY
ESR on UV Irradiated Lead Halides at 80 °K
Vol. **23**, No. 1, p. 137–145 (1967).
See also 4, p. 43; 10.2, p. 140; *19*, p. *330*.

A. A. KAMINSKII, Y. E. LAPSKER, and B. P. SOBOLEV
Induced Emission of $NaCaCeF_6:Nd^{3+}$ at Room Temperature
Vol. **23**, No. 1, p. K5–K7 (1967).
See also 20.1, p. 350; *20.2*, p. *361*; 20.3, p. 367; 22.8, p. 552.

B. I. KULIEV and G. M. NEDLIN
The Types of Magnetic Ordering in Crystals Having the Structure of $RbNiF_3$
Vol. **24**, No. 1, p. K25–K28 (1967).
See also 1.2, p. 11; 3.2, p. 35; *18*, p. *293*; 22.8, p. 552.

L. GHIŢĂ and E. ŢENESCU
On the Exciton Spectrum of PbI_2 on Different Substrates
Vol. **24**, No. 2, p. K175–K178 (1967).
See also 13.2, p. 193; *20.1*, p. *351*.

M. V. DMITRUK, A. A. KAMINSKII, V. V. OSIKO, and T. A. TEVOSYAN
Induced Emission of Hexagonal LaF_3-SrF_2-Nd^{3+} Crystals at Room-Temperature
Vol. **25**, No. 2, p. K75–K78 (1968).
See also 20.1, p. 352; *20.2*, p. *361*; 20.3, p. 368; 22.8, p. 552.

R. FATEHALLY, N. P. SASTRY, and R. NAGARAJAN
Exchange and Spin-Orbit Coupling in $KFeF_3$
Vol. **26**, No. 1, p. 91–96 (1968).
See also *13*, p. *169*; 18.4, p. 323; 22.8, p. 553.

F. VAN DER WOUDE
On the Interpretation of Broadened and Shifted Lines in Mössbauer Spectra
Vol. **26**, No. 1, p. K37–K38 (1968).
See also *6.1*, p. *73*; 18.4, p. 324.

A. V. TULUB and K. PATZER
On the Theory of Cooperative Absorption (Emission) of Light in Crystals
Vol. **26**, No. 2, p. 693–700 (1968).
See also 13.4, p. 212; *20.1*, p. *353*; *20.3*, p. *369*.

G. BALDINI, I. POLLINI, and G. SPINOLO
Optical Properties of α- and β-$TiCl_3$
Vol. **27**, No. 1, p. 95–100 (1968).
See also 13.4, p. 213; *20.1*, p. *353*.

J. S. TIWARI, A. MEHRA, and K. G. SRIVASTAVA
Electronic Absorption Spectrum of K_2CoF_4
Vol. **28**, No. 1, p. 61–63 (1968).
See also 5, p. 53; 13, p. 171; *20.1*, p. *355*; 22.8, p. 554.

S. HAUSSÜHL
Elastic and Thermoelastic Behaviour of MgF_2 and MnF_2
Vol. **28**, No. 1, p. 127–130 (1968).
See also 3.2, p. 36; *12*, p. *158*; 22.5.3, p. 534.

M. R. TUBBS
The Variation of Exciton Peak Position with Lattice Parameter in Metal Iodide Crystals
Vol. **28**, No. 2, p. K135–K137 (1968).
See also 13.2, p. 195; *20.1*, p. *356*; 22.5.2, p. 528; 22.5.3, p. 534.

W. W. HOLLOWAY, JR. and M. KESTIGIAN
Temperature Dependence of the Broad-Band Fluorescence of $RbMnF_3$
Vol. **28**, No. 2, p. K159–K162 (1968).
See also *20.3*, p. *370*; 22.8, p. 554.

E. MOHLER
Linear Electro-Optic Effect of Excitons in CuCl
Vol. **29**, No. 1, p. K55–K57 (1968).
See also 13.2, p. 195; *14.3*, p. *243*; *20*, p. *342*.

S. P. TANEJA and S. P. PURI
Crystal Field Calculation of Principal Magnetic Susceptibilities and Anisotropy Variation with Temperature of Anhydrous Neodymium Chloride
Vol. **30**, No. 1, p. 193–197 (1968).
See also 3, p. 28; *18.1*, p. *298*.

R. V. PISAREV, I. G. SINY, N. N. NESTEROVA, G. A. SMOLENSKY, and P. P. SYRNIKOV
Faraday Rotation and Magnetic Dichroism in $RbNi(Co)F_3$
Vol. **30**, No. 1, p. 367–372 (1968).
See also 13, p. 173; *18.3*, p. *320*; *20*, p. *342*; 22.8, p. 556.

22.5.1 Silver Halides

Review Articles

P. Süptitz and J. Teltow
Transport of Matter in Simple Ionic Crystals (Cubic Halides)
Vol. **23**, No. 1, p. 9–56 (1967).
See also *9*, p. *87*; *10*, p. *93*; 14.4, p. 252; 15, p. 270; 22.5.2, p. 518; 22.5.3, p. 532.

K. N. Subramanian
Cyclic Stressing in Crystals of Rock-Salt Structure
Vol. **28**, No. 1, p. 9–37 (1968).
See also *10*, p. *93*; 10.1, p. 114; 21, p. 273; 21.1, p. 390; 22.5.2, p. 518; 22.6, p. 535.

Original Papers and Short Notes

R. B. Laibowitz and H. S. Sack
Luminescence and Photoconductivity in Silver Chloride with Laser Excitation
Vol. **17**, No. 1, p. 353–360 (1966).
See also 6, p. 57; 13.4, p. 202; *16*, p. *277*; *20.3*, p. *364*.

H. D. Koswig
"Optical Bleaching" of Ni^{2+}-Doped AgCl Crystals
Vol. **17**, No. 1, p. K83–K87 (1966).
See also 10.2, p. 137; *20.1*, p. *345*.

M. Höhne and M. Stasiw
ESR in Se- and Te-Doped AgBr Single Crystals
Vol. **17**, No. 1, p. K119–K121 (1966).
See also *19*, p. *326*.

H. Böttger
Temperature Dependence of the Optical Absorption of Transition Element Ions in Silver Chloride
Vol. **17**, No. 2, p. 855–868 (1966).
See also 13.4, p. 203; *20.1*, p. *346*.

D. Schwarz
Extrinsic Absorption of Chalcogen Doped AgBr Crystals in the Case of Variable Cation-Impurity Concentration
Vol. **18**, No. 2, p. 787–794 (1966).
See also 10, p. 97; *20.1*, p. *346*.

M. Höhne and M. Stasiw
Te_2^{3-} and Se_2^{3-} Centres in AgBr. (I): The *g*-, Superhyperfine, and Hyperfine Tensors
Vol. **20**, No. 2, p. 657–666 (1967).
See also 10.2, p. 139; 13.4, p. 206; *19*, p. *328*.

M. Höhne and M. Stasiw
Te_2^{3-} and Se_2^{3-} Centres in AgBr. (II): Electronic Structure
Vol. **20**, No. 2, p. 667–673 (1967).
See also 10.2, p. 139; 13.4, p. 206; *19*, p. *328*.

I. A. Akimov, V. M. Bentsa, F. I. Vilesov, and A. N. Terenin
External Photoeffect from Sensitizing Dyes Adsorbed on Semiconductors
Vol. **20**, No. 2, p. 771–776 (1967).
See also 13.3, p. 197; 16, p. 279; *17*, p. *287*; 22.4, p. 495; 22.5, p. 512.

P. Müller
On the Calculation of Disorder in Silver Halides
Vol. **21**, No. 2, p. 693–701 (1967).
See also *10*, p. *100.*

K. Vacek
Electroluminescence of AgCl Crystals Excited by Short Field Pulses
Vol. **23**, No. 1, p. 105–111 (1967).
See also *20.3*, p. *367.*

F. Vávra
A Note on Etching of Dislocations in AgCl Crystals
Vol. **23**, No. 2, p. K121–K122 (1967).
See also *10*, p. *102.*

M. Höhne and M. Stasiw
ESR Studies of Different Se_2^{3-} Centres in Silver Halides
Vol. **24**, No. 2, p. 591–604 (1967).
See also 10, p. 103; *19*, p. *331.*

H. D. Koswig, U. Retter, and W. Ulrici
Jahn-Teller Effect of Transition-Metal Ions in Silver Halides. (I): $Ti^{3+}(3d^1)$ and $Fe^{2+}(3d^6)$
Vol. **24**, No. 2, p. 605–614 (1967).
See also *13.4*, p. *210*; 20.1, p. 351.

M. Höhne and M. Stasiw
Detection of Ag^{2+} in Doped AgCl Crystals by ESR
Vol. **25**, No. 1, p. K55–K57 (1968).
See also 10.2, p. 141; *19*, p. *332.*

A. Buroff and J. Malinowski
Charge Compensation of Trapped Holes in Silver Bromide
Vol. **26**, No. 1, p. 267–276 (1968).
See also 10.2, p. 142; 13.4, p. 212; *16*, p. *282.*

J. Donecker
Phonon Scattering by Point-Defect Aggregates in AgCl Crystals
Vol. **26**, No. 2, p. K131–K134 (1968).
See also *6*, p. *65*; 8, p. 84; 10, p. 106.

W. Ulrici
Study of Trivalent Transition-Metal Ions in Silver Halides. (I): AgCl:V
Vol. **27**, No. 1, p. 333–344 (1968).
See also *10*, p. *107*; 13.4, p. 213; 14.4, p. 254; 20.1, p. 354.

K. Vacek, A. Mysyrowicz, R. Levy, A. Bivaš, and S. Nikitine
Luminescence of AgCl Excited by Ultraviolet Laser
Vol. **27**, No. 1, p. K29–K31 (1968).
See also *20.3*, p. *369.*

M. Georgiev and E. Petrova
Investigation of the Drift of Photoelectrons in AgBr at Moderate Temperatures
Vol. **27**, No. 1, p. K33–K36 (1968).
See also *10.2*, p. *142.*

D. Schwarz
ESR Study of Cu^{2+} in AgCl Crystals Additionally Doped with Se or S
Vol. **27**, No. 1, p. K77–K80 (1968).
See also 10, p. 107; 13.4, p. 213; *19*, p. *333.*

W. Ulrici
Study of Trivalent Transition-Metal Ions in Silver Halides. (II): Chromium and Titanium
Vol. **27**, No. 2, p. 489–500 (1968).
See also *10*, p. *107*; 13.4, p. 213; 14.4, p. 254; 20.1, p. 354.

P. Süptitz and R. Weidmann
Diffusion of Mn^{2+} in AgCl and AgBr Crystals
Vol. **27**, No. 2, p. 631–638 (1968).
See also *9*, p. *92*; 10, p. 108.

K. Hennig
Study of Iron Ions in ^{57}Fe-Doped AgCl Using the Mössbauer Effect
Vol. **27**, No. 2, p. K115–K116 (1968).
See also *6.1*, p. *74*; 10, p. 108.

K. Hennig, Kim Yung, and B. S. Skorchev
On States of Iron Impurity Ions in AgCl and NaCl Investigated by the Mössbauer Method
Vol. **27**, No. 2, p. K161–K164 (1968).
See also 6.1, p. 74; *10*, p. *108*; 22.5.2, p. 527.

Yu. E. Perlin and Sh. N. Gifeisman
The Theory of Shallow F-Centres
Vol. **28**, No. 1, p. 45–60 (1968).
See also 6, p. 66; *10.2*, p. *143*; 13, p. 171; 14.4, p. 254; 18, p. 294; 22.2.1, p. 485; 22.2.2, p. 488; 22.2.3, p. 493; 22.4, p. 495; 22.4.1, p. 503; 22.4.2, p. 508; 22.4.3, p. 511; 22.5.2, p. 527.

M. Höhne and M. Stasiw
ESR Detection of Self-Trapped Holes in AgCl
Vol. **28**, No. 1, p. 247–253 (1968).
See also 10.2, p. 143; *19*, p. *334*; 20.1, p. 355.

M. Georgiev, E. Zagorska, and J. Malinowski
Temperature Dependence of the Diffusion of Photoexcited Holes in AgBr
Vol. **28**, No. 2, p. 725–730 (1968).
See also *10.2*, p. *143*; 16, p. 283.

W. W. Walker and L. J. Demer
Indentation Creep in Crystals
Vol. **29**, No. 2, p. K141–K143 (1968).
See also *10*, p. *111*; 22.5.2, p. 529; 22.5.3, p. 534; 22.6, p. 543.

22.5.2 Alkali Halides

Review Articles

P. Süptitz and J. Teltow
Transport of Matter in Simple Ionic Crystals (Cubic Halides)
Vol. **23**, No. 1, p. 9–56 (1967).
See also *9*, p. *87*; *10*, p. *93*; 14.4, p. 252; 15, p. 270; 22.5.1, p. 515; 22.5.3, p. 532.

K. N. Subramanian
Cyclic Stressing in Crystals of Rock-Salt Structure
Vol. **28**, No. 1, p. 9–37 (1968).
See also *10*, p. *93*; 10.1, p. 114; 21, p. 373; 21.1, p. 390; 22.5.1, p. 515; 22.6, p. 535.

Original Papers and Short Notes

P. Sauer, O. Schirmer, and J. Schneider
Comments on Paraelectric Resonance
Vol. **16**, No. 1, p. 79–88 (1966).
See also 10, p. 93; *13.4*, p. *201*.

V. Topa and S. Haragea
Colour Centres in X-Irradiated NaCl:Ag and NaCl:(Ag + Ca) Crystals
Vol. **16**, No. 1, p. 137–145 (1966).
See also *10.2*, p. *137*.

B. I. Smirnov and B. A. Efimov
Development of Glide Bands during Plastic Deformation of LiF Crystals
Vol. **16**, No. 1, p. 191–196 (1966).
See also *10*, p. *93*.

W. Dreybrodt and D. Silber
EPR Measurements on Diatomic Halogen Centers in Alkali Halides
Vol. **16**, No. 1, p. 215–223 (1966).
See also 10.2, p. 137; 13.4, p. 201; *19*, p. *326*.

M. L. Mukherjee and H. N. Bose
Nonradiative Destruction of F-Centres in X-Irradiated KCl Doped with Anionic Impurities
Vol. **16**, No. 2, p. 591–598 (1966).
See also *10.2*, p. *137*; 20.3, p. 364.

M. Musa
Some Results Concerning the Co^{++} Site Symmetry in Alkali Halide Crystals
Vol. **16**, No. 2, p. 771–776 (1966).
See also *10*, p. *94*; 20.1, p. 345.

O. Adametz
Conversion of U-Centres in KCl Crystals into OH-Centres
Vol. **16**, No. 2, p. K161–K164 (1966).
See also *10.2*, p. *137*.

Y. Kazumata, T. Kamikawa, and K. Ozawa
Zero-Phonon Lines of Color Centers in LiF and NaF
Vol. **17**, No. 1, p. 131–138 (1966).
See also *6*, p. *56*; *10.2*, p. *137*.

F. Van de Wiele and G. Jacobs
Optical and Thermal Properties of Color Centers in CsBr Single Crystals Irradiated at Liquid-Helium Temperature
Vol. **17**, No. 1, p. 177–184 (1966).
See also *10.2*, p. *137*.

J. H. Beaumont and P. W. M. Jacobs
Effect of Gaseous Impurities on the Structure-Sensitive Conductance of Potassium Chloride Single Crystals
Vol. **17**, No. 1, p. K45–K48 (1966).
See also 10, p. 95; *14.4*, p. *252*.

T. Kamikawa, Y. Kazumata, and K. Ozawa
New Infrared Absorption in Neutron-Irradiated Lithium Fluoride
Vol. **17**, No. 2, p. 645–655 (1966).
See also *10.2*, p. *137*; 11, p. 146; 20.1, p. 346.

F. Rueda and W. Dekeyser
Charge of Dislocations in Pure and γ-Irradiated LiF
Vol. **17**, No. 2, p. 739–744 (1966).
See also *10*, p. *95*; 10.2, p. 138.

G. P. Upit and S. A. Varchenya
Mircohardness of Alkali Halide Crystals
Vol. **17**, No. 2, p. 831–835 (1966).
See also *12*, p. *156*.

G. Hauret and M. Girard-Nottin
Study of Light Scattering in Potassium Chloride Crystals Doped with Calcium
Vol. **17**, No. 2, p. 881–888 (1966).
See also *10*, p. *95*; 20, p. 338.

W. Dreybrodt and K. Fussgaenger
Interpretation of the Temperature Dependence of Ag^+ Dipole Strengths in Alkali Halides
Vol. **18**, No. 1, p. 133–144 (1966).
See also 13.4, p. 203; *20.1*, p. *346*.

R. C. Lincoln, K. M. Koliwad, and P. B. Ghate
Elastic Constants of Some NaCl Type Alkali Halides
Vol. **18** No. 1, p. 265–277 (1966).
See also 3.2, p. 34; *12*, p. *156*.

W. Degrieck and G. Jacobs
Properties of the B-Centre in Additively Coloured Ag^+ Doped KCl Crystals
Vol. **18**, No. 1, p. 279–284 (1966).
See also *10.2*, p. *138*.

Hj. Matzke
Diffusion of Tritium from Lithium Fluoride
Vol. **18**, No. 1, p. 317–324 (1966).
See also 11, p. 147; *17*, p. *286*.

U. Gerhardt and E. Mohler
Piezooptical Experiments on the Excitons in KBr and KI
Vol. **18**, No. 1, p. K45–K48 (1966).
See also *13.2*, p. *192*; 20.1, p. 346.

P. PETRESCU and V. GHEORDĂNESCU
Anisotropy of the Photoemission from Some Coloured Alkali Halides
Vol. 18, No. 2, p. 511–515 (1966).
See also 10.2, p. 138; *17*, p. *287*.

L. BEN-DOR, A. GLASNER, and S. ZOLOTOV
Colour Centres in KCl Doped with Zn^{+2}, Cd^{+2}, and Hg^{+2}
Vol. 18, No. 2, p. 593–599 (1966).
See also *10.2*, p. *138*.

D. SCHMID
The *g*-Factor of F-Centres in Alkali Halide Crystals
Vol. 18, No. 2, p. 653–666 (1966).
See also 10.2, p. 138; *19*, p. *327*.

K. K. SHVARTS, A. YA. VITOL, A. V. PODIN, D. O. KALNIN, and YU. A. EKMANIS
Radiation Effects in Pile-Irradiated LiF Crystals
Vol. 18, No. 2, p. 897–909 (1966).
See also 10.2, p. 138; *11*, p. *147*; 19, p. 327; 20.1, p. 346.

P. V. SASTRY
On the Photo- and Thermo-Chemical Reactions in Additively Colored Sodium Chloride
Vol. 18, No. 2, p. K127–K130 (1966).
See also *10.2*, p. *138*.

R. LAIHO, P. KETOLAINEN, and E. MÄNTYSALO
Ultrasonic Modulation of F-Band in Additively Colored KBr
Vol. 18, No. 2, p. K143–K144 (1966).
See also 7, p. 76; *10.2*, p. *138*.

E. D. ALUKER and I. P. MEZINA
The Temperature Dependence of Radioluminescence Yield in Alkali Halide Crystal Phosphors
Vol. 19, No. 1, p. 35–40 (1967).
See also 13.4, p. 204; *20.3*, p. *365*.

R. P. HARRISON
The Effect of Cadmium Additions on the Fracture Strength of Sodium Chloride Single Crystals
Vol. 19, No. 1, p. 353–357 (1967).
See also *10*, p. *97*; 12, p. 156.

YU. S. BOYARSKAYA, R. P. ZHITARU, and V. G. KOLOSKOVA
Effect of Different Lattice Defects on the Mobility of Dislocations in Alkali Halide Crystals
Vol. 19, No. 1, p. 441–451 (1967).
See also *10*, p. *97*; 11, p. 147.

YU. A. LOGATCHOV
Model Potential Approximation for Z_1-Centres in Ionic Crystals
Vol. 19, No. 1, p. K35–K38 (1967).
See also 10.2, p. 138; *13.4*, p. *204*.

W. BARTH and B. FRITZ
Local Mode Absorption by U-Centres in Alkali-Halide Mixed Crystals
Vol. 19, No. 2, p. 515–524 (1967).
See also *10.2*, p. *138*.

V. B. Pariiskii
Effect of Stress State on the Shape of the Dislocation Rosette near an Indentation in Some Alkali Halide Crystals
Vol. **19**, No. 2, p. 525—532 (1967).
See also *10*, p. *97*.

S. N. Komnik, V. Z. Bengus, and E. D. Lyak
On the Kinetics of Plastic Deformation of Some Alkali Halide Crystals
Vol. **19**, No. 2, p. 533—541 (1967).
See also *10*, p. *98*.

O. E. Facey and P. W. M. Jacobs
Photochemical Properties of SH^- Ions in KCl and KBr Crystals
Vol. **19**, No. 2, p. 565—576 (1967).
See also *10.2*, p. *139*.

R. Böttcher, W. Windsch, and W. Lüdke
EPR Investigations on Mn^{2+}-Doped LiF Crystals
Vol. **20**, No. 1, p. 121—126 (1967).
See also 9, p. 88; 13.4, p. 205; *19*, p. *328*.

W. Dreybrodt and D. Silber
Electron Spin Resonance of Tl^{++} Centres in KCl Crystals
Vol. **20**, No. 1, p. 337—346 (1967).
See also 13.4, p. 205; *19*, p. *328*.

F. T. Goldstein
F-Center Formation by Fundamental Absorption in KI
Vol. **20**, No. 1, p. 379—393 (1967).
See also *10.2*, p. *139*; 20.3, p. 365.

R. E. DeWames and W. F. Hall
A Relation between the Energy Loss Spectrum of Incoherently Scattered Electrons and the Phonon Density of States in Crystals
Vol. **20**, No. 1, p. K11—K15 (1967).
See also *6*, p. *60*.

H. P. Utech
Interpretation of de Grindberg's Experiments on Lateral Surface Fringes in Sodium Chloride
Vol. **20**, No. 1, p. K41—K43 (1967).
See also *1.3*, p. *15*.

P. Schmeling
Non-Ideal Diffusion of Argon in Potassium Chloride
Vol. **20**, No. 2, p. 597—603 (1967).
See also *9*, p. *89*; 11, p. 148.

W. Blum and B. Ilschner
Creep Behaviour of NaCl Single Crystals
Vol. **20**, No. 2, p. 629—642 (1967).
See also *10*, p. *99*.

I. V. Abarenkov and I. M. Antonova
The Model Potential for Systems with Closed Shells
Vol. **20**, No. 2, p. 643—651 (1967).
See also *13.1*, p. *177*.

H. Bilz, R. Zeyher, and R. K. Wehner
Anharmonic Sideband Effects of the U-Center Local Mode
Vol. **20**, No. 2, p. K167—K171 (1967).
See also *6*, p. *60*; 10.2, p. 139; 20.1, p. 348.

W. Dreybrodt
Temperature Dependence of Hyperfine Structure Tensors of Diatomic Halogen Centres in Alkali Halide Crystals
Vol. **21**, No. 1, p. 99—112 (1967).
See also 13.4, p. 206; *19*, p. *329*.

M. Järvinen and O. Inkinen
An X-Ray Diffraction Study of Rubidium Chloride
Vol. **21**, No. 1, p. 127—135 (1967).
See also 5, p. 51; *13*, p. *165*.

O. Budke and J. Krempaský
Measurement of Thermal Parameters of Thin Films and Plane-Parallel Plates Located on a Thermally Conductive Base
Vol. **21**, No. 1, p. 175—182 (1967).
See also *8*, p. *82*; 22.2.3, p. 491.

B. Dutta, G. Holtmeyer, and J. Kalus
Study of the Anisotropic Slowing-Down of Low-Energy Magnesium Atoms in an NaCl Single Crystal at Various Temperatures
Vol. **21**, No. 1, p. 291—295 (1967).
See also 6.1, p. 72; *11*, p. *148*.

V. Nüsslein and U. Schröder
Calculations of Dispersion Curves and Specific Heat for LiF and NaCl Using the Breathing Shell Model
Vol. **21**, No. 1, p. 309—314 (1967).
See also *6*, p. *61*; 8, p. 82.

A. Behr, H. Peisl, and W. Waidelich
X-Ray Production of Vacancies in KCl at Low Temperatures
Vol. **21**, No. 1, p. K9—K10 (1967).
See also 10, p. 100; *11*, p. *148*.

M. Karras and E. Suoninen
Energy Distribution of Photoelectrons Created by Al K_α X-Rays in Alkali Halide Targets
Vol. **21**, No. 2, p. 487—493 (1967).
See also *17*, p. *287*.

J. Hesse
Yield Stress and Dislocation Density in Calcium-Doped NaCl Single Crystals
Vol. **21**, No. 2, p. 495—505 (1967).
See also *10*, p. *100*.

K. M. Koliwad, P. B. Ghate, and A. L. Ruoff
Pressure Derivatives of the Elastic Constants of NaBr and KF
Vol. **21**, No. 2, p. 507—516 (1967).
See also 7, p. 76; *12*, p. *157*.

H. Kühnert
Polarization of F-Center Luminescence in KCl due to Electric Fields
Vol. **21**, No. 2, p. K171—K173 (1967).
See also 10.2, p. 139; *20.3*, p. *366*.

R. SITTIG
Stark Effect Measurements of the OH^- Absorption Band in KCl
Vol. **21**, No. 2, p. K175–K177 (1967).
See also 10.2, p. 139; 14.4.1, p. 259; *20.1*, p. *349*.

E. W. CLAFFY
Impurity Absorption Bands in Thermoluminescent LiF
Vol. **22**, No. 1, p. 71–76 (1967).
See also 13.4, p. 208; *20.1*, p. *349*; 20.3, p. 366.

K. LAL
Dielectric Properties of Interstitial Impurity Cations in Alkali Halide Crystals
Vol. **22**, No. 1, p. 141–146 (1967).
See also 10, p. 101; *14.4*, p. *253*.

W. DEGRIECK and G. JACOBS
Luminescence and Photoconductivity of the B-Centre in Additively Coloured KCl:Ag Crystals
Vol. **22**, No. 1, p. 177–184 (1967).
See also 10.2, p. 139; *16*, p. *279*; *20.3*, p. *366*.

G. WOLFRAM and V. WITT
Kinetics of the Thermal Mutual Conversion of F and Z_2 Color Centers in KCl:Sr Crystals
Vol. **22**, No. 1, p. 245–253 (1967).
See also *10.2*, p. *139*.

F. AGULLÓ-LÓPEZ
Role of γ-Radiation Intensity on the First Stage of theRoom-Temperature F-Colouring in NaCl
Vol. **22**, No. 2, p. 483–490 (1967).
See also *10.2*, p. *140*; 11, p. 149.

S. C. JAIN and G. D. SOOTHA
Electrical and Optical Properties of Titanium-Doped Lithium Fluoride Crystals
Vol. **22**, No. 2, p. 505–516 (1967).
See also 13.4, p. 208; *14.4*, p. *253*; *20.1*, p. *349*.

J. H. CRAWFORD, JR., W. A. SIBLEY, and E. SONDER
The Influence of Electron-Trapping Impurity on Defect Creation and Bleaching in KCl Irradiated at 78 °K
Vol. **23**, No. 1, p. 301–306 (1967).
See also *10.2*, p. *140*; 11, p. 150; 13.4, p. 209.

S. WANG
Optical Absorption of an Interstitial Alkali-Earth Cation in an Alkali-Halide Crystal
Vol. **23** No. 1, p. 387–399 (1967).
See also 10.2, p. 140; 13.2, p. 193; *13.4*, p. *209*.

S. WANG
Electronic States of the M_A-Centers in KCl
Vol. **23**, No. 1, p. 401–412 (1967).
See also 10.2, p. 140; *13.4*, p. *209*.

R. VOSZKA, T. HORVÁTH, and A. WATTERICH
A Possible Model for V-Centres Stable at Room Temperature in X-Irradiated KCl(Ca) Crystals
Vol. **23**, No. 1, p. K71–K73 (1967).
See also *10.2*, p. *140*.

H. PEISL, H. SPALT, and W. WAIDELICH
X-Ray Diffuse Scattering in γ-Irradiated LiF
Vol. **23**, No. 1, p. K75—K78 (1967).
See also *10*, p. *102*; 10.2, p. 140.

E. YU. GUTMANAS and E. M. NADGORNYI
Estimation of the Peierls Stress in Alkali Halides
Vol **23**, No. 2, p. 571—576 (1967).
See also *10*, p. *102*.

Z. P. CHANG, G. R. BARSCH, and D. L. MILLER
Pressure Dependence of Elastic Constants of Cesium Halides
Vol. **23**, No. 2, p. 577—586 (1967).
See also *12*, p. *157*.

R. G. J. GRISAR, K. P. REINERS, K. F. RENK, and L. GENZEL
Impurity-Induced Far-Infrared Absorption in the Phonon Gap Regions of KI and KBr
Vol. **23**, No. 2, p. 613—620 (1967).
See also *6*, p. *63*; *20,1*, p. *350*.

P. PETRESCU
L-Bands Detected in Photoemission Spectra of NaCl, LiCl, and RbBr
Vol. **23**, No. 2, p. K105—K107 (1967).
See also *10.2*, p. *140*; 17, p. 287.

D. BÄUERLE and B. FRITZ
Infrared Vibrational Absorption by U-Centers in NaI
Vol. **24**, No. 1, p. 207—215 (1967).
See also *6*, p. *63*; 10.2, p. 141; *20.1*, p. *351*.

G. M. BARTENEV and E. G. KORYAK-DORONENKO
The Theoretical Strength of NaCl Crystals under Tension along Its Body and Face Diagonals
Vol. **24**, No. 2, p. 443—449 (1967).
See also *10*, p. *103*; 12, p. 158.

J. B. PAGE, JR. and D. STRAUCH
Shell Model Treatment of the Vibrations of Crystals Containing Substitutional Defects. (I): U-Center Localized Modes
Vol. **24**, No. 2, p. 469—479 (1967).
See also 3.2, p. 35; *6*, p. *64*; 10.2, p. 141.

T. H. CHEUNG and S. WANG
Theoretical Studies on Z_1-Bands in Alkali Halide Crystals
Vol. **24**, No. 2, p. 509—523 (1967).
See also 10.2, p. 141; *13.4*, p. *210*.

F. FRÖHLICH and G. HENSEL
Low Temperature Conductivity of Ca-Doped KCl Crystals Influenced by Segregation Processes
Vol. **24**, No. 2, p. 535—541 (1967).
See also 10, p. 103; *14.4*, p. *253*.

F. FRÖHLICH and O. ADAMETZ
Conversion of OH-Centres in KCl Crystals into U-Centres by Electrolytic Coloration
Vol. **24**, No. 2, p. 583—590 (1967).
See also *10.2*, p. *141*.

A. Glasner, R. Reisfeld, and A. Linenberg
Diffusion of Tl^+ in Doped and Pretreated KCl
Vol. **24**, No. 2, p. 695–706 (1967).
See also *9*, p. *90*.

J. Arends
Defect Formation Energies and Migration Energy Barriers in Cesium Halides
Vol. **24**, No. 2, p. K129–K131 (1967).
See also *10*, p. *103*.

H. L. Arora, S. Kim, and S. Wang
Optical Absorption of the Impurity Ag and Models for the E-Band in an Alkali Halide
Vol. **25**, No. 1, p. 223–232 (1968).
See also *10.2*, p. *141*; 13.4, p. 211.

A. H. Scholz
Computer Calculations for Vacancies in Alkali Halides with NaCl Structure
Vol. **25**, No. 1, p. 285–301 (1968).
See also *10*, p. *104*.

F. Fröhlich and P. Seifert
Potential Differences in NaCl Crystals due to Plastic Deformation by Indentation Experiments between RT and 20 °K
Vol. **25**, No. 1, p. 303–310 (1968).
See also *10*, p. *104*.

G. Giuliani and E. Reguzzoni
X-Ray Production and Thermal Annealing of Frenkel Pairs in KCl
Vol. **25**, No. 1, p. 437–448 (1968).
See also *10.2*, p. *141*; *11*, p. *151*.

G. Baldini, A. Jean, and G. Spinolo
Lifetime of the Excited Cu^+ in Alkali Halides
Vol. **25**, No. 2, p. 557–561 (1968).
See also 13.4, p. 211; 20.1, p. 351; *20.3*, p. *368*.

Y. Kazumata
Atomic ^3H Center in Neutron-Irradiated LiF
Vol. **25**, No. 2, p. 563–566 (1968).
See also *10.2*, p. *142*; 11, p. 151; 19, p. 332.

L. Pelsers and G. Jacobs
Photoconductivity of CsBr Single Crystals Containing F- and M-Centres
Vol. **25**, No. 2, p. 621–628 (1968).
See also 10.2, p. 142; *16*, p. *281*.

R. Kelly, Č. Jech, and Hj. Matzke
Diffusion in Inert-Gas Bombarded KCl, KBr, and KI
Vol. **25**, No. 2, p. 641–653 (1968).
See also *9*, p. *91*; 11, p. 151.

R. Gauthier, D. Bois, P. Pinard, and F. Davoine
Ion and Electron Production by Interaction of a Coherent Light Beam with Substance
Vol. **25**, No. 2, p. 691–696 (1968).
See also *17*, p. *288*; 21.1, p. 403; 21.1.1, p. 420; 22.1.1, p. 464; 22.2.1, p. 485; 22.4.1, p. 502; 22.6, p. 540.

J. HINGSAMMER and H. JODL
Z_1- and Z_2-Centres in Mixed Crystals of Alkali Halides Doped with Sr^{++}
Vol. **26**, No. 1, p. 261—266 (1968).
See also *10.2*, p. *142*; 22.8, p. 553.

K.-H. MATUCHA
Electron-Microscopic Study of Slip Lines on Deformed NaCl
Vol. **26**, No. 1, p. 291—310 (1968).
See also *10*, p. *105*.

R. A. ZHITNIKOV, V. B. KOLTSOV, and N. I. MELNIKOV
Paramagnetic Resonance of Copper Atoms Stabilized in Alkali-Halide Crystals
Vol. **26**, No. 1, p. 371—380 (1968).
See also 10, p. 105; *19*, p. *333*.

R. P. HARRISON
The Effect of Particle Size on the Fracture Strength of Sodium Chloride Single Crystals
Vol. **26**, No. 1, p. K65—K67 (1968).
See also *4*, p. *45*; 10, p. 105; 12, p. 158.

A. B. KUNZ
Notes on a Valence Electron Model of Atoms
Vol. **26**, No. 2, p. 447—452 (1968).
See also *13.4*, p. *212*.

J. ARENDS and H. NIJBOER
The Ionic Conductivity of CsCl Doped with K^+, Rb^+, and Br^-
Vol. **26**, No. 2, p. 537—542 (1968).
See also 10, p. 105; *14.4*, p. *254*.

U. MESSERSCHMIDT
Influence of Impurities on Dislocation Velocities Controlled by Jog-Dragging
Vol. **26**, No. 2, p. 701—708 (1968).
See also *10*, p. *106*.

YU. A. LOGATCHOV
Two-Electron Correlation in Impurity Centres in Model Potential Approximation
Vol. **26**, No. 2, p. 765—768 (1968).
See also *13.4*, p. *212*; 20.3, p. 369.

K. MEYER and E. LINKE
Triboinduced Electron-Emission and Luminescence from Alkali Halides
Vol. **26**, No. 2, p. K89—K92 (1968).
See also *17*, p. *289*; *20.3*, p. *369*.

R. A. SHATAS and J. D. STETTLER
Mollwo-Ivey Relation in Aggregate Defect Centers
Vol **26**, No. 2, p. K155—K157 (1968).
See also 10.2, p. 142; *13.4*, p. *212*; 22.6, p. 541.

F. W. FELIX
Rare-Gas Diffusion in Neutron-Irradiated Rubidium Halides
Vol. **27**, No. 2, p. 529—534 (1968).
See also *9*, p. *91*; 10, p. 107; 11, p. 152.

K. MAIER and W. GEBHARDT
Stress-Induced Inhomogenous Effects of the K-Absorption in RbCl
Vol. **27**, No. 2, p. 713—721 (1968).
See also *10.2*, p. *143*; 13.4, p. 214.

R. P. Harrison
The Yield Strength of Sodium Chloride Containing Manganese
Vol. 27, No. 2, p. K153—K156 (1968).
See also *10*, p. *108*.

K. Hennig, Kim Yung, and B. S. Skorchev
On States of Iron Impurity Ions in AgCl and NaCl Investigated by the Mössbauer Method
Vol. 27, No. 2, p. K161—K164 (1968).
See also 6.1, p. 74; *10*, p. *108*; 22.5.1, p. 517.

R. Balzer, H. Peisl, and W. Waidelich
Evidence for Frenkel Defect Recombination during Thermal Annealing of Low-Temperature X-Irradiated KCl
Vol. 27, No. 2, K165—K167 (1968).
See also *10*, p. *108*; 11, p. 153.

Yu. E. Perlin and Sh. N. Gifeisman
The Theory of Shallow F-Centres
Vol. 28, No. 1, p. 45—60 (1968).
See also 6, p. 66; *10.2*, p. *143*; 13, p. 171; 14.4, p. 254; 18, p. 294; 22.2.1, p. 485, 22.2.2, p. 488; 22.2.3, p. 493; 22.4, p. 495; 22.4.1, p. 503; 22.4.2, p. 508; 22.4.3; p. 511; 22.5.1, p. 517.

K. J. Rao and C. N. R. Rao
Schottky Defects in Alkali Halides: Their Creation, Interaction, and Migration
Vol. 28, No. 1, p. 157—169 (1968).
See also 3.2, p. 36; *10*, p. *108*.

R. Balzer, H. Peisl, and W. Waidelich
Lattice Disorder in Alkali Halides X-Irradiated at Room Temperature
Vol. 28, No. 1, p. 207—210 (1968).
See also *10*, p. *108*; 11, p. 153.

B. G. Dick
Calculation of Properties of Vacancy-Centered Z_1-Center Models in Alkali Halides
Vol. 28, No. 1, p. 223—232 (1968).
See also *10.2*, p. *143*; *13.4*, p. *214*.

T. V. Samoilova, B. I. Smirnov, and E. E. Chernov
Electron Microscopy Observation of Dislocation Multiples in Slip Bands of LiF Crystals
Vol. 28, No. 1, p. K1—K3 (1968).
See also *10*, p. *109*.

G. Wolfram
Photoconductivity of the Z_2-Center in KCl:Sr Crystals
Vol. 28, No. 1, p. K11—K13 (1968).
See also 10.2, p. 143; 13.4, p. 214; *16*, p. *283*.

T. Okada, H. Kawai, S. Mukai, and T. Suita
Hardening of KCl Crystals by F_A-Centers
Vol. 28, No. 1, p. K19—K21 (1968).
See also *10*, p. *109*; *10.2*, p. *143*.

P. O. Szomor
Elastic Constants and Tensile Strength of NaCl-Type Alkali Halides
Vol. 28, No. 2, p. 529—537 (1968).
See also *10*, p. *109*; *12*, p. *158*.

H. Böttger
Multiphonon Absorption of Crystals with Point Defects in a Strong Radiation Field
Vol. **28**, No. 2, p. 611–621 (1968).
See also *6*, p. *67*; *10*, p. *109*.

O. G. Vendik
Calculation of the Point Lattice Potential of a Bounded Crystal
Vol. **28**, No. 2, p. 789–796 (1968).
See also *6*, p. *67*.

T. A. Kudykina, K. B. Tolpygo, and S. M. Zubkova
Exciton Theory in Crystals with Complicated Band Structure with Electron-Phonon Interaction
Vol. **28**, No. 2, p. 807–814 (1968).
See also 13.1, p. 188; *13.2* p. *194*; 22, p. 453.

M. R. Tubbs
The Variation of Exciton Peak Position with Lattice Parameter in Metal Iodide Crystals
Vol. **28**, No. 2, p. K135–K137 (1968).
See also 13.2, p. 195; *20.1*, p. *356*; 22.5, p. 514; 22.5.3, p. 534.

A. B. Kunz
Electronic Energy Bands for Rubidium Chloride and the Face-Centered Cubic Alkali Bromides
Vol. **29**, No. 1, p. 115–120 (1968).
See also *13.1*, p. *188*.

M. P. Fontana, H. Blume, and W. J. van Sciver
Properties of Exciton States in NaI. (I): The Intrinsic Photoluminescence
Vol. **29**, No. 1, p. 159–166 (1968).
See also *13.2*, p. *195*; 20.3, p. 370.

C. Sánchez and F. Agulló-López
Transient Effects in the Room-Temperature F-Colouring of NaCl Irradiated with X- or γ-Rays
Vol. **29**, No. 1, p. 217–230 (1968).
See also *10.2*, p. *143*; 11, p. 153.

P. Petrescu
On the L-Bands and the First Exciton Bands in the Photoemission Spectra of Alkali Iodides
Vol. **29**, No. 1, p. 333–340 (1968).
See also 10.2, p. 143; 13.2, p. 195; *17*, p. *289*.

A. N. Basu and S. Sengupta
A Deformable Shell Model for the Alkali Halides
Vol. **29**, No. 1, p. 367–375 (1968).
See also *6*, p. *67*.

W. Frank
The Critical Shear Stress of Alkali Halides Doped with Divalent Cations. (I): Non-Associated Impurity–Cation Vacancy Pairs
Vol. **29**, No. 1, p. 391–402 (1968).
See also *10*, p. *110*.

P. Petrescu
On the Photoemission Spectrum of LiF:Mg
Vol. **29**, No. 1, p. K1–K2 (1968).
See also 10.2, p. 144; *17*, p. *289*.

P. Petrescu
Evidence for L-Bands in Photoemission Spectra of KF and KCl
Vol. **29**, No. 1, p. K3–K5 (1968).
See also 10.2, p. 144; *17*, p. *289*.

Š. Kupča
Hardness of NaCl Single Crystals Prepared with Various Concentrations of $BaCl_2$ in the Melt
Vol. **29**, No. 1, p. K15–K17 (1968).
See also *10*, p. *110*; 14.4, p. 254.

H. Weijma and B. R. Knottnerus
Ionic Conductivity of Pure and Ba^{2+}-Doped CsCl
Vol. **29**, No. 1, p. K43–K45 (1968).
See also 1.2, p. 12; 10, p. 110; *14.4*, p. *254*.

B. von Guérard, H. Peisl, and W. Waidelich
Equilibrium Vacancy Concentration in KCl
Vol. **29**, No. 1, p. K59–K61 (1968).
See also *10*, p. *110*.

R. Kersten
Hyperfine and Quadrupole Interactions of the F-Centre in KCl as Measured by ENDOR
Vol. **29**, No. 2, p. 575–586 (1968).
See also 10.2, p. 144; *19*, p. *335*.

B. G. Dick
One-Phonon Relaxation of Hydroxyl Ions in Alkali Halide Crystals
Vol. **29**, No. 2, p. 587–596 (1968).
See also 6, p. 68; *10.2*, p. *144*.

D. Bäuerle and B. Fritz
Vibrational Spectra of U_1-Centres in KI
Vol. **29**, No. 2, p. 639–643 (1968).
See also *6*, p. *68*; 10.2, p. 144; *20.1*, p. *357*.

W. Frank
The Critical Shear Stress of Alkali Halides Doped with Divalent Cations. (II): Associated Impurity-Vacancy Dipoles and Impurity-Rich Precipitations
Vol. **29**, No. 2, p. 767–780 (1968).
See also *10*, p. *111*.

A. Mehra
Optical Absorption of Mn^{2+}-Doped Alkali Halides
Vol. **29**, No. 2, p. 847–857 (1968).
See also 6, p. 69; *10*, p. *111*; 13.4, p. 215; *20.1*, p. *357*.

E. V. R. Sastry and T. M. Srinivasan
Growth and Bleach of F-Centers in KCl with Interstitial Cobalt
Vol. **29**, No. 2, p. K107–K109 (1968).
See also 10, p. 111; *10.2*, p. *144*.

W. W. Walker and L. J. Demer
Indentation Creep in Crystals
Vol. **29**, No. 2, p. K141–K143 (1968).
See also *10*, p. *111*; 22.5.1, p. 517; 22.5.3, p. 534; 22.6, p. 543.

K. H. Timmesfeld
Anharmonic Broadening of Resonant Modes
Vol. **30**, No. 1, p. 73–85 (1968).
See also 3.2, p. 36; *6*, p. *69*; *20.1*, p. *358*.

M. L. Reynolds, W. E. Hagston, and G. F. J. Garlick
Absorption and Infra-Red Emission of Nickel and Cobalt Doped Alkali Halides
Vol. **30**, No. 1, p. 97–103 (1968).
See also 13.4, p. 216; *20.1*, p. *358*; *20.3*, p. *371*.

A. J. Watts and J. E. Cousins
Temperature Dependent Spin-Lattice Relaxation Times for Conduction Electrons in Lithium Particles in n-Irradiated Lithium Fluoride
Vol. **30**, No. 1, p. 105–112 (1968).
See also *19*, p. *335*; 21.2, p. 425.

P. Feltham
A Stochastic Model of Creep
Vol. **30**, No. 1, p. 135–146 (1968).
See also *10*, p. *111*; *10.1*, p. *134*; 21, p. 388.

J. Ingels and G. Jacobs
Faraday Rotation of the U_2 Centre in KCl, KBr, and KI
Vol. **30**, No. 1, p. 163–172 (1968).
See also 13.4, p. 216; *18*, p. *295*; *20*, p. *342*.

R. L. Gilbert
Empirical Relationships for the Emission Energies and Lifetimes of the F-Center
Vol. **30**, No. 1, p. 173–182 (1968).
See also *10.2*, p. *144*; 13.4, p. 216; 20.3, p. 371.

N. Itoh
Degradation of Low-Energy Electrons in Alkali Halides
Vol. **30**, No. 1, p. 199–207 (1968).
See also 10, p. 112; *11*, p. *154*.

S. Haussühl and H. Hesse
Quadratic Electrooptical Effect (Kerr Effect) in Alkali Halides of NaCl Type
Vol. **30**, No. 1, p. 209–214 (1968).
See also 14.4, p. 254; *20*, p. *342*.

H. H. Nickle
Comment on a Valence Electron Model of Atoms
Vol. **30**, No. 1, p. K65–K67 (1968).
See also *13.4*, p. *217*.

T. Okada and K. Ozawa
Anomalous Ductility of LiF Crystals at Low Temperatures
Vol. **30**, No. 1, p. K69–K71 (1968).
See also *10*, p. *112*; 11, p. 155.

N. E. Byer and H. S. Sack
Ultrasonic Velocity and Attenuation in Alkali Halides Containing CN^-. (I): Experimental Results
Vol. **30**, No. 2, p. 569–577 (1968).
See also *7*, p. *78*; 10, p. 112; 13.4, p. 217.

N. E. BYER and H. S. SACK
Ultrasonic Velocity and Attenuation in Alkali Halides Containing CN^-. (II): Theoretical Interpretation
Vol. **30**, No. 2, p. 579—586 (1968).
See also 7, p. *78*; 10, p. 112; **13.4**, p. 217.

F. GRANZER, G. WAGNER, and J. EISENBLÄTTER
Atomistic Calculation of Core Structure, Core Energy, and Peierls Stress of an Edge Dislocation in NaCl
Vol. **30**, No. 2, p. 587—600 (1968).
See also *10*, p. *112*.

G. BALDINI and M. GUZZI
Zero-Phonon Transitions of Eu^{++}- and Sm^{++}-Doped Alkali Halides
Vol. **30**, No. 2, p. 601—612 (1968).
See also *6*, p. *70*; 10, p. 112; 20.1, p. 359; 20.3, p. 372.

P. G. BARANOV, R. A. ZHITNIKOV, and N. I. MELNIKOV
Paramagnetic Resonance of Silver Atoms Stabilized in Alkali Halide Crystals
Vol. **30**, No. 2, p. 851—858 (1968).
See also 10.2, p. 144; *19*, p. *336*.

P. G. BARANOV, R. A. ZHITNIKOV, and N. I. MELNIKOV
Theoretical Study of Properties of Copper and Silver Atoms Stabilized in Alkali Halide Crystals
Vol. **30**, No. 2, p. 859—866 (1968).
See also 3.2, p. 36; 10.2, p. 144; 13.4, p. 218; *19*, p. *336*.

H. BETHGE and U. MESSERSCHMIDT
Nonconservative Processes during Dislocation Glide at Elevated Temperatures
Vol. **30**, No. 2, p. K87—K89 (1968).
See also *10*, p. *113*.

K.-J. BERG and K.-P. MEYER
Thermoluminescence of Z_1-Centers in Ca Doped KCl Crystals after X-Irradiation
Vol. **30**, No. 2, p. K161—K164 (1968).
See also *10.2*, p. *144*; *20.3*, p. *372*.

G. HENSEL and J. ZILLER
Precipation Influenced Low Temperature Conductivity of Ca Doped NaCl Crystals
Vol. **30**, No. 2, p. K165—K168 (1968).
See also 10, p. 113; *14.4*, p. *254*.

22.5.3 Alkali-Earth Halides

Review Articles

P. Süptitz and J. Teltow
Transport of Matter in Simple Ionic Crystals (Cubic Halides)
Vol. **23**, No. 1, p. 9–56 (1967).
See also *9*, p. *87*; *10*, p. *93*; 14.4, p. 252; 15, p. 270; 22.5.1, p. 515; 22.5.2, p. 518.

Original Papers and Short Notes

D. J. Barber
Electron Microscopy of Irradiation-Induced Defect Clusters in Magnesium Fluoride
Vol. **16**, No. 2, p. 531–547 (1966).
See also 4, p. 37; *10*, p. *94*; *11*, p. *145*.

V. V. Ratnam
On the Electrical and Magnetic Properties of Coloured and Uncoloured Calcium Fluoride Crystals
Vol. **16**, No. 2, p. 549–557 (1966).
See also *10.2*, p. *137*; 13.4, p. 201; 14.4, p. 252; 18.1, p. 296.

V. V. Ratnam
On the Colour Centres and X-Ray Luminescence of Calcium Fluoride Crystals
Vol. **16**, No. 2, p. 559–567 (1966).
See also *10.2*, p. *137*; *20.3*, p. *364*.

V. V. Eremenko, E. V. Matyushkin, and S. V. Petrov
Study of Energy Transfer from 3d to 4f Electrons in Antiferromagnetic $MnF_2:Eu^{3+}$ Crystals
Vol. **18**, No. 2, p. 683–686 (1966).
See also *13.4*, p. *204*; 18.4, p. 321; *20.3*, p. *364*.

R. J. Fleming
Potential Energy for F-Centre Electrons in Calcium Fluoride
Vol. **18**, No. 2, p. K139–K142 (1966).
See also *10.2*, p. *138*; 13.4, p. 204.

P. Görlich, H. Karras, E. Meisel, and H. Mothes
IR Absorption of Uncoloured and Additively Coloured Synthetic CaF_2 Crystals and CaF_2/Na Crystals
Vol. **19**, No. 1, p. K23–K25 (1967).
See also *10.2*, p. *138*.

P. Schmeling
Diffusion of Xenon in Barium Fluoride
Vol. **20**, No. 1, p. 127–129 (1967).
See also *9*, p. *88*.

A. A. Kaminskii
The Nature of "Aging" of CaF_2–YF_3–Nd^{3+} Crystals (Type I) under Stimulated Emission Conditions
Vol. **20**, No. 1, p. K51–K54 (1967).
See also *20.2*, p. *361*; 22.5, p. 512; 22.8, p. 549.

P. Asadi
On the Existence of a Spontaneous Polarization in Fluorite Single Crystals Doped with Lime
Vol. **20**, No. 1, p. K55–K58 (1967).
See also 10, p. 98; *14.4.1*, p. *258*.

P. Asadi
On the Electrical Conductivity of Fluorite Single Crystals Doped with Uranium Dioxide
Vol. **20**, No. 1, p. K59–K61 (1967).
See also 10, p. 99; *14.4*, p. *252*.

P. Feltham
F-Centres in Alkaline Earth Fluorides
Vol. **20**, No. 2, p. 675–680 (1967).
See also *10.2*, p. *139*; 13.4, p. 206.

H. W. den Hartog and J. Arends
F-Centers in SrF_2
Vol. **22**, No. 1, p. 131–140 (1967).
See also *10.2* p. *139*; 13.4, p. 208; 19, p. 329.

A. Mukerji, G. A. Tanton, and J. E. Williams
X-Ray Induced F-Centers in CaF_2
Vol. **22**, No. 1, p. K19–K22 (1967).
See also *10.2*, p. *140*; 11, p. 149.

J. E. Caffyn, T. L. Goodfellow, E. M. Olymbios, and D. J. Perrin
Dielectric and Optical Absorption Measurements on CaF_2 and CaF_2:Y
Vol. **22**, No. 2, p. 549–557 (1967).
See also 10, p. 101; 13.4, p. 209; *14.4*, p. *253*; *20.1*, p. *350*.

P. Görlich, H. Karras, G. Kötitz, and P. Ullmann
Thermoluminescence Investigations on Doped Akaline Earth Fluorides
Vol. **23**, No. 1, p. 313–323 (1967).
See also 13.4, p. 209; 20.1, p. 350; *20.3*, p. *367*.

H. W. den Hartog and J. Arends
Electronic Structure of F-Centers in Alkaline Earth Fluorides
Vol. **23**, No. 2, p. 713–719 (1967).
See also 10.2, p. 140; *13.4*, p. *210*; 19, p. 330.

H. W. den Hartog
Fourth Shell Isotropic H.F.S.-Constants of F-Centers in CaF_2
Vol. **24**, No. 2, p. K117–K120 (1967).
See also 10.2, p. 141; *13.4*, p. *211*; 19, p. 331.

P. Görlich, H. Karras, Ch. Symanowski, and P. Ullmann
The Colour Centre Absorption of X-Ray Coloured Alkaline Earth Fluoride Crystals
Vol. **25**, No. 1, p. 93–101 (1968).
See also *10.2*, p. *141*.

P. Görlich, H. Karras, G. Kötitz, and R. Rauch
Phonon-Assisted Colour Centre Fluorescence of Additively Coloured Alkali Earth Fluoride Crystals
Vol. **25**, No. 1, p. K15–K18 (1968).
See also 6, p. 65; *10.2*, p. *141*; *20.3*, p. *368*.

P. Feltham and R. Ghosh
On Colour Centres in Plastically Deformed Alkaline Earth Fluorides
Vol. **25**, No. 2, p. K71–K73 (1968).
See also 10, p. 104; *10.2*, p. *142*; 19, p. 332.

H. W. den Hartog
F-Centers in BaF_2
Vol. **25**, No. 2, p. K111–K114 (1968).
See also *10.2*, p. *142*; 13.4, p. 211; 19, p. 333.

YU. K. VORONKO and V. A. SYCHUGOV
The Stimulated Emission of Er^{3+} Ions in CaF_2 at $\lambda_1 = 8456$ Å and $\lambda_2 = 8548$ Å
Vol. **25**, No. 2, p. K119–K121 (1968).
See also 20.1, p. 352; *20.2*, p. *361*; 20.3, p. 368.

Y. KAZUMATA
Self-Trapped Holes in Neutron Irradiated CaF_2 and BaF_2 Crystals
Vol. **26**, No. 2, p. K119–K122 (1968).
See also 10.2, p. 142; 11, p. 152; *19*, p. *333*.

P. GÖRLICH, H. KARRAS, G. KÖTITZ, and R. RAUCH
Polarized Luminescence of X-Irradiated CaF_2:Y and SrF_2:Y Crystals and the Structure of Their Luminescence Centres
Vol. **27**, No. 1, p. 109–115 (1968).
See also 10, p. 106; *20.3*, p. *369*.

R. J. FLEMING
Activation Energies and Emission Spectra of the Thermoluminescence of CaF_2 Crystals Gamma-Irradiated at 77 °K
Vol. **27**, No. 1, p. K57–K61 (1968).
See also 13.4, p. 213; *20.3*, p. *369*.

S. D. McLAUGHLAN and H. W. EVANS
Production of Colloidal Calcium by Electron Irradiation of CaF_2 Crystals
Vol. **27**, No. 2, p. 695–700 (1968).
See also 10.2, p. 142; *11*, p. *152*; 19, p. 333; 20.1, p. 354.

S. HAUSSÜHL
Elastic and Thermoelastic Behaviour of MgF_2 and MnF_2
Vol. **28**, No. 1, p. 127–130 (1968).
See also 3.2, p. 36; *12*, p. *158*; 22.5, p. 514.

M. R. TUBBS
The Variation of Exciton Peak Position with Lattice Parameter in Metal Iodide Crystals
Vol. **28**, No. 2, p. K135–K137 (1968).
See also 13.2, p. 195; *20.1*, p. *356*; 22.5, p. 514; 22.5.2, p. 528.

R. A. SHATAS, G. A. TANTON, and A. MUKERJI
Mollwo-Ivey Relation of H- and F-Bands in Alkaline Earth Fluorides
Vol. **28**, No. 2, p. K143–K145 (1968).
See also *10.2*, p. *143*.

R. J. FLEMING
Evaluation of 1s–2p Optical Transition Energy for F-Centres in CaF_2
Vol. **29**, No. 1, p. K77–K80 (1968).
See also *10.2*, p. *144*; 13.4, p. 215.

W. W. WALKER and L. J. DEMER
Indentation Creep in Crystals
Vol. **29**, No. 2, p. K141–K143 (1968).
See also *10*, p. *111*; 22.5.1, p. 517; 22.5.2, p. 529; 22.6, p. 543.

A. A. KAMINSKII and L. LI
The Photoreduction $Nd^{3+} \rightarrow Nd^{2+}$ in SrF_2 (Type I) Crystals under Stimulated Emission Conditions ("Aging")
Vol. **30**, No. 1, p. K77–K79 (1968).
See also *20.2*, p. *362*.

22.6 Oxides (CdO, ZnO See 22.4; Spinels See 18.3 or 14.4.2)

Review Articles

L. Eckertová
Field Cathodes with Thin Insulating Films
Vol. **18**, No. 1, p. 3–40 (1966).
See also 13, p. 162; 13.3, p. 196; 13.4, p. 200; 14.4.1, p. 255; *17*, p. *286*.

K. N. Subramanian
Cyclic Stressing in Crystals of Rock-Salt Structure
Vol. **28**, No. 1, p. 9–37 (1968).
See also *10*, p. *93*; 10.1, p. 114; 21, p. 373; 21.1, p. 390; 22.5.1, p. 515; 22.5.2, p. 518.

Original Papers and Short Notes

L. I. van Torne
Pumping Induced Imperfections in Glass:Nd^{3+} Lasers
Vol. **16**, No. 1, p. 171–182 (1966).
See also 2, p. 22; 4, p. 37; *20.2* p. *360*.

J. van Landuyt
Determination of the Displacement Vector at the Anti-Phase Boundaries in Rutile by Contrast Experiments in the Electron Microscope
Vol. **16**, No. 2, p. 585–590 (1966).
See also *4*, p. *38*; 10, p. 94.

G. Kluge
Acoustic Activity
Vol. **17**, No. 1, p. 109–118 (1966).
See also 6, p. 56; *7*, p. *76*.

F. van der Woude
Mössbauer Effect in α-Fe_2O_3
Vol. **17**, No. 1, p. 417–432 (1966).
See also *6.1*, p. *71*; 18.4, p. 321.

G. Dorda and J. Vrba
Long-Time Effect of External Fields on Surface Silicon Oxide Conductivity
Vol. **17**, No. 1, p. K127–K130 (1966).
See also *14.3*, p. *232*; 22.1.2, p. 469.

N. Singh and M. D. Karkhanavala
Studies on the Electrical Properties of Uranium Oxides. (II): Non-stoichiometric U_3O_8
Vol. **17**, No. 2, p. 501–508 (1966).
See also 13, p. 163; *14.3*, p. *232*.

J. Devreese, R. De Coninck, and H. Pollak
On the Conduction Mechanism in Uranium Dioxide
Vol. **17**, No. 2, p. 825–829 (1966).
See also 13, p. 163; *14.3*, p. *233*; 15, p. 270.

W. Neumann
Thermal Conductivity of Yellow PbO between 0 and 500 °C
Vol. **18**, No. 1, p. 151–156 (1966).
See also 6, p. 57; *8*, p. *80*.

HJ. MATZKE
Radiation Damage and Xenon Release in Quartz and Fused Silica Following Ion Bombardment
Vol. **18**, No. 1, p. 285–295 (1966).
See also 1.2, p. 9; 2, p. 22; *11*, p. *147*.

R. GEVERS, J. VAN LANDUYT, and S. AMELINCKX
On a Simple Derivation of the Amplitudes of the Electron Beams Transmitted and Scattered by a Crystal Containing Planar Interfaces — Images of Subgrain Boundaries
Vol. **18**, No. 1, p. 325–342 (1966).
See also *4*, p. *39*.

J. VAN LANDUYT, R. GEVERS, and S. AMELINCKX
The Fine Structure of Spots in Electron Diffraction Resulting from the Presence of Planar Interfaces and Dislocations. (II): Observations on Crystals Containing Stacking Faults
Vol. **18**, No. 1, p. 363–378 (1966).
See also 4, p. 39; *10*, p. *96*; *10.1*, p. *118*; 21.1.1, p. 414.

S. A. MOSKALENKO and P. I. KHADSHI
Infrared Absorption by Excitons due to Photoionization and Intraband Lattice Scattering
Vol. **18**, No. 1, p. 379–390 (1966).
See also 6, p. 57; *13.2*, p. *191*; 20.1, p. 346; 22.1.1, p. 460; 22.2.2, p. 487; 22.2.3, p. 490.

W. HASE and W. MEISEL
Mössbauer Effect of Fe^{57} in Sesquioxides of the D 5_3 Structure
Vol. **18**, No. 1, p. K41–K43 (1966).
See also *6.1*, p. *71*.

V. L. VINETSKII and G. A. KHOLODAR
On the Electric Conductivity of Semiconductors Caused by the Ionization of Thermal Lattice Defects
Vol. **19**, No. 1, p. 41–49 (1967).
See also 10, p. 97; 13.4, p. 204; *14.3*, p. *234*.

R. DE BATIST, R. GEVERS, and M. VERSCHUEREN
Magnon Contribution to the Low-Temperature Specific Heat of UO_2
Vol. **19**, No. 1, p. 77–88 (1967).
See also *8*, p. *80*; 18.4, p. 322.

A. M. GEORGE and M. D. KARKHANAVALA
Studies of the Electrical Properties of Uranium Oxides. (III): Thermoelectric Power of α-U_3O_8
Vol. **19**, No. 1, p. 117–121 (1967).
See also 14.3, p. 234; *15*, p. *271*.

A. C. MCLAREN, J. A. RETCHFORD, D. T. GRIGGS, and J. M. CHRISTIE
Transmission Electron Microscope Study of Brazil Twins and Dislocations Experimentally Produced in Natural Quartz
Vol. **19**, No. 2, p. 631–644 (1967).
See also 4, p. 40; *10*, p. *98*.

R. C. RAU and G. A. CHASE
Clustering of Defects in Neutron-Irradiated Beryllium Oxide
Vol. **19**, No. 2, p. 645–654 (1967).
See also *10*, p. *98*; 11, p. 148.

J. Spyridelis, P. Delavignette, and S. Amelinckx
On the Superstructures of Ta_2O_5 and Nb_2O_5
Vol. **19**, No. 2, p. 683–704 (1967).
See also *1*, p. *7*; 4, p. 40; 10, p. 98.

M. P. Verma and B. Dayal
Lattice Dynamics of MgO
Vol. **19**, No. 2, p. 751–756 (1967).
See also *6*, p. *59*.

A. M. Stoneham
Interference Effects in Spin Lattice Relaxation by Two-Phonon Processes
Vol. **19**, No. 2, p. 787–792 (1967).
See also 6, p. 59; *13*, p. *165*.

C. Cherki and R. Coelho
On Charge Storage in Anodic Tantal Oxide Layers
Vol. **19**, No. 2, p. K91–K94 (1967).
See also 14.3.1, p. 245; *14.4.1*, p. *257*.

B. Reppich
Plastic Deformation of Iron-II-Oxide
Vol. **20**, No. 1, p. 69–82 (1967).
See also *10*, p. *98*.

L. L. Hench and D. A. Jenkins
AC Conductivity of a Glass Semiconductor
Vol. **20**, No. 1, p. 327–330 (1967).
See also 2, p. 23; *14.3*, p. *235*; 22.8, p. 549.

Z. Bodó and I. Hevesi
Optical Absorption near the Absorption Edge in V_2O_5 Single Crystals
Vol. **20**, No. 1, p. K45–K49 (1967).
See also 13.1, p. 177; *20.1*, p. *347*.

J. Spyridelis, J. Stoimenos, and N. Economou
Optical and Photoconductive Phenomena in Cuprous Oxide
Vol. **20**, No. 2, p. 623–628 (1967).
See also 13.2, p. 192; 16, p. 279; *20.1*, p. *347*.

M. Jaroš and J. Zítková
Remark on the Theory of the ESR Spectrum of the d^1-Configuration in Rutile (TiO_2)
Vol. **20**, No. 2, p. K99–K101 (1967).
See also 13.4, p. 206; *19*, p. *328*.

T. Sakata, K. Sakata, and I. Nishida
Study of Phase Transition in NbO_2
Vol. **20**, No. 2, p. K155–K157 (1967).
See also *1.2*, p. *10*.

M. Ali and P. Nagels
Evaluation of the Debye Temperature of Thorium Dioxide
Vol. **21**, No. 1, p. 113–116 (1967).
See also *6*, p. *60*; 8, p. 82.

M. Rosenberg, C. Tănăsoiu, and V. Florescu
The Influence of the Magnetic Field on the Domain Structure of Ferrimagnetic Oxides with High Anisotropy Field
Vol. **21**, No. 1, p. 197–204 (1967).
See also 4, p. 42; *18.3*, p. *317*.

R. Gevers, J. van Landuyt, and S. Amelinckx
Two Beam Kinematical Theory for the Diffraction of Electrons by Crystals with Stacking Faults
Vol. **21**, No. 1, p. 393–411 (1967).
See also *10*, p. *99*; *10.1*, p. *122*; 21.1.1, p. 417.

R. C. Rau and K. Lacefield
Microhardness of Irradiated Beryllium Oxide
Vol. **21**, No. 1, p. K69–K71 (1967).
See also 11, p. 148; *12*, p. *157*.

Č. Jech
Ion-Bombardment Enhanced Solubility in Solids
Vol. **21**, No. 2, p. 481–486 (1967).
See also *11*, p. *149*; 22, p. 445.

K. H. Gundlach and G. Heldmann
The Effect of the $E(k)$-Relation on Tunneling through Asymmetric Barriers
Vol. **21**, No. 2, p. 575–579 (1967).
See also 13.1, p. 178; *14.3.1*, p. *245*; 14.3.2, p. 249.

H. Pink
The Determination of Chromium Distribution in Laser-Rubies with Neutron Activation
Vol. **21**, No. 2, p. K111–K113 (1967).
See also *10*, p. *101*; *20.2*, p. *361*.

I. Dézsi, Gy. Erlaki, and L. Keszthelyi
Mössbauer Effect Studies on the $x(Rh_2O_3)1-x(Fe_2O_3)$ System
Vol. **21**, No. 2, p. K121–K125 (1967).
See also *6.1*, p. *72*; 18.4, p. 322; 22.8, p. 551.

Ž. Dimitrijević, S. Kraśnicki, H. Rżany, J. Todorović, A. Wanic, H. Curien, and A. Milojević
Anisotropy of Magnon Dispersion Relation in Haematite
Vol. **21**, No. 2, p. K163–K166 (1967).
See also *13*, p. *166*; *18*, p. *292*.

A. M. de Goer and B. Dreyfus
γ-Irradiation Effect on the Thermal Conductivity of Al_2O_3 with Cr or Mn Impurities at Low Temperature
Vol. **22**, No. 1, p. 77–82 (1967).
See also *8*, p. *83*; 10, p. 101; *11*, p. *149*.

G. E. Arkhangelskii, Z. L. Morgenshtern, and V. B. Neustruev
Colour Centres in Ruby Crystals
Vol. **22**, No. 1, p. 289–295 (1967).
See also *10.2*, p. *140*; 20.1, p. 349; 20.2, p. 361; 20.3, p. 366.

K. Jeges
Electroluminescence of SnO_2 Single Crystals
Vol. **22**, No. 1, p. K7–K9 (1967).
See also *20.3*, p. *366*.

K. Tompa, F. Tóth, and G. Grüner
Susceptibility of MnO Measured by the NMR Method
Vol. **22**, No. 1, p. K11–K13 (1967).
See also 18.1, p. 296; 18.4, p. 322; *19*, p. *329*.

L. N. Cojocaru
Effects of Fast Neutrons on Electric Properties of NiO Single Crystals
Vol. **22**, No. 2, p. 361—364 (1967).
See also *11*, p. *149*; 13.4, p. 208; 14.3, p. 237; 15, p. 272.

P. W. Graves
Single Crystal Growth of Calcium Pyroniobate
Vol. **22**, No. 2, p. 499—504 (1967).
See also *3.1*, p. *32*; 22.8, p. 551.

M. V. Coleman and D. J. D. Thomas
The Structure of Silicon Oxide Films
Vol. **22**, No. 2, p. 593—602 (1967).
See also *1.4*, p. *19*; 22.1.2, p. 471.

S. Ikonopisov, L. Avramov, and S. Veleva
On the Electroluminescence of Indium Oxide Layers in Contact with an Electrolyte
Vol. **23**, No. 1, p. K61—K65 (1967).
See also *20.3*, p. *367*.

M. Stuart
Electrode-Limited to Bulk-Limited Conduction in Silicon Oxide Films
Vol. **23**, No. 2, p. 595—597 (1967).
See also *14.3.1*, p. *245*; *14.4.1*, p. *260*.

A. G. Khachaturyan and M. P. Usikov
A New Type of X-Ray and Electron Diffuse Scattering and Corresponding Strain Contrast on Electron Microscopic Images
Vol. **23**, No. 2, p. 745—754 (1967).
See also *4*, p. *43*; 21.1, p. 400.

S. V. Vonsovskii, V. I. Cherepanov, A. N. Men, and A. E. Nikiforov
Group-Theoretical Classification of Electronic States in Crystals Composed of Pairs and Complexes of Bound Impurity Ions
Vol. **24**, No. 1, p. 51—63 (1967).
See also 3.2, p. 35; *13.4*, p. *210*.

H. Böttger
Splitting of ESR Lines by Spin-Phonon Interaction for Ions with Spin 1
Vol. **24**, No. 1, p. 65—75 (1967).
See also 6, p. 63; 13.4, p. 210; *19*, p. *331*.

J. Antula
A Method of Measuring Effective Electron Mass in Thin Insulating Films
Vol. **24**, No. 1, p. 95—98 (1967).
See also *13.1*, p. *182*; 14.3.2, p. 249; 14.4.1, p. 260.

A. G. Revesz
Noncrystalline Structure and Electronic Conduction of Silicon Dioxide Films
Vol. **24**, No. 1, p. 115—126 (1967).
See also *1.4*, p. *19*; 2, p. 23; *14.3.1*, p. *246*; 14.3.2, p. 249.

L. E. Murr
Study of Erbium Thin Film Oxidation in the Electron Microscope
Vol. **24**, No. 1, p. 135—148 (1967).
See also *1.4*, p. *19*; 5, p. 52.

A. J. Springthorpe
The Rhombohedral Distortion in NiO
Vol. **24**, No. 1, p. K3—K4 (1967).
See also 4, p. 43; *18.4*, p. *323*.

M. Rouzeyre
Contribution to the Study of Transport Phenomena in Aluminium-Alumina-Metal Structures
Vol. **24**, No. 2, p. 399–411 (1967).
See also *14.3.2*, p. *250*; 16, p. 281; 21, p. 383.

J. M. Trooster and A. Dymanus
Mössbauer Effect in $Ga_{2-x}Fe_xO_3$ and Related Compounds
Vol. **24**, No. 2, p. 487–499 (1967).
See also *6.1*, p. *73*; 18.3, p. 318; 22.8, 552.

W. Ludwig and B. Korneffel
Photoemission Studies on Thin Metal–Insulator–Metal Sandwiches
Vol. **24**, No. 2, p. K137–K140 (1967).
See also *14.3.2*, p. *250*; *17*, p. *288*; 21, p. 383.

L. K. Keys
ESR of Ti_2O_3
Vol. **25**, No. 1, p. K5–K6 (1968).
See also *19*, p. *332*.

J. Grabmaier and C. Watson
Plastic Deformation in Single Crystals of Mg–Al-Spinel: Slip Parallel to the $\{1\bar{1}1\}$ Plane
Vol. **25**, No. 1, p. K7–K10 (1968).
See also *10*, p. *104*.

K. N. Shrivastava
Temperature Dependence of Paramagnetic Resonance Transition Probabilities: Mn^{2+} in MgO
Vol. **25**, No. 1, p. K47–K50 (1968).
See also 10, p. 104; *19*, p. *332*.

J. G. Allpress, J. V. Sanders, and A. D. Wadsley
Electron Microscopy of High-Temperature Nb_2O_5 and Related Phases
Vol. **25**, No. 2, p. 541–550 (1968).
See also 3, p. 27; *5*, p. *53*.

L. E. Murr
Investigation of Grain Boundary Energetics in Er_2O_3 Thin Foils by Transmission Electron Microscopy
Vol. **25**, No. 2, p. 629–640 (1968).
See also *1.4*, p. *20*; *4*, p. *44*.

R. Gauthier, D. Bois, P. Pinard, and F. Davoine
Ion and Electron Production by Interaction of a Coherent Light Beam with Substance
Vol. **25**, No. 2, p. 691–696 (1968).
See also *17*, p. *288*; 21.1, p. 403; 21.1.1, p. 420; 22.1.1, p. 464; 22.2.1, p. 485; 22.4.1, p. 502; 22.5.2, p. 525.

H. P. Geserich
Optical and Electrical Measurements on Thallium(III)Oxide Films
Vol. **25**, No. 2, p. 741–751 (1968).
See also *13.1*, p. *184*; 14.3.1, p. 246; 20.1, p. 352.

D. K. Smith and H. W. Newkirk
Irradiation Decoration of Defects in Single-Crystal BeO
Vol. **25**, No. 2, p. K79–K82 (1968).
See also *10*, p. *104*; 11, p. 151.

R. G. Bessent and P. Feltham
Formation of F-Centres in Alkaline Earth Oxides by Plastic Deformation
Vol. **25**, No. 2, p. K107—K110 (1968).
See also 10, p. 105; *10.2*, p. *142*; 19, p. 333.

H. Böttger
Jahn-Teller Splitting of an ESR Line
Vol. **26**, No. 2, p. 681—692 (1968).
See also 6, p. 65; *13.4*, p. *212*; 19, p. 333.

B. Henderson and R. D. King
Colour Centres in Neutron Irradiated Calcium Oxide
Vol. **26**, No. 2, p. K147—K149 (1968).
See also *10.2*, p. *142*; 11, p. 152.

R. A. Shatas and J. D. Stettler
Mollwo-Ivey Relation in Aggregate Defect Centers
Vol. **26**, No. 2, p. K155—K157 (1968).
See also 10.2, p. 142; *13.4*, p. *212*; 22.5.2, p. 526.

D. Michell and A. P. Smith
The Nature of Oxide Layers on Single Crystals of Cadmium and Magnesium
Vol. **27**, No. 1, p. 291—300 (1968).
See also 1.3, p. 15; *1.4*, p. *20*; 10, p. 107.

V. N. Bogomolov and D. N. Mirlin
Optical Absorption by Polarons in Rutile (TiO_2) Single Crystals
Vol. **27**, No. 1, p. 443—453 (1968).
See also 13, p. 170; *20.1*, p. *354*.

T. M. Fitzgerald and B. D. Silverman
Ultrasonic Attenuation in Unirradiated and Neutron Irradiated Quartz
Vol. **27**, No. 2, p. 473—478 (1968).
See also 6, p. 66; *7*, p. *77*; 11, p. 152.

M. I. Klinger
On the Experimental Observation of Small Polarons in Semiconductors
Vol. **27**, No. 2, p. 479—488 (1968).
See also *13*, p. *170*; 14.3, p. 241; 20.1, p. 354.

O. Brümmer and G. Dräger
On the Interpretation of the Fine Structure of X-Ray K Absorption Spectra Far from the Edge
Vol. **27**, No. 2, p. 513—516 (1968).
See also 2, p. 24; *20*, p. *341*; 22.1.1, p. 465; 22.8, p. 554.

H. K. Müller
Electrical and Optical Properties of Sputtered In_2O_3 Films. (I): Electrical Properties and Intrinsic Absorption
Vol. **27**, No. 2, p. 723—731 (1968).
See also 13.4, p. 214; *14.3.1*, p. *247*; 20.1, p. 355.

H. K. Müller
Electrical and Optical Properties of Sputtered In_2O_3 Films. (II): Optical Properties in the Near Infrared
Vol. **27**, No. 2, p. 733—740 (1968).
See also 14.3.1, p. 247; *20.1*, p. *355*.

V. I. CHEREPANOV and A. E. NIKIFOROV
Consideration of Polarization and Covalency Corrections to the Quasi-Coulomb Shift of Optical Lines of Ion Pairs in Crystals
Vol. 28, No. 1, p. 65–73 (1968).
See also *10*, p. *108*; 13.4, p. 214; 20.3, p. 370.

L. MERTEN
Directional Dispersion of Extraordinary Optical Phonons of α-Quartz
Vol. 28, No. 1, p. 111–119 (1968).
See also *6*, p. *66*; 20.1, p. 355.

K. ŽĎÁNSKÝ
Vibrational Resonance Mode and the Temperature Dependence of Hyperfine Coupling of Mn^{2+} in BaO
Vol. 28, No. 1, p. 181–186 (1968).
See also 10, p. 108; 13.4, p. 214; *19*, p. *334*.

S. S. SHEININ and C. D. CANN
Effect of Absorption on Multi-Beam Intensity Profiles of Wedge Crystals
Vol. 28, No. 1, p. 187–192 (1968).
See also *5*, p. *53*; 21, p. 386.

M. ALI, M. FRIDMAN, M. DENAYER, and P. NAGELS
Electrical Conduction in Li-Doped MnO Single Crystals
Vol. 28, No. 1, p. 193–206 (1968).
See also 13.1, p. 187; 13.4, p. 214; *14.3*, p. *242*; 15, p. 274.

YU. A. SHERSTKOV, V. I. NEPSHA, V. A. VAGENIN, A. E. NIKIFOROV, and A. I. KROTKII
Effect of Applied Electric Field on EPR Spectra of Crystal Impurity Centres
Vol. 28, No. 1, p. 269–275 (1968).
See also 10, p. 109; 13, p. 171; *19*, p. *334*.

J. ANTULA
Negative Space-Charge Effect on Schottky Current in Thin Insulating Films
Vol. 28, No. 1, p. 395–400 (1968).
See also 14.3.1, p. 247; *14.4.1*, p. *263*.

K. E. MANN and L. V. HOLROYD
The EPR Spectra of Gd^{3+} in Single Crystals of BaO
Vol. 28, No. 1, p. K27–K29 (1968).
See also *19*, p. *334*.

C. SCHENKER
Ferro–Antiferromagnetic Coupling and Magnetic After-Effect in Multiple Co–CoO and Ni–NiO Films
Vol. 28, No. 2, p. 507–517 (1968).
See also *18.2*, p. *313*; *18.4*, p. *324*; 21.1, p. 406.

I. F. CHANG, S. S. MITRA, J. N. PLENDL, and L. C. MANSUR
Long-Wavelength Longitudinal Phonons of Multi-Mode Crystals
Vol. 28, No. 2, p. 663–673 (1968).
See also *6*, p. *67*; 20.1, p. 356; 22.4.1, p. 504; 22.4.2, p. 508; 22.8, p. 554.

S. N. SHESTATSKII and V. V. SOBOLEV
Electroreflectance Spectrum of Cuprous Oxide Crystals
Vol. 28, No. 2, p. K131–K133 (1968).
See also 13.1, p. 188; 13.2, p. 194; *14.4.1*, p. *264*; *20*, p. *342*; 20.1, p. 356.

J. H. GIESKE and G. R. BARSCH
Pressure Dependence of the Elastic Constants of Single Crystalline Aluminium Oxide
Vol. 29, No. 1, p. 121–131 (1968).
See also *12*, p. *158*.

M. Luukkala
On Coherent Phonon Fusion in Quartz Transducers
Vol. **29**, No. 1, p. 377–382 (1968).
See also 6, p. 67; *7*, p. *78*.

B. Chalupa, R. Michalec, V. Petržílka, J. Tichý, and J. Zelenka
Diffraction of Neutrons on a Vibrating Quartz Crystal
Vol. **29**, No. 1, p. K51–K54 (1968).
See also *5*, p. *54*; 14.4, p. 254.

K. Platzöder
Temperature Effects on the Vacuum Ultraviolet Reflectance of α-Quartz
Vol. **29**, No. 1, p. K63–K64 (1968).
See also 13.1, p. 189; 13.2, p. 195; *20.1*, p. *357*.

A. R. Ruffa
The Valence Bond Approximation in Crystals — Application to an Analysis of the Ultraviolet Spectrum in Quartz
Vol. **29**, No. 2, p. 605–616 (1968).
See also 3.2, p. 36; *13.1*, p. *189*; 20.1, p. 357.

K. N. Shrivastava
Phonon-Induced Hyperfine Coupling of V^{2+} Ion in Cubic Crystalline Fields
Vol. **29**, No 2, p. 737–742 (1968).
See also *6*, p. *68*; 19, p. 335.

G. E. Arkhangelskii, Z. L. Morgenshtern, and V. B. Neustruev
On the Nature of the Colour Centres of Ruby
Vol. **29**, No. 2, p. 831–836 (1968).
See also *10.2*, p. *144*; 13.4, p. 215; *20.3*, p. *371*.

A. Jostsons and P. G. McDougall
Fault Structures in Ti_2O
Vol. **29**, No. 2, p. 873–889 (1968).
See also *4*, p. *47*; 10, p. 111.

S. E. Bronisz and D. L. Douglass
Grain Growth in Thin Films of ThO_2, NpO_2, and PuO_2
Vol. **29**, No. 2, p. K95–K97 (1968).
See also *1.4*, p. *21*; 3.1, p. 33.

L. N. Cojocaru
Effects of Fission Fragments on Electric Properties of the UO_2-SiO_2 System
Vol. **29**, No. 2, p. K119–K122 (1968).
See also *11*, p. *154*; 14.3, p. 244; 15, p. 274; 22.8, p. 556.

W. W. Walker and L. J. Demer
Indentation Creep in Crystals
Vol. **29**, No. 2, p. K141–K143 (1968).
See also *10*, p. *111*; 22.5.1, p. 517; 22.5.2, p. 529; 22.5.3, p. 534.

G. Lecoy and L. Gouskov
Noise in Thin Films Metal–Oxide–Metal Al–Al_2O_3–Al
Vol. **30**, No. 1, p. 9–17 (1968).
See also *14.3.2*, p. *251*.

R. KELLY
On the Problem of Depth Distributions in Oxides Following High-Dose Labelings
Vol. **30**, No. 1, p. 37–48 (1968).
See also *11*, p. *154*.

L. MERTEN
Polariton Dispersion in Biaxial and Uniaxial Crystals
Vol. **30**, No. 2, p. 449–454 (1968).
See also *6*, p. *70*; *20*, p. *343*.

R. DE RIDDER, J. VAN LANDUYT, R. GEVERS, and S. AMELINCKX
The Fine Structure of Spots in Electron Diffraction Resulting from the Presence of Planar Interfaces and Dislocations. (IV): Wedge Crystals
Vol. **30**, No. 2, p. 797–815 (1968).
See also *4*, p. *48*; 10, p. 113; 10.1, p. 136; 21.1.1, p. 423.

H. BÖTTGER
On the Spin-Phonon Interaction in External Electric Field
Vol. **30**, No. 2, p. K119–K123 (1968).
See also 6, p. 70; 13, p. 173; *19*, p. *336*.

22.7 Semiconducting Intermetallic Compounds

Review Articles

G. Harbeke
On the Band Structure of Anisotropic Crystals
Vol. **27**, No. 1, p. 9–44 (1968).
See also *13.1*, p. *174*; 13.2, p. 191; 14.3, p. 231; 20.1, p. 344; 22, p. 439; 22.1, p. 456; 22.1.1, p. 458; 22.1.2, p. 468; 22.1.3, p. 476; 22.4.1, p. 497; 22.4.2, p. 506; 22.4.3, p. 509; 22.5, p. 512; 22.8, p. 547.

Original Papers and Short Notes

K. H. J. Buschow and J. F. Fast
Magnetic and Structural Characteristics of Some Equiatomic Rare-Earth Germanides
Vol. **16**, No. 2, p. 467–473 (1966).
See also *3*, p. *26*; 5, p. 49; 18.1, p. 296; 18.2, p. 300; 18.3, p. 316; 18.4, p. 321.

H. Hora and R. Kantlehner
The Anomalous Vectorial Photoemission from a Cesium–Antimony Layer
Vol. **17**, No. 1, p. 59–66 (1966).
See also 13.1, p. 175; *17*, p. *286*.

M. Hagino and T. Takahashi
Photoelectric Properties of Cs-Sb Films Having Various Compositions
Vol. **17**, No. 2, p. 609–618 (1966).
See also 10, p. 95; 14.3.1, p. 245; 16, p. 277; *17*, p. *286*.

F. Borsa, R. G. Barnes, and R. A. Reese
Nuclear Magnetic Resonance and Mössbauer Effect Study of Sn^{119} in Rare Earth–Tin Intermetallic Compounds
Vol. **19**, No. 1, p. 359–371 (1967).
See also *6.1*, p. *71*; *19*, p. *328*.

L. Żdanowicz
Some Optical Properties of Thin Evaporated Cd_3As_2 Films
Vol. **20**, No. 2, p. 473–480 (1967).
See also *20.1*, p. *347*.

W. P. R. George, C. H. L. Goodman, H. F. Sterling, and R. W. Warren
A Possible New Group of Semiconducting Compounds
Vol. **21**, No. 1, p. 205–210 (1967).
See also 3.1, p. 31; 3.2, p. 35; *14.3*, p. *236*; 21.7, p. 438.

N. Sexer
Classical Galvanomagnetic and Thermomagnetic Effects in Cadmium Arsenide
Vol. **21**, No. 1, p. 225–234 (1967).
See also 13.1, p. 178; *14.3*, p. *236*; 15, p. 272.

K. H. J. Buschow and J. F. Fast
Crystal Structure and Magnetic Properties of Some Rare Earth Germanides
Vol. **21**, No. 2, p. 593–600 (1967).
See also *5*, p. *51*; 18.1, p. 296; 18.3, p. 318; 18.4, p. 322.

A. Stella, A. D. Brothers, R. H. Hopkins, and D. W. Lynch
Pressure Coefficient of the Band Gap in Mg_2Si, Mg_2Ge, and Mg_2Sn
Vol. **23**, No. 2, p. 697–702 (1967).
See also *13.1*, p. *182*; 14.3, p. 238; 20.1, p. 350.

E. O. Osmanov, Yu. V. Rud, and M. E. Stryalkovskii
High-Temperature Modifications of the Semiconductor Compounds $CdSnAs_2$ and $CdGeAs_2$
Vol. **26**, No. 1, p. 85—90 (1968).
See also *1.2*, p. *12*; 2, p. 24; 8, p. 84; 12, p. 158; 14.3, p. 240; 22.8, p. 553.

G. A. Sikharulidze and Yu. I. Ukhanov
Absorption of Polarized Light in $CdSnAs_2$ Crystals
Vol. **26**, No. 1, p. K33—K36 (1968).
See also 13.1, p. 184; *20.1*, p. *353*; 22.8, p. 553.

K. Hirschberg and K. Deutscher
Determination of the Emission Depth of Photoelectrons from Cs–Sb Films
Vol. **26**, No. 2, p. 527—535 (1968).
See also 14.3.1, p. 246; *17*, p. *288*; 20.1, p. 353.

I. K. Andronik, E. K. Arushanov, O. V. Emelyanenko, and D. N. Nasledov
Negative Magnetoresistance in p-CdSb
Vol. **27**, No. 1, p. 45—49 (1968).
See also 13.4, p. 213; *14.3*, p. *241*; 22, p. 451.

L. I. Anatychuk, V. D. Iskra, O. J. Luste, I. M. Rarenko, and L. I. Zarubin
Anisotropy of Electroconductivity in CdSb
Vol. **27**, No. 1, p. 101—107 (1968).
See also 13.1, p. 185; *14.3*, p. *241*; 22, p. 451.

K. Deutscher and K. Hirschberg
Thickness Dependence of the Quantum Yield of Cesium-Antimony Films
Vol. **27**, No. 1, p. 145—150 (1968).
See also 14.3.1, p. 246; *17*, p. *289*; 20.1, p. 354.

U. Birkholz and J. Schelm
Mechanism of Electrical Conduction in β-$FeSi_2$
Vol. **27**, No. 1, p. 413—425 (1968).
See also 13, p. 170; 13.1, p. 186; *14.3*, p. *241*; 15, p. 274; 22, p. 452.

H. Hora and G. B. Kabiersch
Combined Infrared Photoemission from Cs_3Sb
Vol. **27**, No. 2, p. 593—600 (1968).
See also *17*, p. *289*.

G. M. de'Munari, F. Giusiano, and G. Mambriani
A Phase Transition and the Photoemissive Quantum Yield at Low Temperatures in Cs_3Sb Films
Vol. **29**, No. 1, p. 341—356 (1968).
See also *1.2*, p. *12*; 3, p. 27; 13.1, p. 188; 17, p. 289.

U. Birkholz, H. Finkenrath, J. Naegele, and N. Uhle
Infrared Reflectivity of Semiconducting $FeSi_2$
Vol. **30**, No. 1, p. K81—K84 (1968).
See also *20.1*, p. *358*; 22, p. 455.

E. M. Averbakh, V. V. Sobolev, N. N. Syrbu, and Ya. A. Ugai
The Optical Reflectivity Spectra of CdSb and ZnSb Single Crystals
Vol. **30**, No. 2, p. K145—K147 (1968).
See also 13.1, p. 190; *20.1*, p. *359*.

22.8 Tertiary and Higher Compounds (Carbonates, Phosphates, Silicates, and Mica See 22)

Review Articles

A. S. Borshchevskii, N. A. Goryunova, F. P. Kesamanly, and D. N. Nasledov
Semiconducting $A^{II}B^{IV}C_2^V$ Compounds
Vol. **21**, No. 1, p. 9–55 (1967).
See also 3.1, p. 29; 3.2, p. 34; 5, p. 49; 8, p. 79; 10, p. 93; 13.1, p. 174; *14.3*, p. *231*; 15, p. 270; 16, p. 276; 17, p. 286; 18, p. 291; 20.1, p. 344.

G. Harbeke
On the Band Structure of Anisotropic Crystals
Vol. **27**, No. 1, p. 9–44 (1968).
See also *13.1*, p. *174*; 13.2, p. 191; 14.3, p. 231; 20.1, p. 344; 22, p. 439; 22.1, p. 456; 22.1.1, p. 458; 22.1.2, p. 468; 22.1.3, p. 476; 22.4.1, p. 497; 22.4.2, p. 506; 22.4.3, p. 509; 22.5, p. 512; 22.7, p. 545.

Original Papers and Short Notes

R. E. Newnham, R. P. Santoro, P. F. Seal, and G. R. Stallings
Antiferromagnetism in $Mn_3B_2O_6$, $Co_3B_2O_6$, and $Ni_3B_2O_6$
Vol. **16**, No. 1, p. K17–K19 (1966).
See also *18.4*, p. *321*.

Z. Šroubek
On the Theory of the Superexchange Interaction in Ionic Crystals: $KNiF_3$
Vol. **16**, No. 2, p. 405–412 (1966).
See also *3.2*, p. *34*; 13, p. 163; 18.4, p. 321; 22.5, p. 512.

M. V. Kurik and N. I. Vitrikhovskii
Optical Properties and Energy Structure of CdS–CdTe Crystals
Vol. **16**, No. 2, p. K139–K142 (1966).
See also *13.1*, p. *175*; *20.1*, p. *345*; 22.4.1, p. 497; 22.4.3, p. 509.

G. A. Bogomolova, A. A. Kaminskii, and V. A. Timofeeva
Nd^{3+} Optical Centres in $Y_3Al_3O_{12}$ Crystals
Vol. **16**, No. 2, p. K165–K166 (1966).
See also 10, p. 94; *20.1*, p. *345*; 20.2, p. 360.

S. W. Kennedy, G. F. Taylor, and J. H. Patterson
Misfit and Hysteresis at Structure Transformations in Rubidium and Cesium Nitrates; and Thermal Expansions
Vol. **16**, No. 2, p. K175–K178 (1966).
See also *1.2*, p. *9*; 3, p. 26; 8, p. 79.

A. A. Abdurakhmanova and M. I. Aliev
Electrical Properties of GaSb-Ga_2Te_3 and GaSb-GaTe Solid Solutions
Vol. **17**, No. 1, p. 375–380 (1966).
See also 13.1, p. 175; *14.3*, p. *232*; 15, p. 270.

N. A. Gasanova, G. A. Akhundov, and M. A. Nizametdinova
Optical Reflection and Absorption of GaS_xSe_{1-x} Single Crystals
Vol. **17**, No. 1, p. K131–K134 (1966).
See also *20.1*, p. *345*.

P. Bennema
Technique for Measuring the Rate of Growth of Crystals from Solutions in Dependence on the Degree of Supersaturation at Low Supersaturations
Vol. **17**, No. 2, p. 555–562 (1966).
See also *3.1*, p. *30*.

P. Bennema
Crystal Growth Measurements on Potassium Aluminium Alum and Sodium Chlorate from Slightly Supersaturated Solutions
Vol. **17**, No. 2, p. 563–570 (1966).
See also *3.1*, p. *30*.

F. Kubec
Crystal Field Calculation in the $CdWO_4$ Lattice
Vol. **17**, No. 2, p. 571–574 (1966).
See also 3.2, p. 34; *6*, p. *57*; 13.4, p. 202.

C. Verié
On the Band Structure of HgTe–CdTe. (I): Electric Measurements
Vol. **17**, No. 2, p. 889–901 (1966).
See also *13.1*, p. *176*; 14.3, p. 233; 22.4.3, p. 509.

C. Willaime, P. Delavignette, R. Gevers, and S. Amelinckx
The Effect of the Presence of Dislocations on the Shape of Diffraction Spots in Electron Diffraction
Vol. **17**, No. 2, p. K173–K174 (1966).
See also *10*, p. *96*.

H. C. Bolton, L. A. Bursill, A. C. McLaren, and R. G. Turner
On the Origin of the Colour of Labradorite
Vol. **18**, No. 1, p. 221–230 (1966).
See also 4, p. 38; *20.1*, p. *346*.

V. A. Kutasov and I. A. Smirnov
Effect of Excess Antistructural Bi on the Lattice Thermal Conductivity of Bi_2Te_3, Bi_2Te_3-Sb_2Te_3, and Bi_2Te_3–Bi_2S_3 Containing Iodine
Vol. **18**, No. 1, p. 479–488 (1966).
See also *8*, p. *80*; *10*, p. *96*; 14,3, p. 233; 15, p. 271; 22, p. 441.

K. S. Aleksandrov, L. M. Reshchikova, and B. V. Beznosikov
Behaviour of the Elastic Constants of $KMnF_3$ Single Crystals near the Transition of Puckering
Vol. **18**, No. 1, p. K17–K20 (1966).
See also 1.2, p. 10; *12*, p. *156*; 22.5, p. 512.

G. B. Abdullaev, A. U. Malsagov, and V. M. Glazov
Thermoelectric Power of $A^IB^{III}C_2^{VI}$ Type Compounds in the Solid and Liquid State
Vol. **18**, No. 1, p. K29–K31 (1966).
See also 2, p. 22; *15*, p. *271*.

B. P. Kozyrev and T. Boţilă
Some Photoelectrical Properties of KRS-5 Single Crystals
Vol. **18**, No. 1, p. K57–K61 (1966).
See also 13.4, p. 203; *16*. p. *278*; 22.5, p. 512.

B. Tuck
Photoluminescence of $GaAs_{0.7}P_{0.3}$
Vol. **18**, No. 2, p. 541–545 (1966).
See also 13.4, p. 204; *20.3*, p. *364*; 22.2.1, p. 483; 22.2.2, p. 488.

I. P. Molodyan, D. N. Nasledov, S. I. Radautsan, and V. G. Sidorov
The Effective Mass of Electrons in $(InSb)_x\cdot(InTe)_{1-x}$ Crystals
Vol. **18**, No. 2, p. 677–682 (1966).
See also *13.1*, p. *176*; 14.3, p. 234; 15, p. 271.

G. D. Guseinov, M. Z. Ismailov, and A. G. Talybov
On the New Semiconducting Compound $HgTlS_2$
Vol. **18**, No. 2, p. 929–935 (1966).
See also *5*, p. *50*; 8, p. 80; 13.4, p. 204; 14.3, p. 234; 16, p. 278.

P. BENNEMA, R. KERN, and B. SIMON
Interpretation of Deviating Points in the Relation between the Rate of Crystal Growth and the Relative Supersaturation
Vol. **19**, No. 1, p. 211–216 (1967).
See also *3.1*, p. *31*.

C. RAMASASTRY and C. V. RAMAIAH
Dielectric Constant of Sodium Chlorate Crystals at Microwave Frequencies
Vol. **19**, No. 1, p. K15–K17 (1967).
See also *14.4*, p. *252*.

R. MĂNĂILĂ and P. PĂUŞESCU
The Tetragonal to Cubic Transition in the System $Cd_xMg_{1-x}Mn_2O_4$
Vol. **19**, No. 1, p. K19–K22 (1967).
See also *1.2*, p. *10*.

E. I. GAVRILITSA and S. I. RADAUTSAN
Influence of Ordering on the Properties of Solid Solutions of the System $(HgSe)_{3x}$-$(In_2Se_3)_{1-x}$
Vol. **19**, No. 2, p. 609–611 (1967).
See also *1.2*, p. *10*; 8, p. 81; 14.3, p. 235.

R. E. NEWNHAM and R. P. SANTORO
Magnetic and Optical Properties of Dioptase
Vol. **19**, No. 2, p. K87–K90 (1967).
See also *13*, p. *165*; 18, p. 291; *20*, p. *338*.

R. R. GAŁĄZKA and L. SOSNOWSKI
Conduction Band Structure of $Cd_{0.1}Hg_{0.9}Te$
Vol. **20**, No. 1, p. 113–120 (1967).
See also *13.1*; p. *177*; 14.3, p. 235; 22.4.3, p. 509.

L. L. HENCH and D. A. JENKINS
AC Conductivity of a Glass Semiconductor
Vol. **20**, No. 1, p. 327–330 (1967).
See also 2, p. 23; *14.3*, p. *235*; 22.6, p. 537.

S. Y. CHUANG and B. G. HOGG
Positron Annihilation in Alnico and $(La_{0.7}Pb_{0.3})MnO_3$
Vol. **20**, No. 1, p. 331–335 (1967).
See also 11, p. 148; *13*, p. *165*; 20, p. 338; 21.1.1, p. 415.

A. A. KAMINSKII
The Nature of "Aging" of CaF_2-YF_3-Nd^{3+} Crystals (Type I) under Stimulated Emission Conditions
Vol. **20**, No. 1, p. K51–K54 (1967).
See also *20.2*, p. *361*; 22.5, p. 512; 22.5.3, p. 532.

G. BLASSE and A. BRIL
Some Obervations on the Cr^{3+} Fluorescence in the Huntite Structure
Vol. **20**, No. 2, p. 551–556 (1967).
See also 13.4, p. 205; 20.1, p. 347; *20.3*, p. *365*.

G. REMAUT, R. GEVERS, and S. AMELINCKX
Wavy Fringes at Domain Boundaries in Barium Titanate Obersved in the Electron Microscope
Vol. **20**, No. 2, p. 613–621 (1967).
See also *4*, p. *41*; 14.4.2, p. 266.

V. A. BOKOV, N. A. GRIGORYAN, and M. F. BRYZHINA
X-Ray Diffraction and Magnetic Studies of Solid Solutions $Bi_{1-x}Ca_xMnO_3$
Vol. **20**, No. 2, p. 745–754 (1967).
See also *1*, p. *7*; 1.2, p. 10; 18.2, p. 306; 18.3, p. 317; 18.4, p. 322.

V. M. JUDIN and A. B. SHERMAN
Magnetic Properties of $NaNiF_3$
Vol. **20**, No. 2, p. 759—766 (1967).
See also *18.4*, p. *322*; 22.5, p. 512.

A. A. ABDURAKHMANOVA and M. I. ALIEV
On the Mechanism of Thermal Conductivity in $(GaSb)_{3x}$–$(Ga_2Te_3)_{1-x}$ and $(GaSb)_x$–$(GaTe)_{1-x}$ Solid Solutions
Vol. **20**, No. 2, p. 777—781 (1967).
See also 6, p. 60; *8*, p. *82*.

P. ASADI
X-Ray Topography in Coloured and Uncoloured Zones of Antozonite–Fluorite
Vol. **20**, No. 2, p. K71—K72 (1967).
See also *4*, p. *41*; 10.2, p. 139.

P. ASADI
On the Existence of Piezoelectric Textures in Single Crystals of Antozonite–Fluorite
Vol. **20**, No. 2, p. K73—K75 (1967).
See also *4*, p. *41*; 14.4, p. 252.

A. WEISS and K. ZOHNER
Nuclear Quadrupole Interaction of ^{23}Na and ^{11}B and the Crystal Structure of $NaBF_4$
Vol. **21**, No. 1, p. 257—270 (1967).
See also 3.2, p. 35; 5, p. 51; *19*, p. *329*; 22.5, p. 513.

J. HAŃDEREK
Electric Conductivity and Dielectric Absorption Phenomena in Lead Zirconate Ceramics
Vol. **21**, No. 1, p. 323—329 (1967).
See also *14.4*, p. *253*; *14.4.2*, p. *267*.

U. GONSER and R. W. GRANT
Determination of Spin Directions and Electric Field Gradient Axes in Vivianite by Polarized Recoil-Free γ-Rays
Vol. **21**, No. 1, p. 331—342 (1967).
See also *6.1*, p. *72*; 13, p. 165; 18.1, p. 296; 18.4, p. 322.

A. A. KAMINSKII, YU. K. VORONKO, and V. V. OSIKO
Mixed Systems on the Basis of Fluorides as New Laser Materials for Quantum Electronics. The Optical and Emission Parameters
Vol. **21**, No. 1, p. K17—K21 (1967).
See also 20.1, p. 348; *20.2*, p. *361*; 20.3, p. 365; 22.5, p. 513.

A. A. KAMINSKII, KH. S. BAGDASAROV, and L. M. BELYEV
Analysis of the Optical Spectra and Stimulated Radiation of $Y_3Al_5O_{12}$-Nd^{3+} Crystals
Vol. **21**, No. 1, p. K23—K29 (1967).
See also 13.4, p. 207; *20.1*, p. *348*; 20.2, p. 361; *20.3*, p. *365*.

A. BUREWICZ, D. KULGAWCZUK, and A. SZYTUŁA
Internal Magnetic Field in α-FeOOD
Vol. **21**, No. 1, p. K85—K86 (1967).
See also *6.1*, p. *72*; 18.4, p. 322.

F. KELEMEN, A. NÉDA, D. NICULESCU, and E. CRUCEANU
On the Thermal Conductivity of CdTe and Some $CdTe_{1-x}S_x$ and $CdTe_{1-x}Se_x$ Solid Solutions
Vol. **21**, No. 2, p. 557—565 (1967).
See also *8*, p. *82*; 22.4.1, p. 500; 22.4.2, p. 506; 22.4.3, p. 509.

M. PEŘINOVÁ
Origin of Perfect Dislocations and Glide Elements in $NaNO_3$ Single Crystals
Vol. **21**, No. 2, p. K103—K106 (1967).
See also *10*, p. *101*.

I. DÉZSI, GY. ERLAKI, and L. KESZTHELYI
Mössbauer Effect Studies on the $x(Rh_2O_3)1-x(Fe_2O_3)$ System
Vol. 21, No. 2, p. K121–K125 (1967).
See also *6.1*, p. *72*; 18.4, p. 322; 22.6, p. 538.

E. V. MATYUSHKIN, L. S. KUKUSHKIN, and V. V. EREMENKO
Peculiarities of the $RbMnF_3:Nd^{3+}$ Crystals Decay Kinetics due to the Migration of Excitation Energy
Vol. 22, No. 1, p. 65–69 (1967).
See also 13.4, p. 208; *20.3*, p. *366*; 22.5, p. 513.

S. A. ZEINALOV and M. I. ALIEV
Investigation of the Thermal Conductivity of InSb-GaSb Solid Solutions
Vol. 22, No. 1, p. 153–156 (1967).
See also *8*, p. *83*; 22.2.3, p. 491.

A. LODDING
Thermotransport and the Associated Isotope Effect in Solids
Vol. 22, No. 1, p. 157–161 (1967).
See also *9*, p. *89*; 21.2, p. 424.

B. P. KOTRUBENKO, V. N. LANGE, and M. M. MARKUS
Chemical Bond and Structure of the Ternary Compound InAsTe
Vol. 22, No. 1, p. K15–K17 (1967).
See also *3.2*, p. *35*; *5*, p. *52*.

R. V. DAMLE
Lead Zirconate by Coprecipitation
Vol. 22, No. 1, p. K63–K65 (1967).
See also *14.4.2*, p. *267*.

B. RAY and P. M. SPENCER
Phase Diagram of the Alloy System HgTe-CdTe
Vol. 22, No. 2, p. 371–372 (1967).
See also *1*, p. *7*; *1.2*, p. *11*; 22.4.3, p. 509.

G. P. HUFFMAN, P. R. ERRINGTON, and R. M. FISHER
Mössbauer Study of the Fe-Mn Carbides $(Fe_{1-x}Mn_x)_3C$ and $(Fe_{1.1}Mn_{3.9})C_2$
Vol. 22, No. 2, p. 473–481 (1967).
See also *6.1*, p. *72*; 18.2, p. 308.

P. W. GRAVES
Single Crystal Growth of Calcium Pyroniobate
Vol. 22, No. 2, p. 499–504 (1967).
See also *3.1*, p. *32*; 22.6, p. 539.

H.-G. KAHLE, H. KALBFLEISCH, and E. STEIN
Paramagnetic Relaxation and Observation of a Phonon Bottleneck in Terbium Ethyl Sulfate
Vol. 22, No. 2, p. 537–540 (1967).
See also *6*, p. *62*; *18.1*, p. *296*; 19, p. 330.

I. DÉZSI, L. KESZTHELYI, D. KULGAWCZUK, B. MOLNÁR, and N. A. EISSA
Mössbauer Study of β- and δ-FeOOH and Their Disintegration Products
Vol. 22, No. 2, p. 617–629 (1967).
See also 1.2, p. 11; *6.1*, p. *72*; 18.3, p. 318; 18.4, p. 323.

G. D. GUSEINOV, A. M. RAMAZANZADE, E. M. KERIMOVA, and M. Z. ISMAILOV
About a Group of Three-Component Compounds being Analogous to Binary Semiconductors of the $A^{III}B^{VI}$ Type
Vol. 22, No. 2, p. K117–K122 (1967).
See also *14.3*, p. *237*.

S. Lungu
Point Defects in Neutron-Damaged Silica Glass
Vol. **23**, No. 1, p. 147–156 (1967).
See also 2, p. 23; 10, p. 102; *11*, p. *150*.

T. L. Tansley
Spectral Response of p-n Heterojunctions
Vol. **23**, No. 1, p. 241–252 (1967).
See also *14.3.2*, p. *249*; 14.4.1, p. 259; 16, p. 280; 20.1, p. 350; 22.2.1, p. 484.

F. F. Y. Wang and M. Kestigian
Magnetic Susceptibilities of Sodium Neodymium Tungstate and Sodium Terbium Tungstate
Vol. **23**, No. 1, p. 289–294 (1967).
See also 18.1, p. 296; *18.4*, p. *323*.

J. R. Peverley and P. H. E. Meijer
Entropy and Susceptibility of a Pure Dipole-Dipole Substance
Vol. **23**, No. 1, p. 353–360 (1967).
See also *8*, p. *83*; *18*, p. *293*.

A. A. Kaminskii, Y. E. Lapsker, and B. P. Sobolev
Induced Emission of $NaCaCeF_6$:Nd^{3+} at Room Temperature
Vol. **23**, No. 1, p. K5–K7 (1967).
See also 20.1, p. 350; *20.2*, p. *361*; 20.3, p. 367; 22.5, p. 513.

R. R. Gałązka and T. Zakrzewski
Heavy Hole Effective Mass of $Cd_{0.1}Hg_{0.9}Te$
Vol. **23**, No. 1, p. K39–K43 (1967).
See also *13.1*, p. *181*; 14.3, p. 238; 15, p. 272; 22.4.3, p. 510.

J. Przystawa
The Néel Temperatures and Stability of Magnetic Phases in the Disturbed Body-Centered Tetragonal Lattice. Uranium Compounds
Vol. **24**, No. 1, p. 313–322 (1967).
See also *18.2*, p. *309*; *18.4*, p. *323*; 22, p. 448.

V. A. Ivanova, D. Sh. Abdinov, and G. M. Aliev
Some Structural, Electrical, and Thermal Parameters of VCr_2Se_4, $FeCr_2Se_4$, and $NiCr_2Se_4$ Ternary Compounds
Vol. **24**, No. 1, p. K23–K24 (1967).
See also 5, p. 52; 8, p. 83; *14.3*, p. *239*; 15, p. 273.

B. I. Kuliev and G. M. Nedlin
The Types of Magnetic Ordering in Crystals Having the Structure of $RbNiF_3$
Vol. **24**, No. 1, p. K25–K28 (1967).
See also 1.2, p. 11; 3.2, p. 35; *18*, p. *293*; 22.5, p. 513.

J. M. Trooster and A. Dymanus
Mössbauer Effect in $Ga_{2-x}Fe_xO_3$ and Related Compounds
Vol. **24**, No. 2, p. 487–499 (1967).
See also *6.1*, p. *73*; 18.3, p. 318; 22.6, p. 450.

W. Ludwig and G. Voigt
Voltage-Controlled Negative Differential Resistance in Zinc Indium Sulfide Crystals
Vol. **24**, No. 2, p. K161–K164 (1967).
See also *14.4.1*, p. *261*.

M. V. Dmitruk, A. A. Kaminskii, V. V. Osiko, and T. A. Tevosyan
Induced Emission of Hexagonal LaF_3-SrF_2-Nd^{3+} Crystals at Room-Temperature
Vol. **25**, No. 2, p. K75–K78 (1968).
See also 20.1, 352; p. *20.2*, p. *361*; 20.3, p. 368; 22.5, p. 513.

V. Vyšín and V. Janků
Note on the Interpretation of the Rotatory Dispersion of Potassium Dithionate
Vol. **25**, No. 2, p. K103–K104 (1968).
See also *20.1*, p. *352*.

E. O. Osmanov, Yu. V. Rud, and M. E. Stryalkovskii
High-Temperature Modifications of the Semiconductor Compounds $CdSnAs_2$ and $CdGeAs_2$
Vol. **26**, No. 1, p. 85–90 (1968).
See also *1.2*, p. *12*; 2, p. 24; 8, p. 84; 12, p. 158; 14.3, p. 240; 22.7, p. 546.

R. Fatehally, N. P. Sastry, and R. Nagarajan
Exchange and Spin-Orbit Coupling in $KFeF_3$
Vol. **26**, No. 1, p. 91–96 (1968).
See also *13*, p. *169*; 18.4, p. 323; 22.5, p. 513.

J. Hingsammer and H. Jodl
Z_1- and Z_2-Centres in Mixed Crystals of Alkali Halides Doped with Sr^{++}
Vol. **26**, No. 1, p. 261–266 (1968).
See also *10.2*, p. *142*; 22.5.2, p. 526.

V. F. Pissarenko and P. I. Bykovskii
Luminescence of Sm^{2+} in $CaSO_4$
Vol. **26**, No. 1, p. K13–K15 (1968).
See also *20.3*, p. *368*.

G. A. Sikharulidze and Yu. I. Ukhanov
Absorption of Polarized Light in $CdSnAs_2$ Crystals
Vol. **26**, No. 1, p. K33–K36 (1968).
See also 13.1 p. 184; *20.1*, p. *353*; 22.7, p. 546.

D. S. Kulgawczuk, Z. Obuszko, and A. Szytuła
Susceptibility and Magnetization of β- and δ-FeOOH
Vol. **26**, No. 1, p. K83–K85 (1968).
See also *18.3*, p. *319*; 18.4, p. 324.

A. Szytuła, A. Burewicz, Ž. Dimitrijević, S. Kraśnicki, H. Rżany, J. Todorović, A. Wanic, and W. Wolski
Neutron Diffraction Studies of α-FeOOH
Vol. **26**, No. 2, p. 429–434 (1968).
See also *4*, p. *45*; 18.3, p. 319; 18.4, p. 324.

B. Heinrich, J. Zítková, and J. Kaczér
The Temperature Dependence of the Paramagnetic Susceptibility of Ni-I Boracite
Vol. **26**, No. 2, p. 443–446 (1968).
See also *18.1*, p. *297*; 18.4, p. 324.

A. E. Lord, Jr.
Sound Wave Velocity Change due to the Magnon-Phonon Interaction
Vol. **26**, No. 2, p. 717–724 (1968).
See also 6, p. 65; *7*, p. *77*; 13, p. 169; 18.2, p. 311.

Kh. S. Bagdasarov, A. A. Kaminskii, V. S. Krylov, and V. I. Popov
Room-Temperature Induced Emission of Tetragonal YVO_4 Crystals Containing Nd^{3+}
Vol. **27**, No. 1, p. K1–K3 (1968).
See also *20.2*, p. *362*.

B. T. Kolomiets, T. N. Mamontova, and V. V. Negreskul
On the Recombination Radiation of Vitreous Semiconductors
Vol. **27**, No. 1, p. K15–K16 (1968).
See also *2*, p. *24*; *20.3*, p. *369*.

M. YA. VALAKH, N. I. VITRIKHOVSKII, and M. P. LISITSA
The Electron Effective Mass in CdS_xSe_{1-x} Crystals
Vol. 27, No. 1, p. K81–K83 (1968).
See also *13.1*, p. *186*; 20.1, p. 354; 22.4.1, p. 502; 22.4.2, p. 507.

O. BRÜMMER and G. DRÄGER
On the Interpretation of the Fine Structure of X-Ray K Absorption Spectra Far from the Edge
Vol. 27, No. 2, p. 513–516 (1968).
See also 2, p. 24; *20*, p. *341*; 22.1.1, p. 465; 22.6, p. 541.

A. GRAJA
Production of the Second Harmonic of Light in Ammonium Pentaborate and Other Powdered Piezoelectric Crystals
Vol. 27, No. 2, p. K93–K97 (1968).
See also *20*, p. *341*; *20.2*, p. *362*.

J. S. TIWARI, A. MEHRA, and K. G. SRIVASTAVA
Electronic Absorption Spectrum of K_2CoF_4
Vol. 28, No. 1, p. 61–63 (1968).
See also 5, p. 53; 13, p. 171; *20.1*, p. *355*; 22.5, p. 514.

L. KRATĚNA, K. ŽĎÁNSKÝ, and V. ŠIK
Paramagnetic Resonance Spectrum of Gadolinium in $CdWO_4$ Single Crystals
Vol. 28, No. 1, p. 175–179 (1968).
See also *19*, p. *334*.

F. KELEMEN, A. NÉDA, E. CRUCEANU, and D. NICULESCU
On the Thermal Conductivity of $Zn_xHg_{1-x}Te$ Solid Solutions
Vol. 28, No. 1, p. 421–428 (1968).
See also 6, p. 66; *8*, p. *85*; 22.4.3, p. 511.

N. A. GORYUNOVA, S. M. RYVKIN, G. P. SHPENIKOV, I. I. TICHINA, and V. G. FEDOTOV
Investigations of Some Properties of Vitreous and Crystalline $CdGeP_2$
Vol. 28, No. 2, p. 489–494 (1968).
See also 2, p. 24; 14.3, p. 242; *20.3*, p. *370*.

G. K. BORN, R. J. GRASSER, and A. O. SCHARMANN
EPR and Thermoluminescence Measurements on $CaWO_4$ Crystals
Vol. 28, No. 2, p. 583–588 (1968).
See also 10.2, p. 143; *11*, p. *153*; 19, p. 334; 20.3, p. 370.

I. F. CHANG, S. S. MITRA, J. N. PLENDL, and L. C. MANSUR
Long-Wavelength Longitudinal Phonons of Multi-Mode Crystals
Vol. 28, No. 2, p. 663–673 (1968).
See also *6*, p. *67*; 20.1, p. 356; 22.4.1, p. 504; 22.4.2, p. 508; 22.6, p. 542.

V. A. BOKOV, N. A. GRIGORYAN, M. F. BRYZHINA, and V. V. TIKHONOV
Effect of Lattice Distortions on the Magnetic Behaviour of Perovskite-Type Manganites
Vol. 28, No. 2, p. 835–847 (1968).
See also 5, p. 54; *18.2*, p. *313*; *18.4*, p. *324*.

W. W. HOLLOWAY, JR. and M. KESTIGIAN
Temperature Dependence of the Broad-Band Fluorescence of $RbMnF_3$
Vol. 28, No. 2, p. K159–K162 (1968).
See also *20.3*, p. *370*; 22.5, p. 514.

N. N. Sirota and E. A. Vasilev
Magnetic Properties and Structure of $MnAs_xSb_{1-x}$ Solid Solutions
Vol. **28**, No. 2, p. K175—K177 (1968).
See also *18.1*, p. *297*; 18.2, p. 313.

K. Pfeuffer
Determination of the Electron Spin-Lattice Relaxation Time T_1 in the Direct Process
Vol. **29**, No. 1, p. 171—177 (1968).
See also 6, p. 67; *18.1*, p. *297*; 19, p. 335.

L. Ahmad
Change in Electron Affinity of a Semiconductor due to Different Forms of Chemisorption
Vol. **29**, No. 1, p. 179—187 (1968).
See also *13.3*, p. *199*.

A. A. Vaipolin, N. A. Goryunova, L. I. Kleshchinskii, G. V. Loshakova, and E. O. Osmanov
The Structure and Properties of the Semiconducting Compound $ZnSnP_2$
Vol. **29**, No. 1, p. 435—442 (1968).
See also 4, p. 47; *5*, p. *54*; 6, p. 68; 10, p. 110; 14.3, p. 243.

P. Reimers and W. Ruppel
The Preparation of CdS-CdSe Graded Single Crystals
Vol. **29**, No. 1, p. K31—K33 (1968).
See also *3.1*, p. *33*; 5, p. 54; 20.1, p. 357; 22.4.1, p. 504; 22.4.2, p. 508.

D. Bonnet, H. Germer, and H. Rabenhorst
Photoconduction in $Cd(S_xSe_{1-x})$ Films with Graded Composition
Vol. **29**, No. 1, p. K65—K67 (1968).
See also *16*, p. *284*; 22.4.1, p. 504; 22.4.2, p. 508.

V. Plakhty and W. Cochran
X-Ray Study of the Phase Transition and Lattice Vibrations of Lanthanum Aluminate
Vol. **29**, No. 1, p. K81—K83 (1968).
See also *1.2*, p. *12*; *6*, p. *68*.

R. G. Howell and D. J. Newman
High Resolution Study of Crystal Field Lines in the Far Infra-Red
Vol. **29**, No. 2, p. 697—706 (1968).
See also 6, p. 68; 18.1, p. 297; *20.1*, p. *357*.

M. Sauvage
Observation of Sources and Reactions between Partial Twinning Dislocations on X-Ray Topographs
Vol. **29**, No. 2, p. 725—736 (1968).
See also *10*, p. *111*.

J. E. Lewis, H. Rodot, and P. Haen
The Low-Temperature Thermoelectric Power and Thermal Conductivity of GeTe and of Some GeTe-MnTe Alloys
Vol. **29**, No. 2, p. 743—754 (1968).
See also 6, p. 68; *8*, p. *86*; 13, p. 173; *15*, p. *274*; 22, p. 454.

V. Beckmann, W. Bruckner, W. Fuchs, G. Ritter, and H. Wegener
The Measurement of the Anisotropy Constant of Antiferromagnetic $FePO_4$ by Means of the Mössbauer Effect
Vol. **29**, No. 2, p. 781—791 (1968).
See also 6.1, p. 74; *18.4*, p. *325*.

M. Averous, G. Bougnot, and J. Calas
Conduction Bands of GaSb-InSb Mixed Crystals with a Low Concentration of InSb
Vol. **29**, No. 2, p. 807—812 (1968).
See also *13.1*, p. *189*; 14.3, p. 244; 22.2.3, p. 493.

E. Z. Dziuba, D. Niculescu, and N. Niculescu
Energy Band Structure of $Zn_xHg_{1-x}Te$ Solid Solutions
Vol. **29**, No. 2, p. 813—818 (1968).
See also *13.1*, p. *189*; 14.3, p. 244; 20.1, p. 357; 22.4.3, p. 511.

P. P. Salhotra, E. C. Subbarao, and P. Venkateswarlu
Polymorphism of Rubidium Nitrate
Vol. **29**, No. 2, p. 859—864 (1968).
See also *1.2*, p. *13*; *3*, p. *28*; 8, p. 86.

J. Piekoszewski, J. Suwalski, and S. Ligenza
Mössbauer Effect Study in Chalcopyrite
Vol. **29**, No. 2, p. K99—K101 (1968).
See also *6.1*, p. *74*.

L. N. Cojocaru
Effects of Fission Fragments on Electric Properties of the UO_2-SiO_2 System
Vol. **29**, No. 2, p. K119—K122 (1968).
See also *11*, p. *154*; 14.3, p. 244; 15, p. 274; 22.6, p. 543.

S. Haussühl
Piezoelectrical and Electrical Behaviour of Lithiumjodate
Vol. **29**, No. 2, p. K159—K162 (1968).
See also *12*, p. *158*; *14.3*, p. *244*.

A. Murasik, W. Suski, R. Troć, and J. Leciejewicz
The Magnetic Properties of Uranium Oxyselenide
Vol. **30**, No. 1, p. 61—66 (1968).
See also *18.4*, p. *325*.

M. L. Reynolds, W. E. Hagston, and G. F. J. Garlick
Experimental Observations of the Absorption, Infra-Red Emission, and Excitation Spectra of Cr^{3+} Ions in $ZnWO_4$
Vol. **30**, No. 1, p. 113—117 (1968).
See also *13.4*, p. *216*; 20.1, p. 358; 20.3, p. 371.

I. Dézsi, N. A. Eissa, L. Keszthelyi, B. Molnár, and D. L. Nagy
Mössbauer Study of $SnCl_2$ and $Dy(ClO_4)_3$ in Ice
Vol. **30**, No. 1, p. 215—218 (1968).
See also 2, p. 24; *6.1*, p. *74*.

F. J. Schedewie
Paramagnetic Relaxation in Neodymium-Ethylsulfate: Phonon Re-Absorption and Spectral Spin Diffusion
Vol. **30**, No. 1, p. 251—259 (1968).
See also 6, p. 69; 13, p. 173; *18.1*, p. *298*; 19, p. 336.

R. V. Pisarev, I. G. Siny, N. N. Nesterova, G. A. Smolensky, and P. P. Syrnikov
Faraday Rotation and Magnetic Dichroism in $RbNi(Co)F_3$
Vol. **30**, No. 1, p. 367—372 (1968).
See also 13, p. 173; *18.3*, p. *320*; *20*, p. *342*; 22.5, p. 514.

V. V. Boldyrev, G. B. London, and V. K. Zhuravlev
Study of Semiconducting Properties of Potassium Permanganate
Vol. **30**, No. 1, p. K13—K15 (1968).
See also 13.4, p. 217; *14.3*, p. *244*; *15*, p. *275*.

G. Albanese and M. Musci
Mössbauer Resonant Absorption in $B8_2$ Germanium Compounds
Vol. **30**, No. 1, p. K37—K40 (1968).
See also *6.1*, p. *75*.

D. V. Chepur, D. M. Bercha, I. D. Turyanitsa, and V. Yu. Slivka
Peculiarities of the Energy Spectrum and Edge Absorption in the Chain Compounds $A^VB^{VI}C^{VII}$
Vol. **30**, No. 2, p. 461—468 (1968).
See also *13.1*, p. *189*; 16, p. 284; 20.1, p. 358.

H. Imagawa
ESR Studies of Cupric Ion in Various Oxide Glasses
Vol. **30**, No. 2, p. 469—478 (1968).
See also 2, p. 25; 3.2, p. 36; *19*, p. *336*.

M. L. Reynolds, W. E. Hagston, and G. F. J. Garlick
The Theory of the Absorption and Infrared Emission of Cr^{3+} Ions in $ZnWO_4$
Vol. **30**, No. 2, p. 507—514 (1968).
See also *13.4*, p. *217*; 20.1, p. 359; 20.3, p. 371.

M. L. Reynolds, W. E. Hagston, and G. F. J. Garlick
The Optical Absorption Spectra of Co and Ni in $ZnWO_4$
Vol. **30**, No. 2, p. 735—739 (1968).
See also *13.4*, p. *217*; *20.1*, p. *359*.

A. M. White, E. W. Williams, and P. Porteous
Infra-Red and Visible Photoluminescence in $In_{1-x}Ga_xP$
Vol. **30**, No. 2, p. K125—K126 (1968).
See also *20.3*, p. *372*; 22.2.2, p. 488.

22.9 Organic Semiconductors

Original Papers and Short Notes

N. Riehl and P. Thoma
Delayed Fluorescence and Thermoluminescence of Anthracene Single Crystals at Very Low Temperatures
Vol. **16**, No. 1, p. 159–170 (1966).
See also 13.4, p. 201; *20.3*, p. *363*.

G. Gallus and H. C. Wolf
Direct Measurement of the Diffusion Length of Singulet Excitons in Solid Phenanthrene
Vol. **16**, No. 1, p. 277–280 (1966).
See also *13.2*, p. *191*; 20.3, p. 363.

W. L. McCubbin
Symmetry Orbitals and UV Selection Rules for the Polymethylene Chain
Vol. **16**, No. 1, p. 289–293 (1966).
See also *3*, p. *26*; *13.1*, p. *174*; 20, p. 337.

P. Thoma and G. Vaubel
Electrical Properties of Anthracene Single Crystals at Very Low Temperatures
Vol. **16**, No. 2, p. 663–674 (1966).
See also 13.4, p. 201; 14.3, p. 231; 14.4.1, p. 256; *16*, p. *276*.

M. G. Asch, M. C. Félix, and M. R. Ongaro
Electrization of Polyethylene and Polystyrene after γ-Irradiation
Vol. **17**, No. 1, p. 49–57 (1966).
See also 11, p. 146; *13.3*, p. *196*.

J. F. Petroff and A. Authier
Fringes of Equal Deformation in X-Ray Topographs
Vol. **17**, No. 1, p. K3–K5 (1966).
See also *10*, p. *95*.

G. Schoffa
Electron Spin Resonance of Radiation-Produced Radicals in 2,6-Butylhydroxytoluene Single Crystals
Vol. **17**, No. 2, p. 683–689 (1966).
See also 11, p. 146; *19*, p. *327*.

V. L. Broude, E. I. Rashba, and E. F. Sheka
A New Approach to the Vibronic Spectra of Molecular Crystals
Vol. **19**, No. 1, p. 395–406 (1967).
See also *6*, p. *58*; 20.1, p. 347.

R. Stockmeyer and H. H. Stiller
Lattice Dynamics of Solid Adamantane
Vol. **19**, No. 2, p. 781–785 (1967).
See also *6*, p. *59*.

P. Devaux and M. Schott
Thermally Stimulated Currents without Optical Excitation. Application to Copper Phtalocyanine
Vol. **20**, No. 1, p. 301–309 (1967).
See also *13.4*, p. *205*; 14.4.1, p. 257.

G. S. Pawley
A Model for the Lattice Dynamics of Naphthalene and Anthracene
Vol. **20**, No. 1, p. 347–360 (1967).
See also *6*, p. *59*.

P. M. Robinson and H. G. Scott
Tilt Boundaries in Deformed Anthracene Single Crystals
Vol. **20**, No. 2, p. 461–471 (1967).
See also 4, p. 41; *10*, p. *99*.

C. Hamann
On the Electric and Thermoelectric Properties of Copper Phthalocyanine Single Crystals
Vol. **20**, No. 2, p. 481–491 (1967).
See also 13.4, p. 205; *14.3*, p. *236*; 14.4.1, p. 258; 15, p. 271.

E. Majerníková
Faraday Effect and Spin-Orbit Interaction of the Lowest Exciton States in Molecular Crystals
Vol. **22**, No. 1, p. 113–122 (1967).
See also 13.2, p. 192; *18*, p. *292*; *20*, p. *339*.

H. L. Jetter and H. C. Wolf
On the Davydov Splitting of the First Exited Singulet-State of Anthracene Crystals
Vol. **22**, No. 1, p. K39–K43 (1967).
See also 12, p. 157; 13.2, p. 193; *20.1*, p. *349*.

A. Dall'Olio, G. Dascola, and V. Varacca
EPR Evidence of Exchange-Coupled Ion Pairs and of Isolated Ions in Some Chlorosubstituted Cupric Acetates
Vol. **22**, No. 2, p. 365–370 (1967).
See also 13.4, p. 208; *19*, p. *330*.

A. Georgieva and D. Nenow
Spontaneous Solid-Vapour Transformation of Negative Crystals in Diphenyl
Vol. **22**, No. 2, p. 415–421 (1967).
See also 3.1, p. 32; *4*, p. *42*.

H. P. Müller, P. Thoma, and G. Vaubel
The Phosphorescence of Anthracene Single Crystals and Its Spectrum
Vol. **23**, No. 1, p. 253–262 (1967).
See also 13, p. 167; *20.3*, p. *367*.

B. Hilczer, B. Błaszczyk, and S. Goderska
Effect of Storage Temperature on the Time Dependence of Charge Density in Carnauba Wax and Methyl Polymethacrylate Electrets
Vol. **23**, No. 1, p. 419–422 (1967).
See also *14.4.1*, p. *259*.

D. Haarer, D. Schmid, and H. C. Wolf
Electron Spin Resonance of Triplet Excitons in Anthracene
Vol. **23**, No. 2, p. 633–638 (1967).
See also *13.2*, p. *193*; *19*, p. *330*.

G. K. Afanaseva, K. S. Aleksandrov, and A. I. Kitaigorodskii
Elastic Constants of Anthracene
Vol. **24**, No. 1, p. K61–K63 (1967).
See also *12*, p. *157*.

H. Götz, H. Schulze, and K. Werner
Overhauser Effect in Semiconducting Crystalline TCNQ-Complexes
Vol. **24**, No. 1, p. K95—K96 (1967).
See also *19*, p. *331*.

C. Hamann
On the Trap Distribution in Thin Films of Copper Phthalocyanine
Vol. **26**, No. 1, p. 311—318 (1968).
See also *13.4*, p. *212*; 14.3.1, p. 246; 14.4.1, p. 262.

G. Vaubel and H. Baessler
Determination of the Energy of Conduction States in Anthracene Crystals by Photoemission of Electrons from Sodium
Vol. **26**, No. 2, p. 599—606 (1968).
See also *13.1*, p. *185*; 17, p. 288.

H. Baessler, N. Riehl, and G. Vaubel
Calculation of the Width of the Second Conduction Band in Anthracene Crystals from Photoemission Data
Vol. **26**, No. 2, p. 607—610 (1968).
See also *13.1*, p. *185*; 17, p. 289.

D. Michell, P. M. Robinson, and A. P. Smith
Direct Observation of Dislocations in Anthracene
Vol. **26**, No. 2, p. K93—K95 (1968).
See also *10*, p. *106*.

P. Devaux and M. Schott
Field Effect on Photocarriers in Copper Phthalocyanine Single Crystals
Vol. **27**, No. 1, p. 201—208 (1968).
See also 13.2, p. 194; *13.3*, p. *198*; *16*, p. *282*.

R. Stockmeyer and H. Stiller
Phonons, Torsons, and Rotational Diffusion in Adamantane
Vol. **27**, No. 1, p. 269—280 (1968).
See also *6*, p. *65*.

V. M. Agranovich and Yu. V. Konobeev
Diffusion of Free Excitons in Molecular Crystals
Vol. **27**, No. 1, p. 435—442 (1968).
See also *13.2*, p. *194*; 13.4, p. 213.

V. K. Jain and P. K. C. Pillai
Characteristics of Plastic Electrets
Vol. **28**, No. 1, p. 415—419 (1968).
See also 2, p. 24; *14.4*, p. *254*.

S. E. Gukasyan and V. S. Shpinel
Mössbauer Effect in Antimony-Organic Compounds
Vol. **29**, No. 1, p. 49—52 (1968).
See also *6.1*, p. *74*.

V. Hádek
Electron and Hole Conductivity of Donor-Acceptor Complexes of Oligoanilinic Substances and Their Thermoelectric Properties
Vol. **30**, No. 1, p. 275—282 (1968).
See also *13.4*, p. *216*; *14.3*, p. *244*; 15, p. 275.

E. L. Frankevich and B. M. Rumyantsev
Charge-Transfer Excitons and Delayed Fluorescence of Anthracene
Vol. **30**, No. 1, p. 329—340 (1968).
See also 13.2, p. 195; *20.3*, p. *371*.

23. Solidified Gases

Original Papers and Short Notes

V. G. Manzhelii, V. G. Gavrilko, and E. I. Voitovich
Thermal Expansion of Solidified Rare Gases
Vol. **17**, No. 1, p. K139–K141 (1966).
See also *8*, p. *79*.

N. P. Gupta and B. Dayal
Effect of Zero-Point Energy on the Specific Heats of Solidified Neon and Xenon
Vol. **18**, No. 2, p. 731–736 (1966).
See also 3.2, p. 34; *6*, p. *58*; *8*, p. *80*.

B. Dorner and H. H. Stiller
Internal Dynamics of the Low-Temperature Phases of CH_4 Molecular Crystals
Vol. **18**, No. 2, p. 795–806 (1966).
See also 1.2, p. 10; *6*, p. *58*.

N. P. Gupta and B. Dayal
Equation of State of Solid Argon and Krypton
Vol. **20**, No. 1, p. 321–325 (1967).
See also *8*, p. *81*.

N. P. Gupta and B. Dayal
Thermal Expansion and Compressibility of Solid Neon and Xenon
Vol. **21**, No. 2, p. 661–664 (1967).
See also 6, p. 61; *8*, p. *82*; *12*, p. *157*.

J.-Y. Roncin and K. Moorjani
On the Absorption Spectrum of Gaseous and Solid Xenon
Vol. **23**, No. 1, p. K1–K3 (1967).
See also 13.2, p. 193; *20.1*, p. *350*.

M. Bitter, W. Gissler, and T. Springer
Lattice Dynamics of Solid Helium at 2.9 °K and 125 atm by Neutron Scattering
Vol. **23**, No. 2, p. K155–K158 (1967).
See also *6*, p. *63*.

I. N. Krupskii and V. G. Manzhely
Thermal Conductivity of Solid Argon
Vol. **24**, No. 1, p. K53–K56 (1967).
See also *8*, p. *84*.

M. I. Bagatskii, V. A. Kucheryavy, V. G. Manzhelii, and V. A. Popov
Thermal Capacity of Solid Nitrogen
Vol. **26**, No. 2, p. 453–460 (1968).
See also *8*, p. *84*.

I. N. Krupskii, V. G. Manzhely, and L. A. Koloskova
Thermal Conductivity of Solid Ammonia
Vol. **27**, No. 1, p. 263–268 (1968).
See also 6, p. 65; *8*, p. *84*.

Yu. G. Litvinenko, V. V. Eremenko, and T. I. Garber
Antiferromagnetic Ordering Effect on the Light Absorption Spectrum by Crystalline Oxygen
Vol. **30**, No. 1, p. 49–60 (1968).
See also 13.1, p. 189; *18.4*, p. *325*; 20.1, p. 358.

R. K. Gupta and N. P. Gupta
Anharmonic Potential Parameters of the Buckingham Potential for Solid Neon
Vol. **30**, No. 1, p. K41—K43 (1968).
See also *6*, p. *69*.

Author Index

I. V. Abarenkov	20,	643	(1967)	*13.1*, p. *177*; 22.5.2, p. 521
D. Sh. Abdinov	16,	K31	(1966)	2, p. 22; 10, p. 94; *20.1*, p. *344*; 22.1.3, p. 476
	18,	K153	(1966)	*10*, p. *97*; 14.3, p. 234; 22.1.3, p. 477
	24,	K23	(1967)	5, p. 52; 8, p. 83; *14.3*, p. *239*; 15, p. 273; 22.8, p. 552
	24,	K145	(1967)	5, p. 52; *8*, p. *84*; 14.3, p. 239; *15*, p. *273*; 22, p. 449
	28,	K47	(1968)	6, p. 66; *8*, p. *85*; 22, p. 453
S. G. Abdinova	22,	741	(1967)	13.4, p. 209; *14.3*, p. *237*; 22.2.3, p. 491
G. B. Abdullaev	16,	205	(1966)	13.4, p. 201; *14.4.1*, p. *255*; 22, p. 439
	16,	209	(1966)	*14.3.2*, p. *248*; 16, p. 276; 22, p. 439
	16,	K31	(1966)	2, p. 22; 10, p. 94; *20.1*, p. *344*; 22.1.3, p. 476
	16,	K113	(1966)	10, p. 94; *19*, p. *326*; 22.1.3, p. 476
	17,	593	(1966)	13.1, p. 175; *14.4.1*, p. *256*; 22, p. 440
	17,	777	(1966)	6, p. 57; *8*, p. *79*; 22.2.3, p. 490
	18,	K29	(1966)	2, p. 22; *15*, p. *271*; 22.8, p. 548
	18,	K153	(1966)	*10*, p. *97*; 14.3, p. 234; 22.1.3, p. 477
	20,	421	(1967)	*14.3*, p. *235*; *14.4.1*, p. *258*; 22, p. 444
	21,	423	(1967)	13.4, p. 206; *14.3*, p. *236*; 14.4.1, p. 258; 22.1.2, p. 471
	21,	461	(1967)	3.1, p. 31; *14.3*, p. *236*; 15, p. 272; 22, p. 445
	25,	75	(1968)	13.4, p. 211; *16*, p. *281*; 22, p. 449
	26,	65	(1968)	1.2, p. 12; *14.3*, p. *240*; 15, p. 273; 22, p. 450
R. Z. Abdulov	24,	385	(1967)	*18.2*, p. *309*; *18.4*, p. *323*; 21.1.1, p. 419
A. A. Abdurakhma-	17,	375	(1966)	13.1, p. 175; *14.3*, p. *232*; 15, p. 270; 22.8, p. 547
nova	20,	777	(1967)	6, p. 60; *8*, p. *82*; 22.8, p. 550
Sh. Sh. Abelskii	22,	309	(1967)	13, p. 166; *14.1*, p. *223*; 18.2, p. 308
A. A. Adamenko	21,	K81	(1967)	10.1, p. 122; *15*, p. *272*; 21.1, p. 398
O. Adametz	16,	K161	(1966)	*10.2*, p. *137*; 22.5.2, p. 518
	24,	583	(1967)	*10.2*, p. *141*; 22.5.2, p. 524
A. R. Adams	23,	K99	(1967)	*8*, p. *83*; 13, p. 167; 15, p. 273; 22.1.3, p. 477
V. E. Adamyan	18,	873	(1966)	*18.2*, p. *304*; *18.4*, p. *321*; 22, p. 442
Y. Adda	23,	223	(1967)	*9*, p. *90*; 10.1, p. 125; 21.6, p. 434
E. Ady	24,	K65	(1967)	*14.3*, p. *239*; 22.2.1, p. 484
A. M. Afanasev	28,	287	(1968)	5, p. 54; *6*, p. *66*
G. K. Afanaseva	24,	K61	(1967)	*12*, p. *157*; 22.9, p. 559
R. P. Agarwala	30,	319	(1968)	*9*, p. *92*; 21, p. 388; 21.1, p. 408; 21.6, p. 437
V. M. Agranovich	27,	435	(1968)	*13.2*, p. *194*; 13.4, p. 213; 22.9, p. 560
J. P. Agrawal	21,	679	(1967)	13.4, p. 207; 20.1, p. 348; *20.3*, p. *366*; 22.4.1, p. 500
F. Agulló-López	22,	483	(1967)	*10.2*, p. *140*; 11, p. 149; 22.5.2, p. 523
	29,	217	(1968)	*10.2*, p. *143*; 11, p. 153; 22.5.2, p. 528

A. Aharoni	**16,**	3	(1966)	4, p. 37; *18.2*, p. *299*; 21.1, p. 390; 21.1.1, p. 410
L. Ahmad	**29,**	179	(1968)	*13.3*, p. *199*; 22.8, p. 555
G. A. Akhundov	**16,**	209	(1966)	*14.3.2*, p. *248*; 16, p. 276; 22, p. 439
	16,	K15	(1966)	*20.1*, p. *344*; 22, p. 439
	17,	K115	(1966)	*20.1*, p. *345*; 22, p. 440
	17,	K131	(1966)	*20.1*, p. *345*; 22.8, p. 547
	17,	K225	(1966)	*20.3*, p. *364*; 22, p. 441
	17,	K237	(1966)	13.4, p. 203; *14.3*, p. *233*; 15, p. 271; 22, p. 441
	28,	K47	(1968)	6, p. 66; *8*, p. *85*; 22, p. 453
M. R. Akhundov	**16,**	209	(1966)	*14.3.2*, p. *248*; 16, p. 276; 22, p. 439
N. I. Akimenko	**23,**	93	(1967)	*13.1*, p. *180*; 14.3, p. 238; 18, p. 293; 22, p. 447
I. A. Akimov	**17,**	K95	(1966)	13.4, p. 202; *15*, p. *270*; 16, p. 277; 22, p. 440
	20,	771	(1967)	13.3, p. 197; 16, p. 279; *17*, p. *287*; 22.4, p. 495; 22.5, p. 512; 22.5.1, p. 515
F. M. El-Akkad see under letter "E"				
M. M. Akselrod	**27,**	249	(1968)	*13.1*, p. *186*; 14.3, p. 241; 15, p. 274; 22.2.3, p. 492
I. G. Aksyanov	**17,**	K225	(1966)	*20.3*, p. *364*; 22, p. 441
G. Albanese	**30,**	K37	(1968)	*6.1*, p. *75*; 22.8, p. 557
G. Albrecht	**18,**	K119	(1966)	8, p. 80; *13*, p. *164*; 22, p. 442
	23,	K17	(1967)	18.1, p. 297; *19*, p. *330*; 21.1, p. 400
A. Aldea	**22,**	377	(1967)	13.4, p. 208; *14.3*, p. *237*; 22, p. 446
K. S. Aleksandrov	**18,**	K17	(1966)	1.2, p. 10; *12*, p. *156*; 22.5, p. 512; 22.8, p. 548
	24,	K61	(1967)	*12*, p. *157*; 22.9, p. 559
H. Alexander	**26,**	725	(1968)	*10*, p. *106*; 22.1.1, p. 464
	27,	391	(1968)	*10*, p. *107*; 22.1.1, p. 465
G. F. Alfrey	**27,**	541	(1968)	4, p. 46; 10, p. 107; *20.3*, p. *370*; 22.4.1, p. 502
M. Ali	**21,**	113	(1967)	*6*, p. *60*; 8, p. 82; 22.6, p. 537
	28,	193	(1968)	13.1, p. 187; 13.4, p. 214; *14.3*, p. *242*; 15, p. 274; 22.6, p. 542
G. M. Aliev	**16,**	K31	(1966)	2, p. 22; 10, p. 94; *20.1*, p. *344*; 22.1.3, p. 476
	18,	K153	(1966)	*10*, p. *97*; 14.3, p. 234; 22.1.3, p. 477
	24,	K23	(1967)	5, p. 52; 8, p. 83; *14.3*, p. *239*; 15, p. 273; 22.8, p. 552
	24,	K145	(1967)	5, p. 52; *8*, p. *84*; 14.3, p. 239; *15*, p. *273*; 22, p. 449
	29,	K47	(1968)	2, p. 24; *14.3*, p. *243*; 22.1.3, p. 480
M. I. Aliev	**17,**	375	(1966)	13.1, p. 175; *14.3*, p. *232*; 15, p. 270; 22.8, p. 547
	17,	777	(1966)	6, p. 57; *8*, p. *79*; 22.2.3, p. 490
	20,	777	(1967)	6, p. 60; *8*, p. *82*; 22.8, p. 550
	21,	643	(1967)	6, p. 61; *8*, p. *82*; 10, p. 100; 22.1.1, p. 462; 22.1.2, p. 471
	22,	153	(1967)	*8*, p. *83*; 22.2.3, p. 491; 22.8, p. 551
	22,	741	(1967)	13.4, p. 209; *14.3*, p. *237*; 22.2.3, p. 491
S. A. Aliev	**17,**	105	(1966)	13.4, p. 202; *14.3*, p. *232*; 22.2.2, p. 487
M. Kh. Alieva	**25,**	75	(1968)	13.4, p. 211; *16*, p. *281*; 22, p. 449
Z. A. Aliyarova	**21,**	461	(1967)	3.1, p. 31; *14.3*, p. *236*; 15, p. 272; 22, p. 445
	26,	65	(1968)	1.2, p. 12; *14.3*, p. *240*; 15, p. 273; 22, p. 450
Z. I. Alizade	**29,**	K19	(1968)	*14.1*, p. *226*; *18.2*, p. *314*; 21.1, p. 407
R. S. Allgaier	**21,**	343	(1967)	2, p. 23; *13.4*, p. *206*; 21, p. 378; 22, p. 445
J. G. Allpress	**25,**	541	(1968)	3, p. 27; *5*, p. *53*; 22.6, p. 540
L. A. Almasov	**30,**	767	(1968)	*13*, p. *173*; *20*, p. *343*; 22.1.1, p. 467
I. Alstrup	**28,**	555	(1968)	*13.1*, p. *187*; 22.1.2, p. 474
M. Altwein	**29,**	203	(1968)	13.1, p. 188; 14.3.1, p. 247; *20.1*, p. *356*; 22.4, p. 496

E. D. Aluker	19,	35	(1967)	13.4, p. 204; *20.3*, p. *365*; 22.5.2, p. 520
S. Amelinckx	17,	221	(1966)	*10*, p. *94*; 22.1, p. 456
	17,	K173	(1966)	*10*, p. *96*; 22.8, p. 548
	18,	167	(1966)	*10.1*, p. *117*; 22.1.1, p. 413
	18,	325	(1966)	*4*, p. *39*; 22.6, p. 536
	18,	343	(1966)	4, p. 39; *10*, p. *96*; *10.1*, p. *118*
	18,	363	(1966)	4, p. 39; *10*, p. *96*; *10.1*, p. *118*; 21.1.1, p. 414; 22.6, p. 536
	19,	683	(1967)	*1*, p. *7*; 4, p. 40; 10, p. 98; 22.6, p. 537
	20,	613	(1967)	*4*, p. *41*; 14.4.2, p. 266; 22.8, p. 549
	21,	393	(1967)	*10*, p. *99*; *10.1*, p. *122*; 21.1.1, p. 417; 22.6, p. 538
	23,	549	(1967)	*4*, p. *43*; *10.1*, p. *125*; 21.1.1, p. 419
	24,	195	(1967)	4, p. 43; *10*, p. *103*; *10.1*, p. *126*; 21.1, p. 401
	24,	K121	(1967)	*10*, p. *103*; *10.1*, p. *127*
	26,	577	(1968)	*4*, p. *46*; 10, p. 105
	28,	305	(1968)	*1.1*, p. *8*; 4, p. 46; 21, p. 386; 21.1, p. 405
	30,	797	(1968)	*4*, p. *48*; 10, p. 113; 10.1, p. 136; 21.1.1, p. 423; 22.6, p. 544
M. Amin	27,	741	(1968)	1.2, p. 12; *8*, p. *85*; 14.4.2, p. 268
Kh. I. Amirkhanov	20,	K119	(1967)	*15*, p. *272*; 22.2.1, p. 483
A. T. Amos	19,	587	(1967)	*13.3*, p. *197*; 13.4, p. 205
E. Amzallag	30,	407	(1968)	*13.1*, p. *174*; 18, p. 291; 20, p. 337; 22.1.1, p. 458; 22.1.2, p. 468; 22.2.1, p. 482; 22.2.2, p. 487; 22.2.3, p. 489; 22.4, p. 495; 22.4.1, p. 497; 22.4.2, p. 506; 22.4.3, p. 509
T. Anagnostopoulos	29,	K179	(1968)	*10.1*, p. *134*; *11*, p. *154*; 18.2, p. 315; 21.1.1, p. 423
M. S. Anand	30,	319	(1968)	*9*, p. *92*; 21, p. 388; 21.1, p. 408; 21.6, p. 437
T. R. Anantharaman	20,	59	(1967)	*4*, p. *41*
L. I. Anatychuk	27,	101	(1968)	13.1, p. 185; *14.3*, p. *241*; 22, p. 451; 22.7, p. 546
A. Anderman	30,	283	(1968)	3.2, p. 36; *10.1*, p. *135*; 21.1, p. 408
E. E. Anderson	25,	K31	(1968)	5, p. 53; *18*, p. *293*
R. L. Anderson	17,	K79	(1966)	*20.1*, p. *345*; 22.1.1, p. 459
W. Andrä	17,	191	(1966)	10.1, p. 115; *18.2*, p. *302*; 21.1, p. 392
R. Andreytchin	21,	K87	(1967)	2, p. 23; 13.4, p. 207; *20.1*, p. *348*; 22, p. 445
M. V. Andriyashik	28,	277	(1968)	13.1, p. 187; 13.2, p. 194; *20.1*, p. *355*; 22, p. 452
I. K. Andronik	27,	45	(1968)	13.4, p. 213; *14.3*, p. *241*; 22, p. 451; 22.7, p. 546
C. Anghel	18,	K13	(1966)	*10.1*, p. *118*; *18.2*, p. *304*; 21.1, p. 394
	19,	613	(1967)	*18.2*, p. *305*
	21,	601	(1967)	14.1, p. 222; *18.2*, p. *307*; 21.1.1, p. 417
R. I. Anishchenko	18,	923	(1966)	*3.2*, p. *34*; 13, p. 164; 21, p. 376; 21.1, p. 394; 21.1.1, p. 414; 21.2, p. 424; 21.3, p. 426; 21.4, p. 428; 21.6, p. 432; 21.7, p. 438; 22.1.1, p. 460
V. Ansorge	30,	K31	(1968)	*18.1*, p. *298*; *18.2*, p. *315*; 22, p. 454
K. H. Anthony	22,	667	(1967)	*13.1*, p. *180*
A. S. Antonov	16,	761	(1966)	*14.3.2*, p. *248*; *14.4*, p. *252*; 22.1.2, p. 469
V. V. Antonov-Romanovskii	19,	417	(1967)	*20.3*, p. *365*; 22, p. 442
	26,	173	(1968)	*13.4*, p. *211*; 22, p. 450
	30,	341	(1968)	*13.4*, p. *216*; 22, p. 454
I. M. Antonova	20,	643	(1967)	*13.1*, p. *177*; 22.5.2, p. 521
J. Antula	24,	89	(1967)	*14.3.2*, p. *249*; 14.4.1, p. 260; 22, p. 448
	24,	95	(1967)	*13.1*, p. *182*; 14.3.2, p. 249; 14.4.1, p. 260; 22.6, p. 539
	28,	395	(1968)	14.3.1, p. 247; *14.4.1*, p. *263*; 22.6, p. 542
S. Anzai	17,	697	(1966)	*1.2*, p. *9*; 12, p. 156; 14.3, p. 233; 22, p. 441
G. Appelt	27,	657	(1968)	13.1, p. 186; *17*, p. *289*; 20.1, p. 354; 21.1, p. 405

F. SECCO D'ARAGONA see under letter "S" (SECCO)

S. ARAJS	**16,**	475	(1966)	*18.2*, p. *300*; 21.1.1, p. 410
	18,	505	(1966)	*14.1*, p. *221*; 21, p. 376
	21,	191	(1967)	*18.2*, p. *306*; 21.1, p. 397; 21.1.1, p. 417
	22,	737	(1967)	14.1, p. 223; *18.4*, p. *323*; 21.1.1, p. 418
	26,	409	(1968)	*14.1*, p. *225*; 21.1.1, p. 420
	28,	171	(1968)	*14.1*, p. *225*; 21.1, p. 405; 21.6, p. 436
D. G. ARASLI	**21,**	643	(1967)	6, p. 61; *8*, p. *82*; 10, p. 100; 22.1.1, p. 462; 22.1.2, p. 471
N. K. ARCHIPOVA	**16,**	729	(1966)	*9*, p. *87*; 21.1, p. 391; 21.6, p. 430
H. AREND	**20,**	653	(1967)	14.4.2, p. 267; *19*, p. *328*
	24,	K69	(1967)	*14.4.2*, p. *267*
	29,	K149	(1968)	10, p. 111; *14.4.2*, p. *269*
J. ARENDS	**22,**	131	(1967)	*10.2*, p. *139*; 13.4, p. 208; 19, p. 329; 22.5.3, p. 533
	23,	137	(1967)	4, p. 43; 10.2, p. 140; *19*, p. *330*; 22.5, p. 513
	23,	713	(1967)	10.2, p. 140; *13.4*, p. *210*; 19, p. 330; 22.5.3, p. 533
	24,	K129	(1967)	*10*, p. *103*; 22.5.2, p. 525
	26,	537	(1968)	10, p. 105; *14.4*, p. *254*; 22.5.2, p. 526
G. E. ARKHANGELSKII	**22,**	289	(1967)	*10.2*, p. *140*; 20.1, p. 349; 20.2, p. 361; 20.3, p. 366; 22.6, p. 538
	29,	831	(1968)	*10.2*, p. *144*; 13.4, p. 215; *20.3*, p. *371*; 22.6, p. 543
G. ARLT	**25,**	323	(1968)	*14.3*, p. *240*; 22.2.1, p. 485; 22.2.3, p. 492
M. ARNDT	**19,**	K79	(1967)	*10*, p. *98*; 22.1.2, p. 470
J.-U. ARNOLD	**28,**	K93	(1968)	*3*, p. *27*; 5, p. 54; 8, p. 85; 22.1.3, p. 479
H. L. ARORA	**25,**	223	(1968)	*10.2*, p. *141*; 13.4, p. 211; 22.5.2, p. 525
A. ARROTT	**26,**	743	(1968)	10.1, p. 130; *18.2*, p. *311*; 21.1.1, p. 421
R. J. ARSENAULT	**17,**	27	(1966)	*10.1*, p. *115*; 11, p. 145; 21.1, p. 391
I. ARTZNER see J. ARTZNER				
J. ARTZNER	**27,**	K5	(1968)	*18.2*, p. *311*; 21.1, p. 405
	29,	K117	(1968)	*18.2*, p. *314*; 21.1, p. 408; 21.1.1, p. 423
	30,	K91	(1968)	*18.2*, p. *315*; 21.1, p. 409
E. K. ARUSHANOV	**27,**	45	(1968)	13.4, p. 213; *14.3*, p. *241*; 22, p. 451; 22.7, p. 546
P. ASADI	**20,**	K55	(1967)	10, p. 98; *14.4.1*, p. *258*; 22.5.3, p. 532
	20,	K59	(1967)	10, p. 99; *14.4*, p. *252*; 22.5.3, p. 533
	20,	K71	(1967)	*4*, p. *41*; 10.2, p. 139; 22.8, p. 550
	20,	K73	(1967)	*4*, p. *41*; 14.4, p. 252; 22.8, p. 550
G. A. ASADOV	**21,**	461	(1967)	3.1, p. 31; *14.3*, p. *236*; 15, p. 272; 22, p. 445
	26,	65	(1968)	1.2, p. 12; *14.3*, p. *240*; 15, p. 273; 22, p. 450
M. G. ASCH	**17,**	49	(1966)	11, p. 146; *13.3*, p. *196*; 22.9, p. 558
M. ASCHE	**18,**	749	(1966)	10, p. 97; 12, p. 156; *14.3*, p. *234*; 22.1.1, p. 460
E. ASCHER	**26,**	277	(1968)	4, p. 45; *14.4.2*, p. *268*; 20, p. 340
L. ASK	**26,**	K143	(1968)	2, p. 24; 12, p. 158; *14.3.2*, p. *250*; 22.2.2, p. 488
B. M. ASKEROV	**18,**	667	(1966)	6, p. 58; 13.1, p. 176; *14.3*, p. *233*; 22.2, p. 481; 22.2.3, p. 490
	21,	K155	(1967)	*13.1*, p. *179*; 14.3, p. 237; 15, p. 272; 22, p. 446
	28,	783	(1968)	13.1, p. 187; *14.3*, p. *243*; 22, p. 453
CH. M. ASKEROV	**29,**	K47	(1968)	2, p. 24; *14.3*, p. *243*; 22.1.3, p. 480
J. ASKILL	**16,**	K63	(1966)	*9*, p. *87*; 21.1, p. 390
	23,	263	(1967)	*9*, p. *90*; 10, p. 102; 21, p. 381; 21.1, p. 400; 21.1.1, p. 419

J. Askill	**23,**	K21	(1967)	*9*, p. *90*; 21, p. 381
V. M. Asnin	**20,**	755	(1967)	*13.2*, p. *192*; *20.1*, p. *347*; 22.1.1, p. 461
	29,	443	(1968)	13.2, p. 195; 13.4, p. 215; 14.3, p. 243; *20.1*, p. *356*; 22.1.1, p. 466
D. E. Aspnes	**23,**	K79	(1967)	13.1, p. 181; 14.4.1, p. 260; *20*, p. *340*; 22, p. 447
D. Ast	**17,**	269	(1966)	*10.1*, p. *115*; 11, p. 146; 21.1, p. 392
S. M. Atakishiev	**28,**	K47	(1968)	6, p. 66; *8*, p. *85*; 22, p. 453
M. Attardo see M. J. Attardo				
M. J. Attardo	**16,**	449	(1966)	4, p. 37; *11*, p. *145*; 17, p. 286; 21, p. 374
	27,	383	(1968)	10.1, p. 131; *11*, p. *152*; 21.6, p. 436
J. Auleytner	**20,**	K77	(1967)	*1.4*, p. *18*; 22.1.1, p. 461; 22.2.1, p. 483
	24,	K107	(1968)	*10*, p. *103*; *13.2*, p. *193*; 20.1, p. 351; 22.4.1, p. 501
J. Auth	**27,**	653	(1968)	13.4, p. 214; *14.3.2*, p. *250*; 22.1.2, p. 474
A. Authier	**17,**	K3	(1966)	*10*, p. *95*; 22.9, p. 558
	21,	413	(1967)	*10*, p. *100*; 22.1.2, p. 471
	26,	469	(1968)	*4*, p. *45*
	27,	77	(1968)	*10*, p. *106*; 22.1.2, p. 473
B. Autin	**22,**	K135	(1967)	*16*, p. *280*; 22.4.1, p. 500
	24,	501	(1967)	*7*, p. *77*; *14.4.1*, p. *261*; 22, p. 448
E. M. Averbakh	**30,**	K145	(1968)	13.1, p. 190; *20.1*, p. *359*; 22.7, p. 546
M. Averous	**21,**	665	(1967)	13.1, p. 178; 13.4, p. 207; *14.3*, p. *236*; 22.2.3, p. 491
	29,	807	(1968)	*13.1*, p. *189*; 14.3, p. 244; 22.2.3, p. 493; 22.8, p. 556
L. Avramov	**23,**	K61	(1967)	*20.3*, p. *367*; 22.6, p. 539
A. Axmann	**19,**	721	(1967)	*6*, p. *59*; 22.1.3, p. 477
R. Babušková	**20,**	K29	(1967)	*20.1*, p. *347*; 22.2, p. 481
Z. Bachan	**27,**	K25	(1968)	13.1, p. 186; *18*, p. *294*; *20*, p. *341*; 22.1.2, p. 473
H. Bachert	**19,**	K59	(1967)	*20.2*, p. *360*; 22.2.1, p. 483
	29,	K175	(1968)	*20.2*, p. *362*; 22.2.1, p. 486
D. J. Bacon	**23,**	527	(1967)	*10*, p. *102*; *10.1*, p. *125*
	29,	543	(1968)	*10*, p. *110*
H. Baessler	**26,**	599	(1968)	*13.1*, p. *185*; 17, p. 288; 22.9, p. 560
	26,	607	(1968)	*13.1*, p. *185*; 17, p. 289; 22.9, p. 560
M. I. Bagatskii	**26,**	453	(1968)	*8*, p. *84*; 23, p. 561
Kh. S. Bagdasarov	**21,**	K23	(1967)	13.4, p. 207; *20.1*, p. *348*; 20.2, p. 361; *20.3*, p. *365*; 22.8, p. 550
	27,	K1	(1968)	*20.2*, p. *362*; 22.8, p. 553
A. G. Bagirov	**17,**	K225	(1966)	*20.3*, p. *364*; 22, p. 441
H. A. Bahr	**28,**	K23	(1968)	*10*, p. *109*; *10.1*, p. *132*
D. J. Bailey	**22,**	607	(1967)	*10.1*, p. *124*; 21.1, p. 399; 21.1.1, p. 418
F. Bailly	**25,**	317	(1968)	*10*, p. *104*; 22.4.3, p. 510
M. Ya. Bakirov	**30,**	K73	(1968)	*14.4.1*, p. *264*; 22.1.3, p. 480
H. Bakker	**28,**	569	(1968)	*9*, p. *92*; 21.1, p. 406
M. Balarin	**17,**	91	(1966)	*1.2*, p. *9*; 10, p. 94; 11, p. 146; 14.4.2, p. 266
	22,	123	(1967)	*10*, p. *101*
G. Baldini	**25,**	557	(1968)	13.4, p. 211; 20.1, p. 351; *20.3*, p. *368*; 22.5.2, p. 525
	27,	95	(1968)	13.4, p. 213; *20.1*, p. *353*; 22.5, p. 514
	30,	601	(1968)	*6*, p. *70*; 10, p. 112, 20.1, p. 359; 20.3, p. 372; 22.5.2, p. 531

T. O. Baldwin	**25**, 71	(1968)	5, p. 52; *6*, p. *64*; 21.1, p. 402
F. Balibar	**21**, 413	(1967)	*10*, p. *100*; 22.1.2, p. 471
M. Balkanski	**30**, 407	(1968)	*13.1*, p. *174*; 18, p. 291; 20, p. 337; 22.1.1, p. 458; 22.1.2, p. 468; 22.2.1, p. 482; 22.2.2, p. 487; 22.2.3, p. 489; 22.4, p. 495; 22.4.1, p. 497; 22.4.2, p. 506; 22.4.3, p. 509
S. Ballarò	**28**, K109	(1968)	13.1, p. 188; *14.4.1*, p. *264*; *20*, p. *341*; 20.1, p. 356; 22.1.1, p. 466
R. W. Balluffi	**17**, 531	(1966)	*10.1*, p. *116*; 21.6, p. 431
	25, 163	(1968)	4, p. 44; *9*, p. *91*; 21, p. 384
	27, 195	(1968)	*10.1*, p. *130*; *11*, p. *152*; 21.6, p. 436
A. Balzarotti	**28**, K109	(1968)	13.1, p. 188; *14.4.1*, p. *264*; *20*, p. *341*; 20.1, p. 356; 22.1.1, p. 466
R. Balzer	**27**, K165	(1968)	*10*, p. *108*; 11, p. 153; 22.5.2, p. 527
	28, 207	(1968)	*10*, p. *108*; 11, p. 153; 22.5.2, p. 527
M. Bancie-Grillot	**16**, 517	(1966)	13.4, p. 201; *20.3*, p. *363*; 22.4.1, p. 497
J. Bara	**17**, K53	(1966)	*6.1*, p. *71*; 21, p. 374
P. G. Baranov	**30**, 851	(1968)	10.2, p. 144; *19*, p. *336*; 22.5.2, p. 531
	30, 859	(1968)	3.2, p. 36; 10.2, p. 144; 13.4, p. 218; *19*, p. *336*; 22.5.2, p. 531
B. Baranowski	**20**, K37	(1967)	*14*, p. *219*
	30, K33	(1968)	10.1, p. 135; *14.1*, p. *227*; 21, p. 388
D. J. Barber	**16**, 531	(1966)	4, p. 37; *10*, p. *94*; *11*, p. *145*; 22.5.3, p. 532
R. G. Barnes	**19**, 359	(1967)	*6.1*, p. *71*; *19*, p. *328*; 22.7, p. 545
D. M. Barnett	**21**, 239	(1967)	4, p. 42; *10*, p. *99*
G. R. Barsch	**19**, 129	(1967)	*12*, p. *156*
	19, 139	(1967)	*12*, p. *156*; 21, p. 376; 22, p. 442
	23, 577	(1967)	*12*, p. *157*; 22.5.2, p. 524
	29, 121	(1968)	*12*, p. *158*; 22.6, p. 542
G. M. Bartenev	**24**, 443	(1967)	*10*, p. *103*; 12, p. 158; 22.5.2, p. 524
W. Barth	**19**, 515	(1967)	*10.2*, p. *138*; 22.5.2, p. 520
J. Barthel	**22**, K151	(1967)	*3.1*, p. *32*; 14.1, p. 223; 21, p. 381
V. K. Bashenov	**21**, K91	(1967)	*13.4*, p. *207*; 19, p. 329; 22.2.1, p. 483
	25, K51	(1968)	*19*, p. *332*; 22.2, p. 481
R. I. Bashirov	**20**, K119	(1967)	*15*, p. *272*; 22.2.1, p. 483
Sh. Sh. Bashkirov	**26**, 253	(1968)	*6.1*, p. *73*; 13, p. 169; 18, p. 294
A. N. Basu	**29**, 367	(1968)	*6*, p. *67*; 22.5.2, p. 528
R. De Batist	**19**, 77	(1967)	*8*, p. *80*; 18.4, p. 322; 22.6, p. 536
B. W. Batterman	**23**, K147	(1967)	*5*, p. *52*; 22.1.2, p. 472
R. Baubinas	**24**, K91	(1967)	3.1, p. 33; *14.3*, p. *239*; *16*, p. *280*; 22.4.2, p. 507
P. Bauduin	**17**, 517	(1966)	*14.4.1*, p. *256*; *16*, p. *277*; 22.4.1, p. 498
	22, K135	(1967)	*16*, p. *280*; 22.4.1, p. 500
Ch. L. Bauer	**18**, 465	(1966)	*10*, p. *96*
	22, 199	(1967)	1.2, p. 11; *10.1*, p. *124*; 14.2, p. 229; 18, p. 292; 21, p. 380
E. Bauer	**18**, 441	(1966)	*1.4*, p. *17*; *3.1*, p. *30*; 21, p. 376; 21.6, p. 431
G. Bauer	**28**, K51	(1968)	13, p. 171; *14.4.1*, p. *263*; 22.1.1, p. 465
D. Bäuerle	**24**, 207	(1967)	*6*, p. *63*; 10.2, p. 141; *20.1*, p. *351*; 22.5.2, p. 524
	29, 639	(1968)	*6*, p. *68*; 10.2, p. 144; *20.1*, p. *357*; 22.5.2, p. 529
N. P. Baum	**29**, 107	(1968)	13.1, p. 188; *14.1*, p. *226*; *15*, p. *274*; 18.4, p. 324; 21.1, p. 407
F. Baumann	**23**, K99	(1967)	*8*, p. *83*; 13, p. 167; 15, p. 273; 22.1.3, p. 477
C. Baumberger	**22**, K67	(1967)	*14.4.2*, p. *267*
H. Bäurich	**16**, K39	(1966)	*18.2*, p. *300*; 21.1.1, p. 410

H. Bäurich	**23,** K137	(1967)	*18.2*, p. *308*
R. I. Bazyura	**20,** 213	(1967)	*9*, p. *88*; 21.6, p. 432; 22, p. 443; 22.1.3, p. 477
A. R. Beattie	**19,** 577	(1967)	*13.1*, p. *177*; *16*, p. *278*; 22.2.1, p. 483; 22.2.3, p. 490
	20, K135	(1967)	*13*, p. *165*; 14.3, p. 236; 22.2.3, p. 490
J. H. Beaumont	**17,** K45	(1966)	10, p. 95; *14.4*, p. *252*; 22.5.2, p. 519
D. Becker	**30,** 219	(1968)	10.1, p. 135; *11*, p. *155*; 14.1, p. 227; 18.2, p. 315; 21.1, p. 408
V. Beckmann	**29,** 781	(1968)	6.1, p. 74; *18.4*, p. *325*; 22.8, p. 555
A. Behr	**21,** K9	(1967)	10, p. 100; *11*, p. *148*; 22.5.2, p. 522
G. L. Belenky	**24,** 543	(1967)	10, p. 103; *13.4*, p. *210*; 16, p. 281; 20.3, p. 368; 22.4.1, p. 501
R. J. Beltz	**26,** 621	(1968)	*10*, p. *106*; *10.1*, p. *129*
L. M. Belyev	**21,** K23	(1967)	13.4, p. 207; *20.1*, p. *348*; 20.2, p. 361; *20.3*, p. *365*; 22.8, p. 550
L. Ben-Dor	**18,** 593	(1966)	*10.2*, p. *138*; 22.5.2, p. 520
V. Z. Bengus	**19,** 533	(1967)	*10*, p. *98*; 22.5.2, p. 521
P. Bennema	**17,** 555	(1966)	*3.1*, p. *30*; 22.8, p. 547
	17, 563	(1966)	*3.1*, p. *30*; 22.8, p. 548
	19, 211	(1967)	*3.1*, p. *31*; 22.8, p. 549
C. Benoit à la Guillaume	**17,** 875	(1966)	*20.2*, p. *360*; 22.4.1, p. 498
	23, 295	(1967)	*13.4*, p. *209*; 20.3, p. 367; 22.1.2, p. 472
V. M. Bentsa	**20,** 771	(1967)	13.3, p. 197; 16, p. 279; *17*, p. *287*; 22.4, p. 495; 22.5, p. 512; 22.5.1, p. 515
D. M. Bercha	**21,** 769	(1967)	13.1, p. 178; *20.1*, p. *348*; 22, p. 446
	30, 461	(1968)	*13.1*, p. *189*; 16, p. 284; 20.1, p. 358; 22.8, p. 557
K. Berchtold	**20,** 255	(1967)	*14.3.2*, p. *249*; 16, p. 278; 22, p. 444; 22.1.2, p. 470
E. P. Berdnikov	**29,** K145	(1968)	*10*, p. *111*; 12, p. 158; 15, p. 274; 22.1.2, p. 474
C. B. van den Berg	**29,** 837	(1968)	1.2, p. 13; *6.1*, p. *74*; 18.4, p. 325; 22, p. 454
K.-J. Berg	**30,** K161	(1968)	*10.2*, p. *144*; *20.3*, p. *372*; 22.5.2, p. 531
U. Berg	**23,** K87	(1967)	*13.1*, p. *181*; 20, p. 340; 21.3, p. 426
H. Berger	**28,** K97	(1968)	1.4, p. 21; *14.3.1*, p. *247*; 22.4.1, p. 504
D. Bergner	**18,** 67	(1966)	*9*, p. *87*; 21, p. 375
	18, 75	(1966)	*9*, p. *87*; 21, p. 375
I. M. Bernstein	**23,** 539	(1967)	*10.1*, p. *125*; 21.2, p. 425
I. B. Bersuker	**16,** 63	(1966)	10, p. 93; *13.4*, p. *200*; 22, p. 439
U. Bertelsen	**22,** 59	(1967)	*6.1*, p. *72*; 18.4, p. 322; 22.5, p. 513
K.-H. Berthel	**22,** K151	(1967)	*3.1*, p. *32*; 14.1, p. 223, 21, p. 381
V. A. Besfamilnaya	**24,** 757	(1967)	6, p. 64; 13, p. 168; *16*, p. *281*; 22.1.1, p. 464
R. G. Bessent	**25,** K107	(1968)	10, p. 105; *10.2*, p. *142*; 19, p. 333; 22.6, p. 541
H. Bethge	**30,** K87	(1968)	*10*, p. *113*; 22.5.2, p. 531
A. van den Beukel	**19,** 177	(1967)	*1.1*, p. *8*; 10.1, p. 120; 21.1, p. 395; 21.6, p. 432
	23, 165	(1967)	*9*, p. *90*; *10.1*, p. *125*; 21, p. 381
	28, 565	(1968)	*8*, p. *85*; 14.4.2, p. 269
M. Bevis	**20,** 197	(1967)	3, p. 27; *4*, p. *41*; 21.1.1, p. 415
	26, K25	(1968)	3, p. 27; *4*, p. *45*; 21.1.1, p. 420
H. Beyer	**26,** 359	(1968)	4, p. 45; 10.1, p. 129; *11*, p. *151*; 14.1, p. 225; 21.6, p. 435
B. V. Beznosikov	**18,** K17	(1966)	1.2, p. 10; *12*, p. *156*; 22.5, p. 512; 22.8, p. 548
D. L. Bhattacharya	**27,** 427	(1968)	13, p. 170; *18.1*, p. *297*; 21, p. 386; 21.2, p. 425
Z. Biegański	**17,** 837	(1966)	1.2, p. 9; *8*, p. *80*; 18.2, p. 303; 22, p. 441
J. Biersack	**27,** 139	(1968)	*9*, p. *91*; 21, p. 385; 22, p. 451
B. F. Bilenkii	**28,** K5	(1968)	*1.4*, p. *21*; *20.1*, p. *355*; 22, p. 452
H. Bilger	**18,** 207	(1966)	10.1, p. 117; *18.2*, p. *303*; 21.1.1, p. 414

H. Bilger	**18,**	637	(1966)	*10.1*, p. *119*; 21.1.1, p. 414
	18,	K135	(1966)	*10.1*, p. *120*; 18.2, p. 304; 21.1.1, p. 414
	22,	683	(1967)	*10.1*, p. *124*; 14.1, p. 223; 21.1, p. 399
	29,	K179	(1968)	*10.1*, p. *134*; *11*, p. *154*; 18.2, p. 315; 21.1.1, p. 423
H. Bilz	**20,**	K167	(1967)	*6*, p. *60*; 10.2, p. 139; 20.1, p. 348; 22.5.2, p. 522
F. Biragnet	**16,**	295	(1966)	*1.4*, p. *16*; *18.2*, p. *299*; 21.1.1, p. 410
	16,	569	(1966)	4, p. 37; *18.2*, p. *300*; 21.1.1, p. 411
U. Birkholz	**27,**	413	(1968)	13, p. 170; 13.1, p. 186; *14.3*, p. *241*; 15, p. 274; 22, p. 452; 22.7, p. 546
	27,	K69	(1968)	*15*, p. *274*; 16, p. 282; 22.4.1, p. 502
	30,	K81	(1968)	*20.1*, p. *358*; 22, p. 455; 22.7, p. 546
H. K. Birnbaum	**27,**	701	(1968)	*10.1*, p. *131*; 21.1.1, p. 421
M. Bitter	**23,**	K155	(1967)	*6*, p. *63*; 23, p. 561
W. Bitterlich	**28,**	365	(1968)	*1.2*, p. *12*; 18.2, p. 312; 18.4, p. 324
A. Bivas	**22,**	K155	(1967)	6, p. 62; 13.2, p. 193; *20.3*, p. *367*; 22.4.2, p. 507
	27,	K29	(1968)	*20.3*, p. *369*; 22.5.1, p. 516
G. Björkman	**19,**	863	(1967)	*6*, p. *59*; 21.2, p. 424
D. A. Blackburn	**17,**	3	(1966)	*10.1*, p. *115*; 21, p. 374
	23,	177	(1967)	*10.1*, p. *125*; 21, p. 381; 21.6, p. 434
	23,	223	(1967)	*9*, p. *90*; 10.1, p. 125; 21.6, p. 434
M. J. Blackburn	**25,**	K1	(1968)	*10.1*, p. *128*; 21, p. 384
A. Ya. Blank	**22,**	47	(1967)	13.1, p. 179; *19*, p. *329*; 21, p. 380
K. C. A. Blasdale	**18,**	491	(1966)	4, p. 39; *10.1*, p. *118*; 21, p. 376
	21,	649	(1967)	*10.1*, p. *123*; 21, p. 379
G. Blasse	**20,**	551	(1967)	13.4, p. 205; 20.1, p. 347; *20.3*, p. *365*; 22.8, p. 549
B. Błaszczyk	**23,**	419	(1967)	*14.4.1*, p. *259*; 22.9, p. 559
I. A. Blech	**29,**	653	(1968)	*4*, p. *47*; *10*, p. *111*; 22.1.2, p. 474; 22.2.1, p. 486
H. Bleher	**23,**	K13	(1967)	13, p. 167; *14.4.1*, p. *260*; 16, p. 280; 22.1.1, p. 463
J. Blinowski	**20,**	K107	(1967)	10, p. 99; *18*, p. *292*; *20*, p. *339*; 22.1.1, p. 461
J. Blok	**19,**	K107	(1967)	10, p. 98; *14.3*, p. *235*; 22.1.1, p. 461
	27,	219	(1968)	10, p. 106; *13.4*, p. *213*; 14.3, p. 241; 22.1.1, p. 465
	27,	225	(1968)	10, p. 107; *14.3*, p. *241*; 22.1.1, p. 465
K. Blötekjaer	**28,**	711	(1968)	*14.3*, p. *242*; 14.4.1, p. 263; 22.2.3, p. 493
W. Blum	**20,**	629	(1967)	*10*, p. *99*; 22.5.2, p. 521
H. Blume	**29,**	159	(1968)	*13.2*, p. *195*; 20.3, p. 370; 22.5.2, p. 528
A. I. Bobrysheva	**16,**	337	(1966)	6, p. 56; *13.2*, p. *191*; 22, p. 439
D. Boccon-Gibod	**30,**	525	(1968)	7, p. 78; *14.4.1*, p. *264*; *20.1*, p. *359*; 22.2.1, p. 486
Z. Bodó	**20,**	K45	(1967)	13.1, p. 177; *20.1*, p. *347*; 22.6, p. 537
K. W. Böer	**16,**	697	(1966)	*1.4*, p. *16*; 13.4, p. 201; 16, p. 276; 22.4.1, p. 497
	16,	K5	(1966)	13.3, p. 196; *16*, p. *276*; 22.4.1, p. 497
	17,	307	(1966)	*14.4.1*, p. *256*; 22, p. 440
	19,	203	(1967)	*13.4*, p. *204*; 14.3, p. 234; 16, p. 278; 22.4.1, p. 499
	21,	K145	(1967)	*14.4.1*, p. *259*; 22.4.1, p. 500
	28,	355	(1968)	*14.4.1*, p. *263*; 16, p. 282; 22.4.1, p. 503
	30,	291	(1968)	*14.4.1*, p. *264*; 16, p. 284; 20, p. 342; 22.4.1, p. 505
Ir. Boesono	**19,**	107	(1967)	10.1, p. 120; *12.1*, p. *159*; 21.1.1, p. 414
O. V. Bogdankevich	**19,**	K5	(1967)	*20.2*, p. *360*; 22.4.2, p. 506
	29,	715	(1968)	13.4, p. 215; *20.2*, p. *362*; 22.2.1, p. 486
R. Bogdanović	**22,**	603	(1967)	*13.4*, p. *209*; 22.1, p. 457
Yu. A. Bogod	**21,**	797	(1967)	*14.1*, p. *223*; 21.7, p. 438

Yu. A. Bogod	**24,**	K49	(1967)	*14.1*, p. *224*; 21.7, p. 438
	28,	K155	(1968)	*14.1*, p. *226*; 21.7, p. 438
V. N. Bogomolov	**27,**	443	(1968)	13, p. 170; *20.1*, p. *354*; 22.6, p. 541
G. A. Bogomolova	**16,**	K165	(1966)	10, p. 94; *20.1*, p. *345*; 20.2, p. 360; 22.8, p. 547
G. Bohn	**20,**	277	(1967)	*14.1*, p. *222*; 21, p. 377
I. I. Boiko	**16,**	43	(1966)	12, p. 156; *14.3*, p. *231*; 14.4.1, p. 255; 22.1.1, p. 459
	17,	389	(1966)	12, p. 156; *16*, p. *277*; 22.1.1, p. 459
D. Bois	**25,**	691	(1968)	*17*, p. *288*; 21.1, p. 403; 21.1.1, p. 420; 22.1.1, p. 464; 22.2.1, p. 485; 22.4.1, p. 502; 22.5.2, p. 525; 22.6, p. 540
V. A. Bokov	**20,**	745	(1967)	*1*, p. *7*; 1.2, p. 10; 18.2, p. 306; 18.3, p. 317; 18.4, p. 322; 22.8, p. 549
	28,	835	(1968)	5, p. 54; *18.2*, p. *313*; *18.4*, p. *324*; 22.8, p. 554
V. V. Boldyrev	**30,**	K13	(1968)	13.4, p. 217; *14.3*, p. *244*; *15*, p. *275*; 22.8, p. 556
G. F. Bolling	**26,**	743	(1968)	10.1, p. 130; *18.2*, p. *311*; 21.1.1, p. 421
W. Bollmann	**24,**	K141	(1967)	*10*, p. *104*; *10.1*, p. *127*
B. I. Boltaks	**19,**	705	(1967)	*9*, p. *88*; 22.2.1, p. 483
H. C. Bolton	**18,**	221	(1966)	4, p. 38; *20.1*, p. *346*; 22.8, p. 548
V. L. Bonch-Bruevich	**16,**	197	(1966)	*14.4.1*, p. *255*; 22.1.1, p. 459
	22,	267	(1967)	13, p. 166; 14.3, p. 237; *14.4.1*, p. *259*; 22, p. 446; 22.1.1, p. 462; 22.2.1, p. 484
	23,	761	(1967)	13, p. 168; 14.3, p. 238; *14.4.1*, p. *260*; 22.1.1, p. 463
	28,	687	(1968)	13, p. 172; 13.4, p. 215; *14.4.1*, p. *263*; 22.1.1, p. 466
	29,	9	(1968)	6, p. 56; 13, p. 162; 13.2, p. 191; *13.4*, p. *200*; *14.3*, p. *231*; 20, p. 337; 20.3, p. 363; 22.1.1, p. 458; 22.1.2, p. 468
V. N. Bondarenko	**24,**	543	(1967)	10, p. 103; *13.4*, p. *210*; 16, p. 281; 20.3, p. 368; 22.4.1, p. 501
D. Bonnet	**29,**	K65	(1968)	*16*, p. *284*; 22.4.1, p. 504; 22.4.2, p. 508; 22.8, p. 555
A. Bonnot	**21,**	525	(1967)	7, p. 76; *14.4.1*, p. *258*; *20.3*, p. *366*; 22.2.1, p. 484
H. P. Bonzel	**20,**	493	(1967)	*9*, p. *88*; 10, p. 99; 22.1.2, p. 471
	25,	209	(1968)	*9*, p. *91*; 21.1, p. 402
A. N. Borets	**21,**	769	(1967)	13.1, p. 178; *20.1*, p. *348*; 22, p. 446
H. Borik	**29,**	133	(1968)	*6*, p. *67*; 20.1, p. 356; 21.1.1, p. 466; 22.1.2, p. 474
N. A. Borisov	**29,**	715	(1968)	13.4, p. 215; *20.2*, p. *362*; 22.2.1, p. 486
G. K. Born	**28,**	583	(1968)	10.2, p. 143; *11*, p. *153*; 19, p. 334; 20.3, p. 370; 22.8, p. 554
F. Borsa	**19,**	359	(1967)	*6.1*, p. *71*; *19*, p. *328*; 22.7, p. 545
D. Borşan	**23,**	K113	(1967)	1.2, p. 11; *1.4*, p. *19*; 22.4.3, p. 510
A. S. Borshchevskii	**21,**	9	(1967)	3.1, p. 29; 3.2, p. 34; 5, p. 49; 8, p. 79; 10, p. 93; 13.1, p. 174; *14.3*, p. *231*; 15, p. 270; 16, p. 276; 17, p. 286; 18, p. 291; 20.1, p. 344; 22.8, p. 547
D. N. Bose	**24,**	K165	(1967)	*13.3*, p. *198*; 14.3, p. 239; 22.3, p. 494
	27,	325	(1968)	11, p. 152; *13.3*, p. *198*; 22.1.1, p. 465
	30,	K57	(1968)	*13.4*, p. *217*; 16, p. 284; 20.3, p. 371; 22.4.1, p. 505
H. N. Bose	**16,**	591	(1966)	*10.2*, p. *137*; 20.3, p. 364; 22.5.2, p. 518
J. Bosse	**27,**	301	(1968)	*6.1*, p. *73*; 13, p. 170
T. Boţilă	**18,**	K57	(1966)	13.4, p. 203; *16*, p. *278*; 22.5, p. 512; 22.8, p. 548
	30,	K149	(1968)	13.1, p. 190; *20.1*, p. *359*; 22, p. 455
R. Böttcher	**20,**	121	(1967)	9, p. 88; 13.4, p. 205; *19*, p. *328*; 22.5.2, p. 521
	26,	K21	(1968)	*19*, p. *333*; 22.4.1, p. 502

H. Böttger	**17**, 855 (1966)	13.4, p. 203; *20.1*, p. *346*; 22.5.1, p. 515
	23, 325 (1967)	13.4, p. 209; *19*, p. *330*
	24, 65 (1967)	6, p. 63; 13.4, p. 210; *19*, p. *331*; 22.6, p. 539
	26, 681 (1968)	6, p. 65; *13.4*, p. *212*; 19, p. 333; 22.6, p. 541
	28, 611 (1968)	*6*, p. *67*; *10*, p. *109*; 22.5.2, p. 528
	30, K119 (1968)	6, p. 70; 13, p. 173; *19*, p. *336*; 22.6, p. 544
G. Bougnot	**21**, 665 (1967)	13.1, p. 178; 13.4, p. 207; *14.3*, p. *236*; 22.2.3, p. 491
	29, 807 (1968)	*13.1*, p. *189*; 14.3, p. 244; 22.2.3, p. 493; 22.8, p. 556
J. Bougnot	**21**, K31 (1967)	*9*, p. *89*; 10, p. 100; 14.3, p. 236; 22.2.3, p. 491
	26, K127 (1968)	*9*, p. *91*; 22.2.3, p. 492
A. Bourret	**17**, K37 (1966)	*10.1*, p. *116*; 18.2, p. 302; 21.1.1, p. 412
	23, 207 (1967)	*18.2*, p. *308*; 21.1.1, p. 418
	24, K173 (1967)	10.1, p. 127; *11*, p. *151*; 21.1, p. 402
	29, 283 (1968)	4, p. 47; *10.1*, p. *133*; 11, p. 154; 21.1, p. 407
P. Bourtayre	**16**, 517 (1966)	13.4, p. 201; *20.3*, p. *363*; 22.4.1, p. 497
D. I. Bower	**29**, 617 (1968)	6, p. 68; 8, p. 86; *12.1*, p. *160*; 13, p. 172; 18.2, p. 314; 21.1, p. 407; 21.1.1, p. 422
Yu. S. Boyarskaya	**19**, 441 (1967)	*10*, p. *97*; 11, p. 147; 22.5.2, p. 520
A. J. F. Boyle	**22**, K131 (1967)	*6*, p. *62*; 8, p. 83; 12, p. 157; 22.1, p. 457
R. Boyn	**29**, 307 (1968)	10, p. 110; *13.4*, p. *215*; *20.1*, p. *356*; 22.4.1, p. 504
	30, 305 (1968)	13.4, p. 216; *20.1*, p. *358*; *20.3*, p. *371*; 22.4, p. 496
H. P. van de Braak	**19**, 385 (1967)	13, p. 164; *14.1*, p. *222*; 18.2, p. 305; 21.4, p. 428
	24, 733 (1967)	*10.1*, p. *127*; *18.2*, p. *310*
V. A. M. Brabers	**29**, K73 (1968)	18.3, p. 320; *20.1*, p. *357*
R. H. Bragg	**17**, K25 (1966)	1.2, p. 9; *3*, p. *26*; 4, p. 38; 22.1, p. 456
	19, K99 (1967)	1.2, p. 10; *3*, p. *26*; 4, p. 40; 22.1, p. 457
A. Bragiński	**17**, K211 (1966)	*18.3*, p. *316*
	22, K127 (1967)	*18.3*, p. *318*
P. Brand	**28**, 741 (1968)	*13*, p. *172*
W. A. Brantley	**18**, 465 (1966)	*10*, p. *96*
M. Brauer	**27**, 151 (1968)	*14.4.1*, p. *262*; 18, p. 294; 22.1.1, p. 465
P. Bräunlich	**18**, 307 (1966)	*13.4*, p. *203*; 14.3, p. 233; 16, p. 278; 20.3, p. 364; 22.4, p. 495
	21, 383 (1967)	13.4, p. 206; *14.4.1*, p. *258*; 16, p. 279; 20.3, p. 365; 22.4.1, p. 499
	22, 391 (1967)	13.4, p. 208; *16*, p. *280*; *20.3*, p. *366*; 22, p. 447
A. Bril	**20**, 551 (1967)	13.4, p. 205; 20.1, p. 347; *20.3*, p. *365*; 22.8, p. 549
J. L. Brimhall	**27**, K89 (1968)	*10.1*, p. *131*; 11, p. 152; 21, p. 386
A. D. Britov	**26**, K139 (1968)	6, p. 65; *13.2*, p. *194*; *20.2*, p. *361*; 22.4.2, p. 507
J. O. Brittain	**20**, 443 (1967)	*4*, p. *41*; 10.1, p. 122; 21.1, p. 396
	27, 185 (1968)	*10.1*, p. *130*; 21, p. 385; 21.1, p. 404
W. Brodkorb	**16**, 225 (1966)	18.1, p. 296; *18.2*, p. *299*
	16, 577 (1966)	18.1, p. 296; *18.2*, p. *300*; 21.1.1, p. 411
	23, K35 (1967)	*18.4*, p. *323*
	26, 191 (1968)	6.1, p. 73; *18.2*, p. *310*
M. M. Brodsky	**18**, 863 (1966)	*4*, p. *39*; 21.7, p. 438
J. G. Broerman	**25**, 757 (1968)	*13.1*, p. *184*; 22.1.1, p. 464
S. E. Bronisz	**29**, K95 (1968)	*1.4*, p. *21*; 3.1, p. 33; 22.6, p. 543
H. Bross	**20**, 277 (1967)	*14.1*, p. *222*; 21, p. 377
	22, 667 (1967)	*13.1*, p. *180*
A. D. Brothers	**23**, 697 (1967)	*13.1*, p. *182*; 14.3, p. 238; 20.1, p. 350; 22.7, p. 545
F. R. Brotzen	**22**, 9 (1967)	13.3, p. 196; *17*, p. *286*; 21, p. 373

V. L. Broude	**19**,	395	(1967)	*6*, p. *58*; 20.1, p. 347; 22.9, p. 558
F. Brouers	**19**,	867	(1967)	*13*, p. *165*
	22,	213	(1967)	*13*, p. *166*; *20*, p. *339*; 21, p. 380; 21.2, p. 424
H. A. Brown	**25**,	K11	(1968)	*18.2*, p. *310*
	27,	K17	(1968)	*18.2*, p. *312*
H. L. Brown	**18**,	K21	(1966)	*12*, p. *156*; 22.3, p. 494
E. L. Broyda	**27**,	249	(1968)	*13.1*, p. *186*; 14.3, p. 241; 15, p. 274; 22.2.3, p. 492
W. Bruckner	**29**,	781	(1968)	6.1, p. 74; *18.4*, p. *325*; 22.8, p. 555
W. Brückner	**23**,	475	(1967)	*5*, p. *52*; 21.1.1, p. 419
	29,	211	(1968)	*18.2*, p. *314*; *18.4*, p. *325*; 21.1.1, p. 422
R. Brugière	**24**,	K77	(1967)	*10.1*, p. *127*; 11, p. 150; 14.1, p. 224; 21, p. 383
	30,	K139	(1968)	*10.1*, p. *136*; 11, p. 155; 14.1, p. 227; 21.1, p. 409
O. Brümmer	**27**,	513	(1968)	2, p. 24; *20*, p. *341*; 22.1.1, p. 465; 22.6, p. 541; 22.8, p. 554
F. J. Bryant	**21**,	315	(1967)	*11*, p. *148*; *20.3*, p. *365*; 22.4.3, p. 509
M. F. Bryzhina	**20**,	745	(1967)	*1*, p. *7*; 1.2, p. 10; 18.2, p. 306; 18.3, p. 317; 18.4, p. 322; 22.8, p. 549
	28,	835	(1968)	5, p. 54; *18.2*, p. *313*; *18.4*, p. *324*; 22.8, p. 554
V. M. Buchanov	**17**,	K187	(1966)	*17*, p. *286*; 21.1, p. 393
F. Buchy	**17**,	517	(1966)	*14.4.1*, p. *256*; *16*, p. *277*; 22.4.1, p. 498
I. S. Buda	**23**,	229	(1967)	13, p. 167; 14.3, p. 238; *15*, p. *272*; 22.1.1, p. 463
O. Budke	**21**,	175	(1967)	*8*, p. *82*; 22.2.3, p. 491; 22.5.2, p. 522
F. P. Bullen	**27**,	501	(1968)	*10.1*, p. *131*; 21.1, p. 405
H.-J. Bunge	**26**,	167	(1968)	1.1, p. 8; *4*, p. *45*; 21.1.1, p. 420
I. Bunget	**20**,	K163	(1967)	14.4, p. 253; *18.3*, p. *317*
	21,	K131	(1967)	*14.3*, p. *236*; 15, p. 271; 18.3, p. 318
	28,	K39	(1968)	*6.1*, p. *74*; 18.3, p. 320
V. A. Buravikhin	**16**,	651	(1966)	12.1, p. 159; *18.2*, p. *301*; 21.1.1, p. 411
	16,	657	(1966)	4, p. 38; 12.1, p. 159; *18.2*, p. *301*; 21.1.1, p. 411
A. Burewicz	**17**,	K195	(1966)	6.1, p. 71; *18.4*, p. *321*; 22, p. 441
	21,	K85	(1967)	*6.1*, p. *72*; 18.4, p. 322; 22.8, p. 550
	26,	429	(1968)	*4*, p. *45*; 18.3, p. 319; 18.4, p. 324; 22.8, p. 553
H. Burkhard	**27**,	K43	(1968)	*18.2*, p. *312*; *20*, p. *341*; 20.1, p. 354; 21.1.1, p. 421
R. A. Burley	**29**,	551	(1968)	2, p. 24; *10*, p. *110*; 20.1, p. 357; 22.1.3, p. 480
R. A. Burmeister, Jr.	**24**,	683	(1967)	13.4, p. 211; *14.3*, p. *239*; 22.4.2, p. 507
A. Buroff	**26**,	267	(1968)	10.2, p. 142; 13.4, p. 212; *16*, p. *282*; 22.5.1, p. 516
L. A. Bursill	**18**,	221	(1966)	4, p. 38; *20.1*, p. *346*; 22.8, p. 548
K. H. J. Buschow	**16**,	467	(1966)	*3*, p. *26*; 5, p. 49; 18.1, p. 296; 18.2, p. 300; 18.3, p. 316; 18.4, p. 321; 22.7, p. 545
	21,	593	(1967)	*5*, p. *51*; 18.1, p. 296; 18.3, p. 318; 18.4, p. 322; 22.7, p. 545
	24,	715	(1967)	1.2, p. 11; *18.2*, p. *309*; 21, p. 383
	29,	189	(1968)	13, p. 172; 18.1, p. 297; *19*, p. *335*; 21, p. 387, 21.4, p. 428
	29,	825	(1968)	*18.3*, p. *320*; 21.1, p. 407; 21.4, p. 428
	30,	759	(1968)	13, p. 173; 13.1, p. 190; 18.1, p. 298; *19*, p. *336*; 21, p. 389; 21.4, p. 428
N. E. Byer	**30**,	569	(1968)	*7*, p. *78*; 10, p. 112; 13.4, p. 217; 22.5.2, p. 530
	30,	579	(1968)	*7*, p. *78*; 10, p. 112; 13.4, p. 217; 22.5.2, p. 531
P. I. Bykovskii	**26**,	K13	(1968)	*20.3*, p. *368*; 22.8, p. 553
P. Byszewski	**23**,	K53	(1967)	13, p. 167; *18*, p. *293*; *20*, p. *340*; 22.1.1, p. 463; 22.2.1, p. 484
	27,	K25	(1968)	13.1, p. 186; *18*, p. *294*; *20*, p. *341*; 22.1.2, p. 473
	27,	K125	(1968)	*16*, p. *282*; 22.1.1, p. 465

J. E. Caffyn	**22,**	549	(1967)	10, p. 101; 13.4, p. 209; *14.4*, p. *253*; *20.1*, p. *350*; 22.5.3, p. 533
O. Cahen	**22,**	K135	(1967)	*16*, p. *280*; 22.4.1, p. 500
J. Calas	**29,**	807	(1968)	*13.1*, p. *189*; 14.3, p. 244; 22.2.3, p. 493; 22.8, p. 556
J. T. Calow	**28,**	295	(1968)	*3.1*, p. *33*; *14.3.1*, p. *247*; 14.3.2, p. 251; 22.4.2, p. 508
C. D. Cann	**17,**	K143	(1966)	*10*, p. *95*; 22.1.2, p. 469
	28,	187	(1968)	*5*, p. *53*; 21, p. 386; 22.6, p. 542
	30,	791	(1968)	*5*, p. *55*; 22.1.2, p. 475
G. Carter	**25,**	449	(1968)	13.3, p. 198; *14.3.1*, p. *246*; 22.4.1, p. 501
W. J. Caspers	**18,**	761	(1966)	*6.1*, p. *71*
	19,	385	(1967)	13, p. 164; *14.1*, p. *222*; 18.2, p. 305; 21.4, p. 428
	24,	733	(1967)	*10.1*, p. *127*; *18.2*, p. *310*
	25,	721	(1968)	*13*, p. *169*
T. N. Casselman	**20,**	605	(1967)	*6*, p. *60*; 13, p. 165; 18, p. 292; *19*, p. *328*; 22, p. 444
J. Celý	**30,**	671	(1968)	*6.1*, p. *75*; 10, p. 112; 10.1, p. 135
S. Ceresara	**16,**	439	(1966)	*9*, p. *87*; 21, p. 373
	27,	517	(1968)	*9*, p. *91*; 21, p. 386
G. A. Chalikyan	**17,**	K49	(1966)	13.1, p. 175; *16*, p. *277*; 22.2.2, p. 487
B. Chalupa	**29,**	K51	(1968)	*5*, p. *54*; 14.4, p. 254; 22.6, p. 543
H. Chang	**17,**	83	(1966)	*18.2*, p. *301*; 21.1.1, p. 411
I. F. Chang	**28,**	663	(1968)	*6*, p. *67*; 20.1, p. 356; 22.4.1, p. 504; 22.4.2, p. 508; 22.6, p. 542; 22.8, p. 554
R. Chang	**18,**	99	(1966)	*10.1*, p. *117*; 21.1.1, p. 413
Z. P. Chang	**19,**	139	(1967)	*12*, p. *156*; 21, p. 376; 22, p. 442
	23,	577	(1967)	*12*, p. *157*; 22.5.2, p. 524
G. Chanussot	**29,**	K149	(1968)	10, p. 111, *14.4.2*, p. *269*
M. S. R. Chari	**19,**	169	(1967)	8, p. 81; *14.1*, p. *222*; 21.1.1, p. 415
	21,	K127	(1967)	*8*, p. *82*; 21.6, p. 433
	29,	K69	(1968)	6, p. 68; *8*, p. *85*; 13, p. 172; 21.6, p. 437
P. Charsley	**19,**	K63	(1967)	*8*, p. *81*; 21, p. 377; 21.1, p. 396
	21,	357	(1967)	4, p. 42; 6, p. 61; *8*, p. *82*; 21, p. 379; 21.1, p. 397
	25,	531	(1968)	*6*, p. *65*; 8, p. 84; *10.1*, p. *128*; 21, p. 384; 21.1, p. 402
G. A. Chase	**19,**	645	(1967)	*10*, p. *98*; 11, p. 148; 22.6, p. 536
P. de Chatel	**26,**	319	(1968)	*13*, p. *169*; 18, p. 294
V. E. Chelnokov	**21,**	423	(1967)	13.4, p. 206; *14.3*, p. *236*; 14.4.1, p. 258; 22.1.2, p. 471
B. Chelustka	**22,**	K95	(1967)	*11*, p. *149*; 14.3, p. 237; 22.2.3, p. 491
C. W. Chen	**26,**	611	(1968)	10.1, p. 129; *14.1*, p. *225*; 21, p. 385; 21.6, p. 435
D. V. Chepur	**21,**	437	(1967)	*13.4*, p. *206*; 14.4.1, p. 258; 22, p. 445
	30,	461	(1968)	*13.1*, p. *189*; 16, p. 284; 20.1, p. 358; 22.8, p. 557
A. E. Cheredni-chenko	**28,**	K85	(1968)	13.2, p. 194; 13.3, p. 198; *16*, p. *283*; 22.4.1, p. 504
V. I. Cherepanov	**24,**	51	(1967)	3.2, p. 35; *13.4*, p. *210*; 22.6, p. 539
	28,	65	(1968)	*10*, p. *108*; 13.4, p. 214; 20.3, p. 370; 22.6, p. 542
C. Cherki	**19,**	K91	(1967)	14.3.1, p. 245; *14.4.1*, p. *257*; 22.6, p. 537
N. A. Chernoplekov	**20,**	767	(1967)	*6*, p. *60*; 8, p. 81; 21.1, p. 397
E. E. Chernov	**28,**	K1	(1968)	*10*, p. *109*; 22.5.2, p. 527
T. H. Cheung	**24,**	509	(1967)	10.2, p. 141; *13.4*, p. *210*; 22.5.2, p. 524
K.-P. Chik	**16,**	685	(1966)	*10.1*, p. *114*; 21.6, p. 430
	23,	113	(1967)	*4*, p. *43*

K.-P. Chik	**29**, 455	(1968)	*9*, p. *87*; *10*, p. *93*; 11, p. 145; 13.4, p. 200; 19, p. 326; 22.1.1, p. 458; 22.1.2, p. 468
V. S. Chincholkar	**29**, 669	(1968)	*14.4.2*, p. *269*
E. V. Chisler	**17**, 163	(1966)	5, p. 49; *6*, p. *56*; 14.4.2, p. 266; *20.1*, p. *345*
	17, 173	(1966)	1.2, p. 9; *6*, p. *57*; 14.4.2, p. 266; *20.1*, p. *345*
N. S. Chistyakov	**17**, 489	(1966)	*18.2*, p. *303*; 21.1.1, p. 413
Y. T. Chou	**17**, 509	(1966)	*10*, p. *95*
	20, 285	(1967)	*10*, p. *98*
	21, 239	(1967)	4, p. 42; *10*, p. *99*
J. M. Christie	**19**, 631	(1967)	4, p. 40; *10*, p. *98*; 22.6, p. 536
V. Christoph	**26**, K17	(1968)	*14.1*, p. *225*; 18.2, p. 311
S. G. Christov	**17**, 11	(1966)	*17*, p. *286*; 21, p. 374
	21, 159	(1967)	13.3, p. 197; *17*; p. *287*; 22, p. 444
S. Y. Chuang	**20**, 331	(1967)	11, p. 148; *13*, p. *165*; 20, p. 338; 21.1.1, p. 415; 22.8, p. 549
L. K. Chubova	**28**, K155	(1968)	*14.1*, p. *226*; 21.7, p. 438
F. N. Chukhovskii	**28**, 287	(1968)	5, p. 54; *6*, p. *66*
A. Cingolani	**28**, K151	(1968)	*20.3*, p. *370*; 22.4.1, p. 504
T. Claeson	**25**, K95	(1968)	*14.2*, p. *230*
E. W. Claffy	**22**, 71	(1967)	13.4, p. 208; *20.1*, p. *349*; 20.3, p. 366; 22.5.2, p. 523
L. M. Clarebrough	**21**, 369	(1967)	*10.1*, p. *122*; 21.1, p. 397
E. Claridge	**29**, 617	(1968)	6, p. 68; 8, p. 86; *12.1*, p. *160*; 13, p. 172; 18.2, p. 314; 21.1, p. 407; 21.1.1, p. 422
J. Claudel	**26**, 241	(1968)	6, p. 65; *20.1*, p. *352*; 22.4.2, p. 507; 22.4.3, p. 510
K. Clausecker	**24**, 721	(1967)	*14.1*, p. *224*; 21, p. 383
A. Clement	**21**, K97	(1967)	*10.1*, p. *123*; 21.1, p. 398; 21.1.1, p. 418
N. Clement	**21**, K97	(1967)	*10.1*, p. *123*; 21.1, p. 398; 21.1.1, p. 418
G. Clerc	**16**, 569	(1966)	4, p. 37; *18.2*, p. *300*; 21.1.1, p. 411
R. B. Clough	**17**, K175	(1966)	*10*, p. *96*; *10.1*, p. *117*
D. G. Coates	**27**, K11	(1968)	*4*, p. *46*; 5, p. 53; 22.2.1, p. 485
W. Cochran	**25**, 273	(1968)	*6*, p. *64*; 14.4.2, p. 268
	29, K81	(1968)	*1.2*, p. *12*; *6*, p. *68*; 22.8, p. 555
	30, K157	(1968)	1.2, p. 13; *6*, p. *70*; 14.4.2, p. 269
R. Coelho	**19**, K91	(1967)	14.3.1, p. 245; *14.4.1*, p. *257*; 22.6, p. 537
W. A. Coghlan	**25**, 679	(1968)	*10*, p. *104*; *10.1*, p. *128*
L. N. Cojocaru	**22**, 361	(1967)	*11*, p. *149*; 13.4, p. 208; 14.3, p. 237; 15, p. 272; 22.6, p. 539
	29, K119	(1968)	*11*, p. *154*; 14.3, p. 244; 15, p. 274; 22.6, p. 543; 22.8, p. 556
R. Colella	**18**, 157	(1966)	*5*, p. *50*; 22.1.1, p. 460; 22.1.2, p. 469
M. V. Coleman	**22**, 593	(1967)	*1.4*, p. *19*; 22.1.2, p. 471; 22.6, p. 539
	24, K111	(1967)	*1.4*, p. *19*; 2, p. 23; 22.1.2, p. 472
	25, 241	(1968)	*1.4*, p. *19*; 2, p. 23; 22, p. 449
R. De Coninck	**17**, 825	(1966)	13, p. 163; *14.3*, p. *233*; 15, p. 270; 22.6, p. 535
F. Constantinescu	**21**, K151	(1967)	*19*, p. *329*
R. Conte	**28**, 475	(1968)	*10.1*, p. *132*; 21.1.1, p. 422
A. Corciovei	**16**, 329	(1966)	1.4, p. 16; *18.2*, p. *300*
P. C. J. Coremans	**19**, 177	(1967)	*1.1*, p. *8*; 10.1, p. 120; 21.1, p. 395; 21.6, p. 432
P. Costa	**22**, 349	(1967)	*10*, p. *101*; 22.3, p. 494
G. Costache	**16**, 329	(1966)	1.4, p. 16; *18.2*, p. *300*
G. Coste	**20**, 361	(1967)	13.1, p. 177; *15*, p. *271*; 22, p. 444
P. Coulomb	**21**, K97	(1967)	*10.1*, p. *123*; 21.1, p. 398; 21.1.1, p. 418
J. E. Cousins	**30**, 105	(1968)	*19*, p. *335*; 21.2, p. 425; 22.5.2, p. 530

S. McK. Cousland	**27,**	501	(1968)	*10.1*, p. *131*; 21.1, p. 405
T. A. T. Cowell	**24,**	K37	(1967)	10.2, p. 141; *13.4*, p. *210*; 14.3, p. 239; 20.3, p. 367; 22.4.1, p. 501
P. Crăciun	**29,**	761	(1968)	13, p. 173; *17*, p. *290*; 21, p. 388
D. J. Craik	**16,**	321	(1966)	4, p. 37; *18.2*, p. *300*; 21.1.1, p. 410
	28,	407	(1968)	1.4, p. 20; *18.2*, p. *313*; 21.1, p. 405; 21.1.1, p. 422
K. R. Cramer	**20,**	K21	(1967)	*9*, p. *88*
	26,	97	(1968)	*9*, p. *91*
J. H. Crawford, Jr.	**23,**	301	(1967)	*10.2*, p. *140*; 11, p. 150; 13.4, p. 209; 22.5.2, p. 523
J. L. Crawford	**27,**	559	(1968)	1.4, p. 20; *10*, p. *108*; *10.1*, p. *131*
A. G. Crocker	**16,**	K25	(1966)	3, p. 26; *5*, p. *49*
N. Croitoru	**16,**	K143	(1966)	2, p. 22; 14.3.1, p. 245; *15*, p. *270*; 22.1.1, p. 459
	23,	621	(1967)	2, p. 23; 13.1, p. 182; 14.3.1, p. 246; *15*, p. *273*; 22.1.2, p. 472
	23,	627	(1967)	2, p. 23; *16*, p. *280*; 22.1.1, p. 463
	24,	K17	(1967)	*14.3.2*, p. *250*; *14.4.1*, p. *261*; 22.1.1, p. 463
E. Cruceanu	**21,**	557	(1967)	*8*; p. *82*; 22.4.1, p. 500; 22.4.2, p. 506; 22.4.3, p. 509; 22.8, p. 550
	28,	421	(1968)	6, p. 66; *8*, p. *85*; 22.4.3, p. 511; 22.8, p. 554
	30,	K149	(1968)	13.1, p. 190; *20.1*, p. *359*; 22, p. 455
L. Cser	**20,**	581	(1967)	*6.1*, p. *71*; 18, p. 292; 21.1.1, p. 416
	20,	591	(1967)	*6.1*, p. *72*; 18, p. 292; 21.1.1, p. 416
	27,	131	(1968)	*6.1*, p. *73*; 18.3, p. 319
W. L. McCubbin see under letter "M"				
J. A. Cundall	**16,**	613	(1966)	4, p. 38; *18.2*, p. *300*; 21.1.1, p. 411
S. F. Da Cunha see under letter "D"				
H. Curien	**21,**	K163	(1967)	*13*, p. *166*; *18*, p. *292*; 22.6, p. 538
A. Czachor	**20,**	K17	(1967)	5, p. 50; *12.1*, p. *159*; 21, p. 377; 21.1, p. 396; 21.3, p. 426
	29,	423	(1968)	*6*, p. *68*; 21.3, p. 427
J. Czerwonko	**30,**	723	(1968)	*13.1*, p. *190*; 18.1, p. 298; 21, p. 389; 21.1, p. 409; 21.6, p. 437

S. F. Da Cunha	**26,**	K127	(1968)	*9*, p. *91*; 22.2.3, p. 492
S. Däbritz	**29,**	K89	(1968)	3.1, p. 33; *4*, p. *47*; 22.4.1, p. 505
I. V. Dakhovskii	**23,**	229	(1967)	13, p. 167; 14.3, p. 238; *15*, p. *272*; 22.1.1, p. 463
J. R. Dale	**16,**	351	(1966)	3.1, p. 29; 5, p. 49; 13.4, p. 200; *14.3.2*, p. *248*; 16, p. 276; 22.1.1, p. 458; 22.1.2, p. 468; 22.2.1, p. 482; 22.2.2, p. 487; 22.2.3, p. 489
A. Dall'Olio	**22,**	365	(1967)	13.4, p. 208; *19*, p. *330*; 22.9, p. 559
R. V. Damle	**22,**	K63	(1967)	*14.4.2*, p. *267*; 22.8, p. 551
M. Damodar-Pai	**21,**	819	(1967)	13.4, p. 207; *16*, p. *279*; 22.4.1, p. 500
N. van Dang	**20,**	557	(1967)	*18.2*, p. *306*; 21.1.1, p. 416
B. K. Daniels	**27,**	535	(1968)	*10*, p. *107*; *20.3*, p. *369*; 22.4.1, p. 502
G. Dascola	**22,**	365	(1967)	13.4, p. 208; *19*, p. *330*; 22.9, p. 559
J. G. Dash	**19,**	K31	(1967)	*6.1*, p. *71*
S. Dash	**26,**	159	(1968)	*10*, p. *105*; 22.1.2, p. 473
Vu Huy Dat see under letter "V" (Vu)				
L. I. Datsenko	**30,**	231	(1968)	*10*, p. *112*; 20, p. 342; 22.1.1, p. 466
D. Dautreppe	**16,**	281	(1966)	*10.1*, p. *114*; 14.1, p. 220; 18.2, p. 299; 21.1, p. 390
	24,	K173	(1967)	10.1, p. 127; *11*, p. *151*; 21.1, p. 402
	29,	283	(1968)	4, p. 47; *10.1*, p. *133*; 11, p. 154; 21.1, p. 407

A. T. DAVIDSON	**30,**	741	(1968)	3, p. 28; 10, p. 113; *14.4.1*, p. *265*; 20.1, p. 359; 22, p. 455
J. L. DAVIDSON	**23,**	K139	(1967)	*10.1*, p. *126*; 21.1, p. 400
	26,	345	(1968)	*10.1*, p. *129*; 21.1, p. 404
T. A. DAVIS	**26,**	285	(1968)	*20.1*, p. *352*; 22, p. 450
T. L. DAVIS	**26,**	621	(1968)	*10*, p. *106*; *10.1*, p. *129*
F. DAVOINE	**25,**	691	(1968)	*17*, p. *288*; 21.1, p. 403; 21.1.1, p. 420; 22.1.1, p. 464; 22.2.1, p. 485; 22.4.1, p. 502; 22.5.2, p. 525; 22.6, p. 540
A. S. DAVYDOV	**20,**	143	(1967)	*6*, p. *59*; *13.2*, p. *192*; 22, p. 443
	20,	153	(1967)	*13.2*, p. *192*; *20.1*, p. *347*; 22, p. 443
	24,	373	(1967)	*14.4*, p. *253*; 22, p. 448
	27,	51	(1968)	*20.1*, p. *353*; 22, p. 451
	30,	357	(1968)	6, p. 69; *13.4*, p. *216*; 20.1, p. 358; 20.3, p. 371; 22, p. 454
G. V. DAVYDOV	**18,**	41	(1966)	*5*, p. *50*
	18,	K145	(1966)	*5*, p. *50*
B. DAYAL	**16,**	513	(1966)	*6*, p. *56*; 8, p. 79; 21, p. 374
	18,	731	(1966)	3.2, p. 34; *6*, p. *58*; *8*, p. *80*; 23, p. 561
	19,	729	(1967)	*6*, p. *59*; 21.1, p. 396
	19,	751	(1967)	*6*, p. *59*; 22.6, p. 537
	20,	321	(1967)	*8*, p. *81*; 23, p. 561
	21,	661	(1967)	6, p. 61; *8*, p. *82*; *12*, p. *157*; 23, p. 561
	23,	K93	(1967)	5, p. 52; *8*, p. *83*; 22.4.2, p. 507
J.-M. DEBEVER	**17,**	875	(1966)	*20.2*, p. *360*; 22.4.1, p. 498
S. J. DECHTER	**18,**	505	(1966)	*14.1*, p. *221*; 21, p. 376
P. H. DEDERICHS	**23,**	377	(1967)	6, p. 62; *10*, p. *102*; *10.1*, p. *125*
W. DEGRIECK	**18,**	279	(1966)	*10.2*, p. *138*; 22.5.2, p. 519
	22,	177	(1967)	10.2, p. 139; *16*; p. *279*; *20.3*, p. *366*; 22.5.2, p. 523
W. DEKEYSER	**17,**	739	(1966)	*10*, p. *95*; 10.2, p. 138; 22.5.2, p. 519
I. J. DEKHTYAR see I. Y. DEKHTYAR				
I. Y. DEKHTYAR	**21,**	K81	(1967)	10.1, p. 122, *15*; p. *272*; 21.1, p. 398
	26,	K151	(1968)	6.1, p. 73; *10.1*, p. *130*; 14.1, p. 225; 18.1, p. 297; 21.6, p. 436
A. J. DEKKER	**18,**	761	(1966)	*6.1*, p. *71*
P. DEKKER	**29,**	K73	(1968)	18.3, p. 320; *20.1*, p. *357*
L. DELAEY	**24,**	K103	(1967)	*4*, p. *44*; 21.1, p. 402
	25,	697	(1968)	1.1, p. 8; *4*, p. *45*; 21.1, p. 403
J. DELAPLACE	**29,**	819	(1968)	10.1, p. 134; *11*, p. *154*; 14.1, p. 226; 21.3, p. 427
	30,	119	(1968)	10.1, p. 134; *11*, p. *154*; 14.1, p. 226; 21.3, p. 427
P. DELAVIGNETTE	**17,**	221	(1966)	*10*, p. *94*; 22.1, p. 456
	17,	K173	(1966)	*10*, p. *96*; 22.8, p. 548
	19,	683	(1967)	*1*, p. *7*; 4, p. 40; 10, p. 98; 22.6, p. 537
	28,	305	(1968)	*1.1*, p. *8*; 4, p. 46; 21, p. 386; 21.1, p. 405
R. T. DELVES	**16,**	621	(1966)	*3.1*, p. *29*
	17,	119	(1966)	*3.1*, p. *29*
	20,	693	(1967)	*3.1*, p. *31*
K. M. DEMCHUK	**27,**	249	(1968)	*13.1*, p. *186*; 14.3, p. 241; 15, p. 274; 22.2.3, p. 492
D. DEMCO	**21,**	K151	(1967)	*19*, p. *329*
L. J. DEMER	**29,**	K141	(1968)	*10*, p. *111*; 22.5.1, p. 517; 22.5.2, p. 529; 22.5.3, p. 534; 22.6, p. 543
A. F. DEMIDENKO	**30,**	147	(1968)	*6*, p. *69*; *8*, p. *86*; 13.1, p. 189; 22.1.1, p. 466; 22.1.2, p. 474; 22.2.1, p. 486; 22.2.2, p. 488; 22.2.3, p. 493

Z. A. Demidenko	**21,** K59 (1967)	6, p. 61; *20.1*, p. *348*; 22, p. 445
J. Demny	**22,** K1 (1967)	*10.1*, p. *124*; 21, p. 380
M. Denayer	**28,** 193 (1968)	13.1, p. 187; 13.4, p. 214; *14.3*, p. *242*; 15, p. 274; 22.6, p. 542
B. Deo	**17,** 77 (1966)	*8*, p. *79*
A. Deruyttere	**28,** K9 (1968)	*10.1*, p. *132*; 21, p. 387; 21.3, p. 427
K. Dettmann	**22,** 423 (1967)	*11*, p. *149*; 21.1, p. 399
	22, 433 (1967)	*11*, p. *149*; 21, p. 380; 21.1, p. 399; 21.6, p. 434
K. Deutscher	**26,** 527 (1968)	14.3.1, p. 246; *17*, p. *288*; 20.1, p. 353; 22.7, p. 546
	27, 145 (1968)	14.3.1, p. 246; *17*, p. *289*; 20.1, p. 354; 22.7, p. 546
P. Devaux	**20,** 301 (1967)	*13.4*, p. *205*; 14.4.1, p. 257; 22.9, p. 558
	27, 201 (1968)	13.2, p. 194; *13.3*, p. *198*; *16*, p. *282*; 22.9, p. 560
A. Dévényi	**16,** K143 (1966)	2, p. 22; 14.3.1, p. 245; *15*, p. *270*; 22.1.1, p. 459
	23, 621 (1967)	2, p. 23; 13.1, p. 182; 14.3.1, p. 246; *15*, p. *273*; 22.1.2, p. 472
	23, 627 (1967)	2, p. 23; *16*, p. *280*; 22.1.1, p. 463
J. Devenyi	**16,** 295 (1966)	*1.4*, p. *16*; 18.2, p. 299; 21.1.1, p. 410
	16, 569 (1966)	4, p. 37; *18.2*, p. *300*; 21.1.1, p. 411
J. Devreese	**17,** 825 (1966)	13, p. 163; *14.3*, p. *233*; 15, p. 270; 22.6, p. 535
R. E. DeWames see under letter "W"		
A. Dezor	**23,** K159 (1967)	14.4.2, p. 267; *19*, p. *331*
I. Dézsi	**21,** K121 (1967)	*6.1*, p. *72*; 18.4, p. 322; 22.6, p. 538; 22.8, p 551
	22, 617 (1967)	1.2, p. 11; *6.1*, p. *72*; 18.3, p. 318; 18.4, p. 323; 22.8, p. 551
	27, 131 (1968)	*6.1*, p. *73*; 18.3, p. 319
	30, 215 (1968)	2, p. 24; *6.1*, p. *74*; 22.8, p. 556
H. P. Dibbs	**19,** K11 (1967)	*3.2*, p. *35*; 14.3, p. 235; 22, p. 442
B. G. Dick	**28,** 223 (1968)	*10.2*, p. *143*; *13.4*, p. *214*; 22.5.2, p. 527
	29, 587 (1968)	6, p. 68; *10.2*, p. *144*; 22.5.2, p. 529
J. Diehl	**16,** K109 (1966)	*10.1*, p. *115*; 11, p. 145; 21.1, p. 391
	17, 43 (1966)	*10.1*, p. *115*; 11, p. 146; 21.1, p. 392
	17, 269 (1966)	*10.1*, p. *115*; 11, p. 146; 21.1, p. 392
G. J. Dienes	**24,** 461 (1967)	*9*, p. *90*; 21.1, p. 401; 21.6, p. 435
V. Dienys	**17,** 769 (1966)	13, p. 163; *14.4.1*, p. *256*; 22.1.1, p. 460
A. M. van Diepen	**24,** 715 (1967)	1.2, p. 11; *18.2*, p. *309*; 21, p. 383
	29, 189 (1968)	13, p. 172; 18.1, p. 297; *19*, p. *335*; 21, p. 387; 21.4, p. 428
	30, 759 (1968)	13, p. 173; 13.1, p. 190; 18.1, p. 298; *19*, p. *336*; 21, p. 389; 21.4, p. 428
H. Diepers	**16,** K109 (1966)	*10.1*, p. *115*; 11, p. 145; 21.1, p. 391
	24, 235 (1967)	10.1, p. 126; *11*, p. *150*; 21.1, p. 401
	24, 623 (1967)	4, p. 44; 10.1, p. 127; *11*, p. *150*; 21.1, p. 402
R. Dietz	**16,** 789 (1966)	*10.1*, p. *114*; 14.1, p. 220; 21.1, p. 391
H.-D. Dietze	**27,** 601 (1968)	*18.2*, p. *312*
	27, 611 (1968)	*18.2*, p. *312*
E. Dieulesaint	**22,** K135 (1967)	*16*, p. *280*; 22.4.1, p. 500
W. Diez	**27,** 139 (1968)	*9*, p. *91*; 21, p. 385; 22, p. 451
K. Differt	**26,** 327 (1968)	*10.1*, p. *129*; 21.1, p. 404
A. A. Dikii	**30,** K107 (1968)	*14.2*, p. *230*
J. Dillinger	**29,** 707 (1968)	6, p. 68; *13.2*, p. *195*; 20.1, p. 357; 22.4.1, p. 505; 22.4.2, p. 508; 22.4.3, p. 511
I. Dima	**22,** K79 (1967)	*16*, p. *280*; 22.4.2, p. 507
	23, K113 (1967)	1.2, p. 11; *1.4*, p. *19*; 22.4.3, p. 510
Ž. Dimitrijević	**21,** K163 (1967)	*13*, p. *166*; *18*, p. *292*; 22.6, p. 538

Ž. Dimitrijević	**22,**	K55	(1967)	*18.3*, p. *318*
	26,	429	(1968)	*4*, p. *45*; 18.3, p. 319; 18.4, p. 324; 22.8, p. 553
H.-J. Dittfeld	**17,**	K245	(1966)	*14.4.1*, p. *257*; 22.1.1, p. 460; 22.1.2, p. 469
M. V. Dmitruk	**25,**	K75	(1968)	20.1, p. 352; *20.2*, p. *361*; 20.3, p. 368; 22.5, p. 513; 22.8, p. 552
L. Dobrosavljević	**20,**	K63	(1967)	*13.1*, p. *177*; 21, p. 378
	23,	509	(1967)	*14.2*, p. *229*; 18, p. 293; 21, p. 381
A. Dobrzańska	**17,**	K109	(1966)	*18.2*, p. *303*; 21.1.1, p. 413
G. Döhler	**19,**	555	(1966)	13.4, p. 205; *14.4.1*, p. *257*; 16, p. 278; 22.1.3, p. 477; 22.4.1, p. 499
	24,	331	(1967)	*14.4.1*, p. *260*; 16, p. 280; 22.1.3, p. 478
	26,	551	(1968)	*13.1*, p. *185*; 13.4, p. 212; 14.4.1, p. 262; 20.1, p. 353; 22, p. 451
	30,	627	(1968)	13.4, p. 217; *14.4.1*, p. *265*; 22.4.1, p. 505
V. Dolocan	**28,**	719	(1968)	*16*, p. *283*; 22, p. 453
J. Donecker	**26,**	K131	(1968)	*6*, p. *65*; 8, p. 84; 10, p. 106; 22.5.1, p. 516
Nguyen van Dong see under letter "N" (Nguyen)				
G. Dorda	**17,**	K127	(1966)	*14.3*, p. *232*; 22.1.2, p. 469; 22.6, p. 535
B. Dorner	**18,**	795	(1966)	1.2, p. 10; *6*, p. *58*; 23, p. 561
A. V. Doroshenko	**16,**	737	(1966)	*18.2*, p. *301*; *18.4*, p. *321*; 21.1, p. 391; 21.1.1, p. 411
K. Dos	**30,**	619	(1968)	*13.2*, p. *195*; 20.1, p. 359; 22.4.1, p. 505
P. G. McDougall see under letter "M"				
D. L. Douglass	**29,**	K95	(1968)	*1.4*, p. *21*; 3.1, p. 33; 22.6, p. 543
Ya. O. Dovhyj	**27,**	K121	(1968)	13.2, p. 194; *20*, p. *341*; 22, p. 452
W. D. Doyle	**17,**	K67	(1966)	1.4, p. 17; *18.2*, p. *302*; 21.1.1, p. 412
C. M. Dozier	**30,**	87	(1968)	9, p. 92; *10.1*, p. *134*; *12.1*, p. *161*; 21, p. 388; 21.6, p. 437
G. Dräger	**27,**	513	(1968)	2, p. 24; *20*, p. *341*; 22.1.1, p. 465; 22.6, p. 541; 22.8, p. 554
I. Drăghici	**26,**	K39	(1968)	3.1, p. 33; *14.3.1*, p. *246*; 22, p. 450
W. Dreybrodt	**16,**	215	(1966)	10.2, p. 137; 13.4, p. 201; *19*, p. *326*; 22.5.2, p. 518
	18,	133	(1966)	13.4, p. 203; *20.1*, p. *346*; 22.5.2, p. 519
	20,	337	(1967)	13.4, p. 205; *19*, p. *328*; 22.5.2, p. 521
	21,	99	(1967)	13.4, p. 206; *19*, p. *329*; 22.5.2, p. 522
B. Dreyfus	**22,**	77	(1967)	*8*, p. *83*; 10, p. 101; *11*, p. *149*; 22.6, p. 538
V. A. Drozdov	**22,**	K109	(1967)	*12*, p. *157*; 16, p. 280; 22.4.1, p. 500
	29,	K145	(1968)	*10*, p. *111*; 12, p. 158; 15, p. 274; 22.1.2, p. 474
M. Dubejko	**16,**	399	(1966)	13, p. 163; *17*, p. *286*; 21, p. 373
H. Dübler	**25,**	109	(1968)	*9*, p. *91*; 10.1, p. 127; 21, p. 383
S. I. Dudkin	**18,**	853	(1966)	6, p. 58; 13.4, p. 204; *20.1*, p. *346*
	20,	427	(1967)	6, p. 60; 13.4, p. 205; *20.1*, p. *347*
G. Duesing	**23,**	481	(1967)	10.1, p. 125; *11*, p. *150*; 14.1, p. 223; 21.6, p.434
W. J. Duffin	**18,**	755	(1966)	*4*, p. *39*; 5, p. 50; 22, p. 441
T. R. Duncan	**18,**	231	(1966)	*10*, p. *96*; *10.1*, p. *117*; 21, p. 375; 21.1, p. 393; 21.5, p. 429; 21.6, p. 431; 22.1, p. 456; 22.1.1, p. 460; 22.1.2, p. 470
	25,	K105	(1968)	*10*, p. *105*; *10.1*, p. *129*; 21, p. 385; 21.1, p. 403; 21.5, p. 429; 21.6, p. 435; 22.1, p. 457; 22.1.1, p. 464; 22.1.2, p. 473
G. R. Dunmyre	**18,**	505	(1966)	*14.1*, p. *221*; 21, p. 376
	21,	191	(1967)	*18.2*, p. *306*; 21.1, p. 397; 21.1.1, p. 417
J. M. Dupouy	**23,**	K109	(1967)	*10.1*, p. *125*; 21.3, p. 426
	28,	K55	(1968)	*10.1*, p. *132*; 21, p. 387

R. Dupree	**24,**	275	(1967)	*19*, p. *331*; 21, p. 382; 21.1, p. 401; 21.2, p. 425
	24,	525	(1967)	*19*, p. *331*; 21.6, p. 435
C. Dupuis	**23,**	509	(1967)	*14.2*, p. *229*; 18, p. 293; 21, p. 381
G. A. Dussel	**21,**	K145	(1967)	*14.4.1*, p. *259*; 22.4.1, p. 500
B. C. Dutta	**21,**	291	(1967)	6.1, p. 72; *11*, p. *148*; 22.5.2, p. 522
F. Dworschak	**21,**	741	(1967)	*11*, p. *149*; 14.1, p. 223; 21, p. 379
	29,	75	(1968)	*10.1*, p. *132*; 11, p. 153; 21.1, p. 406
	29,	81	(1968)	*10.1*, p. *133*; 11, p. 153; 21.1, p. 406
	30,	219	(1968)	10.1, p. 135; *11*, p. *155*; 14.1, p. 227; 18.2, p. 315; 21.1, p. 408
I. M. Dykman	**30,**	767	(1968)	*13*, p. *173*; *20*, p. *343*; 22.1.1, p. 467
A. Dymanus	**24,**	487	(1967)	*6.1*, p. *73*; 18.3, p. 318; 22.6, p. 540, 22.8, p. 552
K. Dyrek	**17,**	K195	(1966)	6.1, p. 71; *18.4*, p. *321*; 22, p. 441
T. D. Dzhafarov	**19,**	705	(1967)	*9*, p. *88*; 22.2.1, p. 483
E. A. Dzhafarova	**21,**	423	(1967)	13.4, p. 206; *14.3*, p. *236*; 14.4.1, p. 258; 22.1.2, p. 471
S. U. Dzhalilov	**20,**	261	(1967)	*2*, p. *22*; 22.1.3, p. 477
J. Dziesiaty	**26,**	K21	(1968)	*19*, p. *333*; 22.4.1, p. 502
E. Z. Dziuba	**29,**	813	(1968)	*13.1*, p. *189*; 14.3, p. 244; 20.1, p. 357; 22.4.3, p. 511; 22.8, p. 556
R. M. Easson	**23,**	K129	(1967)	*14.2*, p. *229*; 21, p. 382
W. Ebeling	**30,**	533	(1968)	*9*, p. *92*
P. Eckelt	**23,**	307	(1967)	*13.1*, p. *181*; 22.4.1, p. 500; 22.4.2, p. 507; 22.4.3, p. 510
L. Eckertová	**18,**	3	(1966)	13, p. 162; 13.3, p. 196; 13.4, p. 200; 14.4.1, p. 255; *17*, p. *286*; 22.6, p. 535
B. Eckstein	**20,**	83	(1967)	*2*, p. *22*; *10.1*, p. *121*
N. Economou	**20,**	623	(1967)	13.2, p. 192; 16, p. 279; *20.1*, p. *347*; 22.6, p. 537
D. V. Eddolls	**17,**	67	(1966)	13.4, p. 202; *14.3*, p. *232*; 22.2.1, p. 482
I. S. Edelman	**17,**	249	(1966)	4, p. 38; *18.2*, p. *302*; 21.1, p. 392; 21.1.1, p. 412
L. R. Edwards	**26,**	611	(1968)	10.1, p. 129; *14.1*, p. *225*; 21, p. 385; 21.6, p. 435
B. A. Efimov	**16,**	191	(1966)	*10*, p. *93*; 22.5.2, p. 518
O. N. Efimov	**17,**	397	(1966)	*5*, p. *49*; 10, p. 94; 22.1.1, p. 459
	22,	297	(1967)	*5*, p. *51*; 6, p. 62; 22.1.1, p. 462
R. F. Egorov	**26,**	391	(1968)	3, p. 27; *13.1*, p. *185*
V. D. Egorov	**18,**	145	(1966)	13.3, p. 197; *16*, p. *278*; 22.1.1, p. 460
	20,	705	(1967)	13.2, p. 192; 20.2, p. 361; *20.3*, p. *365*; 22.4.1, p. 499
	23,	703	(1967)	13.2, p. 193; 20.2, p. 361; *20.3*, p. *367*; 22.4.1, p. 501
P. Ehrhart	**28,**	K35	(1968)	1.4, p. 20; *11*, p. *153*; 21.1, p. 405; 21.6, p. 436
U. Eichhoff	**22,**	K91	(1967)	*19*, p. *330*
F. Eichhorn	**23,**	237	(1967)	*4*, p. *43*; 22.1.2, p. 472
J. Eisenblätter	**30,**	587	(1968)	*10*, p. *112*; 22.5.2, p. 531
N. A. Eissa	**22,**	617	(1967)	1.2, p. 11; *6.1*, p. *72*; 18.3, p. 318; 18.4, p. 323; 22.8, p. 551
	27,	131	(1968)	*6.1*, p. *73*; 18.3, p. 319
	30,	215	(1968)	2, p. 24; *6.1*, p. *74*; 22.8, p. 556
F. M. El-Akkad	**25,**	K115	(1968)	*14.3*, p. *240*; 22, p. 450
Yu. A. Ekmanis	**18,**	897	(1966)	10.2, p. 138; *11*, p. *147*; 19, p. 327; 20.1, p. 346; 22.5.2, p. 520
G. Elbinger	**17,**	151	(1966)	*18.3*, p. *316*
	21,	303	(1967)	*18.3*, p. *317*; 21.1, p. 397

V. F. Elesin	**24,** 757	(1967)	6, p. 64; 13, p. 168; *16*, p. *281*; 22.1.1, p. 464
P. G. Eliseev	**23,** 587	(1967)	13.4, p. 210; *20.3*, p. *367*; 22.2.1, p. 484
L. Elstner	**17,** 139	(1966)	6, p. 56; *14.3*, p. *232*; 22.1.2, p. 469
	22, 541	(1967)	11, p. 149; *13.4*, p. *208*; 14.3, p. 237; 22.1.2, p. 471
O. V. Emelyanenko	**27,** 45	(1968)	13.4, p. 213; *14.3*, p. *241*; 22, p. 451; 22.7, p. 546
R. Enderlein	**19,** 673	(1967)	13.1, p. 177; 14.4.1, p. 257; *20*, p. *338*; 22, p. 443
	20, 295	(1967)	13.1, p. 177; 14.4.1, p. 257; *20*, p. *338*; 22, p. 444
	23, 127	(1967)	13.1, p. 180; 14.4.1, p. 259; *20*, p. *339*; 22, p. 447
	26, 509	(1968)	13.1, p. 185; 13.2, p. 193; *14.4.1*, p. *262*; *20*, p. *340*; 20.1, p. 353; 22, p. 451
	28, 519	(1968)	*14.4.1*, p. *263*; 22, p. 453
V. V. Eremenko	**18,** 683	(1966)	*13.4*, p. *204*; 18.4, p. 321; *20.3*, p. *364*; 22.5.3, p. 532
	18, K123	(1966)	*8*, p. *80*; 14.1, p. 222; 21.4, p. 428
	21, 797	(1967)	*14.1*, p. *223*; 21.7, p. 438
	22, 65	(1967)	13.4, p. 208; *20.3*, p. *366*; 22.5, p. 513; 22.8, p. 551
	28, K155	(1968)	*14.1*, p. *226*; 21.7, p. 438
	30, 49	(1968)	13.1, p. 189; *18.4*, p. *325*; 20.1, p. 358; 23, p. 561
G. L. Eristavi	**29,** 443	(1968)	13.2, p. 195; 13.4, p. 215; 14.3, p. 243; *20.1*, p. *356*; 22.1.1, p. 466
Gy. Erlaki	**21,** K121	(1967)	*6.1*, p. *72*; 18.4, p. 322; 22.6, p. 538; 22.8, p. 551
I. B. Ermolovich	**24,** 543	(1967)	10, p. 103; *13.4*, p. *210*; 16, p. 281; 20.3, p. 368; 22.4.1, p. 501
G. J. Ernst	**19,** 107	(1967)	10.1, p. 120; *12.1*, p. *159*; 21.1.1, p. 414
L. Ernst	**19,** 89	(1967)	*1.3*, p. *14*; 22.1.1, p. 461
	24, 177	(1967)	13.3, p. 198; *17*, p. *287*; 22.1.1, p. 463
P. R. Errington	**22,** 473	(1967)	*6.1*, p. *72*; 18.2, p. 308; 22.8, p. 551
M. E. Ertl	**17,** K63	(1966)	6, p. 57; *8*, p. *79*; 21.7, p. 438
B. Escaig	**28,** 463	(1968)	*10*, p. *109*; *10.1*, p. *132*
	28, 475	(1968)	*10.1*, p. *132*; 21.1.1, p. 422
P. Escudier	**16,** 295	(1966)	*1.4*, p. *16*; *18.2*, p. *299*; 21.1.1, p. 410
U. Essmann	**17,** 725	(1966)	*10.1*, p. *116*; 21.1, p. 392
	18, 813	(1966)	4, p. 39; *14.2*, p. *228*
	20, 95	(1967)	*14.2*, p. *229*; 18, p. 291; 21, p. 377
	25, 373	(1968)	10.1, p. 128; *14.2*, p. *229*; 21, p. 384
	25, 395	(1968)	10.1, p. 128; *14.2*, p. *230*; 21, p. 384
P. C. Euthymiou	**16,** 183	(1966)	*11*, p. *145*; 14.3, p. 231; 22.2.3, p. 489
B. L. Evans	**25,** 417	(1968)	*13.2*, p. *193*; 20.1, p. 351; 22, p. 449
H. W. Evans	**27,** 695	(1968)	10.2, p. 142; *11*, p. *152*; 19, p. 333; 20.1, p. 354; 22.5.3, p. 534
J. H. Evans	**25,** K39	(1968)	*3.1*, p. *33*; *14.1*, p. *224*; 21, p. 384
K. R. Evans	**22,** 195	(1967)	*10.1*, p. *124*; 21.1, p. 398
	22, 607	(1967)	*10.1*, p. *124*; 21.1, p. 399; 21.1.1, p. 418
I. N. Evdokimov	**19,** 407	(1967)	*17*, p. *287*; 21, p. 377; 22.1, p. 457
G. J. O. Eyoku	**19,** 111	(1967)	*14.3*, p. *234*; 22.1.1, p. 461
B. L. Eyre	**25,** K39	(1968)	*3.1*, p. *33*; *14.1*, p. *224*; 21, p. 384

O. E. Facey	**19,** 565	(1967)	*10.2*, p. *139*; 22.5.2, p. 521
V. L. Falko	**22,** 319	(1967)	7, p. 77; 13.1, p. 179; *19*, p. *329*; 21, p. 380
A. Farwer	**28,** 373	(1968)	4, p. 47; 10, p. 109; *20.1*, p. *355*; 22.4.1, p. 503
J. F. Fast	**16,** 467	(1966)	*3*, p. *26*; 5, p. 49; 18.1, p. 296; 18.2, p. 300; 18.3, p. 316; 18.4, p. 321; 22.7, p. 545

J. F. Fast	**21**,	593	(1967)	*5*, p. *51*; 18.1, p. 296; 18.3, p. 318; 18.4, p. 322; 22.7, p. 545
	24,	715	(1967)	1.2, p. 11; *18.2*, p. *309*; 21, p. 383
	29,	825	(1968)	*18.3*, p. *320*; 21.1, p. 407; 21.4, p. 428
R. Fatehally	**26**,	91	(1968)	*13*, p. *169*; 18.4, p. 323; 22.5, p. 513; 22.8, p. 553
A. B. Fazakas	**28**,	385	(1968)	*14.4.1*, p. *263*; 22, p. 452
	28,	699	(1968)	*13.3*, p. *199*; 14.4.1, p. 263; 22, p. 453
R. G. Fedchenko	**26**,	K151	(1968)	6.1, p. 73; *10.1*, p. *130*; 14.1, p. 225; 18.1, p. 297; 21.6, p. 436
T. Federighi	**16**,	439	(1966)	*9*, p. *87*; 21, p. 373
V. P. Fedorov	**30**,	485	(1968)	*13.2*, p. *195*; 20.1, p. 358; 20.3, p. 371; 22.4.2, p. 508
V. Fedoseev	**27**,	751	(1968)	6, p. 66; *13.2*, p. *194*; 22, p. 452
S. P. Fedotov	**21**,	K91	(1967)	*13.4*, p. *207*; 19, p. 329; 22.2.1, p. 483
	25,	K51	(1968)	*19*, p. *332*; 22.2, p. 481
V. G. Fedotov	**28**,	489	(1968)	2, p. 24; 14.3, p. 242; *20.3*, p. *370*; 22.8, p. 554
V. A. Feigin	**28**,	83	(1968)	8, p. 85; 13, p. 171; *18.2*, p. *312*
J. W. Feitknecht	**16**,	697	(1966)	*1.4*, p. *16*; 13.4, p. 201; 16, p. 276; 22.4.1, p. 497
Ya. S. Feiziev	**24**,	K157	(1967)	14.3, p. 239; *15*, p. *273*; 22, p. 449
	28,	K113	(1968)	*8*, p. *85*; 22, p. 453
E. Feldtkeller	**27**,	161	(1968)	4, p. 46; *18.2*, p. *311*; 21.1, p. 404; 21.1.1, p. 421
F. W. Felix	**27**,	529	(1968)	*9*, p. *91*; 10, p. 107; 11, p. 152; 22.5.2, p. 526
M. C. Félix	**17**,	49	(1966)	11, p. 146; *13.3*, p. *196*; 22.9, p. 558
H. Fellenzer	**23**,	171	(1967)	*13*, p. *167*; 13.3, p. 198; *20*, p. *340*; 21, p. 381
P. Feltham	**20**,	675	(1967)	*10.2*, p. *139*; 13.4, p. 206; 22.5.3, p. 533
	25,	K71	(1968)	10, p. 104; *10.2*, p. *142*; 19, p. 332; 22.5.3, p. 533
	25,	K107	(1968)	10, p. 105; *10.2*, p. *142*; 19, p. 333; 22.6, p. 541
	30,	135	(1968)	*10*, p. *111*; *10.1*, p. *134*; 21, p. 388; 22.5.2, p. 530
A. R. Ferchmin	**21**,	K179	(1967)	*13.1*, p. *179*; *18.2*, p. *307*
	26,	K107	(1968)	*13.1*, p. *185*; *18.2*, p. *311*
R. P. Ferrier	**28**,	335	(1968)	1.4, p. 20; *18.2*, p. *312*; 21.1.1, p. 422
L. Fiermans	**21**,	627	(1967)	*4*, p. *42*; 22.1.2, p. 471
	22,	463	(1967)	*4*, p. *43*; 22.1.2, p. 471
T. Figielski	**20**,	K1	(1967)	13.4, p. 205; *16*, p. *279*; 22.1.1, p. 461
I. Filiński	**18**,	K37	(1966)	*13.2*, p. *192*; 20.3, p. 364; 22.4.2, p. 506
A. S. Filipchenko	**17**,	K113	(1966)	*13.1*, p. *175*; 20.1, p. 345; 22.2.3, p. 490
	19,	435	(1967)	*14.3*, p. *235*; 15, p. 271; 22.2.3, p. 490
V. A. Finkel	**23**,	K97	(1967)	*8*, p. *83*; 21.4, p. 428
H. Finkenrath	**29**,	203	(1968)	13.1, p. 188; 14.3.1, p. 247; *20.1*, p. *356*; 22.4, p. 496
	30,	K81	(1968)	*20.1*, p. *358*; 22, p. 455; 22.7, p. 546
H. J. Fischbeck	**17**,	197	(1966)	*13.1*, p. *175*; 18, p. 291; 21, p. 374
	22,	235	(1967)	*13.1*, p. *179*; 18, p. 292
	22,	649	(1967)	*13.1*, p. *180*; *18*, p. *292*
	27,	345	(1968)	13.1, p. 186; 14.1, p. 225; *19*, p. *333*; 21, p. 386
	30,	779	(1968)	*13.1*, p. *190*; 18, p. 295
R. Fischer (Karlsruhe)	**17**,	335	(1966)	*14.4.1*, p. *256*; 16, p. 277; 22.4.1, p. 498
R. Fischer (Berlin)	**19**,	757	(1967)	*20*, p. *338*
	19,	K103	(1967)	*20*, p. *338*
R. M. Fisher	**22**,	473	(1967)	*6.1*, p. *72*; 18.2, p. 308; 22.8, p. 551
A. G. Fitzgerald	**25**,	263	(1968)	*1.2*, p. *11*; *4*, p. *44*; 22, p. 449
E. R. Fitzgerald	**24**,	37	(1967)	6, p. 63; *7*, p. *77*
T. M. Fitzgerald	**27**,	473	(1968)	6, p. 66; *7*, p. *77*; 11, p. 152; 22.6, p. 541
W. F. Flanagan	**22**,	195	(1967)	*10.1*, p. *124*; 21.1, p. 398
	22,	607	(1967)	*10.1*, p. *124*; 21.1, p. 399; 21.1.1, p. 418

R. J. Fleming	**18,**	K139	(1966)	*10.2*, p. *138*; 13.4, p. 204; 22.5.3, p. 532
	27,	K57	(1968)	13.4, p. 213; *20.3*, p. *369*; 22.5.3, p. 534
	29,	K77	(1968)	*10.2*, p. *144*; 13.4, p. 215; 22.5.3, p. 534
H. Flietner	**18,**	415	(1966)	*13*, p. *164*; *20*, p. *338*; 22.1.1, p. 460
	18,	K5	(1966)	*13.3*, p. *197*; 14.3, p. 233; 22.1.2, p. 470
P. A. Flinn	**24,**	263	(1967)	1.1, p. 8; *4*, p. *43*; 21.1, p. 401; 21.6, p. 434
	26,	K51	(1968)	*6.1*, p. *73*; 21.1.1, p. 420
P. Flögel	**29,**	K89	(1968)	3.1, p. 33; *4*, p. *47*; 22.4.1, p. 505
V. Florescu	**21,**	197	(1967)	4, p. 42; *18.3*, p. *317*; 22.6, p. 537
M. P. Fontana	**29,**	159	(1968)	*13.2*, p. *195*; 20.3, p. 370; 22.5.2, p. 528
F. Forlani	**17,**	657	(1966)	*13.3*, p. *196*; *14.3.2*, p. *248*; 22.1.2, p. 469
E. Fortin	**29,**	K153	(1968)	*16*, p. *284*; 22.2.1, p. 486
C. T. Forwood	**24,**	525	(1967)	*19*, p. *331*; 21.6, p. 435
	29,	99	(1968)	*10.1*, p. *133*; 21.1, p. 407; 21.6, p. 437
A. Fourdeux	**24,**	195	(1967)	4, p. 43; *10*, p. *103*; *10.1*, p. *126*; 21.1, p. 401
W. B. Fowler	**26,**	639	(1968)	*13.4*, p. *212*
Z. Frait	**16,**	K11	(1966)	*18.2*, p. *300*; *19*, p. *326*; 21.1.1, p. 410
D. Fraitová	**23,**	501	(1967)	*13*, p. *167*; *19*, p. *330*; 21.1.1, p. 419
H. Frank	**18,**	401	(1966)	*14.3.1*, p. *245*; 14.4.1, p. 257; 22.1.1, p. 460; 22.1.2, p. 470; 22.2.1, p. 483
W. Frank	**18,**	459	(1966)	*10.1*, p. *118*; 11, p. 147; 21, p. 376; 21.1, p. 394
	19,	239	(1967)	*10.1*, p. *120*; 21.1.1, p. 415
	26,	197	(1968)	*10.1*, p. *129*; 21.1, p. 403
	26,	671	(1968)	*10.1*, p. *130*; *11*, p. *151*; 21.1, p. 404
	29,	391	(1968)	*10*, p. *110*; 22.5.2, p. 528
	29,	767	(1968)	*10*, p. *111*; 22.5.2, p. 529
E. L. Frankevich	**30,**	329	(1968)	13.2, p. 195; *20.3*, p. *371*; 22.9, p. 560
W. Franz	**19,**	597	(1967)	*13*, p. *164*; 14.3, p. 235; 22.1.1, p. 461
	22,	K139	(1967)	*14.3.2*, p. *249*; 22, p. 447
V. Frei	**22,**	381	(1967)	*13.1*, p. *180*; 22, p. 446
D. M. Freik	**28,**	K5	(1968)	*1.4*, p. *21*; *20.1*, p. *355*; 22, p. 452
R. A. French	**21,**	K35	(1967)	13.1, p. 178; *14.2*, p. *229*; 21, p. 379
R. Freud	**20,**	K151	(1967)	*14.2*, p. *229*
	25,	K127	(1968)	*14.2*, p. *230*
H. Freyhardt	**18,**	807	(1966)	*14.2*, p. *228*
A. A. Fridman	**24,**	45	(1967)	*6.1*, p. *73*; 18.2, p. 308; 21.1.1, p. 419
M. Fridman	**28,**	193	(1968)	13.1, p. 187; 13.4, p. 214; *14.3*, p. *242*; 15, p. 274; 22.6, p. 542
A. Friedman	**28,**	385	(1968)	*14.4.1*, p. *263*; 22, p. 452
H. Friedrich	**26,**	107	(1968)	*16*, p. *281*; 22.4.1, p. 502
	27,	237	(1968)	*13.4*, p. *213*; 16, p. 282; 22, p. 451
B. Fritz	**19,**	515	(1967)	*10.2*, p. *138*; 22.5.2, p. 520
	24,	207	(1967)	*6*, p. *63*; 10.2, p. 141; *20.1*, p. *351*; 22.5.2, p. 524
	29,	639	(1968)	*6*, p. *68*; 10.2, p. 144; *20.1*, p. *357*; 22.5.2, p. 529
F. Fröhlich	**24,**	535	(1967)	10, p. 103; *14.4*, p. *253*; 22.5.2, p. 524
	24,	583	(1967)	*10.2*, p. *141*; 22.5.2, p. 524
	25,	303	(1968)	*10*, p. *104*; 22.5.2, p. 525
W. Fuchs	**29,**	781	(1968)	6.1, p. 74; *18.4*, p. *325*; 22.8, p. 555
W. Fuhs	**27,**	171	(1968)	13.4, p. 213; *16*, p. *282*; 22.1.3, p. 479
M. Ya. Fuks	**17,**	543	(1966)	*1.4*, p. *17*; 3.1, p. 30; 10.1, p. 116; *18.2*, p. *303*; 21.1, p. 392; 21.1.1, p. 413
G. N. Fursey	**21,**	789	(1967)	1.3, p. 15; *13.3*, p. *197*; 17, p. 287; 22.1.1, p. 462
	22,	39	(1967)	*17*, p. *287*; 22.1.1, p. 462
K. Fussgaenger	**18,**	133	(1966)	13.4, p. 203; *20.1*, p. *346*; 22.5.2, p. 519

I. Gaal	**29,**	K163	(1968)	*4*, p. *48*; 14.1, p. 226; 21, p. 388
H. Gabriel	**23,**	195	(1967)	*6.1*, p. *72*; 13, p. 167
	27,	301	(1968)	*6.1*, p. *73*; 13, p. 170
F. B. Gadiev	**18,**	K153	(1966)	*10*, p. *97*; 14.3, p. 234; 22.1.3, p. 477
M. M. Gadzhialiev	**20,**	K119	(1967)	*15*, p. *272*; 22.2.1, p. 483
F. B. Gadzhiev	**29,**	K47	(1968)	2, p. 24; *14.3*, p. *243*; 22.1.3, p. 480
P. F. Gaidanskii	**28,**	83	(1968)	8, p. 85; 13, p. 171; *18.2*, p. *312*
R. R. Gałązka	**20,**	113	(1967)	*13.1*, p. *177*; 14.3, p. 235; 22.4.3, p. 509; 22.8, p. 549
	23,	K39	(1967)	*13.1*, p. *181*; 14.3, p. 238; 15, p. 272; 22.4.3, p. 510; 22.8, p. 552
	28,	K121	(1968)	*5*, p. *54*; *8*, p. *85*; 22.4.3, p. 511
A. A. Galkin	**26,**	K55	(1968)	*14.2*, p. *230*
	30,	K107	(1968)	*14.2*, p. *230*
H. le Gall	**17,**	K41	(1966)	*13*, p. *163*; *18.3*, p. *316*
	28,	495	(1968)	13, p. 171; 18, p. 294; *19*, p. *334*
P. C. J. Gallagher	**16,**	95	(1966)	*10.1*, p. *114*; 21.6, p. 430
J. M. Galligan	**16,**	449	(1966)	4, p. 37; *11*, p. *145*; 17, p. 286; 21, p. 374
	23,	K139	(1967)	*10.1*, p. *126*; 21.1, p. 400
	26,	345	(1968)	*10.1*, p. *129*; 21.1, p. 404
	27,	383	(1968)	10.1, p. 131; *11*, p. *152*; 21.6, p. 436
E. E. Galloni	**29,**	K91	(1968)	*1.4*, p. *21*; 21.1, p. 407
J. A. Galloway	**22,**	491	(1967)	*13.1*, p. *180*; 22.1.1, p. 462
	23,	97	(1967)	*13.1*, p. *180*; 22.1.1, p. 463
G. Gallus	**16,**	277	(1966)	*13.2*, p. *191*; 20.3, p. 363; 22.9, p. 558
H. Gamari-Seale	**29,**	323	(1968)	*5*, p. *54*; 18.3, p. 320
F. A. Gambarov	**26,**	K113	(1968)	*2*, p. *24*; 10, p. 106; *12*, p. *158*; 22.1.3, p. 479
B. Garben	**29,**	K27	(1968)	13, p. 172; *18*, p. *294*; *20*, p. *342*; 22.1.3, p. 480
T. I. Garber	**30,**	49	(1968)	13.1, p. 189; *18.4*, p. *325*; 20.1, p. 358; 23, p. 561
A. B. Gardner	**30,**	93	(1968)	*9*, p. *92*; 10.1, p. 134; 21, p. 388; 21.6, p. 437
G. F. J. Garlick	**30,**	97	(1968)	13.4, p. 216; *20.1*, p. *358*; *20.3*, p. *371*; 22.5.2, p. 530
	30,	113	(1968)	*13.4*, p. *216*; 20.1, p. 358; 20.3, p. 371; 22.8, p. 556
	30,	507	(1968)	*13.4*, p. *217*; 20.1, p. 359; 20.3, p. 371; 22.8, p. 557
	30,	735	(1968)	*13.4*, p. *217*; *20.1*, p. *359*; 22.8, p. 557
H. Gärtner	**17,**	K241	(1966)	1.2, p. 9; *1.4*, p. *17*; 21.1, p. 393
N. A. Gasanova	**17,**	K115	(1966)	*20.1*, p. *345*; 22, p. 440
	17,	K131	(1966)	*20.1*, p. *345*; 22.8, p. 547
F. M. Gashimzade	**18,**	667	(1966)	6, p. 58; 13.1, p. 176; *14.3*, p. *233*; 22.2, p. 481; 22.2.3, p. 490
	21,	K155	(1967)	*13.1*, p. *179*; 14.3, p. 237; 15, p. 272; 22, p. 446
	28,	783	(1968)	13.1, p. 187; *14.3*, p. *243*; 22, p. 453
P. Gassenmeier	**30,**	833	(1968)	*10.1*, p. *136*; 21, p. 389; 21.1, p. 409
H. C. Gatos	**30,**	613	(1968)	*14.3*, p. *244*; 22.2.3, p. 493
R. Gauthier	**25,**	691	(1968)	*17*, p. *288*; 21.1, p. 403; 21.1.1, p. 420; 22.1.1, p. 464; 22.2.1, p. 485; 22.4.1, p. 502; 22.5.2, p. 525; 22.6, p. 540
E. I. Gavrilitsa	**19,**	609	(1967)	*1.2*, p. *10*; 8, p. 81; 14.3, p. 235; 22.8, p. 549
V. G. Gavrilko	**17,**	K139	(1966)	*8*, p. *79*; 23, p. 561
R. K. L. Gay	**30,**	755	(1968)	7, p. 78; *14.4.1*, p. *265*; 22.4.1, p. 505
W. Gebhardt	**27,**	713	(1968)	*10.2*, p. *143*; 13.4, p. 214; 22.5.2, p. 526
W. G. Gehman	**30,**	283	(1968)	3.2, p. 36; *10.1*, p. *135*; 21.1, p. 408
R. Geick	**24,**	99	(1967)	*6*, p. *63*; 20.1, p. 351; 22.1.3, p. 478
J. Geiger	**24,**	457	(1967)	*13*, p. *168*; 20, p. 340; 21, p. 383
D. Geist	**29,**	167	(1968)	13.4, p. 215; *14.3*, p. *243*; 19, p. 335; 22.1.1, p. 466
A. Gemperle	**19,**	333	(1967)	4, p. 40; *5*, p. *50*; 21.1.1, p. 415

A. GEMPERLE	**26,** 207 (1968)	1.4, p. 20; *18.2*, p. *310*; 21.1, p. 404
	30, 541 (1968)	3, p. 28; *4*, p. *48*; 21.1.1, p. 423
R. GEMPERLE	**26,** 207 (1968)	1.4, p. 20; *18.2*, p. *310*; 21.1, p. 404
J. GEMPERLOVÁ	**30,** 261 (1968)	*10.1*, p. *135*; 21, p. 388; 21.1, p. 408; 21.1.1, p. 423; 21.2, p. 425
H. GENGNAGEL	**29,** 91 (1968)	*18.2*, p. *314*; 21.1.1, p. 422
L. GENZEL	**23,** 613 (1967)	*6*, p. *63*; *20.1*, p. *350*; 22.5.2, p. 524
A. M. GEORGE	**19,** 117 (1967)	14.3, p. 234; *15*, p. *271*; 22.6, p. 536
W. R. P. GEORGE	**21,** 205 (1967)	3.1, p. 31; 3.2, p. 35; *14.3*, p. *236*; 21.7, p. 438; 22.7, p. 545
L. GEORGESCU	**27,** 125 (1968)	*15*, p. *274*; 22, p. 451
M. GEORGIEV	**27,** K33 (1968)	*10.2*, p. *142*; 22.5.1, p. 516
	28, 725 (1968)	*10.2*, p. *143*; 16, p. 283; 22.5.1, p. 517
A. GEORGIEVA	**22,** 415 (1967)	3.1, p. 32; *4*, p. *42*; 22.9, p. 559
A. N. GEORGOBIANI	**21,** K45 (1967)	14.3.2, p. 249; *20.3*, p. *365*; 22, p. 445; 22.4.1, p. 499
	21, K49 (1967)	*14.3.2*, p. *249*; 16, p. 279; 20.3, p. 365; 22.4.1, p. 499
R. GERBER	**16,** 795 (1966)	*18.3*, p. *316*
A. N. GERGELL	**26,** K77 (1968)	14.3.2, p. 250; *20.3*, p. *368*; 22, p. 450; 22.4.1, p. 502
U. GERHARDT	**18,** K45 (1966)	*13.2*, p. *192*; 20.1, p. 346; 22.5.2, p. 519
W. GERICKE	**30,** K61 (1968)	13.3, p. 199; *14.3.2*, p. *251*; 14.4.1, p. 264; 22.4.3, p. 511
H. GERMER	**29,** K65 (1968)	*16*,' p. *284*; 22.4.1, p. 504; 22.4.2, p. 508; 22.8, p. 555
V. GEROLD	**16,** 675 (1966)	4, p. 38; *10.1*, p. *114*; 21.1, p. 391
L. GERWARD	**22,** 659 (1967)	10.1, p. 124; *18.2*, p. *308*; 21.1, p. 399
H. P. GESERICH	**25,** 741 (1968)	*13.1*, p. *184*; 14.3.1, p. 246; 20.1, p. 352; 22.6, p. 540
T. GESZTI	**20,** 165 (1967)	9, p. 88; *10*, p. *98*; 12, p. 157; 14.4, p. 252
G. GETOV	**21,** K87 (1967)	2, p. 23; 13.4, p. 207; *20.1*, p. *348*; 22, p. 445
R. GEVERS	**17,** 221 (1966)	*10*, p. *94*; 22.1, p. 456
	17, K173 (1966)	*10*, p. *96*; 22.8, p. 548
	18, 167 (1966)	*10.1*, p. *117*; 21.1.1, p. 413
	18, 325 (1966)	*4*, p. *39*; 22.6, p. 536
	18, 343 (1966)	4, p. 39; *10*, p. *96*; *10.1*, p. *118*
	18, 363 (1966)	4, p. 39; *10*, p. *96*; *10.1*, p. *118*; 21.1.1, p. 414; 22.6, p. 536
	19, 77 (1967)	*8*, p. *80*; 18.4, p. 322; 22.6, p. 536
	20, 613 (1967)	*4*, p. *41*; 14.4.2, p. 266; 22.8, p. 549
	21, 393 (1967)	*10*, p. *99*; *10.1*, p. *122*; 21.1.1, p. 417; 22.6, p. 538
	23, 549 (1967)	*4*, p. *43*; *10.1*, p. *125*; 21.1.1, p. 419
	24, 195 (1967)	4, p. 43; *10*, p. *103*; *10.1*, p. *126*; 21.1, p. 401
	26, 577 (1968)	*4*, p. *46*; 10, p. 105
	30, 797 (1968)	*4*, p. *48*; 10, p. 113; 10.1, p. 136; 21.1.1, p. 423; 22.6, p. 544
P. B. GHATE	**18,** 265 (1966)	3.2, p. 34; *12*, p. *156*; 22.5.2, p. 519
	21, 507 (1967)	7, p. 76; *12*, p. *157*; 22.5.2, p. 522
V. GHEORDĂNESCU	**18,** 511 (1966)	10.2, p. 138; *17*, p. *287*; 22.5.2, p. 520
L. GHIŢĂ	**24,** K175 (1967)	13.2, p. 193; *20.1*, p. *351*; 22.5, p. 513
R. GHOSH	**25,** K71 (1968)	10, p. 104; *10.2*, p. *142*; 19, p. 332; 22.5.3, p. 533
R. N. GHOSHTAGORE	**19,** 123 (1967)	*9*, p. *88*
	20, K89 (1967)	*9*, p. *89*; 10, p. 99; 22.1.2, p. 471
G. B. GIBBS	**16,** K27 (1966)	*9*, p. *87*; 10, p. 94

G. B. Gibbs	**17**,	K191	(1966)	3.1, p. 30; *10*, p. *96*
	18,	85	(1966)	*9*, p. *87*; 21, p. 375
J. H. Gieske	**29**,	121	(1968)	*12*, p. *158*; 22.6, p. 542
Sh. N. Gifeisman	**28**,	45	(1968)	6, p. 66; *10.2*, p. *143*; 13, p. 171; 14.4, p. 254; 18, p. 294; 22.2.1, p. 485; 22.2.2, p. 488; 22.2.3, p. 493; 22.4, p. 495; 22.4.1, p. 503; 22.4.2, p. 508; 22.4.3, p. 511; 22.5.1, p. 517; 22.5.2, p. 527
R. L. Gilbert	**30**,	173	(1968)	*10.2*, p. *144*; 13.4, p. 216; 20.3, p. 371; 22.5.2, p. 530
W. D. Gill	**18**,	601	(1966)	13.4, p. 204; 14.3, p. 233; *16*, p. *278*; 22, p. 441
J. Ginter	**27**,	K103	(1968)	13.4, p. 214; *14.3*, p. *242*; 22.4.3, p. 510
M. Girard-Nottin	**17**,	881	(1966)	*10*, p. *95*; 20, p. 338; 22.5.2, p. 519
W. Gissler	**19**,	721	(1967)	*6*, p. *59*; 22.1.3, p. 477
	23,	K155	(1967)	*6*, p. *63*; 23, p. 561
G. Giuliani	**25**,	437	(1968)	*10.2*, p. *141*; *11*, p. *151*; 22.5.2, p. 525
F. Giusiano	**29**,	341	(1968)	*1.2*, p. *12*; 3, p. 27; 13.1, p. 188; 17, p. 289; 22.7, p. 546
N. A. Gjostein	**25**,	209	(1968)	*9*, p. *91*; 21.1, p. 402
I. Gladkih	**27**,	131	(1968)	*6.1*, p. *73*; 18.3, p. 319
W. Glaeser	**24**,	K5	(1967)	*14.1*, p. *224*; 18.2, p. 309; 21.1.1, p. 420
H. J. Gläser	**29**,	167	(1968)	13.4, p. 215; *14.3*, p. *243*; 19, p. 335; 22.1.1, p. 466
A. Glasner	**18**,	593	(1966)	*10.2*, p. *138*; 22.5.2, p. 520
	24,	695	(1967)	*9*, p. *90*; 22.5.2, p. 525
A. A. Glazer	**16**,	745	(1966)	4, p. 38; *18.2*, p. *301*; 21.1.1, p. 411
V. M. Glazov	**18**,	K29	(1966)	2, p. 22; *15*, p. *271*; 22.8, p. 548
R. Gleiss	**17**,	761	(1966)	*6*, p. *57*; 8, p. 79; 21.1, p. 393
A. Glodeanu	**19**,	K43	(1967)	*13.4*, p. *205*; 22.1.1, p. 461; 22.1.2, p. 470; 22.2.1, p. 483
P. Gobin	**24**,	K167	(1967)	*10.1*, p. *127*; 21, p. 383
S. Goderska	**23**,	419	(1967)	*14.4.1*, p. *259*; 22.9, p. 559
E. E. Godik	**30**,	K127	(1968)	*16*, p. *285*; 22.1.1, p. 467; 22.1.2, p. 475
K. Godwod	**17**,	869	(1966)	*4*, p. *38*; 20, p. 338; 22.1.2, p. 469
	20,	K77	(1967)	*1.4*, p. *18*; 22.1.1, p. 461; 22.2.1, p. 483
O. Goede	**17**,	911	(1966)	6, p. 57; 13.4, p. 203; *20.3*, p. *364*; 22.4.1, p. 498
	28,	K167	(1968)	10, p. 109; *20.3*, p. *370*; 22.4.1, p. 504
A. M. de Goer	**22**,	77	(1967)	*8*, p. *83*; 10, p. 101; *11*, p. *149*; 22.6, p. 538
Z. Gołacki	**20**,	K1	(1967)	13.4, p. 205; *16*, p. *279*; 22.1.1, p. 461
Yu. A. Goldberg	**29**,	K103	(1968)	*14.3.2*, p. *251*; *14.4.1*, p. *264*; 22.2.1, p. 486
F. T. Goldstein	**20**,	379	(1967)	*10.2*, p. *139*; 20.3, p. 365; 22.5.2, p. 521
M. Golo	**17**,	K187	(1966)	*17*, p. *286*; 21.1, p. 393
U. Gonser	**21**,	331	(1967)	*6.1*, p. *72*; 13, p. 165; 18.1, p. 296; 18.4, p. 322; 22.8, p. 550
T. L. Goodfellow	**22**,	549	(1967)	10, p. 101; 13.4, p. 209; *14.4*, p. *253*; *20.1*, p. *350*; 22.5.3, p. 533
C. H. L. Goodman	**21**,	205	(1967)	3.1, p. 31; 3.2, p. 35; *14.3*, p. *236*; 21.7, p. 438; 22.7, p. 545
A. S. van der Goot	**29**,	825	(1968)	*18.3*, p. *320*; 21.1, p. 407; 21.4, p. 428
H. Goretzki	**20**,	K141	(1967)	*5*, p. *51*; 21, p. 378
P. Görlich	**19**,	K23	(1967)	*10.2*, p. *138*; 22.5.3, p. 532
	23,	313	(1967)	13.4, p. 209; 20.1, p. 350; *20.3*, p. *367*; 22.5.3, p. 533
	25,	93	(1968)	*10.2*, p. *141*; 22.5.3, p. 533
	25,	K15	(1968)	6, p. 65; *10.2*, p. *141*; *20.3*, p. *368*; 22.5.3, p. 533
	27,	109	(1968)	10, p. 106; *20.3*, p. *369*; 22.5.3, p. 534
	28,	K117	(1968)	*17*, p. *289*; 21, p. 387

V. V. Gorskii	28,	K15	(1968)	13.1, p. 187; *20*, p. *341*; 21, p. 387; 21.3, p. 427
	29,	45	(1968)	*6.1*, p. *74*; 13, p. 172; 21.1.1, p. 422
N. A. Goryunova	21,	9	(1967)	3.1, p. 29; 3.2, p. 34; 5, p. 49; 8, p. 79; 10, p. 93; 13.1, p. 174; *14.3*, p. *231*; 15, p. 270; 16, p. 276; 17, p. 286; 18, 291; 20.1, p. 344; 22.8, p. 547
	25,	513	(1968)	*3.1*, p. *33*; 5, p. 53; 14.3, p. 240; 15, p. 273; 20.1, p. 351; 22, p. 450
	28,	489	(1968)	2, p. 24; 14.3, p. 242; *20.3*, p. *370*; 22.8, p. 554
	29,	435	(1968)	4, p. 47; *5*, p. *54*; 6, p. 68; 10, p. 110; 14.3, p. 243; 22.8, p. 555
H. Götz	24,	K95	(1967)	*19*, p. *331*; 22.9, p. 560
L. Gouskov	16,	721	(1966)	*13.4*, p. *202*; 14.3, p. 232; 16, p. 276; 22.1.1, p. 459
	21,	619	(1967)	13.4, p. 207; *14.3*, p. *236*; *16*, p. *279*; 22.1.1, p. 462
	30,	9	(1968)	*14.3.2*, p. *251*; 22.6, p. 543
D. Gouvernelle	22,	K135	(1967)	*16*, p. *280*; 22.4.1, p. 500
J. Grabmaier	25,	K7	(1968)	*10*, p. *104*; 22.6, p. 540
U. Gradmann	27,	313	(1968)	*1.4*, p. *20*; *18.2*, p. *311*; 21.1, p. 405; 21.1.1, p. 421
W. Gräfe	18,	K5	(1966)	*13.3*, p. *197*; 14.3, p. 233; 22.1.2, p. 470
L. J. Graham	18,	99	(1966)	*10.1*, p. *117*; 21.1.1, p. 413
A. Graja	27,	K93	(1968)	*20*, p. *341*; *20.2*, p. *362*; 22.8, p. 554
R. Granger	16,	599	(1966)	6, p. 56; 13, p. 163; *14.4.1*, p. *255*; 22, p. 440; 22.2.3, p. 489
R. W. Grant	21,	331	(1967)	*6.1*, p. *72*; 13, p. 165; 18.1, p. 296; 18.4, p. 322; 22.8, p. 550
E. B. Granovskii	24,	45	(1967)	*6.1*, p. *73*; 18.2, p. 308; 21.1.1, p. 419
F. Granzer	30,	587	(1968)	*10*, p. *112*; 22.5.2, p. 531
R. J. Grasser	28,	583	(1968)	10.2, p. 143; *11*, p. *153*; 19, p. 334; 20.3, p. 370; 22.8, p. 554
V. Grasso	28,	K109	(1968)	13.1, p. 188; *14.4.1*, p. *264*; *20*, p. *341*; 20.1, p. 356; 22.1.1, p. 466
P. W. Graves	22,	499	(1967)	*3.1*, p. *32*; 22.6, p. 539; 22.8, p. 551
I. P. Grebennik	20,	213	(1967)	*9*, p. *88*; 21.6, p. 432; 22, p. 443; 22.1.3, p. 477
A. K. Green	18,	441	(1966)	*1.4*, p. *17*; *3.1*, p. *30*; 21, p. 376; 21.6, p. 431
B. A. Greenberg	17,	673	(1966)	10, p. 95; *19*, p. *327*; 22, p. 441
	20,	K103	(1967)	4, p. 42; 10, p. 99; *19*, p. *328*
F. C. Greisen	25,	753	(1968)	*13.1*, p. *184*; 21, p. 384; 21.1, p. 403
R. M. Gretchishkin	19,	K1	(1967)	4, p. 40; *18.2*, p. *305*; 21.1, p. 395; 21.6, p. 432
R. D. Gretz	23,	453	(1967)	*3.1*, p. *32*; 21.1, p. 400
Z. S. Gribnikov	28,	815	(1968)	*14.4.1*, p. *263*; 22.1.1, p. 466
D. T. Griggs	19,	631	(1967)	4, p. 40; *10*, p. *98*; 22.6, p. 536
R. V. Grigoriev	28,	K85	(1968)	13.2, p. 194; 13.3, p. 198; *16*, p. *283*; 22.4.1, p. 504
R. Grigorovici	16,	K143	(1966)	2, p. 22; 14.3.1, p. 245; *15*, p. *270*; 22.1.1, p. 459
	23,	621	(1967)	2, p. 23; 13.1, p. 182; 14.3.1, p. 246; *15*, p. *273*; 22.1.2, p. 472
	23,	627	(1967)	2, p. 23; *16*, p. *280*; 22.1.1, p. 463
	24,	K17	(1967)	*14.3.2*, p. *250*; *14.4.1*, p. *261*; 22.1.1, p. 463
N. A. Grigoryan	20,	745	(1967)	*1*, p. *7*; 1.2, p. 10; 18.2, p. 306; 18.3, p. 317; 18.4, p. 322; 22.8, p. 549
	28,	835	(1968)	5, p. 54; *18.2*, p. *313*; *18.4*, p. *324*; 22.8, p. 554
H. G̃. Grimmeiss	28,	547	(1968)	*16*, p. *283*; 22.2.2, p. 488
G. Grimwall	19,	863	(1967)	*6*, p. *59*; 21.1, p. 424
A. A. Grinberg	26,	219	(1968)	*7*, p. *77*; *14.4.1*, p. *262*; 22.2.1, p. 485
R. G. J. Grisar	23,	613	(1967)	*6*, p. *63*; *20.1*, p. *350*; 22.5.2, p. 524

P. Groh	**28**, 475 (1968)	*10.1*, p. *132*; 21.1.1, p. 422
E. F. Gross	**24**, K107 (1967)	*10*, p. *103*; *13.2*, p. *193*; 20.1, p. 351; 22.4.1, p. 501
	26, K139 (1968)	6, p. 65; *13.2*, p. *194*; *20.2*, p. *361*; 22.4.2, p. 507
	30, 485 (1968)	*13.2*, p. *195*; 20.1, p. 358; 20.3, p. 371; 22.4.2, p. 508
P. Grosse	**24**, K33 (1967)	*13.1*, p. *183*; 20.1, p. 351; 22.1.3, p. 478
	26, 139 (1968)	*13.1*, p. *184*; 18, p. 293; *20.1*, p. *352*; 22.1.3, p. 478
	27, K149 (1968)	*14.3*, p. *242*; 22.1.3, p. 479
	28, K93 (1968)	*3*, p. *27*; 5, p. 54; 8, p. 85; 22.1.3, p. 479
J. B. Grun	**22**, K155 (1967)	6, p. 62; 13.2, p. 193; *20.3*, p. *367*; 22.4.2, p. 507
G. Grüner	**22**, K11 (1967)	18.1, p. 296; 18.4, p. 322; *19*, p. *329*; 22.6, p. 538
O. Grüter	**19**, 217 (1967)	4, p. 40; *18.2*, p. *304*; 21.1, p. 395
M. Grynberg	**20**, K107 (1967)	10, p. 99; *18*, p. *292*; *20*, p. *339*; 22.1.1, p. 461
	27, 255 (1968)	*13.1*, p. *186*; 20.1, p. 354; 20.3, p. 369; 22.4.2, p. 507
A. I. Gubanov	**17**, 807 (1966)	*2*, p. *22*; *13*, p. *163*; 21, p. 375; 22, p. 441
	17, 815 (1966)	*3.2*, p. *34*; 5, p. 50; 12.1, p. 159; 21.1, p. 393; 21.6, p. 431
B. von Guérard	**29**, K59 (1968)	*10*, p. *110*; 22.5.2, p. 529
P. Guetin	**24**, K29 (1967)	*13.1*, p. *183*; 22.2.1, p. 484
J. P. Guigay	**29**, 799 (1968)	*4*, p. *47*; 18.2, p. 314
C. Benoit à la Guillaume see under letter "B" (Benoit)		
F. Guiu	**19**, 339 (1967)	*10.1*, p. *121*; 21, p. 377
	25, 189 (1968)	*10.1*, p. *127*; 21, p. 384
	25, 203 (1968)	*10.1*, p. *128*; 21, p. 384
S. E. Gukasyan	**29**, 49 (1968)	*6.1*, p. *74*; 22.9, p. 560
H.-E. Gumlich	**17**, 783 (1966)	13.4, p. 203; *20.3*, p. *364*; 22.4.1, p. 498
	24, K13 (1967)	13.4, p. 210; *20.3*, p. *367*; 22.4.1, p. 501
K. H. Gundlach	**21**, 575 (1967)	13.1, p. 178; *14.3.1*, p. *245*; 14.3.2, p. 249; 22.6, p. 538
	30, K135 (1968)	13.1, p. 190; 14.3.1, p. 247; *14.3.2*, p. *251*; 22, p. 455
A. K. Gupta	**16**, K51 (1966)	*8*, p. *79*; *14.2*, p. *228*; 21, p. 373
N. P. Gupta	**18**, 731 (1966)	3.2, p. 34; *6*, p. *58*; *8*, p. *80*; 23, p. 561
	20, 321 (1967)	*8*, p. *81*; 23, p. 561
	21, 661 (1967)	6, p. 61; *8*, p. *82*; *12*, p. *157*; 23, p. 561
	30, K41 (1968)	*6*, p. *69*; 23, p. 562
R. K. Gupta	**30**, K41 (1968)	*6*, p. *69*; 23, p. 562
R. P. Gupta	**18**, 119 (1966)	6, p. 57; *8*, p. *80*; 21.1, p. 393
	20, 291 (1967)	6, p. 59; *8*, p. *81*; 21, p. 377
A. G. Gurevich	**17**, K7 (1966)	18.3, p. 316; *19*, p. *326*
	27, K85 (1968)	7, p. 78; *18.3*, p. *320*
G. J. van Gurp	**17**, K135 (1966)	4, p. 38; *14.2*, p. *228*
	19, 173 (1967)	*14.2*, p. *228*; 21.1, p. 395
G. D. Guseinov	**18**, 911 (1966)	6, p. 58; *8*, p. *80*; 22, p. 442
	18, 929 (1966)	*5*, p. *50*; 8, p. 80; 13.4, p. 204; 14.3, p. 234; 16, p. 278; 22.8, p. 548
	19, K7 (1967)	*8*, p. *81*; 22, p. 442
	22, K117 (1967)	*14.3*, p. *237*; 22.8, p. 551
	23, 461 (1967)	*10*, p. *102*; 22, p. 448
E. S. Guseinova	**16**, 205 (1966)	13.4, p. 201; *14.4.1*, p. *255*; 22, p. 439
	17, 593 (1966)	13.1, p. 175; *14.4.1*, p. *256*; 22, p. 440
	20, 421 (1967)	*14.3*, p. *235*; *14.4.1*, p. *258*; 22, p. 444
G. I. Guseva	**25**, 775 (1968)	14.1, p. 224; 14.3, p. 240; *15*, p. *273*; 21.7, p. 438; 22.2, p. 481
W. Güth	**17**, 691 (1966)	13.1, p. 175; *13.4*, p. *202*; 22.1, p. 456
	23, K13 (1967)	13, p. 167; *14.4.1*, p. *260*; 16, p. 280; 22.1.1, p. 463

A. A. Gutkin	**30**,	827	(1968)	*14.3.2*, p. *251*; *14.4.1*, p. *265*; 16, p. 285; 22.1.2, p. 475
E. Yu. Gutmanas	**23**,	571	(1967)	*10*, p. *102*; 22.5.2, p. 524
E. Gutsche	**17**,	911	(1966)	6, p. 57; 13.4, p. 203; *20.3*, p. *364*; 22.4.1, p. 498
	17,	K11	(1966)	13.3, p. 196; *16*, p. *277*; 22.4.1, p. 498
	19,	823	(1967)	*13.1*, p. *177*; 22.4, p. 495
	21,	57	(1967)	*13.1*, p. *178*; 22.4, p. 495; 22.4.1, p. 499; 22.4.2, p. 506
	22,	229	(1967)	13.1, p. 179; 13.2, p. 192; *14.4.1*, p. *259*; *20*, p. *339*; 20.1, p. 349; 22.4.1, p. 500; 22.4.2, p. 506
I. Gutzow	**21**,	683	(1967)	*3.1*, p. *31*
	24,	349	(1967)	1.2, p. 11; *3.1*, p. *32*; 22, p. 448
P. Guyon	**16**,	525	(1966)	*14.4.2*, p. *266*
P. Guyot	**19**,	K95	(1967)	*10.1*, p. *121*; 21.1.1, p. 415
	28,	349	(1968)	*1.4*, p. *20*; 4, p. 46
M. Guzzi	**30**,	601	(1968)	*6*, p. *70*; 10, p. 112; 20.1, p. 359; 20.3, p. 372; 22.5.2, p. 531
J. Gyulai	**17**,	K49	(1966)	13.1, p. 175; *16*, p. *277*; 22.2.2, p. 487
	29,	K85	(1968)	13.4, p. 216; *14.4.1*, p. *264*; 16, p. 284; 22.2.2, p. 488
D. Haarer	**23**,	633	(1967)	*13.2*, p. *193*; *19*, p. *330*; 22.9, p. 559
P. Haasen	**18**,	807	(1966)	*14.2*, p. *228*
F. Haberey	**26**,	231	(1968)	*14.4*, p. *253*; 18.3, p. 319
H. Haberkorn	**16**,	675	(1966)	4, p. 38; *10.1*, p. *114*; 21.1, p. 391
K. Hacker	**26**,	551	(1968)	*13.1*, p. *185*; 13.4, p. 212; 14.4.1, p. 262; 20.1, p. 353; 22, p. 451
V. Hádek	**30**,	275	(1968)	*13.4*, p. *216*; *14.3*, p. *244*; 15, p. 275; 22.9, p. 560
A. Hadni	**26**,	241	(1968)	6, p. 65; *20.1*, p. *352*; 22.4.2, p. 507; 22.4.3, p. 510
P. Haen	**29**,	743	(1968)	6, p. 68; *8*, p. *86*; 13, p. 173; *15*, p. *274*; 22, p. 454; 22.8, p. 555
M. Hagino	**17**,	609	(1966)	10, p. 95; 14.3.1, p. 245; 16, p. 277; *17*, p. *286*; 22.7, p. 545
W. E. Hagston	**30**,	97	(1968)	13.4, p. 216; *20.1*, p. *358*; *20.3*, p. *371*; 22.5.2, p. 530
	30,	113	(1968)	*13.4*, p. *216*; 20.1, p. 358; 20.3, p. 371; 22.8, p. 556
	30,	507	(1968)	*13.4*, p. *217*; 20.1, p. 359; 20.3, p. 371; 22.8, p. 557
	30,	735	(1968)	*13.4*, p. *217*; *20.1*, p. *359*; 22.8, p. 557
D. Hahn	**27**,	K133	(1968)	*20.3*, p. *370*; 22.4.1, p. 503
W. Haist	**17**,	691	(1966)	13.1, p. 175; *13.4*, p. *202*; 22.1, p. 456
W. F. Hall	**20**,	K11	(1967)	*6*, p. *60*; 22.5.2, p. 521
C. Hamann	**20**,	481	(1967)	13.4, p. 205; *14.3*, p. *236*; 14.4.1, p. 258; 15, p. 271; 22.9, p. 559
	26,	311	(1968)	*13.4*, p. *212*; 14.3.1, p. 246; 14.4.1, p. 262; 22.9, p. 560
C. Hammar	**24**,	531	(1967)	*13*, p. *168*; *14.3*, p. *239*; 22.2.3, p. 492
M. L. Hammond	**17**,	K25	(1966)	1.2, p. 9; *3*, p. *26*; 4, p. 38; 22.1, p. 456
	19,	K99	(1967)	1.2, p. 10; *3*, p. *26*; 4, p. 40; 22.1, p. 457
P. H. Handel	**28**,	699	(1968)	*13.3*, p. *199*; 14.4.1, p. 263; 22, p. 453
	29,	299	(1968)	*14.4.1*, p. *264*; 22, p. 454
J. Haňderek	**21**,	323	(1967)	*14.4*, p. *253*; *14.4.2*, p. *267*; 22.8, p. 550
H.-J. Hänsch	**28**,	255	(1968)	13.2, p. 194; 14.4.1, p. 263; *20.3*, p. *370*; 22.4.1, p. 503

H. Happ	**16,** 711 (1966)	14.4.2, p. 266; *20.1,* p. *344*
	30, 67 (1968)	*6,* p. *69*; 14.4.2, p. 269; *20.1,* p. *358*
S. Haragea	**16,** 137 (1966)	*10.2,* p. *137*; 22.5.2, p. 518
G. Harbeke	**27,** 9 (1968)	*13.1,* p. *174*; 13.2, p. 191; 14.3, p. 231; 20.1, p. 344; 22, p. 439; 22.1, p. 456; 22.1.1, p. 458; 22.1.2, p. 468; 22.1.3, p. 476; 22.4.1, p. 497; 22.4.2, p. 506; 22.4.3, p. 509; 22.5, p. 512; 22.7, p. 545; 22.8, p. 547
O. Härlin	**25,** 359 (1968)	9, p. 91; *10.1,* p. *128*; 21.6, p. 435
P. G. Harper	**26,** 69 (1968)	*13.2,* p. *193*; 20.1, p. 352; 22, p. 450
B. Harris	**18,** 715 (1966)	*10.1,* p. *119*; 21, p. 376
	29, 383 (1968)	*10.1,* p. *133*; 21, p. 387
R. P. Harrison	**19,** 353 (1967)	*10,* p. *97*; 12, p. 156; 22.5.2, p. 520
	26, K65 (1968)	*4,* p. *45*; 10, p. 105; 12, p. 158; 22.5.2, p. 526
	27, K153 (1968)	*10,* p. *108*; 22.5.2, p. 527
M. Hart	**26,** 185 (1968)	3, p. 27; *5,* p. *53*; 22.1.2, p. 473
E. Hartmann	**29,** K11 (1968)	*14.4.2,* p. *269*; 19, p. 335
H. L. Hartnagel	**24,** K133 (1967)	*7,* p. *77*; *14.4.1,* p. *261*; 22, p. 449
	30, 755 (1968)	7, p. 78; *14.4.1,* p. *265*; 22.4.1, p. 505
H. W. den Hartog	**22,** 131 (1967)	*10.2,* p. *139*; 13.4, p. 208; 19, p. 329; 22.5.3, p. 533
	23, 713 (1967)	10.2, p. 140; *13.4,* p. *210*; 19, p. 330; 22.5.3, p. 533
	24, K117 (1967)	10.2, p. 141; *13.4,* p. *211*; 19, p. 331; 22.5.3, p. 533
	25, K111 (1968)	*10.2,* p. *142*; 13.4, p. 211; 19, p. 333; 22.5.3, p. 533
W. Hase	**18,** K41 (1966)	*6.1,* p. *71*; 22.6, p. 536
	21, K11 (1967)	*5,* p. *51*; 22, p. 445
J. Hatz	**28,** 233 (1968)	10, p. 109; 14.3.2, p. 251; *20.2,* p. *362*; 22.2.1, p. 485
W. Haubenreisser	**16,** 577 (1966)	18.1, p. 296; *18.2,* p. *300*; 21.1.1, p. 411
	16, K147 (1966)	*13,* p. *163*; *18,* p. *291*
	20, K131 (1967)	*13.4,* p. *206*; 18.1, p. 296; 21, p. 378
	26, 191 (1968)	6.1, p. 73; *18.2,* p. *310*
	27, K145 (1968)	6, p. 66; 13, p. 170; *18.3,* p. *320*
A. Haug	**30,** 619 (1968)	*13.2,* p. *195*; 20.1, p. 359; 22.4.1, p. 505
G. Hauret	**17,** 881 (1966)	*10,* p. *95*; 20, p. 338; 22.5.2, p. 519
O. Hauser	**18,** 547 (1966)	*1.2,* p. *10*; 8, p. 80; 11, p. 147; 14.4.2, p. 266
F. Häussermann	**18,** 609 (1966)	*10.1,* p. *119*; 21.1, p. 394; 21.6, p. 431
	22, 689 (1967)	*10.1,* p. *124*; 21.1, p. 399
S. Haussühl	**28,** 127 (1968)	3.2, p. 36; *12,* p. *158*; 22.5, p. 514; 22.5.3, p. 534
	29, K159 (1968)	*12,* p. *158*; *14.3,* p. *244*; 22.8, p. 556
	30, 209 (1968)	14.4, p. 254; *20,* p. *342*; 22.5.2, p. 530
Y. Haven	**26,** 653 (1968)	*10.1,* p. *130*; *12.1,* p. *160*; 21, p. 385
A. K. Head	**19,** 185 (1967)	*10.1,* p. *120*; 21, p. 376; 21.1, p. 395; 21.1.1, p. 415
	20, 505 (1967)	*10.1,* p. *122*; 12.1, p. 159; 21.1, p. 397
	20, 521 (1967)	*10.1,* p. *122*; 12.1, p. 159; 21.1, p. 397
W. Heering	**27,** K69 (1968)	*15,* p. *274*; 16, p. 282; 22.4.1, p. 502
D. Heese	**24,** 281 (1967)	6, p. 64; *13,* p. *168*; 14.4.2, p. 267; 20.1, p. 351
B. Heimann	**24,** 651 (1967)	*17,* p. *288*; 18.1, p. 297; 18.2, p. 309; 21.1, p. 402
U. Heinecke	**18,** 569 (1966)	*18.3,* p. *317*
	30, 551 (1968)	*18.3,* p. *320*
B. Heinrich	**16,** K11 (1966)	*18.2,* p. *300*; *19,* p. *326*; 21.1.1, p. 410
	23, 501 (1967)	*13,* p. *167*; *19,* p. *330*; 21.1.1, p. 419
	26, 443 (1968)	*18.1,* p. *297*; 18.4, p. 324; 22.8, p. 553
H. Heinrich	**28,** K51 (1968)	13, p. 171; *14.4.1,* p. *263*; 22.1.1, p. 465
D. Heitkamp	**24,** 341 (1967)	*9,* p. *90*; 10.1, p. 126; 21, p. 382

B. Hejda	24,	659	(1967)	*13.1*, p. *183*
G. Heldmann	21,	575	(1967)	13.1, p. 178; *14.3.1*, p. *245*; 14.3.2, p. 249; 22.6, p. 538
W. Hellenthal	16,	789	(1966)	*10.1*, p. *114*; 14.1, p. 220; 21.1, p. 391
H. Hemmerich	23,	481	(1967)	10.1, p. 125; *11*, p. *150*; 14.1, p. 223; 21.6, p. 434
L. L. Hench	20,	327	(1967)	2, p. 23; *14.3*, p. *235*; 22.6, p. 537; 22.8, p. 549
B. Henderson	26,	K147	(1968)	*10.2*, p. *142*; 11, p. 152; 22.6, p. 541
O. Henkel	17,	K99	(1966)	*18.2*, p. *302*; 18.3, p. 316; 21.1.1, p. 413
	18,	K113	(1966)	*18.2*, p. *304*; 21.1.1, p. 414
	20,	K127	(1967)	*18.2*, p. *306*; 21.1.1, p. 417
	23,	K151	(1967)	*18.2*, p. *308*; 21.1.1, p. 419
	24,	K153	(1967)	*18.2*, p. *310*; 21.1.1, p. 420
H.-P. Hennig	23,	599	(1967)	13.4, p. 210; *14.4.1*, p. *260*; 22.1.1, p. 463
	26,	225	(1968)	13.4, p. 211; *14.4.1*, p. *262*; 22.1.1, p. 464
	29,	K167	(1968)	*14.4.1*, p. *264*; 22.1.1, p. 466
K. Hennig	27,	K115	(1968)	*6.1*, p. *74*; 10, p. 108; 22.5.1, p. 517
	27,	K161	(1968)	6.1, p. 74; *10*, p. *108*; 22.5.1, p. 517; 22.5.2, p. 527
J. O. Henningsen	22,	441	(1967)	*13.1*, p. *180*; *19*, p. *330*; 21.1, p. 399
W. Henrion	20,	K145	(1967)	13.1, p. 178; *13.2*, p. *192*; 20.1, p. 347; 22.1.3, p. 477
	22,	K33	(1967)	13.1, p. 179; 13.2, p. 192; *20.1*, p. *349*; 22.1.3, p. 477
	23,	K67	(1967)	13.1, p. 181; *20.1*, p. *350*; 22.1.3, p. 477; 22.4.1, p. 501
G. Hensel	24,	535	(1967)	10, p. 103; *14.4*, p. *253*; 22.5.2, p. 524
	30,	K165	(1968)	10, p. 113; *14.4*, p. *254*; 22.5.2, p. 531
M. Henzler	19,	833	(1967)	*13.3*, p. *197*; 14.3, p. 235; 22.1.2, p. 470
P. R. Herczfeld	27,	671	(1968)	*16*, p. *282*; 22.4.1, p. 503; 22.4.2, p. 508
	27,	681	(1968)	*13.4*, p. *214*; *16*, p. *282*; 22.4.1, p. 503; 22.4.2, p. 508
Karin Herrmann	25,	655	(1968)	*13.1*, p. *184*; 19, p. 332; 22.1.3, p. 478
K. H. Herrmann	24,	289	(1967)	13.1, p. 182; *14.3*, p. *238*; 22.1.3, p. 478
	29,	193	(1968)	13.1, p. 188; *14.3*, p. *243*; 22.1.3, p. 480
R. Herrmann	21,	703	(1967)	13.1, p. 178; *14.1*, p. *223*; *19*, p. *329*; 21, p. 379
	25,	427	(1968)	*13.1*, p. *183*; 19, p. 331; 21, p. 384
	25,	655	(1968)	*13.1*, p. *184*; 19, p. 332; 22.1.3, p. 478
	25,	661	(1968)	*13.1*, p. *184*; 19, p. 332; 21, p. 384
C. Hervouet	21,	117	(1967)	*14.4.1*, p. *258*; 22.2.1, p. 483
H. Hesse	30,	209	(1968)	14.4, p. 254; *20*, p. *342*; 22.5.2, p. 530
J. Hesse	21,	495	(1967)	*10*, p. *100*; 22.5.2, p. 522
I. Hevesi	20,	K45	(1967)	13.1, p. 177; *20.1*, p. *347*; 22.6, p. 537
J. Heydenreich	22,	93	(1967)	1.4, p. 18; *10*, p. *101*; 22.1.1, p. 462
W. Heye	18,	K107	(1966)	*10.1*, p. *119*; 21.6, p. 432
B. Hilczer	23,	419	(1967)	*14.4.1*, p. *259*; 22.9, p. 559
J. A. Hilder	26,	69	(1968)	*13.2*, p. *193*; 20.1, p. 352; 22, p. 450
G. Hildebrandt	23,	K147	(1967)	*5*, p. *52*; 22.1.2, p. 472
	24,	245	(1967)	*5*, p. *52*
J. Hillairet	30,	119	(1968)	10.1, p. 134; *11*, p. *154*; 14.1, p. 226; 21.3, p. 427
J. Hingsammer	26,	261	(1968)	*10.2*, p. *142*; 22.5.2, p. 526; 22.8, p. 553
K. Hirschberg	26,	527	(1968)	14.3.1, p. 246; *17*, p. *288*; 20.1, p. 353; 22.7, p. 546
	27,	145	(1968)	14.3.1, p. 246; *17*, p. *289*; 20.1, p. 354; 22.7, p. 546
V. Hivert	17,	317	(1966)	*10.1*, p. *116*; 11, p. 146; 14.1, p. 220; 18.2, p. 302; 21.1.1, p. 412
	17,	K33	(1966)	10.1, p. 116; *14.1*, p. *220*; 21, p. 374
	17,	K37	(1966)	*10.1*, p. *116*; 18.2, p. 302; 21.1.1, p. 412

V. Hizhnyakov	**21**,	755	(1967)	13.4, p. 207; 20.1, p. 348; *20.3*, p. *366*; 22, p. 446
	27,	751	(1968)	6, p. 66; *13.2*, p. *194*; 20.1, p. 355; 22, p. 452
P. Hlawiczka	**23**,	K129	(1967)	*14.2*, p. *229*; 21, p. 382
P. S. Ho	**23**,	489	(1967)	*6*, p. *63*; 13.1, p. 182; 14.1, p. 224; 21.2, p. 425
E. F. Hockings	**17**,	K185	(1966)	*3.1*, p. *30*; 22.1.3, p. 476
H. Hoffmann	**17**,	K105	(1966)	1.4, p. 17; *18.2*, p. *303*
J. Hofmann	**30**,	K53	(1968)	*13*, p. *173*, 17, p. 290; *20*, p. *342*; 21.6, p. 437
U. Hofmann	**29**,	91	(1968)	*18.2*, p. *314*; 21.1.1, p. 422
C. A. Hogarth	**26**,	K135	(1968)	*14.3*, p. *241*; 22.1.3, p. 479
B. G. Hogg	**20**,	331	(1967)	11, p. 148; *13*, p. *165*; 20, p. 338; 21.1.1, p. 415; 22.8, p. 549
J. H. C. Hogg	**18**,	755	(1966)	*4*, p. *39*; 5, p. 50; 22, p. 441
M. Höhne	**17**,	K119	(1966)	*19*, p. *326*; 22.5.1, p. 515
	20,	657	(1967)	10.2, p. 139; 13.4, p. 206; *19*, p. *328*; 22.5.1, p. 515
	20,	667	(1967)	10.2, p. 139; 13.4, p. 206; *19*, p. *328*; 22.5.1, p. 515
	24,	591	(1967)	10, p. 103; *19*, p. *331*; 22.5.1, p. 516
	25,	K55	(1968)	10.2, p. 141; *19*, p. *332*; 22.5.1, p. 516
	28,	247	(1968)	10.2, p. 143; *19*, p. *334*; 20.1, p. 355; 22.5.1, p. 517
B. W. Holland	**24**,	275	(1967)	*19*, p. *331*; 21, p. 382; 21.1, p. 401; 21.2, p. 425
	28,	121	(1968)	13, p. 171; *19*, p. *334*; 21, p. 386; 21.2, p. 425
W. W. Holloway, Jr.	**28**,	K159	(1968)	*20.3*, p. *370*; 22.5, p. 514; 22.8, p. 554
L. V. Holroyd	**28**,	K27	(1968)	*19*, p. *334*; 22.6, p. 542
G. Holtmeyer	**21**,	291	(1967)	6.1, p. 72; *11*, p. *148*; 22.5.2, p. 522
	21,	297	(1967)	2, p. 23; 6.1, p. 72; *11*, p. *148*; 21.7, p. 438
A. Holz	**25**,	567	(1968)	4, p. 44, *18.2*, p. *310*; 21.1, p. 403; 21.1.1, p. 420
	26,	751	(1968)	4, p. 46; *18.2*, p. *311*; 21.1.1, p. 421
	28,	755	(1968)	10.1, p. 132; *18.2*, p. *313*; 21.1, p. 406
J. Hölzl	**18**,	779	(1966)	13, p. 164; *17*, p. *287*; 21.2, p. 424
	24,	651	(1967)	*17*, p. *288*; 18.1, p. 297; 18.2, p. 309; 21.1, p. 402
J. Holzman	**27**,	K125	(1968)	*16*, p. *282*; 22.1.1, p. 465
M. Holzmann	**19**,	543	(1967)	*10.1*, p. *121*; 21.1, p. 396
E. D. Hondros	**21**,	375	(1967)	*9*, p. *89*; 10.1, p. 122; 21.1.1, p. 417
R. H. Hopkins	**23**,	697	(1967)	*13.1*, p. *182*; 14.3, p. 238; 20.1, p. 350; 22.7, p. 545
H. Hora	**17**,	59	(1966)	13.1, p. 175; *17*, p. *286*; 22.7, p. 545
	27,	593	(1967)	*17*, p. *289*; 22.7, p. 546
A. J. Horodecki	**26**,	543	(1968)	*20.3*, p. *369*; 22.4.1, p. 502
W. Hörstel	**23**,	639	(1967)	*10*, p. *102*; 14.3, p. 238; 22.1.3, p. 478
T. Horváth	**23**,	K71	(1967)	*10.2*, p. *140*; 22.5.2, p. 523
R. G. Howell	**29**,	697	(1968)	6, p. 68; 18.1, p. 297; *20.1*, p. *357*; 22.8, p. 555
D. H. Howling	**18**,	579	(1966)	*1.1*, p. *8*; *4*, p. *39*; 21.1, p. 394
I. Hrianca	**29**,	761	(1968)	13, p. 173; *17*, p. *290*; 21, p. 388
Ľ. Hrivnák	**23**,	189	(1967)	7, p. *77*; 8, p. 83; 22.2.3, p. 491
A. Hrynkiewicz see A. Z. Hrynkiewicz				
A. Z. Hrynkiewicz	**17**,	K53	(1966)	*6.1*, p. *71*; 21, p. 374
	17,	K195	(1966)	6.1, p. 71; *18.4*, p. *321*; 22, p. 441
H. U. Hrynkiewicz	**17**,	K53	(1966)	*6.1*, p. *71*; 21, p. 374
R. Hsiang-Tao Yeh see under letter "Y" (R. H. T. Yeh)				
A. Hubert	**22**,	709	(1967)	*18.2*, p. *308*; 21.1, p. 399
	24,	669	(1967)	*4*, p. *44*; 14.2, p. 229; 18.2, p. 309; 21.1, p. 402
H. R. Huff	**30**,	613	(1968)	*14.3*, p. *244*; 22.2.3, p. 493
G. P. Huffman	**22**,	473	(1967)	*6.1*, p. *72*; 18.2, p. 308; 22.8, p. 551
R. A. Huggins	**24**,	301	(1967)	*10.1*, p. *126*; 13.4, p. 210; 21.1, p. 401
D. M. Hughes	**25**,	449	(1968)	13.3, p. 198; *14.3.1*, p. *246*; 22.4.1, p. 501
M. Hulin	**21**,	607	(1967)	*6*, p. *61*; *20.1*, p. *348*
	23,	563	(1967)	*13.1*, p. *182*; 22.1.3, p. 478

P. Humble	**20,**	505	(1967)	*10.1*, p. *122*; 12.1, p. 159; 21.1, p. 397
	20,	521	(1967)	*10.1*, p. *122*; 12.1, p. 159; 21.1, p. 397
	21,	733	(1967)	*10.1*, p. *123*; 21, p. 379; 21.1, p. 398; 21.1.1, p. 417; 21.6, p. 433
	29,	99	(1968)	*10.1*, p. *133*; 21.1, p. 407; 21.6, p. 437
	30,	183	(1968)	*10*, p. *112*; *10.1*, p. *135*
R. E. Hummel	**30,**	515	(1968)	*1.2*, p. *13*; *11*, p. *155*; 14.1, p. 227; 21.1, p. 409
F. Huth	**29,**	K35	(1968)	13.4, p. 215; *14.3*, p. *243*; 22.2.1, p. 485

Vu Huy Dat see under letter "V" (Vu)

G. I. Ibaeev	**30,**	K73	(1968)	*14.4.1*, p. *264*; 22.1.3, p. 480
K. Ibel	**29,**	403	(1968)	*11*, p. *154*; 21.6, p. 437
N. I. Ibragimov	**16,**	K113	(1966)	10, p. 94; *19*, p. *326*; 22.1.3, p. 476
E. Igras	**19,**	K67	(1967)	*1.3*, p. *14*; 4, p. 40; 10, p. 98; 22.1.2, p. 470
	20,	K5	(1967)	1.3, p. 14; *9*, p. *88*; 22.1.2, p. 471
	27,	69	(1968)	*1.3*, p. *15*; *14.3.2*, p. *250*; 22.1.2, p. 473
S. Ikonopisov	**23,**	K61	(1967)	*20.3*, p. *367*; 22.6, p. 539
B. Ilschner	**20,**	629	(1967)	*10*, p. *99*; 22.5.2, p. 521
H. Imagawa	**30,**	469	(1968)	2, p. 25; 3.2, p. 36; *19*, p. *336*; 22.8, p. 557
H. Imgrund	**24,**	K5	(1967)	*14.1*, p. *224*; 18.2, p. 309; 21.1.1, p. 420
J. Ingels	**30,**	163	(1968)	13.4, p. 216; *18*, p. *295*; *20*, p. *342*; 22.5.2, p. 530
O. Inkinen	**21,**	127	(1967)	5, p. 51; *13*; p. *165*; 22.5.2, p. 522
I. V. Ioffe	**19,**	51	(1967)	*6*, p. *58*; 18.2, p. 304; 18.4, p. 321; 19, p. 327; 22, p. 442
	24,	K9	(1967)	*7*, p. *77*; *19*, p. *331*
Ya. A. Iosilevskii	**16,**	633	(1966)	*6.1*, p. *67*; 22, p. 440
	22,	255	(1967)	*5*, p. *51*; 6, p. 62; 10, p. 101
	24,	749	(1967)	6, p. 64; *6.1*, p. *73*
I. P. Ipatova	**23,**	467	(1967)	*6*, p. *63*; 13.4, p. 209; *20.3*, p. *367*; 22.1.2, p. 472
Yu. P. Irkhin	**22,**	309	(1967)	13, p. 166; *14.1*, p. *223*; 18.2, p. 308
G. Ischenko	**27,**	639	(1968)	*6.1*, p. *74*; 22.1.1, p. 465
K. Isebeck	**17,**	259	(1966)	10.1, p. 115; *11*, p. *146*; 14.1, p. 220; 21, p. 374
	18,	427	(1966)	*10.1*, p. *118*; 11, p. 147; 14.1, p. 221; 21, p. 375
K. Ishikawa	**21,**	137	(1967)	*6*, p. *60*
H.-J. Isken	**17,**	619	(1966)	*1.4*, p. *17*; *3.1*, p. *30*; 21.6, p. 431
Z. A. Iskender-Zade	**21,**	423	(1967)	13.4, p. 206; *14.3*, p. *236*; 14.4.1, p. 258; 22.1.2, p. 471
V. D. Iskra	**27,**	101	(1968)	13.1, p. 185; *14.3*, p. *241*; 22, p. 451; 22.7, p. 546
F. I. Ismailov	**17,**	K237	(1966)	13.4, p. 203; *14.3*, p. *233*; 15, p. 271; 22, p. 441
M. Z. Ismailov	**18,**	929	(1966)	*5*, p. *50*; 8, p. 80; 13.4, p. 204; 14.3, p. 234; 16, p. 278; 22.8, p. 548
	19,	K7	(1967)	*8*, p. *81*; 22, p. 442
	22,	K117	(1967)	*14.3*, p. *237*; 22.8, p. 551
T. P. Isotova	**17,**	489	(1966)	*18.2*, p. *303*; 21.1.1, p. 413
J. P. Issi	**17,**	K217	(1966)	*14.1*, p. *221*; 21.7, p. 438
N. Itoh	**30,**	199	(1968)	10, p. 112; *11*, p. *154*; 22.5.2, p. 530
N. R. Ivanov	**22,**	279	(1967)	1.2, p. 11; *14.4.2*, p. *267*
V. G. Ivanov	**21,**	789	(1967)	1.3, p. 15; *13.3*, p. *197*; 17, p. 287; 22.1.1, p. 462
	22,	39	(1967)	*17*, p. *287*; 22.1.1, p. 462
V. A. Ivanova	**24,**	K23	(1967)	5, p. 52; 8, p. 83; *14.3*, p. *239*; 15, p. 273; 22.8, p. 552
	24,	K145	(1967)	5, p. 52; *8*, p. *84*; 14.3, p. 239; *15*, p. *273*; 22, p. 449
I. Ivanov-Omskii	**27,**	K169	(1968)	*13.1*, p. *186*; 15, p. 274; 22.4.3, p. 511
L. K. Ives	**27,**	117	(1968)	*10.1*, p. *130*; 21.1, p. 404; 21.6, p. 436

D. Iwanov	**23,**	663	(1967)	*3.1*, p. *32*; 21, p. 382; 21.1, p. 400
G. Jacobs	**17,**	177	(1966)	*10.2*, p. *137*; 22.5.2, p. 519
	18,	279	(1966)	*10.2*, p. *138*; 22.5.2, p. 519
	22,	177	(1967)	10.2, p. 139; *16*, p. *279*; *20.3*, p. *366*; 22.5.2, p. 523
	25,	621	(1968)	10.2, p. 142; *16*, p. *281*; 22.5.2, p. 525
	30,	163	(1968)	13.4, p. 216; *18*, p. *295*; *20*; p. *342*; 22.5.2, p. 530
P. W. M. Jacobs	**17,**	K45	(1966)	10, p. 95; *14.4*, p. *252*; 22.5.2, p. 519
	19,	565	(1967)	*10.2*, p. *139*; 22.5.2, p. 521
L. Jahn	**19,**	K75	(1967)	*18.3*, p. *317*
E. Jahne	**19,**	823	(1967)	*13.1*, p. *177*; 22.4, p. 495
	21,	57	(1967)	*13.1*, p. *178*; 22.4, p. 495; 22.4.1, p. 499; 22.4.2, p. 506
H. Jahrreiss	**17,**	619	(1966)	*1.4*, p. *17*; *3.1*, p. *30*; 21.6, p. 431
E. Jäger	**20,**	433	(1967)	3.2, p. 35; *13*, p. *165*; 18.3, p. 317
	25,	K43	(1968)	5, p. 53; *13*, p. *169*; 18.3, p. 319
S. C. Jain	**22,**	505	(1967)	13.4, p. 208; *14.4*, p. *253*; *20.1*, p. *349*; 22.5.2, p. 523
V. K. Jain	**28,**	415	(1968)	2, p. 24; *14.4*, p. *254*; 22.9, p. 560
R. von Jan	**17,**	361	(1966)	*10.1*, p. *116*; 11, p. 146; 21, p. 374; 21.1, p. 392; 21.6, p. 430
	29,	755	(1968)	*11*, p. *154*
G. Jäniche	**28,**	K97	(1968)	1.4, p. 21; *14.3.1*, p. *247*; 22.4.1, p. 504
T. Janicki	**30,**	827	(1968)	*14.3.2*, p. *251*; *14.4.1*, p. *265*; 16, p. 285; 22.1.2, p. 475
V. Janků	**25,**	K103	(1968)	*20.1*, p. *352*; 22.8, p. 553
M. Jaroš	**20,**	K99	(1967)	13.4, p. 206; *19*, p. *328*; 22.6, p. 537
	27,	K21	(1968)	*3.2*, p. *36*; 19, p. 333; 22.4.3, p. 510
M. Järvinen	**21,**	127	(1967)	5, p. 51; *13*, p. *165*; 22.5.2, p. 522
J. Jaumann	**27,**	K43	(1968)	*18.2*, p. *312*; *20*; p. *341*; 20.1, p. 354; 21.1.1, p. 421
A. Jean	**25,**	557	(1968)	13.4, p. 211; 20.1, p. 351; *20.3*, p. *368*; 22.5.2, p. 525
Č. Jech	**21,**	481	(1967)	*11*, p. *149*; 22, p. 445; 22.6, p. 538
	25,	641	(1968)	*9*, p. *91*; 11, p. 151; 22.5.2, p. 525
	27,	573	(1968)	1.4, p. 20; *11*, p. *152*; 22, p. 452; 22.1.2, p. 473
K. Jeges	**22,**	K7	(1967)	*20.3*, p. *366*; 22.6, p. 538
R. J. Jelitto	**28,**	365	(1968)	*1.2*, p. *12*; 18.2, p. 312; 18.4, p. 324
H. Jena	**27,**	639	(1968)	*6.1*, p. *74*; 22.1.1, p. 465
D. A. Jenkins	**20,**	327	(1967)	2, p. 23; *14.3*, p. *235*; 22.6, p. 537; 22.8, p. 549
W. A. Jesser	**19,**	95	(1967)	1.4, p. 18; *10*, p. *97*
	21,	533	(1967)	*3.1*, p. *31*; 10, p. 100; 10.1, p. 123; 21, p. 379; 22, p. 445
H. L. Jetter	**22,**	K39	(1967)	12, p. 157; 13.2, p. 193; *20.1*, p. *349*; 22.9, p. 559
B. Jiménez	**19,**	805	(1966)	*14.4.2*, p. *266*
H. Jodl	**26,**	261	(1968)	*10.2*, p. *142*; 22.5.2, p. 526; 22.8, p. 553
K. Johansen	**28,**	555	(1968)	*13.1*, p. *187*; 22.1.2, p. 474
Ch. A. Johnson	**20,**	59	(1967)	*4*, p. *41*
B. K. Jones	**24,**	297	(1967)	*7*, p. *77*; 13.1, p. 182; 21, p. 382; 21.1, p. 401; 21.2, p. 425
G. Jones	**20,**	K135	(1967)	*13*, p. *165*; 14.3, p. 236; 22.2.3, p. 490
G. A. Jones	**19,**	811	(1967)	*1.4*, p. *18*; 18.2, p. 305; 21.1.1, p. 415
	20,	K95	(1967)	*1.4*, p. *18*; *18.2*, p. *306*; 21.1.1, p. 416
M. L. Joshi	**26,**	159	(1968)	*10*, p. *105*; 22.1.2, p. 473
T. Jøssang	**27,**	579	(1968)	*10*, p. *108*
A. Jostsons	**29,**	873	(1968)	*4*, p. *47*; 10, p. 111; 22.6, p. 543
B. Jouffrey	**22,**	349	(1967)	*10*, p. *101*; 22.3, p. 494

J. C. Jousset	**29**,	K127	(1968)	*11*, p. *154*; 14.1, p. 226; 21.5, p. 429
V. M. Judin	**20**,	759	(1967)	*18.4*, p. *322*; 22.5, p. 512; 22.8, p. 550
M. Jung	**18**,	743	(1966)	13.4, p. 204; *14.3.2*, p. *248*; 22.2.2, p. 488
G. Jungk	**18**,	297	(1966)	*13.4*, p. *203*; 14.3, p. 233; 16, p. 278; 22.1.2, p. 470
	25,	K59	(1968)	*13.1*, p. *184*; 16, p. 281; 22.1.2, p. 473
	27,	237	(1968)	*13.4*, p. *213*; 16, p. 282; 22, p. 451
G. B. Kabiersch	**27**,	593	(1968)	*17*, p. *289*; 22.7, p. 546
J. Kaczér	**21**,	635	(1967)	4, p. 42; *18.2*, p. *307*; 21.1.1, p. 417
	26,	443	(1968)	*18.1*, p. *297*; 18.4, p. 324; 22.8, p. 553
S. Kadečková	**16**,	89	(1966)	*10.1*, p. *114*; 21.1.1, p. 410
M. S. Kagan	**24**,	551	(1967)	*14.4.1*, p. *261*; 22.1.1, p. 464
Yu. Kagan	**28**,	287	(1968)	5, p. 54; *6*, p. *66*
H.-G. Kahle	**22**,	537	(1967)	*6*, p. *62*; *18.1*, p. *296*; 19, p. 330; 22.8, p. 551
W. Kahle	**28**,	K97	(1968)	1.4, p. 21; *14.3.1*, p. *247*; 22.4.1, p. 504
D. K. Kaipnazarov	**21**,	805	(1967)	10, p. 100; 11, p. 149; *12*, p. *157*; 22.1.2, p. 471
S. G. Kalashnikov	**16**,	197	(1966)	*14.4.1*, p. *255*; 22.1.1, p. 459
	23,	K9	(1967)	*7*, p. *77*; *14.4.1*, p. *260*; 22.4.1, p. 501
	24,	551	(1967)	*14.4.1*, p. *261*; 22.1.1, p. 464
V. P. Kalashnikov	**16**,	K35	(1966)	*13*, p. *163*; *14.3*, p. *231*; 22, p. 439
	21,	775	(1967)	*6*, p. *61*; 13, p. 166; 14.4, p. 253; 18, p. 292; 22, p. 446
	28,	105	(1968)	6, p. 66; *8*, p. *85*; *14.3*, p. *242*; 22, p. 452
H. Kalbfleisch	**22**,	537	(1967)	*6*, p. *62*; *18.1*, p. *296*; 19, p. 330; 22.8, p. 551
B. Kalińska	**23**,	K53	(1967)	13, p. 167; *18*, p. *293*; *20*, p. *340*; 22.1.1, p. 463; 22.2.1, p. 484
D. O. Kalnin	**18**,	897	(1966)	10.2, p. 138; *11*, p. *147*; 19, p. 327; 20.1, p. 346; 22.5.2, p. 520
J. Kalus	**21**,	291	(1967)	6.1, p. 72; *11*, p. *148*; 22.5.2, p. 522
	21,	297	(1967)	2, p. 23; 6.1, p. 72; *11*, p. *148*; 21.7, p. 438
Z. Kamberská	**17**,	411	(1966)	1.4, p. 17; *18.2*, p. *302*; 21.1, p. 392
V. Kamberský	**17**,	411	(1966)	1.4, p. 17; *18.2*, p. *302*; 21.1, p. 392
	23,	501	(1967)	*13*, p. *167*; *19*, p. *330*; 21.1.1, p. 419
E. Kamieniecki	**16**,	57	(1966)	6, p. 56; *13*, p. *162*; *14.3*, p. *231*; 22.1.1, p. 459; 22.1.2, p. 468
	18,	K105	(1966)	6, p. 58; *13*, p. *164*; *14.3*, p. *234*; 22, p. 442
T. Kamikawa	**17**,	131	(1966)	*6*, p. *56*; *10.2*, p. *137*; 22.5.2, p. 518
	17,	645	(1966)	*10.2*, p. *137*; 11, p. 146; 20.1, p. 346; 22.5.2, p. 519
A. A. Kaminskii	**16**,	K165	(1966)	10, p. 94; *20.1*, p. *345*; 20.2, p. 360; 22.8, p. 547
	20,	K51	(1967)	*20.2*, p. *361*; 22.5, p. 512; 22.5.3, p. 532; 22.8, p. 549
	21,	K17	(1967)	20.1, p. 348; *20.2*, p. *361*; 20.3, p. 365; 22.5, p. 513; 22.8, p. 550
	21,	K23	(1967)	13.4, p. 207; *20.1*, p. *348*; 20.2, p. 361; *20.3*, p. *365*; 22.8, p. 550
	23,	K5	(1967)	20.1, p. 350; *20.2*, p. *361*; 20.3, p. 367; 22.5, p. 513; 22.8, p. 552
	25,	K75	(1968)	20.1, p. 352; *20.2*, p. *361*; 20.3, p. 368; 22.5, p. 513; 22.8, p. 552
	27,	K1	(1968)	*20.2*, p. *362*; 22.8, p. 553
	30,	K77	(1968)	*20.2*, p. *362*; 22.5.3, p. 534

W. Kamprath	**22**,	541 (1967)	11, p. 149; *13.4*, p. *208*; 14.3, p. 237; 22.1.2, p. 471
E. A. Kaner	**22**,	47 (1967)	13.1, p. 179; *19*, p. *329*; 21, p. 380
	22,	319 (1967)	7, p. 77; 13.1, p. 179; *19*, p. *329*; 21, p. 380
	22,	333 (1967)	*13*, p. *166*; 13.1, p. 179; *20*, p. *339*; 21, p. 380
O. Kanert	**30**,	127 (1968)	*10.1*, p. *134*; 19, p. 335; 21, p. 388; 21.1, p. 408
D. G. Kannenberg	**16**,	697 (1966)	*1.4*, p. *16*; 13.4, p. 201; 16, p. 276; 22.4.1, p. 497
R. Kantlehner	**17**,	59 (1966)	13.1, p. 175; *17*, p. *286*; 22.7, p. 545
P. M. Karageorgy-Alkalaev	**26**,	419 (1968)	*13.4*, p. *212*; 20.3, p. 368; 22, p. 450
M. Karapandžić	**17**,	K53 (1966)	*6.1*, p. *71*; 21, p. 374
M. D. Karkhanavala	**17**,	501 (1966)	13, p. 163; *14.3*, p. *232*; 22.6, p. 535
	19,	117 (1967)	14.3, p. 234; *15*, p. *271*; 22.6, p. 536
H. Karras	**19**,	K23 (1967)	*10.2*, p. *138*; 22.5.3, p. 532
	23,	313 (1967)	13.4, p. 209; 20.1, p. 350; *20.3*, p. *367*, 22.5.3, p. 533
	25,	93 (1968)	*10.2*, p. *141*; 22.5.3, p. 533
	25,	K15 (1968)	6, p. 65; *10.2*, p. *141*; *20.3*, p. *368*; 22.5.3, p. 533
	27,	109 (1968)	10, p. 106; *20.3*, p. *369*; 22.5.3, p. 534
M. Karras	**21**,	487 (1967)	*17*, p. *287*; 22.5.2, p. 522
D. Kasperkovitz	**28**,	K51 (1968)	13, p. 171; *14.4.1*, p. *263*; 22.1.1, p. 465
A. A. Kastalskii	**26**,	219 (1968)	*7*, p. *77*; *14.4.1*, p. *262*; 22.2.1, p. 485
L. Kaun	**26**,	485 (1968)	*10.1*, p. *129*; 21, p. 385
G. Kavalyauskene	**26**,	K5 (1968)	2, p. 24; *17*, p. *288*; 22.1.3, p. 479
H. Kawai	**28**,	K19 (1968)	*10*, p. *109*; *10.2*, p. *143*; 22.5.2, p. 527
N. Kawai	**24**,	K83 (1967)	*18.3*, p. *318*
S. Kawaji	**30**,	613 (1968)	*14.3*, p. *244*; 22.2.3, p. 493
Y. Kawasaki	**26**,	77 (1968)	*20.1*, p. *352*; 22.4.3, p. 510
A. Kawski	**26**,	543 (1968)	*20.3*, p. *369*; 22.4.1, p. 502
A. L. Kazakov	**24**,	K9 (1967)	*7*, p. *77*; *19*, p. *331*
V. G. Kazakov	**16**,	651 (1966)	12.1, p. 159; *18.2*, p. *301*; 21.1.1, p. 411
Y. Kazumata	**17**,	131 (1966)	*6*, p. *56*; *10.2*, p. *137*; 22.5.2, p. 518
	17,	645 (1966)	*10.2*, p. *137*; 11, p. 146; 20.1, p. 346; 22.5.2, p. 519
	25,	563 (1968)	*10.2*, p. *142*; 11, p. 151; 19, p. 332; 22.5.2, p. 525
	26,	K119 (1968)	10.2, p. 142; 11, p. 152; *19*, p. *333*; 22.5.3, p. 534
F. J. Kedves	**21**,	K183 (1967)	*10.1*, p. *124*; 14.1, p. 223; 21.6, p. 433
D. W. Keefer	**24**,	217 (1967)	*11*, p. *150*; 14.1, p. 224; 21.1.1, p. 419
	28,	675 (1968)	10.1, p. 132; *11*, p. *153*; 14.1, p. 226; 21, p. 387
R. Keezer	**26**,	K5 (1968)	2, p. 24; *17*; p. *288*; 22.1.3, p. 479
W. Keilig	**17**,	151 (1966)	*18.3*, p. *316*
A. Keiper	**19**,	K59 (1967)	*20.2*, p. *360*; 22.2.1, p. 483
R. Keiper	**19**,	673 (1967)	13.1, p. 177; 14.4.1, p. 257; *20*, p. *338*; 22, p. 443
	23,	127 (1967)	13.1, p. 180; 14.4.1, p. 259; *20*, p. *339*; 22, p. 447
V. V. Kelarev	**24**,	385 (1967)	*18.2*, p. *309*; *18.4*, p. *323*; 21.1.1, p. 419
F. Kelemen	**21**,	557 (1967)	*8*, p. *82*; 22.4.1, p. 500; 22.4.2, p. 506; 22.4.3, p. 509; 22.8, p. 550
	28,	421 (1968)	6, p. 66; *8*, p. *85*; 22.4.3, p. 511; 22.8, p. 554
R. Kelly	**21**,	451 (1967)	*9*, p. *89*; 11, p. 148
	25,	641 (1968)	*9*, p. *91*; 11, p. 151; 22.5.2, p. 525
	30,	37 (1968)	*11*, p. *154*; 22.6, p. 544
N. Kempe	**18**,	415 (1966)	*13*, p. *164*; *20*, p. *338*; 22.1.1, p. 460
C. P. Kempter	**18**,	K21 (1966)	*12*, p. *156*; 22.3, p. 494
	18,	K117 (1966)	*5*, p. *50*; 21.1, p. 394; 21.6, p. 432
C. A. Kennedy	**19**,	203 (1967)	*13.4*, p. *204*; 14.3, p. 234; 16, p. 278; 22.4.1, p. 499

S. W. Kennedy	**16,** K175	(1966)	*1.2*, p. *9*; 3, p. 26; 8, p. 79; 22.8, p. 547
E. M. Kerimova	**19,** K7	(1967)	*8*, p. *81*; 22, p. 442
	22, K117	(1967)	*14.3*, p. *237*; 22.8, p. 551
T. G. Kerimova	**16,** K15	(1966)	*20.1*, p. *344*; 22, p. 439
	16, K31	(1966)	2, p. 22; 10, p. 94; *20.1*, p. *344*; 22.1.3, p. 476
R. Kern	**19,** 211	(1967)	*3.1*, p. *31*; 22.8, p. 549
R. H. Kernohan	**17,** K233	(1966)	*14.2*, p. *228*
R. Kersten	**29,** 575	(1968)	10.2, p. 144; *19*, p. *335*; 22.5.2, p. 529
F. P. Kesamanly	**17,** 105	(1966)	13.4, p. 202; *14.3*, p. *232*; 22.2.2, p. 487
	21, 9	(1967)	3.1, p. 29; 3.2, p. 34; 5, p. 49; 8, p. 79; 10, p. 93; 13.1, p. 174; *14.3*, p. *231*; 15, p. 270; 16, p. 276; 17, p. 286; 18, p. 291; 20.1, p. 344; 22.8, p. 547
	27, K169	(1968)	*13.1*, p. *186*; 15, p. 274; 22.4.3, p. 511
D. Kessel	**17,** 211	(1966)	*14.4*, p. *252*; 22, p. 440
F. R. Kessler	**23,** K25	(1967)	13.1, p. 181; *16*, p. *280*; *20.1*, p. *350*; 22.1.3, p. 477
M. Kestigian	**23,** 289	(1967)	18.1, p. 296; *18.4*, p. *323*; 22.8, p. 552
	25, 119	(1968)	18.2, p. 310; *18.3*, p. *319*; 18.4, p. 323
	28, K159	(1968)	*20.3*, p. *370*; 22.5, p. 514; 22.8, p. 554
L. Keszthelyi	**21,** K121	(1967)	*6.1*, p. *72*; 18.4, p. 322; 22.6, p. 538; 22.8, p. 551
	22, 617	(1967)	1.2, p. 11; *6.1*, p. *72*; 18.3, p. 318; 18.4, p. 323; 22.8, p. 551
	27, 131	(1968)	*6.1*, p. *73*; 18.3, p. 319
	30, 215	(1968)	2, p. 24; *6.1*, p. *74*; 22.8, p. 556
P. Ketolainen	**18,** K143	(1966)	7, p. 76; *10.2*, p. *138*; 22.5.2, p. 520
J. D. Keys	**19,** K11	(1967)	*3.2*, p. *35*; 14.3, p. 235; 22, p. 442
L. K. Keys	**25,** K5	(1968)	*19*, p. *332*; 22.6, p. 540
A. G. Khachaturyan	**23,** 745	(1967)	*4*, p. *43*; 21.1, p. 400; 22.6, p. 539
P. I. Khadshi	**18,** 379	(1966)	6, p. 57; *13.2*, p. *191*; 20.1, p. 346; 22.1.1, p. 460; 22.2.2, p. 487; 22.2.3, p. 490; 22.6, p. 536
Kh. M. Khalilov	**26,** K113	(1968)	*2*, p. *24*; 10, p. 106; *12*, p. *158*; 22.1.3, p. 479
R. Yu. Khansevarov	**22,** K95	(1967)	*11*, p. *149*; 14.3, p. 237; 22.2.3, p. 491
	22, K103	(1967)	10, p. 101; *11*, p. *150*; 14.3, p. 237; 22.1.1, p. 462
L. P. Khiznichenko	**21,** 805	(1967)	10, p. 100; 11, p. 149; *12*, p. *157*; 22.1.2, p. 471
	21, 811	(1967)	10, p. 100; *12*, p. *157*; 22.1.2, p. 471
G. A. Kholodar	**19,** 41	(1967)	10, p 97; 13.4, p. 204; *14.3*, p. *234*; 22.6, p. 536
V. S. Khristosenko	**16,** 651	(1966)	12.1, p. 159; *18.2*, p. *301*; 21.1.1, p. 411
E. Kierzek-Pecold	**19,** 623	(1967)	14.3, p. 235; *18*, p. *291*; *20*, p. *338*; 22, p. 443
	19, K55	(1967)	13.1, p. 176; *14.4.1*, p. *257*; *20*, p. *338*; 22, p. 443
	22, K147	(1967)	*20.1*, p. *350*; 22.1, p. 457
	27, K107	(1968)	*20.1*, p. *355*; 22.1, p. 457
	29, K183	(1968)	13.1, p. 189; *20.1*, p. *357*; 22, p. 454
S. Kikuchi	**24,** K83	(1967)	*18.3*, p. *318*
H. Kilian	**27,** 639	(1968)	*6.1*, p. *74*; 22.1.1, p. 465
S. Kim	**25,** 223	(1968)	*10.2*, p. *141*; 13.4, p. 211; 22.5.2, p. 525
Kim Yung see under letter "Y" (Yung)			
J. Kinel	**27,** K37	(1968)	10.1, p. 131; *12.1*, p. *160*; *18.2*, p. *312*; 21.1.1, p. 421
	29, K39	(1968)	*18.2*, p. *314*; 21.1.1, p. 422
A. P. King	**16,** 613	(1966)	4, p. 38; *18.2*, p. *300*; 21.1.1, p. 411
R. King	**18,** 491	(1966)	4, p. 39; *10.1*, p. *118*; 21, p. 376
R. D. King	**26,** K147	(1968)	*10.2*, p. *142*; 11, p. 152; 22.6, p. 541
L. V. Kirenskii	**17,** 243	(1966)	1.4, p. 16; *3.1*, p. *29*; 21.1, p. 392; 21.1.1, p. 412
	17, 249	(1966)	4, p. 38; *18.2*, p. *302*; 21.1, p. 392; 21.1.1, p. 412
	17, 489	(1966)	*18.2*, p. *303*; 21.1.1, p. 413

K. Kirov	**17**,	581	(1966)	*13.4*, p. *202*; 14.4.1, p. 256; 16, p. 277; 22.4.1, p. 498
R. Kishore	**26**,	133	(1968)	*14.1*, p. *225*; 21.1, p. 403; 21.2, p. 425; 21.6, p. 435
	28,	695	(1968)	*14.1*, p. *226*; 21, p. 387
	30,	817	(1968)	*6*, p. *70*; 8, p. 86; *13*, p. *173*; 14.1, p. 227; 21, p. 389
A. I. Kitaigorodskii	**24**,	K61	(1967)	*12*, p. *157*; 22.9, p. 559
S. A. Kizhaev	**30**,	871	(1968)	18.3, p. 320; *19*, p. *336*
K. Kleinhenz	**29**,	627	(1968)	1.4, p. 21; *10*, p. *110*; 22.1, p. 457
K. Kleinstück	**16**,	127	(1966)	*6.1*, p. *71*; 18.3, p. 316
	23,	237	(1967)	*4*, p. *43*; 22.1.2, p. 472
	23,	475	(1967)	*5*, p. *52*; 21.1.1, p. 419
	25,	607	(1968)	*6.1*, p. *73*; 18.3, p. 319
	29,	211	(1968)	*18.2*, p. *314*; *18.4*, p. *325*; 21.1.1, p. 422
M. Kleman	**23**,	207	(1967)	*18.2*, p. *308*; 21.1.1, p. 418
L. I. Kleshchinskii	**29**,	435	(1968)	4, p. 47; *5*, p. *54*; 6, p. 68; 10, p. 110; 14.3, p. 243; 22.8, p. 555
M. Klesnil	**21**,	717	(1967)	*10.1*, p. *123*; 21.1, p. 398
	27,	545	(1968)	1.3, p. 15; *10.1*, p. *131*; 21.1, p. 405
E. Klier	**28**,	K147	(1968)	*16*, p. *283*; 22.4.2, p. 508; 22.4.3, p. 511
M. I. Klinger	**27**,	479	(1968)	*13*, p. *170*; 14.3, p. 241; 20.1, p. 354; 22.6, p. 541
A. A. Klochichin	**23**,	467	(1967)	*6*, p. *63*; 13.4, p. 209; *20.3*, p. *367*; 22.1.2, p. 472
S. M. Klotsman	**16**,	729	(1966)	*9*, p. *87*; 21.1, p. 391; 21.6, p. 430
	18,	847	(1966)	*9*, p. *88*; 14.1, p. 222; 21.1, p. 394; 21.6, p. 432
G. Kluge	**17**,	109	(1966)	6, p. 56; *7*, p. *76*; 22.6, p. 535
E. Klugmann	**16**,	K117	(1966)	*18.2*, p. *301*; 21.1, p. 391
	28,	K67	(1968)	*18.2*, p. *313*; 21.1, p. 406
Th. Klupsch	**16**,	K147	(1966)	*13*, p. *163*; *18*, p. *291*
V. V. Klyushin	**24**,	385	(1967)	*18.2*, p. *309*; *18.4*, p. *323*; 21.1.1, p. 419
J. E. Knappett	**21**,	K99	(1967)	1.4, p. 18; *3.1*, p. *31*; 14.3.1, p. 245; 22.2.1, p. 484
G. Kneer	**17**,	323	(1966)	*18.2*, p. *302*; 21.1, p. 392; 21.1.1, p. 412
Z. Knésl	**23**,	185	(1967)	*10*, p. *102*
J. S. Knol	**23**,	K83	(1967)	*19*, p. *330*; 21.2, p. 424
B. R. Knottnerus	**29**,	K43	(1968)	1.2, p. 12; 10, p. 110; *14.4*, p. *254*; 22.5.2, p. 529
J. M. Knudsen	**22**,	59	(1967)	*6.1*, p. *72*; 18.4, p. 322; 22.5, p. 513
J. Kobayashi	**21**,	151	(1967)	4, p. 42; *14.4.2*, p. *267*
	26,	277	(1968)	4, p. 45; *14.4.2*, p. *268*; 20, p. 340
L. Ya. Kobelev	**28**,	K81	(1968)	*13*, p. *171*
V. V. Kobelev	**17**,	381	(1966)	*18.2*, p. *302*; 21.1.1, p. 412
J. Kočík	**19**,	333	(1967)	4, p. 40; *5*, p. *50*; 21.1.1, p. 415
J. Kodeš	**17**,	K19	(1966)	*14.3.2*, p. *248*; 22.1.3, p. 476
R. Koepp	**27**,	K117	(1968)	*14.4.1*, p. *262*; 16, p. 282; 22.4.1, p. 503
	30,	K5	(1968)	*14.4.1*, p. *264*; *16*, p. *284*; 22.4.1, p. 505
J. W. Koger	**30**,	515	(1968)	*1.2*, p. *13*; *11*, p. *155*; 14.1, p. 227; 21.1, p. 409
K. M. Koliwad	**18**,	265	(1966)	3.2, p. 34; *12*, p. *156*; 22.5.2, p. 519
	21,	507	(1967)	7, p. 76; *12*, p. *157*; 22.5.2, p. 522
N. P. Kolmakova	**25**,	787	(1968)	*18.2*, p. *310*; *18.4*, *323*; 19, p. 332
	28,	83	(1968)	8, p. 85; 13, p. 171; *18.2*, p. *312*
J. Kołodziejczak	**16**,	K55	(1966)	*13.1*, p. *174*; 22.2.1, p. 482; 22.2.3, p. 489
	17,	701	(1966)	13.1, p. 176; 14.4.1, p. 256; *18*, p. *291*; *20*, p. *337*; 22, p. 441
	19,	231	(1967)	*14.3*, p. *235*
	19,	373	(1967)	13.1, p. 176; 14.4.1, p. 257; *15*, p. *271*; 22, p. 442
	19,	623	(1967)	14.3, p. 235; *18*, p. *291*; *20*, p. *338*; 22, p. 443
	19,	K51	(1967)	*13.1*, p. *176*; 14.3, p. 235; 15, p. 271; 22, p. 443
	19,	K55	(1967)	13.1, p. 176; *14.4.1*, p. *257*; *20*, p. *338*; 22, p. 443

J. Kołodziejczak	**22**, K147	(1967)	*20.1*, p. *350*; 22.1, p. 457
	23, 423	(1967)	6, p. 62; 13.1, p. 181; *14.3*, p. *238*; 14.4.1, p. 260; 22, p. 447; 22.2.3, p. 491
	23, K53	(1967)	13, p. 167; *18*, p. *293*; *20*, p. *340*; 22.1.1, p. 463; 22.2.1, p. 484
	23, K57	(1967)	*13*, p. *167*; 13.1, p. 181; 14.3, p. 238; *20*, p. *340*; 22, p. 447
	24, 323	(1967)	13.1, p. 183; 14.3, p. 238; *19*, p. *331*; 22, p. 448
	27, K25	(1968)	13.1, p. 186; *18*, p. *294*; *20*, p. *341*; 22.1.2, p. 473
	27, K107	(1968)	*20.1*, p. *355*; 22.1, p. 457
	27, K125	(1968)	*16*, p. *282*; 22.1.1, p. 465
	29, 645	(1968)	13.1, p. 189; *18*, p. *295*; *20*, p. *342*; 22, p. 454
	29, K183	(1968)	13.1, p. 189; *20.1*, p. *357*; 22, p. 454
B. T. Kolomiets	**27**, K15	(1968)	*2*, p. *24*; *20.3*, p. *369*; 22.8, p. 553
	27, K169	(1968)	*13.1*, p. *186*; 15, p. 274; 22.4.3, p. 511
L. A. Koloskova	**27**, 263	(1968)	6, p. 65; *8*, p. *84*; 23, p. 561
V. G. Koloskova	**19**, 441	(1967)	*10*, p. *97*; 11, p. 147; 22.5.2, p. 520
O. S. Kolotov	**16**, K121	(1966)	*18.2*, p. *301*; 21.1.1, p. 411
V. B. Koltsov	**26**, 371	(1968)	10, p. 105; *19*, p. *333*; 22.5.2, p. 526
S. N. Komnik	**19**, 533	(1967)	*10*, p. *98*; 22.5.2, p. 521
Č. Koňák	**29**, 203	(1968)	13.1, p. 188; 14.3.1, p. 247; *20.1*, p. *356*; 22.4, p. 496
	29, 707	(1968)	6, p. 68; *13.2*, p. *195*; 20.1, p. 357; 22.4.1, p. 505; 22.4.2, p. 508; 22.4.3, p. 511
V. K. Koniukhov	**21**, K107	(1967)	*20.1*, p. *349*; 22.4.1, p. 500
Yu. V. Konobeev	**27**, 435	(1968)	*13.2*, p. *194*; 13.4, p. 213; 22.9, p. 560
P. P. Konorov	**30**, 845	(1968)	14.3, p. 244; *14.4.1*, p. *265*; 22.1.1, p. 467
P. Konsin	**21**, K39	(1967)	*1.2*, p. *10*; 6, p. 61; *13*, p. *166*; 14.4.2, p. 267
	28, 731	(1968)	*1.2*, p. *12*; *6*, p. *67*; 14.4.2, p. 269; 22, p. 453
H. Konwent	**28**, 39	(1968)	*14.4.2*, p. *269*
G. Koopmans	**27**, 219	(1968)	10, p. 106; *13.4*, p. *213*; 14.3, p. 241; 22.1.1, p. 465
	27, 225	(1968)	10, p. 107; *14.3*, p. *241*; 22.1.1, p. 465
V. A. Kopchik	**27**, 741	(1968)	1.2, p. 12; *8*, p. *85*; 14.4.2, p. 268
H. Koppe	**17**, K229	(1966)	*14.2*, p. *228*
T. J. Koppenaal	**17**, 27	(1966)	*10.1*, p. *115*; 11, p. 145; 21.1, p. 391
D. Kordes	**26**, K103	(1968)	*5*, p. *53*; 10, p. 106; 22.3, p. 494
P. Kordoš	**22**, K59	(1967)	*9*, p. *89*; 10, p. 101; 14.3, p. 237; 22.2.3, p. 491
B. Korneffel	**24**, K137	(1967)	*14.3.2*, *250*; *17*, p. *288*; 21, p. 383; 22.6, p. 540
E. G. Koryak-Doronenko	**24**, 443	(1967)	*10*, p. *103*; 12, p. 158; 22.5.2, p. 524
F. Kosek	**24**, K69	(1967)	*14.4.2*, p. *267*
A. I. Kosinskaya	**23**, 57	(1967)	13.1, p. 180; 18, p. 293; *20*, p. *339*; 22.1.1, p. 463
D. Koskimaki	**25**, K67	(1968)	1.1, p. 8; *1.2*, p. *11*; 21, p. 385; 21.1, p. 403
E. Köster	**19**, 153	(1967)	10.1, p. 120; *18.2*, p. *304*; 21.1, p. 395
	19, 655	(1967)	10.1, p. 121; *18.2*, p. *305*; 21.1, p. 396
	30, 455	(1968)	10.1, p. 135; *18.2*, p. *315*; 21.1, p. 409
H. D. Koswig	**17**, K57	(1966)	*13.4*, p. *202*; 20.1, p. 345; 22, p. 440
	17, K83	(1966)	10.2, p. 137; *20.1*, p. *345*; 22.5.1, p. 515
	24, 605	(1967)	*13.4*, p. *210*; 20.1, p. 351; 22.5.1, p. 516
A. Köthe	**21**, K73	(1967)	*10.1*, p. *122*; 14.1, p. 222; 21, p. 379
G. Kötitz	**23**, 313	(1967)	13.4, p. 209; 20.1, p. 350; *20.3*, p. *367*; 22.5.3, p. 533
	25, K15	(1968)	6, p. 65; *10.2*, p. *141*; *20.3*, p. *368*; 22.5.3, p. 533
	27, 109	(1968)	10, p. 106; *20.3*, p. *369*; 22.5.3, p. 534
B. P. Kotrubenko	**22**, K15	(1967)	*3.2*, p. *35*; *5*, p. *52*; 22.8, p. 551
I. Kovács	**17**, K123	(1966)	*1.1*, p. *8*; 14.1, p. 220; 21, p. 375

E. Kovács-Csetényi	**17**, K123	(1966)	*1.1*, p. *8*; 14.1, p. 220; 21, p. 375
G. G. Kovalevskaya	**23**, 755	(1967)	*6*, p. *63*; 13.1, p. 182; *16*, p. *280*; 22.2.2, p. 488
	30, 441	(1968)	13.4, p. 217; *14.3*, p. *244*; 22.2.2, p. 488
J. Kovár	**23**, 189	(1967)	*7*, p. *77*; 8, p. 83; 22.2.3, p. 491
E. K. Kovev	**17**, 397	(1966)	*5*, p. *49*; 10, p. 94; 22.1.1, p. 459
R. Kowalczyk	**17**, 869	(1966)	*4*, p. *38*; 20, p. 338; 22.1.2, p. 469
	25, 233	(1968)	13.1, p. 183; *18*, p. *293*; *20*, p. *340*; 20.1, p. 351; 22.4.3, p. 510
N. P. Kowalenko	**30**, 533	(1968)	*9*, p. *92*
L. Kozłowski	**27**, K37	(1968)	10.1, p. 131; *12.1*, p. *160*; *18.2*, p. *312*; 21.1.1, p. 421
A. A. Kozma	**17**, 543	(1966)	*1.4*, p. *17*; 3.1, p. 30; 10.1, p. 116; *18.2*, p. *303*; 21.1, p. 392; 21.1.1, p. 413
B. P. Kozyrev	**18**, K57	(1966)	13.4, p. 203; *16*, p. *278*; 22.5, p. 512; 22.8, p. 548
B. Krahl-Urban	**27**, K149	(1968)	*14.3*, p. *242*; 22.1.3, p. 479
A. A. Kralina	**24**, K107	(1967)	*10*, p. *103*; *13.2*, p. *193*; 20.1, p. 351; 22.4.1, p. 501
B. Kramer	**26**, 151	(1968)	*13.1*, p. *184*; 22.1.3, p. 479
E. J. Kramer	**22**, 199	(1967)	1.2, p. 11; *10.1*, p. *124*; 14.2, p. 229; 18, p. 292; 21, p. 380
A. I. Krasilnikov	**19**, K5	(1967)	*20.2*, p. *360*; 22.4.2, p. 506
	23, 587	(1967)	13.4, p. 210; *20.3*, p. *367*; 22.2.1, p. 484
S. Kraśnicki	**21**, K163	(1967)	*13*, p. *166*; *18*, p. *292*; 22.6, p. 538
	22, K55	(1967)	*18.3*, p. *318*
	26, 429	(1968)	*4*, p. *45*; 18.3, p. 319; 18.4, p. 324; 22.8, p. 553
L. Kratěna	**28**, 175	(1968)	*19*, p. *334*; 22.8, p. 554
P. Kratochvíl	**17**, K91	(1966)	*10.1*, p. *116*; 21, p. 374
A. F. Kravchenko	**17**, 479	(1966)	*13.1*, p. *175*; 14.3, p. 232; 22.2.1, p. 482
J. Krejčí	**27**, 545	(1968)	1.3, p. 15; *10.1*, p. *131*; 21.1, p. 405
J. Krempaský	**21**, 175	(1967)	*8*, p. *82*; 22.2.3, p. 491; 22.5.2, p. 522
G. Krenzke	**25**, K131	(1968)	13.3, p. 198; *14.4.1*, p. *261*; 22.1.2, p. 473
W. Kress	**29**, 133	(1968)	*6*, p. *67*; 20.1, p. 356; 22.1.1, p. 466; 22.1.2, p. 474
G. Kretzschmar	**23**, 639	(1967)	*10*, p. *102*; 14.3, p. 238; 22.1.3, p. 478
E. W. Kreutz	**27**, K111	(1968)	*11*, p. *153*; 13.3, p. 198; 14.3, p. 242; 22.2.3, p. 492
F. J. Krings	**24**, 163	(1967)	*10.1*, p. *126*; 18.2, p. 309; 21.1.1, p. 419
P. Krishna	**17**, 437	(1966)	*3*, p. *26*; *5*, p. *49*; 10, p. 93; 22.3, p. 494; 22.4.1, p. 497; 22.5, p. 512
R. Krishnan	**18**, K53	(1966)	*18.3*, p. *317*; 19, p. 327
	21, K5	(1967)	*10*, p. *100*; 18.3, p. 317; 19, p. 329
	26, K47	(1968)	*3.1*, p. *33*; 18.3, p. 319; *19*, p. *333*
P. Krispin	**25**, 147	(1968)	13.4, p. 211; *14.4.1*, p. *261*; 16, p. 281; 22.4.1, p. 501
N. Kristoffel	**21**, K39	(1967)	*1.2*, p. *10*; 6, p. 61; *13*, p. *166*; 14.4.2, p. 267
	28, 731	(1968)	*1.2*, p. *12*; *6*, p. *67*; 14.4.2, p. 269; 22, p. 453
M. A. Krivoglaz	**29**, 53	(1968)	*13*, p. *172*; 18.2, p. 313; 19, p. 335
	29, 61	(1968)	*6.1*, p. *74*; 13, p. 172; 18.2, p. 314; 18.4, p. 324
H. Krogh	**22**, 59	(1967)	*6.1*, p. *72*; 18.4, p. 322; 22.5, p. 513
K.-E. Kröll	**24**, 707	(1967)	*14.4.1*, p. *261*; 22, p. 449
P. F. Kromer	**21**, 805	(1967)	10, p. 100; 11, p. 149; *12*, p. *157*; 22.1.2, p. 471
	21, 811	(1967)	10, p. 100; *12*, p. *157*; 22.1.2, p. 471
E. Kröner	**17**, 211	(1966)	*14.4*, p. *252*; 22, p. 440
H. Kronmüller	**16**, 253	(1966)	13, p. 162; *14.1*, p. *220*; *18.2*, p. *299*; 21, p. 373
	22, 683	(1967)	*10.1*, p. *124*; 14.1, p. 223; 21.1, p. 399
	27, 371	(1968)	10.1, p. 131; *14.2*, p. *230*
	28, 755	(1968)	10.1, p. 132; *18.2*, p. *313*; 21.1, p. 406
	30, K101	(1968)	10.1, p. 136; *18.2*, p. *315*; 21.1, p. 409

A. I. Krotkii	**28,**	269	(1968)	10, p. 109; 13, p. 171; *19*, p. *334*; 22.6, p. 542
F. Kroupa	**18,**	703	(1966)	*10.1*, p. *119*; 21.1.1, p. 414
	23,	K143	(1967)	*10.1*, p. *126*; 21, p. 382
I. V. Krukova	**29,**	715	(1968)	13.4, p. 215; *20.2*, p. *362*; 22.2.1, p. 486
P. I. Krukover	**16,**	651	(1966)	12.1, p. 159; *18.2*, p. *301*; 21.1.1, p. 411
I. N. Krupskii	**24,**	K53	(1967)	*8*, p. *84*; 23, p. 561
	27,	263	(1968)	6, p. 65; *8*, p. *84*; 23, p. 561
V. S. Krylov	**27,**	K1	(1968)	*20.2*, p. *362*; 22.8, p. 553
F. Kubec	**17,**	571	(1966)	3.2, p. 34; *6*, p. *57*; 13.4, p. 202; 22.8, p. 548
	20,	653	(1967)	14.4.2, p. 267; *19*, p. *328*
V. A. Kucheryavy	**26,**	453	(1968)	*8*, p. *84*; 23, p. 561
T. A. Kudykina	**28,**	807	(1968)	13.1, p. 188; *13.2*, p. *194*; 22, p. 453; 22.5.2, p. 528
D. Kuhlmann-Wilsdorf	**18,**	231	(1966)	*10*, p. *96*; *10.1*, p. *117*; 21, p. 375; 21.1, p. 393; 21.5, p. 429; 21.6, p. 431; 22.1, p. 456; 22.1.1, p. 460; 22.1.2, p. 470
	19,	95	(1967)	1.4, p. 18; *10*, p. *97*
	21,	533	(1967)	*3.1*, p. *31*; 10, p. 100; 10.1, p. 123; 21, p. 379; 22, p. 445
	21,	545	(1967)	*10.1*, p. *123*; 21, p. 379; 21.1, p. 398; 21.6, p. 433
	25,	K105	(1968)	*10*, p. *105*; *10.1*, p. *129*; 21, p. 385; 21.1, p. 403; 21.5, p. 429; 21.6, p. 435; 22.1, p. 457; 22.1.1, p. 464; 22.1.2, p. 473
H. Kühnert	**21,**	K171	(1967)	10.2, p. 139; *20.3*, p. *366*; 22.5.2, p. 522
A. A. Kukharskii	**23,**	447	(1967)	*20.1*, p. *350*; 22.1.2, p. 472
L. S. Kukushkin	**22,**	65	(1967)	13.4, p. 208; *20.3*, p. *366*; 22.5, p. 513; 22.8, p. 551
L. A. Kulevskii	**21,**	K107	(1967)	*20.1*, p. *349*; 22.4.1, p. 500
D. Kulgawczuk see D. S. Kulgawczuk				
D. S. Kulgawczuk	**17,**	K195	(1966)	6.1, p. 71; *18.4*, p. *321*; 22, p. 441
	21,	K85	(1967)	*6.1*, p. *72*; 18.4, p. 322; 22.8, p. 550
	22,	617	(1967)	1.2, p. 11; *6.1*, p. *72*; 18.3, p. 318; 18.4, p. 323; 22.8, p. 551
	26,	K83	(1968)	*18.3*, p. *319*; 18.4, p. 324; 22.8, p. 553
	27,	131	(1968)	*6.1*, p. *73*; 18.3, p. 319
B. B. Kuliev	**26,**	K113	(1968)	*2*, p. *24*; 10, p. 106; *12*, p. *158*; 22.1.3, p. 479
B. I. Kuliev	**24,**	K25	(1967)	1.2, p. 11; 3.2, p. 35; *18*, p. *293*; 22.5, p. 513; 22.8, p. 552
G. G. Kulikov	**17,**	K113	(1966)	*13.1*, p. *175*; 20.1, p. 345; 22.2.3, p. 490
M. A. Kumakhof	**30,**	379	(1968)	3.2, p. 36; *11*, p. *155*; 21.1, p. 408; 22.1.2, p. 474
U. Kümmel	**25,**	711	(1968)	10, p. 104; *20.3*, p. *368*; 22.4.1, p. 502
A. B. Kunz	**26,**	447	(1968)	*13.4*, p. *212*; 22.5.2, p. 526
	29,	115	(1968)	*13.1*, p. *188*; 22.5.2, p. 528
K. M. Kunz	**18,**	441	(1966)	*1.4*, p. *17*; *3.1*, p. *30*; 21, p. 376; 21.6, p. 431
G. Kunze	**20,**	179	(1967)	1, p. 7; *4*, p. *41*
Š. Kupča	**29,**	K15	(1968)	*10*, p. *110*; 14.4, p. 254; 22.5.2, p. 529
L. N. Kurbatov	**26,**	K139	(1968)	6, p. 65; *13.2*, p. *194*; *20.2*, p. *361*; 22.4.2, p. 507
V. G. Kurdyumov	**22,**	K75	(1967)	*10.1*, p. *125*; 21.1.1, p. 418
M. V. Kurik	**16,**	K139	(1966)	*13.1*, p. *175*; *20.1*, p. *345*; 22.4.1, p. 497; 22.4.3, p. 509; 22.8, p. 547
M. Kuriyama	**24,**	743	(1967)	*5*, p. *52*
	25,	667	(1968)	*4*, p. *45*; 18.2, p. 310; 21.1.1, p. 420
E. Z. Kurmaev	**24,**	K43	(1967)	*13.1*, p. *183*; *14.2*, p. *229*; 21, p. 383
I. A. Kurova	**24,**	757	(1967)	6, p. 64; 13, p. 168; *16*, p. *281*; 22.1.1, p. 464
V. I. Kutaitsev	**20,**	767	(1967)	*6*, p. *60*; 8, p. 81; 21.1, p. 397
V. A. Kutasov	**18,**	479	(1966)	*8*, p. *80*; *10*, p. *96*; 14.3, p. 233; 15, p. 271; 22, p. 441; 22.8, p. 548

L. B. Kvashnina	**29**,	53 (1968)	*13*, p. *172*; 18.2, p. 313; 19, p. 335
	29,	61 (1968)	*6.1*, p. *74*; 13, p. 172; 18.2, p. 314; 18.4, p. 324
Vu dinh Ky see under letter "V" (Vu)			
R. Labusch	**19**,	715 (1967)	12.1, p. 159; *14.2*, p. *228*
K. Lacefield	**21**,	K69 (1967)	11, p. 148; *12*, p. *157*; 22.6, p. 538
H. Lachowicz	**16**,	313 (1966)	4, p. 37; *18.2*, p. *299*; 21.1.1, p. 410
R. L. Ladd	**30**,	K1 (1968)	*4*, p. *48*; 10.1, p. 135; 21.1, p. 408
J. Łagowski	**26**,	K69 (1968)	*10*, p. *105*; *13.3*, p. *198*; 22.1.1, p. 464
T. S. Lagunova	**17**,	105 (1966)	13.4, p. 202; *14.3*, p. *232*; 22.2.2, p. 487
R. B. Laibowitz	**17**,	353 (1966)	6, p. 57; 13.4, p. 202; *16*, p. *277*; *20.3*, p. *364*; 22.5.1, p. 515
R. Laiho	**18**,	K143 (1966)	7, p. 76; *10.2*, p. *138*; 22.5.2, p. 520
D. I. Lainer	**26**,	K123 (1968)	*14.3*, p. *241*; 15, p. 273; 22, p. 451
J. Lajzerowicz	**16**,	525 (1966)	*14.4.2*, p. *266*
K. Lal	**22**,	141 (1967)	10, p. 101; *14.4*, p. *253*; 22.5.2, p. 523
R. Lallement	**21**,	145 (1967)	*11*, p. *148*; 12, p. 157; 14.1, p. 222; 21.5, p. 429
D. I. Lalović	**28**,	635 (1968)	*13*, p. *171*; *18.2*, p. *313*
E. G. Landsberg	**29**,	9 (1968)	6, p. 56; 13, p. 162; 13.2, p. 191; *13.4*, p. *200*; *14.3*, p. *231*; 20, p. 337; 20.3, p. 363; 22.1.1, p. 458; 22.1.2, p. 468
P. T. Landsberg	**19**,	777 (1967)	13.4, p. 205; *20.2*, p. *360*; 22, p. 443
J. van Landuyt	**16**,	585 (1966)	*4*, p. *38*; 10, p. 94; 22.6, p. 535
	18,	167 (1966)	*10.1*, p. *117*; 21.1.1, p. 413
	18,	325 (1966)	*4*, p. *39*; 22.6, p. 536
	18,	343 (1966)	4, p. 39; *10*, p. *96*; *10.1*, p. *118*
	18,	363 (1966)	4, p. 39; *10*, p. *96*; *10.1*, p. *118*; 21.1.1, p. 414; 22.6, p. 536
	21,	393 (1967)	*10*, p. *99*; *10.1*, p. *122*; 21.1.1, p. 417; 22.6, p. 538
	23,	549 (1967)	*4*, p. *43*; *10.1*, p. *125*; 21.1.1, p. 419
	26,	577 (1968)	*4*, p. *46*; 10, p. 105
	30,	797 (1968)	*4*, p. *48*; 10, p. 113; 10.1, p. 136; 21.1.1, p. 423; 22.6, p. 544
H. Lange	**22**,	229 (1967)	13.1, p. 179; 13.2, p. 192; *14.4.1*, p. *259*; *20*, p. *339*; 20.1, p. 349; 22.4.1, p. 500; 22.4.2, p. 506
	23,	K67 (1967)	13.1, p. 181; *20.1*, p. *350*; 22.1.3, p. 477; 22.4.1, p. 501
V. N. Lange	**22**,	K15 (1967)	*3.2*, p. *35*; *5*, p. *52*; 22.8, p. 551
W. Lange	**18**,	67 (1966)	*9*, p. *87*; 21, p. 375
	18,	75 (1966)	*9*, p. *87*; 21, p. 375
M. J. van Langen	**19**,	107 (1967)	10.1, p. 120; *12.1*, p. *159*; 21.1.1, p. 414
H. P. D. Lanyon	**17**,	K185 (1966)	*3.1*, p. *30*; 22.1.3, p. 476
A. C. McLaren see under letter "M"			
Y. E. Lapsker	**23**,	K5 (1967)	20.1, p. 350; *20.2*, p. *361*; 20.3, p. 367; 22.5, p. 513; 22.8, p. 552
S. D. McLaughlan see under letter "M"			
E. P. Lautenschlager	**20**,	443 (1967)	*4*, p. *41*; 10.1, p. 122; 21.1, p. 396
G. Lautz	**21**,	K139 (1967)	*13*, p. *166*; 14.4.1, p. 259; *20*, p. 339; 22.1.1, p. 462
E. F. Lavrentev	**29**,	569 (1968)	*10.1*, p. *133*; 21.1, p. 407
B. M. Lavrushin	**29**,	715 (1968)	13.4, p. 215; *20.2*, p. *362*; 22.2.1, p. 486
T. J. Law	**16**,	117 (1966)	*1.4*, p. *16*; *3.1*, p. *29*; 21.6, p. 430
	16,	479 (1966)	*1.4*, p. *16*; 3.1, p. 29; 21.6, p. 430
	25,	139 (1968)	*1.4*, p. *19*; *3.1*, p. *33*; 10.1, p. 127; 21.6, p. 435
C. Lea	**25**,	613 (1968)	14.1, p. 224; *17*, p. *288*; 21.5, p. 429

H. J. Leamy	**24,**	149	(1967)	*10*, p. *103*; *10.1*, p. *126*; 21.1.1, p. 419
A. D. W. Leaver	**25,**	531	(1968)	*6*, p. *65*; 8, p. 84; *10.1*, p. *128*; 21, p. 384; 21.1, p. 402
B. M. Lebed	**17,**	K7	(1966)	18.3, p. 316; *19*, p. *326*
	27,	K85	(1968)	7, p. 78; *18.3*, p. *320*
	30,	K111	(1968)	*18.3*, p. *320*; 19, p. 336
J. Leciejewicz	**22,**	517	(1967)	5, p. 52; *18.4*, p. *323*; 22, p. 447
	23,	K123	(1967)	*18.4*, p. *323*; 22, p. 448
	30,	61	(1968)	*18.4*, p. *325*; 22.8, p. 556
	30,	157	(1968)	4, p. 48; *18.4*, p. *325*; 22, p. 454
G. Lecoy	**16,**	721	(1966)	*13.4*, p. *202*; 14.3, p. 232; 16, p. 276; 22.1.1, p. 459
	21,	619	(1967)	13.4, p. 207; *14.3*, p. *236*; *16*, p. *279*; 22.1.1, p. 462
	30,	9	(1968)	*14.3.2*, p. *251*; 22.6, p. 543
E. M. Ledovskaya	**25,**	63	(1968)	*13*, p. *168*
	30,	867	(1968)	*13*, p. *173*
M. Lefeld-	**17,**	869	(1966)	*4*, p. *38*; 20, p. 338; 22.1.2, p. 469
Sosnowska	**30,**	K23	(1968)	*5*, p. *55*; 22.1.2, p. 474
T. Leffers	**25,**	337	(1968)	*10.1*, p. *128*; 21.1, p. 402
H. Le Gall see under letter "G" (Gall)				
S. Legvold	**26,**	611	(1968)	10.1, p. 129; *14.1*, p. *225*; 21, p. 385; 21.6, p. 435
C. Lehmann see Chr. Lehmann				
Chr. Lehmann	**16,**	507	(1966)	1.3, p. 14; *11*, p. *145*; 21.1, p. 390
	30,	219	(1968)	10.1, p. 135; *11*, p. *155*; 14.1, p. 227; 18.2, p. 315; 21, p. 408
G. Leibfried	**22,**	423	(1967)	*11*, p. *149*; 21.1, p. 399
	22,	433	(1967)	*11*, p. *149*; 21, p. 380; 21.1, p. 399; 21.6, p. 434
A. Yu. Leiderman	**26,**	419	(1968)	*13.4*, p. *212*; 20.3, p. 368; 22, p. 450
S. Lele	**20,**	59	(1967)	*4*, p. *41*
B. Lemaire	**17,**	K41	(1966)	*13*, p. *163*; *18.3*, p. *316*
H. Lemke	**16,**	413	(1966)	13.4, p. 201; *14.4.1*, p. *255*; 22, p. 440
	16,	427	(1966)	13.4, p. 201; *14.4.1*, p. *255*; 22.1.2, p. 469
	24,	127	(1967)	*14.4.1*, p. *260*; 22.1.2, p. 472
	25,	K131	(1968)	13.3, p. 198; *14.4.1*, p. *261*; 22.1.2, p. 473
M. C. Lemmens	**19,**	107	(1967)	10.1, p. 120; *12.1*, p. *159*; 21.1.1, p. 414
S. Le Montagner see under letter "M" (Montagner)				
D. Lepski	**23,**	335	(1967)	6, p. 62; *9*, p. *90*
A. G. Lesnik	**17,**	745	(1966)	3.1, p. 30; 10.1, p. 116; *18.2*, p. *303*; 21.1.1, p. 413
	17,	753	(1966)	10.1, p. 117; *18.2*, p. *303*; 19, p. 327; 21.1.1, p. 413
T. N. Letova	**16,**	K121	(1966)	*18.2*, p. *301*; 21.1.1, p. 411
A. Le Traon see under letter "T" (Traon)				
F. Le Traon see under letter "T" (Traon)				
J. Levasseur	**21,**	K1	(1967)	*9*, p. *89*; 21.1, p. 397; 21.1.1, p. 417
J. L. Leveque	**29,**	K179	(1968)	*10.1*, p. *134*; *11*, p. *154*; 18.2, p. 315; 21.1.1, p. 423
A. Levialdi	**28,**	K151	(1968)	*20.3*, p. *370*; 22.4.1, p. 504
G. I. Levin	**17,**	745	(1966)	3.1, p. 30; 10.1, p. 116; *18.2*, p. *303*; 21.1.1, p. 413
R. D. Levine	**19,**	587	(1967)	*13.3*, p. *197*; 13.4, p. 205
M. E. Levinshtein	**28,**	827	(1968)	*14.4.1*, p. *263*; 22, p. 453
	29,	K103	(1968)	*14.3.2*, p. *251*; *14.4.1*, p. *264*; 22.2.1, p. 486
M. E. Levinstein see M. E. Levinshtein				
R. Levy	**22,**	K155	(1967)	6, p. 62; 13.2, p. 193; *20.3*, p. *367*; 22.4.2, p. 507
	27,	K29	(1968)	*20.3*, p. *369*; 22.5.1, p. 516
J. E. Lewis	**29,**	743	(1968)	6, p. 68; *8*, p. *86*; 13, p. 173; *15*, p. *274*; 22, p. 454; 22.8, p. 555
M. H. Lewis	**26,**	571	(1968)	*4*, p. *45*; 10, p. 105; *14.4.2*, p. *268*
S. J. Lewis	**26.**	K51	(1968)	*6.1*, p. *73*; 21.1.1, p. 420

J. C. M. Li	**18**,	517	(1966)	*10*, p. *96*; *10.1*, p. *118*
	18,	527	(1966)	*10*, p. *96*; *10.1*, p. *118*; 21, p. 376; 22, p. 441; 22.1.2, p. 470
	23,	539	(1967)	*10.1*, p. *125*; 21.2, p. 425
L. Li	**30**,	K77	(1968)	*20.2*, p. *362*; 22.5.3, p. 534
S. Libovický	**20**,	K85	(1967)	*4*, p. *42*; 21.1.1, p. 416
I. Licea	**18**,	K9	(1966)	13, p. 164; *14.3*, p. *233*; 22.2.3, p. 490
	22,	147	(1967)	13, p. 166; *14.3.1*, p. *245*; 22, p. 446
	25,	461	(1968)	*13*, p. *168*; *14.3*, p. *240*; 22.2, p. 481; 22.4, p. 495
	26,	115	(1968)	*13*, p. *169*; *14.3*, p. *240*; 22.2.3, p. 492
	27,	K137	(1968)	*13*, p. *170*; 13.1, p. 186; *14.3*, p. *242*; 22.2.3, p. 492
	27,	K143	(1968)	*13*, p. *170*; *14.3*, p. *242*; 14.4.1, p. 263; 22.2.3, p. 492
K. F. Lider	**18**,	K1	(1966)	11, p. 147; *13.2*, p. *191*; *20.3*, p. *364*; 22.4.1, p. 499
M. Lietz	**17**,	597	(1966)	5, p. 50; *13.1*, p. *175*; 20.1, p. 346; 22.4.1, p. 498
S. Ligenza	**29**,	K99	(1968)	*6.1*, p. *74*; 22.8, p. 556
R. C. Lincoln	**18**,	265	(1966)	3.2, p. 34; *12*, p. *156*; 22.5.2, p. 519
U. Lindner	**20**,	K131	(1967)	*13.4*, p. *206*; 18.1, p. 296; 21, p. 378
P. F. Lindquist	**17**,	K25	(1966)	1.2, p. 9; *3*, p. *26*; 4, p. 38; 22.1, p. 456
	19,	K99	(1966)	1.2, p. 10; *3*, p. *26*; 4, p. 40; 22.1, p. 457
A. Linenberg	**24**,	695	(1967)	*9*, p. *90*; 22.5.2, p. 525
R. M. F. Linford	**28**,	K63	(1968)	*14.2*, p. *230*
R. Link	**26**,	K159	(1968)	13.4, p. 212; *14.3*, p. *241*; 22.1.3, p. 479
E. Linke	**26**,	K89	(1968)	*17*, p. *289*; *20.3*, p. *369*; 22.5.2, p. 526
I. S. McLintock see under letter "M"				
D. Linzen	**26**,	K43	(1968)	*18.3*, p. *319*
A. A. Lipnik	**28**,	75	(1968)	6, p. 66; *7*, p. *78*; 13.1, p. 187; 13.2, p. 194; 22, p. 452
H. A. Lipsitt	**19**,	K99	(1967)	1.2, p. 10; *3*, p. *26*; 4, p. 40; 22.1, p. 457
M. P. Lisitsa	**27**,	K81	(1968)	*13.1*, p. *186*; 20.1, p. 354; 22.4.1, p. 502; 22.4.2, p. 507; 22.8, p. 554
V. G. Litovchenko	**16**,	751	(1966)	13.3, p. 196; *16*, p. *276*; 22.1.1, p. 459; 22.1.2, p. 469
Yu. G. Litvinenko	**30**,	49	(1968)	13.1, p. 189; *18.4*, p. *325*; 20.1, p. 358; 23, p. 561
J. Litwin	**20**,	K77	(1967)	*1.4*, p. *18*; 22.1.1, p. 461; 22.2.1, p. 483
G. C. T. Liu	**18**,	517	(1966)	*10*, p. *96*; *10.1*, p. *118*
	18,	527	(1966)	*10*, p. *96*; *10.1*, p. *118*; 21, p. 376; 22, p. 441; 22.1.2, p. 470
C. Llinares	**16**,	721	(1966)	*13.4*, p. *202*; 14.3, p. 232; 16, p. 276; 22.1.1, p. 459
	21,	619	(1967)	13.4, p. 207; *14.3*, p. *236*; *16*, p. *279*; 22.1.1, p. 462
	21,	K31	(1967)	*9*, p. *89*; 10, p. 100; 14.3, p. 236; 22.2.3, p. 491
A. Lodding	**22**,	157	(1967)	*9*, p. *89*; 21.2, p. 424; 22.8, p. 551
Yu. A. Logatchov	**19**,	K35	(1967)	10.2, p. 138; *13.4*, p. *204*; 22.5.2, p. 520
	25,	763	(1968)	*13.4*, p. *211*; 20.3, p. 368; 22, p. 450
	26,	765	(1968)	*13.4*, p. *212*; 20.3, p. 369; 22.5.2, p. 526
G. M. Loginov	**18**,	873	(1966)	*18.2*, p. *304*; *18.4*, p. *321*; 22, p. 442
R. K. Lomnes	**25**,	583	(1968)	*15*, p. *273*; *16*, p. *281*; 22.4.1, p. 501
G. I. Lomova	**28**,	815	(1968)	*14.4.1*, p. *263*; 22.1.1, p. 466
G. B. London	**30**,	K13	(1968)	13.4, p. 217; *14.3*, p. *244*; *15*, p. *275*; 22.8, p. 556
J. Loos	**23**,	721	(1967)	13, p. 168; *19*, p. *331*
A. E. Lord, Jr.	**23**,	521	(1967)	6, p. 63; *13*, p. *167*; 18.2, p. 308
	26,	717	(1968)	6, p. 65; *7*, p. *77*; 13, p. 169; 18.2, p. 311; 22.8, p. 553
M. H. Loretto	**20**,	505	(1967)	*10.1*, p. *122*; 12.1, p. 159; 21.1, p. 397
	20,	521	(1967)	*10.1*, p. *122*; 12.1, p. 159; 21.1, p. 397

G. V. LOSHAKOVA	**29,** 435	(1968)	4, p. 47; *5*, p. *54*; 6, p. 68; 10, p. 110; 14.3, p. 243; 22.8, p. 555
A. F. LUBCHENKO	**18,** 853	(1966)	6, p. 58; 13.4, p. 204; *20.1*, p. *346*
	20, 427	(1967)	6, p. 60; 13.4, p. 205; *20.1*, p. *347*
	21, 785	(1967)	*13.4*, p. *207*; *20.1*, p. *349*
	27, K73	(1968)	10, p. 107; 13.4, p. 213; *20.1*, p. *354*; 22, p. 452
P. LUCASSON	**24,** K77	(1967)	*10.1*, p. *127*; 11, p. 150; 14.1, p. 224; 21, p. 383
	30, K139	(1968)	*10.1*, p. *136*; 11, p. 155; 14.1, p. 227; 21.1, p. 409
R. LÜCK	**18,** 49	(1966)	13.1, p. 176; *14.1*, p. *221*; 21, p. 375
	18, 59	(1966)	13.1, p. 176; *14.1*, p. *221*; 21, p. 375
	21, 671	(1967)	*14.1*, p. *222*; 21.1, p. 398
R. LUDEKE	**23,** 413	(1967)	1.4, p. 19; *3.1*, p. *32*; 13.1, p. 181; *20.1*, p. *350*; 22.4.2, p. 507
W. LÜDKE	**20,** 121	(1967)	9, p. 88; 13.4, p. 205; *19*, p. *328*; 22.5.2, p. 521
M. LUDWIG	**28,** K117	(1968)	*17*, p. *289*; 21, p. 387
W. LUDWIG (Gießen)	**19,** 313	(1967)	*6*, p. *58*; 10, p. 97
W. LUDWIG (Berlin)	**24,** K137	(1967)	*14.3.2*, p. *250*; *17*, p. *288*; 21, p. 383; 22.6, p. 540
	24, K149	(1967)	*14.3.2*, p. *250*; *16*, p. *281*; 22, p. 449; 22.4.1, p. 501
	24, K161	(1967)	*14.4.1*, p. *261*; 22.8, p. 552
	30, K61	(1968)	13.3, p. 199; *14.3.2*, p. *251*; 14.4.1, p. 264; 22.4.3, p. 511
A. LUFT	**26,** 485	(1968)	*10.1*, p. *129*; 21, p. 385
W. VAN DER LUGT	**19,** 327	(1967)	2, p. 22; *19*, p. *327*; 21, p. 377
	23, K83	(1967)	*19*, p. *330*; 21.2, p. 424
P. LUKÁČ	**19,** K47	(1967)	*10.1*, p. *121*; 21, p. 377
P. LUKÁŠ	**21,** 717	(1967)	*10.1*, p. *123*; 21.1, p. 398
	27, 545	(1968)	1.3, p. 15; *10.1*, p. *131*; 21.1, p. 405
L. I. LUKASHENKO	**17,** 543	(1966)	*1.4*, p. *17*; 3.1, p. 30; 10.1, p. 116; *18.2*, p. *303*; 21.1, p. 392; 21.1.1, p. 413
S. LUNGU	**23,** 147	(1967)	2, p. 23; 10, p. 102; *11*, p. *150*; 22.8, p. 552
M. LUPULESCU	**18,** K13	(1966)	*10.1*, p. *118*; *18.2*, p. *304*; 21.1, p. 394
	19, 613	(1967)	*18.2*, p. *305*
	21, 601	(1967)	14.1, p. 222; *18.2*, p. *307*; 21.1.1, p. 417
O. J. LUSTE	**27,** 101	(1968)	13.1, p. 185; *14.3*, p. *241*; 22, p. 451; 22.7, p. 546
H. LUTSCH	**26,** K159	(1968)	13.4, p. 212; *14.3*, p. *241*; 22.1.3, p. 479
M. LUUKKALA	**25,** K99	(1968)	*6*, p. *65*; 21, p. 385
	29, 377	(1968)	6, p. 67; *7*, p. *78*; 22.6, p. 543
A. LUYCKX	**17,** K217	(1966)	*14.1*, p. *221*; 21.7, p. 438
E. D. LYAK	**19,** 533	(1967)	*10*, p. *98*; 22.5.2, p. 521
V. G. LYAPIN	**23,** 429	(1967)	*13.1*, p. *181*; 22, p. 447
D. W. LYNCH	**23,** 697	(1967)	*13.1*, p. *182*; 14.3, p. 238; 20.1, p. 350; 22.7, p. 545
M. H. VAN MAAREN	**24,** K125	(1967)	*9*, p. *90*; 10, p. 103; 22.2.3, p. 492
R. MACH	**24,** K149	(1967)	*14.3.2*, p. *250*; *16*, p. *281*; 22, p. 449; 22.4.1, p. 501
J. MACHÁČKOVÁ	**26,** 435	(1968)	18.3, p. 319; *19*, p. *333*
E. MACHERAUCH	**19,** 793	(1967)	*10.1*, p. *121*; 21.1, p. 396
P. MAČKUS	**25,** 331	(1968)	2, p. 23; 13.4, p. 211; 14.3.1, p. 246; *14.4.1*, p. *261*; 22.1.3, p. 478
E. G. MADATOVA	**26,** K151	(1968)	6.1, p. 73; *10.1*, p. *130*; 14.1, p. 225; 18.1, p. 297; 21.6, p. 436
J. MAEGE	**30,** K153	(1968)	13.4, p. 218; *14.3.1*, p. *247*; 15, p. 275; 16, p. 285; *20.1*, p. *359*; 22, p. 455
T. G. MAGARRAMOV	**29,** K19	(1968)	*14.1*, p. *226*; *18.2*, p. *314*; 21.1, p. 407
K. MAIER	**27,** 713	(1968)	*10.2*, p. *143*; 13.4, p. 214; 22.5.2, p. 526

E. Majerníková	**22,**	113	(1967)	13.2, p. 192; *18*, p. *292*; *20*, p. *339*; 22.9, p. 559
E. F. Makarov	**24,**	45	(1967)	*6.1*, p. *73*; 18.2, p. 308; 21.1.1, p. 419
	25,	607	(1968)	*6.1*, p. *73*; 18.3, p. 319
B. Makiej	**25,**	K127	(1968)	*14.2*, p. *230*
M. J. Małachowski	**19,**	K27	(1967)	*14.3.2*, p. *249*
K. Malén	**26,**	621	(1968)	*10*, p. *106*; *10.1*, p. *129*
C. Malgrange	**30,**	K23	(1968)	*5*, p. *55*; 22.1.2, p. 474
J. Malinowski	**26,**	267	(1968)	10.2, p. 142; 13.4, p. 212; *16*, p. *282*; 22.5.1, p. 516
	28,	725	(1968)	*10.2*, p. *143*; 16, p. 283; 22.5.1, p. 517
A. U. Malsagov	**18,**	K29	(1966)	2, p. 22; *15*, p. *271*; 22.8, p. 548
G. Mambriani	**29,**	341	(1968)	*1.2*, p. *12*; 3, p. 27; 13.1, p. 188; 17, p. 289; 22.7, p. 546
Sh. V. Mamedov	**16,**	K113	(1966)	10, p. 94; *19*, p. *326*; 22.1.3, p. 476
A. Z. Mamedova	**25,**	75	(1968)	13.4, p. 211; *16*, p. *281*; 22, p. 449
B. P. Mamko	**30,**	703	(1968)	13.1, p. 189; *20*, p. *343*; 21.1, p. 409; 22.1.1, p. 467; 22.1.3, p. 480
T. N. Mamontova	**27,**	K15	(1968)	*2*, p. *24*; *20.3*, p. *369*; 22.8, p. 553
J. Man	**19,**	543	(1967)	*10.1*, p. *121*; 21.1, p. 396
R. Mǎnǎilǎ	**19,**	K19	(1967)	*1.2*, p. *10*; 22.8, p. 549
V. S. Mandel	**30,**	K111	(1968)	*18.3*, p. *320*; 19, p. 336
M. A. Manko	**23,**	587	(1967)	13.4, p. 210; *20.3*, p. *367*; 22.2.1, p. 484
E. Mann	**24,**	721	(1967)	*14.1*, p. *224*; 21, p. 383
K. E. Mann	**28,**	K27	(1968)	*19*, p. *334*; 22.6, p. 542
L. C. Mansur	**28,**	663	(1968)	*6*, p. *67*; 20.1, p. 356; 22.4.1, p. 504; 22.4.2, p. 508; 22.6, p. 542; 22.8, p. 554
E. Mäntysalo	**18,**	K143	(1966)	7, p. 76; *10.2*, p. *138*; 22.5.2, p. 520
G. M. McManus see under letters "Mc"				
M. Manzel	**26,**	K43	(1968)	*18.3*, p. *319*
V. G. Manzhelii	**17,**	K139	(1966)	*8*, p. *79*; 23, p. 561
	24,	K53	(1967)	*8*, p. *84*; 23, p. 561
	26,	453	(1968)	*8*, p. *84*; 23, p. 561
	27,	263	(1968)	6, p. 65; *8*, p. *84*; 23, p. 561
V. G. Manzhely see V. G. Manzhelii				
A. Marais	**24,**	635	(1967)	*18.3*, p. *319*
M. J. Marcinkowski	**24,**	149	(1967)	*10*, p. *103*; *10.1*, p. *126*; 21.1.1, p. 419
	25,	K67	(1968)	1.1, p. 8; *1.2*, p. *11*; 21, p. 385; 21.1, p. 403
M. Marina	**24,**	K17	(1967)	*14.3.2*, p. *250*; *14.4.1*, p. *261*; 22.1.1, p. 463
K. Marinova	**27,**	K117	(1968)	*14.4.1*, p. *262*; 16, p. 282; 22.4.1, p. 503
H. Markert	**20,**	K67	(1967)	*7*, p. *76*; 18.2, p. 306; 18.3, p. 317; 21.1.1, p. 416
M. M. Markus	**22,**	K15	(1967)	*3.2*, p. *35*; *5*, p. *52*; 22.8, p. 551
M. Marques	**24,**	481	(1967)	*13.1*, p. *183*; 14.3, p. 239; 22.2.3, p. 491
F. W. Marsh	**25,**	K91	(1968)	*10.1*, p. *128*; 21.1, p. 403
C. Marti	**28,**	97	(1968)	13, p. 171; *19*, p. *334*; 22.4.1, p. 503
G. Martin	**23,**	223	(1967)	*9*, p. *90*; 10.1, p. 125; 21.6, p. 434
J. L. Martin	**22,**	349	(1967)	*10*, p. *101*; 22.3, p. 494
A. A. Maryakhin	**20,**	737	(1967)	*13.1*, p. *178*; 21, p. 378
	23,	K133	(1967)	*13.1*, p. *182*; 21, p. 382
K. Maschke	**28,**	577	(1968)	*13.1*, p. *187*; 22.4, p. 496
S. I. Masharov	**21,**	747	(1967)	*14.1*, p. *223*; 18.4, p. 322; 21, p. 379
	27,	455	(1968)	*6*, p. *65*; 8, p. 84; 13, p. 170; 14.1, p. 225; 15, p. 274; 21, p. 386
V. E. Mashenko	**26,**	K139	(1968)	6, p. 65; *13.2*, p. *194*; *20.2*, p. *361*; 22.4.2, p. 507
V. S. Mashkevich	**21,**	K59	(1967)	6, p. 61; *20.1*, p. *348*; 22, p. 445
E. S. Mashkova	**19,**	407	(1967)	*17*, p. *287*; 21, p. 377; 22.1, p. 457
	19,	425	(1967)	*6*, p. *58*; 21.1, p. 395

T. V. MASHOVETS	**22,**	K95	(1967)	*11*, p. *149*; 14.3, p. 237; 22.2.3, p. 491
	22,	K103	(1967)	10, p. 101; *11*, p. *150*; 14.3, p. 237; 22.1.1, p. 462
O. MASSENET	**16,**	301	(1966)	*18.2*, p. *299*; *18.4*, p. *321*; 21.1, p. 390; 21.1.1, p. 410
	16,	569	(1966)	4, p. 37; *18.2*, p. *300*; 21.1.1, p. 411
B. MASTEL	**27,**	K89	(1968)	*10.1*, p. *131*; 11, p. 152; 21, p. 386
J. MATHON	**30,**	K131	(1968)	*18.2*, p. *315*; 21.1, p. 409; 21.1.1, p. 423
T. MATLAK	**17,**	K53	(1966)	*6.1*, p. *71*; 21, p. 374
F. MATOSSI †	**24,**	K65	(1967)	*14.3*, p. *239*; 22.2.1, p. 484
	28,	373	(1968)	4, p. 47; 10, p. 109; *20.1*, p. *355*; 22.4.1, p. 503
K.-H. MATUCHA	**26,**	291	(1968)	*10*, p. *105*; 22.5.2, p. 526
M. MATYÁŠ	**22,**	K141	(1967)	*9*, p. *90*; 21.2, p. 424
E. V. MATYUSHKIN	**18,**	683	(1966)	*13.4*, p. *204*; 18.4, p. 321; *20.3*, p. *364*; 22.5.3, p. 532
	22,	65	(1967)	13.4, p. 208; *20.3*, p. *366*; 22.5, p. 513; 22.8, p. 551
HJ. MATZKE	**18,**	285	(1966)	1.2, p. 9; 2, p. 22; *11*, p. *147*; 22.6, p. 536
	18,	317	(1966)	11, p. 147; *17*, p. *286*; 22.5.2, p. 519
	25,	641	(1968)	*9*, p. *91*; 11, p. 151; 22.5.2, p. 525
E. MAURER	**19,**	805	(1967)	*14.4.2*, p. *266*
GH. MAXIM	**22,**	K113	(1967)	*18.3*, p. *318*
	23,	K125	(1967)	*18.3*, p. *318*
	30,	K49	(1968)	*18.3*, p. *320*
H. MAYER	**18,**	779	(1966)	13, p. 164; *17*, p. *287*; 21.2, p. 424
W. L. MCCUBBIN	**16,**	289	(1966)	*3*, p. *26*; *13.1*, p. *174*; 20, p. 337; 22.9, p. 558
P. G. MCDOUGALL	**29,**	873	(1968)	*4*, p. *47*; 10, p. 111; 22.6, p. 543
S. MCK. COUSLAND see under letter "C" (COUSLAND)				
A. C. MCLAREN	**18,**	221	(1966)	4, p. 38; *20.1*, p. *346*; 22.8, p. 548
	19,	631	(1967)	4, p. 40; *10*, p. *98*; 22.6, p. 536
S. D. MCLAUGHLAN	**27,**	695	(1968)	10.2, p. 142; *11*, p. *152*; 19, p. 333; 20.1, p. 354; 22.5.3, p. 534
I. S. MCLINTOCK	**29,**	K157	(1968)	*19*, p. *335*
G. M. MCMANUS	**25,**	667	(1968)	*4*, p. *45*; 18.2, p. 310; 21.1.1, p. 420
S. B. MCRICKARD	**18,**	89	(1966)	*10.1*, p. *117*; 21.1.1, p. 413
D. B. MEADOWCROFT	**27,**	535	(1968)	*10*, p. *107*; *20.3*, p. *369*; 22.4.1, p. 502
	27,	541	(1968)	4, p. 46; 10, p. 107; *20.3*, p. *370*; 22.4.1, p. 502
H. MECHETTI	**25,**	K65	(1968)	*14.2*, p. *230*
	28,	683	(1968)	*14.2*, p. *230*
C. H. B. MEE	**25,**	613	(1968)	14.1, p. 224; *17*, p. *288*; 21.5, p. 429
A. MEESSEN	**26,**	125	(1968)	*17*, p. *288*; 20.1, p. 352; 21, p. 385
A. MEHRA	**28,**	61	(1968)	5, p. 53; 13, p. 171; *20.1*, p. *355*; 22.5, p. 514; 22.8, p. 554
	29,	847	(1968)	6, p. 69; *10*, p. *111*; 13.4, p. 215; *20.1*, p. *357*; 22.5.2, p. 529
H. MEHRER	**29,**	231	(1968)	*9*, p. *92*; *10.1*, p. *133*; 21.6, p. 437
E. S. MEIERAN	**29,**	653	(1968)	*4*, p. *47*; *10*, p. *111*; 22.1.2, p. 474; 22.2.1, p. 486
P. H. E. MEIJER	**23,**	353	(1967)	*8*, p. *83*; *18*, p. *293*; 22.8, p. 552
E. MEISEL	**19,**	K23	(1967)	*10.2*, p. *138*; 22.5.3, p. 532
W. MEISEL	**16,**	127	(1966)	*6.1*, p. *71*; 18.3, p. 316
	18,	K41	(1966)	*6.1*, p. *71*; 22.6, p. 536
D. MEISSNER	**23,**	481	(1967)	10.1, p. 125; *11*, p. *150*; 14.1, p. 223; 21.6, p. 434
A. SH. MEKHTIEV	**27,**	K169	(1968)	*13.1*, p. *186*; 15, p. 274; 22.4.3, p. 511
S. I. MEKHTIEVA	**16,**	K31	(1966)	2, p. 22; 10, p. 94; *20.1*, p. *344*; 22.1.3, p. 476
	18,	K153	(1966)	*10*, p. *97*; 14.3, p. 234; 22.1.3, p. 477
H. MELL	**24,**	183	(1967)	13.1, p. 182; *14.3*, p. *238*; 22.1.3, p. 478
J. MELNGAILIS	**16,**	247	(1966)	*10.1*, p. *114*; 12.1, p. 159; 21.1, p. 390

N. I. Melnikov	**26,** 371	(1968)	10, p. 105; *19*, p. *333*; 22.5.2, p. 526
	30, 851	(1968)	10.2, p. 144; *19*, p. *336*; 22.5.2, p. 531
	30, 859	(1968)	3.2, p. 36; 10.2, p. 144; 13.4, p. 218; *19*, p. *336*; 22.5.2, p. 531
A. N. Men	**24,** 51	(1967)	3.2, p. 35; *13.4*, p. *210*; 22.6, p. 539
J. Mendiola	**19,** 805	(1967)	*14.4.2*, p. *266*
H. Menniger	**25,** K59	(1968)	*13.1*, p. *184*; 16, p. 281; 22.1.2, p. 473
A. Meňovský	**30,** K31	(1968)	*18.1*, p. *298*; *18.2*, p. *315*; 22, p. 454
T. Merceron	**24,** 635	(1967)	*18.3*, p. *319*
K. L. Merkle	**18,** 173	(1966)	10.1, p. 117; *11*, p. *146*; 21.1, p. 393; 21.6, p. 431
A. Merlini	**18,** 157	(1966)	*5*, p. *50*; 22.1.1, p. 460; 22.1.2, p. 469
L. Merten	**25,** 125	(1968)	*6*, p. *64*; 14.4.2, p. 268
	28, 111	(1968)	*6*, p. *66*; 20.1, p. 355; 22.6, p. 542
	30, 449	(1968)	*6*, p. *70*; *20*, p. *343*; 22.6, p. 544
J. Mertsching	**16,** 267	(1966)	*7*, p. *76*; 13, p. 162; 21, p. 373
	17, 799	(1966)	*13*, p. *163*; *20*, p. *337*
	21, K167	(1967)	*14*, p. *219*
	26, 9	(1968)	*7*, p. *76*; 13, p. 162; 19, p. 326; 20, p. 337; 21, p. 373
	27, 345	(1968)	13.1, p. 186; 14.1, p. 225; *19*, p. *333*; 21, p. 386
M. Meshii	**27,** 185	(1968)	*10.1*, p. *130*; 21, p. 385; 21.1, p. 404
A. M. Meshkov	**17,** K95	(1966)	13.4, p. 202; *15*, p. *270*; 16, p. 277; 22, p. 440
U. Messerschmidt	**26,** 701	(1968)	*10*, p. *106*; 22.5.2, p. 526
	30, K87	(1968)	*10*, p. *113*; 22.5.2, p. 531
E. A. Metzbower	**20,** 681	(1967)	3.2, p. 35; *6*, p. *60*; 8, p. 81; 12.1, p. 160; 21.1, p. 397; 21.3, p. 426
	25, 403	(1968)	*3.2*, p. *36*; 6, p. 64; 8, p. 84; 12.1, p. 160; 21.1, p. 402
W. Metzdorf	**17,** 285	(1966)	1.4, p. 16; *18.2*, p. *302*; 21.1.1, p. 412
K. Meyer	**22,** K123	(1967)	*20.3*, p. *366*; 22.4.1, p. 500
	26, K89	(1968)	*17*, p. *289*; *20.3*, p. *369*; 22.5.2, p. 526
K.-P. Meyer	**30,** K161	(1968)	*10.2*, p. *144*; *20.3*, p. *372*; 22.5.2, p. 531
I. P. Mezina	**19,** 35	(1967)	13.4, p. 204; *20.3*, p. *365*; 22.5.2, p. 520
R. Michalec	**29,** K51	(1968)	*5*, p. *54*; 14.4, p. 254; 22.6, p. 543
C. Michalk	**27,** K51	(1968)	*18.3*, p. *320*
L. Michalowsky	**16,** 795	(1966)	*18.3*, p. *316*
D. Michell	**21,** 369	(1967)	*10.1*, p. *122*; 21.1, p. 397
	26, K93	(1968)	*10*, p. *106*; 22.9, p. 560
	27, 291	(1968)	1.3, p. 15; *1.4*, p. *20*; 10, p. 107; 22.6, p. 541
Y. Mihai	**21,** K65	(1967)	2, p. 23; 13.4, p. 207; *20.1*, p. *348*; 22.1.3, p. 477
K. Mika	**21,** 279	(1967)	*6*, p. *61*
M. P. Mikhailova	**19,** 429	(1967)	*14.3.2*, p. *249*; 16, p. 278; 22.2.3, p. 490
A. G. Mikolaichuk	**28,** K5	(1968)	*1.4*, p. *21*; *20.1*, p. *355*; 22, p. 452
D. L. Miller	**23,** 577	(1967)	*12*, p. *157*; 22.5.2, p. 524
A. D. Milne	**26,** 185	(1968)	3, p. 27; *5*, p. *53*; 22.1.2, p. 473
	26, 469	(1968)	*4*, p. *45*
A. Milojevič	**21,** K163	(1967)	*13*, p. *166*; *18*, p. *292*; 22.6, p. 538
S. Milošević	**20,** K63	(1967)	*13.1*, p. *177*; 21, p. 378
J. Mimkes	**27,** K133	(1968)	*20.3*, p. *370*; 22.4.1, p. 503
M. A. Mindlina	**25,** K83	(1968)	13.1, p. 184; *20*, p. *340*; 21.1, p. 403
	30, 703	(1968)	13.1, p. 189; *20*, p. *343*; 21.1, p. 409; 22.1.1, p. 467; 22.1.3, p. 480
C. Minier-Cassayre	**16,** 281	(1966)	*10.1*, p. *114*; 14.1, p. 220; 18.2, p. 299; 21.1, p. 390
	17, 317	(1966)	*10.1*, p. *116*; 11, p. 146; 14.1, p. 220; 18.2, p. 302; 21.1.1, p. 412
	17, 793	(1966)	*10.1*, p. *117*; *11*, p. *146*; 14.1, p. 220; 21.1, p. 393

C. Minier-Cassayre	**17**, K33	(1966)	10.1, p. 116; *14.1*, p. *220*; 21, p. 374
N. Minnaja	**17**, 657	(1966)	*13.3*, p. *196*; *14.3.2*, p. *248*; 22.1.2, p. 469
P. Miosga	**19**, 597	(1966)	*13*, p. *164*; 14.3, p. 235; 22.1.1, p. 461
D. N. Mirlin	**27**, 443	(1968)	13, p. 170; *20.1*, p. *354*; 22.6, p. 541
S. A. Mironov	**27**, K85	(1968)	7, p. 77; *18.3*, p. *320*
D. D. Mishin	**19**, K1	(1967)	4, p. 40; *18.2*, p. *305*; 21.1, p. 395; 21.6, p. 432
	20, K9	(1967)	*18.2*, p. *305*; 21.1, p. 396; 21.6, p. 433
	20, K123	(1967)	*18.2*, p. *306*; 21.1.1, p. 416
	27, K49	(1968)	*10.1*, p. *131*; 18.2, p. 312; 21.1.1, p. 421
	30, 31	(1968)	*10.1*, p. *134*; *18.2*, p. *315*; 21.1.1, p. 423
U. K. Mishra	**24**, K87	(1967)	*13.1*, p. *183*; 18, p. 293; 20, p. 340; 22.1.2, p. 472; 22.2.1, p. 485
W. H. Mitchell	**17**, 605	(1966)	*14.3*, p. *232*; *16*, p. *277*; 22.2.3, p. 490
S. S. Mitra	**28**, 663	(1968)	*6*, p. *67*; 20.1, p. 356; 22.4.1, p. 504; 22.4.2, p. 508; 22.6, p. 542; 22.8, p. 554
A. I. Mitsek	**25**, 787	(1968)	*18.2*, p. *310*; *18.4*, p. *323*; 19, p. 332
	28, 83	(1968)	8, p. 85; 13, p. 171; *18.2*, p. *312*
N. N. Mochalkin	**26**, K139	(1968)	6, p. 65; *13.2*, p. *194*; *20.2*, p. *361*; 22.4.2, p. 507
E. Mohler	**18**, K45	(1966)	*13.2*, p. *192*; 20.1, p. 346; 22.5.2, p. 519
	22, K49	(1967)	13.1, p. 179; *20.1*, p. *349*; 22.1.3, p. 477
	29, K55	(1968)	13.2, p. 195; *14.3*, p. *243*; *20*, p. *342*; 22.5, p. 514
V. F. Moiseev	**18**, 881	(1966)	*10.1*, p. *119*; 21.1.1, p. 414
V. A. Molchanov	**19**, 407	(1967)	*17*, p. *287*; 21, p. 377; 22.1, p. 457
	19, 425	(1967)	*6*, p. *58*; 21.1, p. 395
S. B. van der Molen	**19**, 327	(1967)	2, p. 22; *19*, p. *327*; 21, p. 377
B. Molnár	**22**, 617	(1967)	1.2, p. 11; *6.1*, p. *72*; 18.3, p. 318; 18.4, p. 323; 22.8, p. 551
	30, 215	(1968)	2, p. 24; *6.1*, p. *74*; 22.8, p. 556
I. P. Molodyan	**18**, 677	(1966)	*13.1*, p. *176*; 14.3, p. 234; 15, p. 271; 22.8, p. 548
S. Le Montagner	**23**, 709	(1967)	*14.4.2*, p. *267*
E. Monteil	**21**, K31	(1967)	*9*, p. *89*; 10, p. 100; 14.3, p. 236; 22.2.3, p. 491
R. Montmory	**16**, 301	(1966)	*18.2*, p. *299*; *18.4*, p. *321*; 21.1, p. 390; 21.1.1, p. 410
	16, 569	(1966)	4, p. 37; *18.2*, p. *300*; 21.1.1, p. 411
J. T. Moore	**25**, K105	(1968)	*10*, p. *105*; *10.1*, p. *129*; 21, p. 385; 21.1, p. 403; 21.5, p. 429; 21.6, p. 435; 22.1, p. 457; 22.1.1, p. 464; 22.1.2, p. 473
K. Moorjani	**23**, K1	(1967)	13.2, p. 193; *20.1*, p. *350*; 23, p. 561
D. J. Morgan	**22**, 491	(1967)	*13.1*, p. *180*; 22.1.1, p. 462
	23, 97	(1967)	*13.1*, p. *180*; 22.1.1, p. 463
D. P. Morgan	**24**, 9	(1967)	6, p. 56; *13*, p. *162*; 14.1, p. 220; 14.3, p. 231; 19, p. 326; *20*, p. *337*; 21, p. 373; 21.1, p. 390; 21.2, p. 424; 21.6, p. 430; 22.1.1, p. 458; 22.2.1, p. 482; 22.2.3, p. 489
G. J. Morgan	**30**, 247	(1968)	2, p. 25; 6, p. 69; *13*, p. *173*
Z. L. Morgenshtern	**22**, 289	(1967)	*10.2*, p. *140*; 20.1, p. 349; 20.2, p. 361; 20.3, p. 366; 22.6, p. 538
	29, 831	(1968)	*10.2*, p. *144*; 13.4, p. 215; *20.3*, p. *371*; 22.6, p. 543
L. M. Morgulis	**22**, K75	(1967)	*10.1*, p. *125*; 21.1.1, p. 418
J. W. Moroń	**16**, K171	(1966)	*18.2*, p. *301*; 21.1.1, p. 411
	17, K109	(1966)	*18.2*, p. *303*; 21.1.1, p. 413
	27, K37	(1968)	10.1, p. 131; *12.1*, p. *160*; *18.2*, p. *312*; 21.1.1, p. 421
	29, K39	(1968)	*18.2*, p. *314*; 21.1.1, p. 422
A. I. Morosov	**23**, K9	(1967)	7, p. 77; *14.4.1*, p. *260*; 22.4.1, p. 501
A. J. Morton	**17**, 101	(1966)	*1.2*, p. *9*; 3, p. 26; 21.1.1, p. 412

M. E. de Morton	**26**, K73 (1968)	*10.1*, p. *129*; 12.1, p. 160; 21, p. 385
P. Moser	**16**, 281 (1966)	*10.1*, p. *114*; 14.1, p. 220; 18.2, p. 299; 21.1, p. 390
	17, 317 (1966)	*10.1*, p. *116*; 11, p. 146; 14.1, p. 220; 18.2, p. 302; 21.1.1, p. 412
	17, 793 (1966)	*10.1*, p. *117*; *11*, p. *146*; 14.1, p. 220; 21.1, p. 393
P. Moser	**17**, K33 (1966)	10.1, p. 116; *14.1*, p. *220*; 21, p. 374
	17, K37 (1966)	*10.1*, p. *116*; 18.2, p. 302; 21.1.1, p. 412
	29, K179 (1968)	*10.1*, p. *134*; *11*, p. *154*; 18.2, p. 315; 21.1.1, p. 423
R. Moser	**17**, 783 (1966)	13.4, p. 203; *20.3*, p. *364*; 22.4.1, p. 498
	24, K13 (1967)	13.4, p. 210; *20.3*, p. *367*; 22.4.1, p. 501
S. A. Moskalenko	**18**, 379 (1966)	6, p. 57; *13.2*, p. *191*; 20.1, p. 346; 22.1.1, p. 460; 22.2.2, p. 487; 22.2.3, p. 490; 22.6, p. 536
H. Mothes	**19**, K23 (1967)	*10.2*, p. *138*; 22.5.3, p. 532
N. F. Mott	**21**, 343 (1967)	2, p. 23; *13.4*, p. *206*; 21, p. 378; 22, p. 445
K. Motzke	**16**, 795 (1966)	*18.3*, p. *316*
A. Moureau	**17**, K217 (1966)	*14.1*, p. *221*; 21.7, p. 438
L. A. Mozgovaya	**22**, K109 (1967)	*12*, p. *157*; 16, p. 280; 22.4.1, p. 500
	29, K145 (1968)	*10*, p. *111*; 12, p. 158; 15, p. 274; 22.1.2, p. 474
H. Mughrabi	**19**, 251 (1967)	*10.1*, p. *120*; 21.1, p. 395
H. Mühe	**21**, K115 (1967)	*18.2*, p. *307*; 21, p. 380; 21.1.1, p. 418
S. Mukai	**28**, K19 (1968)	*10*, p. *109*; *10.2*, p. *143*; 22.5.2, p. 527
A. Mukerji	**22**, K19 (1967)	*10.2*, p. *140*; 11, p. 149; 22.5.3, p. 533
	28, K143 (1968)	*10.2*, p. *143*; 22.5.3, p. 534
B. Mukherjee	**22**, K131 (1967)	*6*, p. *62*; 8, p. 83; 12, p. 157; 22.1, p. 457
M. L. Mukherjee	**16**, 591 (1966)	*10.2*, p. *137*; 20.3, p. 364; 22.5.2, p. 518
G. O. Müller	**20**, K173 (1967)	*6*, p. *60*; 7, p. 76; 22.4.1, p. 499
	23, 703 (1967)	13.2, p. 193; 20.2, p. 361; *20.3*, p. *367*; 22.4.1, p. 501
	24, 127 (1967)	*14.4.1*, p. *260*; 22.1.2, p. 472
	25, K131 (1968)	13.3, p. 198; *14.4.1*, p. *261*; 22.1.2, p. 473
H. K. Müller	**27**, 723 (1968)	13.4, p. 214; *14.3.1*, p. *247*; 20.1, p. 355; 22.6, p. 541
	27, 733 (1968)	14.3.1, p. 247; *20.1*, p. *355*; 22.6, p. 541
H. P. Müller	**23**, 253 (1967)	13, p. 167; *20.3*, p. *367*; 22.9, p. 559
J. Müller	**27**, 313 (1968)	*1.4*, p. *20*; *18.2*, p. *311*; 21.1, p. 405; 21.1.1, p. 421
K.-H. Müller	**26**, K109 (1968)	*13*, p. *169*; *18.2*, p. *311*; 21.1.1, p. 421
L. Müller	**21**, K65 (1967)	2, p. 23; 13.4, p. 207; *20.1*, p. *348*; 22.1.3, p. 477
P. Müller	**21**, 693 (1967)	*10*, p. *100*; 22.5.1, p. 516
R. Müller	**18**, 427 (1966)	*10.1*, p. *118*; 11, p. 147; 14.1, p. 221; 21, p. 375
G. M. de'Munari	**29**, 341 (1968)	*1.2*, p. *12*; 3, p. 27; 13.1, p. 188; 17, p. 289; 22.7, p. 546
C. Muntoni	**28**, K43 (1968)	*1.2*, p. *12*; 13.1, p. 187; 20.1, p. 356; 22.4.1, p. 503
A. Murasik	**22**, 517 (1967)	5, p. 52; *18.4*, p. *323*; 22, p. 447
	23, K123 (1967)	*18.4*, p. *323*; 22, p. 448
	30, 61 (1968)	*18.4*, p. *325*; 22.8, p. 556
	30, 157 (1968)	4, p. 48; *18.4*, p. *325*; 22, p. 454
R. J. Murphy	**27**, 559 (1968)	1.4, p. 20; *10*, p. *108*; *10.1*, p. *131*
L. E. Murr	**19**, 7 (1967)	*1.4*, p. *16*; 4, p. 37; 10, p. 93; 21.1, p. 390; 21.1.1, p. 410
	24, 135 (1967)	*1.4*, p. *19*; 5, p. 52; 22.6, p. 539
	25, 629 (1968)	*1.4*, p. *20*; *4*, p. *44*; 22.6, p. 540
M. Musa	**16**, 771 (1966)	*10*, p. *94*; 20.1, p. 345; 22.5.2, p. 518
M. Musci	**30**, K37 (1968)	*6.1*, p. *75*; 22.8, p. 557
H. E. Müser	**24**, 109 (1967)	*14.4.2*, p. *267*
E. N. Myasnikov	**20**, 153 (1967)	*13.2*, p. *192*; *20.1*, p. *347*; 22, p. 443

A. Mysyrowicz	**22**,	K155	(1967)	6, p. 62; 13.2, p. 193; *20.3*, p. *367*; 22.4.2, p. 507
	27,	K29	(1968)	*20.3*, p. *369*; 22.5.1, p. 516
V. P. Naberezhnykh	**20**,	737	(1967)	*13.1*, p. *178*; 21, p. 378
	29,	K23	(1968)	*13.1*, p. *188*; 21, p. 387
E. M. Nadgornyi	**23**,	571	(1967)	*10*, p. *102*; 22.5.2, p. 524
A. J. Nadolny	**16**,	K133	(1966)	*16*, p. *276*; 22.1, p. 456
J. Naegele	**30**,	K81	(1968)	*20.1*, p. *358*; 22, p. 455; 22.7, p. 546
R. Nagarajan	**26**,	91	(1968)	*13*, p. *169*; 18.4, p. 323; 22.5, p. 513; 22.8, p. 553
P. Nagels	**21**,	113	(1967)	*6*, p. *60*; 8, p. 82; 22.6, p. 537
	28,	193	(1968)	13.1, p. 187; 13.4, p. 214; *14.3*, p. *242*; 15, p. 274; 22.6, p. 542
D. L. Nagy	**30**,	215	(1968)	2, p. 24; *6.1*, p. *74*; 22.8, p. 556
E. Nagy	**23**,	K49	(1967)	*1.2*, p. *11*; 16, p. 280; 21.1, p. 400
M. C. Naik	**30**,	319	(1968)	*9*, p. *92*; 21, p. 388; 21.1, p. 408; 21.6, p. 437
Chr. Nanev	**16**,	777	(1966)	*1.3*, p. *14*; 3.1, p. 29; 21, p. 374; 21.1, p. 391
	23,	663	(1967)	*3.1*, p. *32*; 21, p. 382; 21.1, p. 400
Ya. N. Nasirov	**24**,	K157	(1967)	14.3, p. 239; *15*, p. *273*; 22, p. 449
	28,	K113	(1968)	*8*, p. *85*; 22, p. 453
D. N. Nasledov	**17**,	105	(1966)	13.4, p. 202; *14.3*, p. *232*; 22.2.2, p. 487
	17,	K113	(1966)	*13.1*, p. *175*; 20.1, p. 345; 22.2.3, p. 490
	18,	677	(1966)	*13.1*, p. *176*; 14.3, p. 234; 15, p. 271; 22.8, p. 548
	19,	429	(1967)	*14.3.2*, p. *249*; 16, p. 278; 22.2.3, p. 490
	19,	435	(1967)	*14.3*, p. *235*; 15, p. 271; 22.2.3, p. 490
	21,	9	(1967)	3.1, p. 29; 3.2, p. 34; 5, p. 49; 8, p. 79; 10, p. 93; 13.1, p. 174; *14.3*, p. *231*; 15, p. 270; 16, p. 276; 17, p. 286; 18, p. 291; 20.1, p. 344; 22.8, p. 547
	23,	755	(1967)	*6*, p. *63*; 13.1, p. 182; *16*, p. *280*; 22.2.2, p. 488
	27,	45	(1968)	13.4, p. 213; *14.3*, p. *241*; 22, p. 451; 22.7, p. 546
	29,	K103	(1968)	*14.3.2*, p. *251*; *14.4.1*, p. *264*; 22.2.1, p. 486
N. S. Natarajan	**21**,	K127	(1967)	*8*, p. *82*; 21.6, p. 433
	29,	K69	(1968)	6, p. 68; *8*, p. *85*; 13, p. 172; 21.6, p. 437
Yu. P. Naumov	**30**,	485	(1968)	*13.2*, p. *195*; 20.1, p. 358; 20.3, p. 371; 22.4.2, p. 508
V. Naundorf	**29**,	K123	(1968)	10.1, p. 134; *11*, p. *154*; 12.1, p. 161; 21.1, p. 408
E. Nebauer	**29**,	269	(1968)	*9*, p. *92*; 10, p. 110, 22.4.1, p. 504
	29,	K89	(1968)	3.1, p. 33; *4*, p. *47*; 22.4.1, p. 505
A. Néda	**21**,	557	(1967)	*8*, p. *82*; 22.4.1, p. 500; 22.4.2, p. 506; 22.4.3, p. 509; 22.8, p. 550
	28,	421	(1968)	6, p. 66; *8*, p. *85*; 22.4.3, p. 511; 22.8, p. 554
G. M. Nedlin	**24**,	K25	(1967)	1.2, p. 11; 3.2, p. 35; *18*, p. *293*; 22.5, p. 513; 22.8, p. 552
V. M. Nedostup	**17**,	745	(1966)	3.1, p. 30; 10.1, p. 116; *18.2*, p. *303*; 21.1.1, p. 413
H. H. Neely	**24**,	217	(1967)	*11*, p. *150*; 14.1, p. 224; 21.1.1, p. 419
	28,	675	(1968)	10.1, p. 132; *11*, p. *153*; 14.1, p. 226; 21, p. 387
V. V. Negreskul	**27**,	K15	(1968)	*2*, p. *24*; *20.3*, p. *369*; 22.8, p. 553
E. Nembach	**18**,	807	(1966)	*14.2*, p. *228*
S. A. Nemnonov	**24**,	K43	(1967)	*13.1*, p. *183*; *14.2*, p. *229*; 21, p. 383
V. V. Nemoshkalenko	**25**,	K83	(1968)	13.1, p. 184; *20*, p. *340*; 21.1, p. 403
	28,	K15	(1968)	13.1, p. 187; *20*, p. *341*; 21, p. 387; 21.3, p. 427
	29,	45	(1968)	*6.1*, p. *74*; 13, p. 172; 21.1.1, p. 422
	30,	703	(1968)	13.1, p. 189; *20*, p. *343*; 21.1, p. 409; 22.1.1, p. 467; 22.1.3, p. 480
D. Nenow	**17**,	575	(1966)	*3.1*, p. *30*
	22,	415	(1967)	3.1, p. 32; *4*, p. *42*; 22.9, p. 559

V. I. NEPSHA	28, 269	(1968)	10, p. 109; 13, p. 171; *19*, p. *334*; 22.6, p. 542
N. N. NESTEROVA	30, 367	(1968)	13, p. 173; *18.3*, p. *320*; *20*, p. *342*; 22.5, p. 514; 22.8, p. 556
R. E. NETTLETON	24, 561	(1967)	*13*, p. *168*; 14.4.2, p. 268
H. NEUHÄUSER	17, 35	(1966)	*10.1*, p. *115*; 11, p. 145; 21.1, p. 391
	25, 593	(1968)	*10.1*, p. *128*; 11, p. 151; 21.1, p. 403
	27, 281	(1968)	*10.1*, p. *130*; 11, p. 152; 21.1, p. 404
E. NEUMANN	17, 783	(1966)	13.4, p. 203; *20.3*, p. *364*; 22.4.1, p. 498
	24, K13	(1967)	13.4, p. 210; *20.3*, p. *367*; 22.4.1, p. 501
H. NEUMANN	20, K33	(1967)	*17*, p. *287*; 22, p. 444
P. NEUMANN	17, K71	(1966)	3, p. 26; *5*, p. *49*
W. NEUMANN	18, 151	(1966)	6, p. 57; *8*, p. *80*; 22.6, p. 535
V. B. NEUSTRUEV	22, 289	(1967)	*10.2*, p. *140*; 20.1, p. 349; 20.2, p. 361; 20.3, p. 366; 22.6, p. 538
	29, 831	(1968)	*10.2*, p. *144*; 13.4, p. 215; *20.3*, p. *371*; 22.6, p. 543
C. W. A. NEWEY	29, 357	(1968)	*10.1*, p. *133*; 21.1, p. 407
H. W. NEWKIRK	25, K79	(1968)	*10*, p. *104*; 11, p. 151; 22.6, p. 540
D. J. NEWMAN	29, 697	(1968)	6, p. 68; 18.1, p. 297; *20.1*, p. *357*; 22.8, p. 555
R. E. NEWNHAM	16, K17	(1966)	*18.4*, p. *321*; 22.8, p. 547
	19, K87	(1967)	*13*, p. *165*; 18, p. 291; *20*, p. *338*; 22.8, p. 549
PHAM NGU TUNG see under letter "P" (PHAM)			
NGUYEN VAN DONG	30, 557	(1968)	2, p. 25; *14.3*, p. *244*; 14.4, p. 254; *15*, p. *275*; 22, p. 455
H. H. NICKLE	30, K65	(1968)	*13.4*, p. *217*; 22.5.2, p. 530
J. C. NICOUD	29, 819	(1968)	10.1, p. 134; *11*, p. *154*; 14.1, p. 226; 21.3. p.427
	30, 119	(1968)	10.1, p. 134; *11*, p. *154*; 14.1, p. 226; 21.3, p. 427
D. NICULESCU	21, 557	(1967)	*8*, p. *82*; 22.4.1, p. 500; 22.4.2, p. 506; 22.4.3, p. 509; 22.8, p. 550
	28, 421	(1968)	6, p. 66; *8*, p. *85*; 22.4.3, p. 511; 22.8, p. 554
	29, 813	(1968)	*13.1*, p. *189*; 14.3, p. 244; 20.1, p. 357; 22.4.3, p. 511; 22.8, p. 556
N. NICULESCU	29, 813	(1968)	*13.1*, p. *189*; 14.3, p. 244; 20.1, p. 357; 22.4.3, p. 511; 22.8, p. 556
J. NIEMIEC	17, K89	(1966)	*14.2*, p. *228*
S. D. NIGAM	28, K71	(1968)	3, p. 27; *12*, p. *158*; *12.1*, p. *160*
J. NIHOUL	17, 295	(1966)	*10*, p. *94*
H. NIJBOER	26, 537	(1968)	10, p. 105; *14.4*, p. *254*; 22.5.2, p. 526
A. E. NIKIFOROV	24, 51	(1967)	3.2, p. 35; *13.4*, p. *210*; 22.6, p. 539
	28, 65	(1968)	*10*, p. *108*; 13.4, p. 214; 20.3, p. 370; 22.6, p. 542
	28, 269	(1968)	10, p. 109; 13, p. 171; *19*, p. *334*; 22.6, p. 542
M. NIKIFOROVA	21, K87	(1967)	2, p. 23; 13.4, p. 207; *20.1*, p. *348*; 22, p. 445
S. NIKITINE	22, K155	(1967)	6, p. 62; 13.2, p. 193; *20.3*, p. *367*; 22.4.2, p. 507
	27, K29	(1968)	*20.3*, p. *369*; 22.5.1, p. 516
G. S. NIKOLSKII	18, K123	(1966)	*8*, p. *80*; 14.1, p. 222; 21.4, p. 428
V. K. NIKULIN	16, K125	(1966)	*3.2*, p. *34*; 5, p. 49; 21.1, p. 391; 21.6, p. 430
	17, 815	(1966)	*3.2*, p. *34*; 5, p. 50; 12.1, p. 159; 21.1, p. 393; 21.6, p. 431
	28, 801	(1968)	*3.2*, p. *36*; 13.1, p. 187; 21.1, p. 406; 21.6, p. 437
I. NISHIDA	20, K155	(1967)	*1.2*, p. *10*; 22.6, p. 537
O. NISHIDA	26, 77	(1968)	*20.1*, p. *352*; 22.4.3, p. 510
P. S. NISIN	26, K151	(1968)	6.1, p. 73; *10.1*, p. *130*; 14.1, p. 225; 18.1, p. 297; 21.6, p. 436
M. V. NITSOVICH	16, 459	(1966)	6, p. 56; 13.1, p. 174; *15*, p. *270*; 22, p. 440
V. M. NITSOVICH	16, 459	(1966)	6, p. 56; 13.1, p. 174; *15*, p. *270*; 22, p. 440
W. D. NIX	25, 679	(1968)	*10*, p. *104*; *10.1*, p. *128*

M. A. Nizametdinova	**17,** K115	(1966)	*20.1*, p. *345*; 22, p. 440
	17, K131	(1966)	*20.1*, p. *345*; 22.8, p. 547
	19, K111	(1967)	13.1, p. 177; *20.1*, p. *347*; 22, p. 443
H. Norden	**22,** 631	(1967)	1.4, p. 19; *3.1*, p. *32*; 21.6, p. 434
T. V. Nordstrom	**24,** K121	(1967)	*10*, p. *103*; *10.1*, p. *127*
P. Novák	**23,** K45	(1967)	*13.4*, p. *209*
B. V. Novikov	**18,** K1	(1966)	11, p. 147; *13.2*, p. *191*; *20.3*, p. *364*; 22.4.1, p. 499
	28, K85	(1968)	13.2, p. 194; 13.3, p. 198; *16*, p. *283*; 22.4.1, p. 504
O. N. Novruzov	**28,** K113	(1968)	*8*, p. *85*; 22, p. 453
A. S. Nowick	**24,** 461	(1967)	*9*, p. *90*; 21.1, p. 401; 21.6, p. 435
R. H. Nussbaum	**19,** K31	(1967)	*6.1*, p. *71*
V. Nüsslein	**21,** 309	(1967)	*6*, p. *61*; 8, p. 82; 22.5.2, p. 522
S. Oberländer	**25,** 345	(1968)	*14.3*, p. *240*; 14.3.2, p. 250; 22, p. 449
D. Obrikat	**22,** K123	(1967)	*20.3*, p. *366*; 22.4.1, p. 500
Z. Obuszko	**17,** K195	(1966)	6.1, p. 71; *18.4*, p. *321*; 22, p. 441
	26, K83	(1968)	*18.3*, p. *319*; 18.4, p. 324; 22.8, p. 553
J. L. Oddou	**16,** 281	(1966)	*10.1*, p. *114*; 14.1, p. 220; 18.2, p. 299; 21.1, p. 390
	17, 793	(1966)	*10.1*, p. *117*; *11*, p. *146*; 14.1, p. 220; 21.1, p. 393
D. D. Odintsov	**19,** 407	(1967)	*17*, p. *287*; 21, p. 377; 22.1, p. 457
G. Oelgart	**24,** 289	(1967)	13.1, p. 182; *14.3*, p. *238*; 22.1.3, p. 478
T. O. Ogurtani	**24,** 301	(1967)	*10.1*, p. *126*; 13.4, p. 210; 21.1, p. 401
T. Okada	**28,** K19	(1968)	*10*, p. *109*; *10.2*, p. *143*; 22.5.2, p. 527
	30, K69	(1968)	*10*, p. *112*; 11, p. 155; 22.5.2, p. 530
P. R. Okamoto	**25,** 81	(1968)	*3*, p. *27*; 5, p. 53
V. I. Okulov	**21,** 89	(1967)	*14*, p. *219*; 14.4, p. 253
G. Olofsson	**28,** 547	(1968)	*16*, p. *283*; 22.2.2, p. 488
S. Olszewski	**16,** 399	(1966)	13, p. 163; *17*; p. *286*; 21, p. 373
E. M. Olymbios	**22,** 549	(1967)	10, p. 101; 13.4, p. 209; *14.4*, p. *253*; *20.1*, p. 350; 22.5.3, p. 533
M. R. Ongaro	**17,** 49	(1966)	11, p. 146; *13.3*, p. *196*; 22.9, p. 558
V. A. Onishchuk	**24,** 373	(1967)	*14.4*, p. *253*; 22, p. 448
H. Oppermann	**22,** K151	(1967)	*3.1*, p. *32*; 14.1, p. 223; 21, p. 381
V. M. Orĺov	**25,** 513	(1968)	*3.1*, p. *33*; 5, p. 53; 14.3, p. 240; 15, p. 273; 20.1, p. 351; 22, p. 450
N. N. Ormont	**24,** 757	(1967)	6, p. 64; 13, p. 168; *16*, p. *281*; 22.1.1, p. 464
J. C. Orr	**29,** K157	(1968)	*19*, p. *335*
B. N. Oshcherin	**25,** K123	(1968)	*10*, p. *105*; *10.1*, p. *129*; 21, p. 385; 22, p. 450
V. V. Osiko	**21,** K17	(1967)	20.1, p. 348; *20.2*, p. *361*; 20.3, p. 365; 22.5, p. 513; 22.8, p. 550
	25, K75	(1968)	20.1, p. 352; *20.2*, p. *361*; 20.3, p. 368; 22.5, p. 513; 22.8, p. 552
Yu. A. Osipiyan	**30,** 19	(1968)	*10*, p. *111*; 22.4, p. 496
E. O. Osmanov	**26,** 85	(1968)	*1.2*, p. *12*; 2, p. 24; 8, p. 84; 12, p. 158; 14.3, p. 240; 22.7, p. 546; 22.8, p. 553
	29, 435	(1968)	4, p. 47; *5*, p. *54*; 6, p. 68; 10, p. 110; 14.3, p. 243; 22.8, p. 555
J. Ostanevich	**20,** 581	(1967)	*6.1*, p. *71*; 18, p. 292; 21.1.1, p. 416
	20, 591	(1967)	*6.1*, p. *72*; 18, p. 292; 21.1.1, p. 416
V. V. Ostroborodova	**24,** 757	(1967)	6, p. 64; 13, p. 168; *16*, p. *281*; 22.1.1, p. 464
J. W. Ostrowski	**16,** K133	(1966)	*16*, p. *276*; 22.1, p. 456
	20, K159	(1967)	*16*, p. *279*; 22.1, p. 457

E. Otenyazov	**21**, 805 (1967)	10, p. 100; 11, p. 149; *12*, p. *157*; 22.1.2, p. 471
K. Otsuka	**22**, 559 (1967)	1.4, p. 19; *3.1*, p. *32*; 21.1, p. 399
	22, 579 (1967)	1.4, p. 19; *3.1*, p. *32*; 21.1, p. 399
H. M. Otte	**16**, K25 (1966)	3, p. 26; *5*, p. *49*
	19, K99 (1967)	1.2, p. 10; *3*, p. *26*; 4, p. 40; 22.1, p. 457
A. Otto	**22**, 401 (1967)	*13*, p. *166*; 13.3, p. 197; 20, p. 339
	23, 171 (1967)	*13*, p. *167*; 13.3, p. 198; *20*, p. *340*; 21, p. 381
	26, K99 (1968)	*13*, p. *168*; *20*, p. *341*; 21.6, p. 436
H. Overhof	**26**, 461 (1968)	*3*, p. *27*; 13.1, p. 185; 18, p. 294
M. A. Ovsyannikov	**17**, 243 (1966)	1.4, p. 16; *3.1*, p. *29*; 21.1, p. 392; 21.1.1, p. 412
K. Owen	**17**, K79 (1966)	*20.1*, p. *345*; 22.1.1, p. 459
S. J. T. Owen	**21**, K99 (1967)	1.4, p. 18; *3.1*, p. *31*; 14.3.1, p. 245; 22.2.1, p. 484
	28, 295 (1968)	*3.1*, p. *33*; *14.3.1*, p. *247*; 14.3.2, p. 251; 22.4.2, p. 508
M. Owner-Petersen	**28**, 211 (1968)	13.1, p. 187; *19*, p. *334*; 22.1.2, p. 474
K. Ozawa	**17**, 131 (1966)	*6*, p. *56*; *10.2*, p. *137*; 22.5.2, p. 518
	17, 645 (1966)	*10.2*, p. *137*; 11, p. 146; 20.1, p. 346; 22.5.2, p. 519
	17, 697 (1966)	*1.2*, p. *9*; 12, p. 156; 14.3, p. 233; 22, p. 441
	30, K69 (1968)	*10*, p. *112*; 11, p. 155; 22.5.2, p. 530
T. Paakkari	**28**, 539 (1968)	*5*, p. *54*; 21, p. 387
D. Paccard	**16**, 301 (1966)	*18.2*, p. *299*; *18.4*, p. *321*; 21.1, p. 390; 21.1.1, p. 410
Yu. B. Paderno	**24**, K11 (1967)	*18.1*, p. *297*; 22, p. 448
	24, K79 (1967)	*18.1*, p. *297*; 22, p. 448
J. B. Page, Jr.	**24**, 469 (1967)	3.2, p. 35; *6*, p. *64*; 10.2, p. 141; 22.5.2, p. 524
E. Pagiola	**17**, 657 (1966)	*13.3*, p. *196*; *14.3.2*, p. *248*; 22.1.2, p. 469
H. Pagnia	**24**, K97 (1967)	14.3.1, p. 246; *16*, p. *281*; 22.2.3, p. 491
	27, K111 (1968)	*11*, p. *153*; 13.3, p. 198; 14.3, p. 242; 22.2.3, p. 492
	28, K89 (1968)	13.3, p. 198; *16*, p. *283*; 22.2.3, p. 493
M. D. Pai	**27**, 671 (1968)	*16*, p. *282*; 22.4.1, p. 503; 22.4.2, p. 508
A. Pajączkowska	**20**, K159 (1967)	*16*, p. *279*; 22.1, p. 457
V. I. Pakhomov	**27**, K129 (1968)	14.4.2, p. 269; *19*, p. *334*
L. Pál	**20**, 581 (1967)	*6.1*, p. *71*; 18, p. 292; 21.1.1, p. 416
	20, 591 (1967)	*6.1*, p. *72*; 18, p. 292; 21.1.1, p. 416
S. Pal	**18**, 119 (1966)	6, p. 57; *8*, p. *80*; 21.1, p. 393
	23, 361 (1967)	*6*, p. *62*; 21, p. 381; 21.1, p. 400; 21.1.1, p. 419; 21.2, p. 424; 21.6, p. 434
L. S. Palatnik	**17**, 543 (1966)	*1.4*, p. *17*; 3.1, p. 30; 10.1, p. 116; *18.2*, p. *303*; 21.1, p. 392; 21.1.1, p. 413
N. A. Pangarov	**20**, 365 (1967)	*3.1*, p. *31*; 3.2, p. 35; 21.6, p. 432
	20, 371 (1967)	*3.1*, p. *31*; 4, p. 41; 21, p. 377
J. M. Pankratz	**26**, K97 (1968)	4, p. 46; *10*, p. *106*; 22.1.2, p. 473
G. Kh. Panova	**20**, 767 (1967)	*6*, p. *60*; 8, p. 81; 21.1, p. 397
A. C. Papadakis	**21**, K77 (1967)	*14.4*, p. *253*; 14.4.1, p. 258; 22, p. 445
P. Papamantellos	**29**, 323 (1968)	*5*, p. *54*; 18.3, p. 320
V. V. Paranjape	**18**, K103 (1966)	6, p. 58; *13*, p. *164*; *14.3*, p. *234*; 22, p. 442
V. B. Pariiskii	**19**, 525 (1967)	*10*, p. *97*; 22.5.2, p. 521
Y. S. Park	**29**, K7 (1968)	13.4, p. 215; *16*, p. *284*; 22.4.1, p. 504
W. F. Parks	**27**, K17 (1968)	*18.2*, p. *312*
H. Pascard	**16**, K153 (1966)	*18.2*, p. *301*; 21.1.1, p. 411
N. Paschoff	**22**, 83 (1967)	*1.4*, p. *18*; *3.1*, p. *31*; 22.1.1, p. 462
	22, 93 (1967)	1.4, p. 18; *10*, p. *101*; 22.1.1, p. 462

R. T. Pascoe	**29**,	357	(1968)	*10.1*, p. *133*; 21.1, p. 407
J. Pastrňák	**20**,	K29	(1967)	*20.1*, p. *347*; 22.2, p. 481
	22,	407	(1967)	*14.4.1*, p. *259*; 22, p. 447
	26,	591	(1968)	13.1, p. 185; *20.1*, p. *353*; 22.2, p. 481
K. N. Pathak	**17**,	77	(1966)	*8*, p. *79*
J. H. Patterson	**16**,	K175	(1966)	*1.2*, p. *9*; 3, p. 26; 8, p. 79; 22.8, p. 547
K. Patzer	**26**,	693	(1968)	13.4, p. 212; *20.1*, p. *353*; *20.3*, p. *369*; 22.5, p. 513
P. Paufler	**24**,	77	(1967)	*10.1*, p. *126*; 21.1, p. 401; 21.3, p. 426.
A. R. Paul	**30**,	319	(1968)	*9*, p. *92*; 21, p. 388; 21.1, p. 408; 21.6, p. 437
W. Paul	**23**,	413	(1967)	1.4, p. 19; *3.1*, p. *32*; 13.1, p. 181; *20.1*, p. *350*; 22.4.2, p. 507
M. Paulus	**22**,	K87	(1967)	*18.3*, p. *318*
M. Paunov	**22**,	103	(1967)	*3.1*, p. *31*; 21, p. 380
P. Păuşescu	**19**,	K19	(1967)	*1.2*, p. *10*; 22.8, p. 549
G. S. Pawley	**20**,	347	(1967)	*6*, p. *59*; 22.9, p. 559
R. F. Peart	**20**,	545	(1967)	*9*, p. *88*; 12.1, p. 160; 21, p. 378; 21.1.1, p. 416
	23,	263	(1967)	*9*, p. *90*; 10, p. 102; 21, p. 381; 21.1, p. 400; 21.1.1, p. 419
A. N. Pechenov	**19**,	K5	(1967)	*20.2*, p. *360*; 22.4.2, p. 506
B. Pegel	**22**,	223	(1967)	6, p. 61; *9*, p. *89*
	22,	K45	(1967)	6, p. 62; *9*, p. *89*
	23,	335	(1967)	6, p. 62; *9*, p. *90*
	28,	603	(1968)	*10.1*, p. *132*; 21, p. 387; 21.1, p. 406; 21.1.1, p. 422
	29,	K133	(1968)	*6*, p. *69*; 10.1, p. 134
H. Peibst	**20**,	K173	(1967)	*6*, p. *60*; 7, p. 76; 22.4.1, p. 499
H. Peisl	**21**,	K9	(1967)	10, p. 100; *11*, p. *148*; 22.5.2, p. 522
	23,	K75	(1967)	*10*, p. *102*; 10.2, p. 140; 22.5.2, p. 524
	27,	K165	(1968)	*10*, p. *108*; 11, p. 153; 22.5.2, p. 527
	28,	207	(1968)	*10*, p. *108*; 11, p. 153; 22.5.2, p. 527
	29,	K59	(1968)	*10*, p. *110*; 22.5.2, p. 529
J. Pelleg	**22**,	K83	(1967)	*9*, p. *90*; 21.1, p. 400
L. Pelsers	**25**,	621	(1968)	10.2, p. 142; *16*, p. *281*; 22.5.2, p. 525
C. M. Penchina	**29**,	K7	(1968)	13.4, p. 215; *16*, p. *284*; 22.4.1, p. 504
P. Peretto	**16**,	281	(1966)	*10.1*, p. *114*; 14.1, p. 220; 18.2, p. 299; 21.1, p. 390
J. Perez	**24**,	K167	(1967)	*10.1*, p. *127*; 21, p. 383
M. Peřinová	**21**,	K103	(1967)	*10*, p. *101*; 22.8, p. 550
Yu. E. Perlin	**28**,	45	(1968)	6, p. 66; *10.2*, p. *143*; 13, p. 171; 14.4, p. 254; 18, p. 294; 22.2.1, p. 485; 22.2.2, p. 488; 22.2.3, p. 493; 22.4, p. 495; 22.4.1, p. 503; 22.4.2, p. 508; 22.4.3, p. 511; 22.5.1, p. 517; 22.5.2, p. 527
S. A. Permogorov	**18**,	K1	(1966)	11, p. 147; *13.2*, p. *191*; *20.3*, p. *364*; 22.4.1, p. 499
D. J. Perrin	**22**,	549	(1967)	10, p. 101; 13.4, p. 209; *14.4*, p. *253*; *20.1*, p. *350*; 22.5.3, p. 533
A. J. Perry	**24**,	K141	(1967)	*10*, p. *104*; *10.1*, p. *127*
B. Persson	**22**,	631	(1967)	1.4, p. 19; *3.1*, p. *32*; 21.6, p. 434
R. Perthel	**17**,	151	(1966)	*18.3*, p. *316*
	20,	433	(1967)	3.2, p. 35; *13*, p. *165*; 18.3, p. 317
	29,	211	(1968)	*18.2*, p. *314*; *18.4*, p. *325*; 21.1.1, p. 422
G. Petö	**23**,	K49	(1967)	*1.2*, p. *11*; 16, p. 280; 21.1, p. 400
G. A. Petrakovskii	**30**,	K111	(1968)	*18.3*, p. *320*; 19, p. 336
P. Petrescu	**18**,	511	(1966)	10.2, p. 138; *17*, p. *287*; 22.5.2, p. 520
	23,	K105	(1967)	*10.2*, p. *140*; 17, p. 287; 22.5.2, p. 524
	29,	333	(1968)	10.2, p. 143; 13.2, p. 195; *17*, p. *289*; 22.5.2, p. 528
	29,	K1	(1968)	10.2, p. 144; *17*, p. *289*; 22.5.2, p. 528
	29,	K3	(1968)	10.2, p. 144; *17*; p. *289*; 22.5.2, p. 529

V. Petrescu	**22**, K113	(1967)	*18.3*, p. *318*
	23, K125	(1967)	*18.3*, p. *318*
	30, K49	(1968)	*18.3*, p. *320*
J. F. Petroff	**17**, K3	(1966)	*10*, p. *95*; 22.9, p. 558
M. P. Petrov	**30**, 871	(1968)	18.3, p. 320; *19*, p. *336*
S. V. Petrov	**18**, 683	(1966)	*13.4*, p. *204*; 18.4, p. 321; *20.3*, p. *364*; 22.5.3, p. 532
E. Petrova	**27**, K33	(1968)	*10.2*, p. *142*; 22.5.1, p. 516
V. Petržílka	**29**, K51	(1968)	*5*, p. *54*; 14.4, p. 254; 22.6, p. 543
Z. Petykiewicz	**21**, 183	(1967)	*1.4*, p. *18*; 18.2, p. 306; 21.1.1, p. 417
K. Peuker	**30**, 479	(1968)	6, p. 70; 13.4, p. 217; *20.3*, p. *371*; 22, p. 455
J. R. Peverley	**23**, 353	(1967)	*8*, p. *83*; *18*, p. *293*; 22.8, p. 552
K.-H. Pfeffer	**19**, 735	(1967)	*10.1*, p. *121*; *18.2*, p. *305*; 21.1, p. 396; 21.1.1, p. 415
	20, 395	(1967)	10.1, p. 121; *18.2*, p. *305*; 21.1, p. 396; 21.1.1, p. 416
	21, 837	(1967)	10.1, p. 123; *18.2*, p. *307*; 21.1, p. 398
	21, 857	(1967)	10.1, p. 123; *18.2*, p. *307*; 21.1, p. 398; 21.1.1, p. 418
J. Pfeifer	**17**, K15	(1966)	1.3, p. 14; *3.1*, p. *29*; 22.1.1, p. 459
K. Pfeuffer	**29**, 171	(1968)	6, p. 67; *18.1*, p. *297*; 19, p. 335; 22.8, p. 555
H. Pfleiderer	**29**, 597	(1968)	*16*, p. *284*; 22.2.3, p. 493
Pham Ngu Tung	**30**, 557	(1968)	2, p. 25; *14.3*, p. *244*; 14.4, p. 254; *15*, p. *275*; 22, p. 455
J. Philibert	**21**, K1	(1967)	*9*, p. *89*; 21.1, p. 397; 21.1.1, p. 417
M. Picard	**23**, 563	(1967)	*13.1*, p. *182*; 22.1.3, p. 478
R. Pichon	**17**, K33	(1966)	10.1, p. 116; *14.1*, p. *220*; 21, p. 374
W. Picht	**26**, K163	(1968)	*16*, p. *282*; 20.1, p. 353; 20.3, p. 369; 22.4.1, p. 502
H. Pick	**19**, 313	(1967)	*6*, p. *58*; 10, p. 97
P. B. Pickar	**21**, 271	(1967)	5, p. 51; 6, p. 61; 13.1, p. 178; *20.1*, p. *348*; 22.3, p. 494
	29, 153	(1968)	5, p. 54; 13.1, p. 188; 16, p. 283; *20.1*, p. *356*; 22, p. 453
J. Piekoszewski	**29**, K99	(1968)	*6.1*, p. *74*; 22.8, p. 556
F. Pieragostini	**16**, 439	(1966)	*9*, p. *87*; 21, p. 373
B. Pietrass	**24**, 571	(1967)	*6*, p. *64*
J. R. Pilbrow	**20**, 225	(1967)	*19*, p. *328*; 22, p. 444; 22.5, p. 512
	20, 237	(1967)	*19*, p. *328*; 22, p. 444; 22.5, p. 512
B. Pilch	**16**, K171	(1966)	*18.2*, p. *301*; 21.1.1, p. 411
M. H. Pilkuhn	**25**, 9	(1968)	13.4, p. 200; 14.3.2, p. 248; *14.4.1*, p. *255*; *20.2* p. *360*; 20.3, p. 363; 22.2.1, p. 482; 22.2.2, p. 487; 22.2.3, p. 489
P. K. C. Pillai	**28**, 415	(1968)	2, p. 24; *14.4*, p. *254*; 22.9, p. 560
P. Pinard	**25**, 691	(1968)	*17*, p. *288*; 21.1, p. 403; 21.1.1, p. 420; 22.1.1, p. 464; 22.2.1, p. 485; 22.4.1, p. 502; 22.5.2, p. 525; 22.6, p. 540
I. I. Pinchuk	**20**, 537	(1967)	3, p. 27; *15*, p. *271*; 21.7, p. 438
A. Pindor	**20**, K17	(1967)	5, p. 50; *12.1*, p. *159*; 21, p. 377; 21.1, p. 396; 21.3, p. 426
B. Ya. Pines	**20**, 213	(1967)	*9*, p. *88*; 21.6, p. 432; 22, p. 443; 22.1.3, p. 477
H. Pink	**21**, K111	(1967)	*10*, p. *101*; *20.2*, p. *361*; 22.6, p. 538
I. Z. Pinsker	**23**, 587	(1967)	13.4, p. 210; *20.3*, p. *367*; 22.2.1, p. 484
R. V. Pisarev	**30**, 367	(1968)	13, p. 173; *18.3*, p. *320*; *20*, p. *342*; 22.5, p. 514; 22.8, p. 556
V. F. Pissarenko	**26**, K13	(1968)	*20.3*, p. *368*; 22.8, p. 553

B. Pistoulet	**24,**	481	(1967)	*13.1*, p. *183*; 14.3, p. 239; 22.2.3, p. 491
V. Plakhty	**29,**	K81	(1968)	*1.2*, p. *12*; *6*, p. *68*; 22.8, p. 555
J. van der Planken	**28,**	K9	(1968)	*10.1*, p. *132*; 21, p. 387; 21.3, p. 427
K. Platzöder	**29,**	K63	(1968)	13.1, p. 189; 13.2, p. 195; *20.1*, p. *357*; 22.6, p. 543
C. N. Plăviţu	**16,**	69	(1966)	8, p. 79; 13, p. 162; *14.4*, p. *252*; 15, p. 270; 22, p. 439
	24,	361	(1967)	6, p. 64; 13, p. 168; 14.3, p. 239; *15*, p. *273*; 22, p. 448
J. N. Plendl	**28,**	663	(1968)	*6*, p. *67*; 20.1, p. 356; 22.4.1, p. 504; 22.4.2, p. 508; 22.6, p. 542; 22.8, p. 554
A. V. Podin	**18,**	897	(1966)	10.2, p. 138; *11*, p. *147*; 19, p. 327; 20.1, p. 346; 22.5.2, p. 520
B. Pödör	**16,**	K167	(1966)	10, p. 94; *14.3*, p. *232*; 22.1.1, p. 459
V. A. Pogozhev	**16,**	K121	(1966)	*18.2*, p. *301*; 21.1.1, p. 411
S. Pokrzywnicki	**24,**	K11	(1967)	*18.1*, p. *297*; 22, p. 448
	24,	K73	(1967)	*18.1*, p. *297*; 22, p. 448
J. Polák	**21,**	581	(1967)	*10.1*, p. *123*; 21.6, p. 433
	28,	773	(1968)	*10.1*, p. *132*; 14.1, p. 226; 21.6, p. 436
M. Polcarová	**21,**	635	(1967)	4, p. 42; *18.2*, p. *307*; 21.1.1, p. 417
A. Policec	**19,**	613	(1967)	*18.2*, p. *305*
	21,	601	(1967)	14.1, p. 222; *18.2*, p. *307*; 21.1.1, p. 417
	27,	K5	(1968)	*18.2*, p. *311*; 21.1, p. 405
H. Pollak	**17,**	825	(1966)	13, p. 163; *14.3*, p. *233*; 15, p. 270; 22.6, p. 535
I. Pollini	**27,**	95	(1968)	13.4, p. 213; *20.1*, p. *353*; 22.5, p. 514
F. Polly	**22,**	K123	(1967)	*20.3*, p. *366*; 22.4.1, p. 500
W. Pompe	**28,**	K23	(1968)	*10*, p. *109*; *10.1*, p. *132*
V. A. Popov	**26,**	453	(1968)	*8*, p. *84*; 23, p. 561
V. I. Popov	**16,**	657	(1966)	4, p. 38; 12.1, p. 159; *18.2*, p. *301*; 21.1.1, p. 411
	27,	K1	(1968)	*20.2*, p. *362*; 22.8, p. 553
Yu. V. Popov	**30,**	349	(1968)	13.1, p. 189; *20.1*, p. *358*; 22, p. 454
M. Porsch	**23,**	365	(1967)	6, p. 62; *13*, p. *167*; 13.1, p. 181; 22, p. 447
	27,	359	(1968)	13.1, p. 186; 14.4.1, p. 262; *20*, p. *341*; 22, p. 451
P. Porteous	**30,**	K125	(1968)	*20.3*, p. *372*; 22.2.2, p. 488; 22.8, p. 557
A. P. Potapov	**16,**	745	(1966)	4, p. 38; *18.2*, p. *301*; 21.1.1, p. 411
J. Pottharst	**24,**	109	(1967)	*14.4.2*, p. *267*
V. A. Povitskii	**24,**	45	(1967)	*6.1*, p. *73*; 18.2, p. 308; 21.1.1, p. 419
	25,	607	(1968)	*6.1*, p. *73*; 18.3, p. 319
J. Pozhela	**17,**	769	(1966)	13, p. 163; *14.4.1*, p. *256*; 22.1.1, p. 460
J. K. Pozhela	**27,**	757	(1968)	13, p. 170; 14.4.1, p. 262; *15*, p. *274*; 22.1.1, p. 465
I. Pracka	**22,**	K147	(1967)	*20.1*, p. *350*; 22.1, p. 457
	27,	K107	(1968)	*20.1*, p. *355*; 22.1, p. 457
	29,	K183	(1968)	13.1, p. 189; *20.1*, p. *357*; 22, p. 454
E. Praveczki	**18,**	K33	(1966)	*18.2*, p. *304*
	28,	K105	(1968)	13, p. 172; *18.2*, p. *313*
V. A. Presnov	**21,**	K91	(1967)	*13.4*, p. *207*; 19, p. 329; 22.2.1, p. 483
	25,	K51	(1968)	*19*, p. *332*; 22.2, p. 481
A. M. Prokhorov	**21,**	K107	(1967)	*20.1*, p. *349*; 22.4.1, p. 500
V. V. Proklov	**23,**	K9	(1967)	*7*, p. *77*; *14.4.1*, p. *260*; 22.4.1, p. 501
V. Prosser	**16,**	K21	(1966)	13, p. 162; *18*, p. *291*; *20*, p. *337*; 22.4.3, p. 509
	29,	707	(1968)	6, p. 68; *13.2*, p. *195*; 20.1, p. 357; 22.4.1, p. 505; 22.4.2, p. 508; 22.4.3, p. 511
M. Z. Przegalińska-Mieszkowska	**16,**	K133	(1966)	*16*, p. *276*; 22.1, p. 456
J. Przybyła	**29,**	K39	(1968)	*18.2*, p. *314*; 21.1.1, p. 422
J. Przystawa	**17,**	155	(1966)	*18.4*, p. *321*

J. Przystawa	**20,**	451	(1967)	3.2, p. 35; *18.2*, p. *306*; *18.4*, p. *322*; 22, p. 444
	24,	313	(1967)	*18.2*, p. *309*; *18.4*, p. *323*; 22, p. 448; 22.8, p. 552
	30,	K115	(1968)	3, p. 28; *18.4*, p. *325*; 22, p. 455
M. Psoda	**28,**	K121	(1968)	*5*, p. *54*; *8*, p. *85*; 22.4.3, p. 511
I. B. Puchalska	**17,**	K1	(1966)	1.4, p. 17; *18.2*, p. *302*; 21.1.1, p. 412
	21,	183	(1967)	*1.4*, p. *18*; 18.2, p. 306; 21.1.1, p. 417
	21,	K57	(1967)	4, p. 42; *18.2*, p. *307*; 21.1.1, p. 417
	28,	335	(1968)	1.4, p. 20; *18.2*, p. *312*; 21.1.1, p. 422
J. Puhlmann	**29,**	193	(1968)	13.1, p. 188; *14.3*, p. *243*; 22.1.3, p. 480
J. Punzel	**24,**	K1	(1967)	*1.3*, p. *15*; 11, p. 150; 22.1.2, p. 472
S. P. Puri	**30,**	193	(1968)	3, p. 28; *18.1*, p. *298*; 22.5, p. 514
R. K. Purohit	**24,**	K57	(1967)	*14.3.2*, p. *250*; 22.2.1, p. 484; 22.2.2, p. 488
H. Puszkarski	**22,**	355	(1967)	*13*, p. *166*; 13.3, p. 197; 19, p. 329; 21.1, p. 399
E. H. Putley	**17,**	605	(1966)	*14.3*, p. *232*; *16*, p. *277*; 22.2.3, p. 490
K. E. Puttick	**18,**	491	(1966)	4, p. 39; *10.1*, p. *118*; 21, p. 376
G. P. Pynko	**17,**	243	(1966)	1.4, p. 16; *3.1*, p. *29*; 21.1, p. 392; 21.1.1, p. 412
	17,	249	(1966)	4, p. 38; *18.2*, p. *302*; 21.1, p. 392; 21.1.1, p. 412
V. G. Pynko	**17,**	243	(1966)	1.4, p. 16; *3.1*, p. *29*; 21.1, p. 392; 21.1.1, p. 412
	17,	249	(1966)	4, p. 38; *18.2*, p. *302*; 21.1, p. 392; 21.1.1, p. 412
P. Quadflieg	**25,**	323	(1968)	*14.3*, p. *240*; 22.2.1, p. 485; 22.2.3, p. 492
Y. Quéré	**30,**	713	(1968)	*10.1*, p. *135*; *11*, p. *155*; 21, p. 389
P. L. Quinn	**17,**	307	(1966)	*14.4.1*, p. *256*; 22, p. 440
S. Raab	**19,**	K59	(1967)	*20.2*, p. *360*; 22.2.1, p. 483
	29,	K175	(1968)	*20.2*, p. *362*; 22.2.1, p. 486
H. Rabenhorst	**29,**	K65	(1968)	*16*, p. *284*; 22.4.1, p. 504; 22.4.2, p. 508; 22.8, p. 555
S. I. Radautsan	**18,**	677	(1966)	*13.1*, p. *176*; 14.3, p. 234; 15, p. 271; 22.8, p. 548
	19,	609	(1967)	*1.2*, p. *10*; 8, p. 81; 14.3, p. 235; 22.8, p. 549
S. Radelaar	**27,**	K63	(1968)	*1.1*, p. *8*; 9, p. 91; 10.1, p. 131; 21.6, p. 436
V. Radhakrishnan	**18,**	113	(1966)	*14.2*, p. *228*
	20,	783	(1967)	*14.2*, p. *229*; 21, p. 378
F. Raga	**22,**	K155	(1967)	6, p. 62; 13.2, p. 193; *20.3*, p. *367*; 22.4.2, p. 507
J. S. Rajput	**16,**	K51	(1966)	*8*, p. *79*; *14.2*, p. *228*; 21, p. 373
C. V. Ramaiah	**19,**	K15	(1967)	*14.4*, p. *252*; 22.8, p. 549
C. Ramasastry	**19,**	K15	(1967)	*14.4*, p. *252*; 22.8, p. 549
A. M. Ramazanzade	**22,**	K117	(1967)	*14.3*, p. *237*; 22.8, p. 551
	23,	461	(1967)	*10*, p. *102*; 22, p. 448
R. Ramji Rao see under "Rao"				
R. T. Ramsey	**25,**	103	(1968)	3.2, p. 36; *12.1*, p. *160*; 21, p. 383; 21.3, p. 426
K. Rander	**27,**	301	(1968)	*6.1*, p. *73*; 13, p. 170
C. N. R. Rao	**28,**	157	(1968)	3.2, p. 36; *10*, p. *108*; 22.5.2, p. 527
K. J. Rao	**28,**	157	(1968)	3.2, p. 36; *10*, p. *108*; 22.5.2, p. 527
R. Ramji Rao	**29,**	865	(1968)	6, p. 69; *7*, p. *78*; *8*, p. *86*
I. M. Rarenko	**27,**	101	(1968)	13.1, p. 185; *14.3*, p. *241*; 22, p. 451; 22.7, p. 546
J. Rasek	**17,**	K109	(1966)	*18.2*, p. *303*; 21.1.1, p. 413
E. I. Rashba	**16,**	43	(1966)	12, p. 156; *14.3*, p. *231*; 14.4.1, p. 255; 22.1.1, p. 459
	19,	395	(1967)	*6*, p. *58*; 20.1, p. 347; 22.9, p. 558
A. I. Rasulov	**18,**	911	(1966)	6, p. 58; *8*, p. *80*; 22, p. 442
	19,	K7	(1967)	*8*, p. *81*; 22, p. 442
O. N. Rasumov	**29,**	45	(1968)	*6.1*, p. *74*; 13, p. 172; 21.1.1, p. 422

V. V. Ratnam	**16**,	549	(1966)	*10.2*, p. *137*; 13.4, p. 201; 14.4, p. 252; 18.1, p. 296; 22.5.3, p. 532
	16,	559	(1966)	*10.2*, p. *137*; *20.3*, p. *364*; 22.5.3, p. 532
R. Rattke	**22**,	123	(1967)	*10*, p. *101*
F. Rau	**17**,	259	(1966)	10.1, p. 115; *11*, p. *146*; 14.1, p. 220; 21, p. 374
R. C. Rau	**19**,	645	(1967)	*10*, p. *98*; 11, p. 148; 22.6, p. 536
	21,	K69	(1967)	11, p. 148; *12*, p. *157*; 22.6, p. 538
	30,	K1	(1968)	*4*, p. *48*; 10.1, p. 135; 21.1, p. 408
A. Räuber	**18**,	125	(1966)	13.4, p. 203; *19*, p. *327*; 22.4.1, p. 498
R. Rauch	**25**,	K15	(1968)	6, p. 65; *10.2*, p. *141*; *20.3*, p. *368*; 22.5.3, p. 533
	27,	109	(1968)	10, p. 106; *20.3*, p. *369*; 22.5.3, p. 534
A. G. Ravlik	**17**,	543	(1966)	*1.4*, p. *17*; 3.1, p. 30; 10.1, p. 116; *18.2*, p. *303*; 21.1, p. 392; 21.1.1, p. 413
B. Ray	**22**,	371	(1967)	*1*, p. *7*; *1.2*, p. *11*; 22.4.3, p. 509; 22.8, p. 551
B. S. Razbirin	**18**,	K37	(1966)	*13.2*, p. *192*; 20.3, p. 364; 22.4.2, p. 506
	24,	K107	(1967)	*10*, p. *103*; *13.2*, p. *193*; 20.1, p. 351; 22.4.1, p. 501
	26,	K139	(1968)	6, p. 65; *13.2*, p. *194*; *20.2*, p. *361*; 22.4.2, p. 507
	30,	485	(1968)	*13.2*, p. *195*; 20.1, p. 358; 20.3, p. 371; 22.4.2, p. 508
R. A. Reese	**19**,	359	(1967)	*6.1*, p. *71*; *19*, p. *328*; 22.7, p. 545
P. Regnier	**23**,	K109	(1967)	*10.1*, p. *125*; 21.3, p. 426
	28,	K55	(1968)	*10.1*, p. *132*; 21, p. 387
E. Reguzzoni	**25**,	437	(1968)	*10.2*, p. *141*; *11*, p. *151*; 22.5.2, p. 525
H. Rehme	**18**,	K101	(1966)	*4*, p. *39*; 14.4.2, p. 266
	26,	K1	(1968)	4, p. 45; *14.4.1*, p. *262*; 14.4.2, p. 268
H. G. Reik	**24**,	281	(1967)	6, p. 64; *13*, p. *168*; 14.4.2, p. 267; 20.1, p. 351
P. Reimers	**22**,	K27	(1967)	*9*, p. *89*; 21.6, p. 433
	29,	K31	(1968)	*3.1*, p. *33*; 5, p. 54; 20.1, p. 357; 22.4.1, p. 504; 22.4.2, p. 508; 22.8, p. 555
K. P. Reiners	**23**,	613	(1967)	*6*, p. *63*; *20.1*, p. *350*; 22.5.2, p. 524
R. Reisfeld	**24**,	695	(1967)	*9*, p. *90*; 22.5.2, p. 525
G. Remaut	**20**,	613	(1967)	*4*, p. *41*; 14.4.2, p. 266; 22.8, p. 549
K. F. Renk	**23**,	613	(1967)	*6*, p. *63*; *20.1*, p. *350*; 22.5.2, p. 524
B. Reppich	**20**,	69	(1967)	*10*, p. *98*; 22.6, p. 537
K. K. Repshas	**27**,	757	(1968)	13, p. 170; 14.4.1, p. 262; *15*, p. *274*; 22.1.1, p. 465
B. I. Reser	**26**,	391	(1968)	3, p. 27; *13.1*, p. *185*
L. M. Reshchikova	**18**,	K17	(1966)	1.2, p. 10; *12*, p. *156*; 22.5, p. 512; 22.8, p. 548
J. A. Retchford	**19**,	631	(1967)	4, p. 40; *10*, p. *98*; 22.6, p. 536
U. Retter	**24**,	605	(1967)	*13.4*, p. *210*; 20.1, p. 351; 22.5.1, p. 516
A. G. Revesz	**19**,	193	(1967)	*1.3*, p. *14*; 13.3, p. 197; 14.3.2, p. 248; 22.1.2, p. 470
	24,	115	(1968)	*1.4*, p. *19*; 2, p. 23; *14.3.1*, p. *246*; 14.3.2, p. 249; 22.6, p. 539
O. P. Revokatov	**22**,	K91	(1967)	*19*, p. *330*
M. L. Reynolds	**30**,	97	(1968)	13.4, p. 216; *20.1*, p. *358*; *20.3*, p. *371*; 22.5.2, p. 530
	30,	113	(1968)	*13.4*, p. *216*; 20.1, p. 358; 20.3, p. 371; 22.8, p. 556
	30,	507	(1968)	*13.4*, p. *217*; 20.1, p. 359; 20.3, p. 371; 22.8, p. 557
	30,	735	(1968)	*13.4*, p. *217*; *20.1*, p. *359*; 22.8, p. 557
I. S. Rez	**27**,	K129	(1968)	14.4.2, p. 269; *19*, p. *334*
N. Rezlescu	**23**,	K117	(1967)	*18.3*, p. *318*
R. G. Rhodes	**28**,	K63	(1968)	*14.2*, p. *230*
K. J. Richards	**20**,	K21	(1967)	*9*, p. *88*
	26,	97	(1968)	*9*, p. *91*
R. H. Richman	**26**,	743	(1968)	10.1, p. 130; *18.2*, p. *311*; 21.1.1, p. 421
G. Richter	**18**,	K149	(1966)	13.1, p. 176; *14.3*, p. *234*; 16, p. 278; 22.2.3, p. 490
	30,	K97	(1968)	*20.3*, p. *372*; 22.1.1, p. 467

J. Richter	**26,**	485	(1968)	*10.1*, p. *129*; 21, p. 385
S. B. McRickard see under letter "M"				
Th. Ricker	**30,**	K93	(1968)	4, p. 48; *18.2*, p. *315*; 21.1, p. 409; 21.1.1, p. 423
R. de Ridder	**30,**	797	(1968)	*4*, p. *48*; 10, p. 113; 10.1, p. 136; 21.1.1, p. 423; 22.6, p. 544
K. T. Rie	**30,**	219	(1968)	10.1, p. 135; *11*, p. *155*; 14.1, p. 227; 18.2, p. 315; 21.1, p. 408
N. Riehl	**16,**	159	(1966)	13.4, p. 201; *20.3*, p. *363*; 22.9, p. 558
	26,	607	(1968)	*13.1*, p. *185*; 17, p. 289; 22.9, p. 560
G. Ritter	**29,**	781	(1968)	6.1, p. 74; *18.4*, p. *325*; 22.8, p. 555
A. Rizzo	**28,**	K151	(1968)	*20.3*, p. *370*; 22.4.1, p. 504
D. Robaschik	**16,**	237	(1966)	4, p. 37; *18.2*, p. *299*; 18.3, p. 316
J. L. Robert	**24,**	481	(1967)	*13.1*, p. *183*; 14.3, p. 239; 22.2.3, p. 491
F. E. Roberts	**22,**	373	(1967)	*1*, p. *7*; 22.2.1, p. 484
G. G. Roberts	**27,**	209	(1968)	*14.4.1*, p. *262*; 22.1.3, p. 479
P. M. Robinson	**20,**	461	(1967)	4, p. 41; *10*, p. *99*; 22.9, p. 559
	26,	K93	(1968)	*10*, p. *106*; 22.9, p. 560
K. P. Rodionov	**27,**	249	(1968)	*13.1*, p. *186*; 14.3, p. 241; 15, p. 274; 22.2.3, p. 492
H. Rodot	**29,**	743	(1968)	6, p. 68; *8*, p. *86*; 13, p. 173; *15*, p. *274*; 22, p. 454; 22.8, p. 555
B. Roessler	**20,**	713	(1967)	4, p. 41; *18.2*, p. *306*; 21.1.1, p. 416
	24,	263	(1967)	1.1, p. 8; *4*, p. *43*; 21.1, p. 401; 21.6, p. 434
A. A. Rogachev	**20,**	755	(1967)	*13.2*, p. *192*; *20.1*, p. *347*; 22.1.1, p. 461
	29,	443	(1968)	13.2, p. 195; 13.4, p. 215; 14.3, p. 243; *20.1*, p. *356*; 22.1.1, p. 466
H. J. Rohde	**16,**	K179	(1966)	13.4, p. 202; *14.4.1*, p. *256*; 22.4.1, p. 498
	23,	277	(1967)	13.4, p. 209; *14.4.1*, p. *259*; 22.4.1, p. 500
P. Rohner	**30,**	619	(1968)	*13.2*, p. *195*; 20.1, p. 359; 22.4.1, p. 505
V. A. Romanov	**16,**	43	(1966)	12, p. 156; *14.3*, p. *231*; 14.4.1, p. 255; 22.1.1, p. 459
	17,	389	(1966)	12, p. 156; *16*, p. *277*; 22.1.1, p. 459
	28,	797	(1968)	14.3.1, p. 247; *16*, p. *283*; 22, p. 453
	28,	815	(1968)	*14.4.1*, p. *263*; 22.1.1, p. 466
R. Romestain	**28,**	97	(1968)	13, p. 171; *19*, p. *334*; 22.4.1, p. 503
J.-Y. Roncin	**23,**	K1	(1967)	13.2, p. 193; *20.1*, p. *350*; 23, p. 561
J.-P. Ropé	**21,**	517	(1967)	7, p. 76; *14.4.1*, p. *258*; 22.2.1, p. 484
M. F. Rose	**17,**	K199	(1966)	*12.1*, p. *159*; 21, p. 375; 21.1, p. 393; 21.6, p. 431
	21,	235	(1967)	6, p. 60; 8, p. 82; *12.1*, p. *160*; 21, p. 378; 21.1, p. 397; 21.6, p. 433
	25,	103	(1968)	3.2, p. 36; *12.1*, p. *160*; 21, p. 383; 21.3, p. 426
J. Röseler	**23,**	365	(1967)	6, p. 62; *13*, p. *167*; 13.1, p. 181; 22, p. 447
	25,	311	(1968)	*6*, p. *64*; *13*, p. *168*
M. Rosenberg	**20,**	K163	(1967)	14.4, p. 253; *18.3*, p. *317*
	21,	197	(1967)	4, p. 42; *18.3*, p. *317*; 22.6, p. 537
	21,	K131	(1967)	*14.3*, p. *236*; 15, p. 272; 18.3, p. 318
Yu. B. Rosenfeld	**28,**	K179	(1968)	6, p. 67; *20.1*, p. *356*
L. Roskovcová	**20,**	K29	(1967)	*20.1*, p. *347*; 22.2, p. 481
	26,	591	(1968)	13.1, p. 185; *20.1*, p. *353*; 22.2, p. 481
U. Rössler	**17,**	597	(1966)	5, p. 50; *13.1*, p. *175*; 20.1, p. 346; 22.4.1, p. 498
	26,	461	(1968)	*3*, p. *27*; 13.1, p. 185; 18, p. 294
	28,	577	(1968)	*13.1*, p. *187*; 22.4, p. 496
G. Roth	**29,**	K123	(1968)	10.1, p. 134; *11*, p. *154*; 12.1, p. 161; 21.1, p. 408
B. F. Rothenstein	**18,**	K13	(1966)	*10.1*, p. *118*; *18.2*, p. *304*; 21.1, p. 394
	19,	613	(1967)	*18.2*, p. *305*
	21,	601	(1967)	14.1, p. 222; *18.2*, p. *307*; 21.1.1, p. 417

B. F. Rothenstein	**27,**	K5	(1968)	*18.2*, p. *311*; 21.1, p. 405
	29,	K117	(1968)	*18.2*, p. *314*; 21.1, p. 408; 21.1.1, p. 423
	30,	K91	(1968)	*18.2*, p. *315*; 21.1, p. 409
L. Rothkirch	**26,**	709	(1968)	13.1, p. 185; *14.3*, p. *241*; 22.1.3, p. 479
J. L. Routbort	**22,**	203	(1967)	*10.1*, p. *124*; 21, p. 380
M. Rouzeyre	**24,**	399	(1967)	*14.3.2*, p. *250*; 16, p. 281; 21, p. 383; 22.6, p. 540
P. Rowlands	**26,**	K25	(1968)	3, p. 27; *4*, p. *45*; 21.1.1, p. 420
D. K. Roy	**28,**	K75	(1968)	6, p. 67; *14.3.2*, p. *251*; 22.1.1, p. 465
	29,	K111	(1968)	*13.4*, p. *216*; *14.3.2*, p. *251*; 22.1.1, p. 466
Yu. V. Rud	**26,**	85	(1968)	*1.2*, p. *12*; 2, p. 24; 8, p. 84; 12, p. 158; 14.3, p. 240; 22.7, p. 546; 22.8, p. 553
M. L. Rudee	**26,**	K97	(1968)	4, p. 46; *10*, p. *106*; 22.1.2, p. 473
A. F. Rudolph	**19,**	K79	(1967)	*10*, p. *98*; 22.1.2, p. 470
F. Rueda	**17,**	739	(1966)	*10*, p. *95*; 10.2, p. 138; 22.5.2, p. 519
E. Ruedl	**28,**	305	(1968)	*1.1*, p. *8*; 4, p. 46; 21, p. 386; 21.1, p. 405
A. W. Ruff, Jr.	**27,**	117	(1968)	*10.1*, p. *130*; 21.1, p. 404; 21.6, p. 436
A. R. Ruffa	**29,**	605	(1968)	3.2, p. 36; *13.1*, p. *189*; 20.1, p. 357; 22.6, p. 543
M. Rühle	**16,**	K105	(1966)	*10.1*, p. *115*; 11, p. 145; 21.1, p. 391
	19,	263	(1967)	*10.1*, p. *120*; 11, p. 147; 21.1, p. 395
	19,	279	(1967)	*10.1*, p. *120*; 11, p. 147; 21.1, p. 395
	22,	689	(1967)	*10.1*, p. *124*; 21.1, p. 399
	23,	113	(1967)	*4*, p. *43*
	26,	661	(1968)	*10.1*, p. *130*; *11*, p. *151*; 21.1, p. 404; 21.6, p. 435
	26,	671	(1968)	*10.1*, p. *130*; *11*, p. *151*; 21.1, p. 404
D. Rühlicke	**16,**	K1	(1966)	*10.1*, p. *114*; 21.3, p. 426
B. M. Rumyantsev	**30,**	329	(1968)	13.2, p. 195; *20.3*, p. *371*; 22.9, p. 560
P. Runow	**29,**	627	(1968)	1.4, p. 21; *10*, p. *110*; 22.1, p. 457
A. L. Ruoff	**21,**	507	(1967)	7, p. 76; *12*, p. *157*; 22.5.2, p. 522
	23,	489	(1967)	*6*, p. *63*; 13.1, p. 182; 14.1, p. 224; 21.2, p. 425
W. Rupp	**28,**	K35	(1968)	1.4, 20; *11*, p. *153*; 21.1, p. 405; 21.6, p. 436
W. Ruppel	**26,**	K163	(1968)	*16*, p. *282*; 20.1, p. 353; 20.3, p. 369; 22.4.1, p. 502
	29,	K31	(1968)	*3.1*, p. *33*; 5, p. 54; 20.1, p. 357; 22.4.1, p. 504; 22.4.2, p. 508; 22.8, p. 555
W. Ruske	**17,**	185	(1966)	1.4, p. 16; *18.2*, p. *301*; 21.1.1, p. 412
G. I. Rusov	**17,**	249	(1966)	4, p. 38; *18.2*, p. *302*; 21.1, p. 392; 21.1.1, p. 412
M. A. Ruvinskii	**28,**	75	(1968)	6, p. 66; *7*, p. *78*; 13.1, p. 187; 13.2, p. 194; 22, p. 452
S. M. Ryvkin	**28,**	489	(1968)	2, p. 24; 14.3, p. 242; *20.3*, p. *370*; 22.8, p. 554
K. I. Rzaev	**20,**	261	(1967)	*2*, p. *22*; 22.1.3, p. 477
H. Rżany	**17,**	K195	(1966)	6.1, p. 71; *18.4*, p. *321*; 22, p. 441
	21,	K163	(1967)	*13*, p. *166*; *18*, p. *292*; 22.6, p. 538
	26,	429	(1968)	*4*, p. *45*; 18.3, p. 319; 18.4, p. 324; 22.8, p. 553
H. S. Sack	**17,**	353	(1966)	6, p. 57; 13.4, p. 202; *16*, p. *277*; *20.3*, p. *364*; 22.5.1, p. 515
	22,	203	(1967)	*10.1*, p. *124*; 21, p. 380
	30,	569	(1968)	*7*, p. *78*; 10, p. 112; 13.4, p. 217; 22.5.2, p. 530
	30,	579	(1968)	*7*, p. *78*; 10, p. 112; 13.4, p. 217; 22.5.2, p. 531
K. E. Saeger	**21,**	671	(1967)	*14.1*, p. *222*; 21.1, p. 398
	28,	589	(1968)	*14.1*, p. *225*; 21.1, p. 406
G. I. Safaraliev	**17,**	777	(1966)	6, p. 57; *8*, p. *79*; 22.2.3, p. 490
	22,	741	(1967)	13.4, p. 209; *14.3*, p. *237*; 22.2.3, p. 491
J. Sak	**16,**	147	(1966)	*13.1*, p. *174*; *13.2*, p. *191*; 20.1, p. 344; 22, p. 439
	25,	155	(1968)	8, p. 84; *13.1*, p. *183*

J. Sak	**27,**	521	(1968)	6, p. 66; 13.1, p. 186; *13.4*, p. *213*; 22, p. 452
	29,	707	(1968)	6, p. 68; *13.2*, p. *195*; 20.1, p. 357; 22.4.1, p. 505; 22.4.2, p. 508; 22.4.3, p. 511
K. Sakata	**20,**	K155	(1967)	*1.2*, p. *10*; 22.6, p. 537
T. Sakata	**20,**	K155	(1967)	*1.2*, p. *10*; 22.6, p. 537
A. Sakalas	**24,**	K91	(1967)	3.1, p. 33; *14.3*, p. *239*; *16*, p. *280*; 22.4.2, p. 507
M. Yu. Sakhnovskii	**28,**	277	(1968)	13.1, p. 187; 13.2, p. 194; *20.1*, p. *355*; 22, p. 452
N. M. Salanskii	**17,**	489	(1966)	*18.2*, p. *303*; 21.1.1, p. 413
P. P. Salhotra	**29,**	859	(1968)	*1.2*, p. *13*; *3*, p. *28*; 8, p. 86; 22.8, p. 556
O. P. Salita	**29,**	569	(1968)	*10.1*, p. *133*; 21.1, p. 407
J. A. M. Salter	**19,**	K63	(1967)	*8*, p. *81*; 21, p. 377; 21.1, p. 396
	21,	357	(1967)	4, p. 42; 6, p. 61; *8*, p. *82*; 21, p. 379; 21.1, p. 397
	25,	531	(1968)	*6*, p. *65*; 8, p. 84; *10.1*, p. *128*; 21, p. 384; 21.1, p. 402
G. K. Samedov	**26,**	K123	(1968)	*14.3*, p. *241*; 15, p. 273; 22, p. 451
B. N. Samoilov	**20,**	767	(1967)	*6*, p. *60*; 8, p. 81; 21.1, p. 397
T. V. Samoilova	**28,**	K1	(1968)	*10*, p. *109*; 22.5.2, p. 527
A. G. Samoilovich	**16,**	459	(1966)	6, p. 56; 13.1, p. 174; *15*, p. *270*; 22, p. 440
	23,	229	(1967)	13, p. 167; 14.3, p. 238; *15*, p. *272*; 22.1.1, p. 463
M. R. Samuelsen	**28,**	211	(1968)	13.1, p. 187; *19*, p. *334*; 22.1.2, p. 474
C. Sánchez	**29,**	217	(1968)	*10.2*, p. *143*; 11, p. 153; 22.5.2, p. 528
J. V. Sanders	**25,**	541	(1968)	3, p. 27; *5*, p. *53*; 22.6, p. 540
R. L. Sanders	**30,**	93	(1968)	*9*, p. *92*; 10.1, p. 134; 21, p. 388; 21.6, p. 437
L. M. Sandler	**17,**	753	(1966)	10.1, p. 117; *18.2*, p. *303*; 19, p. 327; 21.1.1, p. 413
R. Sandrock	**16,**	487	(1966)	5, p. 49; *13.1*, p. *175*; 22.1.3, p. 476
R. P. Santoro	**16,**	K17	(1966)	*18.4*, p. *321*; 22.8, p. 547
	19,	K87	(1967)	*13*, p. *165*; 18, p. 291; *20*, p. *338*; 22.8, p. 549
O. G. Sarbej	**18,**	749	(1966)	10, p. 97; 12, p. 156; *14.3*, p. *234*; 22.1.1, p. 460
V. S. Sardaryan	**17,**	479	(1966)	*13.1*, p. *175*; 14.3, p. 232; 22.2.1, p. 482
A. S. Sastri	**25,**	K67	(1968)	1.1, p. 8; *1.2*, p. *11*; 21, p. 385; 21.1, p. 403
D.H. Sastry	**29,**	K137	(1968)	*4*, p. *47*; 10.1, p. 134; 14.1, p. 226; 21.1, p. 408; 21.3, p. 427
E. V. R. Sastry	**29,**	K107	(1968)	10, p. 111; *10.2*, p. *144*; 22.5.2, p. 529
N. P. Sastry	**26,**	91	(1968)	*13*, p. *169*; 18.4, p. 323; 22.5, p. 513; 22.8, p. 553
P. V. Sastry	**18,**	K127	(1966)	*10.2*, p. *138*; 22.5.2, p. 520
P. Sauer	**16,**	79	(1966)	10, p. 93; *13.4*, p. *201*; 22.5.2, p. 518
M. Sauvage	**26,**	469	(1968)	*4*, p. *45*
	29,	725	(1968)	*10*, p. *111*; 22.8, p. 555
W. R. Savage	**25,**	131	(1968)	*1.3*, p. *15*; 17, p. 288; 21, p. 383
M. Savelli	**16,**	721	(1966)	*13.4*, p. *202*; 14.3, p. 232; 16, p. 276; 22.1.1, p. 459
A. Sawaoka	**24,**	K83	(1967)	*18.3*, p. *318*
V. K. Saxena	**27,**	427	(1968)	13, p. 170; *18.1*, p. *297*; 21, p. 386; 21.2, p. 425
M. Saxlová	**26,**	671	(1968)	*10.1*, p. *130*; *11*, p. *151*; 21.1, p. 404
S. Schäfer	**19,**	297	(1967)	*10*, p. *97*; 22.1.1, p. 461
A. Scharmann see A. O. Scharmann				
A. O. Scharmann	**18,**	307	(1966)	*13.4*, p. *203*; 14.3, p. 233; 16, p. 278; 20.3, p. 364; 22.4, p. 495
	28,	583	(1968)	10.2, p. 143; *11*, p. *153*; 19, p. 334; 20.3, p. 370; 22.8, p. 554
O. Schärpf	**30,**	637	(1968)	4, p. 48; *18.2*, p. *315*; 21.1, p. 409
F. J. Schedewie	**30,**	251	(1968)	6, p. 69; 13, p. 173; *18.1*, p. *298*; 19, p. 336; 22.8, p. 556
J. Schelm	**27,**	413	(1968)	13, p. 170; 13.1, p. 186; *14.3*, p. *241*; 15, p. 274; 22, p. 452; 22.7, p. 546
E. Schelzke	**17,**	K241	(1966)	1.2, p. 9; *1.4*, p. *17*; 21.1, p. 393

M. Schenk	**17,**	91	(1966)	*1.2,* p. *9*; 10, p. 94; 11, p. 146; 14.4.2, p. 266
	18,	547	(1966)	*1.2,* p. *10*; 8, p. 80; 11, p. 147; 14.4.2, p. 266
W. Schilling	**17,**	259	(1966)	10.1, p. 115; *11,* p. *146*; 14.1, p. 220; 21, p. 374
	18,	427	(1966)	*10.1,* p. *118*; 11, p. 147; 14.1, p. 221; 21, p. 375
	23,	481	(1967)	10.1, p. 125; *11,* p. *150*; 14.1, p. 223; 21.6, p. 434
W. Schilz	**29,**	559	(1968)	*13,* p. *172*; 14.3, p. 244; *20,* p. *342*; 22, p. 454
O. Schirmer	**16,**	79	(1966)	10, p. 93; *13.4,* p. *201*; 22.5.2, p. 518
F. Schlät	**21,**	K73	(1967)	*10.1,* p. *122*; 14.1, p. 222; 21, p. 379
M. Schleifstein	**30,**	K97	(1968)	*20.3,* p. *372*; 22.1.1, p. 467
C. Schlenker	**16,**	301	(1966)	*18.2,* p. *299*; *18.4,* p. *321*; 21.1, p. 390; 21.1.1, p. 410
	28,	507	(1968)	*18.2,* p. *313*; *18.4,* p. *324*; 21.1, p. 406; 22.6, p. 542
P. Schmeling	**20,**	127	(1967)	*9,* p. *88*; 22.5.3, p. 532
	20,	597	(1967)	*9,* p. *89*; 11, p. 148; 22.5.2, p. 521
D. Schmid	**18,**	653	(1966)	10.2, p. 138; *19,* p. *327*; 22.5.2, p. 520
	23,	633	(1967)	*13.2,* p. *193*; *19,* p. *330*; 22.9, p. 559
G. Schmid	**18,**	829	(1966)	*10.1,* p. *119*; 21.1, p. 394
H. Schmid	**26,**	277	(1968)	4, p. 45; *14.4.2,* p. *268*; 20, p. 340
E. Schmidt	**27,**	57	(1968)	13.1, p. 185; *20.1,* p. *353*; 22.1.1, p. 464; 22.1.2, p. 473
J. Schnakenberg	**28,**	623	(1968)	*6,* p. *67*; *13,* p. *171*; 14.3, p. 242; 22, p. 453
J. Schneider	**16,**	79	(1966)	10, p. 93; *13.4,* p. *201*; 22.5.2, p. 518
	18,	125	(1966)	13.4, p. 203; *19,* p. *327*; 22.4.1, p. 498
P. Schnupp	**21,**	567	(1967)	13.1, p. 178; *14.3.1,* p. *245*; 22, p. 446
	25,	455	(1968)	13.1, p. 183; *14.3.1,* p. *246*; 22, p. 449
E. Schnürer	**20,**	K173	(1967)	*6,* p. *60*; 7, p. 76; 22.4.1, p. 499
T. Schober	**27,**	195	(1968)	*10.1,* p. *130*; *11,* p. *152*; 21.6, p. 436
G. Schoffa	**17,**	683	(1966)	11, p. 146; *19,* p. *327*; 22.9, p. 558
A. H. Scholz	**25,**	285	(1968)	*10,* p. *104*; 22.5.2, p. 525
H.-G. Schöpf	**28,**	K23	(1968)	*10,* p. *109*; *10.1,* p. *132*
M. Schott	**20,**	301	(1967)	*13.4,* p. *205*; 14.4.1, p. 257; 22.9, p. 558
	27,	201	(1968)	13.2, p. 194; *13.3,* p. *198*; *16,* p. *282*; 22.9, p. 560
K. Schröder	**27,**	601	(1968)	*18.2,* p. *312*
	27,	611	(1968)	*18.2,* p. *312*
	29,	107	(1968)	13.1, p. 188; *14.1,* p. *226*; *15,* p. *274*; 18.4, p. 324; 21.1, p. 407
U. Schröder	**21,**	309	(1967)	*6,* p. *61*; 8, p. 82; 22.5.2, p. 522
	24,	99	(1967)	*6,* p. *63*; 20.1, p. 351; 22.1.3, p. 478
K. Schroeder	**22,**	423	(1967)	*11,* p. *149*; 21.1, p. 399
	22,	433	(1967)	*11,* p. *149*; 21, p. 380; 21.1, p. 399; 21.6, p. 434
W. Schröter	**21,**	211	(1967)	10, p. 99; 13.4, p. 206; *14.3,* p. *236*; 22.1.1, p. 461
R. Schubert	**16,**	K5	(1966)	13.3, p. 196; *16,* p. *276*; 22.4.1, p. 497
	16,	K157	(1966)	13.3, p. 196; *16,* p. *276*; 22.4.1, p. 497
W. Schülke	**23,**	K87	(1967)	*13.1,* p. *181*; 20, p. 340; 21.3, p. 426
M. Schulz	**21,**	K139	(1967)	*13,* p. *166*; 14.4.1, p. 259; *20,* p. *339*; 22.1.1, p. 462
	25,	521	(1968)	*13,* p. *169*; 14.4.1, p. 261; *20,* p. *340*; 22.1.1, p. 464
D. Schulze	**26,**	485	(1968)	*10.1,* p. *129*; 21, p. 385
G. E. R. Schulze	**23,**	475	(1967)	*5,* p. *52*; 21.1.1, p. 419
	24,	77	(1967)	*10.1,* p. *126*; 21.1, p. 401; 21.3, p. 426
	29,	211	(1968)	*18.2,* p. *314*; *18.4,* p. *325*; 21.1.1, p. 422
H. Schulze	**24,**	K95	(1967)	*19,* p. *331*; 22.9, p. 560
	30,	K9	(1968)	14.4.2, p. 269; *19,* p. *336*
D. Schumacher	**25,**	359	(1968)	9, p. 91; *10.1,* p. *128*; 21.6, p. 435
	29,	819	(1968)	10.1, p. 134; *11,* p. *154*; 14.1, p. 226; 21.3, p. 427
	30,	119	(1968)	10.1, p. 134; *11,* p. *154*; 14.1, p. 226; 21.3, p. 427

H. Schuster	**21,**	741	(1967)	*11*, p. *149*; 14.1, p. 223; 21, p. 379
	29,	75	(1968)	*10.1*, p. *132*; 11, p. 153; 21.1, p. 406
	29,	81	(1968)	*10.1*, p. *133*; 11, p. 153; 21.1, p. 406
	30,	219	(1968)	10.1, p. 135; *11*, p. *155*; 14.1, p. 227; 18.2, p. 315; 21.1, p. 408
W. Schüz	**25,**	253	(1968)	1.4, p. 20; *5*, p. *53*; 22.1.3, p. 478
D. Schwarz	**18,**	787	(1966)	10, p. 97; *20.1*, p. *346*; 22.5.1, p. 515
	27,	K77	(1968)	10, p. 107; 13.4, p. 213; *19*, p. *333*; 22.5.1, p. 516
A. Schweighofer	**25,**	K91	(1968)	*10.1*, p. *128*; 21.1, p. 403
Ch. Schwink	**16,**	K45	(1966)	4, p. 37; *18.2*, p. *300*; 21.1, p. 390
	17,	35	(1966)	*10.1*, p. *115*; 11, p. 145; 21.1, p. 391
	18,	557	(1966)	*10.1*, p. *118*; 21.1, p. 394
	19,	217	(1967)	4, p. 40; *18.2*, p. *304*; 21.1, p. 395
	30,	387	(1968)	*10.1*, p. *135*; 21.1, p. 408
	30,	637	(1968)	4, p. 48; *18.2*, p. *315*; 21.1, p. 409
W. J. van Sciver	**29,**	159	(1968)	*13.2*, p. *195*; 20.3, p. 370; 22.5.2, p. 528
H. G. Scott	**20,**	461	(1967)	4, p. 41; *10*, p. *99*; 22.9, p. 559
B. Sczaniecki	**23,**	K159	(1967)	14.4.2, p. 267; *19*, p. *331*
P. F. Seal	**16,**	K17	(1966)	*18.4*, p. *321*; 22.8, p. 547
F. Secco D'Aragona	**30,**	K1	(1968)	*4*, p. *48*; 10.1, p. 135; 21.1, p. 408
J. Šedivý	**20,**	249	(1967)	*8*, p. *81*; 21.6, p. 432
A. Seeger	**16,**	253	(1966)	13, p. 162; *14.1*, p. *220*; *18.2*, p. *299*; 21, p. 373
	17,	709	(1966)	*10*, p. *95*; *10.1*, p. *116*; 21, p. 375; 21.1, p. 392; 21.3, p. 426; 21.6, p. 431; 22.1.1, p. 460; 22.1.2, p. 469
	18,	189	(1966)	*10.1*, p. *117*; 21.1, p. 393; 21.6, p. 431
	19,	251	(1967)	*10.1*, p. *120*; 21.1, p. 395
	24,	721	(1967)	*14.1*, p. *224*; 21, p. 383
	25,	359	(1968)	9, p. 91; *10.1*, p. *128*; 21.6, p. 435
	27,	371	(1968)	10.1, p. 131; *14.2*, p. *230*
	29,	231	(1968)	*9*, p. *92*; *10.1*, p. *133*; 21.6, p. 437
	29,	455	(1968)	*9*, p. *87*; *10*, p. *93*; 11, p. 145; 13.4, p. 200; 19, p. 326; 22.1.1, p. 458; 22.1.2, p. 468
	30,	K101	(1968)	10.1, p. 136; *18.2*, p. *315*; 21.1, p. 409
G. P. Seidel	**17,**	43	(1966)	*10.1*, p. *115*; 11, p. 146; 21.1, p. 392
	25,	175	(1968)	*10.1*, p. *127*; *11*, p. *151*; 21.1.1, p. 420
D. N. Seidman	**17,**	531	(1966)	*10.1*, p. *116*; 21.6, p. 431
P. Seifert	**25,**	303	(1968)	*10*, p. *104*; 22.5.2, p. 525
M. Selders	**24,**	K33	(1967)	*13.1*, p. *183*; 20.1, p. 351; 22.1.3, p. 478
H. Seliger	**29,**	K27	(1968)	13, p. 172; *18*, p. *294*; *20*, p. *342*; 22.1.3, p. 480
B. Selle	**24,**	K149	(1967)	*14.3.2*, p. *250*; *16*, p. *281*; 22, p. 449; 22.4.1, p. 501
	30,	K153	(1968)	13.4, p. 218; *14.3.1*, p. *247*; 15, p. 275; 16, p. 285; *20.1*, p. *359*; 22, p. 455
G. Ya. Selyutin	**22,**	K23	(1967)	*6.1*, p. *72*; 21, p. 380
	26,	253	(1968)	*6.1*, p. *73*; 13, p. 169; 18, p. 294
D. G. Semak	**21,**	437	(1967)	*13.4*, p. *206*; 14.4.1, p. 258; 22, p. 445
G. K. Semin	**27,**	K129	(1968)	14.4.2, p. 269; *19*, p. *334*
S. Sengupta	**29,**	367	(1968)	*6*, p. *67*; 22.5.2, p. 528
V. V. Serdyuk	**28,**	K31	(1968)	13.4, p. 214; *16*, p. *283*; 22.4.1, p. 503
P. S. Serebrennikov	**21,**	431	(1967)	*14.3.2*, p. *249*; 22, p. 445
A. Serpi	**28,**	K43	(1968)	*1.2*, p. *12*; 13.1, p. 187; 20.1, p. 356; 22.4.1, p. 503
B. Šesták	**16,**	89	(1966)	*10.1*, p. *114*; 21.1.1, p. 410
D. Seyboth	**27,**	639	(1968)	*6.1*, p. *74*; 22.1.1, p. 465
N. Sexer	**21,**	225	(1967)	13.1, p. 178; *14.3*, p. *236*; 15, p. 272; 22.7, p. 545
P. K. Sharma	**18,**	119	(1966)	6, p. 57; *8*, p. *80*; 21.1, p. 393

P. K. Sharma	23, 361	(1967)	*6*, p. *62*; 21, p. 381; 21.1, p. 400; 21.1.1, p. 419; 21.2, p. 429; 21.6, p. 434
	28, 643	(1968)	*6*, p. *67*; *8*, p. *85*; 21, p. 387; 21.1, p. 406; 21.2, p. 425; 21.6, p. 436
S. Sharma	30, 239	(1968)	6, p. 69; *14.3*, p. *244*; 22.4.1, p. 505
R. A. Shatas	26, K155	(1968)	10.2, p. 142; *13.4*, p. *212*; 22.5.2, p. 526; 22.6, p. 541
	28, K143	(1968)	*10.2*, p. *143*; 22.5.3, p. 534
D. Shaw	29, 145	(1968)	10, p. 109; *13.4*, p. *215*; 14.3, p. 243; 22.4.3, p. 511
N. Shaw	17, 605	(1966)	*14.3*, p. *232*; *16*, p. *277*; 22.2.3, p. 490
S. V. Shegolikhina	30, 845	(1968)	14.3, p. 244; *14.4.1*, p. *265*; 22.1.1, p. 467
S. S. Sheinin	17, K143	(1966)	*10*, p. *95*; 22.1.2, p. 469
	21, 247	(1967)	*5*, p. *51*; 21, p. 378
	28, 187	(1968)	*5*, p. *53*; 21, p. 386; 22.6, p. 542
	30, 791	(1968)	*5*, p. *55*; 22.1.2, p. 475
M. K. Sheinkman	24, 543	(1967)	10, p. 103; *13.4*, p. *210*; 16, p. 281; 20.3, p. 368; 22.4.1, p. 501
E. F. Sheka	19, 395	(1967)	*6*, p. *58*; 20.1, p. 347; 22.9, p. 558
A. B. Sherman	20, 759	(1967)	*18.4*, p. *322*; 22.5, p. 512; 22.8, p. 550
Yu. A. Sherstkov	28, 269	(1968)	10, p. 109; 13, p. 171; *19*, p. *334*; 22.6, p. 542
S. N. Shestatskii	28, K131	(1968)	13.1, p. 188; 13.2, p. 194; *14.4.1*, p. *264*; *20*, p. *342*; 20.1, p. 356; 22.6, p. 542
	30, 349	(1968)	13.1, p. 189; *20.1*, p. *358*; 22, p. 454
K. V. Shevlyagin	27, K85	(1968)	7, p. 78; *18.3*, p. *320*
V. B. Shipilo	25, K23	(1968)	*14.3*, p. *240*; 22.2.3, p. 492
	25, K27	(1968)	*14.3*, p. *240*; 22.2.3, p. 492
V. P. Shirokovskii	26, 391	(1968)	3, p. 27; *13.1*, p. *185*
G. P. Shpenikov	28, 489	(1968)	2, p. 24; 14.3, p. 242; *20.3*, p. *370*; 22.8, p. 554
G. P. Shpenkov	25, 513	(1968)	*3.1*, p. *33*; 5, p. 53; 14.3, p. 240; 15, p. 273; 20.1, p. 351; 22, p. 450
V. S. Shpinel	29, 49	(1968)	*6.1*, p. *74*; 22.9, p. 560
K. N. Shrivastava	25, K47	(1968)	10, p. 104; *19*, p. *332*; 22.6, p. 540
	29, 737	(1968)	*6*, p. *68*; 19, p. 335; 22.6, p. 543
M. M. Shukla	16, 513	(1966)	*6*, p. *56*; 8, p. 79; 21, p. 374
	19, 729	(1967)	*6*, p. *59*; 21.1, p. 396
M. S. Shur	17, 163	(1966)	5, p. 49; *6*, p. *56*; 14.4.2, p. 266; *20.1*, p. *345*
	17, 173	(1966)	1.2, p. 9; *6*, p. *57*; 14.4.2, p. 266; *20.1*, p. *345*
	28, 827	(1968)	*14.4.1*, p. *263*; 22, p. 453
Ya. S. Shur	16, 745	(1966)	4, p. 38; *18.2*, p. *301*; 21.1.1, p. 411
S. D. Shutov	30, 349	(1968)	13.1, p. 189; *20.1*, p. *358*; 22, p. 454
L. A. Shuvalov	22, 279	(1967)	1.2, p. 11; *14.4.4*, p. *267*
K. K. Shvarts	18, 897	(1966)	10.2, p. 138; *11*, p. *147*; 19, p. 327; 20.1, p. 346; 22.5.2, p. 520
V. N. Shvetsov	30, 845	(1968)	14.3, p. 244; *14.4.1*, p. *265*; 22.1.1, p. 467
W. A. Sibley	23, 301	(1967)	*10.2*, p. *140*; 11, p. 150; 13.4, p. 209; 22.5.2, p. 523
H. Šíchová	20, 249	(1967)	*8*, p. *81*; 21.6, p. 432
S. K. Sidorov	16, 737	(1966)	*18.2*, p. *301*; *18.4*, p. *321*; 21.1, p. 391; 21.1.1, p. 411
	24, 385	(1967)	*18.2*, p. *309*; *18.4*, p. *323*; 21.1.1, p. 419
V. G. Sidorov	18, 677	(1966)	*13.1*, p. *176*; 14.3, p. 234; 15, p. 271; 22.8, p. 548
D. Siebert	24, K65	(1967)	*14.3*, p. *239*; 22.2.1, p. 484
R. Siems	30, 645	(1968)	6, p. 70; *10*, p. *112*; *10.1*, p. *135*
Z. Siermiński	17, 211	(1966)	*18.3*, p. *316*
J. A. Sigler	21, 545	(1967)	*10.1*, p. *123*; 21, p. 379; 21.1, p. 398; 21.6, p. 433
H. Sigmund	20, 255	(1967)	*14.3.2*, p. *249*; 16, p. 278; 22, p. 444; 22.1.2, p. 470

P. Sigmund	**16**, 507	(1966)	1.3, p. 14; *11*, p. *145*; 21.1, p. 390
V. Šik	**28**, 175	(1968)	*19*, p. *334*; 22.8, p. 554
G. A. Sikharulidze	**26**, K33	(1968)	13.1, p. 184; *20.1*, p. *353*; 22.7, p. 546; 22.8, p. 553
A. Sikora	**25**, K127	(1968)	*14.2*, p. *230*
S. Sikorski	**25**, 345	(1968)	*14.3*, p. *240*; 14.3.2, p. 250; 22, p. 449
D. Silber	**16**, 215	(1966)	10.2, p. 137; 13.4, p. 201; *19*, p. *326*; 22.5.2, p. 518
	20, 337	(1967)	13.4, p. 205; *19*, p. *328*; 22.5.2, p. 521
J. M. Silcock	**20**, 725	(1967)	*4*, p. *41*; 21, p. 378
J. Silcox	**19**, 57	(1967)	*1.4*, p. *17*; 4, p. 39; 21, p. 376
	19, 63	(1967)	*1.4*, p. *18*; 4, p. 40; 21.1, p. 395; 21.6, p. 432
B. D. Silverman	**27**, 473	(1968)	6, p. 66; *7*, p. *77*; 11, p. 152; 22.6, p. 541
P. Simidtchieva	**21**, K87	(1967)	2, p. 23; 13.4, p. 207; *20.1*, p. *348*; 22, p. 445
C. Şimoi	**29**, 761	(1968)	13, p. 173; *17*, p. *290*; 21, p. 388
B. Simon	**19**, 211	(1967)	*3.1*, p. *31*; 22.8, p. 549
W. Simonet	**22**, K87	(1967)	*18.3*, p. *318*
A. K. Singh	**28**, 643	(1968)	*6*, p. *67*; *8*, p. *85*; 21, p. 387; 21.1, p. 406; 21.2, p. 425; 21.6, p. 436
H. P. Singh	**23**, K93	(1967)	5, p. 52; *8*, p. *83*; 22.4.2, p. 507
N. Singh	**17**, 501	(1966)	13, p. 163; *14.3*, p. *232*; 22.6, p. 535
I. G. Siny	**30**, 367	(1968)	13, p. 173; *18.3*, p. *320*; *20*, p. *342*; 22.5, p. 514; 22.8, p. 556
D. Sippel	**23**, 237	(1967)	*4*, p. *43*; 22.1.2, p. 472
N. N. Sirota	**25**, K23	(1968)	*14.3*, p. *240*; 22.2.3, p. 492
	25, K27	(1968)	*14.3*, p. *240*; 22.2.3, p. 492
	28, K175	(1968)	*18.1*, p. *297*; 18.2, p. 313; 22.8, p. 555
R. Sittig	**21**, K175	(1967)	10.2, p. 139; 14.4.1, p. 259; *20.1*, p. *349*; 22.5.2, p. 523
N. I. Sivkov	**17**, 243	(1966)	1.4, p. 16; *3.1*, p. *29*; 21.1, p. 392; 21.1.1, p. 412
	17, 249	(1966)	4, p. 38; *18.2*, p. *302*; 21.1, p. 392; 21.1.1, p. 412
S. J. Sivonen	**29**, K171	(1968)	14.1, p. 226; *18.2*, p. *314*; 21.1.1, p. 423; 21.4, p. 428
R. Sizmann	**26**, 359	(1968)	4, p. 45; 10.1, p. 129; *11*, p. *151*; 14.1, p. 225; 21.6, p. 435
	28, K35	(1968)	1.4, p. 20; *11*, p. *153*; 21.1, p. 405; 21.6, p. 436
	29, 403	(1968)	*11*, p. *154*; 21.6, p. 437
V. G. Skobov	**22**, 333	(1967)	*13*, p. *166*; 13.1, p. 179; *20*, p. *339*; 21, p. 380
B. S. Skorchev	**27**, K161	(1968)	6.1, p. 74; *10*, p. *108*; 22.5.1, p. 517; 22.5.2, p. 527
M. Ya. Skorokhod	**30**, 231	(1968)	*10*, p. *112*; 20, p. 342; 22.1.1, p. 466
T. Skośkiewicz	**30**, K33	(1968)	10.1, p. 135; *14.1*, p. *227*; 21, p. 388
V. A. Skripkin	**27**, K169	(1968)	*13.1*, p. *186*; 15, p. 274; 22.4.3, p. 511
St. Slădaru	**30**, K149	(1968)	13.1, p. 190; *20.1*, p. *359*; 22, p. 455
L. M. Slifkin	**30**, 93	(1968)	*9*, p. *92*; 10.1, p. 134; 21, p. 388; 21.6, p. 437
V. Yu. Slivka	**30**, 461	(1968)	*13.1*, p. *189*; 16, p. 284; 20.1, p. 358; 22.8, p. 557
S. V. Slobodchikov	**19**, 429	(1967)	*14.3.2*, p. *249*; 16, p. 278; 22.2.3, p. 490
	23, 755	(1967)	*6*, p. *63*; 13.1, p. 182; *16*, p. *280*; 22.2.2, p. 488
	30, 441	(1968)	13.4, p. 217; *14.3*, p. *244*; 22.2.2, p. 488
Z. Smetana	**25**, K87	(1968)	10, p. 104; 18.3, p. 319; *19*, p. *332*
A. Smilga	**24**, K91	(1967)	3.1, p. 33; *14.3*, p. *239*; *16*, p. *280*; 22.4.2, p. 507
	25, 331	(1968)	2, p. 23; 13.4, p. 211; 14.3.1, p. 246; *14.4.1*, p. *261*; 22.1.3, p. 478
B. I. Smirnov	**16**, 191	(1966)	*10*, p. *93*; 22.5.2, p. 518
	28, K1	(1968)	*10*, p. *109*; 22.5.2, p. 527
I. A. Smirnov	**18**, 479	(1966)	*8*, p. *80*; *10*, p. *96*; 14.3, p. 233; 15, p. 271; 22, p. 441; 22.8, p. 548
Y. N. Smirnov	**23**, K97	(1967)	*8*, p. *83*; 21.4, p. 428

I. S. Smirnova **30,** 19 (1968) *10*, p. *111*; 22.4, p. 496

A. P. Smith **21,** 369 (1967) *10.1*, p. *122*; 21.1, p. 397

26, K93 (1968) *10*, p. *106*; 22.9, p. 560

27, 291 (1968) 1.3, p. 15; *1.4*, p. *20*; 10, p. 107; 22.6, p. 541

D. K. Smith **25,** K79 (1968) *10*, p. *104*; 11, p. 151; 22.6, p. 540

G. Smith **19** 577 (1967) *13.1*, p. *177*; *16*, p. *278*; 22.2.1, p. 483; 22.2.3, p. 490

20, K135 (1967) *13*, p. *165*; 14.3, p. 236; 22.2.3, p. 490

M. J. A. Smith **24,** 525 (1967) *19*, p. *331*; 21.6, p. 435

E. M. Smokotin **30,** K111 (1968) *18.3*, p. *320*; 19, p. 336

B. Smola **17,** K91 (1966) *10.1*, p. *116*; 21, p. 374

G. A. Smolenskii **18,** 873 (1966) *18.2*, p. *304*; *18.4*, p. *321*; 22, p. 442

30, 367 (1968) 13, p. 173; *18.3*, p. *320*; *20*, p. *342*; 22.5, p. 514; 22.8, p. 556

30, 871 (1968) 18.3, p. 320; *19*, p. *336*

G. A. Smolensky see G. A. Smolenskii

L. Śniadower **25,** 233 (1968) 13.1, p. 183; *18*, p. *293*; *20*, p. *340*; 20.1, p. 351; 22.4.3, p. 510

28, K121 (1968) *5*, p. *54*; *8*, p. *85*; 22.4.3, p. 511

28, K125 (1968) 13.1, p. 188; *18*, p. *294*; *20.1*, p. *356*; 22.4.3, p. 511

O. V. Snitko **24,** 543 (1967) 10, p. 103; *13.4*, p. *210*; 16, p. 281; 20.3, p. 368; 22.4.1, p. 501

A. Sobaszek **26,** K59 (1968) *9*, p. *91*; 21.6, p. 435

B. P. Sobolev **23,** K5 (1967) 20.1, p. 350; *20.2*, p. *361*; 20.3, p. 367; 22.5, p. 513; 22.8, p. 552

V. V. Sobolev **28,** K131 (1968) 13.1, p. 188; 13.2, p. 194; *14.4.1*, p. *264*; *20*, p. *342*; 20.1, p. 356; 22.6, p. 542

30, 349 (1968) 13.1, p. 189; *20.1*, p. *358*; 22, p. 454

30, K145 (1968) 13.1, p. 190; *20.1*, p. *359*; 22.7, p. 546

K. Šobra **17,** K19 (1966) *14.3.2*, p. *248*; 22.1.3, p. 476

M. Sodha **30,** 239 (1968) 6, p. 69; *14.3*, p. *244*; 22.4.1, p. 505

V. I. Sokolova **25,** 513 (1968) *3.1*, p. *33*; 5, p. 53; 14.3, p. 240; 15, p. 273; 20.1, p. 351; 22, p. 450

I. L. Sokolskaya **21,** 789 (1967) 1.3, p. 15; *13.3*, p. *197*; 17, p. 287; 22.1.1, p. 462

22, 39 (1967) *17*, p. *287*; 22.1.1, p. 462

V. P. Soldatov **18,** 863 (1966) *4*, p. *39*; 21.7, p. 438

J. Sólyom **25,** 473 (1968) *8*, p. *84*; 18, p. 293

E. Sonder **23,** 301 (1967) *10.2*, p. *140*; 11, p. 150; 13.4, p. 209; 22.5.2, p. 523

K. Sonnenberg **17,** 259 (1966) 10.1, p. 115; *11*, p. *146*; 14.1, p. 220; 21, p. 374

G. D. Sootha **22,** 505 (1967) 13.4, p. 208; *14.4*, p. *253*; *20.1*, p. *349*; 22.5.2, p. 523

M. Sorohan **26,** K9 (1968) *11*, p. *151*; *18.2*, p. *310*; 21.1, p. 404

28, K59 (1968) *18.2*, p. *313*; 21.1, p. 406

V. Soshka **19,** 425 (1967) *6*, p. *58*; 21.1, p. 395

A. Sosin **28,** 675 (1968) 10.1, p. 132; *11*, p. *153*; 14.1, p. 226; 21, p. 387

L. Sosnowski **20,** 113 (1967) *13.1*, p. *177*; 14.3, p. 235; 22.4.3, p. 509; 22.8, p. 549

J. M. Spaeth **20,** 225 (1967) *19*, p. *328*; 22, p. 444; 22.5, p. 512

20, 237 (1967) *19*, p. *328*; 22, p. 444; 22.5, p. 512

H. Spalt **23,** K75 (1967) *10*, p. *102*; 10.2, p. 140; 22.5.2, p. 524

H. N. Spector **20,** 605 (1967) *6*, p. *60*; 13, p. 165; 18, p. 292; *19*, p. *328*; 22, p. 444

22, 185 (1967) 13.1, p. 179; 14.4.1, p. 259; 18, p. 292; *20.1*, p. *349*; 22, p. 446

M. O. Speidel **22,** K71 (1967) *10.1*, p. *124*; 21, p. 381

P. M. Spencer	22, 371	(1967)	*1*, p. *7*; *1.2*, p. *11*; 22.4.3, p. 509; 22.8, p. 551
F. Spiegelberg	17, K11	(1966)	13.3, p. 196; *16*, p. *277*; 22.4.1, p. 498
	30, 659	(1968)	6, p. 70; *13.2*, p. *195*; 20.1, p. 359; 20.3, p. 372; 22.4.1, p. 505
G. Spinolo	25, 557	(1968)	13.4, p. 211; 20.1, p. 351; *20.3*, p. *368*; 22.5.2, p. 525
	27, 95	(1968)	13.4, p. 213; *20.1*, p. *353*; 22.5, p. 514
I. Spînulescu-Carnaru	18, 769	(1966)	*1.4*, p. *17*; 3.1, p. 30; 5, p. 50; 22.4.3, p. 509
	23, 157	(1967)	*1.4*, p. *19*; 4, p. 43; 22.4.3, p. 510
	26, K39	(1968)	3.1, p. 33; *14.3.1*, p. *246*; 22, p. 450
H. Spreen	16, K45	(1966)	4, p. 37; *18.2*, p. *300*; 21.1, p. 390
	24, 413	(1967)	4, p. 44; *18.2*, p. *309*; 21.1, p. 401
	24, 431	(1967)	4, p. 44; *18.2*, p. *309*; 21.1, p. 401
T. Springer	23, K155	(1967)	*6*, p. *63*; 23, p. 561
A. J. Springthorpe	24, K3	(1967)	4, p. 43; *18.4*, p. *323*; 22.6, p. 539
B. Sprušil	19, K83	(1967)	*14.1*, p. *222*; 21.6, p. 432
	24, K41	(1967)	*10.1*, p. *126*; 14.1, p. 224; 21.6, p. 434
J. Spyridelis	19, 683	(1967)	*1*, p. *7*; 4, p. 40; 10, p. 98; 22.6, p. 537
	20, 623	(1967)	13.2, p. 192; 16, p. 279; *20.1*, p. *347*; 22.6, p. 537
R. Srinivasan	29, 865	(1968)	6, p. 69; *7*, p. *78*; *8*, p. *86*
T. M. Srinivasan	29, K107	(1968)	10, p. 111; *10.2*, p. *144*; 22.5.2, p. 529
T. P. Srinivasan	28, K71	(1968)	3, p. 27; *12*, p. *158*; *12.1*, p. *160*
K. G. Srivastava	28, 61	(1968)	5, p. 53; 13, p. 171; *20.1*, p. *355*; 22.5, p. 514; 22.8, p. 554
Z. Šroubek	16, 405	(1966)	*3.2*, p. *34*; 13, p. 163; 18.4, p. 321; 22.5, p. 512; 22.8, p. 547
M. D. Stafleu	23, 675	(1967)	*13.1*, p. *182*; 18, p. 293; 21, p. 382
	23, 683	(1967)	*13.1*, p. *182*; 21, p. 382
I. M. Stakhyra	21, 769	(1967)	13.1, p. 178; *20.1*, p. *348*; 22, p. 446
B. Staliński	17, 837	(1966)	1.2, p. 9; *8*, p. *80*; 18.2, p. 303; 22, p. 441
	24, K73	(1967)	*18.1*, p. *297*; 22, p. 448
G. R. Stallings	16, K17	(1966)	*18.4*, p. *321*; 22.8, p. 547
L. Stals	17, 295	(1966)	*10*, p. *94*
J. Stankowski	23, K159	(1967)	14.4.2, p. 267; *19*, p. *331*
	24, 451	(1967)	10, p. 103; 14.4.2, p. 268; *19*, p. *331*
J. T. Stanley	27, 701	(1968)	*10.1*, p. *131*; 21.1.1, p. 421
S. S. Starobinets	17, K7	(1966)	18.3, p. 316; *19*, p. *326*
V. I. Startsev	18, 863	(1966)	*4*, p. *39*; 21.7, p. 438
P. M. Staryk	23, 93	(1967)	*13.1*, p. *180*; 14.3, p. 238; 18, p. 293; 22, p. 447
M. Stasiw	17, K119	(1966)	*19*, p. *326*; 22.5.1, p. 515
	20, 657	(1967)	10.2, p. 139; 13.4, p. 206; *19*, p. *328*; 22.5.1, p. 515
	20, 667	(1967)	10.2, p. 139; 13.4, p. 206; *19*, p. *328*; 22.5.1, p. 515
	24, 591	(1967)	10, p. 103; *19*, p. *331*; 22.5.1, p. 516
	25, K55	(1968)	10.2, p. 141; *19*, p. *332*; 22.5.1, p. 516
	28, 247	(1968)	10.2, p. 143; *19*, p. *334*; 20.1, p. 355; 22.5.1, p. 517
V. I. Steblin	21, K45	(1967)	14.3.2, p. 249; *20.3*, p. *365*; 22, p. 445; 22.4.1, p. 499
	21, K49	(1967)	*14.3.2*, p. *249*; 16, p. 279; 20.3, p. 365; 22.4.1, p. 499
K. Steenbeck	17, 191	(1966)	10.1, p. 115; *18.2*, p. *302*; 21.1, p. 392
	18, K131	(1966)	*18.2*, p. *304*; 21.1, p. 394
E. Stein	22, 537	(1967)	*6*, p. *62*; *18.1*, p. *296*; 19, p. 330; 22.8, p. 551
E. Steinbeiss	16, 499	(1966)	*18.3*, p. *316*
	16, 795	(1966)	*18.3*, p. *316*
	18, 241	(1966)	*18.3*, p. *317*

J. Steinert	**21**,	K13	(1967)	10.1, p. 122; *18.2*, p. *307*; 21, p. 379; 21.1.1, p. 417
W. Steinmann	**28**,	437	(1968)	*13*, p. *162*; 13.3, p. 196; *20*, p. *337*; 20.1, p. 344; 21, p. 373; 21.2, p. 424; 21.6, p. 430
	30,	K53	(1968)	*13*, p. *173*; 17, p. 290; *20*, p. *342*; 21.6, p. 437
A. Stella	**23**,	697	(1967)	*13.1*, p. *182*; 14.3, p. 238; 20.1, p. 350; 22.7, p. 545
E. Sterk	**27**,	131	(1968)	*6.1*, p. *73*; 18.3, p. 319
H. F. Sterling	**21**,	205	(1967)	3.1, p. 31; 3.2, p. 35; *14.3*, p. *236*; 21.7, p. 438; 22.7, p. 545
J. D. Stettler	**26**,	K155	(1968)	10.2, p. 142; *13.4*, p. *212*; 22.5.2, p. 526; 22.6, p. 541
D. A. Stevenson	**24**,	683	(1967)	13.4, p. 211; *14.3*, p. *239*; 22.4.2, p. 507
H. Stiller see H. H. Stiller				
H. H. Stiller	**18**,	795	(1966)	1.2, p. 10; *6*, p. *58*; 23, p. 561
	19,	781	(1967)	*6*, p. *59*; 22.9, p. 558
	27,	269	(1968)	*6*, p. *65*; 22.9, p. 560
F. Stöckmann	**17**,	335	(1966)	*14.4.1*, p. *256*; 16, p. 277; 22.4.1, p. 498
	27,	K69	(1968)	*15*, p. *274*; 16, p. 282; 22.4.1, p. 502
	28,	K101	(1968)	*14.3*, p. *243*; 22.1.3, p. 479
R. Stockmeyer	**19**,	781	(1967)	*6*, p. *59*; 22.9, p. 558
	27,	269	(1968)	*6*, p. *65*; 22.9, p. 560
J. Stoimenos	**20**,	623	(1967)	13.2, p. 192; 16, p. 279; *20.1*, p. *347*; 22.6, p. 537
M.-P. Stoll	**22**,	163	(1967)	*18.2*, p. *307*; 20, p. 339; 21.1.1, p. 418
H. Stolz	**18**,	251	(1966)	*13*, p. *164*
	21,	77	(1967)	13, p. 165; *14*, p. *219*; 20, p. 339
	23,	729	(1967)	*14*, p. *219*
A. M. Stoneham	**19**,	787	(1967)	6, p. 59; *13*, p. *165*; 22.6, p. 537
S. Stonkus	**22**,	K3	(1967)	13.4, p. 208; *16*, p. *279*; 22.4.2, p. 506
W. Stössel	**30**,	311	(1968)	*11*, p. *155*; 13.4, p. 216; *16*, p. *284*; 22.4.1, p. 505
M. J. Stowell	**16**,	117	(1966)	*1.4*, p. *16*; *3.1*, p. *29*; 21.6, p. 430
	16,	479	(1966)	*1.4*, p. *16*; 3.1, p. 29; 21.6, p. 430
	25,	139	(1968)	*1.4*, p. *19*; *3.1*, p. *33*; 10.1, p. 127; 21.6, p. 435
H. Stramska	**17**,	701	(1966)	13.1, p. 176; 14.4.1, p. 256; *18*, p. *291*; *20*, p. *337*; 22, p. 441
	27,	K25	(1968)	13.1, p. 186; *18*, p. *294*; *20*, p. *341*; 22.1.2, p. 473
R. Straubel	**16**,	237	(1966)	4, p. 37; *18.2*, p. *299*; 18.3, p. 316
	20,	267	(1967)	*18.2*, p. *305*
D. Strauch	**24**,	469	(1967)	3.2, p. 35; *6*, p. *64*; 10.2, p. 141; 22.5.2, p. 524
	30,	495	(1968)	*6*, p. *70*; 10, p. 112; *20.1*, p. *359*; 22.1, p. 457; 22.1.1, p. 466; 22.1.2, p. 475
G. B. Street	**18**,	601	(1966)	13.4, p. 204; 14.3, p. 233; *16*, p. *278*; 22, p. 441
H. W. Streitwolf	**21**,	K167	(1967)	*14*, p. *219*
H. Streller	**17**,	K245	(1966)	*14.4.1*, p. *257*; 22.1.1, p. 460; 22.1.2, p. 469
P. Strimer	**26**,	241	(1968)	6, p. 65; *20.1*, p. *352*; 22.4.2, p. 507; 22.4.3, p. 510
R. Struikmans	**19**,	K107	(1967)	10, p. 98; *14.3*, p. *235*; 22.1.1, p. 461
	27,	225	(1968)	10, p. 107; *14.3*, p. *241*; 22.1.1, p. 465
B. A. Strukov	**27**,	741	(1968)	1.2, p. 12; *8*, p. *85*; 14.4.2, p. 268
M. E. Stryalkovskii	**26**,	85	(1968)	*1.2*, p. *12*; 2, p. 24; 8, p. 84; 12, p. 158; 14.3, p. 240; 22.7, p. 546; 22.8, p. 553
M. Stuart	**23**,	595	(1967)	*14.3.1*, p. *245*; *14.4.1*, p. *260*; 22.6, p. 539
J. Stuke	**17**,	335	(1966)	*14.4.1*, p. *256*; 16, p. 277; 22.4.1, p. 498
	17,	343	(1966)	13.1, p. 175; *14.4.1*, p. *256*; *20*, p. *337*; 22.1.3, p. 476
	22,	K49	(1967)	13.1, p. 179; *20.1*, p. *349*; 22.1.3, p. 477
	23,	K99	(1967)	*8*, p. *83*; 13, p. 167; 15, p. 273; 22.1.3, p. 477
	24,	99	(1967)	*6*, p. *63*; 20.1, p. 351; 22.1.3, p. 478

J. Stuke	**24,**	183	(1967)	13.1, p. 182; *14.3*, p. *238*; 22.1.3, p. 478
	27,	171	(1968)	13.4, p. 213; *16*, p. *282*; 22.1.3, p. 479
	29,	203	(1968)	13.1, p. 188; 14.3.1, p. 247; *20.1*, p. *356*; 22.4, p. 496
A. V. Subashiev	**23,**	467	(1967)	*6*, p. *63*; 13.4, p. 209; *20.3*, p. *367*; 22.1.2, p. 472
V. K. Subashiev	**17,**	K49	(1966)	13.1, p. 175; *16*, p. *277*; 22.2.2, p. 487
	23,	447	(1967)	*20.1*, p. *350*; 22.1.2, p. 472
E. C. Subbarao	**29,**	859	(1968)	*1.2*, p. *13*; *3*, p. *28*; 8, p. 86; 22.8, p. 556
K. N. Subramanian	**28,**	9	(1968)	*10*, p. *93*; 10.1, p. 114; 21, p. 373; 21.1, p. 390; 22.5.1, p. 515; 22.5.2, p. 518; 22.6, p. 535
T. Suita	**28,**	K19	(1968)	*10*, p. *109*; *10.2*, p. *143*; 22.5.2, p. 527
R. V. Sukhanova	**17,**	243	(1966)	1.4, p. 16; *3.1*, p. *29*; 21.1, p. 392; 21.1.1, p. 412
	17,	249	(1966)	4, p. 38; *18.2*, p. *302*; 21.1, p. 392; 21.1.1, p. 412
A. Sukiennicki	**17,**	903	(1966)	*1.3*, p. *14*; *3.2*, p. *34*; 21.1.1, p. 413
	17,	K1	(1966)	1.4, p. 17; *18.2*, p. *302*; 21.1.1, p. 412
	21,	183	(1967)	*1.4*, p. *18*; 18.2, p. 306; 21.1.1, p. 417
	21,	K57	(1967)	4, p. 42; *18.2*, p. *307*; 21.1.1, p. 417
	25,	K19	(1968)	*18.2*, p. *310*; 21.1, p. 402; 21.1.1, p. 420
	29,	417	(1968)	4, p. 47; *18.2*, p. *314*; 21.1.1, p. 422
E. Suoninen see E. J. Suoninen				
E. J. Suoninen	**21,**	487	(1967)	*17*, p. *287*; 22.5.2, p. 522
	29,	K171	(1968)	14.1, p. 226; *18.2*, p. *314*; 21.1.1, p. 423; 21.4, p. 428
P. Suortti	**28,**	539	(1968)	*5*, p. *54*; 21, p. 387
P. Süptitz	**23,**	9	(1967)	*9*, p. *87*; *10*, p. *93*; 14.4, p. 252; 15, p. 270; 22.5.1, p. 515; 22.5.2, p. 518; 22.5.3, p. 532
	27,	631	(1968)	*9*, p. *92*; 10, p. 108; 22.5.1, p. 517
W. Suski	**20,**	451	(1967)	3.2, p. 35; *18.2*, p. *306*; *18.4*, p. *322*; 22, p. 444
	30,	61	(1968)	*18.4*, p. *325*; 22.8, p. 556
E. Sutter	**23,**	K25	(1967)	13.1, p. 181; *16*, p. *280*; *20.1*, p. *350*; 22.1.3, p. 477
J. Suwalski	**29,**	K99	(1968)	*6.1*, p. *74*; 22.8, p. 556
I. V. Svechkarev	**23,**	K133	(1967)	*13.1*, p. *182*; 21, p. 382
V. M. Svistunov	**26,**	K55	(1968)	*14.2*, p. *230*
	30,	K107	(1968)	*14.2*, p. *230*
L. Švob	**22,**	K141	(1967)	*9*, p. *90*; 21.2, p. 424
J. B. Swan	**23,**	171	(1967)	*13*, p. *167*; 13.3, p. 198; *20*, p. *340*; 21, p. 381
M. L. Swanson	**23,**	649	(1967)	10.1, p. 125; *11*, p. *150*; 14.1, p. 224; 21, p. 381; 21.6, p. 434
N. Swindells	**20,**	197	(1967)	3, p. 27; *4*, p. *41*; 21.1.1, p. 415
V. A. Sychugov	**25,**	K119	(1968)	20.1, p. 352; *20.2*, p. *361*; 20.3, p. 368; 22.5.3, p. 534
Ch. Symanowski	**25,**	93	(1968)	*10.2*, p. *141*; 22.5.3, p. 533
N. N. Syrbu	**30,**	K145	(1968)	13.1, p. 190; *20.1*, p. *359*; 22.7, p. 546
P. P. Syrnikov	**30,**	367	(1968)	13, p. 173; *18.3*, p. *320*; *20*, p. *342*; 22.5, p. 514; 22.8, p. 556
L. A. Sysoev	**26,**	107	(1968)	*16*, p. *281*; 22.4.1, p. 502
V. I. Syutkina	**21,**	465	(1967)	1.1, p. 8; 4, p. 42; *10.1*, p. *123*; 21.1, p. 398; 21.6, p. 433
I. Szabó	**26,**	319	(1968)	*13*, p. *169*; 18, p. 294
L. Szepessy	**26,**	K127	(1968)	*9*, p. *91*; 22.2.3, p. 492
P. O. Szomor	**28,**	529	(1968)	*10*, p. *109*; *12*, p. *158*; 22.5.2, p. 527
W. Szymańska	**23,**	69	(1967)	*13.1*, p. *180*; 15, p. 272; 22.4.3, p. 510
	27,	K103	(1968)	13.4, p. 214; *14.3*, p. *242*; 22.4.3, p. 510
J. Szymaszek	**20,**	K37	(1967)	*14*, p. *219*
A. Szytuła	**17,**	K195	(1966)	6.1, p. 71; *18.4*, p. *321*; 22, p. 441

A. Szytuła	**21**, K85	(1967)	*6.1*, p. *72*; 18.4, p. 322; 22.8, p. 550
	26, 429	(1968)	*4*, p. *45*; 18.3, p. 319; 18.4, p. 324; 22.8, p. 553
	26, K83	(1968)	*18.3*, p. *319*; 18.4, p. 324; 22.8, p. 553
B. G. Tagiev	**16**, 205	(1966)	13.4, p. 201; *14.4.1*, p. *255*; 22, p. 439
	17, 593	(1966)	13.1, p. 175; *14.4.1*, p. *256*; 22, p. 440
	20, 421	(1967)	*14.3*, p. *235*; *14.4.1*, p. *258*; 22, p. 444
R. I. Tagirov	**16**, 745	(1966)	4, p. 38; *18.2*, p. *301*; 21.1.1, p. 411
E. Taglauer	**29**, 259	(1968)	4, p. 47; *10.1*, p. *133*; 21.1, p. 407
S. Takácz	**21**, 709	(1967)	*14.2*, p. *229*
T. Takahashi	**17**, 609	(1966)	10, p. 95; 14.3.1, p. 245; 16, p. 277; *17*, p. *286*; 22.7, p. 545
A. G. Talybov	**18**, 929	(1966)	*5*, p. *50*; 8, p. 80; 13.4, p. 204; 14.3, p. 234; 16, p. 278; 22.8, p. 548
C. Tănăsoiu	**21**, 197	(1967)	4, p. 42; *18.3*, p. *317*; 22.6, p. 537
S. P. Taneja	**30**, 193	(1968)	3, p. 28; *18.1*, p. *298*; 22.5, p. 514
A. L. Tanenbaum	**16**, 183	(1966)	*11*, p. *145*; 14.3, p. 231; 22.2.3, p. 489
L. E. Tanner	**30**, 685	(1968)	*1.2*, p. *13*; 3.1, p. 33; 4, p. 48; 21, p. 388
T. L. Tansley	**18**, 105	(1966)	13, p. 164; *14.3.2*, p. *248*; 22.2.1, p. 483
	23, 241	(1967)	*14.3.2*, p. *249*; 14.4.1, p. 259; 16, p. 280; 20.1, p. 350; 22.2.1, p. 484; 22.8, p. 552
	24, 615	(1967)	*14.3.2*, p. *250*; 14.4.1, p. 261; 16, p. 281; 22.2.1, p. 485; 22.2.2, p. 488
G. A. Tanton	**22**, K19	(1967)	*10.2*, p. *140*; 11, p. 149; 22.5.3, p. 533
	28, K143	(1968)	*10.2*, p. *143*; 22.5.3, p. 534
V. V. Tarassov	**20**, 37	(1967)	2, p. 22; *6*, p. *59*; 8, p. 81; 22, p. 443
	20, 45	(1967)	2, p. 22; *6*, p. *59*; 8, p. 81; 22, p. 443
	30, 147	(1968)	*6*, p. *69*; *8*, p. *86*; 13.1, p. 189; 22.1.1, p. 466; 22.1.2, p. 474; 22.2.1, p. 486; 22.2.2, p. 488; 22.2.3, p. 493
J. Tavernier	**22**, K135	(1967)	*16*, p. *280*; 22.4.1, p. 500
B. Tavger	**22**, 31	(1967)	13.1, p. 179; *14.3.1*, p. *245*; 15, p. 272; 22, p. 446
G. F. Taylor	**16**, K175	(1966)	*1.2*, p. *9*; 3, p. 26; 8, p. 79; 22.8, p. 547
H. Teichler	**23**, 341	(1967)	*10*, p. *102*; 22.1.1, p. 463; 22.1.2, p. 472
I. Tehver	**21**, 755	(1967)	13.4, p. 207; 20.1, p. 348; *20.3*, p. *366*; 22, p. 446
R. V. Telesnin	**16**, K121	(1966)	*18.2*, p. *301*; 21.1.1, p. 411
J. Teltow	**23**, 9	(1967)	*9*, p. *87*; *10*, p. *93*; 14.4, p. 252; 15, p. 270; 22.5.1, p. 515; 22.5.2, p. 518; 22.5.3, p. 532
E. Ţenescu	**24**, 175	(1967)	13.2, p. 193; *20.1*, p. *351*; 22.5, p. 513
A. N. Terenin	**17**, K95	(1966)	13.4, p. 202; *15*, p. *270*; 16, p. 277; 22, p. 440
	20, 771	(1967)	13.3, p. 197; 16, p. 279; *17*, p. *287*; 22.4, p. 495; 22.5, p. 512; 22.5.1, p. 515
T. A. Tevosyan	**25**, K75	(1968)	20.1, p. 352; *20.2*, p. *361*; 20.3, p. 368; 22.5, p. 513; 22.8, p. 552
H. Thiel	**20**, K173	(1967)	*6*, p. *60*; 7, p. 76; 22.4.1, p. 499
R. C. Thiel	**29**, 837	(1968)	1.2, p. 13; *6.1*, p. *74*; 18.4, p. 325; 22, p. 454
W. Thielemann	**26**, K29	(1968)	13.1, p. 184; 18, p. 294; *20.1*, p. *352*; 22.2.1, p. 485
K. Thiessen	**26**, 107	(1968)	*16*, p. *281*; 22.4.1, p. 502
P. Thoma	**16**, 159	(1966)	13.4, p. 201; *20.3*, p. *363*; 22.9, p. 558
	16, 663	(1966)	13.4, p. 201; 14.3, p. 231; 14.4.1, p. 256; *16*, p. *276*; 22.9, p. 558
	23, 253	(1967)	13, p. 167; *20.3*, p. *367*; 22.9, p. 559
D. J. D. Thomas	**20**, 131	(1967)	*1.4*, p. *18*; 22, p. 443
	22, 593	(1967)	*1.4*, p. *19*; 22.1.2, p. 471; 22.6, p. 539

D. J. D. THOMAS	**24,** K111	(1967)	*1.4,* p. *19*; 2, p. 23; 22.1.2, p. 472
	25, 241	(1968)	*1.4,* p. *19*; 2, p. 23; 22, p. 449
G. THOMAS	**25,** 81	(1968)	*3,* p. *27*; 5, p. 53
	25, 263	(1968)	*1.2,* p. *11*; *4,* p. *44*; 22, p. 449
L. K. THOMAS	**28,** 401	(1968)	13.1, p. 187; *20.3,* p. *370*; 21, p. 386
P. THOMAS	**26,** 151	(1968)	*13.1,* p. *184*; 22.1.3, p. 479
D. A. THOMPSON	**17,** 83	(1966)	*18.2,* p. *301*; 21.1.1, p. 411
I. I. TICHINA	**28,** 489	(1968)	2, p. 24; 14.3, p. 242; *20.3,* p. *370*; 22.8, p. 554
J. TICHÝ	**29,** K51	(1968)	*5,* p. *54*; 14.4, p. 254; 22.6, p. 543
V. V. TIKHONOV	**28,** 835	(1968)	5, p. 54; *18.2,* p. *313*; *18.4,* p. *324*; 22.8, p. 554
H. D. TILLER	**29,** 153	(1968)	5, p. 54; 13.1, p. 188; 16, p. 283; *20.1,* p. *356*; 22, p. 453
K. H. TIMMESFELD	**30,** 73	(1968)	3.2, p. 36; *6,* p. *69*; *20.1,* p. *358*; 22.5.2, p. 530
A. N. TIMOFEEV	**16,** 729	(1966)	*9,* p. *87*; 21.1, p. 391; 21.6, p. 430
	18, 847	(1966)	*9,* p. *88*; 14.1, p. 222; 21.1, p. 394; 21.6, p. 432
V. B. TIMOFEEV	**28,** 277	(1968)	13.1, p. 187; 13.2, p. 194; *20.1,* p. *355*; 22, p. 452
V. A. TIMOFEEVA	**16,** K165	(1966)	10, p. 94; *20.1,* p. *345*; 20.2, p. 360; 22.8, p. 547
P. TISCHER	**17,** 259	(1966)	10.1, p. 115; *11,* p. *146*; 14.1, p. 220; 21, p. 374
T. C. TISONE	**20,** 443	(1967)	*4,* p. *41*; 10.1, p. 122; 21.1, p. 396
	27, 185	(1968)	*10.1,* p. *130*; 21, p. 385; 21.1, p. 404
J. S. TIWARI	**28,** 61	(1968)	5, p. 53; 13, p. 171; *20.1,* p. *355*; 22.5, p. 514; 22.8, p. 554
J. TOBISCH	**21,** K11	(1967)	*5,* p. *51*; 22, p. 445
J. TODOROVIĆ	**21,** K163	(1967)	*13,* p. *166*; *18,* p. *292*; 22.6, p. 538
	22, K55	(1967)	*18.3,* p. *318*
	26, 429	(1968)	*4,* p. *45*; 18.3, p. 319; 18.4, p. 324; 22.8, p. 553
K. B. TOLPYGO	**23,** 429	(1967)	*13.1,* p. *181*; 22, p. 447
	28, 807	(1968)	13.1, p. 188; *13.2,* p. *194*; 22, p. 453; 22.5.2, p. 528
M. TOMÁŠEK	**16,** 389	(1966)	*13.3,* p. *196*; 22, p. 440
K. TOMPA	**18,** 391	(1966)	*19,* p. *327*; 21.1, p. 393
	22, K11	(1967)	18.1, p. 296; 18.4, p. 322; *19,* p. *329*; 22.6, p. 538
V. TOPA	**16,** 137	(1966)	*10.2,* p. *137*; 22.5.2, p. 518
L. I. VAN TORNE	**16,** 171	(1966)	2, p. 22; 4, p. 37; *20.2,* p. *360*; 22.6, p. 535
	18, 737	(1966)	10.1, p. 119; *14.1,* p. *221*; 21.3, p. 426
	19, 855	(1967)	10.1, p. 121; *13,* p. *165*; 14.1, p. 222; 21, p. 377
S. TOSCHEV	**21,** 683	(1967)	*3.1,* p. *31*
	24, 349	(1967)	1.2, p. 11; *3.1,* p. *32*; 22, p. 448
B. S. TOŠIĆ	**27,** 623	(1968)	*13,* p. *170*; *18.2,* p. *312*
	28, 635	(1968)	*13,* p. *171*; *18.2,* p. *313*
F. TÓTH	**22,** K11	(1967)	18.1, p. 296; 18.4, p. 322; *19,* p. *329*; 22.6, p. 538
J. TÓTH	**27,** K47	(1968)	*14.1,* p. *225*; 21, p. 386
J. TOUŠEK	**28,** K147	(1968)	*16,* p. *283*; 22.4.2, p. 508; 22.4.3, p. 511
K. D. TOVSTYUK	**21,** 769	(1967)	13.1, p. 178; *20.1,* p. *348*; 22, p. 446
	23, 75	(1967)	13.1, p. 180; *14.3,* p. *237*; 22, p. 447
	23, 93	(1967)	*13.1,* p. *180*; 14.3, p. 238; 18, p. 293; 22, p. 447
I. SH. TRAKHTENBERG	**16,** 729	(1966)	*9,* p. *87*; 21.1, p. 391; 21.6, p. 430
	18, 847	(1966)	*9,* p. *88*; 14.1, p. 222; 21.1, p. 394; 21.6, p. 432
A. LE TRAON	**23,** 709	(1967)	*14.4.2,* p. *267*
F. LE TRAON	**23,** 709	(1967)	*14.4.2,* p. *267*
H. TRAUB	**30,** 387	(1968)	*10.1,* p. *135*; 21.1, p. 408
H. TRÄUBLE	**18,** 813	(1966)	4, p. 39; *14.2,* p. *228*
	20, 95	(1967)	*14.2,* p. *229*; 18, p. 291; 21, p. 377
	25, 373	(1968)	10.1, p. 128; *14.2,* p. *229*; 21, p. 384
	25, 395	(1968)	10.1, p. 128; *14.2,* p. *230*; 21, p. 384
R. H. TREDGOLD	**26,** K5	(1968)	2, p. 24; *17,* p. *288*; 22.1.3, p. 479

V. I. Trefilov	**18,**	881	(1966)	*10.1*, p. *119*; 21.1.1, p. 414
H. Treptow	**30,**	K61	(1968)	13.3, p. 199; *14.3.2*, p. *251*; 14.4.1, p. 264; 22.4.3, p. 511
J. Treusch	**16,**	487	(1966)	5, p. 49; *13.1*, p. *175*; 22.1.3, p. 476
	19,	603	(1967)	*13.1*, p. *177*
E. D. Trifonov	**26,**	519	(1968)	13.4, p. 212; *20.1*, p. *353*; *20.3*, p. *368*; 22, p. 451
	30,	479	(1968)	6, p. 70; 13.4, p. 217; *20.3*, p. *371*; 22, p. 455
R. Trivedi	**17,**	221	(1966)	*10*, p. *94*; 22.1, p. 456
R. Troć	**17,**	837	(1966)	1.2, p. 9; *8*, p. *80*; 18.2, p. 303; 22, p. 441
	22,	517	(1967)	5, p. 52; *18.4*, p. *323*; 22, p. 447
	23,	K123	(1967)	*18.4*, p. *323*; 22, p. 448
	30,	61	(1968)	*18.4*, p. *325*; 22.8, p. 556
	30,	157	(1968)	4, p. 48; *18.4*, p. *325*; 22, p. 454
E. Trojnar	**17,**	K89	(1966)	*14.2*, p. *228*
J. M. Trooster	**24,**	487	(1967)	*6.1*, p. *73*; 18.3, p. 318; 22.6, p. 540; 22.8, p. 552
A. S. Troshin	**26,**	519	(1968)	13.4, p. 212; *20.1*, p. *353*; *20.3*, p. *368*; 22, p. 451
O. V. Troshin	**28,**	K81	(1968)	*13*, p. *171*
J. Trylski	**21,**	K53	(1967)	*13.4*, p. *207*; 22.4, p. 495; 22.4.1, p. 499; 22.4.2, p. 506
M. B. Trzhaskovskaya	**28,**	801	(1968)	*3.2*, p. *36*; 13.1, p. 187; 21.1, p. 406; 21.6, p. 437
I. M. Tsidilkovski	**27,**	249	(1968)	*13.1*, p. *186*; 14.3, p. 241; 15, p. 274; 22.2.3, p. 492
I. S. T. Tsong	**29,**	617	(1968)	6, p. 68; 8, p. 86; *12.1*, p. *160*; 13, p. 172; 18.2, p. 314; 21.1, p. 407; 21.1.1, p. 422
B. S. Tsukerblat	**28,**	K179	(1968)	6, p. 67; *20.1*, p. *356*
E. V. Tsvetkova	**25,**	513	(1968)	*3.1*, p. *33*; 5, p. 53; 14.3, p. 240; 15, p. 273; 20.1, p. 351; 22, p. 450
M. R. Tubbs	**28,**	K135	(1968)	13.2, p. 195; *20.1*, p. *356*; 22.5, p. 514; 22.5.2, p. 528; 22.5.3, p. 534
B. Tuck	**18,**	541	(1966)	13.4, p. 204; *20.3*, p. *364*; 22.2.1, p. 483; 22.2.2, p. 488; 22.8, p. 548
	29,	793	(1968)	13.4, p. 215; *20.3*, p. *370*; 22.2.1, p. 486
A. V. Tulub	**26,**	693	(1968)	13.4, p. 212; *20.1*, p. *353*; *20.3*, p. *369*; 22.5, p. 513
Pham Ngu Tung see under letter "P" (Pham)				
T. O. Tuomi	**28,**	329	(1968)	*11*, p. *153*; 14.3, p. 242; 16, p. 282; 22.1.3, p. 479
R. G. Turner	**18,**	221	(1966)	4, p. 38; *20.1*, p. *346*; 22.8, p. 548
T. J. Turner	**26,**	645	(1968)	*10.1*, p. *130*; *12.1*, p. *160*; 21.6, p. 435
	30,	87	(1968)	9, p. 92; *10.1*, p. *134*; *12.1*, p. *161*; 21, p. 388; 21.6, p. 437
I. D. Turyanitsa	**30,**	461	(1968)	*13.1*, p. *189*; 16, p. 284; 20.1, p. 358; 22.8, p. 557
V. A. Tyagai	**24,**	543	(1967)	10, p. 103; *13.4*, p. *210*; 16, p. 281; 20.3, p. 368; 22.4.1, p. 501
T. Tymosz	**21,**	183	(1967)	*1.4*, p. *18*; 18.2, p. 306; 21.1.1, p. 417
Ya. A. Ugai	**30,**	K145	(1968)	13.1, p. 190; *20.1*, p. *359*; 22.7, p. 546
N. Uhle	**30,**	K81	(1968)	*20.1*, p. *358*; 22, p. 455; 22.7, p. 546
Yu. I. Ukhanov	**17,**	K113	(1966)	*13.1*, p. *175*; 20.1, p. 345; 22.2.3, p. 490
	26,	K33	(1968)	13.1, p. 184; *20.1*, p. *353*; 22.7, p. 546; 22.8, p. 553
H. A. Ullmaier	**17,**	631	(1966)	*14.2*, p. *228*
	17,	K233	(1966)	*14.2*, p. *228*
P. Ullmann	**23,**	313	(1967)	13.4, p. 209; 20.1, p. 350; *20.3*, p. *367*; 22.5.3, p. 533
	25,	93	(1968)	*10.2*, p. *141*; 22.5.3, p. 533
H.-J. Ullrich	**20,**	K113	(1967)	*5*, p. *51*; 21, p. 378; 22, p. 444
	29,	K89	(1968)	3.1, p. 33; *4*, p. *47*; 22.4.1, p. 505

W. Ulrici	**24**, 605 (1967)	*13.4*, p. *210*; 20.1, p. 351; 22.5.1, p. 516
	27, 333 (1968)	*10*, p. *107*; 13.4, p. 213; 14.4, p. 254; 20.1, p. 354; 22.5.1, p. 516
	27, 489 (1968)	*10*, p. *107*; 13.4, p. 213; 14.4, p. 254; 20.1, p. 354; 22.5.1, p. 517
H.-G. Unruh	**29**, 669 (1968)	*14.4.2*, p. *269*
G. P. Upit	**17**, 831 (1966)	*12*, p. *156*; 22.5.2, p. 519
L. Uray	**29**, K163 (1968)	*4*, p. *48*; 14.1, p. 226; 21, p. 388
A. Urbelis	**30**, K5 (1968)	*14.4.1*, p. *264*; *16*, p. *284*; 22.4.1, p. 505
S. I. Uritsky	**23**, 57 (1967)	13.1, p. 180; 18, p. 293; *20*, p. *339*; 22.1.1, p. 463
M. P. Usikov	**23**, 745 (1967)	*4*, p. *43*; 21.1, p. 400; 22.6, p. 539
H. P. Utech	**20**, K41 (1967)	*1.3*, p. *15*; 22.5.2, p. 521
L. M. Utevskii	**22**, K75 (1967)	*10.1*, p. *125*; 21.1.1, p. 418
K. Vacek	**23**, 105 (1967)	*20.3*, p. *367*; 22.5.1, p. 516
	27, K29 (1968)	*20.3*, p. *369*; 22.5.1, p. 516
V. A. Vagenin	**28**, 269 (1968)	10, p. 109; 13, p. 171; *19*, p. *334*; 22.6, p. 542
A. A. Vaipolin	**29**, 435 (1968)	4, p. 47; *5*, p. *54*; 6, p. 68; 10, p. 110; 14.3, p. 243; 22.8, p. 555
J. Vaitkus	**22**, K3 (1967)	13.4, p. 208; *16*, p. *279*; 22.4.2, p. 506
M. Ya. Valakh	**27**, K81 (1968)	*13.1*, p. *186*; 20.1, p. 354; 22.4.1, p. 502; 22.4.2, p. 507; 22.8, p. 554
L. Valenta	**17**, 903 (1966)	*1.3*, p. *14*; *3.2*, p. *34*; 21.1.1, p. 413
	26, 191 (1968)	6.1, p. 73; *18.2*, p. *310*
M. Vaněček	**28**, K147 (1968)	*16*, p. *283*; 22.4.2, p. 508; 22.4.3, p. 511
R. Vanselow	**21**, 69 (1967)	*1.3*, p. *15*; 10.1, p. 122; 21.6, p. 433
V. Varacca	**22**, 365 (1967)	13.4, p. 208; *19*, p. *330*; 22.9, p. 559
S. A. Varchenya	**17**, 831 (1966)	*12*, p. *156*; 22.5.2, p. 519
C. Varotto	**18**, K49 (1966)	10.1, p. 118; *14.1*, p. *221*; 21.1.1, p. 414
Y. P. Varshni	**19**, 459 (1967)	13.1, p. 174; 13.2, p. 191; 13.4, p. 200; 20.1, p. 344; 20.2, p. 360; *20.3*, p. *363*; 22.1, p. 456; 22.1.1, p. 458; 22.1.2, p. 468; 22.1.3, p. 476; 22.2.1, p. 482; 22.2.2, p. 487; 22.2.3, p. 489
	20, 9 (1967)	13.1, p. 174; 13.2, p. 191; 13.4, p. 200; 20.1, p. 344; 20.2, p. 360; *20.3*, p. *363*; 22.1, p. 456; 22.1.1, p. 458; 22.1.2, p. 468; 22.1.3, p. 476; 22.2.1, p. 482; 22.2.2, p. 487; 22.2.3, p. 489
V. M. Vasetskii	**18**, 749 (1966)	10, p. 97; 12, p. 156; *14.3*, p. *234*; 22.1.1, p. 460
E. A. Vasilev	**28**, K175 (1968)	*18.1*, p. *297*; 18.2, p. 313; 22.8, p. 555
G. Vasiliu	**22**, K79 (1967)	*16*, p. *280*; 22.4.2, p. 507
A. S. Vasilkovskii	**30**, 231 (1968)	*10*, p. *112*; 20, p. 342; 22.1.1, p. 466
C. R. Vassel	**17**, K123 (1966)	*1.1*, p. *8*; 14.1, p. 220; 21, p. 375
K. I. Vasu	**29**, K137 (1968)	*4*, p. *47*; 10.1, p. 134; 14.1, p. 226; 21.1, p. 408; 21.3, p. 427
G. Vaubel	**16**, 663 (1966)	13.4, p. 201; 14.3, p. 231; 14.4.1, p. 256; *16*, p. *276*; 22.9, p. 558
	23, 253 (1967)	13, p. 167; *20.3*, p. *367*; 22.9, p. 559
	26, 599 (1968)	*13.1*, p. *185*; 17, p. 288; 22.9, p. 560
	26, 607 (1968)	*13.1*, p. *185*; 17, p. 289; 22.9, p. 560
D. Vaughan	**20**, 725 (1967)	*4*, p. *41*; 21, p. 378
F. Vávra	**23**, K121 (1967)	*10*, p. *102*; 22.5.1, p. 516
K. Vedam	**26**, 285 (1968)	*20.1*, p. *352*; 22, p. 450
B. G. Vekhter	**16**, 63 (1966)	10, p. 93; *13.4*, p. *200*; 22, p. 439
	28, K179 (1968)	6, p. 67; *20.1*, p. *356*

S. VELEVA	**23**, K61	(1967)	*20.3*, p. *367*; 22.6, p. 539
B. VELICKÝ	**16**, 147	(1966)	*13.1*, p. *174*; *13.2*, p. *191*; 20.1, p. 344; 22, p. 439
	19, K39	(1967)	6, p. 59; 13, p. 164; 13.1, p. 176; *20.1*, p. *347*; 22, p. 442
O. G. VENDIK	**27**, K99	(1968)	*14.4.2*, p. *268*
	28, 789	(1968)	*6*, p. *67*; 22.5.2, p. 528
P. VENKATESWARLU	**29**, 859	(1968)	*1.2*, p. *13*; *3*, p. *28*; 8, p. 86; 22.8, p. 556
J. VENNIK	**21**, 627	(1967)	*4*, p. *42*; 22.1.2, p. 471
	22, 463	(1967)	*4*, p. *43*; 22.1.2, p. 471
C. VERIÉ	**17**, 889	(1966)	*13.1*, p. *176*; 14.3, p. 233; 22.4.3, p. 509; 22.8, p. 548
AJIT RAM VERMA	**17**, 437	(1966)	*3*, p. *26*; *5*, p. *49*; 10, p. 93; 22.3, p. 494; 22.4.1, p. 497; 22.5, p. 512
M. P. VERMA	**19**, 751	(1967)	*6*, p. *59*; 22.6, p. 537
O. R. VERNICH	**17**, K237	(1966)	13.4, p. 203; *14.3*, p. *233*; 15, p. 271; 22, p. 441
M. VERSCHUEREN	**19**, 77	(1967)	*8*, p. *80*; 18.4, p. 322; 22.6, p. 536
G. VERTOGEN	**25**, 721	(1968)	*13*, p. *169*
	25, 729	(1968)	*13*, p. *169*; 18.2, p. 310
J. F. VERWEY	**23**, 137	(1967)	4, p. 43; 10.2, p. 140; *19*, p. *330*; 22.5, p. 513
L. VESCAN	**24**, K17	(1967)	*14.3.2*, p. *250*; *14.4.1*, p. *261*; 22.1.1, p. 463
D. VESELÝ	**29**, 675	(1968)	*10.1*, p. *133*; 21, p. 388
	29, 685	(1968)	*10.1*, p. *133*; 21, p. 388
M. VICTORIA	**28**, 131	(1968)	4, p. 46; *10.1*, p. *131*; 21.1, p. 405; 21.1.1, p. 422
A. E. VIDOZ	**18**, K49	(1966)	10.1, p. 118; *14.1*, p. *221*; 21.1.1, p. 414
	28, 131	(1968)	4, p. 46; *10.1*, p. *131*; 21.1, p. 405; 21.1.1, p. 422
	28, 145	(1968)	4, p. 46; *10.1*, p. *131*; 21.1, p. 405; 21.1.1, p. 422; 21.6, p. 436
P. VIGIER	**17**, 317	(1966)	*10.1*, p. *116*; 11, p. 146; 14.1, p. 220; 18.2, p. 302; 21.1.1, p. 412
	17, K37	(1966)	*10.1*, p. *116*; 18.2, p. 302; 21.1.1, p. 412
F. I. VILESOV	**20**, 771	(1967)	13.3, p. 197; 16, p. 279; *17*, p. *287*; 22.4, p. 495; 22.5, p. 512; 22.5.1, p. 515
J. VILLAIN	**19**, 767	(1967)	*13*, p. *165*; *18*, p. *291*
	26, 501	(1968)	*18.2*, p. *311*; *18.4*, p. *324*
V. L. VINETSKII	**19**, 41	(1967)	10, p. 97; 13.4, p. 204; *14.3*, p. *234*; 22.6, p. 536
J. VIŠČAKAS	**22**, K3	(1967)	13.4, p. 208; *16*, p. *279*; 22.4.2, p. 506
	24, K91	(1967)	3.1, p. 33; *14.3*, p. *239*; *16*, p. *280*; 22.4.2, p. 507
	25, 331	(1968)	2, p. 23; 13.4, p. 211; 14.3.1, p. 246; *14.4.1*, p. *261*; 22.1.3, p. 478
R. VISOCEKAS	**28**, 97	(1968)	13, p. 171; *19*, p. *334*; 22.4.1, p. 503
V. VÍTEK	**18**, 687	(1966)	*10.1*, p. *119*; 21, p. 376
	18, 703	(1966)	*10.1*, p. *119*; 21.1.1, p. 414
	22, 453	(1967)	*10.1*, p. *124*; 21, p. 381; 21.1.1, p. 418
A. YA. VITOL	**18**, 897	(1966)	10.2, p. 138; *11*, p. *147*; 19, p. 327; 20.1, p. 346; 22.5.2, p. 520
N. A. VITOVSKII	**22**, K95	(1967)	*11*, p. *149*; 14.3, p. 237; 22.2.3, p. 491
	22, K103	(1967)	10, p. 101; *11*, p. *150*; 14.3, p. 237; 22.1.1, p. 462
N. I. VITRIKHOVSKII	**16**, K139	(1966)	*13.1*, p. *175*; *20.1*, p. *345*; 22.4.1, p. 497; 22.4.3, p. 509; 22.8, p. 547
	27, K81	(1968)	*13.1*, p. *186*; 20.1, p. 354; 22.4.1, p. 502; 22.4.2, p. 507; 22.8, p. 554
B. VLACH	**19**, 543	(1967)	*10.1*, p. *121*; 21.1, p. 396
V. L. VLADIMIROVA	**29**, 569	(1968)	*10.1*, p. *133*; 21.1, p. 407
N. A. VLASENKO	**20**, 311	(1967)	14.4.1, p. 257; *20.3*, p. *365*; 22.4.1, p. 499
	26, K77	(1968)	14.3.2, p. 250; *20.3*, p. *368*; 22, p. 450; 22.4.1, p. 502

K. M. van Vliet	**21,** 819	(1967)	13.4, p. 207; *16*, p. *279*; 22.4.1, p. 500
	27, 671	(1968)	*16*, p. *282*; 22.4.1, p. 503; 22.4.2, p. 508
	27, 681	(1968)	*13.4*, p. *214*; *16*, p. *282*; 22.4.1, p. 503; 22.4.2, p. 508
G. Vogl	**29,** 819	(1968)	10.1, p. 134; *11*, p. *154*; 14.1, p. 226; 21.3, p. 427
	30, 119	(1968)	10.1, p. 134; *11*, p. *154*; 14.1, p. 226; 21.3, p. 427
H. Vogt	**16,** 711	(1966)	14.4.2, p. 266; *20.1*, p. *344*
	30, 67	(1968)	*6*, p. *69*; 14.4.2, p. 269; *20.1*, p. *358*
O. Vöhringer	**19,** 793	(1967)	*10.1*, p. *121*; 21.1, p. 396
F. Voigt	**23,** 607	(1967)	13, p. 168; 18.3, p. 318; *19*, p. *330*
G. Voigt	**24,** K161	(1967)	*14.4.1*, p. *261*; 22.8, p. 552
J. Voigt	**17,** K11	(1966)	13.3, p. 196; *16*, p. *277*; 22.4.1, p. 498
	30, 659	(1968)	6, p. 70; *13.2*, p. *195*; 20.1, p. 359; 20.3, p. 372; 22.4.1, p. 505
E. I. Voitovich	**17,** K139	(1966)	*8*, p. *79*; 23, p. 561
T. E. Volin	**25,** 163	(1968)	4, p. 44; *9*, p. *91*; 21, p. 384
G. Völkel	**17,** K75	(1966)	10, p. 95; 14.4.2, p. 266; *19*, p. *326*
	25, K35	(1968)	14.4.2, p. 268; *19*, p. *332*
	28, K163	(1968)	14.4.2, p. 269; *19*, p. *335*
	30, K9	(1968)	14.4.2, p. 269; *19*, p. *336*
S. Völlmar	**28,** K171	(1968)	*18.2*, p. *313*
	30, K27	(1968)	*18.2*, p. *315*; 21.1.1, p. 423
A. N. Voloshinskii	**22,** 309	(1967)	13, p. 166; *14.1*, p. *223*; 18.2, p. 308
S. V. Vonsovskii	**24,** 51	(1967)	3.2, p. 35; *13.4*, p. *210*; 22.6, p. 539
M. Voos	**23,** 295	(1967)	*13.4*, p. *209*; 20.3, p. 367; 22.1.2, p. 472
V. V. Vorobjov	**23,** K97	(1967)	*8*, p. *83*; 21.4, p. 428
Yu. K. Voronko	**21,** K17	(1967)	20.1, p. 348; *20.2*, p. *361*; 20.3, p. 365; 22.5, p. 513; 22.8, p. 550
	25, K119	(1968)	20.1, p. 352; *20.2*, p. *361*; 20.3, p. 368; 22.5.3, p. 534
P. I. Voronyuk	**23,** 75	(1967)	13.1, p. 180; *14.3*, p. *237*; 22, p. 447
P. Voss	**17,** K21	(1966)	*14.4.1*, p. *256*; 22.4.1, p. 498
	28, 355	(1968)	*14.4.1*, p. *263*; 16, p. 282; 22.4.1, p. 503
	30, 291	(1968)	*14.4.1*, p. *264*; 16, p. 284; 20, p. 342; 22.4.1, p. 505
P. Vostrý	**19,** K83	(1967)	*14.1*, p. *222*; 21.6, p. 432
	24, K41	(1967)	*10.1*, p. *126*; 14.1, p. 224; 21.6, p. 434
R. Voszka	**23,** K71	(1967)	*10.2*, p. *140*; 22.5.2, p. 523
J. Vrba	**17,** K127	(1966)	*14.3*, p. *232*; 22.1.2, p. 469; 22.6, p. 535
G. de Vries	**19,** 107	(1967)	10.1, p. 120; *12.1*, p. *159*; 21.1.1, p. 414
M. M. A. Vrijhoef	**19,** 177	(1967)	*1.1*, p. *8*; 10.1, p. 120; 21.1, p. 395; 21.6, p. 432
A. R. de Vroomen	**23,** 675	(1967)	*13.1*, p. *182*; 18, p. 293; 21, p. 382
	23, 683	(1967)	*13.1*, p. *182*; 21, p. 382
Vu dinh Ky	**17,** K203	(1966)	14.1, p. 221; 15, p. 270; *18.2*, p. *303*; 21.1.1, p. 413
	17, K207	(1966)	14.1, p. 221; 15, p. 270; *18.2*, p. *303*; 21.1.1, p. 413
	21, K159	(1967)	*18.2*, p. *307*; 21.1, p. 398; 21.1.1, p. 418
	22, 729	(1967)	13, p. 167; *14.1*, p. *223*; *15*, p. *272*; 18.2, p. 308; 21.1, p. 400; 21.1.1, p. 418
	26, 565	(1968)	*14.1*, p. *225*; *18.2*, p. *311*; 21.1, p. 404; 21.1.1, p. 421
R. Vu Huy Dat	**22,** K67	(1967)	*14.4.2*, p. *267*
	27, K41	(1968)	*14.4.2*, p. *268*
H. R. Vydyanath	**29,** K137	(1968)	*4*, p. *47*; 10.1, p. 134; 14.1, p. 226; 21.1, p. 408; 21.3, p. 427
V. Vyšín	**25,** K103	(1968)	*20.1*, p. *352*; 22.8, p. 553

T. Wächter	**30,**	K91	(1968)	*18.2*, p. *315*; 21.1, p. 409
R. H. Wade	**19,**	57	(1967)	*1.4*, p. *17*; 4, p. 39; 21, p. 376
	19,	63	(1967)	*1.4*, p. *18*; 4, p. 40; 21.1, p. 395; 21.6, p. 432
	19,	847	(1967)	4, p. 40; *18.2*, p. *305*; 21.1, p. 396; 21.1.1, p. 415
	29,	799	(1968)	*4*, p. *47*; 18.2, p. 314
A. D. Wadsley	**25,**	541	(1968)	3, p. 27; *5*, p. *53*; 22.6, p. 540
H. Wagenblast	**26,**	409	(1968)	*14.1*, p. *225*; 21.1.1, p. 420
G. Wagner	**30,**	587	(1968)	*10*, p. *112*; 22.5.2, p. 531
W. Waidelich	**21,**	K9	(1967)	10, p. 100; *11*, p. *148*; 22.5.2, p. 522
	23,	K75	(1967)	*10*, p. *102*; 10.2, p. 140; 22.5.2, p. 524
	27,	K111	(1968)	*11*, p. *153*; 13.3, p. 198; 14.3, p. 242; 22.2.3, p. 492
	27,	K165	(1968)	*10*, p. *108*; 11, p. 153; 22.5.2, p. 527
	28,	207	(1968)	*10*, p. *108*; 11, p. 153; 22.5.2, p. 527
	29,	K59	(1968)	*10*, p. *110*; 22.5.2, p. 529
W. W. Walker	**29,**	K141	(1968)	*10*, p. *111*; 22.5.1, p. 517; 22.5.2, p. 529; 22.5.3, p. 534; 22.6, p. 543
A. K. Walton	**19,**	111	(1967)	*14.3*, p. *234*; 22.1.1, p. 461
	24,	K87	(1967)	*13.1*, p. *183*; 18, p. 293; 20, p. 340; 22.1.2, p. 472; 22.2.1, p. 485
F. Walz	**29,**	245	(1968)	10.1, p. 133; 11, p. 153; *18.2*, p. *314*; 21.1, p. 407
R. E. DeWames	**20,**	K11	(1967)	*6*, p. *60*; 22.5.2, p. 521
F. F. Y. Wang	**23,**	289	(1967)	18.1, p. 296; *18.4*, p. *323*; 22.8, p. 552
	25,	119	(1968)	18.2, p. 310; *18.3*, p. *319*; 18.4, p. 323
S. Wang	**23,**	387	(1967)	10.2, p. 140; 13.2, p. 193; *13.4*, p. *209*; 22.5.2, p. 523
	23,	401	(1967)	10.2, p. 140; *13.4*, p. *209*; 22.5.2, p. 523
	24,	509	(1967)	10.2, p. 141; *13.4*, p. *210*; 22.5.2, p. 524
	25,	223	(1968)	*10.2*, p. *141*; 13.4, p. 211; 22.5.2, p. 525
A. Wanic	**17,**	K195	(1966)	6.1, p. 71; *18.4*, p. *321*; 22, p. 441
	21,	K163	(1967)	*13*, p. *166*; *18*, p. *292*; 22.6, p. 538
	22,	K55	(1967)	*18.3*, p. *318*
	26,	429	(1968)	*4*, p. *45*; 18.3, p. 319; 18.4, p. 324; 22.8, p. 553
S. Waplak	**23,**	K159	(1967)	14.4.2, p. 267; *19*, p. *331*
T. Warmiński	**19,**	K67	(1967)	*1.3*, p. *14*; 4, p. 40; 10, p. 98; 22.1.2, p. 470
	20,	K5	(1967)	1.3, p. 14; *9*, p. *88*; 22.1.2, p. 471
	27,	69	(1968)	*1.3*, p. *15*; *14.3.2*, p. *250*; 22.1.2, p. 473
R. W. Warren	**21,**	205	(1967)	3.1, p. 31; 3.2, p. 35; *14.3*, p. *236*; 21.7, p. 438; 22.7, p. 545
S. Wartewig	**17,**	K221	(1966)	14.4.2, p. 266; *19*, p. *327*
G. Wassermann	**18,**	K107	(1966)	*10.1*, p. *119*; 21.6, p. 432
C. Watson	**25,**	K7	(1968)	*10*, p. *104*; 22.6, p. 540
A. Watterich	**23,**	K71	(1967)	*10.2*, p. *140*; 22.5.2, p. 523
A. J. Watts	**30,**	105	(1968)	*19*, p. *335*; 21.2, p. 425; 22.5.2, p. 530
C. M. Wayman	**17,**	101	(1966)	*1.2*, p. *9*; 3, p. 26; 21.1.1, p. 412
	22,	559	(1967)	1.4, p. 19; *3.1*, p. *32*; 21.1, p. 399
	22,	579	(1967)	1.4, p. 19; *3.1*, p. *32*; 21.1, p. 399
P. W. Webb	**28,**	295	(1968)	*3.1*, p. *33*; *14.3.1*, p. *247*; 14.3.2, p. 251; 22.4.2, p. 508
E.-H. Weber	**17,**	843	(1966)	13.3, p. 197; 13.4, p. 203; 14.4.1, p. 257; *16*, p. *277*; 22.4, p. 495
	28,	649	(1968)	13.3, p. 199; *16*, p. *283*; 22.4.1, p. 504
G. Weber	**23,**	K17	(1967)	18.1, p. 297; *19*, p. *330*; 21.1, p. 400
H. H. Weber	**23,**	703	(1967)	13.2, p. 193; 20.2, p. 361; *20.3*, p. *367*; 22.4.1, p. 501
P. Weber	**17,**	185	(1966)	1.4, p. 16; *18.2*, p. *301*; 21.1.1, p. 412

E. Webster	**21,**	315	(1967)	*11*, p. *148*; *20.3*, p. *365*; 22.4.3, p. 509
J. H. P. van Weeren	**19,**	K107	(1967)	10, p. 98; *14.3*, p. *235*; 22.1.1, p. 461
	27,	219	(1968)	10, p. 106; *13.4*, p. *213*; 14.3, p. 241; 22.1.1, p. 465
	27,	225	(1968)	10, p. 107; *14.3*, p. *241*; 22.1.1, p. 465
H. Wegener	**29,**	781	(1968)	6.1, p. 74; *18.4*, p. *325*; 22.8, p. 555
R. Wehner see R. K. Wehner				
R. K. Wehner	**17,**	K179	(1966)	3.2, p. 34; *6*, p. *57*
	20,	K167	(1967)	*6*, p. *60*; 10.2, p. 139; 20.1, p. 348; 22.5.2, p. 522
	22,	527	(1967)	*6*, p. *62*
	29,	133	(1968)	*6*, p. *67*; 20.1, p. 356; 22.1.1, p. 466; 22.1.2, p. 474
D. T. Y. Wei	**29,**	K7	(1968)	13.4, p. 215; *16*, p. *284*; 22.4.1, p. 504
F. L. Weichman	**25,**	583	(1968)	*15*, p. *273*; *16*, p. *281*; 22.4.1, p. 501
R. Weidmann	**27,**	631	(1968)	*9*, p. *92*; 10, p. 108; 22.5.1, p. 517
H. Weijma	**29,**	K43	(1968)	1.2, p. 12; 10, p. 110; *14.4*, p. *254*; 22.5.2, p. 529
G. Weise	**22,**	K151	(1967)	*3.1*, p. *32*; 14.1, p. 223; 21, p. 381
G. Weiser	**17,**	343	(1966)	13.1, p. 175; *14.4.1*, p. *256*; *20*, p. *337*; 22.1.3, p. 476
A. Weiss	**21,**	257	(1967)	3.2, p. 35; 5, p. 51; *19*, p. *329*; 22.5, p. 513; 22.8, p. 550
P. Weissglas	**24,**	531	(1967)	*13*, p. *168*; *14.3*, p. *239*; 22.2.3, p. 492
	28,	711	(1968)	*14.3*, p. *242*; 14.4.1, p. 263; 22.2.3, p. 493
W. Weller	**30,**	373	(1968)	*14.2*, p. *230*
H. Wenzl	**17,**	259	(1966)	10.1, p. 115; *11*, p. *146*; 14.1, p. 220; 21, p. 374
	18,	427	(1966)	*10.1*, p. *118*; 11, p. 147; 14.1, p. 221; 21, p. 375
K. Werner	**24,**	K95	(1967)	*19*, p. *331*; 22.9, p. 560
K. Wetzig	**19,**	K71	(1967)	*1.3*, p. *14*; 4, p. 40; 21, p. 377; 21.1.1, p. 415
	27,	K7	(1968)	1.2, p. 12; *4*, p. *46*; 21, p. 386; 21.1.1, p. 421
H. Wever	**18,**	K25	(1966)	*9*, p. *88*
	24,	K5	(1967)	*14.1*, p. *224*; 18.2, p. 309; 21.1.1, p. 420
	25,	109	(1968)	*9*, p. *91*; 10.1, p. 127; 21, p. 383
R. C. Whelan	**29,**	145	(1968)	10, p. 109; *13.4*, p. *215*; 14.3, p. 243; 22.4.3, p. 511
A. M. White	**30,**	K125	(1968)	*20.3*, p. *372*; 22.2.2, p. 488; 22.8, p. 557
B. J. Wicks	**26,**	571	(1968)	*4*, p. *45*; 10, p. 105; *14.4.2*, p. *268*
H. E. Wiehl	**17,**	285	(1966)	1.4, p. 16; *18.2*, p. *302*; 21.1.1, p. 412
F. Van de Wiele	**17,**	177	(1966)	*10.2*, p. *137*; 22.5.2, p. 519
H. Wienhold	**30,**	K17	(1968)	13.3, p. 199; *14.3.2*, p. *251*; 22.1.2, p. 474
E. Wiesner	**16,**	127	(1966)	*6.1*, p. *71*; 18.3, p. 316
	25,	607	(1968)	*6.1*, p. *73*; 18.3, p. 319
D. A. Wigley	**21,**	145	(1967)	*11*, p. *148*; 12, p. 157; 14.1, p. 222; 21.5, p. 429
H. P. J. Wijn	**26,**	231	(1967)	*14.4*, p. *253*; 18.3, p. 319
H. W. de Wijn	**24,**	715	(1967)	1.2, p. 11; *18.2*, p. *309*; 21, p. 383
	29,	189	(1968)	13, p. 172; 18.1, p. 297; *19*, p. *335*; 21, p. 387; 21.4, p. 428
	30,	759	(1968)	13, p. 173; 13.1, p. 190; 18.1, p. 298; *19*, p. *336*; 21, p. 389; 21.4, p. 428
H. B. Wild	**26,**	K135	(1968)	*14.3*, p. *241*; 22.1.3, p. 479
W. E. Wilhelm	**16,**	K59	(1966)	*14.4.1*, p. *255*; 22, p. 439
M. Wilkens	**16,**	K105	(1966)	*10.1*, p. *115*; 11, p. 145; 21.1, p. 391
	18,	609	(1966)	*10.1*, p. *119*; 21.1, p. 394; 21.6, p. 431
	22,	689	(1967)	*10.1*, p. *124*; 21.1, p. 399
	23,	113	(1967)	*4*, p. *43*
	30,	833	(1968)	*10.1*, p. *136*; 21, p. 389; 21.1, p. 409
C. Willaime	**17,**	K173	(1966)	*10*, p. *96*; 22.8, p. 548
E. W. Williams	**30,**	K125	(1968)	*20.3*, p. *372*; 22.2.2, p. 488; 22.8, p. 557

F. Williams	**25,**	493	(1968)	10, p. 93; *13.4*, p. *200*; 20.3, p. 363; 22.1, p. 456; 22.1.1, p. 458; 22.1.2, p. 468; 22.2.2, p. 487; 22.4.1, p. 497
G. P. Williams, Jr.	**26,**	645	(1968)	*10.1*, p. *130*; *12.1*, p. *160*; 21.6, p. 435
	30,	87	(1968)	9, p. 92; *10.1*, p. *134*; *12.1*, p. *161*; 21, p. 388; 21.6, p. 437
J. C. Williams	**25,**	K1	(1968)	*10.1*, p. *128*; 21, p. 384
J. E. Williams	**22,**	K19	(1967)	*10.2*, p. *140*; 11, p. 149; 22.5.3, p. 533
R. H. Williams	**26,**	K5	(1968)	2, p. 24; *17*, p. *288*; 22.1.3, p. 479
W. Windsch	**17,**	K75	(1966)	10, p. 95; 14.4.2, p. 266; *19*, p. *326*
	17,	K221	(1966)	14.4.2, p. 266; *19*, p. *327*
	20,	121	(1967)	9, p. 88; 13.4, p. 205; *19*, p. *328*; 22.5.2, p. 521
	28,	K163	(1968)	14.4.2, p. 269; *19*, p. *335*
	29,	K11	(1968)	*14.4.2*, p. *269*; 19, p. 335
K. Winzer	**26,**	139	(1968)	*13.1*, p. *184*; 18, p. 293; *20.1*, p. *352*; 22.1.3, p. 478
R. De Wit	**20,**	567	(1967)	*10*, p. *99*
	20,	575	(1967)	*10*, p. *99*
V. Witt	**22,**	245	(1967)	*10.2*, p. *139*; 22.5.2, p. 523
G. Wobser	**17,**	709	(1966)	*10*, p. *95*; *10.1*, p. *116*; 21, p. 375; 21.1, p. 392; 21.3, p. 426; 21.6, p. 431; 22.1.1, p. 460; 22.1.2, p. 469
	18,	189	(1966)	*10.1*, p. *117*; 21.1, p. 393; 21.6, p. 431
E. P. Wohlfarth	**30,**	131	(1968)	*18.2*, p. *315*; 21.1, p. 409; 21.1.1, p. 423
A. Wojakowski	**16,**	K129	(1966)	*14.3*, p. *232*; 15, p. 270; 20.1, p. 345; 22, p. 440
L. Wojtczak	**23,**	K163	(1967)	6, p. 63; *8*, p. *83*; 21, p. 382; 21.1, p. 400; 21.6, p. 434
	25,	K19	(1968)	*18.2*, p. *310*; 21.1, p. 402; 21.1.1, p. 420
B. Wojtowicz-Natanson	**18,**	K37	(1966)	*13.2*, p. *192*; 20.3, p. 364; 22.4.2, p. 506
G. Wolf	**18,**	K119	(1966)	8, p. 80; *13*, p. *164*; 22, p. 442
H. C. Wolf	**16,**	277	(1966)	*13.2*, p. *191*; 20.3, p. 363; 22.9, p. 558
	22,	K39	(1967)	12, p. 157; 13.2, p. 193; *20.1*, p. *349*; 22.9, p. 559
	23,	633	(1967)	*13.2*, p. *193*; *19*, p. *330*; 22.9, p. 559
M. Wolf	**26,**	K109	(1968)	*13*, p. *169*; *18.2*, p. *311*; 21.1.1, p. 421
G. Wolfram	**22,**	245	(1967)	*10.2*, p. *139*; 22.5.2, p. 523
	28,	K11	(1968)	10.2, p. 143; 13.4, p. 214; *16*, p. *283*; 22.5.2, p. 527
H. Wollenberger	**21,**	741	(1967)	*11*, p. *149*; 14.1, p. 223; 21, p. 379
	29,	75	(1968)	*10.1*, p. *132*; 11, p. 153; 21.1, p. 406
	29,	81	(1968)	*10.1*, p. *133*; 11, p. 153; 21.1, p. 406
	30,	219	(1968)	10.1, p. 135; *11*, p. *155*; 14.1, p. 227; 18.2, p. 315; 21.1, p. 408
Z. Wołoszyn	**20,**	K77	(1967)	*1.4*, p. *18*; 22.1.1, p. 461; 22.2.1, p. 483
W. Wolski	**26,**	429	(1968)	*4*, p. *45*; 18.3, p. 319; 18.4, p. 324; 22.8, p. 553
M. J. Wood	**16,**	321	(1966)	4, p. 37; *18.2*, p. *300*; 21.1.1, p. 410
	28,	407	(1968)	1.4, p. 20; *18.2*, p. *313*; 21.1, p. 405; 21.1.1, p. 422
J. Woods	**24,**	K37	(1967)	10.2, p. 141; *13.4*, p. *210*; 14.3, p. 239; 20.3, p. 367; 22.4.1, p. 501
F. J. Worzala	**20,**	K81	(1967)	*10.1*, p. *122*; 11, p. 148; 21.1, p. 397
F. van der Woude	**17,**	417	(1966)	*6.1*, p. *71*; 18.4, p. 321; 22.6, p. 535
	26,	K37	(1968)	*6.1*, p. *73*; 18.4, p. 324; 22.5, p. 513
H. A. Wriedt	**16,**	475	(1966)	*18.2*, p. *300*; 21.1.1, p. 410
T. W. Wright	**24,**	37	(1967)	6, p. 63; *7*, p. *77*
J. Wurm	**21,**	741	(1967)	*11*, p. *149*; 14.1, p. 223; 21, p. 379
	29,	75	(1968)	*10.1*, p. *132*; 11, p. 153; 21.1, p. 406
	29,	81	(1968)	*10.1*, p. *133*; 11, p. 153; 21.1, p. 406

J. Wurm	**30,**	219	(1968)	10.1, p. 135; *11*, p. *155*; 14.1, p. 227; 18.2, p. 315; 21.1, p. 408
M. Wuttig	**27,**	701	(1968)	*10.1*, p. *131*; 21.1.1, p. 421
A. S. Yakimova	**28,**	277	(1968)	13.1, p. 187; 13.2, p. 194; *20.1*, p. *355*; 22, p. 452
E. S. Yakovleva	**21,**	465	(1967)	1.1, p. 8; 4, p. 42; *10.1*, p. *123*; 21.1, p. 398; 21.6, p. 433
S. Yamada	**26,**	77	(1968)	*20.1*, p. *352*; 22.4.3, p. 510
R. H. T. Yeh	**25,**	551	(1968)	*7*, p. *77*; *14.2*, p. *230*
	25,	K65	(1968)	*14.2*, p. *230*
	28,	683	(1968)	*14.2*, p. *230*
	30,	K45	(1968)	1.2, p. 13; *14.4.2*, p. *269*
A. Yelon	**16,**	295	(1966)	*1.4*, p. *16*; *18.2*, p. *299*; 21.1.1, p. 410
	16,	301	(1966)	*18.2*, p. *299*; *18.4*, p. *321*; 21.1, p. 396; 21.1.1, p. 410
	16,	569	(1966)	4, p. 37; *18.2*, p. *300*; 21.1.1, p. 411
A. D. Yoffe	**30,**	741	(1968)	3, p. 28; 10, p. 113; *14.4.1*, p. *265*; 20.1, p. 359; 22, p. 455
H. Yoshinaga	**18,**	625	(1966)	10.1, p. 119; *14.1*, p. *221*; 21.1, p. 394; 21.6, p. 431
P. A. Young	**25,**	417	(1968)	*13.2*, p. *193*; 20.1, p. 351; 22, p. 449
Kim Yung	**27,**	K161	(1968)	6.1, p. 74; *10*, p. *108*; 22.5.1, p. 517; 22.5.2, p. 527
V. E. Yurasova	**17,**	K187	(1966)	*17*, p. *286*; 21.1, p. 393
D. Yusupova	**21,**	805	(1967)	10, p. 100; 11, p. 149; *12*, p. *157*; 22.1.2, p. 471
E. Zagorska	**28,**	725	(1968)	*10.2*, p. *143*; 16, p. 283; 22.5.1, p. 517
T. Zakrzewski	**23,**	K39	(1967)	*13.1*, p. *181*; 14.3, p. 238; 15, p. 272; 22.4.3, p. 510; 22.8, p. 552
R. B. Žakula	**27,**	623	(1968)	*13*, p. *170*; *18.2*, p. *312*
	28,	635	(1968)	*13*, p. *171*; *18.2*, p. *313*
F. Žaloudek	**16,**	K21	(1966)	13, p. 162; *18*, p. *291*; *20*, p. *337*; 22.4.3, p. 509
E. H. Zamanova	**26,**	65	(1968)	1.2, p. 12; *14.3*, p. *240*; 15, p. 273; 22, p. 450
A. Zaręba	**21,**	K135	(1967)	13.4, p. 208; *16*, p. *279*; 22.1, p. 457
L. I. Zarubin	**27,**	101	(1968)	13.1, p. 185; *14.3*, p. *241*; 22, p. 451; 22.7, p. 546
A. Żawadowski	**25,**	473	(1968)	*8*, p. *84*; 18, p. 293
L. Żdanovicz	**20,**	473	(1967)	*20.1*, p. *347*; 22.7, p. 545
W. Żdanowicz	**16,**	K129	(1966)	*14.3*, p. *232*; 15, p. 270; 20.1, p. 345; 22, p. 440
K. Žďánský	**20,**	653	(1967)	14.4.2, p. 267; *19*, p. *328*
	28,	175	(1968)	*19*, p. *334*; 22.8, p. 554
	28,	181	(1968)	10, p. 108; 13.4, p. 214; *19*, p. *334*; 22.6, p. 542
S. A. Zeinalov	**22,**	153	(1967)	*8*, p. *83*; 22.2.3, p. 491; 22.8, p. 551
J. Zelenka	**29,**	K51	(1968)	*5*, p. *54*; 14.4, p. 254; 22.6, p. 543
M. G. Zemlyanov	**20,**	767	(1967)	*6*, p. *60*; 8, p. 81; 21.1, p. 397
A. Zetzsche	**22,**	123	(1967)	*10*, p. *123*
R. Zeyfang	**24,**	221	(1967)	6, p. 63; *8*, p. *83*; 10.1, p. 126; 21.1, p. 401
R. Zeyher	**20,**	K167	(1967)	*6*, p. *60*; 10.2, p. 139; 20.1, p. 348; 22.5.2, p. 522
I. P. Zhadko	**16,**	43	(1966)	12, p. 156; *14.3*, p. *231*; 14.4.1, p. 255; 22.1.1, p. 459
	17,	389	(1966)	12, p. 156; *16*, p. *277*; 22.1.1, p. 459
	28,	797	(1968)	14.3.1, p. 247; *16*; p. *283*; 22, p. 453
N. G. Zhdanova	**24,**	551	(1967)	*14.4.1*, p. *261*; 22.1.1, p. 464
R. P. Zhitaru	**19,**	441	(1967)	*10*, p. *97*; 11, p. 147; 22.5.2, p. 520

R. A. Zhitnikov	**26**, 371	(1968)	10, p. 105; *19*, p. *333*; 22.5.2, p. 526
	30, 851	(1968)	10.2, p. 144; *19*; p. *336*; 22.5.2, p. 531
	30, 859	(1968)	3.2, p. 36; 10.2, p. 144; 13.4, p. 218; *19*, p. *336*; 22.5.2, p. 531
A. P. Zhukov	**27**, K129	(1968)	14.4.2, p. 269; *19*, p. *334*
V. K. Zhuravlev	**30**, K13	(1968)	13.4, p. 217; *14.3*, p. *244*; *15*, p. *275*; 22.8, p. 556
T. Ch. Zhuvarly	**16**, K113	(1966)	10, p. 94; *19*, p. *326*; 22.1.3, p. 476
V. P. Zhuze	**18**, 873	(1966)	*18.2*, p. *304*; *18.4*, p. *321*; 22, p. 442
A. Zia	**25**, 273	(1968)	*6*, p. *64*; 14.4.2, p. 268
E. Zielińska-Rohozińska	**28**, K139	(1968)	*4*, p. *47*; 5, p. 54; 22.1.2, p. 474
J. Ziller	**30**, K165	(1968)	10, p. 113; *14.4*, p. *254*; 22.5.2, p. 531
F. Zimmer	**29**, K89	(1968)	3.1, p. 33; *4*, p. *47*; 22.4.1, p. 505
G. Zimmerer	**22**, K49	(1967)	13.1, p. 179; *20.1*, p. *349*; 22.1.3, p. 477
	29, 203	(1968)	13.1, p. 188; 14.3.1, p. 247; *20.1*, p. *356*; 22.4, p. 496
R. Zimmerman	**29**, K91	(1968)	*1.4*, p. *21*; 21.1, p. 407
B. H. Zimmermann	**27**, 639	(1968)	*6.1*, p. *74*; 22.1.1, p. 465
W. Zimmermann	**28**, K101	(1968)	*14.3*, p. *243*; 22.1.3, p. 479
	30, 311	(1968)	*11*, p. *155*; 13.4, p. 216; *16*; p. *284*; 22.4.1, p. 505
W. Zinn	**17**, 323	(1966)	*18.2*, p. *302*; 21.1, p. 392; 21.1.1, p. 412
J. S. Ziomek	**21**, 271	(1967)	5, p. 51; 6, p. 61; 13.1, p. 178; *20.1*, p. *348*; 22.3, p. 494
J. Zítková	**20**, K99	(1967)	13.4, p. 206; *19*, p. *328*; 22.6, p. 537
	26, 443	(1968)	*18.1*, p. *297*; 18.4, p. 324; 22.8, p. 553
A. M. Zlobin	**28**, 105	(1968)	6, p. 66; *8*, p. *85*; *14.3*, p. *242*; *22*, p. 452
K. Zohner	**21**, 257	(1967)	3.2, p. 35; 5, p. 51; *19*, p. *329*; 22.5, p. 513; 22.8, p. 550
G. Zollfrank	**25**, 711	(1968)	10, p. 104; *20.3*, p. *368*; 22.4.1, p. 502
V. F. Zolotaryov	**21**, 437	(1967)	*13.4*, p. *206*; 14.4.1, p. 258; 22, p. 445
S. Zolotov	**18**, 593	(1966)	*10.2*, p. *138*; 22.5.2, p. 520
B. C. van Zorge	**18**, 761	(1966)	*6.1*, p. *71*
V. V. Zotov	**28**, K31	(1968)	13.4, p. 214; *16*, p. *283*; 22.4.1, p. 503
L. G. Zotova	**21**, 805	(1967)	10, p. 100; 11, p. 149; *12*, p. *157*; 22.1.2, p. 471
G. V. Zozulya	**20**, 213	(1967)	*9*, p. *88*; 21.6, p. 432; 22, p. 443; 22.1.3, p. 477
L. Zsoldos	**20**, K25	(1967)	1.2, p. 10; 5, p. 51; *18.2*, p. *306*; *18.4*, p. *322*; 21.1.1, p. 416
S. M. Zubkova	**28**, 807	(1968)	13.1, p. 188; *13.2*, p. *194*; 22, p. 453; 22.5.2, p. 528
V. A. Zuev	**16**, 751	(1966)	13.3, p. 196; *16*; p. *276*; 22.1.1, p. 459; 22.1.2, p. 469
S. Żukotyński	**16**, K55	(1966)	*13.1*, p. *174*; 22.2.1, p. 482; 22.2.3, p. 489
	19, K51	(1966)	*13.1*, p. *176*; 14.3, p. 235; 15, p. 271; 22, p. 443
	25, 583	(1968)	*15*, p. *273*; *16*, p. *281*; 22.4.1, p. 501
M. Zvára	**16**, K21	(1966)	13, p. 162; *18*, p. *291*; *20*, p. *337*; 22.4.3, p. 509
	27, K157	(1968)	13, p. 170; 18, p. 294; *20*, p. *341*; 22.2.1, p. 485
	29, 707	(1968)	6, p. 68; *13.2*, p. *195*; 20.1, p. 357; 22.4.1, p. 505; 22.4.2, p. 508; 22.4.3, p. 511
M. M. Zverev	**19**, K5	(1967)	*20.2*, p. *360*; 22.4.2, p. 506
A. Zygmunt	**22**, 517	(1967)	5, p. 52; *18.4*, p. *322*; 22, p. 447
	23, K123	(1967)	*18.4*, p. *323*; 22, p. 448
S. A. Zynio	**20**, 311	(1967)	14.4.1, p. 257; *20.3*, p. *365*; 22.4.1, p. 499
P. S. Zyryanov	**21**, 89	(1967)	*14*, p. *219*; 14.4, p. 253
	25, 775	(1968)	14.1, p. 224; 14.3, p. 240; *15*, p. *273*; 21.7, p. 438; 22.2, p. 481

Classification Scheme

1. Structure of Solids
   1.1 Alloys. Metallurgy
   1.2 Solid-State Phase Transformations
   1.3 Surfaces
   1.4 Films
2. Non-Crystalline State
3. Crystallography
   3.1 Crystal Growth
   3.2 Interatomic Forces
4. Microstructure of Solids
5. Perfectly Periodic Structures
6. Lattice Mechanics. Phonons
   6.1 Mössbauer Investigations
7. Acoustic Properties of Solids
8. Thermal Properties of Solids
9. Diffusion in Solids
10. Defect Properties of Solids (Irradiation Defects See 11)
    10.1 Defect Properties of Metals
    10.2 Photochemical Reactions. Colour Centres
11. Irradiation Effects in Solids
12. Mechanical Properties of Solids (Plastic Deformations See 10)
    12.1 Mechanical Properties of Metals (Plastic Deformations See 10.1)
13. Electron States in Solids
    13.1 Band Structure. Fermi Surfaces
    13.2 Excitons
    13.3 Surface States
    13.4 Impurity and Defect States
14. Electrical Properties of Solids. Transport Phenomena
    14.1 Metals. Conductors
    14.2 Superconductivity. Superconducting Materials and Devices
    14.3 Semiconductors
        14.3.1 Semiconducting Films
        14.3.2 Semiconducting Devices. Junctions (Contact Problems See 14.4.1)
    14.4 Dielectrics
        14.4.1 High Field Phenomena, Space Charge Effects, Inhomogeneities, Injected Carriers (Electroluminescence see 20.3; Junctions see 14.3.2)
        14.4.2 Ferroelectric Materials and Phenomena
15. Thermoelectric and Thermomagnetic Properties of Solids
16. Photoconductivity. Photovoltaic Effects
17. Emission of Electrons and Ions from Solids

18. Magnetic Properties of Solids
    18.1 Paramagnetic Properties
    18.2 Ferromagnetic Properties
    18.3 Ferrimagnetic Properties. Ferrites
    18.4 Antiferromagnetic Properties
19. Magnetic Resonance in Solids
20. Optical Properties of Solids
    20.1 Spectra. Optical Constants (Colour Centres See 10.2; X-Ray Spectra See 20)
    20.2 Lasers
    20.3 Luminescence of Solids (X-Ray Spectra See 20)
21. Metals. Alloys
    21.1 Transition Metals (Group Sc to Zn) and Their Alloys
        21.1.1 Fe and Fe Alloys
    21.2 Alkali Metals
    21.3 Alkali-Earth Metals
    21.4 Rare-Earth Metals
    21.5 Actinides
    21.6 Noble Metals
    21.7 Semi-Metals (Semiconducting Alloys See 22.7)
22. Semiconductors and Ionic Crystals
    22.1 Elements
        22.1.1 Germanium
        22.1.2 Silicon
        22.1.3 Group VI Elements
    22.2 III—V Compounds
        22.2.1 Arsenides
        22.2.2 Phosphides
        22.2.3 Antimonides
    22.3 Carbides
    22.4 II—VI Compounds
        22.4.1 Sulphides
        22.4.2 Selenides
        22.4.3 Tellurides
    22.5 Halides
        22.5.1 Silver Halides
        22.5.2 Alkali-Halides
        22.5.3 Alkali-Earth Halides
    22.6 Oxides (CdO, ZnO See 22.4; Spinels See 18.3 or 14.4.2)
    22.7 Semiconducting Intermetallic Compounds
    22.8 Tertiary and Higher Compounds (Carbonates, Phosphates, Silicates, and Mica See 22)
    22.9 Organic Semiconductors
23. Solidified Gases

For comments on the use of the classification scheme see p. 4.

18. Magnetic Properties of Solids
    - 18.1 Paramagnetic Properties
    - 18.2 Ferromagnetic Properties
    - 18.3 Ferrimagnetic Properties. Ferrites
    - 18.4 Antiferromagnetic Properties

19. Magnetic Resonance in Solids

20. Optical Properties of Solids
    - 20.1 Spectra. Optical Constants (Colour Centres See 10.2; X-Ray Spectra See 20)
    - 20.2 Lasers
    - 20.3 Luminescence of Solids (X-Ray Spectra See 20)

21. Metals. Alloys
    - 21.1 Transition Metals (Group Sc to Zn) and Their Alloys
        - 21.1.1 Fe and Fe Alloys
    - 21.2 Alkali Metals
    - 21.3 Alkali-Earth Metals
    - 21.4 Rare-Earth Metals
    - 21.5 Actinides
    - 21.6 Noble Metals
    - 21.7 Semi-Metals (Semiconducting Alloys See 22.7)

22. Semiconductors and Ionic Crystals
    - 22.1 Elements
        - 22.1.1 Germanium
        - 22.1.2 Silicon
        - 22.1.3 Group VI Elements
    - 22.2 III—V Compounds
        - 22.2.1 Arsenides
        - 22.2.2 Phosphides
        - 22.2.3 Antimonides
    - 22.3 Carbides
    - 22.4 II—VI Compounds
        - 22.4.1 Sulphides
        - 22.4.2 Selenides
        - 22.4.3 Tellurides
    - 22.5 Halides
        - 22.5.1 Silver Halides
        - 22.5.2 Alkali Halides
        - 22.5.3 Alkali-Earth Halides
    - 22.6 Oxides (CdO, ZnO See 22.4; Spinels See 18.3 or 14.4.2)
    - 22.7 Semiconducting Intermetallic Compounds
    - 22.8 Tertiary and Higher Compounds (Carbonates, Phosphates, Silicates, and Mica See 22)
    - 22.9 Organic Semiconductors

23. Solidified Gases

For comments on the use of the classification scheme see p. 4.

Manuscripts and letters for physica status solidi and physica status solidi (a) – applied research

should be sent to

DDR, 102 Berlin, Neue Schönhauser Str. 20

or

Internationaler Buchversand GmbH, 1 Berlin 19, Eschenallee 21.

The object of "physica status solidi" and "physica status solidi (a) — applied research" is to publish rapidly important and topical results concerning fundamental research in the field of solid state physics and aspects of application and experimental techniques of fundamental solid-state research, respectively. The issues of "physica status solidi" and "physica status solidi (a)" will be published on the 1st and the 16th day of each month, respectively. Either series will contain the following categories:

1. Review articles written by arrangement with the editors. They will be remunerated.
2. Original papers selected from those which have not been submitted for publication elsewhere, after being reviewed. (The latest date for acceptance of contributions is the 8th and the 23rd day of each month for "physica status solidi" and "physica status solidi (a)", respectively, 50 days before the delivery of the journal.)
3. Short notes published in the off-set part of the journal. They are, if necessary after being reviewed, selected by the editorial staff. (The latest date for acceptance of manuscripts for this part is the 12th and the 27th day of each month for "physica status solidi" and "physica status solidi (a)", respectively, 16 days before the delivery of tho journal).
4. Pre-printed titles of papers to be published in the next issue of both series.

Editorial Information

Manuscripts should be submitted preferentially in English. Besides, papers in German, French, and Russian are admitted. Authors are asked to observe the following rules:

1. Manuscripts have to be written as concisely as possible. As most, 20 and 3 pages (28 lines per page; abstracts, tables, references, captions of figures included) are admitted for original papers and short notes, respectively. All manuscripts should be typewritten on one side only. Two copies are required. Badly arranged manuscripts are returned.
2. The author's name, initials, and institution should be given.
3. An abstract of about ten lines in English and a second one in German, French, or Russian are required for original papers.
4. Formulae should not be written too small and with pencil. Separate lines for formulae are desirable. Indistinct writing will lead to a return of the manuscript.
5. Drawings should be enclosed on separate sheets of paper; photographs should be submitted, on white paper, only for review articles and original papers. Duplicates are required.
6. Captions for the figures should be submitted on separate sheets of paper, and special instructions with regard to the size of illustrations should be given on these.
7. References should be numbered in order of appearance and should be given, on a separate sheet, in the form used in physica status solidi.
8. Units, symbols, etc. should be used in accordance with the recommendations given by the IUPAP.
9. Each paper should contain a proposal for the subject classification according to the classification scheme appearing on the second and third cover of each issue.

The authors will be sent two galley proofs of review articles and original papers. Proof reading of short notes and pre-printed titles will be done by the editorial staff.

For each original paper, and short note, a total of 75 reprints is available free of charge. The number of reprints free of charge per review article is 25. Requests for additional reprints (against payment) should be given together with the submitted manuscript. **The deadline for orders of additional reprints is the 15th and the 30ies day of each month** for "physica status solidi" and "physica status solidi (a)", respectively, two weeks before the delivery of the journal.

Bei Fragen zur Produktsicherheit wenden Sie sich bitte an:
If you have any questions regarding product safety,
please contact:

Walter de Gruyter GmbH
Genthiner Straße 13
10785 Berlin
productsafety@degruyterbrill.com